Chapter 6

ATCF	after-tax cash flow
BTCF	before-tax cash flow
BV	book value of an asset
MACRS	modified accelerated cost recovery system
MV	market value of an asset; the price that a buyer will pay for a particular type of property
t	effective income-tax rate

Chapter 7

Z_u	the number of input resource units needed to produce output unit number u

Chapter 8

A$	actual (current) dollars
f	general inflation rate
R$	real (constant) dollars

Chapter 9

EUAC	equivalent uniform annual cost

Chapter 13

$E(X)$	mean of a random variable
$f(x)$	probability density function of a continuous random variable
$p(x)$	probability mass function of a discrete random variable
$SD(X)$	standard deviation of a random variable
$V(X)$	variance of a random variable

Engineering Economy

12th Edition

William G. Sullivan
Virginia Polytechnic Institute and State University

Elin M. Wicks
Crowe Associates, L.L.P.

James T. Luxhoj
Rutgers University

Prentice
Hall

Pearson Education, Inc.
Upper Saddle River, New Jersey 07458

Library of Congress Cataloging-in-Publication Data

Sullivan, William G., 1942–
    Engineering economy / William G. Sullivan, Elin M. Wicks, and James Luxhoj.
    p.   cm.
    Includes bibliographical references and index.
    ISBN 0-13-067338-2
1. Engineering economy.   I. Wicks, Elin M.   II. Luxhoj, James.
TA177.4.S85   2003
658.15–dc21   2001059806

Vice President and Editorial Director, ECS: *Marcia Horton*
Acquisitions Editor: *Dorothy Marrero*
Editorial Assistant: *Jessica Romeo*
Vice President and Director of Production and Manufacturing, ESM: *David W. Riccardi*
Executive Managing Editor: *Vince O'Brien*
Managing Editor: *David A. George*
Production Editor: *Scott Disanno*
Director of Creative Services: *Paul Belfanti*
Creative Director: *Carole Anson*
Art Director: *John Christiana*
Cover Designer: *Bruce Kenselaar*
Art Editor: *Xiaohong Zhu*
Manufacturing Manager: *Trudy Pisciotti*
Manufacturing Buyer: *Lisa McDowell*
Marketing Manager: *Holly Stark*

© 2003, 2000, 1997 by Pearson Education, Inc.
Pearson Education, Inc.
Upper Saddle River, New Jersey 07458

Earlier editions entitled *Engineering Economy,* copyright 1993, 1988, 1984, 1979, 1973, 1967, and 1960 by Macmillan Publishing Company. Earlier editions entitled *Introduction to Engineering Economy,* copyright 1953 by Macmillan Publishing Company, and copyright 1942 by E. P. DeGarmo and B. M. Woods.

The author and publisher of this book have used their best efforts in preparing this book. These efforts include the development, research, and testing of the theories and programs to determine their effectiveness. The author and publisher make no warranty of any kind, expressed or implied, with regard to these programs or the documentation contained in this book. The author and publisher shall not be liable in any event for incidental or consequential damages in connection with, or arising out of, the furnishing, performance, or use of these programs.

Printed in the United States of America

10 9 8 7 6 5 4 3 2 1

ISBN 0-13-067338-2

Pearson Education Ltd., *London*
Pearson Education Australia Pty. Limited, *Sydney*
Pearson Education Singapore, Pte. Ltd.
Pearson Education North Asia Ltd., *Hong Kong*
Pearson Education Canada, Inc., *Toronto*
Pearson Educatíon de Mexico, S.A. de C.V.
Pearson Education—Japan, *Tokyo*
Pearson Education Malaysia, Pte. Ltd.
Pearson Education, Inc., *Upper Saddle River, New Jersey*

Contents

CHAPTER 3

Money–Time Relationships and Equivalence 70

PART II

Basic Topics in Engineering Economy 149

CHAPTER 4

Applications of Money–Time Relationships 150

CHAPTER 5

Comparing Alternatives 197

CHAPTER 6

CHAPTER 7

CHAPTER 8

CHAPTER 9

Replacement Analysis 406

CHAPTER 10

Dealing with Uncertainty 447

PART III
Additional Topics in Engineering Economy 483

CHAPTER 11

Evaluating Projects with the Benefit–Cost-Ratio Method 484

CHAPTER 12

Engineering Economy Studies in Investor-Owned Utilities 519

CHAPTER 13

CHAPTER 14

CHAPTER 15

x

Preface

About Engineering Economy

Engineering economy—what is it, and why is it important? The initial reaction of many engineering students to these questions is "Money matters will be handled by someone else. It is not something I need to worry about." In reality, any engineering project must be not only physically realizable, but also economically affordable. For example, a child's tricycle could be built with an aluminum frame or a composite frame. Some may argue that because the composite frame will be stronger and lighter, it is a better choice. However, there is not much of a market for thousand dollar tricycles! One might suggest that this argument is ridiculously simplistic and that common sense would dictate choosing aluminum for the framing material. Although the scenario is an exaggeration, it reinforces the idea that the economic factors of a design weigh heavily in the design process, and that engineering economy is an integral part of that process, regardless of the engineering discipline. *Engineering, without economy, makes no sense at all.*

In broad terms, for an engineering design to be successful, it must be technically sound and produce benefits. These benefits must exceed the costs associated with the design in order for the design to enhance net value. The field of engineering economy is concerned with the systematic evaluation of the benefits and costs of projects involving engineering design and analysis. In other words, engineering economy quantifies the benefits and costs associated with engineering projects to determine whether they make (or save) enough money to warrant their capital investments. Thus, engineering economy requires the application of engineering design and analysis principles to provide goods and services that satisfy the consumer at an affordable cost. As we shall see, engineering economy is as relevant to the design engineer who considers material selection as it is to the chief executive officer who approves capital expenditures for new ventures.

History of the Book

The original *Introduction to Engineering Economy*, authored by Woods and De-Garmo, appeared in 1942. The extensive use of this text for the past 60 years has encouraged the authors to continue building on the original purpose of the book—to teach lucidly the principles of engineering economy. In this spirit, the twelfth edition of *Engineering Economy* has built upon the rich and time-tested teaching materials of earlier editions, and its publication makes it the second-oldest book on the market that deals exclusively with engineering economy.

Twelfth Edition of *Engineering Economy*

New or Enhanced Features to This Edition

- Design economics problems are expanded in Chapter 2.
- Cost estimating has been clarified and given expanded coverage.
- A number of new and updated end-of-chapter problems are included.
- A Web site devoted to electronic media to support an engineering economy course is fully operational (and maintained by Prentice Hall).
- Spreadsheet templates appear throughout the text.
- An extra supplement dealing with development and use of spreadsheets is available.
- An Instructor's Manual containing full solutions to all problems in the book is available.
- Suggestions for using "student portfolios" to facilitate the integrated learning of topics in engineering economy are presented in this Preface.
- "Economic value added" by an engineering project is explained in terms of an after-tax cash-flow analysis.
- The cost of equity and debt capital and the weighted average cost of capital and its relationship to rate of return concepts are explained.
- Replacement Analysis (Chapter 9) has been rewritten to clarify concepts and principles of this important topic.
- Chapter 15, which deals with multiattributed decision making, has been added.

Pedagogy of this Book

This book has two primary purposes: (1) to provide students with a sound understanding of the principles, basic concepts, and methodology of engineering economy; and (2) to help them develop proficiency with these methods and with the process for making rational decisions regarding situations they are likely to encounter in professional practice. Consequently, *Engineering Economy* is intended to serve as a text for classroom instruction *and* as a basic reference for use by practicing engineers in all specialty areas (e.g., chemical, civil, computer, electrical, industrial, and mechanical engineering). The book is also useful to persons engaged in the management of technical activities.

As a textbook, the twelfth edition is written principally for the first formal course in engineering economy. The contents of the book and the accompanying Instructor's Manual and Electronic Spreadsheets Supplement (both available from Prentice Hall) are organized for effective presentation and teaching of the subject matter. A three-credit-hour semester course should be able to cover the majority of topics in this edition, and there is sufficient depth and breadth to enable an instructor to arrange course content to suit individual needs. Representative syllabi for a three-credit and a two-credit semester course in engineering economy are provided in Table P-1. Moreover, because several advanced topics are included, this book can also be used for a second course in engineering economy.

Every chapter and the appendices have been revised and updated to reflect current trends and issues. Also, numerous exercises that involve open-ended problem statements and iterative problem-solving skills are included throughout the book. A large number of the 500+ end-of-chapter exercises are new, and many solved examples representing realistic problems that arise in various engineering disciplines are presented.

An engineering economy course may be classified, for Accreditation Board for Engineering and Technology (ABET) purposes, as part engineering science and part engineering design. It is generally advisable to develop and teach such a course at the upper division level. Here, an engineering economy course incorporates the accumulated knowledge students have acquired in other areas of the curriculum also dealing with iterative problem solving, open-ended exercises, creativity in formulating and evaluating feasible solutions to problems, and consideration of realistic constraints (economic, aesthetic, safety, etc.) in problem solving.

Internet Web Site Supplement

An engineering home page is accessible to instructors and students at:

http://www.prenhall.com/sullivan_engineering

This resource contains numerous teaching and learning aids, such as (1) sample Microsoft PowerPoint® slides for selected chapters in the book; (2) sample test/exam questions; (3) an engineering economy tutorial that includes green engineering examples; (4) electronic spreadsheet templates authored by James A. Alloway Jr.; and (5) case studies developed by engineering students working in interdisciplinary teams.

Our engineering economy home page is an apt resource for transitioning the teaching of engineering economy into the twenty-first century. Now instructors and students can electronically draw upon the Internet to "cut" and "paste" the desired learning supplements to suit their individual needs and interests. We are positive that this feature of the twelfth edition will motivate the curiosity, imagination, and learning of your students in engineering economy.

TABLE P-1 Typical Syllabi for Courses in Engineering Economy

Semester Course (Three Credit Hours)

Chapter	Week of the semester	Topic(s)
1	1	Introduction to Engineering Economy
2 App. A		Cost Concepts and Design Economics
3	2-3	Money–Time Relationships and Equivalence
4	4	Applications of Money–Time Relationships
5	5	Comparing Alternatives
6	6	Depreciation and Income Taxes
	7	**Midterm Examination**
11	8	Evaluating Projects with the Benefit–Cost-Ratio Method
7	9	Cost Estimation Techniques
8	10	Price Changes and Exchange Rates
10	11	Dealing with Uncertainity
9	12	Replacement Analysis
14-15	13-14	Capital Financing and Allocation, Dealing with Multiattributed Decisions
	15	**Final Examination**

Number of class periods: 45

Semester Course (Two Credit Hours)

Chapter(s)	No. of Class Periods	Topic(s)
1	1	Introduction to Engineering Economy
2	4	Cost Concepts, Single Variable Tradeoff Analysis and Present Economy
3	5	Money–Time Relationships and Equivalence
1-3	1	**Test #1**
7	3	Developing Cash Flows and Cost Estimating Techniques
4	2	Applications of Money–Time Relationshps
5	4	Comparing Alternatives
4,5,7	1	**Test #2**
10	2	Dealing With Uncertainty
6	5	Depreciation and Income Taxes
15	1	Dealing with Multiattributed Decisions
All the above	1	**Final Examination**

Number of class periods: 30

Instructional Features

The Instructor's Manual is designed as a comprehensive aid in teaching the text material. Full solutions of all problems at the end of each chapter are presented. Several *comprehensive examples (case studies)* have been included in the twelfth edition. These fairly complex examples and problems provide the instructor with essential material for teaching both the first formal course and a second, more advanced course in engineering economy. They also integrate the principles, basic concepts, and methodologies that are needed by engineers in typical real-world situations, and also serve as a bridge from the classroom to professional practice.

Spreadsheet Supplement

A second supplement entitled *Spreadsheet Modeling to Accompany Engineering Economy, Twelfth Edition* is authored by James A. Alloway, Jr. Electronic spreadsheets are a mainstay in many undergraduate engineering economy courses; the spreadsheet supplement ensures that the twelfth edition of *Engineering Economy* will maintain its leadership position by providing basic templates for all major topics in the text. In addition, it provides a concise summary of formulas and key concepts, which students will find invaluable for review and quick reference.

The greatest advantage is that it is no longer necessary to enter the spreadsheets by hand. The templates can be downloaded and opened directly in Excel for Windows. Most other spreadsheet software packages provide conversion utilities to convert these files into their respective native formats. Users can then modify the basic templates for the specific problem at hand. As a bonus, advanced templates have also been developed for such techniques as Monte Carlo simulation, three-factor simultaneous sensitivity analysis, and integer linear programming.

Engineering Economy Portfolio

In many engineering economy courses, students are required to design, develop, and maintain an "Engineering Economy Portfolio." The purpose of the portfolio is to demonstrate and integrate knowledge of engineering economy beyond the required assignments and tests. This is usually an individual assignment. Professional presentation, clarity, brevity, and creativity are important criteria that will be used to evaluate portfolios. Students are asked to keep the audience (i.e., the grader) in mind when constructing their portfolios.

The portfolio should contain a variety of content. To get credit for content, students must display their knowledge. Simply collecting articles in a folder demonstrates very little. To get credit for collected articles, students should read them and write a brief summary. The summary could explain how the article is relevant to engineering economy, it could critique the article, or it could check or extend any economic calculations in the article. The portfolio should include both the summary and the article itself. Annotating the article by writing comments in the margin is also a good idea. Other suggestions for portfolio content follows (note that students are encouraged to be creative):

- Describe and set up or solve an engineering economy problem from your own discipline (e.g., electrical engineering or building construction).
- Choose a project or problem in society or at your university and apply engineering economic analysis to one or more proposed solutions.
- Develop proposed homework or test problems for engineering economy. Include the complete solution. Additionally, state which course objective(s) this problem demonstrates (include text section).
- Reflect upon and write about your progress in the class. You might include a self-evaluation against the course objectives.
- Include a photo or graphic that illustrates some aspect of engineering economy. Include a caption that explains the relevance of the photo or graphic.
- Include completely worked out practice problems. Use a different color pen to show these were checked against the provided answers.
- Rework missed test problems, including an explanation of each mistake.

(The preceding list could reflect the relative value of the suggested items; that is, items at the top of the list are worth more than items at the bottom of the list.)

Develop an introductory section that explains the purpose and organization of the portfolio. A table of contents and clearly marked sections or headings are highly recommended. Cite the source (i.e., a complete bibliographic entry) of all material other than your own work. Remember, portfolios provide evidence that students know more about engineering economy than what is reflected in the assignments and exams. Focus on quality of evidence, not quantity.

WILLIAM G. SULLIVAN

ELIN M. WICKS

JAMES T. LUXHOJ

*F*undamentals of Engineering Economy

[Engineering] is the art of doing that well with one dollar which any bungler can do with two.

—A. M. Wellington, *The Economic Theory of the Location of Railways*
(New York: John Wiley & Sons, 1887)

1

Introduction to Engineering Economy

*T*he objectives of Chapter 1 are to (1) introduce the subject of engineering economy, (2) discuss its critical role in engineering design and analysis, (3) discuss the basic principles of the subject, and (4) provide an overview of the book.

The following topics are discussed in this chapter:

The importance of this subject in engineering practice
Origins of engineering economy
The principles of engineering economy
Engineering economy and the design process
Accounting and engineering economy studies
Overview of the book

1.1 Introduction

The technological and social environments in which we live continue to change at a rapid rate. In recent decades, advances in science and engineering have made space travel possible, transformed our transportation systems, revolutionized the practice of medicine, and miniaturized electronic circuits so that a computer can be placed on a semiconductor chip. The list of such achievements seems almost endless. In your science and engineering courses, you will learn about some of the physical laws that underlie these accomplishments.

The utilization of scientific and engineering knowledge for our benefit is achieved through the *design* of things we use, such as machines, structures, products, and services. However, these achievements don't occur without a price, monetary or otherwise. Therefore, the purpose of this book is to develop and illustrate the principles and methodology required to answer the basic economic question of any design: Do its benefits exceed its costs?

The Accreditation Board for Engineering and Technology states that engineering "is the profession in which a knowledge of the mathematical and natural sciences gained by study, experience, and practice is applied with judgment to develop ways to utilize, economically, the materials and forces of nature for the benefit of mankind."* In this definition, the economic aspects of engineering are emphasized, as well as the physical aspects. Clearly, it is essential that the economic part of engineering practice be accomplished well.

Engineering Economy involves the systematic evaluation of the economic merits of proposed solutions to engineering problems. To be economically acceptable (i.e., affordable), *solutions to engineering problems* must demonstrate a positive balance of long-term benefits over long-term costs, and they must also

- promote the well-being and survival of an organization,
- embody creative and innovative technology and ideas,
- permit identification and scrutiny of their estimated outcomes, and
- translate profitability to the "bottom line" through a valid and acceptable measure of merit.

Therefore, engineering economy is the dollars-and-cents side of the decisions that engineers make or recommend as they work to position a firm to be profitable in a highly competitive marketplace. Inherent to these decisions are trade-offs among different types of costs and the performance (response time, safety, weight, reliability, etc.) provided by the proposed design or problem solution. *The mission of engineering economy is to balance these trade-offs in the most economical manner.* For instance, if an engineer at Ford Motor Company invents a new transmission lubricant that increases fuel mileage by 10% and extends the life of the transmission by 30,000 miles, how much can the company afford to spend to implement this invention? Engineering economy can provide an answer.

A few more of the myriad situations in which engineering economy plays a crucial role come to mind:

1. Choosing the best design for a high-efficiency gas furnace.
2. Selecting the most suitable robot for a welding operation on an automotive assembly line.
3. Making a recommendation about whether jet airplanes for an overnight delivery service should be purchased or leased.
4. Determining the optimal staffing plan for a computer help desk.

From these illustrations, it should be obvious that engineering economy includes significant technical considerations. Thus, engineering economy involves technical analysis, with emphasis on the economic aspects, and has the objective of assisting decisions. This is true whether the decision maker is an engineer interactively ana-

---

*Accreditation Board of Engineering and Technology, *Criteria for Accrediting Programs in Engineering in the United States* (New York: 1998, ABET, Baltimore, MD).

lyzing alternatives at a computer-aided design workstation or the Chief Executive Officer (CEO) considering a new project. *An engineer who is unprepared to excel at engineering economy is not properly equipped for his or her job.*

1.2 Origins of Engineering Economy

Cost considerations and comparisons are fundamental aspects of engineering practice. This basic point was emphasized in Section 1.1. However, the development of engineering economy methodology, which is now used in nearly all engineering work, is relatively recent. This does not mean that, historically, costs were usually overlooked in engineering decisions. However, the perspective that ultimate economy is a primary concern to the engineer and the availability of sound techniques to address this concern differentiate this aspect of modern engineering practice from that of the past.

A pioneer in the field was Arthur M. Wellington,* a civil engineer, who in the latter part of the nineteenth century specifically addressed the role of economic analysis in engineering projects. His particular area of interest was railroad building in the United States. This early work was followed by other contributions in which the emphasis was on techniques that depended primarily on financial and actuarial mathematics. In 1930, Eugene Grant published the first edition of his textbook.† This was a milestone in the development of engineering economy as we know it today. He placed emphasis on developing an economic point of view in engineering, and (as he stated in the preface) "this point of view involves a realization that quite as definite a body of principles governs the economic aspects of an engineering decision as governs its physical aspects." In 1942, Woods and DeGarmo wrote the first edition of this book, later titled *Engineering Economy.*

1.3 What Are the Principles of Engineering Economy?

The development, study, and application of any discipline must begin with a basic foundation. We define the foundation for engineering economy to be a set of principles, or fundamental concepts, that provide a comprehensive doctrine for developing the methodology.‡ These principles will be mastered by students as

---

* A. M. Wellington, *The Economic Theory of Railway Location,* 2nd ed. (New York: John Wiley & Sons, 1887).

† E. L. Grant, *Principles of Engineering Economy* (New York: The Ronald Press Company, 1930).

‡ The definition of the principles of engineering economy varies somewhat with different authors. Examples of other definitions may be found in the following works:

1. E. L. Grant, W. G. Ireson, and R. S. Leavenworth, *Principles of Engineering Economy,* 8th ed. (New York: John Wiley & Sons, 1990).
2. Report titled "Research Planning Conference for Developing a Research Framework for Engineering Economics," Gerald J. Thuesen (editor), Georgia Institute of Technology, March 1986. The report was the result of National Science Foundation Grant MEA-8501237.

they progress through this book. However, in engineering economic analysis, experience has shown that most errors can be traced to some violation of or lack of adherence to the basic principles. Once a problem or need has been clearly defined, the foundation of the discipline can be discussed in terms of seven principles.

> **P R I N C I P L E 1—DEVELOP THE ALTERNATIVES:**
> The choice (decision) is among alternatives. The alternatives need to be identified and then defined for subsequent analysis.

A decision situation involves making a choice among two or more alternatives. Developing and defining the alternatives for detailed evaluation is important because of the resulting impact on the quality of the decision. Engineers and managers should place a high priority on this responsibility. Creativity and innovation are essential to the process.

One alternative that may be feasible in a decision situation is making no change to the current operation or set of conditions (i.e., doing nothing). If you judge this option feasible, make sure it is considered in the analysis. However, do not focus on the status quo to the detriment of innovative or necessary change.

> **P R I N C I P L E 2—FOCUS ON THE DIFFERENCES:**
> Only the differences in expected future outcomes among the alternatives are relevant to their comparison and should be considered in the decision.

If all prospective outcomes of the feasible alternatives were exactly the same, there would be no basis or need for comparison. We would be indifferent among the alternatives and could make a decision using a random selection.

Obviously, only the differences in the future outcomes of the alternatives are important. Outcomes that are common to all alternatives can be disregarded in the comparison and decision. For example, if your feasible housing alternatives were two residences with the same purchase (or rental) price, price would be inconsequential to your final choice. Instead, the decision would depend on other factors, such as location and annual operating and maintenance expenses. This example illustrates, in a simple way, Principle 2, which emphasizes the basic purpose of an engineering economic analysis: to recommend a future course of action based on the differences among feasible alternatives.

> **P R I N C I P L E 3—USE A CONSISTENT VIEWPOINT:**
> The prospective outcomes of the alternatives, economic and other, should be consistently developed from a defined viewpoint (perspective).

The perspective of the decision maker, which is often that of the owners of the firm, would normally be used. However, it is important that the viewpoint for the particular decision be first defined and then used consistently in the description, analysis, and comparison of the alternatives.

As an example, consider a public organization operating for the purpose of developing a river basin, including the generation and wholesale distribution of electricity from dams on the river system. A program is being planned to upgrade and increase the capacity of the power generators at two sites. What perspective should be used in defining the technical alternatives for the program? The "owners of the firm" in this example means the segment of the public that will pay the cost of the program and their viewpoint should be adopted in this situation.

Now let us look at an example where the viewpoint may not be that of the owners of the firm. Suppose that the company in this example is a private firm and that the problem deals with providing a flexible benefits package for the employees. Also, assume that the feasible alternatives for operating the plan all have the same future costs to the company. The alternatives, however, have differences from the perspective of the employees, and their satisfaction is an important decision criterion. The viewpoint for this analysis and decision should be that of the employees of the company as a group, and the feasible alternatives should be defined from their perspective.

> **PRINCIPLE 4—USE A COMMON UNIT OF MEASURE:**
> Using a common unit of measurement to enumerate as many of the prospective outcomes as possible will simplify the analysis and comparison of the alternatives.

It is desirable to make as many prospective outcomes as possible *commensurable* (directly comparable). For economic consequences, a monetary unit such as dollars is the common measure. You should also try to translate other outcomes (which do not initially appear to be economic) into the monetary unit. This translation, of course, will not be feasible with some of the outcomes, but the additional effort toward this goal will enhance commensurability and make the subsequent analysis and comparison of alternatives easier.

What should you do with the outcomes that are not economic (i.e., the expected consequences that cannot be translated (and estimated) using the monetary unit)? First, if possible, quantify the expected future results using an appropriate unit of measurement for each outcome. If this is not feasible for one or more outcomes, describe these consequences explicitly so that the information is useful to the decision maker in the comparison of the alternatives.

> **PRINCIPLE 5—CONSIDER ALL RELEVANT CRITERIA:**
> Selection of a preferred alternative (decision making) requires the use of a criterion (or several criteria). The decision process should consider both the outcomes enumerated in the monetary unit and those expressed in some other unit of measurement or made explicit in a descriptive manner.

The decision maker will normally select the alternative that will best serve the long-term interests of the owners of the organization. In engineering economic analysis, the primary criterion relates to the long-term financial interests of the owners. This is based on the assumption that available capital will be allocated to provide maximum monetary return to the owners. Often, though, there are other organizational objectives you would like to achieve with your decision, and these should be considered and given weight in the selection of an alternative. These nonmonetary attributes and multiple objectives become the basis for additional criteria in the decision-making process.

P R I N C I P L E 6—MAKE UNCERTAINTY EXPLICIT:
Uncertainty is inherent in projecting (or estimating) the future outcomes of the alternatives and should be recognized in their analysis and comparison.

The analysis of the alternatives involves projecting or estimating the future consequences associated with each of them. The magnitude and the impact of future outcomes of any course of action are uncertain. Even if the alternative involves no change from current operations, the probability is high that today's estimates of, for example, future cash receipts and expenses will not be what eventually occurs. Thus, dealing with uncertainty is an important aspect of engineering economic analysis and is the subject of Chapters 10 and 13.

P R I N C I P L E 7—REVISIT YOUR DECISIONS:
Improved decision making results from an adaptive process; to the extent practicable, the initial projected outcomes of the selected alternative should be subsequently compared with actual results achieved.

A good decision-making process can result in a decision that has an undesirable outcome. Other decisions, even though relatively successful, will have results significantly different from the initial estimates of the consequences. Learning from and adapting based on our experience are essential and are indicators of a good organization.

The evaluation of results versus the initial estimate of outcomes for the selected alternative is often considered impracticable or not worth the effort. Too often, no feedback to the decision-making process occurs. Organizational discipline is needed to ensure that implemented decisions are routinely postevaluated and that the results used to improve future analyses of alternatives and the quality of decision making. The percentage of important decisions in an organization that are not postevaluated should be small. For example, a common mistake made in the comparison of alternatives is the failure to examine adequately the impact of uncertainty in the estimates for selected factors on the decision. Only postevaluations will highlight this type of weakness in the engineering economy studies being done in an organization.

▼ 1.4 Engineering Economy and the Design Process

> An engineering economy study is accomplished using a structured procedure and mathematical modeling techniques. The economic results are then used in a decision situation that involves two or more alternatives and normally includes other engineering knowledge and input.

A sound *engineering economic analysis procedure* incorporates the basic principles discussed in Section 1.3 and involves several steps. We represent the procedure, and will discuss it later in this section, in terms of the *seven steps* listed in the left-hand column of Table 1-1. There are several feedback loops (not shown) within the procedure. For example, within Step 1, information developed in evaluating the problem will be used as feedback to refine the problem definition. As another example, information from the analysis of alternatives (Step 5) may indicate the need to change one or more of them or to develop additional alternatives.

The seven-step procedure is also used to assist decision making within the engineering design process, shown as the right-hand column in Table 1-1. In this case, activities in the design process contribute information to related steps in the economic analysis procedure. The general relationship between the activities in the design process and the steps of the economic analysis procedure is indicated in Table 1-1.

TABLE 1-1 The General Relationship between the Engineering Economic Analysis Procedure and the Engineering Design Process

Engineering Economic Analysis Procedure	Engineering Design Process (see Figure P1.15 on p. 22)
Step	*Activity*
1. Problem recognition, definition, and evaluation.	1. Problem/need definition.
	2. Problem/need formulation and evaluation.
2. Development of the feasible alternatives.	3. Synthesis of possible solutions (alternatives).
3. Development of the outcomes and cash flows for each alternative.	
4. Selection of a criterion (or criteria).	4. Analysis, optimization, and evaluation.
5. Analysis and comparison of the alternatives.	
6. Selection of the preferred alternative.	5. Specification of preferred alternative.
7. Performance monitoring and post-evaluation of results.	6. Communication.

Middendorf* states that "engineering design is an iterative, decision making activity whereby scientific and technological information is used to produce a system, device, or process which is different, in some degree, from what the designer knows to have been done before and which is meant to meet human needs." Also, we want to meet the human needs *economically* as emphasized in the definition of engineering in Section 1.1.

The engineering design process may be repeated in phases to accomplish a total design effort. For example, in the first phase, a full cycle of the process may be undertaken to select a conceptual or preliminary design alternative. Then, in the second phase, the activities are repeated to develop the preferred detailed design based on the selected preliminary design. The seven-step economic analysis procedure would be repeated as required to assist decision making in each phase of the total design effort. This procedure is discussed next.

1.4.1 Problem Definition

It is not adequate simply to think about a perplexing question or situation. Rather, a problem must be well understood and stated in an explicit form before the project team proceeds with the rest of the analysis. The first step of the engineering economic analysis procedure (problem definition) is particularly important, since it provides the basis for the rest of the analysis.

The term *problem* is used here generically. It includes all decision situations for which an engineering economy analysis is required. Recognition of the problem is normally stimulated by internal or external organizational needs or requirements. An operating problem within a company (internal need) or a customer expectation about a product or service (external requirement) are examples.

Once the problem is recognized, its formulation should be viewed from a *systems perspective*. That is, the boundary or extent of the situation needs to be carefully defined, thus establishing the elements of the problem and what constitutes its environment.

Evaluation of the problem includes refinement of needs and requirements, and information from the evaluation phase may change the original formulation of the problem. In fact, redefining the problem until a consensus is reached may be the most important part of the problem-solving process!

1.4.2 Development of Alternatives[†]

The two primary actions in Step 2 of the procedure are (1) searching for potential alternatives and (2) screening them to select a smaller group of feasible alterna-

---

*W. H. Middendorf, *Design of Devices and Systems* (New York: Marcel Dekker, Inc., 1986), p. 2.

[†]This is sometimes called *option development*. This important step is described in detail in A. B. Van Gundy, *Techniques of Structured Problem Solving*, 2nd ed. (New York: Van Nostrand Reinhold Co., 1988). For additional reading, see E. Lumsdaine and M. Lumsdaine, *Creative Problem Solving—An Introductory Course for Engineering Students* (New York: McGraw-Hill Book Co., 1990), and J. L. Adams, *Conceptual Blockbusting—A Guide to Better Ideas* (Reading, MA: Addison-Wesley Publishing Co., 1986).

tives for detailed analysis and comparison in Step 5. The term *feasible* here means that each alternative selected for further analysis is judged, based on preliminary evaluation, to meet or exceed the requirements established for the situation.

1.4.2.1 Searching for Superior Alternatives In the discussion of Principle 1 (Section 1.3), creativity and resourcefulness were emphasized as being absolutely essential to the development of potential alternatives. The difference between good alternatives and great alternatives depends largely on an individual's or group's *problem-solving efficiency.* Such efficiency can be increased in the following ways:

1. Concentrate on redefining one problem at a time in Step 1.
2. Develop many redefinitions for the problem.
3. Avoid making judgments as new problem definitions are created.
4. Attempt to redefine a problem in terms that are dramatically different from the original Step 1 problem definition.
5. Make sure that the *true problem* is well researched and understood.

In searching for superior alternatives or identifying the true problem, several limitations invariably exist, including (1) lack of time and money, (2) preconceptions of what will and what will not work, and (3) lack of knowledge. Consequently, the engineer or project team will be working with less-than-perfect problem solutions in the practice of engineering.

EXAMPLE 1-1

The management team of a small furniture-manufacturing company is under pressure to increase profitability in order to get a much-needed loan from the bank to purchase a more modern pattern-cutting machine. One proposed solution is to sell waste wood chips and shavings to a local charcoal manufacturer instead of using them to fuel space heaters for the company's office and factory areas.

(a) Define the company's problem. Next, reformulate the problem in a variety of creative ways.

(b) Develop at least one potential alternative for your reformulated problems in part (a). (Don't concern yourself with feasibility at this point.)

SOLUTION

(a) The company's problem appears to be that revenues are not sufficiently covering costs. Several reformulations can be posed:

1. The problem is to increase revenues while reducing costs.
2. The problem is to maintain revenues while reducing costs.
3. The problem is an accounting system that provides distorted cost information.
4. The problem is that the new machine is really not needed (and hence there is no need for a bank loan).

(b) Based only on reformulation 1, an alternative is to sell wood chips and shavings as long as increased revenue exceeds extra expenses that may be required to heat the buildings. Another alternative is to discontinue the manufacture of specialty items and concentrate on standardized, high-volume products. Yet another alternative is to pool purchasing, accounting, engineering, and other white-collar support services with other small firms in the area by contracting with a local company involved in providing these services.

1.4.2.2 Developing Investment Alternatives "It takes money to make money," as the old saying goes. Did you know that in the United States the average firm spends over $250,000 in capital on each of its employees? So, to make money, each firm must invest capital to support its important human resource—but in what should an individual firm invest? There are usually hundreds of opportunities for a company to make money. Engineers are at the very heart of creating value for a firm by turning innovative and creative ideas into new or reengineered commercial products and services. Most of these ideas require investment of money, and only a few of all feasible ideas can be developed, due to lack of time, knowledge, or resources.

Consequently, most investment alternatives created by good engineering ideas are drawn from a larger population of equally good problem solutions. But how can this larger set of equally good solutions be tapped into? Interestingly, studies have concluded that designers and problem solvers tend to pursue a few ideas that involve "patching and repairing" an old idea.* Truly new ideas are often excluded from consideration! This section outlines two approaches that have found wide acceptance in industry for developing sound investment alternatives by removing some of the barriers to creative thinking: (1) classical brainstorming and (2) the Nominal Group Technique.

(1) Classical Brainstorming. Classical brainstorming is the most well-known and often-used technique for idea generation. It is based on the fundamental principles of *deferment of judgment* and that *quantity breeds quality.* There are four rules for successful brainstorming:

1. Criticism is ruled out.
2. Freewheeling is welcomed.
3. Quantity is wanted.
4. Combination and improvement are sought.

A. F. Osborn lays out a detailed procedure for successful brainstorming.[†] A classical brainstorming session has the following basic steps:

---

*S. Finger and J. R. Dixon, "A Review of Research in Mechanical Engineering Design. Part I: Descriptive, Prescriptive, and Computer-Based Models of Design Processes," *Research in Engineering Design* (New York: Springer-Verlag, 1990).

[†] A. F. Osborn, *Applied Imagination,* 3rd ed. (New York: Charles Scribner's Sons, 1963). Also refer to P. R. Scholtes, B. L. Joiner and B. J. Streibel, *The Team Handbook,* 2nd ed. (Madison, WI: Oriel Inc., 1996).

1. *Preparation.* The participants are selected, and a preliminary statement of the problem is circulated.
2. *Brainstorming.* A warm-up session with simple unrelated problems is conducted, the relevant problem and the four rules of brainstorming are presented, and ideas are generated and recorded using checklists and other techniques if necessary.
3. *Evaluation.* The ideas are evaluated relative to the problem.

Generally, a brainstorming group should consist of four to seven people, although some suggest larger groups.

(2) Nominal Group Technique. The Nominal Group Technique (NGT), developed by Andre P. Delbecq and Andrew H. Van de Ven,* involves a structured group meeting designed to incorporate individual ideas and judgments into a group consensus. By correctly applying the NGT, it is possible for groups of people (preferably, 5 to 10) to generate investment alternatives or other ideas for improving the competitiveness of the firm. Indeed, the technique can be used to obtain group thinking (consensus) on a wide range of topics. For example, a question that might be given to the group is, "What are the most important problems or opportunities for improvement of. . . ?"

The technique, when properly applied, draws on the creativity of the individual participants, while reducing two undesirable effects of most group meetings: (a) the dominance of one or more participants and (b) the suppression of conflicting ideas. The basic format of an NGT session is as follows:

1. Individual silent generation of ideas
2. Individual round-robin feedback and recording of ideas
3. Group clarification of each idea
4. Individual voting and ranking to prioritize ideas
5. Discussion of group consensus results

The NGT session begins with an explanation of the procedure and a statement of question(s), preferably written by the facilitator.† The group members are then asked to prepare individual listings of alternatives, such as investment ideas or issues that they feel are crucial for the survival and health of the organization. This is known as the silent-generation phase and usually takes only a few minutes to "get the thoughts rolling." After this phase has been completed, the facilitator calls on each participant, in round-robin fashion, to present one idea from his or her list (or further thoughts as the round-robin session is proceeding). Each idea (or opportunity) is then identified in turn and recorded on a flip chart or board by the NGT facilitator, leaving ample space between ideas for comments or clarification. This process continues until all the opportunities have been recorded, clarified, and

---

* A. Van de Ven and A. Delbecq, "The Effectiveness of Nominal, Delphi, and Interactive Group Decision Making Processes," *Academy of Management Journal,* vol. 17, no. 4, December 1974, pp. 605–621.

† A good example of the NGT is given in D. S. Sink, "Using the Nominal Group Technique Effectively," *National Productivity Review,* Spring 1983, pp. 173–184.

displayed for all to see. At this point a voting procedure is used to prioritize the ideas or opportunities. Finally, voting results lead to the development of group consensus on the topic being addressed.

1.4.3 Development of Prospective Outcomes

Step 3 of the engineering economic analysis procedure incorporates Principles 2, 3, and 4 from Section 1.3 and uses the basic *cash-flow approach* employed in engineering economy. A cash flow occurs when money is transferred from one organization or individual to another. Thus, a cash flow represents the economic effects of an alternative in terms of money spent and received.

Consider the concept of an organization having only one "window" to its external environment through which *all* monetary transactions occur—receipts of revenues and payments to suppliers, creditors, and employees. The key to developing the related cash flows for an alternative is estimating what would happen to the revenues and costs, as seen at this window, if the particular alternative were implemented. The *net cash flow* for an alternative is the difference between all cash inflows (receipts or savings) and cash outflows (costs or expenses) during each time period.

In addition to the economic aspects of decision making, *nonmonetary factors* (*attributes*) often play a significant role in the final recommendation. Examples of objectives other than profit maximization or cost minimization that can be important to an organization include the following:

1. Meeting or exceeding customer expectations
2. Safety
3. Improving employee satisfaction
4. Maintaining production flexibility to meet changing demands
5. Meeting or exceeding all environmental requirements
6. Achieving good public relations or being an exemplary member of the community

1.4.4 Selection of a Decision Criterion

The selection of a decision criterion (Step 4 of the analysis procedure) incorporates Principle 5. The decision maker will normally select the alternative that will best serve the long-term interests of the owners of the organization. It is also true that the economic decision criterion should reflect a consistent and proper viewpoint (Principle 3) to be maintained throughout an engineering economy study.

1.4.5 Analysis and Comparison of Alternatives

Analysis of the economic aspects of an engineering problem (Step 5) is largely based on cash-flow estimates for the feasible alternatives selected for detailed study. A substantial effort is normally required to obtain reasonable accurate forecasts of cash flows and other factors in view of, for example, inflationary (or deflationary) pressures, exchange rate movements, and regulatory (legal) mandates that often occur. Clearly, the consideration of future uncertainties (Principle 6) is an essential part of an engineering economy study. When cash flow and other required estimates are eventually determined, alternatives can be compared based on their differences as called for by Principle 2. Usually, these differences will be quantified in terms of a monetary unit such as dollars.

Companion Web Site (http://www.prenhall.com/sullivan_engineering/): The disposal of plastic bags associated with curbside Autumn leaf collection presents an environmental problem in many urban communities. Visit the Web site to view the issues associated with the economic alternative of using biodegradeable bags.

1.4.6 Selection of the Preferred Alternative

When the first five steps of the engineering economic analysis procedure have been done properly, the preferred alternative (Step 6) is simply a result of the total effort. Thus, the soundness of the technical-economic modeling and analysis techniques dictates the quality of the results obtained and the recommended course of action. Step 6 is included in Activity 5 of the engineering design process (specification of the preferred alternative) when done as part of a design effort.

1.4.7 Performance Monitoring and Postevaluation of Results

This final step implements Principle 7 and is accomplished while and after the results achieved from the selected alternative are collected. Monitoring project performance during its operational phase improves the achievement of related goals and objectives and reduces the variability in desired results. Step 7 is also the follow-up step to a previous analysis, comparing actual results achieved with the previously estimated outcomes. The aim is to learn how to do better analyses, and the feedback from postimplementation evaluation is important to the continuing improvement of operations in any organization. Unfortunately, like Step 1, this final step is often not done consistently or well in engineering practice; therefore, it needs particular attention to ensure feedback for use in ongoing and subsequent studies.

EXAMPLE 1-2

Bad news: You have just wrecked your car! You need another car immediately because you have decided that walking, riding a bike, and taking a bus are not acceptable. An automobile wholesaler offers you $2,000 for your wrecked car "as is." Also, your insurance company's claims adjuster estimates that there is $2,000 in damages to your car. Because you have collision insurance with a $1,000 deductibility provision, the insurance company mails you a check for $1,000. The odometer reading on your wrecked car is 58,000 miles.

What should you do? Use the seven-step procedure from Table 1-1 to analyze your situation. Also, identify which principles accompany each step.

SOLUTION

STEP 1—Define the Problem

Your basic problem is that you need transportation. Further evaluation leads to the elimination of walking, riding a bicycle, and taking a bus as feasible alternatives.

STEP 2—Develop Your Alternatives (Principle 1 is used here.)

Your problem has been reduced to either replacing or repairing your automobile. The alternatives would appear to be

1. Sell the wrecked car for $2,000 to the wholesaler and spend this money, the $1,000 insurance check, and all of your $7,000 savings account on a newer car. The total amount paid out of your savings account is $7,000, and the car will have 28,000 miles of prior use.
2. Spend the $1,000 insurance check and $1,000 of savings to fix the car. The total amount paid out of your savings is $1,000, and the car will have 58,000 miles of prior use.
3. Spend the $1,000 insurance check and $1,000 of your savings to fix the car and then sell the car for $4,500. Spend the $4,500 plus $5,500 of additional savings to buy the newer car. The total amount paid out of savings is $6,500, and the car will have 28,000 miles.
4. Give the car to a part-time mechanic, who will repair it for $1,100 ($1,000 insurance and $100 of your savings), but will take an additional month of repair time. You will also have to rent a car for that time at $400/month (paid out of savings). The total amount paid out of savings is $500, and the car will have 58,000 miles on the odometer.
5. Same as Alternative 4, but you then sell the car for $4,500 and use this money plus $5,500 of additional savings to buy the newer car. The total amount paid out of savings is $6,000, and the newer car will have 28,000 miles of prior use.

ASSUMPTIONS:

1. The less reliable repair shop in Alternatives 4 and 5 will not take longer than an additional month to repair the car.
2. Each car will perform at a satisfactory operating condition (as it was originally intended) and will provide the same total mileage before being sold or salvaged.
3. Interest earned on money remaining in savings is negligible.

STEP 3—Estimate the Cash Flows for Each Alternative (Principle 2 should be adhered to in this step.)

1. Alternative 1 varies from all others because the car is not to be repaired at all but merely sold. This eliminates the benefit of the $500 increase in the value of the car when it is repaired and then sold. Also this alternative leaves no money in your savings account. There is a cash flow of $-\$8,000$ to gain a newer car valued at $10,000.

2. Alternative 2 varies from Alternative 1 because it allows the old car to be repaired. Alternative 2 differs from Alternatives 4 and 5 because it utilizes a more expensive ($500 more) and less risky repair facility. It also varies from Alternatives 3 and 5 because the car will be kept. The cash flow is −$2,000 and the repaired car can be sold for $4,500.

3. Alternative 3 gains an additional $500 by repairing the car and selling it to buy the same car as in Alternative 1. The cash flow is −$7,500 to gain the newer car valued at $10,000.

4. Alternative 4 uses the same idea as Alternative 2, but involves a less expensive repair shop. The repair shop is more risky in the quality of its end product, but will only cost $1,100 in repairs and $400 in an additional month's rental of a car. The cash flow is −$1,500 to keep the older car valued at $4,500.

5. Alternative 5 is the same as Alternative 4, but gains an additional $500 by selling the repaired car and purchasing a newer car as in Alternatives 1 and 3. The cash flow is −$7,000 to obtain the newer car valued at $10,000.

STEP 4—Select a Criterion

It is very important to use a consistent viewpoint (Principle 3) and a common unit of measure (Principle 4) in performing this step. The viewpoint in this situation is yours (the owner of the wrecked car).

The value of the car to the owner is its market value (i.e., $10,000 for the newer car and $4,500 for the repaired car). Hence, the dollar is used as the consistent value against which everything is measured. This reduces all decisions to a quantitative level, which can then be reviewed later with qualitative factors that may carry their own dollar value (e.g., how much is low mileage or a reliable repair shop worth?).

STEP 5—Analyze and Compare the Alternatives

Make sure you consider all relevant criteria (Principle 5).

1. Alternative 1 is eliminated, because Alternative 3 gains the same end result and would also provide the car owner with $500 more cash. This is experienced with no change in the risk to the owner. (Car value = $10,000, savings = 0, total worth = $10,000.)

2. Alternative 2 is a good alternative to consider, because it spends the least amount of cash, leaving $6,000 in the bank. Alternative 2 provides the same end result as Alternative 4, but costs $500 more to repair. Therefore, Alternative 2 is eliminated. (Car value = $4,500, savings = $6,000, total worth = $10,500.)

3. Alternative 3 is eliminated, because Alternative 5 also repairs the car but at a lower out-of-savings cost ($500 difference), and both Alternatives 3 and 5 have the same end result of buying the newer car. (Car value = $10,000, savings = $500, total worth = $10,500.)

4. Alternative 4 is a good alternative, because it saves $500 by using a cheaper repair facility, provided that the risk of a poor repair job is judged to be small. (Car value = $4,500, savings = $6,500, total worth = $11,000.)

5. Alternative 5 repairs the car at a lower cost ($500 cheaper) and eliminates the risk of breakdown by selling the car to someone else at an additional $500 gain. (Car value = $10,000, savings = $1,000, total worth = $11,000.)

STEP 6—Select the Best Alternative

When performing this step of the procedure, you should make uncertainty explicit (Principle 6). Among the uncertainties that can be found in this problem, the following are the most relevant to the decision. If the original car is repaired and kept, there is a possibility that it would have a higher frequency of breakdowns (based on personal experience). If a cheaper repair facility is used, the chance of a later breakdown is even greater (based on personal experience). Buying a newer car will use up most of your savings. Also, the newer car purchased may be too expensive, based on the additional price paid (which is at least $6,000/30,000 miles = 20 cents per mile). Finally, the newer car may also have been in an accident and could have a worse repair history than the presently owned car.

Based on the information in all previous steps, *Alternative 5* was actually chosen.

STEP 7—Monitor the Performance of Your Choice

This step goes hand-in-hand with Principle 7 (revisit your decisions). The newer car turned out after being "test driven" for 20,000 miles to be a real beauty. Mileage was great, and no repairs were needed. The systematic process of identifying and analyzing alternative solutions to this problem really paid off!

1.5 Accounting and Engineering Economy Studies

In Section 1.1, we emphasized that engineers and managers use the principles and methodology of engineering economy to assist decision making. Thus, engineering economy studies provide information on which current decisions pertaining to the *future* operation of an organization can be based.

After a decision to invest capital in a project has been made and the money has been invested, those who supply and manage the capital want to know the financial results. Therefore, accounting procedures are established so that financial events relating to the investment can be recorded and summarized and *financial performance* determined. At the same time, through the use of proper financial information, controls can be established and utilized to aid in guiding the operation toward the desired financial goals.

General accounting and *cost accounting* are the procedures that provide these necessary services in a business organization. Thus, accounting data are primarily concerned with *past* and *current* financial events, even though such data are often used to make projections about the future.

General accounting is a source of much of the past financial data needed for estimating future financial conditions. Accounting is also a source of data for analyses of how well the results of a capital investment turned out compared with the results that were predicted in the engineering economic analysis.

Cost accounting, or management accounting, is a subset of accounting that is of particular importance because it is concerned principally with decision making and control in a firm. Consequently, it is the source of some of the cost data that are needed in engineering economy studies. Modern cost accounting may satisfy any or all of the following objectives:

1. To determine the cost of products or services
2. To provide a rational basis for pricing goods or services
3. To provide a means for controlling expenditures
4. To provide information on which operating decisions may be based and the results evaluated

Although the basic objectives of cost accounting are simple, the exact determination of costs usually is not. As a result, some of the procedures used are arbitrary conventions that make it possible to obtain reasonably accurate costs in some situations, but in many others the information is too aggregated and distorted to be relevant in managerial planning and control decisions.

Some of the inaccuracies of traditional cost accounting techniques have been remedied by a relatively recent methodology known as *activity-based accounting*. This methodology is aimed at producing more accurate and timely cost information primarily by (1) carefully tracing overhead to its causal activities and (2) assigning technology costs equitably over the entire product life cycle. Because overhead and technology account for as much as 60 percent of total product cost in many industries, improved cost reporting and control are made possible by being able to trace these two major cost components to the activities, and subsequently products, that truly create them.

An adequate understanding of the origins and meaning of accounting data is necessary to be able to interpret those data for use in engineering economy studies. Thus, a brief discussion of accounting, including activity-based accounting, is provided in Appendix A.

1.6 Overview of the Book

The contents of this book have been organized in three parts, with the chapters in a logical sequence for both teaching and applying the principles and methodology of engineering economy. The three parts of the book, and the chapters in each part, are

1. Part I: Fundamentals of Engineering Economy (Chapters 1–3)
2. Part II: Basic Topics in Engineering Economy (Chapters 4–10)
3. Part III: Additional Topics in Engineering Economy (Chapters 11–15)

In this first chapter, we presented the fundamental concepts of engineering economy in terms of seven basic principles. We next discussed the steps involved in an engineering economic analysis and then related the analysis procedure to the engineering design process. The interface between accounting and engineering economy was also discussed. Thus, in Chapter 1 the basic foundation of the subject has been established.

In Chapter 2, selected cost concepts important in engineering economy studies are presented. Particular emphasis is placed on the economic principles of engineering design. The application of life cycle cost concepts is also discussed, including break-even analysis and present economy studies.

Chapter 3 concentrates on the concepts of money–time relationships and economic equivalence. Specifically, we consider the time value of money in evaluating the future revenues and costs associated with alternative uses of money. Then, in

Chapter 4, the methods commonly utilized to analyze the economic consequences and profitability of an alternative are demonstrated. These methods, and their proper use in the comparison of alternatives, are primary subjects of Chapter 5, which also includes a discussion of the appropriate time period for an analysis. Thus, Chapters 3, 4, and 5 together develop an essential part of the methodology needed for understanding the remainder of the book and for performing engineering economy studies on a before-tax basis.

In Chapter 6, the additional techniques required to accomplish engineering economy studies on an after-tax basis are explained. In the private sector, most engineering economy studies are done on an after-tax basis. Therefore, Chapter 6 adds to the basic part of the methodology developed in Chapters 3, 4, and 5. A part of Chapter 6 is concerned with depreciation under the Modified Accelerated Cost Recovery System authorized by the Tax Reform Act of 1986. However, techniques applicable to assets acquired prior to the effective date of the act are also included. Similarly, the remainder of Chapter 6 deals with performing after-tax analyses.

Chapter 7 considers the critical question of how to develop estimates of future consequences associated with each feasible alternative. The process associated with this step in an engineering economic analysis constitutes a crucial aspect of application and practice. Cost estimation is placed in Chapter 7, instead of earlier in the book, so that the basic methodology of comparing alternatives on both a before-tax and an after-tax basis can be discussed in an integrated manner. Therefore, the development of estimated revenues, costs, and other information for an alternative is given concentrated attention in Chapter 7.

The effects of inflation (or deflation), price changes, and exchange rates are the topic of Chapter 8. The concepts for handling price changes and exchange rates in an engineering economy study are discussed both comprehensively and pragmatically from an application viewpoint.

Often, an organization must analyze whether existing assets should be continued in service or replaced with new assets to meet current and future operating needs. In Chapter 9, techniques for addressing this question are developed and presented. Because the replacement of assets demands significant capital, decisions made in this area are important and require special attention.

Concern over uncertainty and risk is a reality in engineering practice. In Chapter 10, the impact of potential variation between the estimated economic outcomes of an alternative and the results that may occur is considered. Various nonprobabilistic techniques for analyzing the consequences of uncertainty in future estimates of revenues and costs are discussed and illustrated.

In Part III, Chapter 11 is dedicated to the analysis of public projects with the benefit–cost-ratio method of comparison. This widely used method of evaluating alternatives was motivated through the Flood Control Act passed by the U.S. Congress in 1936.

Privately owned, regulated public utilities are an important part of the U.S. economy. In Chapter 12, the unique characteristics of these firms, and the revenue requirements method of accomplishing engineering economy studies related to their operations, are discussed. In Chapter 13, probabilistic techniques for analyzing the consequences of uncertainty in future cash-flow estimates and other factors are explained. Discrete and continuous probability concepts, as well as Monte Carlo simulation techniques, are included in Chapter 13.

Chapter 14 is concerned with the proper identification and analysis of all projects and other needs for capital within an organization. Accordingly, the capital financing and capital allocation process to meet these needs is addressed. This process is crucial to the welfare of an organization, since it affects most operating outcomes, whether in terms of current product quality and service effectiveness or long-term capability to compete in the world marketplace. Finally, Chapter 15 discusses many time-tested methods for including nonmonetary attributes in engineering economy studies.

1.7 Problems

The number(s) in color at the end of a problem refer to the section(s) in that chapter most closely related to the problem.

1-1. List 10 typical situations in the operation of an organization where an engineering economic analysis would significantly assist decision making. You may assume a specific type of organization (e.g., manufacturing firm, medical health center, transportation company, government agency) if it will assist in the development of your answer (state any assumptions). (1.1)

1-2. Explain why the subject of engineering economy is important to the practicing engineer. (1.1–1.4)

1-3. Assume that your employer is a manufacturing firm that produces several different electronic consumer products. What are five nonmonetary factors (attributes) that may be important when a significant change is considered in the design of the current best-selling product? (1.3, 1.4)

1-4. Will the increased use of automation increase the importance of engineering economy studies? Why or why not?

1-5. Explain the meaning of the statement, "The choice (decision) is among alternatives." (1.3)

1-6. Describe the outcomes that should be expected from a feasible alternative. What are the differences between potential alternatives and feasible alternatives? (1.4)

1-7. Define uncertainty. What are some of the basic causes of uncertainty in engineering economy studies? (1.3)

1-8. You have discussed with a co-worker in the engineering department the importance of explicitly defining the viewpoint (perspective) from which future outcomes of a course of action being analyzed are to be developed. Explain what you mean by a viewpoint or perspective. (1.3)

1-9. Two years ago, you were a member of the project team that analyzed whether your company should upgrade some building, equipment, and related facilities to support the expanding operation of the company. The project team analyzed three feasible alternatives, one of which makes no changes in facilities and the remaining two involve significant facility changes. Now you have been selected to lead a postevaluation team. Delineate your technical plan for comparing the estimated consequences (developed two years ago) of implementing the selected alternative with the results that have been achieved. (1.3, 1.4)

1-10. Describe how it might be feasible in an engineering economic analysis to consider the following different situations in terms of the monetary unit: (1.3)

  a. A piece of equipment that is being considered as a replacement for an existing item has greater reliability; that is, the Mean Time Between Failures (MTBF) during operation of the new equipment has been increased 40% in comparison with the present item.

  b. A company manufactures wrought iron patio furniture for the home market. Some changes in material and metal treatment, which involve increased manufacturing costs, are being considered to reduce the rusting problem significantly.

  c. A large foundry operation has been in the same location in a metropolitan area for the past 35 years. Even though it is in compliance with current air pollution regulations, the continuing residential and commercial development of that area is causing an increasing expectation on the part of local residents for improved environmental control by the foundry. The company considers community relations to be important.

1-11. Explain the relationship between engineering economic analysis and engineering design. How does economic analysis assist decision making in the design process? (1.4)

1-12. During your first month as an employee at Greenfield Industries (a large drill-bit manufacturer), you are asked to evaluate alternatives for producing a newly designed drill bit on a turning machine. Your boss' memorandum to you has practically no information about what the alternatives are and what criteria should be used. The same task was posed to a previous employee who could not finish the analysis, but she has given you the following information: An old turning machine valued at $350,000 exists (in the warehouse) that can be modified for the new drill bit. The in-house technicians have given an estimate of $40,000 to modify this machine, and they assure you that they will have the machine ready before the projected start date (although they have never done any modifications of this type). It is hoped that the old turning machine will be able to meet production requirements at full capacity. An outside company, McDonald Inc., made the machine seven years ago and can easily do the same modifications for $60,000. The cooling system used for this machine is not environmentally safe and would require some disposal costs. McDonald Inc. has offered to build a new turning machine with more environmental safeguards and higher capacity for a price of $450,000. McDonald Inc. has promised this machine before the startup date and is willing to pay any late costs. Your company has $100,000 set aside for the startup of the new product line of drill bits.
For this situation,

  a. Define the problem.
  b. List key assumptions.
  c. List alternatives facing Greenfield Industries.
  d. Select a criterion for evaluation of alternatives.
  e. Introduce risk into this situation.

  f. Discuss how nonmonetary considerations may impact the selection.
  g. Describe how a post-audit could be performed.

1-13. *The Almost-Graduating Senior Problem.* Consider this situation faced by many a first-semester senior in civil engineering who is exhausted from extensive job interviewing and penniless from excessive partying. Mary's impulse is to accept immediately a highly attractive job offer to work in her brother's successful manufacturing company. She would then be able to relax for a year or two, save some money, and then return to college to complete her senior year and graduate. Mary is cautious about this impulsive desire, because it may lead to no college degree at all!

  a. Develop at least two formulations for Mary's problem.
  b. Identify feasible solutions for each problem formulation in part (a). *Be creative!*

1-14. While studying for the Engineering Economy final exam, you and two friends find yourselves craving a fresh pizza. You can't spare the time to pick up the pizza and must have it delivered. "Pick-Up-Sticks" offers a 1-1/4-inch-thick (including toppings), 20-inch square pizza with your choice of two toppings for $15 plus 5% sales tax and a $1.50 delivery charge (no sales tax on delivery charge). "Fred's" offers the round, deep-dish Sasquatch which is 20 inches in diameter. It is 1-3/4 inches thick, which includes two toppings, and costs $17.25 plus 5% sales tax and free delivery.

  a. What is the problem in this situation? Please state it in an explicit and precise manner.
  b. Systematically apply the seven principles of engineering economy (pp. 5–7) to the problem you have defined in Part (a).
  c. Assuming that your common unit of measure is $ (i.e., cost), what is the better value for getting a pizza based on the criterion of *minimizing cost per unit of volume*?
  d. What other criteria might be used to select which pizza to purchase?

1-15. Storm doors have been installed on 50% of all homes in Anytown, USA. The remaining 50% of homeowners without storm doors think they may have a problem that a storm door could solve, but they're not sure. Use activities 1, 2, and 3 in the engineering design process (Table

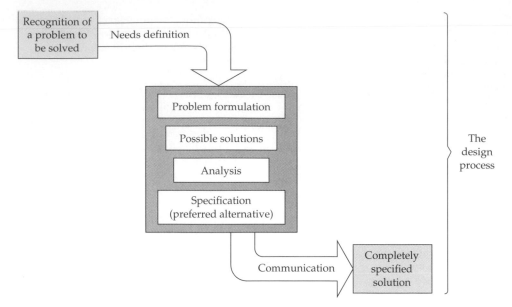

Figure P1-15 Figure for Problem 1-15

1-1) to help these homeowners systematically think through the definition of their need (activity 1), a formal statement of their problem (activity 2), and the generation of alternatives (activity 3).

The design process begins in Figure P1.15 with a statement of need and terminates with the specifications for a means of fulfilling that need.

1-16. *Brain Teaser.* A friend of yours bought a small apartment building for $100,000 in a college town. She spent $10,000 of her own money for the building and obtained a mortgage from a local bank for the remaining $90,000. The *annual* mortgage payment to the bank is $10,500. Your friend also expects that annual maintenance on the building and grounds will be $15,000. There are four apartments (two bedrooms each) in the building that can each be rented for $360 per month.

Refer to the seven-step procedure in Table 1-1 (left-hand side) to answer these questions:

a. Does your friend have a problem? If so, what is it?

b. What are her alternatives (identify at least three)?

c. Estimate the economic consequences and other required data for the alternatives in Part (b).

d. Select a criterion for discriminating among alternatives, and use it to advise your friend on which course of action to pursue.

e. Attempt to analyze and compare the alternatives in view of at least one criterion in addition to cost.

f. What should your friend do based on the information you and she have generated?

g. Develop a plan for your friend to follow in evaluating (after the fact) how good her actual choice was. She may *not* have followed your advice. Be creative in Part (g).

1-17. *In-Class Team Exercise.* Divide your class into groups of four persons each. Spend 15 minutes brainstorming *ethical* matters that may arise during an engineering economy study. Have each group present a two-minute summary of their findings to the entire class.

Cost Concepts and Design Economics

The objectives of Chapter 2 are to (1) describe some of the basic cost terminology and concepts that are used throughout this book and (2) illustrate how they should be used in engineering economic analysis and decision making.

The following topics are discussed in this chapter:

Cost estimating

Fixed, variable, and incremental costs

Recurring and nonrecurring costs

Direct, indirect, and overhead costs

Cash cost versus book cost

Sunk costs and opportunity costs

Life-cycle cost

The general economic environment

The relationship between price and demand

The total revenue function

Breakeven point relationships

Maximizing profit/minimizing cost

Cost-driven design optimization

Present economy studies

▼ 2.1 Introduction

> Designing to meet economic needs and achieving competitive operations in private-
> and public-sector organizations depends on prudently balancing what is technically
> feasible and what is economically acceptable. Unfortunately, there is no shortcut
> method available to reach this balance between technical and economic feasibility.
> Thus, methods of engineering economic analysis should be used to provide results
> that will help to attain an acceptable balance.

The word *cost* (or *expense*) has meanings that vary in usage.* The *concepts* and
other economic principles used in an engineering economy study depend on the
problem situation and on the decision to be made. Consequently, the content of
Chapter 2, which integrates cost concepts, the principles of engineering economy,
and design considerations, is important preparation for the applications covered
in subsequent chapters of the book.

▼ 2.2 Cost Estimating and Cost Terminology

Often, the most difficult, expensive, and time-consuming part of an engineering
economy study is the estimation of costs, revenues, useful lives, residual values,
and other data pertaining to the design of alternatives being analyzed. In this
section, we briefly introduce the role of *cost estimating* in engineering practice. (The
interested reader is referred to Chapter 7 for a more detailed discussion.) Also, we
provide definitions and examples of important cost concepts and reemphasize the
economic perspective in engineering design.

2.2.1 Cost Estimating

The term "cost estimating" is frequently used to describe the process by which
the present and future cost consequences of engineering designs are forecast. A
primary difficulty in estimating for economic analyses is that most prospective
projects are relatively unique; that is, substantially similar design efforts have not
been previously undertaken to meet the same functional requirements and economic constraints. Hence, accurate past data that can be used in estimating costs
and benefits directly, without substantial modification, often do not exist. It may
be possible, however, to develop data on certain past design outcomes that are
related to the outcomes being estimated and to adjust that data based on design
requirements and expected future conditions.

Whenever an engineering economic analysis is performed for a major capital
investment, the cost estimating effort for that analysis should be an integral part of
a comprehensive planning and design process requiring the active participation of
not only engineering designers but also personnel from marketing, manufacturing,
finance, and top management. Results of cost estimating are used for a variety of
purposes, including

---

*For purposes of this book, the words *cost* and *expense* are used interchangeably.

1. Providing information used in setting a selling price for quoting, bidding, or evaluating contracts,

2. Determining whether a proposed product can be made and distributed at a profit (for simplicity, price = cost + profit),

3. Evaluating how much capital can be justified for process changes or other improvements, and

4. Establishing benchmarks for productivity improvement programs.

There are two fundamental approaches to cost estimating: the "top-down" approach and the "bottom-up" approach. The top-down approach basically uses historical data from similar engineering projects to estimate the costs, revenues, and other data for the current project by modifying these data for changes in inflation or deflation, activity level, weight, energy consumption, size, and other factors. This approach is best used early in the estimating process when alternatives are still being developed and refined.

The bottom-up approach is a more detailed method of cost estimating. This method attempts to break down a project into small, manageable units and to estimate their economic consequences. These smaller unit costs are added together with other types of costs to obtain an overall cost estimate. This approach usually works best when the detail concerning the desired output (a product or a service) has been defined and clarified.

EXAMPLE 2-1

A simple example of cost estimating is to forecast the expense of getting a Bachelor of Science (B.S.) from the university you are attending. In our solution, we outline the two basic approaches just discussed for estimating these costs.

SOLUTION

A top-down approach would take the published cost of a four-year degree at the same (or a similar) university and adjust it for inflation and extraordinary items that an incoming student might encounter such as fraternity/sorority membership, scholarships, and tutoring. For example, suppose the published cost of attending your university is $15,750 for the current year. This figure is anticipated to increase at the rate of 6 percent per year and includes full-time tuition and fees, university housing, and a weekly meal plan. Not included are the costs of books, supplies, and other personal expenses. For our initial estimate, these "other" expenses are assumed to remain constant at $5,000 per year.

The total estimated cost for four years can now be computed. We simply need to adjust the published cost for inflation each year and add in the cost of "other" expenses.

Year	Tuition, Fees, Room and Board	"Other" Expenses	Total Estimated Cost for Year
1	$15,750 × 1.06 = $16,995	$5,000	$21,695
2	16,695 × 1.06 = 17,697	5,000	22,697
3	17,697 × 1.06 = 18,759	5,000	23,759
4	18,759 × 1.06 = 19,885	5,000	24,885
		Grand Total	$93,036

In contrast with the top-down approach, a bottom-up approach to the same cost estimate would be to first break down anticipated expenses into the typical categories shown in Figure 2-1 for each of the four years at the university. Tuition and fees can be estimated fairly accurately in each year, as can books and supplies. For example, suppose that the average cost of a college textbook is $80. You can estimate your annual textbook cost by simply multiplying the average cost per book by the number of courses you plan to take. Assume that you plan on taking five courses each semester during the first year. Your estimated textbook costs would be

$$\left(\frac{5 \text{ courses}}{\text{semester}}\right)(2 \text{ semesters})\left(\frac{1 \text{ book}}{\text{course}}\right)\left(\frac{\$80}{\text{book}}\right) = \$800.$$

The other two categories, living expenses and transportation, are probably more dependent on your lifestyle. For example, whether you own and operate an

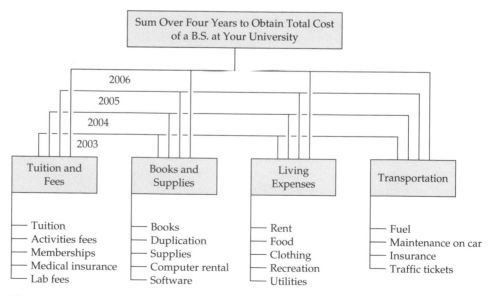

Figure 2-1 **Bottom-Up Approach to Determining the Cost of a College Education**

automobile and live in a "high-end" apartment off-campus can dramatically affect the estimated expenses during your college years. Cost-estimating procedures and techniques are discussed more fully in Chapter 7.

2.2.2 Fixed, Variable, and Incremental Costs

Fixed costs are those unaffected by changes in activity level over a feasible range of operations for the capacity or capability available. Typical fixed costs include insurance and taxes on facilities, general management and administrative salaries, license fees, and interest costs on borrowed capital.

Of course, any cost is subject to change, but fixed costs tend to remain constant over a specific range of operating conditions. When large changes in usage of resources occur, or when plant expansion or shutdown is involved, fixed costs will be affected.

Variable costs are those associated with an operation that vary in total with the quantity of output or other measures of activity level. If you were making an engineering economic analysis of a proposed change to an existing operation, the variable costs would be the primary part of the prospective differences between the present and changed operations as long as the range of activities is not significantly changed. For example, the costs of material and labor used in a product or service are variable costs, because they vary in total with the number of output units, even though the costs per unit stay the same.

An *incremental cost* (or *incremental revenue*) is the additional cost (or revenue) that results from increasing the output of a system by one (or more) units. Incremental cost is often associated with "go–no go" decisions that involve a limited change in output or activity level. For instance, the incremental cost per mile for driving an automobile may be $0.27, but this cost depends on considerations such as total mileage driven during the year (normal operating range), mileage expected for the next major trip, and the age of the automobile. Also, it is common to read of the "incremental cost of producing a barrel of oil" and "incremental cost to the state for educating a student." As these examples indicate, the incremental cost (or revenue) is often quite difficult to determine in practice.

EXAMPLE 2-2

In connection with surfacing a new highway, a contractor has a choice of two sites on which to set up the asphalt mixing plant equipment. The contractor estimates that it will cost $1.15 per cubic yard per mile (yd^3-mile) to haul the asphalt paving material from the mixing plant to the job location. Factors relating to the two mixing sites are as follows (production costs at each site are the same):

Cost Factor	Site A	Site B
Average hauling distance	6 miles	4.3 miles
Monthly rental of site	$1,000	$5,000
Cost to set up and remove equipment	$15,000	$25,000
Hauling expense	$1.15/yd^3-mile	$1.15/yd^3-mile
Flagperson	not required	$96/day

The job requires 50,000 cubic yards of mixed-asphalt-paving material. It is estimated that four months (17 weeks of five working days per week) will be required for the job. Compare the two sites in terms of their fixed, variable, and total costs. Assume that the cost of the return trip is negligible. Which is the better site? For the selected site, how many cubic yards of paving material does the contractor have to deliver before starting to make a profit if paid $8.05 per cubic yard delivered to the job location?

SOLUTION

The fixed and variable costs for this job are indicated in the table shown next. Site rental, setup, and removal costs (and the cost of the flagperson at Site B) would be constant for the total job, but the hauling cost would vary in total amount with the distance and thus with the total output quantity of yd^3-miles.

Cost	Fixed	Variable	Site A		Site B	
Rent	X			= $ 4,000		= $ 20,000
Setup/removal	X			= 15,000		= 25,000
Flagperson	X			= 0	5(17)($96) =	8,160
Hauling		X	6(50,000)($1.15) =	345,000	4.3(50,000)($1.15) =	247,250
			Total:	$364,000		= $300,410

Thus, Site B, which has the larger fixed costs, has the smaller total cost for the job. Note that the extra fixed costs of Site B are being "traded off" for reduced variable costs at this site.

The contractor will begin to make a profit at the point where total revenue equals total cost as a function of the cubic yards of asphalt pavement mix delivered. Based on Site B, we have

$$4.3(\$1.15) = \$4.945 \text{ in variable cost per yd}^3 \text{ delivered}$$

$$\text{Total cost} = \text{total revenue}$$

$$\$53,160 + \$4.945x = \$8.05x$$

$$x = 17,121 \text{ yd}^3 \text{ delivered.}$$

Therefore, by using Site B, the contractor will begin to make a profit on the job after delivering 17,121 cubic yards of material.

EXAMPLE 2-3

Four college students who live in the same geographical area intend to go home for Christmas vacation (a distance of 400 miles each way). One of the students has an automobile and agrees to take the other three if they will pay the cost of operating the automobile for the trip. When they return from the trip, the owner presents each of them with a bill for $102.40, stating that she has kept careful records of the cost of operating the car and that, based on an annual average of 15,000 miles, their cost per mile is $0.384. The three others feel that the charge is too high and ask to see the cost figures on which it is based. The owner shows them the following list:

Cost Element	Cost per Mile
Gasoline	$0.120
Oil and lubrication	0.021
Tires	0.027
Depreciation	0.150
Insurance and taxes	0.024
Repairs	0.030
Garage	0.012
Total	$0.384

The three riders, after reflecting on the situation, form the opinion that only the costs for gasoline, oil and lubrication, tires, and repairs are a function of mileage driven (variable costs) and thus could be caused by the trip. Because these four costs total only $0.198 per mile, and thus total $158.40 for the 800-mile trip, the share for each student would be $158.40/3 = $52.80. Obviously, the opposing views are substantially different. Which, if either, is correct? What are the consequences of the two different viewpoints in this matter, and what should the decision-making criterion be?

SOLUTION

In this instance, assume that the owner of the automobile agreed to accept $52.80 per person from the three riders, based on the variable costs that were purely incremental for the Christmas trip versus the owner's average annual mileage. That is, the $52.80 per person is the "with a trip" cost relative to the "without" alternative.

Now, what would the situation be if the three students, because of the low cost, returned and proposed another 800-mile trip the following weekend? And what if there were several more such trips on subsequent weekends? Quite clearly, what started out to be a small marginal (and temporary) change in operating conditions—from 15,000 miles per year to 15,800 miles—soon would become a normal operating condition of 18,000 or 20,000 miles per year. On this basis, it would not be valid to compute the extra cost per mile as $0.198.

Because the normal operating range would change, the fixed costs would have to be considered. A more valid incremental cost would be obtained by computing the total annual cost if the car were driven, say, 18,000 miles, then subtracting the total cost for 15,000 miles of operation, and thereby determining the cost of the 3,000 additional miles of operation. From this difference the cost per mile for the additional mileage could be obtained. In this instance the total cost for 15,000 miles of driving per year was 15,000 × $0.384 = $5,760. If the cost of service—due to increased depreciation, repairs, and so forth—turned out to be $6,570 for 18,000 miles per year, evidently the cost of the additional 3,000 miles would be $810. Then the corresponding incremental cost per mile due to the increase in the operating range would be $0.27. Therefore, if several weekend trips were expected to become normal operation, the owner would be on more reasonable economic ground to quote an incremental cost of $0.27 per mile for even the first trip.

2.2.3 Recurring and Nonrecurring Costs

These two general cost terms are often used to describe various types of expenditures. *Recurring costs* are those that are repetitive and occur when an organization produces similar goods or services on a continuing basis. Variable costs are also recurring costs, because they repeat with each unit of output. But recurring costs are not limited to variable costs. A fixed cost that is paid on a repeatable basis is a recurring cost. For example, in an organization providing architectural and engineering services, office space rental, which is a fixed cost, is also a recurring cost.

Nonrecurring costs, then, are those which are not repetitive, even though the total expenditure may be cumulative over a relatively short period of time. Typically, nonrecurring costs involve developing or establishing a capability or capacity to operate. For example, the purchase cost for real estate upon which a plant will be built is a nonrecurring cost, as is the cost of constructing the plant itself.

2.2.4 Direct, Indirect, and Standard Costs

These frequently encountered cost terms involve most of the cost elements that also fit into the previous overlapping categories of fixed and variable costs, and recurring and nonrecurring costs. *Direct costs* are costs that can be reasonably measured and allocated to a specific output or work activity. The labor and material costs directly associated with a product, service, or construction activity are direct costs. For example, the materials needed to make a pair of scissors would be a direct cost.

Indirect costs are costs that are difficult to attribute or allocate to a specific output or work activity. The term normally refers to types of costs that would involve too much effort to allocate directly to a specific output. In this usage, they are costs allocated through a selected formula (such as, proportional to direct labor hours, direct labor dollars, or direct material dollars) to the outputs or work activities. For example, the costs of common tools, general supplies, and equipment maintenance in a plant are treated as indirect costs.

Overhead consists of plant operating costs that are not direct labor or direct material costs. In this book, the terms *indirect costs, overhead,* and *burden* are used interchangeably. Examples of overhead include electricity, general repairs, property taxes, and supervision. Administrative and selling expenses are usually added to direct costs and overhead costs to arrive at a unit selling price for a product or service. (Appendix A provides a more detailed discussion of cost accounting principles.)

Various methods are used to allocate overhead costs among products, services, and activities. The most commonly used methods involve allocation in proportion to direct labor costs, direct labor hours, direct materials costs, the sum of direct labor and direct materials costs (referred to as *prime cost* in a manufacturing operation), or machine hours. In each of these methods, it is necessary to know what the total overhead costs have been or are estimated to be for a time period (typically a year) to allocate them to the production (or service delivery) outputs.

Standard costs are representative costs per unit of output that are established in advance of actual production or service delivery. They are developed from anticipated direct labor hours, materials, and overhead categories (with their established costs per unit). Because total overhead costs are associated with a *certain level of production,* this is an important condition that should be remembered when dealing with standard cost data (for example, see Section 2.5.3). Standard costs play an important role in cost control and other management functions. Some typical uses are the following:

1. Estimating future manufacturing costs.
2. Measuring operating performance by comparing actual cost per unit with the standard unit cost.
3. Preparing bids on products or services requested by customers.
4. Establishing the value of work in process and finished inventories.

2.2.5 Cash Cost versus Book Cost

A cost that involves payment of cash is called a *cash cost* (and results in a cash flow) to distinguish it from one that does not involve a cash transaction and is reflected in the accounting system as a *noncash cost.* This noncash cost is often referred to as a *book cost.* Cash costs are estimated from the perspective established for the analysis (Principle 3, Section 1.3) and are the future expenses incurred for the alternatives being analyzed. Book costs are costs that do not involve cash payments, but rather represent the recovery of past expenditures over a fixed period of time. The most common example of book cost is the *depreciation* charged for the use of assets such as plant and equipment. In engineering economic analysis, only those costs that are cash flows or potential cash flows from the defined perspective for the analysis need to be considered. *Depreciation, for example, is not a cash flow* and is important in an analysis only because it affects income taxes, which are cash flows. We discuss the topics of depreciation and income taxes in Chapter 6.

2.2.6 Sunk Cost

A *sunk cost* is one that has occurred in the past and has no relevance to estimates of future costs and revenues related to an alternative course of action. Thus, a sunk cost is common to all alternatives, is not part of the future (prospective) cash flows, and can be disregarded in an engineering economic analysis. For instance, sunk costs are nonrefundable cash outlays, such as earnest money on a house or money spent on a passport.

We need to be able to recognize sunk costs and then handle them properly in an analysis. Specifically, we need to be alert for the possible existence of sunk costs in any situation that involves a past expenditure that cannot be recovered, or capital that has already been invested and cannot be retrieved.

The concept of sunk cost is illustrated in the next simple example. Suppose that Joe College finds a motorcycle he likes and pays $40 as a down payment, which will be applied to the $1,300 purchase price, but which must be forfeited if he decides not to take the cycle. Over the weekend, Joe finds another motorcycle he considers equally desirable for a purchase price of $1,230. For the purpose of deciding which cycle to purchase, the $40 is a sunk cost and thus would not enter into the decision, except that it lowers the remaining cost of the first cycle. The decision then is between paying $1,260 ($1,300 − $40) for the first motorcycle versus $1,230 for the second motorcycle.

In summary, sunk costs result from past decisions and therefore are irrelevant in the analysis and comparison of alternatives that affect the future. Even though it is sometimes emotionally difficult to do, sunk costs should be ignored, except possibly to the extent that their existence assists you to anticipate better what will happen in the future.

EXAMPLE 2-4

A classic example of sunk cost involves the replacement of assets. Suppose that your firm is considering the replacement of a piece of equipment. It originally cost $50,000, is presently shown on the company records with a value of $20,000, and can be sold for an estimated $5,000. For purposes of replacement analysis, the $50,000 is a sunk cost. However, one view is that the sunk cost should be considered as the difference between the value shown in the company records and the present realizable selling price. According to this viewpoint, the sunk cost is $20,000 minus $5,000, or $15,000. Neither the $50,000 nor the $15,000, however, should be considered in an engineering economic analysis, except for the manner in which the $15,000 may affect income taxes, which will be discussed in Chapter 9.

2.2.7 Opportunity Cost

An *opportunity cost* is incurred because of the use of limited resources, such that the opportunity to use those resources to monetary advantage in an alternative use is foregone. Thus, it is the cost of the best rejected (i.e., foregone) opportunity and is often hidden or implied.

For example, suppose that a project involves the use of vacant warehouse space presently owned by a company. The cost for that space to the project should be the income or savings that possible alternative uses of the space may bring to the firm. In other words, the opportunity cost for the warehouse space should be the income derived from the best alternative use of the space. This may be more than or less than the average cost of that space obtained from the accounting records of the company.

Consider also a student who could earn $20,000 for working during a year, but chooses instead to go to school for a year and spend $5,000 to do so. The opportunity cost of going to school for that year is $25,000: $5,000 cash outlay and $20,000 for income foregone. (This figure neglects the influence of income taxes and assumes that the student has no earning capability while in school.)

EXAMPLE 2-5

The concept of an opportunity cost is often encountered in analyzing the replacement of a piece of equipment or other capital asset. Let us reconsider Example 2-4, in which your firm considered the replacement of an existing piece of equipment that originally cost $50,000, is presently shown on the company records with a value of $20,000, but has a present market value of only $5,000. For purposes of an engineering economic analysis of whether to replace the equipment, the present investment in that equipment should be considered as $5,000, because, by keeping the equipment, the firm is giving up the *opportunity* to obtain $5,000 from its disposal. Thus, the $5,000 immediate selling price is really the investment cost of not replacing the equipment and is based on the opportunity cost concept.

2.2.8 Life-Cycle Cost

In engineering practice, the term *life-cycle cost* is often encountered. This term refers to a summation of all the costs, both recurring and nonrecurring, related to a product, structure, system, or service during its life span. The *life cycle* is illustrated in Figure 2-2. The life cycle begins with identification of the economic need or want (the requirement) and ends with retirement and disposal activities. It is a time horizon that must be defined in the context of the specific situation—whether it is a highway bridge, a jet engine for commercial aircraft, or an automated flexible manufacturing cell for a factory. The end of the life cycle may be projected on a functional or an economic basis. For example, the amount of time that a structure or piece of equipment is able to perform economically may be shorter than that permitted by its physical capability. Changes in the design efficiency of a boiler illustrate this situation. The old boiler may be able to produce the steam required, but not economically enough for the intended use.

The life cycle may be divided into two general time periods: the acquisition phase and the operation phase. As shown in Figure 2-2, each of these phases is further subdivided into interrelated but different activity periods.

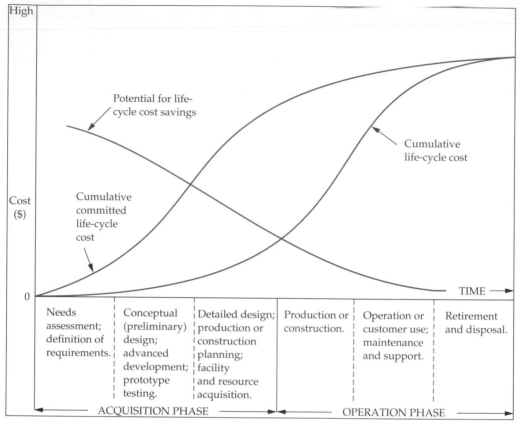

Figure 2-2 **Phases of the Life Cycle and Their Relative Cost**

The acquisition phase begins with an analysis of the economic need or want—the analysis necessary to make explicit the requirement for the product, structure, system, or service. Then, with the requirement explicitly defined, the other activities in the acquisition phase can proceed in a logical sequence. The conceptual design activities translate the defined technical and operational requirements into a preferred preliminary design. Included in these activities are development of the feasible alternatives and engineering economic analyses to assist in selection of the preferred preliminary design. Also, advanced development and prototype-testing activities to support the preliminary design work occur during this period.

The next group of activities in the acquisition phase involves detailed design and planning for production or construction. This step is followed by the activities necessary to prepare, acquire, and make ready for operation the facilities and other resources needed. *Again, engineering economy studies are an essential part of the design process to analyze and compare alternatives and to assist in determining the final detailed design.*

In the operation phase the production, delivery, or construction of the end item(s) or service and their operation or customer use occur. This phase ends with retirement from active operation or use and, often, disposal of the physical assets involved. The priorities for engineering economy studies during the operation phase are (1) achieving efficient and effective support to operations, (2) determining whether (and when) replacement of assets should occur, and (3) projecting the timing of retirement and disposal activities.

Figure 2-2 shows relative cost profiles for the life cycle. The greatest potential for achieving life-cycle cost savings is early in the acquisition phase. How much of the life-cycle costs for a product (for example) can be saved is dependent on many factors. However, effective engineering design and economic analysis during this phase are critical in maximizing potential savings.

One aspect of cost-effective engineering design is the minimizing of the impact of design changes during the steps in the life cycle. In general, the cost of a design change increases by a multiple of approximately 10 with each step, as illustrated in Figure 2-3. Thus, there is a large savings incentive to have an excellent conceptual design on which to base the detailed design and to prevent any changes during the production or construction and operation stages of the life cycle.

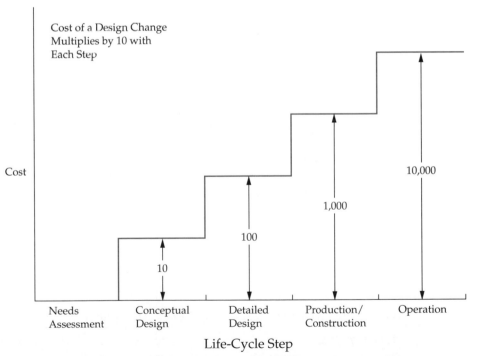

Figure 2-3 Costs of Design Changes Are Significant

> The cumulative committed life-cycle cost curve increases rapidly during the acquisition phase. In general, approximately 80% of life-cycle costs are "locked in" at the end of this phase by the decisions made during requirements analysis and preliminary and detailed design. In contrast, as reflected by the cumulative life-cycle cost curve, only about 20% of actual costs occur during the acquisition phase, with about 80% being incurred during the operation phase.

Thus, one purpose of the life-cycle concept is to make explicit the interrelated effects of costs over the total life span for a product. An objective of the design process is to minimize the life-cycle cost—while meeting other performance requirements—by making the right trade-offs between prospective costs during the acquisition phase and those during the operation phase.

The cost elements of the life cycle that need to be considered will vary with the situation. Because of their common use, however, several basic life-cycle cost categories will now be defined.

The *investment cost* is the capital required for most of the activities in the acquisition phase. In simple cases, such as acquiring specific equipment, an investment cost may be incurred as a single expenditure. On a large, complex construction project, however, a series of expenditures over an extended period could be incurred. This cost is also called a *capital investment*.

EXAMPLE 2-6

Consider a situation in which the equipment and related support for a new Computer-Aided Design/Computer-Aided Manufacturing (CAD/CAM) workstation are being acquired for your engineering department. The applicable cost elements and estimated expenditures are as follows:

Cost Element	Cost
Lease a telephone line for communication	$ 1,100/month
Lease CAD/CAM software (includes installation and debugging)	550/month
Purchase hardware (CAD/CAM workstation)	20,000
Purchase a 57,600-baud modem	250
Purchase a high-speed printer	1,500
Purchase a four-color plotter	10,000
Shipping costs	500
Initial training (in house) to gain proficiency with CAD/CAM software	6,000

What is the investment cost of this CAD/CAM system?

SOLUTION

The investment cost in this example is the sum of all the cost elements except the two monthly lease expenditures—specifically, the sum of the initial costs for the CAD/CAM workstation, modem, printer, and plotter ($31,750); shipping cost

($500); and the initial training cost ($6,000). These cost elements result in a total investment cost of $38,250. The two cost elements that involve lease payments on a monthly basis (telephone line and CAD/CAM software) are part of the recurring costs in the operation phase.

The term *working capital* refers to the funds required for current assets (i.e., other than fixed assets such as equipment, facilities, etc.) that are needed for the startup and support of operational activities. For example, products cannot be made or services delivered without having materials available in inventory. Functions such as maintenance cannot be supported without spare parts, tools, trained personnel, and other resources. Also, cash must be available to pay employee salaries and the other expenses of operation. The amount of working capital needed will vary with the project involved, and some or all of the investment in working capital is usually recovered at the end of a project's life.

Operation and maintenance cost includes many of the recurring annual expense items associated with the operation phase of the life cycle. The direct and indirect costs of operation associated with the five primary resource areas—people, machines, materials, energy, and information—are a major part of the costs in this category.

Disposal cost includes those nonrecurring costs of shutting down the operation and the retirement and disposal of assets at the end of the life cycle. Normally, costs associated with personnel, materials, transportation, and one-time special activities can be expected. These costs will be offset in some instances by receipts from the sale of assets with remaining market value. A classic example of a disposal cost is that associated with cleaning up a site where a chemical processing plant had been located.

2.3 The General Economic Environment

There are numerous general economic concepts that must be taken into account in engineering studies. In broad terms, economics deals with the interactions between people and wealth, and engineering is concerned with the cost-effective use of scientific knowledge to benefit humankind. This section introduces some of these basic economic concepts and indicates how they may be factors for consideration in engineering studies and managerial decisions.

2.3.1 Consumer and Producer Goods and Services

The goods and services that are produced and utilized may be divided conveniently into two classes. *Consumer goods and services* are those products or services that are directly used by people to satisfy their wants. Food, clothing, homes, cars, television sets, haircuts, opera, and medical services are examples. The producers of consumer goods and services must be aware of, and are subject to, the changing wants of the people to whom their products are sold.

Producer goods and services are used to produce consumer goods and services or other producer goods. Machine tools, factory buildings, buses, and farm machinery are examples. In the long run, producer goods serve to satisfy human wants, but only as a means to that end. Thus, the amount of producer goods needed is determined indirectly by the amount of consumer goods or services that are demanded by people. However, because the relationship is much less direct than for consumer goods and services, the demand for and production of producer goods may greatly precede or lag behind the demand for the consumer goods that they will produce.

2.3.2 Measures of Economic Worth

Goods and services are produced and desired because directly or indirectly they have *utility*—the power to satisfy human wants and needs. Thus, they may be used or consumed directly, or they may be used to produce other goods or services that may, in turn, be used directly. Utility is most commonly measured in terms of *value*, expressed in some medium of exchange as the *price* that must be paid to obtain the particular item.

Much of our business activity, including engineering, focuses on increasing the utility (value) of materials and products by changing their form or location. Thus, iron ore, worth only a few dollars per ton, significantly increases in value by being processed, combined with suitable alloying elements, and converted into razor blades. Similarly, snow, worth almost nothing when high in distant mountains, becomes quite valuable when it is delivered in melted form several hundred miles away to dry southern California.

2.3.3 Necessities, Luxuries, and Price Demand

Goods and services may be divided into two types: *necessities* and *luxuries*. Obviously, these terms are relative, because, for most goods and services, what one person considers a necessity may be considered a luxury by another. For example, a person living in one community may find that an automobile is a necessity to get to and from work. If the same person lived and worked in a different city, adequate public transportation might be available, and an automobile would be a luxury. For all goods and services, there is a relationship between the price that must be paid and the quantity that will be demanded or purchased. This general relationship is depicted in Figure 2-4. As the selling price per unit (p) is increased, there will be less demand (D) for the product, and as the selling price is decreased, the demand will increase. The relationship between price and demand can be expressed as the linear function

$$p = a - bD \qquad \text{for } 0 \le D \le \frac{a}{b}, \text{ and } a > 0,\ b > 0, \tag{2-1}$$

where a is the intercept on the price axis and $-b$ is the slope. Thus, b is the amount by which demand increases for each unit decrease in p. Both a and b are constants.

Figure 2-4
General Price–Demand
Relationship. (Note that
price is considered to
be the independent
variable, but is shown
as the vertical axis.
This convention is
commonly used by
economists.)

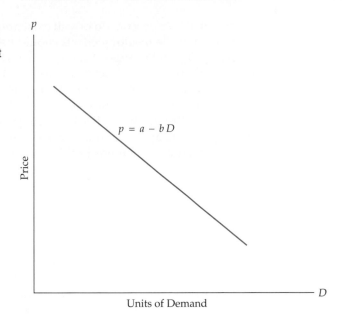

It follows, of course, that

$$D = \frac{a - p}{b} \qquad (b \neq 0). \tag{2-2}$$

Although Figure 2-4 illustrates the general relationship between price and demand, this relationship would probably be different for necessities and luxuries. Consumers can readily forego the consumption of luxuries if the price is greatly increased, but they find it more difficult to reduce their consumption of true necessities. Also, they will use the money saved by not buying luxuries to pay the increased cost of necessities.

2.3.4 Competition

Because economic laws are general statements regarding the interaction of people and wealth, they are affected by the economic environment in which people and wealth exist. Most general economic principles are stated for situations in which *perfect competition* exists.

Perfect competition occurs in a situation in which any given product is supplied by a large number of vendors and there is no restriction on additional suppliers entering the market. Under such conditions, there is assurance of complete freedom on the part of both buyer and seller. Perfect competition may never occur in actual practice, because of a multitude of factors that impose some degree of limitation upon the actions of buyers or sellers, or both. However, with conditions of perfect competition assumed, it is easier to formulate general economic laws.

The existing competitive situation is an important factor in most engineering economy studies. Unless information is available to the contrary, it should be assumed that competitors do or will exist and that they produce a quality product or service, and the resulting effects should be taken into account.

Monopoly is at the opposite pole from perfect competition. A perfect monopoly exists when a unique product or service is only available from a single supplier and that vendor can prevent the entry of all others into the market. Under such conditions, the buyer is at the complete mercy of the supplier in terms of the availability and price of the product. Perfect monopolies rarely occur in practice, because (1) few products are so unique that substitutes cannot be used satisfactorily, and (2) governmental regulations prohibit monopolies if they are unduly restrictive.

2.3.5 The Total Revenue Function

The total revenue, TR, that will result from a business venture during a given period is the product of the selling price per unit, p, and the number of units sold, D. Thus,

$$\text{TR} = \text{price} \times \text{demand} = p \cdot D. \tag{2-3}$$

If the relationship between price and demand as given in Equation (2-1) is used,

$$\text{TR} = (a - bD)D = aD - bD^2 \quad \text{for } 0 \le D \le \frac{a}{b} \text{ and } a > 0, b > 0. \tag{2-4}$$

The relationship between total revenue and demand for the condition expressed in Equation (2-4) may be represented by the curve shown in Figure 2-5. From calculus, the demand, $\hat{D}$, that will produce maximum total revenue can be obtained by solving

$$\frac{d\text{TR}}{dD} = a - 2bD = 0. \tag{2-5}$$

Thus,*

$$\hat{D} = \frac{a}{2b}. \tag{2-6}$$

It must be emphasized that because of cost-volume relationships discussed in the next section, *most businesses would not obtain maximum profits by maximizing revenue.* Accordingly, the cost-volume relationship must be considered and related to revenue, because cost reductions provide a key motivation for many engineering process improvements. If a solution to an engineering problem cannot be justified through cost reductions, the solution may depend on expansion of the revenue side of the profit equation as discussed in Section 2.3.6.

---

*To guarantee that $\hat{D}$ maximizes total revenue, check the second derivative to be sure it is negative:

$$\frac{d^2\text{TR}}{dD^2} = -2b.$$

Also, recall that in cost minimization problems a positively signed second derivative is necessary to guarantee a minimum-value optimal cost solution.

2.3.6 Cost, Volume, and Breakeven Point Relationships

Fixed costs remain constant over a wide range of activities as long as the business does not permanently discontinue operations, but variable costs vary in total with the volume of output (Section 2.2.1). Thus, at any demand D, total cost is

$$C_T = C_F + C_V, \tag{2-7}$$

where C_F and C_V denote fixed and variable costs, respectively. For the linear relationship assumed here,

$$C_V = c_v \cdot D, \tag{2-8}$$

where c_v is the variable cost per unit. In this section, we consider two scenarios for finding breakeven points. In the first scenario demand is a function of price. The second scenario assumes that price and demand are independent of each other.

Scenario 1 When total revenue, as depicted in Figure 2-5, and total cost, as given by Equations (2-7) and (2-8), are combined, the typical results as a function of demand are depicted in Figure 2-6. At *breakeven point* D'_1, total revenue is equal to total cost, and an increase in demand will result in a profit for the operation. Then at optimal demand, D^*, profit is maximized [Equation (2-10)]. At breakeven point D'_2, total revenue and total cost are again equal, but additional volume will result in an operating loss instead of a profit. Obviously, the conditions for which breakeven and maximum profit occur are our primary interest. First, at any volume (demand), D,

Profit (loss) = total revenue − total costs

$$= (aD - bD^2) - (C_F + c_v D)$$

$$= -bD^2 + (a - c_v)D - C_F \quad \text{for } 0 \le D \le \frac{a}{b} \text{ and } a > 0,\ b > 0. \tag{2-9}$$

Figure 2-5
Total Revenue Function
as a Function of
Demand

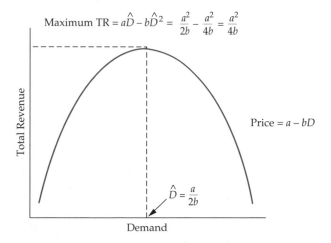

Maximum TR $= a\hat{D} - b\hat{D}^2 = \dfrac{a^2}{2b} - \dfrac{a^2}{4b} = \dfrac{a^2}{4b}$

Total Revenue

Price $= a - bD$

$\hat{D} = \dfrac{a}{2b}$

Demand

Figure 2-6
Combined Cost and
Revenue Functions,
and Breakeven Points,
as Functions of
Volume, and Their
Effect on Typical Profit
(Scenario 1)

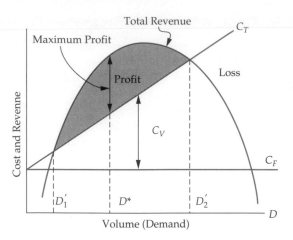

In order for a profit to occur, based on Equation (2-9), and to achieve the typical results depicted in Figure 2-6, two conditions must be met:

1. $(a - c_v) > 0$; that is, the price per unit that will result in no demand has to be greater than the variable cost per unit. (This avoids negative demand.)
2. Total revenue (TR) must exceed total cost (C_T) for the period involved.

If these conditions are met, we can find the optimal demand at which maximum profit will occur by taking the first derivative of Equation (2-9) with respect to D and setting it equal to zero:

$$\frac{d(\text{profit})}{dD} = a - c_v - 2bD = 0.$$

The optimal value of D that maximizes profit is

$$D^* = \frac{a - c_v}{2b}. \qquad (2\text{-}10)$$

To ensure that we have *maximized* profit (rather than minimized it), the sign of the second derivative must be negative. Checking this, we find that

$$\frac{d^2(\text{profit})}{dD^2} = -2b,$$

which will be negative for $b > 0$ (as earlier specified).

An economic breakeven point for an operation occurs when total revenue equals total cost. Then for total revenue and total cost, as used in the development of Equations (2-9) and (2-10) and at any demand D,

$$\text{Total revenue} = \text{total cost} \qquad (\text{breakeven point})$$

$$aD - bD^2 = C_F + c_v D$$

$$-bD^2 + (a - c_v)D - C_F = 0. \qquad (2\text{-}11)$$

Because Equation (2-11) is a quadratic equation with one unknown (D), we can solve for the breakeven points D'_1 and D'_2 (the roots of the equation):*

$$D' = \frac{-(a - c_v) \pm [(a - c_v)^2 - 4(-b)(-C_F)]^{1/2}}{2(-b)}. \qquad (2\text{-}12)$$

With the conditions for a profit satisfied [Equation (2-9)], the quantity in the brackets of the numerator (the discriminant) in Equation (2-12) will be greater than zero. This will ensure that D'_1 and D'_2 have real positive, unequal values.

EXAMPLE 2-7

A company produces an electronic timing switch that is used in consumer and commercial products made by several other manufacturing firms. The fixed cost (C_F) is $73,000 per month, and the variable cost (c_v) is $83 per unit. The selling price per unit is $p = \$180 - 0.02(D)$, based on Equation (2-1). For this situation, (a) determine the optimal volume for this product and confirm that a profit occurs (instead of a loss) at this demand, and (b) find the volumes at which breakeven occurs; that is, what is the domain of profitable demand?

SOLUTION

$$\text{(a)} D^* = \frac{a - c_v}{2b} = \frac{\$180 - \$83}{2(0.02)} = 2{,}425 \text{ units per month} \quad \text{[from Equation (2-10)]}$$

Is $(a - c_v) > 0$?

$$(\$180 - \$83) = \$97, \qquad \text{which is greater than 0.}$$

And is (total revenue − total cost) > 0 for $D^* = 2{,}425$ units per month?

$$[\$180(2{,}425) - 0.02(2{,}425)^2] - [\$73{,}000 + \$83(2{,}425)] = \$44{,}612$$

A demand of $D^* = 2{,}425$ units per month results in a maximum profit of $44,612 per month. Notice that the second derivative is negative (-0.04).

$$\text{(b)} \text{Total revenue} = \text{total cost} \qquad \text{(breakeven point)}$$

$$-bD^2 + (a - c_v)D - C_F = 0 \qquad \text{[from Equation (2-11)]}$$

$$-0.02D^2 + (\$180 - \$83)D - \$73{,}000 = 0$$

$$-0.02D^2 + 97D - 73{,}000 = 0$$

---

*Given $ax^2 + bx + c = 0$, the quadratic equation is $x = \dfrac{-b \pm \sqrt{b^2 - 4ac}}{2a}$.

And, from Equation (2-12),

$$D' = \frac{-97 \pm [(97)^2 - 4(-0.02)(-73,000)]^{0.5}}{2(-0.02)}$$

$$D_1' = \frac{-97 + 59.74}{-0.04} = 932 \text{ units per month}$$

$$D_2' = \frac{-97 - 59.74}{-0.04} = 3,918 \text{ units per month.}$$

Thus, the domain of profitable demand is 932 to 3,918 units per month.

Scenario 2 When the price per unit (p) for a product or service can be represented more simply as being independent of demand [versus being a linear function of demand, as assumed in Equation (2-1)] and is greater than the variable cost per unit (c_v), a single breakeven point results. Then under the assumption that demand is immediately met, total revenue (TR) $= p \cdot D$. If the linear relationship for costs in Equations (2-7) and (2-8) is also used in the model, the typical situation is depicted in Figure 2-7.

Figure 2-7
Typical Breakeven
Chart with Price (p) a
Constant (Scenario 2)

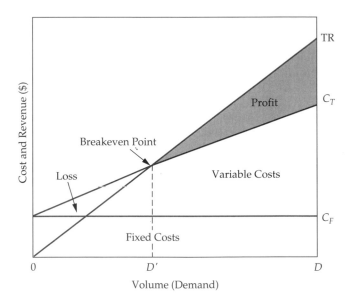

EXAMPLE 2-8

An engineering consulting firm measures its output in a standard service hour unit, which is a function of the personnel grade levels in the professional staff. The variable cost (c_v) is $62 per standard service hour. The charge-out rate [i.e., selling price (p)] is $85.56 per hour. The maximum output of the firm is 160,000 hours per year, and its fixed cost (C_F) is $2,024,000 per year. For this firm, (a) what is the breakeven point in standard service hours and in percentage of total capacity, and (b) what is the percentage reduction in the breakeven point (sensitivity) if fixed costs are reduced 10%; if variable cost per hour is reduced 10%; if both costs are reduced 10%; and if the selling price per unit is increased by 10%?

SOLUTION

(a)

$$\text{Total revenue} = \text{total cost} \qquad \text{(breakeven point)}$$

$$pD' = C_F + c_v D'$$

$$D' = \frac{C_F}{(p - c_v)}, \qquad (2\text{-}13)$$

and

$$D' = \frac{\$2,024,000}{(\$85.56 - \$62)} = 85,908 \text{ hours per year}$$

$$D' = \frac{85,908}{160,000} = 0.537,$$

or 53.7% of capacity.

(b) A 10% reduction in C_F gives

$$D' = \frac{0.9(\$2,024,000)}{(\$85.56 - \$62)} = 77,318 \text{ hours per year}$$

and

$$\frac{85,908 - 77,318}{85,908} = 0.10,$$

or a 10% reduction in D'.
A 10% reduction in c_v gives

$$D' = \frac{\$2,024,000}{[\$85.56 - 0.9(\$62)]} = 68,011 \text{ hours per year}$$

and

$$\frac{85,908 - 68,011}{85,908} = 0.208,$$

or a 20.8% reduction in D'.
A 10% reduction in both C_F and c_v gives

$$D' = \frac{0.9(\$2,024,000)}{[\$85.56 - 0.9(\$62)]} = 61,210 \text{ hours per year}$$

and

$$\frac{85,908 - 61,210}{85,908} = 0.287,$$

or a 28.7% reduction in D'.

A 10% increase in p gives

$$D' = \frac{\$2,024,000}{[1.1(\$85.56) - \$62]} = 63,021 \text{ hours per year}$$

and

$$\frac{85,908 - 63,021}{85,908} = 0.266,$$

or a 26.6% reduction in D'.

Thus, the breakeven point is more sensitive to a reduction in variable cost per hour than to the same percentage reduction in the fixed cost, but reduced costs in both areas should be sought. Furthermore, notice that the breakeven point in this example is highly sensitive to the selling price per unit, p. These results are summarized as follows:

Change in Factor Value(s)	Decrease in Breakeven Point
10% reduction in C_F	10.0%
10% reduction in c_v	20.8
10% reduction in C_F and in c_v	28.7
10% increase in p	26.6

The breakeven point for an operating situation can be determined in units of output, percentage utilization of capacity, or sales volume (demand). In Example 2-8(a), the breakeven point (D') was calculated in units of output (85,908 standard service hours per year), and then, using the total capacity figure (160,000 hours per year), it was also expressed as percentage utilization of capacity (53.7%). In terms of sales volume, the breakeven point in Example 2-8 is $85.56 (85,908) = \$7,350,288$.

> Market competition often creates pressure to lower the breakeven point of an operation; the lower the breakeven point, the less likely that a loss will occur during market fluctuations. Also, if the selling price remains constant (or increases), a larger profit will be achieved at any level of operation above the reduced breakeven point.

2.4 Cost-Driven Design Optimization

As discussed in Section 2.2.8, engineers must maintain a *life-cycle* (i.e., "cradle to grave") viewpoint as they design products, processes, and services. Such a complete perspective ensures that engineers consider initial investment costs; operation and maintenance expenses and other annual expenses in later years; and environmental and social consequences over the life of their designs. In fact, a movement called *Design for the Environment* (DFE), or "green engineering," has prevention of waste, improved materials selection, and reuse and recycling of resources among its goals. Designing for energy conservation, for example, is a subset of green engineering. Another example is the design of an automobile bumper that can be easily recycled. As you can see, *engineering design is an economically driven art*.

Minimum life-cycle cost, consistent with the consideration of other factors, is a goal achieved largely in the early stages of design. The attitude that the engineer can develop something that will work and then think about controlling the cost is fallacious, because, by the time most of the functional requirements are built into a design, many of the best opportunities for reducing costs have been missed. Engineers can accomplish a great deal toward the goal of minimizing life-cycle costs by simply keeping in mind its importance!

Examples of cost minimization through effective design are plentiful in the practice of engineering. Consider the design of a heat exchanger in which tube material and configuration affect cost and dissipation of heat. The problems in this section designated as "cost-driven design optimization" are simple design models intended to illustrate the importance of cost in the design process. These problems show the procedure for determining an optimal design using cost concepts. We will consider discrete and continuous optimization problems that involve a single design variable, X. This variable is also called a *primary cost driver,* and knowledge of its behavior may allow a designer to account for a large portion of total cost behavior.

For cost-driven design optimization problems, the two main tasks are as follows:

1. Determine the optimal value for a certain alternative's design variable. For example, what velocity of an aircraft minimizes the total annual costs of owning and operating the aircraft?
2. Select the best alternative, each with its own unique value for the design variable. For example, what insulation thickness is best for a home in Virginia: R11, R19, R30, or R38?

In general, the cost models developed in these problems consist of three types of costs:

1. Fixed cost(s)
2. Cost(s) that vary *directly* with the design variable
3. Cost(s) that vary *indirectly* with the design variable

A simplified format of a cost model with one design variable is

$$\text{Cost} = aX + \frac{b}{X} + k, \qquad (2\text{-}14)$$

where a is a parameter that represents the directly varying cost(s),

             b is a parameter that represents the indirectly varying cost(s),

             k is a parameter that represents the fixed cost(s), and

             X represents the design variable in question (e.g., weight or velocity).

In a particular problem, the parameters a, b, and k may actually represent the sum of a group of costs in that category, and the design variable may be raised to some power for either directly or indirectly varying costs.*

The following steps outline a general approach for optimizing a design with respect to cost:

1. Identify the design variable that is the primary cost driver (e.g., pipe diameter or insulation thickness).
2. Write an expression for the cost model in terms of the design variable.
3. Set the first derivative of the cost model with respect to the continuous design variable equal to zero. For discrete design variables, compute the value of the cost model for each discrete value over a selected range of potential values.
4. Solve the equation found in Step 3 for the optimum value of the continuous design variable.† For discrete design variables, the optimum value has the minimum cost value found in Step 3. In Example 2-10, we will also use an incremental procedure for selecting the best-valued discrete design variable. (Incremental analysis is a very important topic in Chapter 5.) This method is analogous to taking the first derivative for a continuous design variable and setting it equal to zero to determine an optimal value.
5. For continuous design variables, use the second derivative of the cost model with respect to the design variable to determine whether the optimum value found in Step 4 corresponds to a global maximum or minimum.

EXAMPLE 2-9

The cost of operating a jet-powered commercial (passenger-carrying) airplane varies as the three-halves (3/2) power of its velocity; specifically, $C_O = knv^{3/2}$, where n is the trip length in miles, k is a constant of proportionality, and v is velocity in miles per hour. It is known that at 400 miles per hour the *average* cost of operation is $300 per mile. The company that owns the aircraft wants to minimize the cost of operation, but that cost must be balanced against the cost of the passengers' time (C_C), which has been set at $300,000 per hour.

(a) At what velocity should the trip be planned to minimize the total cost, which is the sum of the cost of operating the airplane and the cost of passengers' time?

(b) How do you know that your answer for the problem in part (a) minimizes the total cost?

---

*A more general model is: $\text{Cost} = k + ax + b_1 x^{e_1} + b_2 x^{e_2} + \ldots$, where $e_1 = -1$ reflects costs that vary inversely with X, $e_2 = 2$ indicates costs that vary as the square of X, and so forth.

†If multiple optima (stationary points) are found in Step 4, finding the global optimum value of the design variable will require a little more effort. One approach is to systematically use each root in the second derivative equation and assign each point as a maximum or a minimum based on the sign of the second derivative. A second approach would be to use each root in the objective function and see which point best satisfies the cost function.

SOLUTION

The equation for total cost (C_T) is

$$C_T = C_O + C_C = knv^{3/2} + (\$300{,}000 \text{ per hour}) \left(\frac{n}{v} \right), \quad \text{where } n/v \text{ has time}$$
$$\text{(hours) as its unit.}$$

Now we solve for the value of k:

$$\frac{C_O}{n} = kv^{3/2}$$

$$\frac{\$300}{\text{mile}} = k \left(400 \frac{\text{miles}}{\text{hour}} \right)^{3/2}$$

$$k = \frac{\$300/\text{mile}}{\left(400 \frac{\text{miles}}{\text{hour}} \right)^{3/2}}$$

$$k = \frac{\$300/\text{mile}}{8000 \left(\frac{\text{miles}^{3/2}}{\text{hours}^{3/2}} \right)}$$

$$k = \$0.0375 \frac{\text{hours}^{3/2}}{\text{miles}^{5/2}}.$$

Thus,

$$C_T = \left(\$0.0375 \frac{\text{hours}^{3/2}}{\text{miles}^{5/2}} \right) (n \text{ miles}) \left(v \frac{\text{miles}}{\text{hour}} \right)^{3/2} + \left(\frac{\$300{,}000}{\text{hour}} \right) \left(\frac{n \text{ miles}}{v \frac{\text{miles}}{\text{hour}}} \right)$$

$$C_T = \$0.0375 n v^{3/2} + \$300{,}000 \left(\frac{n}{v} \right).$$

Next, the first derivative is taken:

$$\frac{dC_T}{dv} = \frac{3}{2}(\$0.0375)nv^{1/2} - \frac{\$300{,}000n}{v^2} = 0.$$

So

$$0.05625 v^{1/2} - \frac{300{,}000}{v^2} = 0$$

$$0.05625 v^{5/2} - 300{,}000 = 0$$

$$v^{5/2} = \frac{300{,}000}{0.05625} = 5{,}333{,}333$$

$$v^* = (5{,}333{,}333)^{0.4} = 490.68 \text{ mph}.$$

Finally, we check the second derivative to confirm a minimum cost solution:

$$\frac{d^2 C_T}{dv^2} = \frac{0.028125}{v^{1/2}} + \frac{600,000}{v^3} \qquad \text{for } v > 0, \text{ and therefore, } \frac{d^2 C_T}{dv^2} > 0.$$

The company concludes that $v = 490.68$ mph minimizes the total cost of this particular airplane's flight.

EXAMPLE 2-10

This example deals with a discrete optimization problem of determining the most economical amount of attic insulation for a large single-story home in Virginia. In general, the heat lost through the roof of a single-story home is

$$\begin{pmatrix} \text{Heat loss} \\ \text{in Btu} \\ \text{per hour} \end{pmatrix} = \begin{pmatrix} \Delta \text{Temperature} \\ \text{in } °F \end{pmatrix} \begin{pmatrix} \text{Area} \\ \text{in} \\ \text{ft}^2 \end{pmatrix} \begin{pmatrix} \text{Conductance in} \\ \text{Btu/hour} \\ \overline{\text{ft}^2 - ° F} \end{pmatrix},$$

or

$$Q = (T_{\text{in}} - T_{\text{out}}) \cdot A \cdot U.$$

In southwest Virginia, the number of heating days per year is approximately 230, and the annual heating degree-days equals 230 $(65°F - 46°F) = 4,370$ degree-days per year. Here $65°F$ is assumed to be the average inside temperature and $46°F$ is the average outside temperature each day.

Consider a 2,400-ft^2 single-story house in Blacksburg. The typical annual space-heating load for this size of a house is 100×10^6 Btu. That is, with no insulation in the attic, we lose about 100×10^6 Btu per year.* Common sense dictates that the "no insulation" alternative is not attractive and is to be avoided.

With insulation in the attic, the amount of heat lost each year will be reduced. The value of energy savings that results from adding insulation and reducing heat loss is dependent on what type of residential heating furnace is installed. For this example, we assume that an electrical resistance furnace is installed by the builder, and its efficiency is near 100%.

Now we're in a position to answer the question: What amount of insulation is most economical? An additional piece of data we need involves the cost of electricity, which is $0.074 per kWh. This can be converted to dollars per 10^6 Btu as follows (1 kWh = 3,413 Btu):

$$\frac{\text{kWh}}{3,413 \text{ Btu}} = 293 \text{ kWh per million Btu}$$

$$\frac{293 \text{ kWh}}{10^6 \text{ Btu}} \left(\frac{\$0.074}{\text{kWh}} \right) \cong \$21.75/10^6 \text{ Btu}.$$

---

* 100×10^6 Btu/yr $\cong \left(\dfrac{4,370 \text{ }°F\text{-days per year}}{1.00 \text{ efficiency}} \right) (2,400 \text{ ft}^2)(24 \text{ hr/day}) \dfrac{0.397 \text{ Btu/hr}}{\text{ft}^2\text{-}°F}$, where 0.397 is the U-factor with no insulation.

The cost of several insulation alternatives and associated space heating loads for this house are given in the following table:

| | Amount of Insulation | | | |
	R11	R19	R30	R38
Investment cost	$600	$900	$1,300	$1,600
Annual heating load (Btu/year)	74×10^6	69.8×10^6	67.2×10^6	66.2×10^6

In view of these data, which amount of attic insulation is most economical? The life of the insulation is estimated to be 25 years.

SOLUTION
Set up a table to examine total life-cycle costs:

	R11	R19	R30	R38
A. Investment cost	$600	$900	$1,300	$1,600
B. Cost of heat loss per year	$1,609.50	$1,518.15	$1,461.60	$1,439.85
C. Cost of heat loss over 25 years	$40,237.50	$37,953.75	$36,540	$35,996.25
D. Total life cycle costs (A + C)	$40,837.50	$38,853.75	$37,840	$37,596.25

Answer: To minimize total life-cycle costs, select R38 insulation.

ANOTHER SOLUTION
Another approach for selecting the best alternative from a discrete set is to examine the incremental (Δ) differences among them (remember Principle 2 on page 5) when the alternatives are ranked from low investment cost to high investment cost. We illustrate this procedure here and return to it in Chapter 5.

We begin by examining total energy savings over 25 years for each added amount of insulation, less the added investment cost associated with each amount of insulation.

The following questions lead to the computation of the relevant trade-offs involved:

1. What savings occur if we decide to insulate with R19 instead of R11?

Δ(R19 − R11) Δ Investment = $300
Δ Savings/year = $91.35 = [−1,518.15 − (−1,609.5)]$
Δ Savings over 25 years = $2,283.75

By putting in R19 rather than R11, the total net savings over 25 years is $1,983.75.

2. What total net savings are realized if we choose R30 instead of R19?

$$\Delta(R30-R19) \quad \Delta \quad \text{Investment} = \$400$$
$$\Delta \quad \text{Savings/year} = \$56.55 = [-1,461.60 - (-1,518.15)]$$
$$\Delta \quad \text{Savings over 25 years} = \$1,413.75$$

The total net savings over 25 years is $1,013.75.

3. Finally, what net savings are achieved if we add the maximum amount of insulation (R38) rather than R30?

$$\Delta(R38-R30) \quad \Delta \quad \text{Investment} = \$300$$
$$\Delta \quad \text{Savings per year} = \$21.75 = [-1,439.85 - (-1,461.60)]$$
$$\Delta \quad \text{Savings over 25 years} = \$543.75$$

The total net savings over 25 years is $243.75.

If we ignore the time value of money (to be discussed in Chapter 3) over the 25-year period and select the amount of attic insulation that gives us positive net savings, our best (most economical) choice would be R38.

CAUTION

This conclusion may change when we consider the time value of money (i.e., an interest rate greater than zero) in Chapter 3. In such a case, it will not necessarily be true that adding more and more insulation is the optimal course of action.

2.5 Present Economy Studies

When alternatives for accomplishing a specific task are being compared over *one year or less* and the influence of time on money can be ignored, engineering economic analyses are referred to as *present economy studies*. Several situations involving present economy studies are illustrated in this section. The rules, or criteria, shown next will be used to select the preferred alternative when defect-free output (yield) is variable *or* constant among the alternatives being considered. In addition, other criteria of acceptability (e.g., compliance with environmental regulations) must also be satisfied.

RULE 1:
When revenues and other economic benefits are present and vary among alternatives, choose the alternative that *maximizes* overall profitability based on the number of defect-free units of a product or service produced.

> **RULE 2:**
> When revenues and other economic benefits are *not* present *or* are constant among all alternatives, consider only the costs and select the alternative that *minimizes* total cost per defect-free unit of product or service output.

Companion Web Site (http://www.prenhall.com/sullivan_engineering/): What happens to the food waste from restaurants? This waste contributes to the landfill problem in many communities. Visit the Web site to view a *present economy study* of an environmentally friendly alternative that converts food waste into pellet feed for livestock.

2.5.1 Total Cost in Material Selection

In many cases, economic selection among materials cannot be based solely on the costs of materials. Frequently, a change in materials will affect the design and processing costs, and shipping costs may also be altered.

EXAMPLE 2-11

A good example of this situation is illustrated by the part in Figure 2-8 for which annual demand is 100,000 units. The part shown is produced on a high-speed turret lathe, using 1112 screw-machine steel costing $0.30 per pound. A study was conducted to determine whether it might be cheaper to use brass screw stock, costing $1.40 per pound. Because the weight of steel required per piece was 0.0353 pounds and that of brass was 0.0384 pounds, the material cost per piece was $0.0106 for steel and $0.0538 for brass. However, when the manufacturing engineering department was consulted, it was found that, although 57.1 defect-free parts per hour were being produced using steel, the output would be 102.9 defect-free parts per hour if brass were used. Which material should be used for this part?

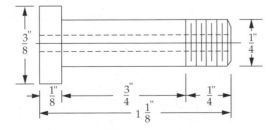

Figure 2-8 Small Screw Machine Product

SOLUTION

The machine attendant was paid $15.00 per hour, and the variable (i.e., traceable) overhead costs for the turret lathe were estimated to be $10.00 per hour. Thus, the total-cost comparison for the two materials was as follows:

	1112 Steel	Brass
Material	$0.30 \times 0.0353 = $0.0106	$1.40 \times 0.0384 = $0.0538
Labor	$15.00/57.1 = 0.2627	$15.00/102.9 = 0.1458
Variable Overhead	$10.00/57.1 = 0.1751	$10.00/102.9 = 0.0972
Total cost per piece	$0.4484	$0.2968
Saving per piece by use of brass = $0.4484 − $0.2968 = $0.1516		

Because 100,000 parts are made each year, revenues are constant across the alternatives. Rule 2 would select brass, and its use will produce a savings of $151.60 per thousand (a total of $15,160 for the year). It is also clear that costs other than the cost of material were of basic importance in the study.

Care should be taken in making economic selections between materials to ensure that any differences in shipping costs, yields, or resulting scrap are taken into account. Commonly, alternative materials do not come in the same stock sizes, such as sheet sizes and bar lengths. This may considerably affect the yield obtained from a given weight of material. Similarly, the resulting scrap may differ for various materials.

In addition to deciding what material a product should be made of, there are often alternative methods or machines that can be used to produce the product, which, in turn, can impact processing costs. Processing times may vary with the machine selected as may the product yield. As illustrated in Example 2-12, these considerations can have important economic implications.

EXAMPLE 2-12

Two currently owned machines are being considered for the production of a part. The capital investment associated with the machines is about the same and can be ignored for purposes of this example. The important differences between the machines are their production capacities (production rate × available production hours) and their reject rates (percentage of parts produced that cannot be sold). Consider the following table:

	Machine *A*	Machine *B*
Production Rate	100 parts/hr	130 parts/hr
Hours Available for Production	7 hr/day	6 hr/day
Percent Parts Rejected	3%	10%

The material cost is $6.00 per part and all defect-free parts produced can be sold for $12 each (rejected parts have negligible scrap value). For either machine, the operator cost is $15.00 per hour and the variable overhead rate for traceable costs is $5.00 per hour.

(a) Assume that the daily demand for this part is large enough that all defect-free parts can be sold. Which machine should be selected?

(b) What would the percent of parts rejected have to be for Machine B to be as profitable as Machine A?

SOLUTION

(a) Rule 1 applies in this situation because total daily revenues (selling price per part times the number of parts sold per day) and total daily costs will vary depending on the machine chosen. Therefore, we should select the machine that will maximize the profit per day:

$$\text{Profit per Day} = \text{Revenue per Day} - \text{Cost per Day}$$

$$= (\text{Production Rate})(\text{Production Hours})(\$12/\text{part})\cdot$$

$$[1 - (\%\ \text{rejected}/100)]$$

$$- (\text{Production Rate})(\text{Production Hours})(\$6/\text{part})$$

$$- (\text{Production Hours})(\$15/\text{hour} + \$5/\text{hour}).$$

Machine A. Profit per Day $= \left(\dfrac{100\ \text{parts}}{\text{hr}}\right)\left(\dfrac{7\ \text{hr}}{\text{day}}\right)\left(\dfrac{\$12}{\text{part}}\right)(1 - 0.03)$

$$- \left(\dfrac{100\ \text{parts}}{\text{hr}}\right)\left(\dfrac{7\ \text{hr}}{\text{day}}\right)\left(\dfrac{\$6}{\text{part}}\right) - \left(\dfrac{7\ \text{hr}}{\text{day}}\right)\left(\dfrac{\$15}{\text{hr}} + \dfrac{\$5}{\text{hr}}\right)$$

$$= \$3,808\ \text{per day}.$$

Machine B: Profit per Day $= \left(\dfrac{130\ \text{parts}}{\text{hr}}\right)\left(\dfrac{6\ \text{hr}}{\text{day}}\right)\left(\dfrac{\$12}{\text{part}}\right)(1 - 0.10)$

$$- \left(\dfrac{130\ \text{parts}}{\text{hr}}\right)\left(\dfrac{6\ \text{hr}}{\text{day}}\right)\left(\dfrac{\$6}{\text{part}}\right) - \left(\dfrac{6\ \text{hr}}{\text{day}}\right)\left(\dfrac{\$15}{\text{hr}} + \dfrac{\$5}{\text{hr}}\right)$$

$$= \$3,624\ \text{per day}.$$

Therefore, *select Machine A* to maximize profit per day.

(b) To find the breakeven percent of parts rejected, X, for Machine B, set the profit per day of Machine A equal to the profit per day of Machine B, and solve for X:

$$\$3,808/\text{day} = \left(\frac{130 \text{ parts}}{\text{hr}}\right)\left(\frac{6 \text{ hr}}{\text{day}}\right)\left(\frac{\$12}{\text{part}}\right)(1 - X)$$

$$- \left(\frac{130 \text{ parts}}{\text{hr}}\right)\left(\frac{6 \text{ hr}}{\text{day}}\right)\left(\frac{\$6}{\text{part}}\right) - \left(\frac{6 \text{ hr}}{\text{day}}\right)\left(\frac{\$15}{\text{hr}} + \frac{\$5}{\text{hr}}\right).$$

Thus, $X = 0.08$, so the percent of parts rejected for Machine B can be no higher than 8%.

2.5.2 Alternative Machine Speeds

Machines can frequently be operated at various speeds, resulting in different rates of product output. However, this usually results in different frequencies of machine downtime to permit servicing or maintaining the machine, such as resharpening or adjusting tooling. Such situations lead to present economy studies to determine the preferred operating speed. We *first* assume that there is an unlimited amount of work to be done in Example 2-13. Secondly, Example 2-14 illustrates how to deal with a fixed (limited) amount of work.

EXAMPLE 2-13

A simple example of alternative machine speeds involves the planing of lumber. Lumber put through the planer increases in value by $0.10 per board foot. When the planer is operated at a cutting speed of 5,000 feet per minute, the blades have to be sharpened after 2 hours of operation, and the lumber can be planed at the rate of *1,000 board-feet per hour.* When the machine is operated at 6,000 feet per minute, the blades have to be sharpened after 1-1/2 hours of operation, and the rate of planing is *1,200 board-feet per hour.* Each time the blades are changed, the machine has to be shut down for 15 minutes. The blades, unsharpened, cost $50 per set and can be sharpened 10 times before having to be discarded. Sharpening costs $10 per set. The crew that operates the planer changes and resets the blades. At what speed should the planer be operated?

SOLUTION

Because the labor cost for the crew would be the same for either speed of operation and because there was no discernible difference in wear upon the planer, these factors did not have to be included in the study.

In problems of this type, the operating time plus the delay time due to the necessity for tool changes constitute a cycle time that determines the output from the machine. The time required for a complete cycle determines the number of cycles that can be accomplished in a period of available time (e.g., one day) and a certain portion of each complete cycle is productive. The actual productive time will be the product of the productive time per cycle and the number of cycles per day.

	Value per Day
At 5,000 feet per minute	
Cycle time = 2 hours + 0.25 hour = 2.25 hours	
Cycles per day = 8 ÷ 2.25 = 3.555	
Value added by planing = 3.555 × 2 × 1,000 × $0.10 =	$711.00*
Cost of resharpening blades = 3.555 × $10 = $35.55	
Cost of blades = 3.555 × $50/10 = 17.18	
Total Cost Cash Flow	−53.33
Net increase in value (profit) per day	$657.67
At 6,000 feet per minute	
Cycle time = 1.5 hours + 0.25 hour = 1.75 hours	
Cycles per day = 8 ÷ 1.75 = 4.57	
Value added by planing = 4.57 × 1.5 × 1,200 × $0.10 =	$822.60*
Cost of resharpening blades = 4.57 × $10 = $45.70	
Cost of blades = 4.57 × $50/10 = 22.85	
Total Cost Cash Flow	−68.55
Net increase in value (profit) per day	$754.05

*The units work out as follows: (cycles/day)(hours/cycle)(board feet/hour)(dollar value/board-foot) = dollars/day

Thus, in Example 2-13 it is more economical according to Rule 1 to operate at 6,000 feet per minute, in spite of the more frequent sharpening of blades that is required.

It should be noted that this analysis assumes that the added production of lumber can be used. If, for example, the maximum production needed is equal to or less than that obtained by the slower machine speed (1,000 × 3.555 cycles × 2 hours − 7,110 board-feet per day), then the value added would be the same for each speed, and the decision then should be based on which speed minimizes total cost.

EXAMPLE 2-14

Example 2-13 assumed that every board-foot of lumber that is planed can be sold. If there is limited demand for the lumber, a correct choice of machining speeds can be made with Rule 2 by *minimizing total cost* per unit of output. Suppose now we want to know the better machining speed when only one job requiring 6,000 board-feet of planing is considered.

SOLUTION
For a fixed planing requirement of 6,000 board-feet, the value added by planing is 6,000 ($0.10) = $600 for either cutting speed. Hence, we want to minimize total cost per board-foot planed.

At 5,000 feet per minute cutting speed, we get

Cycle time = 2.25 hours,

Production per cycle = 2(1,000) = 2,000 board-feet,

Number of cycles = 6,000/2,000 = 3, or 6.75 hours,

Total cost = 3($10/cycle) + 3($50/10) = $45 (cost per board-foot = $0.0075).

At 6,000 feet per minute cutting speed, we get

Cycle time = 1.75 hours,

Production per cycle = 1.5(1,200) = 1,800 board-feet,

Number of cycles = 6,000/1,800 = 3.33, or 5.83 hours,

Total cost = 3.33($10/cycle) + 3.33($50/10) = $50 (cost per board-foot = $0.0083).

For a 6,000-board foot job, select the slower cutting speed (5,000 feet per minute) to minimize cost. During the 0.92 hour of time savings for the 6,000-feet-per-minute cutting speed, we assume that the operator is idle.

2.5.3 Making versus Purchasing (Outsourcing) Studies*

In the short run, say, one year or less, a company may consider producing an item in-house even though the item can be purchased (outsourced) from a supplier at a price lower than the company's standard production costs. (See Section 2.2.4.) This could occur if (1) direct, indirect, and overhead costs are incurred regardless of whether the item is purchased from an outside supplier and (2) the *incremental* cost of producing an item in the shortrun is less than the supplier's price. Therefore, the relevant short-run costs of make vs. purchase decisions are the *incremental costs* incurred and the *opportunity costs* of the resources involved.

Opportunity costs may become significant when in-house manufacture of an item causes other production opportunities to be foregone (often because of insufficient capacity). But in the long run, capital investments in additional manufacturing plant and capacity are often feasible alternatives to outsourcing. (Much of this book is concerned with evaluating the economic worthiness of proposed capital investments.) Because engineering economy often deals with *changes* to existing operations, standard costs may not be too useful in make vs. purchase studies. In fact, if they are used, standard costs can lead to uneconomical decisions. Example 2-15 illustrates the correct procedure to follow in performing make vs. purchase studies based on incremental costs.

---

* Much interest has been shown in outsourcing decisions. For example, see P. Chalos, "Costing, Control, and Strategic Analysis in Outsourcing Decisions," *Journal of Cost Management*, vol. 8, no. 4 (Winter 1995), pp. 31–37.

EXAMPLE 2-15

A manufacturing plant consists of three departments: A, B, and C. Department A occupies 100 square meters in one corner of the plant. Product X is one of several products being produced in Department A. The daily production of Product X is 576 pieces. The cost accounting records show the following average daily production costs for Product X:

Direct labor	(1 operator working 4 hours per day at $22.50/hr, including fringe benefits, plus a part-time foreman at $30 per day)	$120.00
Direct material		86.40
Overhead	(at $0.82 per square meter of floor area)	82.00
	Total cost per day =	$288.40

The department foreman has recently learned about an outside company that sells Product X at $0.35 per piece. Accordingly, the foreman figured a cost per day of $0.35(576) = 201.60, resulting in a daily savings of $288.40 − $201.60 = 86.80. Therefore, a proposal was submitted to the plant manager for shutting down the production line of Product X and buying it from the outside company.

However, after examining each component separately, the plant manager decided not to accept the foreman's proposal based on the unit cost of Product X:

1. *Direct labor*: Because the foreman was supervising the manufacture of other products in Department A in addition to Product X, the only possible savings in labor would occur if the operator working 4 hours per day on Product X were not reassigned after this line is shut down. That is, a maximum savings of $90.00 per day would result.

2. *Materials*: The maximum savings on direct material will be $86.40. However, this figure could be lower if some of the material for Product X is obtained from scrap of another product.

3. *Overhead*: Because other products are made in Department A, no reduction in total floor space requirements will probably occur. Therefore, no reduction in overhead costs will result from discontinuing Product X. It has been estimated that there will be daily savings in the variable overhead costs traceable to Product X of about $3.00 due to a reduction in power costs and in insurance premiums.

SOLUTION

If the manufacture of Product X is discontinued, the firm will save at most $90.00 in direct labor, $86.40 in direct materials, and $3.00 in variable overhead costs, which totals $179.40 per day. This estimate of actual cost savings per day is less than the potential savings indicated by the cost accounting records ($288.40 per day), and it would not exceed the $201.60 to be paid to the outside company if Product X

is purchased. For this reason, the plant manager used Rule 2 and rejected the proposal of the foreman and continued the manufacture of Product X.

In conclusion, Example 2-15 shows how an erroneous decision might be made by using the unit cost of Product X from the cost accounting records without detailed analysis. The fixed cost portion of Product X's unit cost, which is present even if the manufacture of Product X is discontinued, was not properly accounted for in the original analysis by the foreman.

2.5.4 Trade-Offs in Energy Efficiency Studies

Energy efficiency affects the annual expense of operating an electrical device such as a pump or motor. Typically, a more energy-efficient device requires a higher capital investment than does a less energy-efficient device, but the extra capital investment usually produces annual savings in electrical power expenses relative to a second pump or motor that is less energy efficient. This important trade-off between capital investment and annual electric power consumption will be considered in several chapters of this book. Hence, the purpose of Section 2.5.4 is to explain how the annual expense of operating an electrical device is calculated and traded off against capital investment cost.

If an electric pump, for example, can deliver a given horsepower (hp) or kilo-Watt (kW) rating to an industrial application, the *input* energy requirement is determined by dividing the given output by the energy efficiency of the device. The input requirement in hp or kW is then multiplied by the annual hours that the device operates and the unit cost of electric power. You can see that the higher the efficiency of the pump, the lower the annual cost of operating the device is relative to another less-efficient pump.

EXAMPLE 2-16

Two pumps capable of delivering 100 hp to an agricultural application are being evaluated in a present economy study. The selected pump will only be utilized for one year, and it will have no market value at the end of the year. Pertinent data are summarized as follows:

	ABC Pump	XYZ Pump
Purchase Price	$2,900	$6,200
Annual Maintenance	$170	$510
Efficiency	80%	90%

If electric power costs $0.10 per kWh and the pump will be operated 4,000 hours per year, which pump should be chosen? Recall that 1 hp = 0.746 kW.

SOLUTION
The annual expense of electric power for the ABC Pump is

$$(100 \text{ hp}/0.80)(0.746 \text{ kW/hp})(\$0.10/\text{kWh})(4{,}000 \text{ hr/yr}) = \$37{,}300.$$

For the XYZ Pump, the annual expense of electric power is $(100 \text{ hp}/0.90)(0.746 \text{ kW/hp})(\$0.10/\text{kWhr})(4{,}000 \text{ hr/yr}) = \$33{,}156$. Thus, the total annual cost of owning and operating the ABC Pump is \$40,370, while the total cost of owning and operating the XYZ Pump for one year is \$39,866. Consequently, the more energy-efficient XYZ Pump should be selected to minimize total annual cost. Notice the difference in annual energy expense (\$4,144) that results from a 90% efficient pump relative to an 80% efficient pump. This cost reduction more than balances the extra \$3,300 in capital investment and \$340 in annual maintenance required for the XYZ Pump.

2.6 Summary

In this chapter, we have discussed cost estimating, terminology, and concepts important in engineering economy. A listing of important abbreviations and notation, by chapter, is provided in Appendix B. It is important that the meaning and use of various cost terms and concepts be understood in order to communicate effectively with other engineering and management personnel.

Several general economic concepts were discussed and illustrated. First, the ideas of consumer and producer goods and services, measures of economic growth, competition, and necessities and luxuries were covered. Then, some relationships among costs, price, and volume (demand) were discussed. Included were the concepts of optimal volume and breakeven points. Important economic principles of design optimization are also illustrated in this chapter.

The use of present-economy studies in engineering decision making can provide satisfactory results and save considerable analysis effort. When an adequate engineering economic analysis can be accomplished by considering the various monetary consequences that occur in a short time period (usually one year or less), a present-economy study should be used.

2.7 References

Bierman, H., and Smidt, S. *The Capital Budgeting Decision: Economic Analysis of Investment Projects*, 8th ed. (New York: Macmillan Publishing Co., 1993).

Malik, S. A., and Sullivan, W. G. "Impact of Capacity Utilization on Product Mix and Costing Decisions.," *IEEE Transactions on Engineering Management*, vol. 42, no. 2 (May 1995). pp. 171–176.

Schweyer, Herbert E. *Analytic Models for Managerial and Engineering Economics* (New York: Reinhold Publishing Corp., 1964).

▼ 2.8 Problems

The number(s) in color at the end of a problem refer to the section(s) in that chapter most closely related to the problem.

2-1. A company in the process industry produces a chemical compound that is sold to manufacturers for use in the production of certain plastic products. The plant that produces the compound employs approximately 300 people. Develop a list of six different cost elements that would be *fixed* and a similar list of six cost elements that would be *variable*. (2.2)

2-2. Refer to Problem 2-1 and your answer to it. (2.2)

  a. Develop a table that shows the cost elements you defined and classified as fixed and variable. Indicate which of these costs are also *recurring, nonrecurring, direct,* or *indirect.*
  b. Identify one additional cost element for each of the cost categories: recurring, nonrecurring, direct, and indirect.

2-3. Classify each of the following cost items as mostly fixed or variable: (2.2)

    Raw materials
    Direct labor
    Depreciation
    Supplies
    Utilities
    Property taxes
    Administrative salaries
    Payroll taxes
    Insurance (building and equipment)
    Clerical salaries
    Sales commissions
    Rent
    Interest on borrowed money

2-4. In your own words, describe the life-cycle cost concept. Why is the potential for achieving life-cycle cost savings greatest in the acquisition phase of the life cycle? (2.2)

2-5. Explain why perfect competition is an ideal that is difficult to attain in the United States. List several business situations in which perfect competition is approached. (2.3)

2-6. A company produces circuit boards used to update outdated computer equipment. The fixed cost is $42,000 per month, and the variable cost is $53 per circuit board. The selling price per unit is $p = \$150 - 0.02D$. Maximum output of the plant is 4,000 units per month. (2.3)

  a. Determine optimum demand for this product.
  b. What is the maximum profit per month?
  c. At what volumes does breakeven occur?
  d. What is the company's range of profitable demand?

2-7. Suppose we know that $p = 1,000 - D/5$, where p = price in dollars and D = annual demand. The *total* cost per year can be approximated by $\$1,000 + 2D^2$. (2.3)

  a. Determine the value of D that maximizes profit.
  b. Show that in part (a) profit has been maximized rather than minimized.

2-8. A company has established that the relationship between the sales price for one of its products and the quantity sold per month is approximately $D = 780 - 10p$ units. (D is the demand or quantity sold per month, and p is the price in dollars.) The fixed cost is $800 per month, and the variable cost is $30 per unit produced. What number of units, D^*, should be produced per month and sold to maximize net profit? What is the maximum profit per month related to the product? Also, determine D_1' and D_2'. (2.3)

2-9. A company estimates that the relationship between unit price and demand per month for a potential new product is approximated by $p = \$100.00 - \$0.10D$. The company can produce the product by increasing fixed costs $17,500 per month, and the estimated variable cost is $40.00 per unit. What is the optimal demand, D^*, and based on this demand, should the company produce the new product? Why? (2.3)

  a. Work out the complete solution by differential calculus, starting with the formula for profit or loss per month.
  b. Solve graphically for an approximate answer.

2-10. A large wood products company is negotiating a contract to sell plywood overseas. The fixed cost that can be allocated to the production of plywood is $900,000 per month. The variable cost per thousand board feet is $131.50. The price charged will be determined by $p = \$600 - (0.05)D$ per 1,000 board feet. (2.3)

TABLE P2-13 Table for Problem 2-13		
	Site A	Site B
Average hauling distance	4 miles	3 miles
Annual rental fee for solid waste site	$5,000	$100,000
Hauling cost	$1.50/yd^3-mile	$1.50/yd^3-mile

a. For this situation determine the optimal monthly sales volume for this product and calculate the profit (or loss) at the optimal volume.

b. What is domain of profitable demand during a month?

2-11. A company produces and sells a consumer product and thus far has been able to control the volume of the product by varying the selling price. The company is seeking to maximize its net profit. It has been concluded that the relationship between price and demand, per month, is approximately $D = 500 - 5p$, where p is the price per unit in dollars. The fixed cost is $1,000 per month, and the variable cost is $20 per unit. Obtain the answer, both mathematically and graphically, to the following questions: (2.3)

a. What is the optimal number of units that should be produced and sold per month?

b. What is the maximum profit per month?

c. What are the breakeven sales quantities (range of profitable demand volume)?

2-12. A company has determined that the price and the monthly demand of one of its products are related by the equation

$$D = \sqrt{(400 - p)},$$

where p is the price per unit in dollars and D is the monthly demand. The associated fixed costs are $1,125/month and the variable costs are $100/unit. (2.3)

a. How many units should be produced and sold each month to maximize profit?

b. How do you know that the answer to part (a) maximizes profit?

c. Which of the following values of D represents the breakeven point? Why? (i) 10 units. (ii) 15 units. (iii) 20 units. (iv) 25 units.

2-13. A municipal solid-waste site for a city must be located at Site A or Site B. After sorting, some of the solid refuse will be transported to an electric power plant where it will be used as fuel. Data for the hauling of refuse from each site to the power plant are shown in Table P2-13.

a. If the power plant will pay $8.00 per cubic yard of sorted solid waste delivered to the plant, where should the solid-waste site be located? Use the city's viewpoint and assume that 200,000 cubic yards of refuse will be hauled to the plant for one year only. One site must be selected. (2.2)

b. Referring to the electric power plant, the cost Y in dollars per hour to produce electricity is $Y = 12 + 0.3X + 0.27X^2$, where X is in megawatts. Revenue in dollars per hour from the sale of electricity is $15X - 0.2X^2$. Find the value of X that gives maximum profit. (2.3)

2-14. A plant has a capacity of 4,100 hydraulic pumps per month. The fixed cost is $504,000 per month. The variable cost is $166 per pump, and the sales price is $328 per pump (assume that sales equal output volume). What is the breakeven point in number of pumps per month? What percentage reduction will occur in the breakeven point if fixed costs are reduced by 18% and unit variable costs by 6%? (2.3)

2-15. Suppose that the ABC Corporation has a production (and sales) capacity of $1,000,000 per month. Its fixed costs—over a considerable range of volume—are $350,000 per month, and the variable costs are $0.50 per dollar of sales. (2.3)

a. What is the annual breakeven point volume (D')? Develop (graph) the breakeven chart.

b. What would be the effect on D' of decreasing the variable cost per unit by 25% if the fixed costs thereby increased by 10%?

c. What would be the effect on D' if the fixed costs were decreased by 10% and the variable cost per unit were increased by the same percentage?

2-16. A company produces and sells a consumer product and is able to control the demand for the product by varying the selling price. The approximate relationship between price and demand is

$$p = \$38 + \frac{2,700}{D} - \frac{5,000}{D^2}, \quad \text{for } D > 1,$$

where p is the price per unit in dollars and D is the demand per month. The company is seeking to maximize its profit. The fixed cost is $1,000 per month and the variable cost (c_v) is $40 per unit. (2.3)

a. What is the number of units that should be produced and sold each month to maximize profit?

b. Show that your answer to part (a) maximizes profit.

2-17. A local defense contractor is considering the production of fireworks as a way to reduce dependence on the military. The variable cost per unit is $40 D. The fixed cost that can be allocated to the production of fireworks is negligible. The price charged per unit will be determined by the equation $p = \$180 - (5)D$, where D represents demand in units sold per week. (2.3)

a. What is the optimum number of units the defense contractor should produce in order to maximize profit per week?

b. What is the profit if the optimum number of units are produced?

2-18. A plant operation has fixed costs of $2,000,000 per year, and its output capacity is 100,000 electrical appliances per year. The variable cost is $40 per unit, and the product sells for $90 per unit.

a. Construct the economic breakeven chart.

b. Compare annual profit when the plant is operating at 90% of capacity with the plant operation at 100% capacity. Assume that the first 90% of capacity output is sold at $90 per unit and that the remaining 10% of production is sold at $70 per unit. (2.3)

2-19. The fixed cost for a steam line per meter of pipe is $450X + \$50$ per year. The cost for loss of heat from the pipe per meter is $\$4.8/X^{1/2}$ per year. Here X represents the thickness of insulation in meters, and X is a continuous design variable. (2.4)

a. What is the optimum thickness of the insulation?

b. How do you know that your answer in part (a) minimizes total cost per year?

c. What is the basic trade-off being made in this problem?

2-20. A farmer estimates that if he harvests his soybean crop now, he will obtain 1,000 bushels, which he can sell at $3.00 per bushel. However, he estimates that this crop will increase by an additional 1,200 bushels of soybeans for each week he delays harvesting, but the price will drop at a rate of 50 cents per bushel per week; in addition, it is likely that he will experience spoilage of approximately 200 bushels per week for each week he delays harvesting. When should he harvest his crop to obtain the largest net cash return, and how much will be received for his crop at that time? (2.4)

2-21. A recent engineering graduate was given the job of determining the best production rate for a new type of casting in a foundry. After experimenting with many combinations of hourly production rates and total production cost per hour, he summarized his findings in Table I. (See Table P2-21.) The engineer then talked to the

TABLE P2-21	Table for Problem 2-21					
Table I	Total cost/hour	$1,000	$2,600	$3,200	$3,900	$4,700
	Castings produced/hour	100	200	300	400	500
Table II	Selling price/casting	$20.00	$17.00	$16.00	$15.00	$14.50
	Castings produced/hour	100	200	300	400	500

firm's marketing specialist, who provided estimates of selling price per casting as a function of production output (see Table II). There are 8,760 hours in a year. (2.4)

a. What production rate would you recommend to maximize total profits per year?

b. How sensitive is the rate in part (a) to changes in total production cost per hour?

2-22. The cost of operating a large ship (C_O) varies as the square of its velocity (v); specifically, $C_O = knv^2$, where n is the trip length in miles and k is a constant of proportionality. It is known that at 12 miles/hour the *average* cost of operation is $100 per mile. The owner of the ship wants to minimize the cost of operation, but it must be balanced against the cost of the perishable cargo (C_c), which the customer has set at $1,500 per hour. At what velocity should the trip be planned to minimize the total cost (C_T), which is the sum of the cost of operating the ship and the cost of perishable cargo? (2.4)

2-23. Suppose you are going on a long trip to your grandmother's home in Seattle, 3,000 miles away. You have decided to drive your old Ford out there, which gets approximately 18 miles per gallon when cruising at 70 mph. Because grandma is an excellent cook and you can stay and eat at her place as long as you want (for free), you want to get to Seattle as economically as possible. However, you are also worried about your fuel consumption rate at high speeds. You also have cost of food, snacks, and lodging to balance against the cost of fuel.

What is the optimum average speed you should use so as to minimize your total trip cost, C_T? (2.4)

$C_T = C_G + C_{FSS}$, where

$C_G = n \times p_g \times f$ ($C_G = $ cost of gas),

$C_{FSS} = n \times p_{fss} \times v^{-1}$ ($C_{FSS} = $ cost of food, snacks, and lodging),

n: trip length (miles),

p_g: gas price, $1.26/gallon,

p_{fss}: average hourly spending money, $2/hour, (motel, breakfast, snacks, etc., $48 per 24 hours!),

v: average Ford velocity (mph),

$f = kv$, where k is a constant of proportionality and f is the fuel consumption rate in gallons per mile.

2-24.

a. Compare the probable part cost from Machine A and Machine B, assuming that each will make the part to the same specification. Which machine yields the lowest part cost? Assume that the interest rate is negligible.

b. If the cost of labor can be cut in half by using part-time employees, which machine should be recommended?

	Machine A	Machine B
Initial capital investment	$35,000	$150,000
Life	10 years	8 years
Market (salvage) value	$3,500	$15,000
Parts required per year	10,000	10,000
Labor cost per hour	$16	$20
Time to make one part	20 minutes	10 minutes
Maintenance cost per year	$1,000	$3,000

2-25. The following results were obtained after analyzing the operational effectiveness of a production machine at two different speeds:

Speed	Output (pieces per hour)	Time between Tool Grinds (hours)
A	400	15
B	540	10

A set of unsharpened tools costs $1,000 and can be ground 20 times. The cost of each grinding is $25. The time required to change and reset the tools is 1.5 hours, and such changes are made by a tool-setter who is paid $18/hour. The production machine operator is paid $15/hour, including the time that the machine is down for tool sharpening. Variable overhead on the machine is charged at the rate of $25/hour, including tool-changing time. A fixed-size production run will be made (independent of machine speed). (2.5)

a. At what speed should the machine be operated to minimize the total cost per piece? State your assumptions.

b. What is the basic trade-off in this problem?

2-26. Either tool steel or carbon steel can be used for the set of tools on a certain lathe. It is necessary to sharpen the tools periodically. Relevant information for each is shown in Table P2-26.

TABLE P2-26 Table for Problem 2-26

	Carbon Steel	Tool Steel
Output at optimum speed	100 pieces/hour	130 pieces/hour
Time between tool grinds	3 hours	6 hours
Time required to change tools	1 hour	1 hour
Cost of unsharpened tools	$400	$1200
Number of times tools can be ground	10	5

The cost of the lathe operator is $14.00 per hour, including the tool-changing time during which he is idle. The tool changer costs $20.00 per hour for just the time he is changing tools. Variable overhead costs for the lathe are $28.00 per hour, including tool-changing time. Which type of steel should be used to minimize overall cost per piece? (2.5)

2-27. An automatic machine can be operated at three speeds, with the following results:

Speed	Output (pieces per hour)	Time between Tool Grinds(hours)
A	400	15
B	480	12
C	540	10

A set of unsharpened tools costs $500 and can be ground 20 times. The cost of each grinding is $25. The time required to change and reset the tools is 1.5 hours, and such changes are made by a toolsetter who is paid $8.00 per hour. Variable overhead on the machine is charged at the rate of $3.75 per hour, including tool-changing time. At which speed should the machine be operated to minimize total cost per piece? The basic trade-off in this problem is between the rate of output and tool usage. (2.5)

2-28. A company is analyzing a make-versus-purchase situation for a component used in several products, and the engineering department has developed these data:

Option *A*: Purchase 10,000 items per year at a fixed price of $8.50 per item. The cost of placing the order is negligible according to the present cost accounting procedure.

Option *B*: Manufacture 10,000 items per year using available capacity in the factory. Cost estimates are direct materials = $5.00 per item and direct labor = $1.50 per item. Manufacturing overhead is allocated at 200% of direct labor (= $3.00 per item).

a. Based on these data, should the item be purchased or manufactured? (2.5)

b. If manufacturing overhead can be traced directly to this item—thus avoiding the 200% overhead rate—and it amounts to $2.15 per item, what choice should be recommended? (Traceable overhead is possible with an activity-based cost-accounting procedure; is incremental to the manufacture of the part; and consists of such cost elements as employee training, material handling, quality control, supervision, and utilities.) Traceable overhead associated with purchasing this item (vendor certification, benchmarking, etc.) is $0.50 per item.

2-29. In the design of an automobile radiator, an engineer has a choice of using either a brass–copper alloy casting or a plastic molding. Either material provides the same service. However, the brass-copper alloy casting weighs 25 pounds, compared with 20 pounds for the plastic molding. Every pound of extra weight in the automobile has been assigned a penalty of $6 to account for increased fuel consumption during the life cycle of the car. The brass-copper alloy casting costs $3.35 per pound, whereas the plastic molding costs $7.40 per pound. Machining costs per casting are $6.00 for the brass–copper alloy. Which material should the engineer select, and what is the difference in unit costs? (2.5)

2-30. For the production of part R-193, two operations are being considered. The capital investment associated with each operation is identical. Each completed part increases in value by $0.40 per part.

Operation 1 produces 2,000 parts per hour. After each hour, the tooling must be adjusted by the machine operator. This adjustment takes 20 minutes. The machine operator for Operation 1 is paid $20 per hour. (This includes fringe benefits.)

Operation 2 produces 1,750 parts per hour, but the tooling needs to be adjusted by the operator only once every two hours. This adjustment takes 30 minutes. The machine operator for Operation 2 is paid $11 per hour. (This includes fringe benefits.)

Assume an 8-hour work day. Further assume that all parts produced can be sold. (2.5)

a. Should Operation 1 or Operation 2 be recommended? Show all work.

b. What is the basic trade-off in this problem?

2-31. Rework Example 2-12 for the case where the capacity of each machine is further reduced by 25% because of machine failures, materials shortages, and operator errors. In this situation, 30,000 units of good (nondefective) product must be manufactured during the next three months. Assume one shift per day and five work days per week. (2.5)

a. Can the order be delivered on time?

b. If only one machine (*A* or *B*) can be used in part (a), which one should it be?

2-32. Two alternative designs are under consideration for a tapered fastening pin. The fastening pins are sold for $0.70 each. Either design will serve equally well and will involve the same material and manufacturing cost except for the lathe and drill operations.

Design *A* will require 16 hours of lathe time and 4.5 hours of drill time per 1,000 units. Design *B* will require 7 hours of lathe time and 12 hours of drill time per 1,000 units. The variable operating cost of the lathe, including labor, is $18.60 per hour. The variable operating cost of the drill, including labor, is $16.90 per hour. Finally, there is a sunk cost of $5,000 for Design *A* and $9,000 for Design B due to obsolete tooling. (2.5)

a. Which design should be adopted if 125,000 units are sold each year?

b. What is the annual saving over the other design?

2-33. In some countries, motorists are required to drive with their headlights on at all times. General Motors is beginning to equip their cars with daytime running lights. Most people would agree that driving with the headlights on at night is cost-effective with respect to extra fuel consumption and safety considerations. Given the following data and any additional assumptions you feel are necessary, analyze the cost-effectiveness of driving with your headlights on during the day by answering the following questions [cost-effective means that benefits outweigh (exceed) the costs]: (2.5)

75% of driving takes place during the daytime.

2% of fuel consumption is due to accessories (radio, headlights, etc.).

Cost of fuel = $1.15 per gallon.

Average distance traveled per year = 15,000 miles.

Average cost of an accident = $2,500.

Purchase price of headlights = $25.00 per set (2 headlights).

Average time car is in operation per year = 350 hours.

Average life of a headlight = 200 operating hours.

Average fuel consumption = 1 gallon per 30 miles.

a. What are the extra costs associated with driving with your headlights on during the day?

b. What are the benefits associated with driving with your headlights on during the day?

c. What additional assumptions (if any) do you need to complete your analysis?

d. It is cost-effective to drive with your headlights on during the day? Be sure to support your recommendation with the necessary calculations.

2-34. Suppose you are a mechanical engineer faced with the problem of designing a rigid coupling that will be used to join two odd-sized instrument shafts for a special customer order. Only 40 couplings will be produced, and there is no reason to suspect that there will be a repeat order in the near future. The coupling is fairly simple and can be turned from round steel rod stock. The manufacturing engineering department indicates that two machining methods are available. The following table summarizes the data for the metal lathe production and the automatic-screw production alternatives for the rigid coupling.

Comparative Costs of Two Production Processes

	Lathe	Automatic Screw Machine
Production rate	4 pieces/hr	18 pieces/hr
Machine charge	$5/hr	$25/hr
Setup charge (labor)	—	$15
Operating charge (labor)	$15/hr	$12/hr
Material cost	Same	Same
Inspection cost	Same	Same

Since an automatic screw machine is a more complex and versatile device than a turret lathe, it is not surprising that its hourly cost is higher. A skilled machine operator is needed to operate the lathe, whereas a less-skilled machine operator tends the automatic-screw machine. The setup charge for the screw machine is to pay for the services of a highly skilled setup man who initially adjusts its operation. The operator then keeps it supplied with raw material. Raw material and inspection costs would be independent of the method of production. The actual cutting tools for an automatic-screw machine would probably be more expensive than those for a lathe, because the screw machine operates at a higher cutting speed. For this short run (40 units), however, tool wear will be negligible, and this cost can be ignored. (2.5)

a. Compute the cost of producing the coupling by each method.

b. How does cost per part vary with number of items produced? Draw a graph to illustrate your answer.

2-35. One method for developing a mine containing an estimated 100,000 tons of ore will result in the recovery of 62% of the available ore deposit and will cost $23 per ton of material removed. A second method of development will recover only 50% of the ore deposit, but it will cost only $15 per ton of material removed. Subsequent processing of the removed ore recovers 300 pounds of metal from each ton of processed ore and costs $40 per ton of ore processed. The recovered metal can be sold for $0.80 per pound. Which method for developing the mine should be used if your objective is to maximize total profit from the mine? (2.5)

2-36. Ocean water contains 0.9 ounce of gold per ton. Method *A* costs $220 per ton of water processed and will recover 85% of the metal. Method *B* costs $160 per ton of water processed and will recover 65% of the metal. The two methods require the same investment and are capable of producing the same amount of gold each day. If the extracted gold can be sold for $350 per ounce, which method of extraction should be used? Assume that the supply of ocean water is unlimited. Work this problem *on the basis of profit per ounce of gold extracted.* (2.5)

2-37. Which of the following statements are true and which are false? (all sections)

a. Working capital is a variable cost.

b. The greatest potential for cost savings occurs in the operation phase of the life cycle.

c. If the capacity of an operation is significantly changed (e.g., a manufacturing plant), the fixed costs will also change.

d. The initial investment cost for a project is a nonrecurring cost.

e. Variable costs per output unit are a recurring cost.

f. A noncash cost is a cash flow.

g. Goods and services have utility because they have the power to satisfy human wants and needs.

h. The demand for necessities is more inelastic than the demand for luxuries.

i. Indirect costs can normally be allocated to a specific output or work activity.

j. Present economy studies are often done when the time value of money is not a significant factor in the situation.

k. Overhead costs normally include all costs that are not direct costs.

l. Optimal volume (demand) occurs when total costs equal total revenues.

m. Standard costs per unit of output are established in advance of actual production or service delivery.

n. A related sunk cost will normally affect the prospective cash flows associated with a situation.

o. The life cycle needs to be defined within the context of the specific situation.

p. The greatest commitment of costs occurs in the acquisition phase of the life cycle.

2-38. One component of a system's life-cycle cost is the cost of system failure. Failure costs can be reduced by designing a more reliable system. A simplified expression for system life-cycle cost, C, can be written as a function of the system's failure rate:

$$C = \frac{C_I}{\lambda} + C_R \cdot \lambda \cdot t.$$

Here C_I = investment cost (\$ per hour per failure),

C_R = system repair cost,

λ = system failure rate (failures/operating hour),

t = operating hours.

a. Assume that C_I, C_R, and t are constants. Derive an expression for λ, say, λ^*, that optimizes C. (2.4)

b. Does the equation in part (a) correspond to a maximum or minimum value of C? Show all work to support your answer.

c. What trade-off is being made in this problem?

2-39. A bicycle component manufacturer produces hubs for bike wheels. Two processes are possible for manufacturing, and the parameters of each process are as follows:

	Process 1	Process2
Production Rate	35 parts/hr	15 parts/hr
Daily Production Time	4 hr/day	7 hr/day
Percent of Parts Rejected Based on Visual Inspection	20%	9%

Assume that the daily demand for hubs allows all defect-free hubs to be sold. Additionally, tested or rejected hubs cannot be sold.

Find the process that maximizes profit per day if each part is made from \$4 worth of material and can be sold for \$30. Both processes are fully automated, and variable overhead cost is charged at the rate of \$40 per hour. (2.5)

2-40. *Brain Teaser* The student chapter of the American Society of Mechanical Engineers is planning a six-day trip to the national conference in Albany, NY. For transportation, the group will rent a car from either the State Tech Motor Pool or a local car dealer. The Motor Pool charges \$0.26 per mile, has no daily fee, and the motor pool pays for the gas. The car dealer charges \$25 per day and \$0.14 per mile, but the group must pay for the gas. The car's fuel rating is 20 miles per gallon, and the price of gas is estimated to be \$1.20 per gallon. (2.3)

a. At what point, in miles, is the cost of both options equal?

b. The car dealer has offered a special student discount and will give the students 100 free miles per day. What is the new breakeven point?

c. Suppose now that the Motor Pool reduces its all-inclusive rate to \$0.23 per mile and that the car dealer increases its rate to \$25 per day and \$0.21 per mile. In this case, the car dealer wants to encourage student business, so he offers 1,000 free miles for the entire six-day trip. He claims that if more than 882 miles are driven, students will come out ahead with one of his rental cars. If the students anticipate driving 1,600 miles (total), from whom should they rent a car? Is the car dealer's claim entirely correct?

CHAPTER

3

*M*oney–Time Relationships and Equivalence

*T*he objective of this chapter is to describe the return to capital in the form of interest (or profit) and to illustrate how basic equivalence calculations are made with respect to the time value of capital in engineering economy studies.

The following topics are discussed in this chapter:

Return to capital
Origins of interest
Simple interest
Compound interest
The concept of equivalence
Cash-flow diagrams/tables
Interest formulas
Arithmetic sequences of cash flows
Geometric sequences of cash flows
Interest rates that vary with time
Nominal versus effective interest rates
Continuous compounding

3.1 Introduction

The term *capital* refers to wealth in the form of money or property that can be used to produce more wealth. The majority of engineering economy studies involve commitment of capital for extended periods of time, so the effect of time must be considered. In this regard, it is recognized that a dollar today is worth more than a dollar one or more years from now because of the interest (or profit) it can earn. Therefore, money has a *time value.*

3.2 Why Consider Return to Capital?

Capital in the form of money for the people, machines, materials, energy, and other things needed in the operation of an organization may be classified into two basic categories. *Equity capital* is that owned by individuals who have invested their money or property in a business project or venture in the hope of receiving a profit. *Debt capital,* often called *borrowed capital,* is obtained from lenders (e.g., through the sale of bonds) for investment. In return, the lenders receive interest from the borrowers.

Normally, lenders do not receive any other benefits that may accrue from the investment of borrowed capital. They are not owners of the organization and do not participate as fully as the owners in the risks of the project or venture. Thus, lenders' fixed return on the capital loaned, in the form of interest, is more assured (i.e., has less risk) than the receipt of profit by the owners of equity capital. If the project or venture is successful, the return (profit) to the owners of equity capital can be substantially more than the interest received by lenders of capital. However, the owners could lose some or all of their money invested, whereas the lenders still could receive all the interest owed plus repayment of the money borrowed by the firm.

There are fundamental reasons why return to capital in the form of interest and profit is an essential ingredient of engineering economy studies. First, interest and profit pay the providers of capital for forgoing its use during the time the capital is being used. The fact that the supplier can realize a return on capital acts as an *incentive* to accumulate capital by savings, thus postponing immediate consumption in favor of creating wealth in the future. Second, interest and profit are payments for the *risk* the investor takes in permitting another person, or an organization, to use his or her capital.

In typical situations, investors must decide whether the expected return on their capital is sufficient to justify buying into a proposed project or venture. If capital is invested in a project, investors would expect, as a minimum, to receive a return at least equal to the amount they have sacrificed by not using it in some other available opportunity of comparable risk. This interest or profit available from an alternative investment is the *opportunity cost* of using capital in the proposed undertaking. Thus, whether borrowed capital or equity capital is involved, there is a cost for the capital employed in the sense that the project and venture must provide a sufficient return to be financially attractive to suppliers of money or property.

In summary, whenever capital is required in engineering and other business projects and ventures, it is essential that proper consideration be given to its cost (i.e., time value). The remainder of this chapter deals with time value of money principles, which are vitally important to the proper evaluation of engineering projects that form the foundation of a firm's competitiveness, and hence to its very survival.

3.3 The Origins of Interest

Like taxes, interest has existed from earliest recorded human history. Records reveal its existence in Babylon in 2000 B.C. In the earliest instances, interest was paid in money for the use of grain or other commodities that were borrowed; it was also paid in the form of grain or other goods. Many existing interest practices stem from early customs in the borrowing and repayment of grain and other crops.

History also reveals that the idea of interest became so well established that a firm of international bankers existed in 575 B.C., with home offices in Babylon. The firm's income was derived from the high interest rates it charged for the use of its money for financing international trade.

Throughout early recorded history, typical annual rates of interest on loans of money were in the neighborhood of 6 to 25%, although legally sanctioned rates as high as 40% were permitted in some instances. The charging of exorbitant interest rates on loans was termed *usury*, and prohibition of usury is found in the Bible. (See *Exodus* 22: 21–27.)

During the Middle Ages, interest taking on loans of money was generally outlawed on scriptural grounds. In 1536, the Protestant theory of usury was established by John Calvin, and it refuted the notion that interest was unlawful. Consequently, interest taking again became viewed as an essential and legal part of doing business. Eventually, published interest tables became available to the public.

3.4 Simple Interest

When the total interest earned or charged is linearly proportional to the initial amount of the loan (principal), the interest rate, and the number of interest periods for which the principal is committed, the interest and interest rate are said to be *simple*. Simple interest is not used frequently in modern commercial practice.

When simple interest is applicable, the total interest, I, earned or paid may be computed in the formula

$$I = (P)(N)(i), \tag{3-1}$$

where P = principal amount lent or borrowed,

N = number of interest periods (e.g., years),

i = interest rate per interest period.

The total amount repaid at the end of N interest periods is $P + I$. Thus, if $1,000 were loaned for three years at a simple interest rate of 10% per year, the interest earned would be

$$I = \$1,000 \times 3 \times 0.10 = \$300.$$

The total amount owed at the end of three years would be $1,000 + $300 = $1,300. Notice that the cumulative amount of interest owed is a linear function of time until the interest is repaid (usually not until the end of period N).

3.5 Compound Interest

Whenever the interest charge for any interest *period* (a year, for example) is based on the remaining principal amount plus any accumulated interest charges up to the *beginning* of that period, the interest is said to be *compound*. The effect of compounding of interest can be seen in the following table for $1,000 loaned for three periods at an interest rate of 10% compounded each period:

Period	(1) Amount Owed at Beginning of Period	(2) = (1) × 10% Interest Amount for Period	(3) = (1) + (2) Amount Owed at End of Period
1	$1,000	$100	$1,100
2	$1,100	$110	$1,210
3	$1,210	$121	$1,331

As you can see, a total of $1,331 would be due for repayment at the end of the third period. If the length of a period is one year, the $1,331 at the end of three periods (years) can be compared with the $1,300 given earlier for the same problem with simple interest. A graphical comparison of simple interest and compound interest is given in Figure 3-1. The difference is due to the effect of *compounding*, which is essentially the calculation of interest on previously earned interest. This difference would be much greater for larger amounts of money, higher interest rates, or greater numbers of interest periods. Thus, simple interest does consider the time value of money but does not involve compounding of interest. Compound interest is much more common in practice than simple interest and is used throughout the remainder of this book.

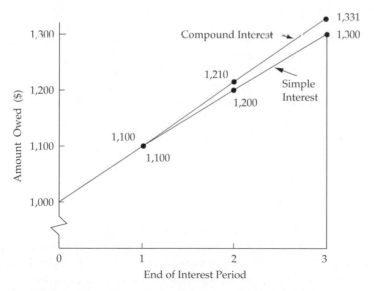

Figure 3-1 Illustration of Simple versus Compound Interest

3.6 The Concept of Equivalence

Alternatives should be compared as far as possible when they produce similar results, serve the same purpose, or accomplish the same function. This is not always possible in some types of economy studies, as we shall see later, but now our attention is directed at answering the question: How can alternatives for providing the same service or accomplishing the same function be compared when interest is involved over extended periods of time? Thus, we should consider the comparison of alternative options, or proposals, by reducing them to an *equivalent basis* that is dependent on (1) the interest rate, (2) the amounts of money involved, (3) the timing of the monetary receipts or expenses, and (4) the manner in which the interest, or profit, on invested capital is paid and the initial capital recovered.

To better understand the mechanics of interest and to expand on the notion of economic equivalence, consider a situation in which we borrow $8,000 and agree to repay it in four years at an interest rate of 10% per year. There are many plans by which the principal of this loan (i.e., $8,000) and the interest on it can be repaid. For simplicity, we have selected four plans to demonstrate the idea of economic equivalence. Here *equivalence* means that all four plans are equally desirable to the borrower. In each plan, the interest rate is 10% per year and the original amount borrowed is $8,000; thus differences among the plans rest with items (3) and (4) above. The four plans are shown in Table 3-1, and it will soon be apparent that all are *equivalent* at an interest rate of 10% per year.

In Plan 1, $2,000 of the loan principal is repaid at the end of each of years one through four. As a result, the interest we repay at the end of a particular year is affected by how much we still owe on the loan at the *beginning* of that year. Our end-of-year payment is just the sum of $2,000 and interest computed on the beginning-of-year amount owed.

Plan 2 indicates that none of the loan principal is repaid until the end of the fourth year. Our interest cost each year is $800, and it is repaid at the end of years one through four. Because interest does not accumulate in either Plan 1 or Plan 2, compounding of interest is not present. Notice that $3,200 in interest is paid in Plan 2, whereas only $2,000 is paid in Plan 1. We had the use of the $8,000 principal for four years in Plan 2 but, on average, had the use of much less than $8,000 in Plan 1.

Plan 3 requires that we repay equal end-of-year amounts of $2,524 each. Later in this chapter (Section 3.9), we will show how the $2,524 per year is computed. For our purposes here, the reader should observe that the four end-of-year payments in Plan 3 completely repay the $8,000 loan principal with interest at 10% per year. Furthermore, compounding of interest occurs in Plan 3.

Finally, Plan 4 shows that no interest and no principal are repaid for the first three years of the loan period. Then at the end of the fourth year, the original loan principal plus the accumulated interest for the four years is repaid in a single lump-sum amount of $11,712.80 (rounded in Table 3-1 to $11,713). Plan 4 involves compound interest. The total amount of interest repaid in Plan 4 is highest of all the plans considered. Not only was the principal repayment in Plan 4 deferred until the end of year four, but we also deferred all interest payment until that time.

If annual interest rates rise above 10% per year during the period of the loan, can you see that Plan 4 causes bankers to turn gray-haired rather quickly?

This brings us back to the notion of economic equivalence. If interest rates remain constant at 10% for the plans shown in Table 3-1, all four plans are equivalent. This assumes that one can freely borrow and lend at the 10% rate. Hence, we would be indifferent about whether the principal is repaid early in the loan's life

TABLE 3-1 Four Plans for Repayment of $8,000 in Four Years with Interest at 10% Per Year

(1) Year	(2) Amount Owed at Beginning of Year	(3) = 10% × (2) Interest Accrued for Year	(4) = (2) + (3) Total Money Owed at End of Year	(5) Principal Payment	(6) = (3) + (5) Total End-of-Year Payment (Cash Flow)
Plan 1: At End of Each Year Pay $2,000 Principal Plus Interest Due					
1	$8,000	$800	$8,800	$2,000	$2,800
2	6,000	600	6,600	2,000	2,600
3	4,000	400	4,400	2,000	2,400
4	2,000	200	2,200	2,000	2,200
	20,000 $-yr	$2,000		$8,000	$10,000
		(total interest)			(total amount repaid)
Plan 2: Pay Interest Due at End of Each Year and Principal at End of Four Years					
1	$8,000	$800	$8,800	$0	$800
2	8,000	800	8,800	0	800
3	8,000	800	8,800	0	800
4	8,000	800	8,800	8,000	8,800
	32,000 $-yr	$3,200		$8,000	$11,200
		(total interest)			(total amount repaid)
Plan 3. Pay in Four Equal End-of-Year Payments					
1	$8,000	$800	$8,800	$1,724	$2,524
2	6,276	628	6,904	1,896	2,524
3	4,380	438	4,818	2,086	2,524
4	2,294	230	2,524	2,294	2,524
	20,960 $-yr	$2,096		$8,000	$10,096
		(total interest)			(total amount repaid)
Plan 4: Pay Principal and Interest in One Payment at End of Four Years					
(here, column 6 ≠ column 3 + column 5)					
1	$8,000	$800	$8,800	$0	$0
2	8,800	880	9,680	0	0
3	9,680	968	10,648	0	0
4	10,648	1,065	11,713	8,000	11,713
	37,130 $-yr	$3,713		$8,000	$11,713
		(total interest)			(total amount repaid)

(e.g., Plans 1 and 3) or repaid at the end of year four (e.g., Plans 2 and 4). *Economic equivalence is established, in general, when we are indifferent between a future payment, or series of future payments, and a present sum of money.*

To see *why* the four plans in Table 3-1 are equivalent at 10%, we could plot the amount owed at the beginning of each year (column 2) versus the year. The area under the resultant bar chart represents the *dollar-years* that the money is owed. For example, the dollar-years for Plan 1 equals 20,000, which is obtained from the following bar chart:

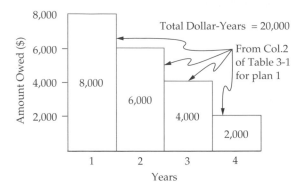

When total dollar-years are calculated for each plan and divided into total interest paid over the four years (the sum of column 3), the ratio is found to be constant:

Plan	Area under Curve (Dollar-Years) (Sum of Column 2 in Table 3-1)	Total Interest Paid (Sum of Column 3 in Table 3-1)	Ratio of Total Interest to Dollar-Years
1	$20,000	$2,000	0.10
2	32,000	3,200	0.10
3	20,960	2,096	0.10
4	37,130	3,713	0.10

Because the ratio is constant at 0.10 for all plans, we can deduce that all repayment methods considered in Table 3-1 are equivalent, even though each involves a different total end-of-year payment in column 6. Dissimilar dollar-years of borrowing, by itself, does not necessarily mean that different loan repayment plans *are* or *are not* equivalent. In summary, equivalence is established when total interest paid, divided by dollar-years of borrowing, is a constant ratio among financing plans (i.e., alternatives).

One last important point to emphasize is that the loan repayment plans of Table 3-1 are equivalent only at an interest rate of 10%. If these plans are evaluated with methods presented later in this chapter at interest rates other than 10%, one plan can be identified that is superior to the other three. For instance, when $8,000 has been lent at 10% interest and subsequently the cost of borrowed money increases to 15%, the *lender* would prefer Plan 1 in order to recover his or her funds quickly so that they might be reinvested elsewhere at higher interest rates.

3.7 Notation and Cash-Flow Diagrams and Tables

The following notation is utilized in formulas for compound interest calculations:

i = effective interest rate per interest period;

N = number of compounding periods;

P = present sum of money; the *equivalent* value of one or more cash flows at a reference point in time called the present;

F = future sum of money; the *equivalent* value of one or more cash flows at a reference point in time called the future;

A = end-of-period cash flows (or *equivalent* end-of-period values) in a uniform series continuing for a specified number of periods, starting at the end of the first period and continuing through the last period.

The use of cash-flow (time) diagrams or tables is strongly recommended for situations in which the analyst needs to clarify or visualize what is involved when flows of money occur at various times. In addition, viewpoint (remember Principle 3?) is an essential feature of cash-flow diagrams.

The difference between total cash inflows (receipts) and cash outflows (expenditures) for a specified period of time (e.g., one year) is the net cash flow for the period. As discussed in Chapter 2, cash flows are important in engineering economy because they form the basis for evaluating alternatives. Indeed, the usefulness of a cash-flow diagram for economic analysis problems is analogous to that of the free-body diagram for mechanics problems.

Figure 3-2 shows a cash-flow diagram for Plan 4 of Table 3-1, and Figure 3-3 depicts the net cash flows of Plan 3. These two figures also illustrate the definition of the preceding symbols and their placement on a cash-flow diagram. Notice that all cash flows have been placed at the end of the year to correspond with the convention used in Table 3-1. In addition, a viewpoint has been specified.

The cash-flow diagram employs several conventions:

1. The horizontal line is a *time scale*, with progression of time moving from left to right. The period (e.g., year, quarter, month) labels can be applied to intervals of time rather than to points on the time scale. Note, for example, that the end of Period 2 is coincident with the beginning of Period 3. When the end-of-period cash-flow convention is used, period numbers are placed at the end of each time interval as illustrated in Figures 3-2 and 3-3.

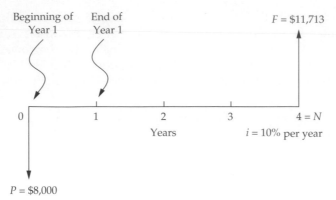

Figure 3-2 Cash-Flow Diagram for Plan 4 of Table 3-1
(Lender's Viewpoint)

2. The arrows signify cash flows and are placed at the end of the period. If a distinction needs to be made, downward arrows represent expenses (negative cash flows or cash outflows) and upward arrows represent receipts (positive cash flows or cash inflows).

3. The cash-flow diagram is dependent on the point of view. For example, the situations shown in Figures 3-2 and 3-3 were based on cash flow as seen by the lender. If the directions of all arrows had been reversed, the problem would have been diagrammed from the borrower's viewpoint.

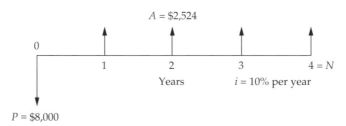

Figure 3-3 Cash-Flow Diagram for Plan 3 of Table 3-1
(Lender's Viewpoint)

EXAMPLE 3-1

Before evaluating the economic merits of a proposed investment, the XYZ Corporation insists that its engineers develop a cash-flow diagram of the proposal. An investment of $10,000 can be made that will produce uniform annual revenue of $5,310 for five years and then have a market (recovery) value of $2,000 at the end of year five. Annual expenses will be $3,000 at the end of each year for operating and maintaining the project. Draw a cash-flow diagram for the five-year life of the project. Use the corporation's viewpoint.

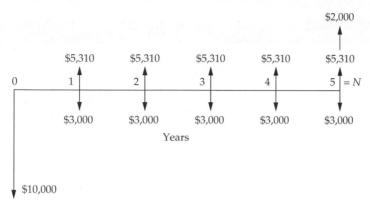

Figure 3-4 Cash-Flow Diagram for Example 3-1

SOLUTION

As shown in Figure 3-4, the initial investment of $10,000 and annual expenses of $3,000 are cash outflows, while annual revenues and the market value are cash inflows.

Notice that the beginning of a given year is the end of the preceding year. For example, the beginning of year two is the end of year one.

Example 3-2 presents a situation in which cash flows are represented in tabular form to facilitate the analysis of plans and designs.

EXAMPLE 3-2

In a company's renovation of a small office building, two feasible alternatives for upgrading the heating, ventilation, and air conditioning (HVAC) system have been identified. Either Alternative *A or* Alternative *B* must be implemented. The costs are as follows:

Alternative *A*   *Rebuild (overhaul) the existing HVAC system*
    · Equipment, labor, and materials to upgrade $18,000
    · Annual cost of electricity   32,000
    · Annual maintenance expenses   2,400

Alternative *B*   *Install a new HVAC system that utilizes existing ductwork*
    · Equipment, labor, and materials to install $60,000
    · Annual cost of electricity   9,000
    · Annual maintenance expenses   16,000
    · Replacement of a major component 4 years hence . . .   9,400

TABLE 3-2 Cash-Flow Table for Example 3-2

End-of-Year	Alternative *A* Net Cash Flow	Alternative *B* Net Cash Flow	Difference (*B* − *A*)	Cumulative Difference
0 (now)	−$18,000	−$60,000	−$42,000	−$42,000
1	−34,400	−25,000	9,400	−32,600
2	−34,400	−25,000	9,400	−23,200
3	−34,400	−25,000	9,400	−13,800
4	−34,400	−25,000 − 9,400	0	−13,800
5	−34,400	−25,000	9,400	−4,400
6	−34,400	−25,000	9,400	5,000
7	−34,400	−25,000	9,400	14,400
8	−34,400 + 2,000	−25,000 + 8,000	15,400	29,800
Total	−$291,200	−$261,400		

At the end of eight years, the estimated market value for Alternative *A* is $2,000, and for Alternative *B* it is $8,000. Assume that both alternatives will provide comparable service (comfort) over an eight-year period, and assume that the major component replaced in Alternative *B* will have no market value at the end of year eight. (1) Use a cash-flow table and end-of-year convention to tabulate the net cash flows for both alternatives. (2) Determine the annual net cash flow difference between the alternatives (*B* − *A*). (3) Compute the cumulative difference through the end of year eight. (The cumulative difference is the sum of differences, *B* − *A*, from year zero through year eight.)

SOLUTION

The cash-flow table (company's viewpoint) for this example is shown in Table 3-2. Based on these results, several points can be made: (1) Doing nothing is not an option—either A or B must be selected; (2) even though positive and negative cash flows are included in the table, on balance we are investigating two "cost-only" alternatives; (3) a decision between the two alternatives can be made just as easily on the *difference* in cash flows (i.e., on the avoidable difference) as it can on the stand-alone net cash flows for Alternatives *A* and *B*; (4) Alternative *B* has cash flows identical to those of Alternative *A*, *except for* the differences shown in the table, so if the avoidable difference can "pay its own way," Alternative *B* is the recommended choice; (5) cash flow changes caused by inflation or other suspected influences could have easily been inserted into the table and included in the analysis; and (6) it takes six *years* for the extra $42,000 investment in Alternative *B* to generate sufficient cumulative savings in annual expenses to justify the higher investment. (This ignores the time value of money.) So, which alternative is better? We'll be able to answer this question later when we consider the time value of money in order to recommend choices between alternatives.

It should be apparent that a cash-flow table clarifies the timing of cash flows, the assumptions that are being made, and the data that are available. A cash-flow table is often useful when the complexity of a situation makes it difficult to show all cash flow amounts on a diagram.

The remainder of Chapter 3 deals with the development and illustration of equivalence (time value of money) principles for assessing the economic attractiveness of investments such as those proposed in Examples 3-1 and 3-2.

Viewpoint: In most examples presented in this chapter, the company's (investor's) viewpoint will be taken.

3.8 Interest Formulas Relating Present and Future Equivalent Values of Single Cash Flows

Figure 3-5 shows a cash-flow diagram involving a present single sum, P, and a future single sum, F, separated by N periods with interest at $i\%$ per period.

Throughout this chapter a *dashed arrow,* such as that shown in Figure 3-5, indicates the quantity to be determined. Two formulas relating a given P and its unknown equivalent F are provided in Equations (3-2) and (3-3).

3.8.1 Finding *F* when Given *P*

If an amount of P dollars is invested at a point in time and $i\%$ is the interest (profit or growth) rate per period, the amount will grow to a future amount of $P + Pi = P(1 + i)$ by the end of one period; by the end of two periods, the amount will grow

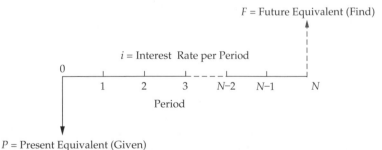

Figure 3-5 General Cash-Flow Diagram Relating Present Equivalent and Future Equivalent of Single Payments

to $P(1+i)(1+i) = P(1+i)^2$; by the end of three periods, the amount will grow to $P(1+i)^2(1+i) = P(1+i)^3$; and by the end of N periods the amount will grow to

$$F = P(1+i)^N. \tag{3-2}$$

EXAMPLE 3-3

Suppose that you borrow $8,000 now, promising to repay the loan principal plus accumulated interest in four years at $i = 10\%$ per year. How much would you repay at the end of four years?

SOLUTION

Year	Amount Owed at Start of Year	Interest Owed for Each Year	Amount Owed at End of Year	Total End-of-Year Payment
1	P $= \$\ 8{,}000$	iP $= \$\ 800$	$P(1+i)$ $= \$\ 8{,}800$	0
2	$P(1+i)$ $= \$\ 8{,}800$	$iP(1+i)$ $= \$\ 880$	$P(1+i)^2$ $= \$\ 9{,}680$	0
3	$P(1+i)^2$ $= \$\ 9{,}680$	$iP(1+i)^2$ $= \$\ 968$	$P(1+i)^3$ $= \$10{,}648$	0
4	$P(1+i)^3$ $= \$10{,}648$	$iP(1+i)^3$ $= \$1{,}065$	$P(1+i)^4$ $= \$11{,}713$	$F = \$11{,}713$

In general, we see that $F = P(1+i)^N$, and the total amount to be repaid is $11,713. This further illustrates Plan 4 in Table 3-1 in terms of notation that we shall be using throughout this book.

The quantity $(1+i)^N$ in Equation (3-2) is commonly called the *single payment compound amount factor*. Numerical values for this factor are given in the second column from the left in the tables of Appendix C for a wide range of values of i and N. In this book, we shall use the functional symbol $(F/P, i\%, N)$ for $(1+i)^N$. Hence, Equation (3-2) can be expressed as

$$F = P(F/P, i\%, N), \tag{3-3}$$

where the factor in parentheses is read "find F given P at $i\%$ interest per period for N interest periods." Note that the sequence of F and P in F/P is the same as in the initial part of Equation (3-3), where the unknown quantity, F, is placed on the left-hand side of the equation. This sequencing of letters is true of all functional symbols used in this book and makes them easy to remember.

Another example of finding F when given P, together with a cash-flow diagram and solution, appears in Table 3-3. Note in Table 3-3 that for each of the six common discrete compound interest circumstances covered, two problem statements are given—(a) *in borrowing–lending terminology* and (b) *in equivalence terminology*—but they both represent the same cash-flow situation. Indeed, there are generally many ways in which a given cash-flow situation can be expressed.

In general, a good way to interpret a relationship such as Equation (3-3) is that the calculated amount, F, at the point in time at which it occurs, *is equivalent to* (i.e., can be traded for) the known value, P, at the point in time at which it occurs, for the given interest or profit rate, i.

3.8.2 Finding *P* when Given *F*

From Equation (3-2), $F = P(1+i)^N$. Solving this for P gives the relationship

$$P = F \left(\frac{1}{1+i} \right)^N = F(1+i)^{-N}. \tag{3-4}$$

The quantity $(1+i)^{-N}$ is called the *single payment present worth factor*. Numerical values for this factor are given in the third column of the tables in Appendix C for a wide range of values of i and N. We shall use the functional symbol $(P/F, i\%, N)$ for this factor. Hence,

$$P = F(P/F, i\%, N). \tag{3-5}$$

EXAMPLE 3-4

An investor (owner) has an option to purchase a tract of land that will be worth $10,000 in six years. If the value of the land increases at 8% each year, how much should the investor be willing to pay now for this property?

SOLUTION

The purchase price can be determined from Equation (3-5) and Table C-11 in Appendix C as follows:

$$P = \$10{,}000(P/F, 8\%, 6)$$

$$P = \$10{,}000(0.6302)$$

$$= \$6{,}302.$$

Another example of this type of problem, together with a cash-flow diagram and solution, is given in Table 3-3.

TABLE 3-3 Discrete Cash-Flow Examples Illustrating Equivalence

Example Problems (all using an interest rate of $i = 10\%$ per year—see Table C-13 of Appendix C)

To Find:	Given:	(a) In Borrowing-Lending Terminology:	(b) In Equivalence Terminology:	Cash-Flow Diagram[a]	Solution
For single cash flows:					
F	P	A firm borrows $1,000 for eight years. How much must it repay in a lump sum at the end of the eighth year?	What is the future equivalent at the end of eight years of $1,000 at the beginning of those eight years?	$P = \$1,000$ $N = 8$ 0 $F = ?$	$F = P(F/P, 10\%, 8)$ $= \$1,000(2.1436)$ $= \$2,143.60$
P	F	A firm wishes to have $2,143.60 eight years from now. What amount should be deposited now to provide for it?	What is the present equivalent of $2,143.60 received eight years from now?	$F = \$2,143.60$ 0 $N = 8$ $P = ?$	$P = F(P/F, 10\%, 8)$ $= \$2,143.60(0.4665)$ $= \$1,000.00$
For uniform series:					
F	A	If eight annual deposits of $187.45 each are placed in an account, how much money has accumulated immediately after the last deposit?	What amount at the end of the eighth year is equivalent to eight end-of-year payments of $187.45 each?	$F = ?$ $1\,2\,3\,4\,5\,6\,7\,8$ $A = \$187.45$	$F = A(F/A, 10\%, 8)$ $= \$187.45(11.4359)$ $= \$2,143.60$

TABLE 3-3 (continued)

P	A	How much should be deposited in a fund now to provide for eight end-of-year withdrawals of $187.45 each?	What is the present equivalent of eight end-of-year payments of $187.45 each? A = $187.45 1 2 3 4 5 6 7 8 P = ?	$P = A(P/A, 10\%, 8)$ $= \$187.45(5.3349)$ $= \$1,000.00$
A	F	What uniform annual amount should be deposited each year in order to accumulate $2,143.60 at the time of the eighth annual deposit?	What uniform payment at the end of eight successive years is equivalent to $2,143.60 at the end of the eighth year? F = $2,143.60 1 2 3 4 5 6 7 8 A = ?	$A = F(A/F, 10\%, 8)$ $= \$2,143.60(0.0874)$ $= \$187.45$
A	P	What is the size of eight equal annual payments to repay a loan of $1,000? The first payment is due one year after receiving the loan.	What uniform payment at the end of eight successive years is equivalent to $1,000 at the beginning of the first year? P = $1,000 1 2 3 4 5 6 7 8 A = ?	$A = P(A/P, 10\%, 8)$ $= \$1,000(0.18745)$ $= \$187.45$

[a] The cash-flow diagram represents the example as stated in borrowing–lending terminology.

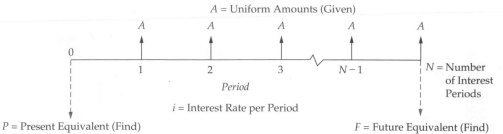

Figure 3-6 General Cash-Flow Diagram Relating Uniform Series (Ordinary Annuity) to Its Present Equivalent and Future Equivalent Values

3.9 Interest Formulas Relating a Uniform Series (Annuity) to Its Present and Future Equivalent Values

Figure 3-6 shows a general cash-flow diagram involving a series of uniform (equal) receipts, each of amount A, occurring at the end of each period for N periods with interest at $i\%$ per period. Such a uniform series is often called an *annuity*. It should be noted that the formulas and tables to be presented are derived such that A occurs at the end of each period, and thus,

1. P (present equivalent value) occurs one interest period before the first A (uniform amount),
2. F (future equivalent value) occurs at the same time as the last A, and N periods after P, and
3. A (annual equivalent value) occurs at the end of periods 1 through N, inclusive.

The timing relationship for P, A, and F can be observed in Figure 3-6. Four formulas relating A to F and P will be developed.

3.9.1 Finding *F* when Given *A*

If a cash flow in the amount of A dollars occurs at the end of each period for N periods and $i\%$ is the interest (profit or growth) rate per period, the future equivalent value, F, at the end of the Nth period is obtained by summing the future equivalents of each of the cash flows. Thus,

$$F = A(F/P, i\%, N-1) + A(F/P, i\%, N-2) + A(F/P, i\%, N-3) + \cdots$$

$$+ A(F/P, i\%, 1) + A(F/P, i\%, 0)$$

$$= A[(1+i)^{N-1} + (1+i)^{N-2} + (1+i)^{N-3} + \cdots + (1+i)^1 + (1+i)^0].$$

The bracketed terms comprise a geometric sequence having a common ratio of $(1+i)^{-1}$. Recall that the sum of the first N terms of a geometric sequence is

$$S_N = \frac{a_1 - ba_N}{1 - b} \quad (b \neq 1),$$

where a_1 is the first term in the sequence, a_N is the last term, and b is the common ratio. If we let $b = (1+i)^{-1}$, $a_1 = (1+i)^{N-1}$, and $a_N = (1+i)^0$, then

$$F = A \left[\frac{(1+i)^{N-1} - \dfrac{1}{(1+i)}}{1 - \dfrac{1}{(1+i)}} \right],$$

which reduces to

$$F = A \left[\frac{(1+i)^N - 1}{i} \right]. \qquad (3\text{-}6)$$

The quantity $\{[(1+i)^N - 1]/i\}$ is called the *uniform series compound amount factor*. It is the starting point for developing the remaining three uniform series interest factors.

Numerical values for the uniform series compound amount factor are given in the fourth column of the tables in Appendix C for a wide range of values of i and N. We shall use the functional symbol $(F/A, i\%, N)$ for this factor. Hence Equation (3-6) can be expressed as

$$F = A(F/A, i\%, N). \qquad (3\text{-}7)$$

Examples of this type of "wealth accumulation" problem based on the $(F/A, i\%, N)$ factor are provided here and in Table 3-3.

EXAMPLE 3-5

(a) Suppose you make 15 equal annual deposits of $1,000 each into a bank account paying 5% interest per year. The first deposit will be made one year from today. How much money can be withdrawn from this bank account immediately after the 15th deposit?

SOLUTION
The value of A is $1,000, N equals 15 years, and $i = 5\%$ per year. Immediately after the 15th payment, the future equivalent amount is

$$F = \$1{,}000(F/A, 5\%, 15)$$

$$= \$1{,}000(21.5786)$$

$$= \$21{,}578.60.$$

Notice in the following cash-flow diagram that the value of F is coincident with the last payment of $1,000.

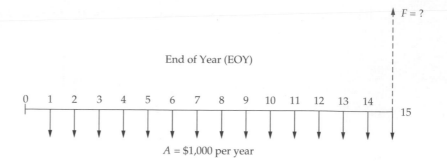

(b) To illustrate further the amazing effects of compound interest, we consider the credibility of this statement: "If you are 20 years of age and save $1.00 each day for the rest of your life, you can become a millionaire." Let's assume that you live to age 80 and that the annual interest rate is 10% ($i = 10\%$). Under these specific conditions, we compute the future compound amount (F) to be

$$F = \$365/\text{yr} \, (F/A, 10\%, 60 \text{ years})$$

$$= \$365 \, (3{,}034.81)$$

$$= \underline{\$1{,}107{,}706.}$$

Thus, the statement is true for the assumptions given! The moral is to *start saving early* and let the "magic" of compounding work on your behalf!

A few words to the wise: Saving money early and preserving resources through frugality (avoiding waste) are extremely important ingredients of *wealth creation* in general. Often, being frugal means postponing the satisfaction of immediate material wants for the creation of a better tomorrow. In this regard, be very *cautious* about spending tomorrow's cash today by undisciplined borrowing (e.g., with credit cards). The ($F/A, i\%, N$) factor demonstrates how *fast* your debt can accumulate!

3.9.2 Finding *P* when Given *A*

From Equation (3-2), $F = P(1 + i)^N$. Substituting for F in Equation (3-6), one determines that

$$P(1 + i)^N = A \left[\frac{(1 + i)^N - 1}{i} \right].$$

Dividing both sides by $(1 + i)^N$, we get

$$P = A \left[\frac{(1 + i)^N - 1}{i(1 + i)^N} \right].$$ (3-8)

Thus, Equation (3-8) is the relation for finding the present equivalent value (as of the beginning of the first period) of a uniform series of end-of-period cash flows of amount A for N periods. The quantity in brackets is called the *uniform series present worth factor*. Numerical values for this factor are given in the fifth column of the tables in Appendix C for a wide range of values of i and N. We shall use the functional symbol $(P/A, i\%, N)$ for this factor. Hence,

$$P = A(P/A, i\%, N).$$ (3-9)

EXAMPLE 3-6

If a certain machine undergoes a major overhaul now, its output can be increased by 20%—which translates into additional cash flow of $20,000 at the end of each year for five years. If $i = 15\%$ per year, how much can we afford to invest to overhaul this machine?

SOLUTION
The increase in cash flow is $20,000 per year, and it continues for five years at 15% annual interest. The upper limit on what we can afford to spend now is

$$P = \$20,000(P/A, 15\%, 5)$$

$$= \$20,000(3.3522)$$

$$= \$67,044.$$

EXAMPLE 3-7

Suppose that your rich uncle has $1,000,000 that he wishes to distribute to his heirs at the rate of $100,000 per year. If the $1,000,000 is deposited in a bank account that earns 6% interest per year, how many years will it take to completely deplete the account? How long will it take if the account earns 8% interest per year instead of 6%?

SOLUTION
From Table C-9, solve for N in the following equation: $\$1,000,000 = \$100,000(P/A, 6\%, N)$; $N = 15.7$ years. When the interest rate is increased to 8% per year, it will take 20.9 years to bring the account balance to zero, which is found by solving this equation: $\$1,000,000/\$100,000 = (P/A, 8\%, N)$.

3.9.3 Finding *A* when Given *F*

Taking Equation (3-6) and solving for A, one finds that

$$A = F\left[\frac{i}{(1+i)^N - 1}\right]. \tag{3-10}$$

Thus, Equation (3-10) is the relation for finding the amount, A, of a uniform series of cash flows occurring at the end of N interest periods that would be equivalent to (have the same value as) its future equivalent value occurring at the end of the last period. The quantity in brackets is called the *sinking fund factor*. Numerical values for this factor are given in the sixth column of the tables in Appendix C for a wide range of values of i and N. We shall use the functional symbol $(A/F, i\%, N)$ for this factor. Hence,

$$A = F(A/F, i\%, N). \tag{3-11}$$

EXAMPLE 3-8

An enterprising student is planning to have personal savings totaling $1,000,000 when she retires at age 65. She is now 20 years old. If the annual interest rate will average 7% over the next 45 years on her savings account, what equal end-of-year amount must she save to accomplish her goal?

SOLUTION

The future amount, F, is $1,000,000. The equal annual amount this student must place in a *sinking fund* that grows to $1,000,000 in 45 years at 7% annual interest (see Table C-10) is

$$A = \$1,000,000 \, (A/F, 7\%, 45)$$

$$= \$1,000,000(0.0035)$$

$$= \$3,500.$$

Another example of this type of problem, together with a cash-flow diagram and solution, is given in Table 3-3.

3.9.4 Finding *A* when Given *P*

Taking Equation (3-8) and solving for A, one finds that

$$A = P\left[\frac{i(1+i)^N}{(1+i)^N - 1}\right].$$ (3-12)

Thus, Equation (3-12) is the relation for finding the amount, A, of a uniform series of cash flows occurring at the end of each of N interest periods that would be equivalent to, or could be traded for, the present equivalent P, occurring at the beginning of the first period. The quantity in brackets is called the *capital recovery factor*.* Numerical values for this factor are given in the seventh column of the tables in Appendix C for a wide range of values of i and N. We shall use the functional symbol $(A/P, i\%, N)$ for this factor. Hence,

$$A = P(A/P, i\%, N).$$ (3-13)

An example that utilizes the equivalence between a present lump-sum loan amount and a series of equal uniform annual payments starting at the end of year one and continuing through year four was provided in Table 3-1 as Plan 3. Equation (3-13) yields the equivalent value of A that repays the $8,000 loan plus 10% interest per year over four years:

$$A = \$8,000(A/P, 10\%, 4) = \$8,000(0.3155) = \$2,524.$$

The entries in columns three and five of Plan 3 in Table 3-1 can now be better understood. Interest owed at the end of year one equals $8,000(0.10), and therefore the principal repaid out of the total end-of-year payment of $2,524 is the difference, $1,724. At the beginning of year two, the amount of principal owed is $8,000 − $1,724 = $6,276. Interest owed at the end of year two is $6,276(0.10) ≅ $628, and the principal repaid at that time is $2,524 − $628 = $1,896. The remaining entries in Plan 3 are obtained by performing these calculations for years three and four.

A graphical summary of Plan 3 is given in Figure 3-7. Here it can be seen that 10% interest is being paid on the beginning-of-year amount owed and that year-end payments of $2,524, consisting of interest and principal, bring the amount owed to $0 at the end of the fourth year. (The exact value of A is $2,523.77 and produces an exact value of $0 at the end of four years.) It is important to note that all the uniform series interest factors in Table 3-3 involve the same concept as the one illustrated in Figure 3-7.

Another example of a problem where we desire to compute an equivalent value for A, from a given value of P and a known interest rate and number of compounding periods, is given in Table 3-3.

---

*The capital recovery factor is more conveniently expressed as $i/[1 - (1+i)^{-N}]$ for computation with a hand-held calculator.

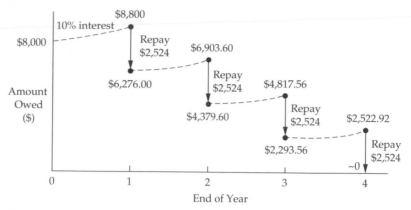

Figure 3-7 Relationship of Cash Flows for Plan 3 of Table 3-1 to Repayment of the $8,000 Loan Principal

For an annual interest rate of 10%, the reader should now be convinced from Table 3-3 that $1,000 at the beginning of year one is equivalent to $187.45 at the end of years one through eight, which is then equivalent to $2,143.60 at the end of year eight.

3.9.5 Interest Factor Relationships: Summary

This section is summarized by presenting equations and graphs of the relationships between an annuity and its present and future equivalent values:

$$(A/P, i\%, N) = \frac{1}{(P/A, i\%, N)}; \tag{3-14}$$

$$(A/F, i\%, N) = \frac{1}{(F/A, i\%, N)}; \tag{3-15}$$

$$(F/A, i\%, N) = (P/A, i\%, N)(F/P, i\%, N); \tag{3-16}$$

$$(P/A, i\%, N) = \sum_{k=1}^{N}(P/F, i\%, k); \tag{3-17}$$

$$(F/A, i\%, N) = \sum_{k=1}^{N}(F/P, i\%, N - k); \tag{3-18}$$

$$(A/F, i\%, N) = (A/P, i\%, N) - i. \tag{3-19}$$

For a fixed value of N, the following graphs help to visualize the preceding equations:

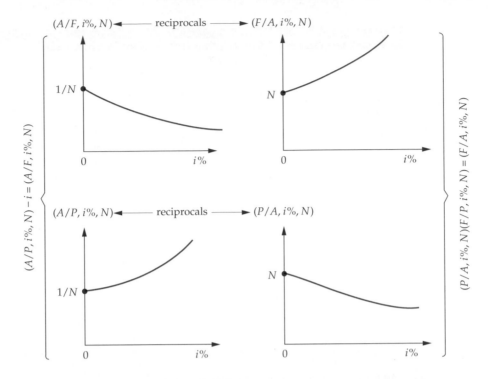

3.10 Interest Formulas for Discrete Compounding and Discrete Cash Flows

Table 3-4 provides a summary of the six most common discrete compound interest factors, utilizing notation of the preceding sections. The formulas are for *discrete compounding*, which means that the interest is compounded at the end of each finite-length period, such as a month or a year. Furthermore, the formulas also assume discrete (i.e., lump-sum) cash flows spaced at the end of equal time intervals on a cash-flow diagram. Discrete compound interest factors are given in Appendix C, where the assumption is made that i remains constant during the N compounding periods.

3.11 Deferred Annuities (Uniform Series)

All annuities (uniform series) discussed to this point involve the first cash flow being made at the end of the first period, and they are called *ordinary annuities*.

TABLE 3-4 Discrete Compounding-Interest Factors and Symbols[a]

To Find:	Given:	Factor by Which to Multiply "Given"[a]	Factor Name	Factor Functional Symbol[b]
For single cash flows:				
F	P	$(1+i)^N$	Single payment compound amount	$(F/P, i\%, N)$
P	F	$\dfrac{1}{(1+i)^N}$	Single payment present worth	$(P/F, i\%, N)$
For uniform series (annuities):				
F	A	$\dfrac{(1+i)^N - 1}{i}$	Uniform series compound amount	$(F/A, i\%, N)$
P	A	$\dfrac{(1+i)^N - 1}{i(1+i)^N}$	Uniform series present worth	$(P/A, i\%, N)$
A	F	$\dfrac{i}{(1+i)^N - 1}$	Sinking fund	$(A/F, i\%, N)$
A	P	$\dfrac{i(1+i)^N}{(1+i)^N - 1}$	Capital recovery	$(A/P, i\%, N)$

[a] i, effective interest rate per interest period; N, number of interest periods; A, uniform series amount (occurs at the end of each interest period); F, future equivalent; P, present equivalent.
[b] The functional symbol system is used throughout this book.

If the cash flow does not begin until some later date, the annuity is known as a *deferred annuity*. If the annuity is deferred J periods ($J < N$), the situation is as portrayed in Figure 3-8, in which the entire framed ordinary annuity has been moved forward from "time present," or "time 0," by J periods. Remember that in an annuity deferred for J periods the first payment is made at the end of period $(J + 1)$, assuming that all periods involved are equal in length.

The present equivalent at the end of period J of an annuity with cash flows of amount A is, from Equation (3-9), $A(P/A, i\%, N - J)$. The present equivalent of the single amount $A(P/A, i\%, N - J)$ as of time 0 will then be

$$A(P/A, i\%, N - J)(P/F, i\%, J).$$

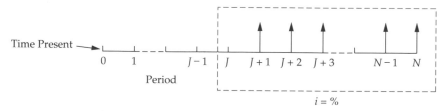

Figure 3-8 General Cash-Flow Representation of a Deferred Annuity (Uniform Series)

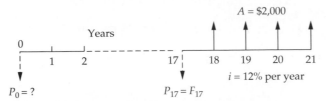

Figure 3-9 Cash-Flow Diagram of the Deferred Annuity Problem in Example 3-9

EXAMPLE 3-9

To illustrate the preceding discussion, suppose that a father, on the day his son is born, wishes to determine what lump amount would have to be paid into an account bearing interest of 12% per year to provide withdrawals of $2,000 on each of the son's 18th, 19th, 20th, and 21st birthdays.

SOLUTION

The problem is represented in Figure 3-9. One should first recognize that an ordinary annuity of four withdrawals of $2,000 each is involved, and that the present equivalent of this annuity occurs at the 17th birthday when a $(P/A, i\%, N - J)$ factor is utilized. In this problem, $N = 21$ and $J = 17$. It is often helpful to use a *subscript* with P or F to denote the respective point in time. Hence,

$$P_{17} = A(P/A, 12\%, 4) = \$2,000(3.0373) = \$6,074.60.$$

Note the dashed arrow in Figure 3-9, denoting P_{17}. Now that P_{17} is known, the next step is to calculate P_0. With respect to P_0, P_{17} is a future equivalent, and hence it could also be denoted F_{17}. Money at a given point in time, such as the end of period 17, is the same regardless of whether it is called a present equivalent or a future equivalent. Hence,

$$P_0 = F_{17}(P/F, 12\%, 17) = \$6,074.60(0.1456) = \$884.46,$$

which is the amount that the father would have to deposit on the day his son is born.

EXAMPLE 3-10

As an addition to the problem in Example 3-9, suppose that the father wishes to determine the equivalent worth of the four $2,000 withdrawals as of the son's 24th birthday. This could mean that four amounts were never withdrawn or possibly that the son took them and immediately redeposited them in an account also earning interest at 12% per year. Using our subscript system, we wish to calculate F_{24} as shown in Figure 3-10.

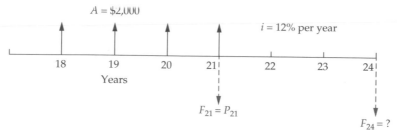

Figure 3-10 Cash-Flow Diagram for the Deferred Annuity Problem in Example 3-10

SOLUTION
One way to work this is to calculate

$$F_{21} = A(F/A, 12\%, 4) = \$2,000(4.7793) = \$9,558.60.$$

To determine F_{24}, F_{21} can now be denoted P_{21}, and

$$F_{24} = P_{21}(F/P, 12\%, 3) = \$9,558.60(1.4049) = \$13,428.88.$$

Another, quicker way to work the problem is to recognize that $P_{17} = \$6,074.60$ and $P_0 = \$884.46$ are each equivalent to the four $2,000 withdrawals. Hence, one can find F_{24} directly, given P_{17} or P_0. Using P_0, we obtain

$$F_{24} = P_0(F/P, 12\%, 24) = \$884.46(15.1786) = \$13,424.86,$$

which closely approximates the previous answer. The two numbers differ by $4.02, which can be attributed to round-off error in the interest factors.

3.12 Equivalence Calculations Involving Multiple Interest Formulas

The reader should now be comfortable with equivalence problems that involve discrete compounding of interest and discrete cash flows. All compounding of interest takes place once per time period (e.g., a year), and to this point cash flows also occur once per time period. This section provides three examples involving two or more equivalence calculations to solve for an unknown quantity. The end-of-year cash-flow convention is used. Again, the interest rate is constant over the N time periods.

EXAMPLE 3-11

Figure 3-11 depicts an example problem with a series of year-end cash flows extending over eight years. The amounts are $100 for the first year, $200 for the second year,

$500 for the third year, and $400 for each year from the fourth through the eighth. These could represent something like the expected maintenance expenditures for a certain piece of equipment or payments into a fund. Note that the payments are shown at the end of each year, which is a standard assumption (convention) for this book and for economic analyses in general unless one has information to the contrary. It is desired to find the (a) present equivalent expenditure, P_0; (b) future equivalent expenditure, F_8; and (c) annual equivalent expenditure, A, of these cash flows if the annual interest rate is 20%.

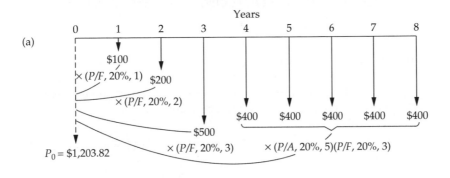

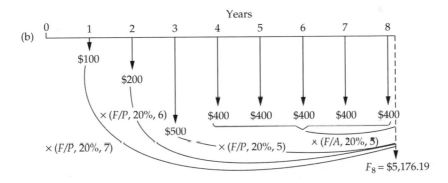

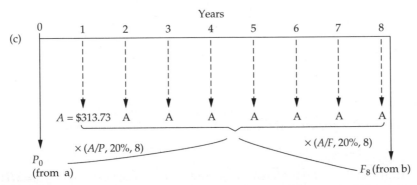

Figure 3-11 Example 3-11 for Calculating the Equivalent *P*, *F*, and *A* Values

SOLUTION

(a) To find the equivalent P_0, one needs to sum the equivalent values of all payments as of the beginning of the first year (time zero). The required movements of money through time are shown graphically in Figure 3-11(a).

$$
\begin{aligned}
P_0 = \ &F_1(P/F, 20\%, 1) & = & \quad \$100(0.8333) & = & \quad \$ \quad 83.33 \\
&+F_2(P/F, 20\%, 2) & + & \quad \$200(0.6944) & & + \quad 138.88 \\
&+F_3(P/F, 20\%, 3) & + & \quad \$500(0.5787) & & + \quad 289.35 \\
&+A(P/A, 20\%, 5) \times (P/F, \ 20\%, 3) & + & \quad \$400(2.9900) \times (0.5787) & + & \quad 692.26 \\
\hline
& & & & & \$1,203.82.
\end{aligned}
$$

(b) To find the equivalent F_8, one can sum the equivalent values of all payments as of the end of the eighth year (time eight). Figure 3-11(b) indicates these movements of money through time. However, since the equivalent P_0 is already known to be $1,203.82, one can directly calculate

$$F_8 = P_0(F/P, 20\%, 8) = \$1,203.82(4.2998) = \$5,176.19.$$

(c) The equivalent A of the irregular cash flows can be calculated directly from either P_0 or F_8 as

$$A = P_0(A/P, 20\%, 8) = \$1,203.82(0.2606) = \$313.73$$

or

$$A = F_8(A/F, 20\%, 8) = \$5,176.19(0.0606) = \$313.73$$

The computation of A from P_0 and F_8 is shown in Figure 3-11(c). Thus, one finds that the irregular series of payments shown in Figure 3-11 is equivalent to $1,203.82 at time zero, $5,176.19 at time eight, or a uniform series of $313.73 at the end of each of the eight years.

EXAMPLE 3-12

Transform the cash flows on the left-hand side of Figure 3-12 to their equivalent cash flows on the right-hand side. That is, take the left-hand quantities as givens and determine the unknown value of Q in terms of H in Figure 3-12. The interest rate is 10% per year. (Notice that "$\Longleftrightarrow$" means "equivalent to.")

SOLUTION

If all cash flows on the left are discounted to year zero, we have $P_0 = 2H(P/A, 10\%, 4) + H(P/A, 10\%, 3)(P/F, 10\%, 5) = 7.8839H$. When cash flows on the right are also discounted to year zero, we can solve for Q in terms of H. [Notice that Q at the end

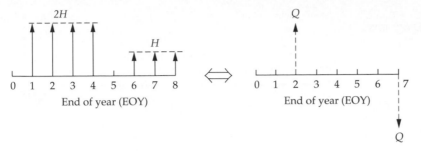

Figure 3-12 Cash-Flow Diagrams for Example 3-12

of year (EOY) two is positive, Q at EOY seven is negative, and the two Q values must be equal in amount.] So

$$7.8839H = Q(P/F, 10\%, 2) - Q(P/F, 10\%, 7),$$

or

$$Q = 25.172H.$$

EXAMPLE 3-13

Suppose you start a savings plan in which you save $500 each year for 15 years. You make your first payment at age 22 and then leave the accumulated sum in the savings plan (and make NO more annual payments) until you reach age 65, at which time you withdraw the total accumulated amount. The average annual interest rate you'll earn on this savings plan is 10%.

A friend of yours (exactly your age) from Minnesota State University waits 10 years to start her savings plan. (That is, she is age 32.) She decides to save $2,000 each year in an account earning interest at the rate of 10% per year. She will make these annual payments until she is 65 years old, at which time she will withdraw the total accumulated amount.

How old will you be when your friend's *accumulated* savings amount (including interest) exceeds yours? State any assumptions you think are necessary.

SOLUTION

Creating cash-flow diagrams for Example 3-13 is an important first step in solving for the unknown number of years, N, until the future equivalent values of both savings plans are equal. The two diagrams are shown in Figure 3-13. The future equivalent (F) of your plan is $500(F/A, 10\%, 15)(F/P, 10\%, N - 36)$, and that of your friend is $F' = \$2,000(F/A, 10\%, N - 31)$. It is clear that N, the age at which

Figure 3-13
Cash-Flow Diagrams
for Example 3-13

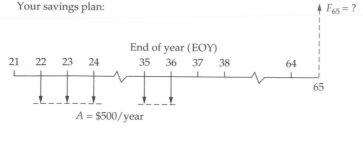

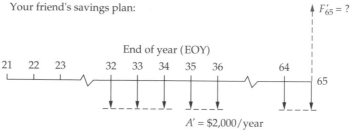

$F = F'$, is greater than 32. Assuming that the interest rate remains constant at 10% per year, the value of N can be determined by trial and error:

N	Your Plan's F	Friend's F'
36	$15,886	$12,210
38	$19,222	$18,974
39	$21,145	$22,872
40	$23,259	$27,159

By the time you reach age 39, your friend's accumulated savings will exceed yours. (If you had deposited $1,000 instead of $500, you would be over 76 years old when your friend's plan surpassed yours. Moral: Start saving early!)

▼ 3.13 Interest Formulas Relating a Uniform Gradient of Cash Flows to Its Annual and Present Equivalents

Some problems involve receipts or expenses that are projected to increase or decrease by a uniform *amount* each period, thus constituting an arithmetic sequence of cash flows. For example, because of leasing a certain type of equipment, maintenance and repair savings relative to purchasing the equipment may increase by a roughly constant amount each period. This situation can be modeled as a *uniform gradient* of cash flows.

Figure 3-14 is a cash-flow diagram of a sequence of end-of-period cash flows increasing by a constant amount, G, in each period. The G is known as the *uniform gradient amount*. Note that the timing of cash flows on which the derived formulas and tabled values are based is as follows:

Figure 3-14
Cash-Flow Diagram for
a Uniform Gradient
Increasing by G
Dollars per Period

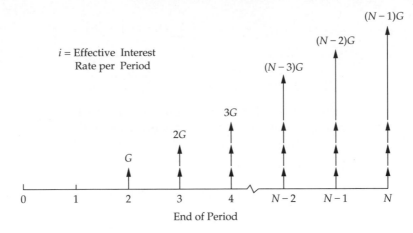

End of Period	Cash Flows
1	0
2	G
3	$2G$
.	.
.	.
.	.
$N-1$	$(N-2)G$
N	$(N-1)G$

Notice that the first cash flow occurs at the end of period two.

3.13.1 Finding *F* when Given *G*

The future equivalent, F, of the arithmetic sequence of cash flows shown in Figure 3-14 is

$$F = G(F/A, i\%, N-1) + G(F/A, i\%, N-2) + \cdots$$
$$+ G(F/A, i\%, 2) + G(F/A, i\%, 1),$$

or

$$F = G\left[\frac{(1+i)^{N-1} - 1}{i} + \frac{(1+i)^{N-2} - 1}{i} + \cdots\right.$$
$$\left. + \frac{(1+i)^2 - 1}{i} + \frac{(1+i)^1 - 1}{i}\right]$$
$$= \frac{G}{i}[(1+i)^{N-1} + (1+i)^{N-2} + \cdots$$
$$+ (1+i)^2 + (1+i)^1 + 1] - \frac{NG}{i}$$

$$= \frac{G}{i}\left[\sum_{k=0}^{N-1}(1+i)^k\right] - \frac{NG}{i}$$

$$= \frac{G}{i}(F/A, i\%, N) - \frac{NG}{i}. \tag{3-20}$$

Instead of working with future equivalent values, it is usually more practical to deal with annual and present equivalents in Figure 3-14.

3.13.2 Finding *A* when Given *G*

From Equation (3-20), it is easy to develop an expression for *A* as follows:

$$A = F(A/F, i, N)$$

$$= \left[\frac{G}{i}(F/A, i, N) - \frac{NG}{i}\right](A/F, i, N)$$

$$= \frac{G}{i} - \frac{NG}{i}(A/F, i, N)$$

$$= \frac{G}{i} - \frac{NG}{i}\left[\frac{i}{(1+i)^N - 1}\right]$$

$$= G\left[\frac{1}{i} - \frac{N}{(1+i)^N - 1}\right]. \tag{3-21}$$

The term in brackets in Equation (3-21) is called the *gradient to uniform series conversion factor*. Numerical values for this factor are given on the right side of Appendix C for a range of *i* and *N* values. We shall use the functional symbol $(A/G, i\%, N)$ for this factor. Thus,

$$A = G(A/G, i\%, N). \tag{3-22}$$

3.13.3 Finding *P* when Given *G*

We may now utilize Equation (3-21) to establish the equivalence between *P* and *G*:

$$P = A(P/A, i\%, N)$$

$$= G\left[\frac{1}{i} - \frac{N}{(1+i)^N - 1}\right]\left[\frac{(1+i)^N - 1}{i(1+i)^N}\right]$$

$$= G\left[\frac{(1+i)^N - 1 - Ni}{i^2(1+i)^N}\right]$$

$$= G\left\{\frac{1}{i}\left[\frac{(1+i)^N - 1}{i(1+i)^N} - \frac{N}{(1+i)^N}\right]\right\}. \tag{3-23}$$

The term in braces in Equation (3-23) is called the *gradient to present equivalent conversion factor.* It can also be expressed as $(1/i)[(P/A, i\%, N) - N(P/F, i\%, N)]$. Numerical values for this factor are given in column 8 of Appendix C for a wide assortment of i and N values. We shall use the functional symbol $(P/G, i\%, N)$ for this factor. Hence,

$$P = G(P/G, i\%, N). \tag{3-24}$$

3.13.4 Computations Using *G*

Be sure to notice that the direct use of gradient conversion factors applies when there is no cash flow at the end of period one, as in Example 3-14. There may be an *A* amount at the end of period one, but it is treated separately, as illustrated in Examples 3-15 and 3-16. A major advantage of using gradient conversion factors (i.e., computational time savings) is realized when *N* becomes large.

EXAMPLE 3-14

As an example of the straightforward use of the gradient conversion factors, suppose that certain end-of-year cash flows are expected to be $1,000 for the *second* year, $2,000 for the third year, and $3,000 for the fourth year and that if interest is 15% per year, it is desired to find the (a) present equivalent value at the beginning of the first year, and (b) uniform annual equivalent value at the end of each of the four years.

SOLUTION

Observe that this schedule of cash flows fits the model of the arithmetic gradient formulas with $G = \$1,000$ and $N = 4$. (See Figure 3-14.) Note that there is no cash flow at the end of the first period.

(a) The present equivalent can be calculated as

$$P_0 = G(P/G, 15\%, 4) = \$1,000(3.79) = \$3,790.$$

(b) The annual equivalent can be calculated from Equation (3-22) as

$$A = G(A/G, 15\%, 4) = \$1,000(1.3263) = \$1,326.30.$$

Of course, once P_0 is known, the value of *A* can be calculated as

$$A = P_0(A/P, 15\%, 4) = \$3,790(0.3503) = \$1,326.30.$$

EXAMPLE 3-15

As a further example of the use of arithmetic gradient formulas, suppose that one has cash flows as follows:

End of Year	Cash Flows ($)
1	−5,000
2	−6,000
3	−7,000
4	−8,000

Also, assume that one wishes to calculate their present equivalent at $i = 15\%$ per year using gradient conversion factors.

SOLUTION

The schedule of cash flows is depicted in the top diagram of Figure 3-15. The bottom two diagrams of Figure 3-15 show how the original schedule can be broken into two separate sets of cash flows, an annuity series of $5,000 *payments*, plus an arithmetic gradient *payment* of $1,000 that fits the general gradient model for which factors are tabled. The summed present equivalents of these two separate sets of payments equal the present equivalent of the original problem. Thus, using the symbols shown in Figure 3-15, we have

$$P_{0T} = P_{0A} + P_{0G}$$

$$= -A(P/A, 15\%, 4) - G(P/G, 15\%, 4)$$

$$= -\$5,000(2.8550) - \$1,000(3.79) = -\$14,275 - 3,790 = -\$18,065.$$

Figure 3-15
Example 3-15
Breakdown of Cash
Flows

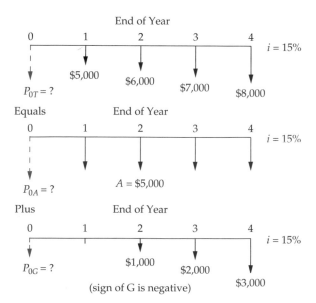

The annual equivalent of the original cash flows could be calculated with the aid of Equation (3-22) as follows:

$$A_T = A + A_G$$

$$= -\$5,000 - \$1,000(A/G, 15\%, 4) = -\$6,326.30.$$

A_T is equivalent to P_{0T} because $-\$6,326.30(P/A, 15\%, 4) = -\$18,061$, which is the same value obtained previously (subject to round-off error).

EXAMPLE 3-16

For another example of the use of arithmetic gradient formulas, suppose that one has cash flows that are timed in exact reverse of the situation depicted in Example 3-15. The top diagram of Figure 3-16 shows the following sequence of cash flows:

End of Year	Cash Flows ($)
1	−8,000
2	−7,000
3	−6,000
4	−5,000

Figure 3-16
Example 3-16
Breakdown of Cash
Flows

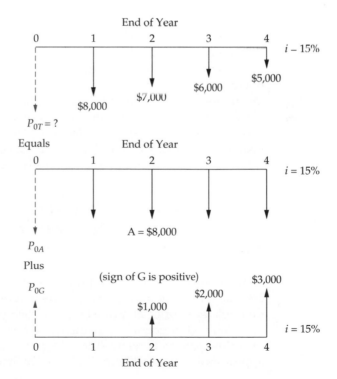

Calculate the present equivalent at $i = 15\%$ per year using arithmetic gradient interest factors.

SOLUTION

The bottom two diagrams of Figure 3-16 show how the uniform gradient can be broken into two separate sets of cash-flow diagrams. It must be remembered that the arithmetic gradient factors in Appendix C are for increasing gradient amounts. So

$$P_{0T} = P_{0A} + P_{0G}$$

$$= -A(P/A, 15\%, 4) + G(P/G, 15\%, 4)$$

$$= -\$8,000(2.8550) + \$1,000(3.79)$$

$$= -\$22,840 + \$3,790 = -\$19,050.$$

Again, the annual equivalent of the original decreasing series of cash flows can be calculated by the same rationale:

$$A = A + A_G$$

$$= -\$8,000 + \$1,000(A/G, 15\%, 4)$$

$$= -\$6,673.70.$$

Note from Examples 3-15 and 3-16 that the present equivalent of $-\$18,065$ for an increasing arithmetic gradient series of payments is different from the present equivalent of $-\$19,050$ for an arithmetic gradient of payments of identical amounts, but reversed timing (decreasing series of payments). This difference would be even greater for higher interest rates and gradient amounts and exemplifies the marked effect of the timing of cash flows on equivalent values. *It is also helpful to observe that the sign of G corresponds to the general slope of the total cash flows over time.* For instance, in Figure 3-15, the slope of the total cash flows is negative (G is negative), whereas in Figure 3-16 the slope is positive (G is positive).

3.14 Interest Formulas Relating a Geometric Sequence of Cash Flows to Its Present and Annual Equivalents

Some economic equivalence problems involve projected cash-flow patterns that are changing at an average *rate, $\overline{f}$,* each period. A fixed amount of a commodity that inflates in price at a constant rate each year is a typical situation that can be modeled with a geometric sequence of cash flows. The resultant end-of-period cash-flow pattern is referred to as a *geometric gradient series* and has the general appearance shown in Figure 3-17. Notice that *the initial cash-flow in this series, A_1,*

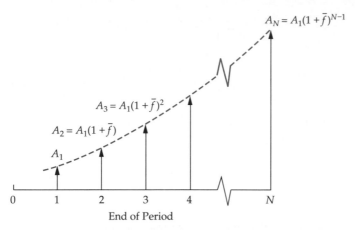

Figure 3-17 **Cash-Flow Diagram for a Geometric Sequence of Cash Flows Increasing at a Constant Rate of $\overline{f}$ per Period**

occurs at the end of period 1 and that $A_k = (A_{k-1})(1 + \overline{f}), 2 \leq k \leq N$. The Nth term in this geometric sequence is $A_N = A_1(1 + \overline{f})^{N-1}$, and the common ratio throughout the sequence is $(A_k - A_{k-1})/A_{k-1} = \overline{f}$. Be sure to notice that $\overline{f}$ can be positive *or* negative.

Each term in Figure 3-17 could be discounted, or compounded, at interest rate i per period to obtain a value of P or F, respectively. However, this becomes quite tedious for large N, so it is convenient to have a single equation instead.

To develop a compact expression for P at interest rate i per period for the cash flows of Figure 3-17, consider the summation

$$P = \sum_{k=1}^{N} A_k(1 + i)^{-k} = \sum_{k=1}^{N} A_1(1 + \overline{f})^{k-1}(1 + i)^{-k},$$

or

$$P = \frac{A_1}{1 + \overline{f}} \sum_{k=1}^{N} \left(\frac{1 + \overline{f}}{1 + i} \right)^{k}. \tag{3-25}$$

When $i \neq \overline{f}$, we can simplify Equation (3-25) by defining a "convenience rate," i_{CR}, as follows:

$$i_{CR} = \frac{1 + i}{1 + \overline{f}} - 1. \tag{3-26}$$

The convenience rate can also be written as $i_{CR} = (i - \overline{f})/(1 + \overline{f})$. In the situation where $i \neq \overline{f}$, Equation (3-25) can thus be rewritten as

$$P = \frac{A_1}{1 + \overline{f}} \sum_{k=1}^{N} \left(\frac{1 + i}{1 + \overline{f}} \right)^{-k}$$

$$= \frac{A_1}{1 + \overline{f}} \sum_{k=1}^{N} (1 + i_{CR})^{-k}$$

$$= \frac{A_1}{1 + \overline{f}} (P/A, i_{CR}\%, N).^{*} \tag{3-27}$$

Equation (3-27) makes use of the fact that

$$(P/A, i_{CR}\%, N) = \sum_{k=1}^{N} (1 + i_{CR})^{-k} = \sum_{k=1}^{N} (P/F, i_{CR}\%, k).$$

When $i = \overline{f}$ and $i_{CR} = 0$, Equation (3-27) reduces to

$$P = \frac{A_1}{1 + \overline{f}} (P/A, 0\%, N) = \frac{N A_1}{1 + \overline{f}}. \tag{3-28}$$

The interested reader can verify Equation (3-28) by applying L'Hôpital's Rule to the $(P/A, i_{CR}\%, N)$ factor in Equation (3-27) and taking the limit as $i_{CR} \to 0$.

Values of i_{CR} used in connection with Equation (3-27) are typically not included in the tables in Appendix C. Because i_{CR} is usually a noninteger interest rate, resorting to the definition of a $(P/A, i_{CR}\%, N)$ factor (see Table 3-4) and substituting terms into it is a satisfactory way to obtain values of these interest factors.

The end-of-period uniform annual equivalent, A, of a geometric gradient series can be determined from Equation (3-27) [or Equation (3-28)] as follows:

$$A = P(A/P, i\%, N). \tag{3-29}$$

The year zero "base" of this annuity, which increases at a constant rate of $\overline{f}\%$ per period, is A_0 and equals

$$A_0 = P(A/P, i_{CR}\%, N). \tag{3-30}$$

The difference between A and A_0 can be seen in Figure 3-18. Finally, the future equivalent of this geometric gradient series is simply

$$F = P(F/P, i\%, N). \tag{3-31}$$

Additional discussion of geometric sequences of cash flows is provided in Chapter 8 (Section 8.3), which deals with price changes and exchange rates.

---

*When $\overline{f}$ exceeds i, i_{CR} is negative, and the preceding summation is valid only when N is finite.

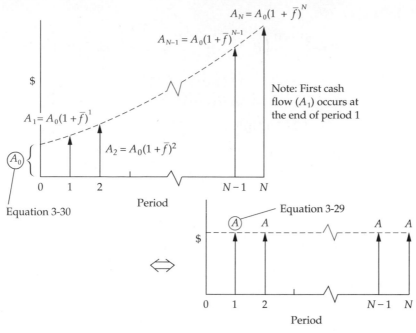

Figure 3-18 Graphical Interpretation of A and A_0 Terms in a Geometric Gradient Series when $\bar{f} > 0$

EXAMPLE 3-17

Consider the end-of-year geometric sequence of cash flows in Figure 3-19 and determine the P, A, A_0, and F equivalent values. The rate of increase is 20% per year after the first year, and the interest rate is 25% per year.

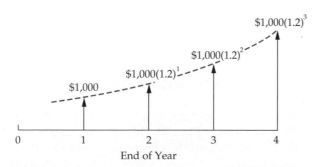

Figure 3-19 Cash-Flow Diagram for Example 3-17

SOLUTION

$$P = \frac{\$1,000}{1.2}\left(P/A, \frac{25\% - 20\%}{1.20}, 4\right) = \$833.33(P/A, 4.167\%, 4)$$

$$= \$833.33\left[\frac{(1.04167)^4 - 1}{0.04167(1.04167)^4}\right]$$

$$= \$833.33(3.6157) = \$3,013.08;$$

$$A = \$3,013.08(A/P, 25\%, 4) = \$1,275.86;$$

$$A_0 = \$3,013.08(A/P, 4.167\%, 4)$$

$$= \$3,103.08\left[\frac{0.04167(1.04167)^4}{(1.04167)^4 - 1}\right] = \$833.34;$$

$$F = \$3,013.08(F/P, 25\%, 4) = \$7,356.15.$$

EXAMPLE 3-18

Suppose that the geometric gradient in Example 3-17 begins with $1,000 at the end of year one and *decreases* by 20% per year after the first year. Determine P, A, A_0, and F under this condition.

SOLUTION

The value of $\overline{f}$ is −20% in this case and $i_{CR} = [(1+i)/(1+\overline{f})] - 1 = (1.25/0.80) - 1 = 0.5625$, or 56.25% per year. The desired quantities are as follows:

$$P = \frac{\$1,000}{0.8}(P/A, 56.25\%, 4) = \$1,250(1.4795)$$

$$= \$1,849.38;$$

$$A = \$1,849.38(A/P, 25\%, 4) = \$783.03;$$

$$A_0 = \$1,849.38(A/P, 56.25\%, 4) = \$1,250.00;$$

$$F = \$1,849.38(F/P, 25\%, 4) = \$4,515.08.$$

▼3.15 Interest Rates That Vary with Time

When the interest rate on a loan can vary with, for example, the Federal Reserve Board's discount rate, it is necessary to take this into account when determining the future equivalent value of the loan. It is becoming common to see interest-rate "escalation riders" on some types of loans. Example 3-19 demonstrates how this situation is treated.

EXAMPLE 3-19

A person has made an arrangement to borrow $1,000 now and another $1,000 two years hence. The entire obligation is to be repaid at the end of four years. If the projected interest rates in years one, two, three, and four are 10%, 12%, 12%, and 14%, respectively, how much will be repaid as a lump-sum amount at the end of four years?

SOLUTION
This problem can be solved by compounding the amount owed at the beginning of each year by the interest rate that applies to each individual year and repeating this process over the four years to obtain the total future equivalent value:

$$F_1 = \$1,000(F/P, 10\%, 1) = \$1,100;$$

$$F_2 = \$1,100(F/P, 12\%, 1) = \$1,232;$$

$$F_3 = (\$1,232 + \$1,000)(F/P, 12\%, 1) = \$2,500;$$

$$F_4 = \$2,500(F/P, 14\%, 1) = \$2,850.$$

To obtain the present equivalent of a series of future cash flows subject to varying interest rates, a procedure similar to the preceding one would be utilized with a sequence of $(P/F, i_k\%, k)$ factors. In general, the present equivalent value of a cash flow occurring at the end of period N can be computed with Equation (3-32), where i_k is the interest rate for the kth period (the symbol $\prod$ means "the product of"):

$$P = \frac{F_N}{\prod_{k=1}^{N}(1 + i_k)}. \tag{3-32}$$

For instance, if $F_4 = \$1,000$ and $i_1 = 10\%$, $i_2 = 12\%$, $i_3 = 13\%$, and $i_4 = 10\%$, then

$$P = \$1,000[(P/F, 10\%, 1)(P/F, 12\%, 1)(P/F, 13\%, 1)(P/F, 10\%, 1)]$$

$$= \$1,000[(0.9091)(0.8929)(0.8850)(0.9091)] = \$653.$$

3.16 Nominal and Effective Interest Rates

Very often the interest period, or time between successive compounding, is less than one year. It has become customary to quote interest rates on an annual basis, followed by the compounding period if different from one year in length. For example, if the interest rate is 6% per interest period and the interest period is six months, it is customary to speak of this rate as "12% compounded semiannually." Here the annual rate of interest is known as the *nominal rate*, 12% in this case. A nominal interest rate is represented by r. But the actual (or effective) annual rate on the principal is not 12%, but something greater, because compounding occurs twice during the year.

Consequently, the frequency at which a nominal interest rate is compounded each year can have a pronounced effect on the dollar amount of total interest earned. For instance, consider a principal amount of $1,000 to be invested for three years at a nominal rate of 12% compounded semiannually. The interest earned during the first six months would be $1,000 × (0.12/2) = $60.

Total principal and interest at the beginning of the second six-month period is

$$P + Pi = \$1,000 + \$60 = \$1,060.$$

The interest earned during the second six months would be

$$\$1,060 \times (0.12/2) = \$63.60.$$

Then total interest earned during the year is

$$\$60.00 + \$63.60 = \$123.60.$$

Finally, the *effective* annual interest rate for the entire year is

$$\frac{\$123.60}{\$1,000} \times 100 = 12.36\%.$$

If this process is repeated for years two and three, the *accumulated* (compounded) *amount of interest* can be plotted as in Figure 3-20. Suppose that the same $1,000 had been invested at 12% compounded *monthly*, which is 1% per month. The accumulated interest over three years that results from monthly compounding is shown in Figure 3-21.

The actual or exact rate of interest earned on the principal during one year is known as the *effective rate*. It should be noted that effective interest rates are always

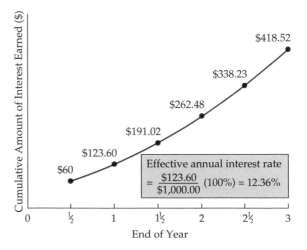

Figure 3-20
$1,000 Compounded at a Semiannual Frequency ($r = 12\%$)

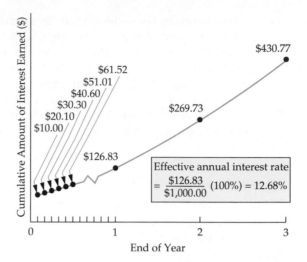

Figure 3-21
$1,000 Compounded at a Monthly Frequency (*r* = 12%)

$430.77

$61.52
$51.01
$40.60
$30.30
$20.10
$10.00

$269.73

$126.83

Effective annual interest rate
$= \dfrac{\$126.83}{\$1,000.00} (100\%) = 12.68\%$

Cumulative Amount of Interest Earned ($)

End of Year

expressed on an annual basis, unless specifically stated otherwise. In this text, the effective interest rate per year is customarily designated by i and the nominal interest rate per year by r. In engineering economy studies in which compounding is annual, $i = r$. The relationship between effective interest, i, and nominal interest, r, is

$$i = (1 + r/M)^M - 1$$
$$= (F/P, r/M, M) - 1, \qquad (3\text{-}33)$$

where M is the number of compounding periods per year. It is now clear from Equation (3-33) why $i > r$ when $M > 1$.

The effective rate of interest is useful for describing the compounding effect of interest earned on interest during one year. Table 3-5 shows effective rates for various nominal rates and compounding periods.

TABLE 3-5 **Effective Interest Rates for Various Nominal Rates and Compounding Frequencies**

Compounding Frequency	Number of Compounding Periods per Year, M	Effective Rate (%) for Nominal Rate of					
		6%	8%	10%	12%	15%	24%
Annually	1	6.00	8.00	10.00	12.00	15.00	24.00
Semiannually	2	6.09	8.16	10.25	12.36	15.56	25.44
Quarterly	4	6.14	8.24	10.38	12.55	15.87	26.25
Bimonthly	6	6.15	8.27	10.43	12.62	15.97	26.53
Monthly	12	6.17	8.30	10.47	12.68	16.08	26.82
Daily	365	6.18	8.33	10.52	12.75	16.18	27.11

Interestingly, the federal truth in lending law now requires a statement regarding the annual percentage rate (APR) being charged in contracts involving borrowed money. The APR is a nominal interest rate and *does not* account for compounding that may occur, or be appropriate, during a year. Before this legislation was passed by Congress in 1969, creditors had no obligation to explain how interest charges were determined or what the true cost of money on a loan was. As a result, borrowers were generally unable to compute their APR and compare different financing plans.

EXAMPLE 3-20

A credit card company charges an interest rate of 1.375% per month on the unpaid balance of all accounts. The annual interest rate, they claim, is 12(1.375%) = 16.5%. What is the effective rate of interest per year being charged by the company?

SOLUTION

Interest tables in Appendix C are based on time periods that may be annual, quarterly, monthly, and so on. Because we have no 1.375% tables (or 16.5% tables), Equation (3-33) must be used to compute the effective rate of interest in this example:

$$i = \left(1 + \frac{0.165}{12}\right)^{12} - 1$$

$$= 0.1781, \text{ or } 17.81\%/\text{year}.$$

Note that $r = 12(1.375\%) = 16.5\%$, which is the APR. It is true that $r = M(r/M)$, as seen in Example 3-20, where r/M is the interest rate per period.

3.17 Interest Problems with Compounding More Often than Once per Year

3.17.1 Single Amounts

If a nominal interest rate is quoted and the number of compounding periods per year and number of years are known, any problem involving future, annual, or present equivalent values can be calculated by straightforward use of Equations (3-3) and (3-33), respectively.

EXAMPLE 3-21

Suppose that a $100 lump-sum amount is invested for 10 years at a nominal interest rate of 6% compounded quarterly. How much is it worth at the end of the 10th year?

SOLUTION
There are four compounding periods per year, or a total of $4 \times 10 = 40$ interest periods. The interest rate per interest period is $6\%/4 = 1.5\%$. When the values are used in Equation (3-3), one finds that

$$F = P(F/P, 1.5\%, 40) = \$100.00(1.015)^{40} = \$100.00(1.814) = \$181.40.$$

Alternatively, the effective interest rate from Equation (3-33) is 6.14%. Therefore, $F = P(F/P, 6.14\%, 10) = \$100.00(1.0614)^{10} = \$181.40.$

3.17.2 Uniform Series and Gradient Series

When there is more than one compounded interest period per year, the formulas and tables for uniform series and gradient series can be used *as long as* there is a cash flow at the end of each interest period, as shown in Figures 3-6 and 3-14 for a uniform annual series and a uniform gradient series, respectively.

EXAMPLE 3-22

Suppose that one has a bank loan for $10,000, which is to be repaid in equal *end-of-month* installments for five years with a nominal interest rate of 12% compounded monthly. What is the amount of each payment?

SOLUTION
The number of installment payments is $5 \times 12 = 60$, and the interest rate per month is $12\%/12 = 1\%$. When these values are used in Equation (3-13), one finds that

$$A = P(A/P, 1\%, 60) = \$10,000(0.0222) = \$222.$$

Notice that there is a cash flow at the end of each month (interest period), including month 60, in this example.

EXAMPLE 3-23

Certain operating savings are expected to be 0 at the end of the first six months, to be $1,000 at the end of the second six months, and to increase by $1,000 at the end of each six-month period thereafter for a total of four years. It is desired to find the equivalent uniform amount, A, at the end of each of the eight six-month periods if the nominal interest rate is 20% compounded semiannually.

SOLUTION
A cash-flow diagram is given in Figure 3-22, and the solution is

$$A = G(A/G, 10\%, 8) = \$1,000(3.0045) = \$3,004.50$$

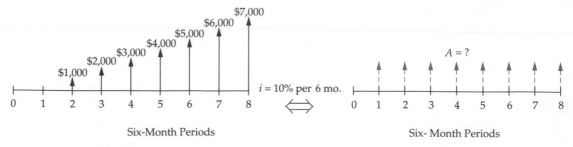

Figure 3-22 Arithmetic Gradient with Compounding More Often Than Once per Year in Example 3-23

The symbol " $\Longleftrightarrow$ " in Figure 3-22 indicates that the left-hand cash-flow diagram is *equivalent to* the right-hand cash-flow diagram when the correct value of A has been determined. In Example 3-23, the interest rate per six-month period is 10%, and cash flows occur every six months.

▼ 3.18 Interest Problems with Cash Flows Less Often than Compounding Periods

In general, if i is the effective interest rate per interest period and there is a uniform cash flow, X, at the *end* of each Kth interest period ($K > 1$), then the equivalent amount, A, at the end of each interest period is

$$A = X(A/F, i\%, K). \tag{3-34}$$

By similar reasoning, if i is the effective interest rate per interest period and there is a uniform cash flow, X, at the *beginning* of each Kth interest period, then the equivalent amount, A, at the end of each interest period is

$$A = X(A/P, i\%, K). \tag{3-35}$$

EXAMPLE 3-24

In cash-flow diagram (a), write an equation to convert the three amounts, X, into their annual equivalent value over N years ($N = 3Z$) when the interest rate is $i\%$ per year. For cash-flow diagram (b), determine the annual equivalent amount over 18 years when $i = 10\%$ per year.

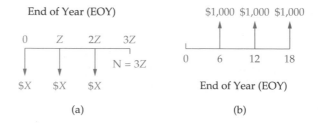

(a)   (b)

SOLUTION

(a) By using Equation (3-35), we see that the value of A corresponding to the three payments of $\$X$ is

$$A = \$X(A/P, i\%, Z).$$

(b) Equation (3-34) enables one to calculate the value of A, which extends from end-of-year 1 through end-of-year 18:

$$A = \$1,000(A/F, 10\%, 6) = \$129.60.$$

EXAMPLE 3-25

Suppose that there exists a series of 10 end-of-year receipts of $\$1,000$ each and that it is desired to compute their equivalent worth as of the *end* of the 10th year if the nominal interest rate is 12% compounded quarterly. The cash flows are depicted in Figure 3-23.

SOLUTION

Interest is 12%/4 = 3% per quarter, but the uniform series cash flows does not occur at the end of each quarter. In such cases, one can make special adaptations to fit the interest formulas to the tables provided. To solve this type of problem, (1) compute an equivalent cash flow for the time interval that corresponds to the stated compounding frequency, *or* (2) determine an effective interest rate for the interval of time separating the cash flows.

The *first* adaptation procedure is to take the number of compounding periods over which a cash flow occurs and convert the cash flow into its equivalent uniform end-of-period series. The upper cash-flow diagrams in Figure 3-24 show this approach applied to the first year (four interest periods) in the example of Figure 3-23. The uniform end-of-quarter amount, equivalent to $\$1,000$ at the end of the year with interest at 3% per quarter, can be calculated by using Equation (3-34):

$$A = F(A/F, 3\%, 4) = \$1,000(0.2390) = \$239.$$

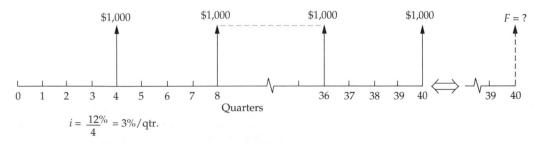

Figure 3-23 **Uniform Series with Cash Flows Less Often than Compounding Periods in Example 3-25**

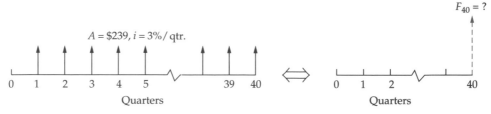

Figure 3-24 First Adaptation to Solve Example 3-25

Thus, $239 at the end of each quarter is equivalent to $1,000 at the end of each year. This is true not only for the first year, but also for each of the 10 years under consideration. Hence the original series of 10 end-of-year cash flows of $1,000 each can be converted to a problem involving 40 end-of-quarter amounts of $239 each, as shown in the lower cash-flow diagrams of Figure 3-24.

The future equivalent at the end of the 10th year (40th quarter) may then be computed as

$$F_{40} = A(F/A, 3\%, 40) = \$239(75.4012) = \$18,021.$$

The *second* procedure for handling cash flows occurring less often than compounding periods is to find the exact interest rate for each time period *separating* cash flows and then to straightforwardly apply the interest formulas and tables for the exact interest rate. For Example 3-25, interest is 3% per quarter and payments occur each year. Hence, the interest rate to be found is the exact rate each year, or the *effective rate* per year. The effective rate per year that corresponds to 3% per quarter (12% nominal) can be found from Equation (3-33):

$$\left(1 + \frac{0.12}{4}\right)^4 - 1 = (F/P, 3\%, 4) - 1 = 0.1255.$$

Hence, the original problem in Figure 3-23 can now be expressed as shown in Figure 3-25. The future equivalent of this series can then be found as

$$F_{10} = A(F/A, 12.55\%, 10) = \$1,000(F/A, 12.55\%, 10) = \$18,022$$

Because interest factors are not commonly tabled for $i = 12.55\%$, one must compute the $(F/A, 12.55\%, 10)$ factor by substituting $i = 0.1255$ and $N = 10$ into its algebraic equivalent, $[(1 + i)^N - 1]/i$. (See Table 3-4.)

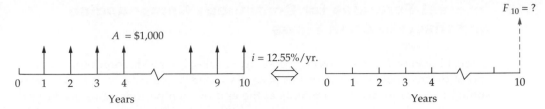

Figure 3-25 Second Adaptation to Solve Example 3-25

The second procedure just illustrated is probably the more popular way to deal with problems in which cash flows occur every K $(K > 1)$ compounding periods. By using the second procedure, we find that the basic question becomes: "How do we find an effective interest rate for the fixed time interval (K compounding periods) separating the cash flows?" We now formalize this procedure by using a more general version of Equation (3-33) to determine an effective interest rate per K compounding periods:

$$i(\text{per } K \text{ compounding periods}) = (1 + r/M)^K - 1. \qquad (3\text{-}36)$$

Here, K = number of compounding periods per fixed time interval separating cash flows;

r = nominal interest rate per year;

M = number of compounding periods per year.

EXAMPLE 3-26

Determine the present equivalent, P, of the following cash-flow diagram:

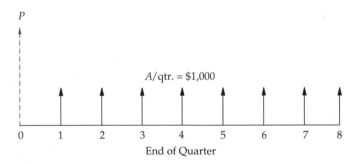

The nominal interest rate is 15% *compounded monthly.* Cash flows occur every three months (once per quarter).

SOLUTION
By using Equation (3-36), we determine the effective interest rate per quarter: $i/\text{qtr.} = (1 + \frac{0.15}{12})^3 - 1 = (1.0125)^3 - 1 = 0.038$, or 3.8%. Then $P = \$1,000 \times (P/A, 3.8\%, 8) = \$6,788.70$. Note that the second procedure can also be easily used with nonuniform cash flows.

▼ 3.19 Interest Formulas for Continuous Compounding and Discrete Cash Flows

In most business transactions and economy studies, interest is compounded at the end of discrete periods, and, as has been discussed previously, cash flows are assumed to occur in discrete amounts at the end of such periods. *This practice will be used throughout the remaining chapters of this book.* However, it is evident that in most enterprises, cash is flowing in and out in an almost continuous stream. Because cash, whenever it's available, can usually be used profitably, this situation creates opportunities for very frequent compounding of the interest earned. So that this condition can be dealt with (modeled) when continuously compounded interest rates are available, the concepts of continuous compounding and continuous cash flow are sometimes used in economy studies. Actually, the effects of these procedures compared to those of discrete compounding are rather small in most cases.

Continuous compounding assumes that cash flows occur at discrete intervals (e.g., once per year), but that compounding is continuous throughout the interval. For example, with a nominal rate of interest per year of r, if the interest is compounded M times per year, one unit of principal will amount to $[1 + (r/M)]^M$ at the end of one year. Letting $M/r = p$, we find that the foregoing expression becomes

$$\left[1 + \frac{1}{p}\right]^{rp} = \left[\left(1 + \frac{1}{p}\right)^p\right]^r. \tag{3-37}$$

Because

$$\lim_{p \to \infty} \left(1 + \frac{1}{p}\right)^p = e^1 = 2.71828\ldots,$$

Equation (3-37) can be written as e^r. Consequently, the *continuously compounded compound amount factor (single cash flow)* at $r\%$ nominal interest for N years is e^{rN}. Using our functional notation, we express this as

$$(F/P, \underline{r}\%, N) = e^{rN}. \tag{3-38}$$

Note that the symbol $\underline{r}$ is directly comparable to that used for discrete compounding and discrete cash flows ($i\%$) except that $\underline{r}\%$ is used to denote the nominal rate *and* the use of continuous compounding.

Since e^{rN} for continuous compounding corresponds to $(1 + i)^N$ for discrete compounding, e^r is equal to $(1 + i)$. Hence, we may correctly conclude that

$$i = e^r - 1. \tag{3-39}$$

By using this relationship, the corresponding values of (P/F), (F/A), and (P/A) for continuous compounding may be obtained from Equations (3-4), (3-6), and (3-8), respectively, by substituting $e^r - 1$ for i in these equations. Thus, for continuous compounding and discrete cash flows,

$$(P/F, \underline{r}\%, N) = \frac{1}{e^{rN}} = e^{-rN}; \tag{3-40}$$

$$(F/A, \underline{r}\%, N) = \frac{e^{rN} - 1}{e^r - 1}; \tag{3-41}$$

$$(P/A, \underline{r}\%, N) = \frac{1 - e^{-rN}}{e^r - 1} = \frac{e^{rN} - 1}{e^{rN}(e^r - 1)}. \tag{3-42}$$

Values for $(A/P, \underline{r}\%, N)$, and $(A/F, \underline{r}\%, N)$ may be derived through their inverse relationships to $(P/A, \underline{r}\%, N)$ and $(F/A, \underline{r}\%, N)$, respectively. Numerous continuous compounding, discrete cash flow interest factors and their uses are summarized in Table 3-6.

TABLE 3-6 Continuous Compounding and Discrete Cash Flows: Interest Factors and Symbols[a]

To Find:	Given:	Factor by Which to Multiply "Given"	Factor Name	Factor Functional Symbol
For single cash flows:				
F	P	e^{rN}	Continuous compounding compound amount (single cash flow)	$(F/P, \underline{r}\%, N)$
P	F	e^{-rN}	Continuous compounding present equivalent (single cash flow)	$(P/F, \underline{r}\%, N)$
For uniform series (annuities):				
F	A	$\dfrac{e^{rN} - 1}{e^r - 1}$	Continuous compounding compound amount (uniform series)	$(F/A, \underline{r}\%, N)$
P	A	$\dfrac{e^{rN} - 1}{e^{rN}(e^r - 1)}$	Continuous compounding present equivalent (uniform series)	$(P/A, \underline{r}\%, N)$
A	F	$\dfrac{e^r - 1}{e^{rN} - 1}$	Continuous compounding sinking fund	$(A/F, \underline{r}\%, N)$
A	P	$\dfrac{e^{rN}(e^r - 1)}{e^{rN} - 1}$	Continuous compounding capital recovery	$(A/P, \underline{r}\%, N)$

[a] $\underline{r}$, nominal annual interest rate, compounded continuously; N, number of periods (years); A, annual equivalent amount (occurs at the end of each year); F, future equivalent; P, present equivalent.

Because continuous compounding is used infrequently in this text, detailed values for $(A/F, \underline{r}\%, N)$ and $(A/P, \underline{r}\%, N)$ are not given in Appendix D. However, the tables in Appendix D do provide values of $(F/P, \underline{r}\%, N)$, $(P/F, \underline{r}\%, N)$, $(F/A, \underline{r}\%, N)$, and $(P/A, \underline{r}\%, N)$ for a limited number of interest rates.

Note that tables of interest and annuity factors for continuous compounding are tabulated in terms of nominal annual rates of interest.

EXAMPLE 3-27

Suppose that one has a present loan of $1,000 and desires to determine what equivalent uniform end-of-year payments, A, could be obtained from it for 10 years if the nominal interest rate is 20% compounded continuously ($M = \infty$).

SOLUTION

Here we utilize the formulation

$$A = P(A/P, \underline{r}\%, N).$$

Since the (A/P) factor is not tabled for continuous compounding, we substitute its inverse (P/A), which is tabled in Appendix D. Thus,

$$A = P \times \frac{1}{(P/A, \underline{20}\%, 10)} = \$1,000 \times \frac{1}{3.9054} = \$256.$$

Note that the answer to the same problem, with discrete annual compounding ($M = 1$), is

$$A = P(A/P, 20\%, 10)$$

$$= \$1,000(0.2385) = \$239.$$

EXAMPLE 3-28

An individual needs $12,000 immediately as a down payment on a new home. Suppose that he can borrow this money from his insurance company. He must repay the loan in equal payments every six months over the next eight years. The nominal interest rate being charged is 7% compounded continuously. What is the amount of each payment?

SOLUTION

The nominal interest rate per six months is 3.5%. Thus, A each six months is $12,000(A/P, \underline{r} = 3.5\%, 16)$. By substituting terms in Equation (3-42) and then using its inverse, we determine the value of A per six months to be $997:

$$A = \$12,000 \left[\frac{1}{(P/A, \ \underline{r} = 3.5\%, 16)} \right] = \frac{\$12,000}{12.038} = \$997.$$

▼ 3.20 Interest Formulas for Continuous Compounding and Continuous Cash Flows

Continuous flow of funds means a series of cash flows occurring at infinitesimally short intervals of time; this corresponds to an annuity having an infinite number of short periods. This model could apply to companies having receipts and expenses that occur frequently during each working day. In such cases, the interest is normally compounded continuously. If the nominal interest rate per year is r and there are p payments per year, which amount to a total of one unit per year, then, by using Equation (3-8), the present equivalent at the beginning of the year (for one year) is

$$P = \frac{1}{p} \left\{ \frac{[1 + (r/p)]^p - 1}{r/p[1 + (r/p)]^p} \right\} = \frac{[1 + (r/p)]^p - 1}{r[1 + (r/p)]^p}. \tag{3-43}$$

The limit of $[1+(r/p)]^p$ as p approaches infinity is e^r. Calling the present equivalent of one unit per year, flowing continuously and with continuous compounding of interest, the *continuous compounding present equivalent factor (continuous, uniform cash flow over one period)*, one finds that

$$(P/\overline{A}, r\%, 1) = \frac{e^r - 1}{re^r}, \tag{3-44}$$

where $\overline{A}$ is the amount flowing uniformly and continuously over one year (here $1).

For $\overline{A}$ flowing each year over N years, as depicted in Figure 3-26,

$$(P/\overline{A}, r\%, N) = \frac{e^{rN} - 1}{re^{rN}}, \tag{3-45}$$

which is the *continuous compounding present equivalent factor (continuous, uniform cash flows)*.

Figure 3-26
General Cash-Flow
Diagram for Continuous
Compounding,
Continuous Cash Flows

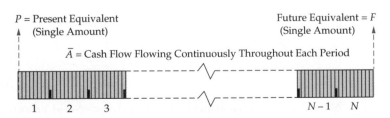

P = Present Equivalent (Single Amount)

Future Equivalent = F (Single Amount)

$\overline{A}$ = Cash Flow Flowing Continuously Throughout Each Period

1 2 3   N – 1   N

r = Nominal Interest Rate Compounded Continuously

Equation (3-44) can also be written as

$$(P/\overline{A}, \underline{r}\%, 1) = e^{-r}\left[\frac{e^r - 1}{r}\right] = (P/F, \underline{r}\%, 1)\left[\frac{e^r - 1}{r}\right].$$

Because the present equivalent of $1 per year, flowing continuously with continuous compounding of interest, is $(P/F, \underline{r}\%, 1)(e^r - 1)/r$, it follows that $(e^r - 1)/r$ must also be the compound amount of $1 per year, flowing continuously with continuous compounding of interest. Consequently, the *continuous compounding compound amount factor (continuous, uniform cash flow over one year)* is

$$(F/\overline{A}, \underline{r}\%, 1) = \frac{e^r - 1}{r}. \tag{3-46}$$

For N years,

$$(F/\overline{A}, \underline{r}\%, N) = \frac{e^{rN} - 1}{r}. \tag{3-47}$$

Equation (3-47) can also be developed by integration in this manner:

$$F = \overline{A}\int_0^N e^{rt}dt = \overline{A}\left(\frac{1}{r}\right)\int_0^N re^{rt}\,dt,$$

or

$$F = \frac{\overline{A}}{r}(e^{rt})\Big|_0^N = \overline{A}\left[\frac{e^{rN} - 1}{r}\right].$$

This is the *continuous compounding compound amount factor (continuous uniform cash flows for N years).*

Values of $(P/\overline{A}, \underline{r}\%, N)$ and $(F/\overline{A}, \underline{r}\%, N)$ are given in the tables in Appendix D for various interest rates. Values for $(\overline{A}/P, \underline{r}\%, \underline{N})$ and $(\overline{A}/F, \underline{r}\%, N)$ can be readily obtained through their inverse relationship to $(P/\overline{A}, \underline{r}\%, N)$ and $(F/\overline{A}, \overline{r}\%, N)$, respectively. A summary of these factors and their use is given in Table 3-7.

EXAMPLE 3-29

What will be the future equivalent amount at the end of five years of a uniform, continuous cash flow, at the rate of $500 per year for five years, with interest compounded continuously at the nominal annual rate of 8%?

SOLUTION
We have

$$F = \overline{A}(F/\overline{A}, \underline{8}\%, 5) = \$500 \times 6.1478 = \$3,074.$$

Note that if this cash flow had been in year-end amounts of $500 with discrete annual compounding of $i = 8\%$, the future equivalent amount would have been

$$F = A(F/A, 8\%, 5) = \$500 \times 5.8666 = \$2,933.$$

TABLE 3-7 Continuous Compounding Continuous Uniform Cash Flows: Interest Factors and Symbols[a]

To Find:	Given:	Factor by Which to Multiply "Given"[a]	Factor Name	Factor Functional Symbol
F	$\overline{A}$	$\dfrac{e^{rN}-1}{r}$	Continuous compounding compound amount (continuous, uniform cash flows)	$(F/\overline{A}, \underline{r}\%, N)$
P	$\overline{A}$	$\dfrac{e^{rN}-1}{re^{rN}}$	Continuous compounding present equivalent (continuous, uniform cash flows)	$(P/\overline{A}, \underline{r}\%, N)$
$\overline{A}$	F	$\dfrac{r}{e^{rN}-1}$	Continuous compounding sinking-fund (continuous, uniform cash flows)	$(\overline{A}/F, \underline{r}\%, N)$
$\overline{A}$	P	$\dfrac{re^{rN}}{e^{rN}-1}$	Continuous compounding capital recovery (continuous, uniform cash flows)	$(\overline{A}/P, \underline{r}\%, N)$

[a] $\underline{r}$, nominal annual interest rate, compounded continuously; N, number of periods (years); $\overline{A}$, amount of money flowing continuously and uniformly during each period; F, future equivalent; P, present equivalent.

If the year-end payments had occurred with 8% nominal interest compounded continuously, the future equivalent would then have been

$$F = A(F/A, \underline{8}\%, 5) = \$500 \times 5.9052 = \$2,953.$$

It is clear that for a given A amount and continuous compounding of a given nominal interest rate, continuous funds flow produces the largest-valued future equivalent amount.

EXAMPLE 3-30

What is the future equivalent of $10,000 per year that flows continuously for 8.5 years if the nominal interest is 10% compounded continuously?

SOLUTION

There are 17 six-month periods in 8.5 years, and the $\underline{r}$ per six months is 5%. The $\overline{A}$ every six months is $5,000, so $F = \$5,000(F/\overline{A}, \underline{5}\%, 17) = \$133,964.50$. This formulation is utilized to enable us to find an interest factor having an integer-valued N. The same answer could have been obtained by resorting to the definition of the $(F/\overline{A}, \underline{r}\%, N)$ factor given in Table 3-7 with $N = 8.5$ years:

$$F = \$10,000 \left[\frac{e^{0.10(8.5)} - 1}{0.10} \right]$$

$$= \$133,964.50.$$

▼ 3.21 Additional Solved Problems

This section contains many solved problems that further illustrate the economic equivalence concepts of Chapter 3.

PROBLEM 1

Given the following information and table, determine the value of each "?":

Loan Principal = $10,000
Interest Rate = 8% per year
Duration of Loan = 3 years

End of Year k	Interest Paid	Principal Repayment
1	$800	?
2	$553.60	$3,326.40
3	?	?

SOLUTION
A uniform annual payment scheme is implied by the table entries. Thus, the total annual payment = $10,000 $(A/P, 8\%, 3)$ = $3,880. At the end of year one, the principal repayment will be $3,880 – $800 = $3,080. At the beginning of year three, the remaining principal to be repaid is $10,000 – $3,080 – $3,326.40 = $3,593.60. Therefore, the interest paid during year three is about 0.08($3,593.60) = $286.40. (Some rounding is present in this problem because of four significant digits in the interest tables.)

PROBLEM 2

Suppose the 8% interest rate in Problem 1 is a nominal interest rate. If compounding occurs monthly, what is the effective annual interest rate?

SOLUTION
Use Equation (3-33) to find

$$i = \left(1 + \frac{0.08}{12} \right)^{12} - 1$$

$$= 0.083 \text{ (8.3\% effective annual interest rate).}$$

PROBLEM 3

Compare the interest earned by $9,000 for five years at 8% simple interest per year with the interest earned by the same amount for five years at 8% compounded annually. Explain why a difference occurs.

SOLUTION

Simple interest:
$$\underline{I} = (P)(N)(i) = \$9,000(0.08)(5) = \underline{\$3,600};$$
$$\text{Total} = \$9,000 + \$3,600 = \$12,600.$$

Compound interest:
$$F = P(F/P, 8\%, 5) = \$9,000(1.4693) = \$13,223.70;$$
$$\text{Total interest} = \$13,223.70 - \$9,000 = \underline{\$4,223.70}.$$

There is a difference in the amount of interest earned because compounding allows interest from previous years to earn interest, whereas simple interest does not.

PROBLEM 4

With the minimum number of interest factors, find the value of X in the following diagram so that the two cash-flow diagrams are equivalent when the interest rate is 10% per year:

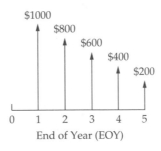

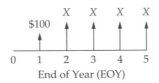

SOLUTION

Use EOY one as the reference point and three interest factors:

$$\$1{,}000 + \$800(P/A, 10\%, 4) - \$200(P/G, 10\%, 4) = 100 + X(P/A, 10\%, 4)$$

$$X = \frac{\$900 + \$800(P/A, 10\%, 4) - \$200(P/G, 10\%, 4)}{(P/A, 10\%, 4)}.$$

PROBLEM 5

Set up an expression for the value of Z on the left-hand cash-flow diagram that establishes equivalence with the right-hand cash-flow diagram. The nominal interest rate is 12% compounded quarterly:

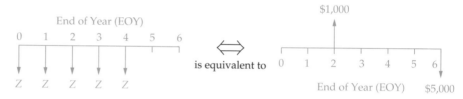

SOLUTION

We have the following:

$$i = (1 + 0.12/4)^4 - 1 \quad \text{[Equation (3-33)]}$$

$$\simeq 0.1255 \ (12.55\%);$$

$$-Z - Z(P/A, 12.55\%, 4) = \$1{,}000(P/F, 12.55\%, 2) - \$5{,}000(P/F, 12.55\%, 6)$$

$$Z = \frac{\$1{,}000(P/F, 12.55\%, 2) - \$5{,}000(P/F, 12.55\%, 6)}{[-1 - (P/A, 12.55\%, 4)]}.$$

PROBLEM 6

A student decides to make semiannual payments of $500 each into a bank account that pays an APR (nominal interest) of 8% compounded weekly. How much money will this student have accumulated in this bank account at the end of 20 years? Assume that only one (the final) withdrawal is made.

SOLUTION

Using Equation (3-36), we see that i per six months (26 weeks) is equal to

$$\left(1 + \frac{0.08}{52}\right)^{26} - 1 = 0.0408 \ (4.08\%).$$

Then F at the end of year 20 will be $F = \$500(F/A, 4.08\%, 40)$, or

$$F = \$500\left[\frac{(1.0408)^{40} - 1}{0.0408}\right] = \$500\left[\frac{4.9510 - 1}{0.0408}\right] = \$48,419.$$

PROBLEM 7

Consider an EOY geometric gradient, which lasts for eight years, whose initial value at EOY one is \$5,000, and $\bar{f} = 6.04\%$ per year thereafter. Find the equivalent uniform gradient amount over the same period if the initial value of the cash flows at the end of year one is \$4,000. Complete the following questions in determining the value of the gradient amount, G. The nominal interest rate is 8% compounded semiannually.

(a) What is i_{CR}?

$$i = \left(1 + \frac{0.08}{2}\right)^2 - 1 = 0.0816$$

$$= 8.16\%;$$

$$i_{CR} = \frac{1 + 0.0816}{1 + 0.0604} - 1 = 0.02\ (2\%).$$

(b) What is P_0 for the geometric gradient series?

$$P_0 = \frac{\$5,000}{1 + 0.0604}(P/A, 2\%, 8)$$

$$= \$34,541.$$

(c) What is P_0' of the uniform (arithmetic) gradient of cash flows?

$$P_0' = \$4,000(P/A, 8.16\%, 8) + G(P/G, 8.16\%, 8).$$

(d) What is the value of G?

Set $P_0 = P_0'$ and solve for G. Answer: $G = \$662.53$.

PROBLEM 8

An individual makes five annual deposits of \$2,000 in a savings account that pays interest at a rate of 4% per year. One year after making the last deposit, the interest rate changes to 6% per year. Five years after the last deposit, the accumulated money is withdrawn from the account. How much is withdrawn?

SOLUTION

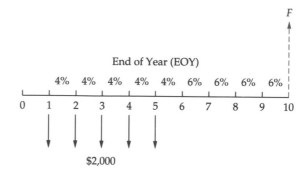

$$F = \$2,000(F/A, 4\%, 5)(F/P, 4\%, 1)(F/P, 6\%, 4) = \$14,223.$$

PROBLEM 9

Some future amount, F, is equivalent to \$2,000 being received every six months over the next 12 years. The nominal interest rate is 20% compounded continuously. What is the value of F?

SOLUTION

$$F = \frac{\$2,000}{6 \text{ mo}}(F/A, \underline{10}\% \text{ per 6 mo, 24 six-month periods})$$

$$= \$190,607.40.$$

PROBLEM 10

What is the value of P that is equivalent to $\overline{A} = \$800/\text{yr}$ (\$800 flowing continuously each year) for 11.2 years? The nominal rate of interest is 10%, continuously compounded.

SOLUTION

$$P = \frac{\$800}{\text{yr.}}(P/\overline{A}, \underline{10}\%/\text{yr}, 11.2 \text{ years})$$

$$= \$800\left(\frac{e^{0.10(11.2)} - 1}{0.10e^{0.10(11.2)}}\right) = \$800\left(\frac{e^{1.12} - 1}{0.10e^{1.12}}\right)$$

$$\cong \$5,390.$$

	A	B
1	i	30.0%
2	N	7
3		
4		
5	$(F/P, i\%, N) =$	6.2749
6	$(P/F, i\%, N) =$	0.1594
7	$(F/A, i\%, N) =$	17.5828
8	$(P/A, i\%, N) =$	2.8021
9	$(A/F, i\%, N) =$	0.0569
10	$(A/P, i\%, N) =$	0.3569
11		
12	$(P/G, i\%, N) =$	5.6218
13	$(A/G, i\%, N) =$	2.0063
14	$(F/G, i\%, N) =$	35.2761

Cell	Name
B1	i
B2	N

Cell	Contents
B5	$(1+i)^{\wedge}N$
B6	$1/(1+i)^{\wedge}N$
B7	$((1+i)^{\wedge}N-1)/i$
B8	$((1+i)^{\wedge}N-1)/(i*(1+i)^{\wedge}N)$
B9	$i/((1+i)^{\wedge}N-1)$
B10	$i*(1+i)^{\wedge}N/((1+i)^{\wedge}N-1)$
B12	$(((1+i)^{\wedge}N-1)/(i*(1+i)^{\wedge}N)-N/(1+i)^{\wedge}N)/i$
B13	$(1/i)-N/((1+i)^{\wedge}N-1)$
B14	$((1+i)^{\wedge}N-1)/i^{\wedge}2-N/i$

Figure 3-27 Spreadsheet for Generating Interest Factor Values with Discrete Compounding

▼ 3.22 Spreadsheet Applications

Appendices C and D tabulate the most common interest factors for a variety of interest rates and a number of compounding periods. However, frequently, we use an interest rate that does not have a corresponding table in the appendices. In this situation, we must resort to using the equations that define the interest factors. This process can be facilitated by using a spreadsheet.

Figure 3-27 shows a spreadsheet (and corresponding cell formulas) that can be used to generate the discrete compounding factor values for a given interest rate (i) and number of compounding periods (N). A similar spreadsheet for generating the factor values in the case of continuous compounding is shown in Figure 3-28.

Figure 3-29 is a spreadsheet model that can compute effective interest rates. Given a nominal interest rate (r) and the number of compounding periods per year (M), the spreadsheet computes the effective annual interest rate. If cash flows occur less frequently than compounding periods—for example, monthly compounding and quarterly cash flows—the spreadsheet can compute the effective interest rate for the time interval separating cash flows.

▼ 3.23 Summary

Chapter 3 has presented the fundamental time value of money relationships that are used throughout the remainder of this book. Considerable emphasis has been placed on the notion of economic equivalence, whether the relevant cash flows

	A	B
1	r	15.0%
2	N	5
3		
4		
5	(F/P, r %, N) =	2.1170
6	(P/F, r %, N) =	0.4724
7	(F/A, r %, N) =	6.9021
8	(P/A, r %, N) =	3.2603
9	(A/F, r %, N) =	0.1449
10	(A/P, r %, N) =	0.3067

Cell	Name
B1	r
B2	N

Cell	Formula
B5	=EXP(r*N)
B6	=EXP(-r*N)
B7	=(EXP(r*N)-1)/(EXP(r)-1)
B8	=(EXP(r*N)-1)/(EXP(r*N)*(EXP(r)-1))
B9	=(EXP(r)-1)/(EXP(r*N)-1)
B10	=(EXP(r*N)*(EXP(r)-1))/(EXP(r*N)-1)

Figure 3-28 Spreadsheet for Generating Interest Factor Values with Continuous Compounding

	A	B	C	D	E
1	Nominal interest rate, r				12%
2	Compounding periods per year, M				12
3	Number of compounding periods				
4	per fixed time interval				
5	separating cash flows, K				4
6					
7	Effective interest rates:				
8	i (annual)				12.68%
9	i (per K compounding periods)				4.06%

Cell	Name
E1	r
E2	M
E5	K

Cell	Contents
E8	=((1 + r/M)^M)-1
E9	=((1 + r/M)^K)-1

Figure 3-29 Spreadsheet for Computing Effective Interest Rates

and interest rates are discrete or continuous. Students should feel comfortable with the material in this chapter before embarking on their journey through subsequent chapters. Important abbreviations and notation in Chapter 3 are listed in Appendix B, which will serve as a handy reference in your use of this book.

3.24 **References**

Au, T., and T. P. Au, *Engineering Economics for Capital Investment Analysis* (Boston: Allyn and Bacon, 1983).

Bussey, L. E., and T. G. Eschenbach, *The Economic Analysis of Industrial Projects* (Englewood Cliffs, NJ: Prentice Hall, 1992).

Thuesen, G. J., and W. J. Fabrycky, *Engineering Economy,* 9th ed. (Upper Saddle River, NJ: Prentice Hall, 2001).

White, J. A., K. E. Case, D. B. Pratt, and M. H. Agee, *Principles of Engineering Economic Analysis,* 4th ed. (New York: John Wiley, 1998).

3.25 Problems

The number(s) in color at the end of a problem refer to the section(s) in that chapter most closely related to the problem.

3-1. What lump-sum amount of interest will be paid on a $10,000 loan that was made on August 1, 2002, and repaid on November 1, 2006, with ordinary simple interest at 10% per year? (3.4)

3-2. Draw a cash-flow diagram for $10,500 being loaned out at an interest rate of 12% per year over a period of six years. How much simple interest would be repaid as a lump-sum amount at the end of the sixth year? (3.4, 3.7)

3-3. What is the future equivalent of $1,000 invested at 8% simple interest per year for $2\frac{1}{2}$ years? (3.4)

    a. $1,157.   **b.** $1,188.   **c.** $1,200.
    d. $1,175.   **e.** $1,150.

3-4. How much interest is *payable each year* on a loan of $2,000 if the interest rate is 10% per year when half of the loan principal will be repaid as a lump sum at the end of four years and the *other half* will be repaid in one lump-sum amount at the end of eight years? How much interest will be paid over the eight-year period? (3.6)

3-5. In Problem 3-4, if the interest had not been paid each year, but had been added to the outstanding principal plus accumulated interest, how much interest would be due to the lender as a lump sum at the end of the eighth year? How much extra interest is being paid here (as compared with Problem 3-4), and what is the reason for the difference? (3.6)

3-6.

    a. Suppose that in Plan 1 of Table 3-1, $4,000 in principal is to be repaid at the end of years two and four only. How much total interest would have been paid by the end of year four? (3.6)

    b. Rework Plan 3 of Table 3-1 when an annual interest rate of 8% is being charged on the loan. How much *principal* is now being repaid in the third year's total end-of-year payment? How much total interest has been paid by the end of the fourth year? (3.6, 3.9)

3-7.

    a. Based on the information, determine the value of each "?" in the following table: (3.6)

$$\text{Loan Principal} = \$10,000$$
$$\text{Interest Rate} = 6\%/\text{yr}$$
$$\text{Duration of Loan} = 3 \text{ yr}$$

EOY k	Interest Paid	Principal Repayment
1	$600	?
2	$411.54	$3,329.46
3	?	?

    b. What is the amount of principal owed at the *beginning* of year three?

    c. Why is the total interest paid in (a) different from $10,000(1.06)^3 - \$10,000 \simeq \$1,910$ that would be repaid according to plan 4 in Table 3-1?

3-8. A future amount of $150,000 is to be accumulated through annual payments, A, over 20 years. The last payment of A, occurs simultaneously with the future amount at the end of year 20. If the interest rate is 9% per year, what is the value of A? (3.9)

3-9. What amount would need to be paid each January 1 into a savings account if at the end of 13 years (13 payments) you desired $10,000? Annual interest is 7%. (*Note:* The last payment will coincide with the time of the $10,000 balance.) (3.9)

3-10. A future amount, F, is equivalent to $1,500 now when eight years separate the amounts and the annual interest rate is 10%. What is the value of F? (3.8)

3-11. A present obligation of $20,000 is to be repaid in uniform annual amounts, each of which includes repayment of the debt (principal) and interest on the debt, over a period of five years. If the interest rate is 12% per year, what is the amount of the annual repayment? (3.9)

3-12. Suppose that the $20,000 in Problem 3-11 is to be repaid at a rate of $4,000 per year plus the interest that is owed based on the beginning-of-year unpaid principal. Compute the total amount of interest repaid in this situation and compare it with that of Problem 3-11. Why are the two amounts different? (3.6)

3-13. A person wishes to accumulate $5,000 over a period of 15 years so that a cash payment can be made for a new roof on a summer cottage. To have this amount when it is needed, annual payments will be made into a savings account that earns 8% interest per year. How much must each annual payment be? Draw a cash-flow diagram. (3.7, 3.9)

3-14. You have just learned that ABC Corporation has an investment opportunity that costs $35,000 and eight years later pays a lump-sum amount of $100,000. The cash-flow diagram is as follows:

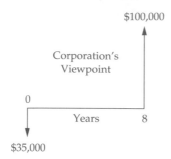

$100,000

Corporation's Viewpoint

0

Years   8

$35,000

What interest rate per year would be earned on this investment? Calculate your answer to the nearest one-tenth of 1%. (3.8)

3-15. It is estimated that a copper mine will produce 10,000 tons of ore during the coming year. Production is expected to increase by 5% per year thereafter in each of the following six years. Profit per ton will be $14 for years one through seven.

  a. Draw a cash-flow diagram for this copper mine operation from the company's viewpoint. (3.7)

  b. If the company can earn 15% per year on its capital, what is the future equivalent of the copper mine's cash flows at the end of year seven? (3.8 or 3.14)

3-16. Mrs. Green has just purchased a new car for $20,000. She makes a down payment of 30% of the negotiated price and then makes payments of $415.90 per month thereafter for 36 months.

Furthermore, she believes the car can be sold for $7,000 at the end of three years. Draw a cash-flow diagram of this situation from Mrs. Green's viewpoint. (3.7)

3-17. If $25,000 is deposited now into a savings account that earns 6% per year, what uniform annual amount could be withdrawn at the end of each year for ten years so that nothing would be left in the account after the 10th withdrawal? (3.9)

3-18. It is estimated that a certain piece of equipment can save $22,000 per year in labor and materials costs. The equipment has an expected life of five years and no market value. If the company must earn a 15% annual return on such investments, how much could be justified now for the purchase of this piece of equipment? Draw a cash-flow diagram from the company's viewpoint. (3.7, 3.9)

3-19. Suppose that installation of Low-Loss thermal windows in your area is expected to save $350 a year on your home heating bill for the next 18 years. If you can earn 8% per year on other investments, how much could you afford to spend now for these windows? (3.9)

3-20. A proposed product modification to avoid production difficulties will require an immediate expenditure of $14,000 to modify certain dies. What annual savings must be realized to recover this expenditure in four years with interest at 10% per year? (3.9)

3-21. You can buy a machine for $100,000 that will produce a net income, after operating expenses, of $10,000 per year. If you plan to keep the machine for four years, what must the market (resale) value be at the end of four years to justify the investment? You must make a 15% annual return on your investment. (3.9)

3-22. Consider the accompanying cash-flow diagram. (See Figure P3-22.) (3.9)

Figure P3-22
Figure for Problem 3-22

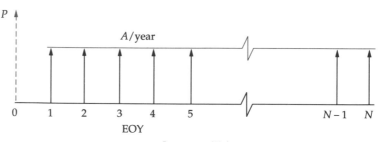

P

A/year

0   1   2   3   4   5   N–1   N

EOY

Interest = i%/year

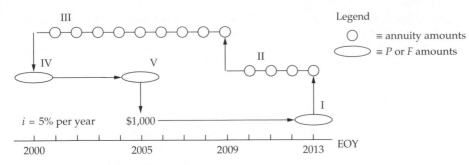

Figure P3-25 Figure for Problem P3-25

a. If $P = \$1,000$, $A = \$200$, and $i\% = 12\%$ per year, then $N = ?$

b. If $P = \$1,000$, $A = \$200$, and $N = 10$ years, then $i = ?$

c. If $A = \$200$, $i\% = 12\%$ per year, and $N = 5$ years, then $P = ?$

d. If $P = \$1,000$, $i\% = 12\%$ per year, and $N = 5$ years, then $A = ?$

3-23. Use the rule of 72 to determine how long it takes to accumulate $10,000 in a savings account when $P = \$5,000$ and $i = 10\%$ per year. (3.8)

> Rule of 72: The time required to double the value of a lump-sum investment that is allowed to compound is approximately
>
> $72 \div$ annual interest rate (as a %)

3-24.

a. Show that the following relationship is true: $(A/P, i, N) = i/[1 - (P/F, i, N)]$. (3.10)

b. Show that $(A/G, 0\%, N) = (N - 1)/2$. (3.14)

c. If an annuity, A, starts at the end of year one and continues each year thereafter forever, what is its P_0 equivalent value when $i = 12\%$ per year? (3.9)

3-25. Using Figure P3-25, find the equivalent values of cash flows I–V to a single $1,000 cash flow at the end of 2005 when the interest rate is 5% per year. (*Hint*: Moving $1,000 from 2005 to I, I to II, and so on, with time value of money calculations should result in $1,000 at EOY 2005.) (3.8)

3-26. Suppose that $10,000 is borrowed now at 15% interest per year. A partial repayment of

$4,000 is made four years from now. The amount that will remain to be paid then is most nearly: (3.8)

a. $7,000, **b.** $8,050, **c.** $8,500,
d. $13,490 **e.** $14,490.

3-27. How much should be deposited each year for 12 years if you wish to withdraw $309 each year for five years, beginning at the end of the 15th year? Let $i = 8\%$ per year. (3.11)

3-28. Suppose that you have $10,000 cash today and can invest it at an interest rate of 10% compounded each year. How many years will it take you to become a millionaire? (3.8)

3-29. Equal end-of-year payments of $263.80 each are being made on a $1,000 loan at 10% effective interest per year. (3.6, 3.9)

a. How many payments are required to repay the entire loan?

b. Immediately after the second payment, what lump-sum amount would completely pay off the loan?

3-30. Maintenance costs for a small bridge with an expected 50-year life are estimated to be $1,000 each year for the first 5 years, followed by a $10,000 expenditure in the year 15 and a $10,000 expenditure in year 30. If $i = 10\%$ per year, what is the equivalent uniform annual cost over the entire 50-year-period? (3.12)

3-31. In 1971, first-class postage for a one-ounce envelope was $0.08. In 2001, a first-class stamp for the same envelope cost $0.34. What compounded *annual* increase in the cost of first-class postage was experienced during the 30 years? (3.8)

3-32. You purchase special equipment that reduces defects by $10,000 per year on an item. This item is sold on contract for the next five years. After the contract expires, the special equipment will save approximately $3,000 per year for five years. You assume that the machine has no market value at the end of ten years. How much can you afford to pay for this equipment now if you require a 20% annual return on your investment? All cash flows are end-of-year amounts. (3.12)

3-33. John Q. wants his estate to be worth $200,000 at the end of 10 years. His net worth is now zero. He can accumulate the desired $200,000 by depositing $14,480 at the end of each year for the next 10 years. At what interest rate per year must his deposits be invested? (3.9)

3-34. What lump sum of money must be deposited into a bank account at the present time so that $500 per month can be withdrawn for five years, with the first withdrawal scheduled for six years from today? The interest rate is 3/4% per month. (*Hint:* Monthly withdrawals begin at the end of month 72.) (3.11)

3-35. Solve for the value of Z in the accompanying figure below so that the top cash-flow diagram is equivalent to the bottom one. Let $i = 8\%$ per year. (3.12)

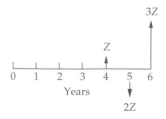

3-36. An individual is borrowing $100,000 at 8% interest compounded annually. The loan is to be repaid in equal annual payments over 30 years. However, just after the eighth payment is made,

the lender allows the borrower to triple the annual payment. The borrower agrees to this increased payment. If the lender is still charging 8% per year, compounded annually, on the unpaid balance of the loan, what is the balance still owed just after the twelfth payment is made? (3.12)

3-37. A woman arranges to repay a $1,000 bank loan in 10 equal payments at a 10% effective annual interest rate. Immediately after her third payment, she borrows another $500, also at 10% per year. When she borrows the $500, she talks the banker into letting her repay the remaining debt of the first loan and the entire amount of the second loan in 12 equal annual payments. The first of these 12 payments would be made one year after she receives the $500. Compute the amount of each of the 12 payments. (3.12)

3-38. A loan of $10,000 is to be repaid over a period of eight years. During the first four years, exactly half of the *loan principal* is to be repaid (along with accumulated compound interest) by a uniform series of payments of A_1 dollars per year. The other half of the loan principal is to be repaid over four years with accumulated interest by a uniform series of payments of A_2 dollars per year. If $i = 9\%$ per year, what are A_1 and A_2? (3.12)

3-39. On January 1, 2002, a person's savings account was worth $200,000. Every month thereafter, this person makes a cash contribution of $676 to the account. If the fund is expected to be worth $400,000 on January 1, 2007, what annual rate of interest is being earned on this fund? (3.17)

3-40. Determine the present equivalent value at time 0 in the accompanying cash-flow diagram (see Figure P3-40) above when $i = 7\%$ per year. Try to minimize the number of interest factors you use. (3.12)

3-41. Transform the cash flows on the left-hand side of the accompanying diagram (see Figure P3-41) to their equivalent amount, F, shown on the right-hand side. The interest rate is 8% per year. (3.12)

3-42. Determine the value of W on the right-hand side of the accompanying diagram (see Figure P3-42) that makes the two cash-flow diagrams equivalent when $i = 12\%$ per year. (3.12)

Figure P3-40
Figure for Problem P3-40

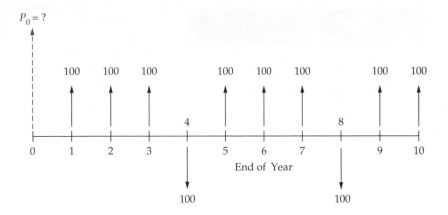

Figure P3-41
Figure for Problem P3-41

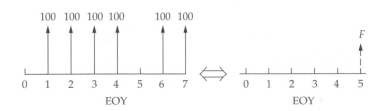

Figure P3-42
Figure for Problem P3-42

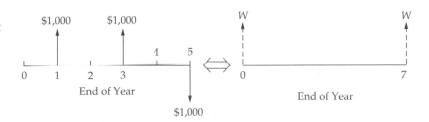

Figure P3-43
Figure for Problem P3-43

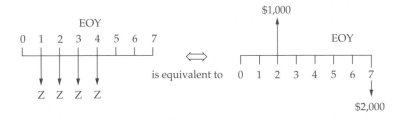

3-43. Determine the value of Z on the left-hand side of the accompanying cash-flow diagram (see Figure P3-43) that establishes equivalence with the right-hand side. The interest rate is 10% per year. (3.12)

3-44. Determine the value of "A" (uniform annual amount in years 1 through 10) in Table P3-44 that is equivalent to the following cash-flow pattern (the interest rate is 10% per year): (3.12)

TABLE P3-44 Cash flow pattern for Problem P3-44

End of Year	Amount
0	$800
1	1,000
2	1,000
3	1,100
4	1,200
5	1,300
6	1,400
7	1,500
8	1,600
9	1,700
10	1,800

3-45. A certain fluidized-bed combustion vessel has an investment cost of $100,000, a life of 10 years, and negligible market (resale) value. Annual costs of materials, maintenance, and electric power for the vessel are expected to total $10,000. A major relining of the combustion vessel will occur during the fifth year at a cost of $30,000. If the interest rate is 15% per year, what is the lump-sum equivalent cost of this project at the present time? (3.12)

3-46. Suppose that $400 is deposited each year into a bank account that pays interest annually ($i =$ 8%). If 12 payments are made into the account, how much would be accumulated in this fund by the end of the 12th year? The first payment occurs at time zero (now). (3.9)

3-47. An expenditure of $20,000 is made to modify a material-handling system in a small job shop. This modification will result in first-year savings of $2,000, a second-year savings of $4,000, and a savings of $5,000 per year thereafter. How many years must the system last if a 18% return on investment is required? The system is tailor made for this job shop and has no market (salvage) value at any time. (3.12)

3-48. Determine the present equivalent and annual equivalent value of the cash-flow pattern shown in Figure P3-48 when $i = 8\%$ per year. (3.13)

3-49. Find the uniform annual amount that is equivalent to a uniform gradient series in which the first year's payment is $500, the second year's payment is $600, the third year's payment is $700, and so on, and there is a total of 20 payments. The annual interest rate is 8%. (3.13)

3-50. Suppose that annual income from a rental property is expected to start at $1,300 per year and decrease at a uniform amount of $50 each year after the first for the 15-year expected life of the property. The investment cost is $8,000 and i is 9% per year. Is this a good investment? Assume that the investment occurs at time zero (now) and that the annual income is first received at the end of year one. (3.13)

3-51. For a repayment schedule that starts at the end of year four at $Z and proceeds for years 4 through 10 at $2Z, $3Z,..., what is the value of Z if the principal of this loan is $10,000 and the interest rate is 7% per year? Use a uniform gradient amount (G) in your solution. (3.13)

3-52. If $10,000 now is equivalent to 4Z at the end of year two, 3Z at the end of year three, 2Z at the end of year four, and Z at the end of year five, what is the value of Z when $i = 8\%$ per year? Use a uniform gradient amount (G) in your solution. (3.13)

3-53. Refer to the accompanying cash-flow diagram (see Figure P3-53), and solve for the unknown quantity in parts (a) through (d) that makes the equivalent value of cash outflows equal to the equivalent value of the cash inflow, F. (3.13)

   a. If $F = \$10,000$, $G = \$600$, and $N = 6$, then $i = ?$
   b. If $F = \$10,000$, $G = \$600$, and $i = 5\%$ per period, then $N = ?$
   c. If $G = \$1,000$, $N = 12$, and $i = 10\%$ per period, then $F = ?$
   d. If $F = \$8,000$, $N = 6$, and $i = 10\%$ per period, then $G = ?$

Figure P3-48
Figure for Problem P3-48

End of Year	0	1	2	3	4	5	6	7
Amount ($)	−1,500	+500	+500	+500	+400	+300	+200	+100

Figure P3-53
Figure for Problem P3-53

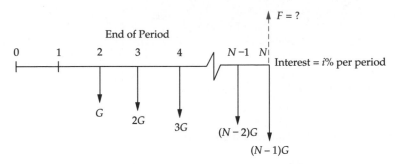

Figure P3-54
Figure for Problem P3-54

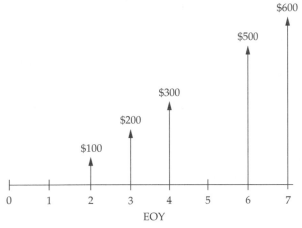

3-54. Solve for P_0 in the accompanying cash-flow diagram, Figure P3-54, by using only two interest factors. The interest rate is 15% per year. (3.13)

3-55. In the accompanying diagram, Figure P3-55, what is the value of K on the left-hand cash-flow diagram that is equivalent to the right-hand cash-flow diagram? Let $i = 12\%$ per year. (3.13)

3-56. For the accompanying cash-flow diagram, Figure P3-56, complete the following equivalence equation: $P_0 = \$100(P/A, 10\%, 4) +$ _____. (This can be completed with one more term.) (3.13)

3-57. Calculate the future equivalent at the end of 1999, at 8% per year, of the following series of cash flows in Figure P3-57 [use a uniform gradient amount (G) in your solution]: (3.13)

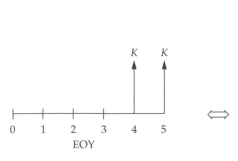

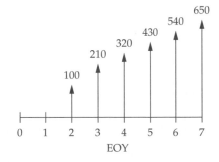

Figure P3-55 **Figure for Problem P3-55**

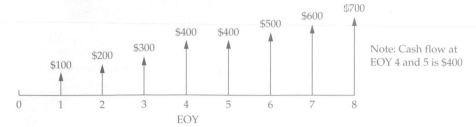

Note: Cash flow at
EOY 4 and 5 is $400

Figure P3-56 Figure for Problem P3-56

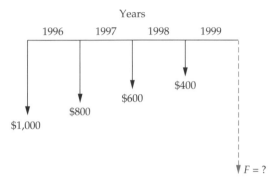

Figure P3-57 Figure for Problem P3-57

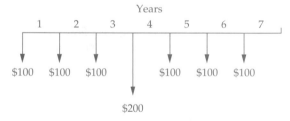

Figure P3-58 Figure for Problem P3-58

3-58. Convert the cash-flow pattern shown in Figure P3-58 to a uniform series of end-of-year costs over a seven-year period and let $i = 9\%$ per year: (3.12)

3-59. Suppose that the parents of a young child decide to make annual deposits into a savings account, with the first deposit being made on the child's fifth birthday and the last deposit being made on the fifteenth birthday. Then starting on the child's eighteenth birthday, the withdrawals shown below will be made. If the effective annual interest rate is 8% during this period of time, what are the annual deposits in years five through fifteen? Use a uniform gradient amount (G) in your solution. (See Figure P3-59.) (3.13)

3-60. Find the value of the unknown quantity in the accompanying cash-flow diagram, Figure P3-60, to establish equivalence of cash inflows and outflows. Let $i = 8\%$ per year. Use a uniform gradient factor in your solution. (3.13)

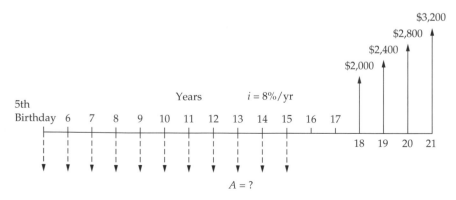

Figure P3-59 Figure for Problem P3-59

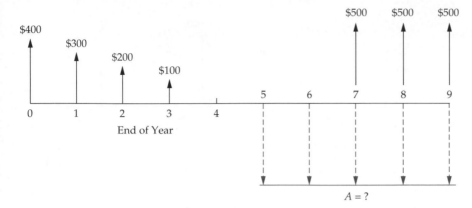

Figure P3-60 Figure for Problem P3-60

3-61. The heat loss through the exterior walls of a certain poultry processing plant is estimated to cost the owner $3,000 next year. A salesman from Superfiber Insulation, Inc., has told you, the plant engineer, that he can reduce the heat loss by 80% with the installation of $18,000 worth of Superfiber now. If the cost of heat loss rises by $200 per year (gradient) after the next year and the owner plans to keep the present building for 15 more years, what would you recommend if the interest rate is 10% per year? (3.13)

3-62. Find the equivalent value of Q in the accompanying cash-flow diagram. (3.13)

3-63. What value of N comes closest to making the left-hand cash-flow diagram of the accompanying figure, Figure P3-63, equivalent to the one on the right? Let $i = 15\%$ per year. Use a uniform gradient amount (G) in your solution. (3.13)

3-64. Find the value of B on the left-hand diagram of Figure P3-64, that makes the two cash-flow diagrams equivalent at $i = 10\%$ per year. (3.13)

3-65. You are the manager of a large crude oil refinery. As part of the refining process, a certain heat exchanger (operated at high temperatures and with abrasive material flowing through it) must be replaced every year. The replacement and downtime cost in the first year is $175,000. This cost is expected to increase due to inflation at a rate of 8% per year for five years, at which time this particular heat exchanger will no longer be needed. If the company's

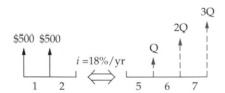

Figure P3-63
Figure for Problem P3-63

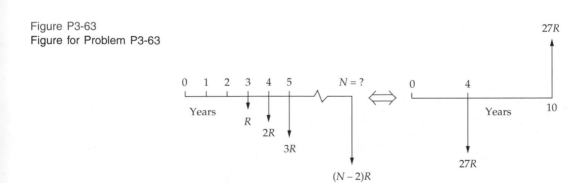

Figure P3-64
Figure for Problem P3-64

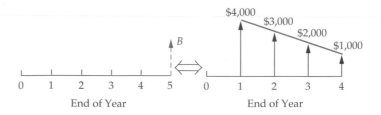

cost of capital is 18% per year, how much could you afford to spend for a higher quality heat exchanger so that these annual replacement and downtime costs could be eliminated? (3.14)

3-66. A geometric gradient that increases at $\overline{f} = 6\%$ per year for 15 years is shown in the accompanying diagram. The annual interest rate is 12%. What is the present equivalent value of this gradient? (3.14)

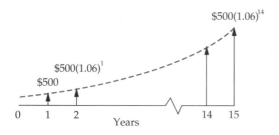

3-67. In a geometric sequence of annual cash flows starting at end of year *zero*, the value of A_0 is $1,304.35 (which *is* a cash flow). The value of the last term in the series, A_{10}, is $5,276.82. What is the equivalent value of A for years one through ten? Let $i = 20\%$ per year. (3.14)

3-68. An electronic device is available that will reduce this year's labor costs by $10,000. The equipment is expected to last for eight years. If labor costs increase at an average rate of 7% per year and the interest rate is 12% per year,

  a. What is the maximum amount that we could justify spending for the device?
  b. What is the uniform annual equivalent value (A) of labor costs over the eight-year period?
  c. What annual year zero amount (A_0) that inflates at 7% per year is equivalent to the answer in part (a)? (3.14)

3-69. Determine the present equivalent (at time zero) of the accompanying geometric sequence of cash flows. Let $i = 15.5\%$ per year and $\overline{f} = 10\%$. (3.14)

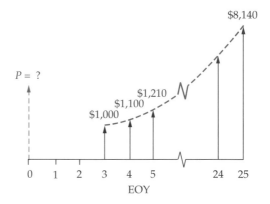

3-70. Rework Problem 3-69 when the cash flow at the end of year 3 is $8,140, and cash flows at the end of years 4 through 25 *decrease* by 10% per year (i.e., $\overline{f} = -10\%$ per year). (3.14)

3-71. For the cash-flow diagram in Figure P3-71, solve for X such that the cash receipt in year zero is equivalent to the cash outflows in years one through six. (3.14)

3-72. An EOY geometric gradient lasts for 10 years, whose initial value at EOY three is $5,000 and $\overline{f} - 6.04\%$ per year thereafter. Find the equivalent uniform gradient amount (G) over the same time period (beginning in year 1 and ending in year 12) if the initial value of the series at EOY one is $4,000. Answer the following questions in determining the value of the gradient amount, G. The interest rate is 8% nominal, compounded semiannually. (3.13, 3.14)

  a. What is i_{CR}?
  b. What is P_0 for the geometric gradient?
  c. What is P_0 of the uniform (arithmetic) gradient?
  d. What is the value of G?

3-73. *Set up an expression* for the unknown quantity, Z, in the cash-flow diagram in Figure P3-73. (3.13, 3.14)

Figure P3-71
Figure for Problem P3-71

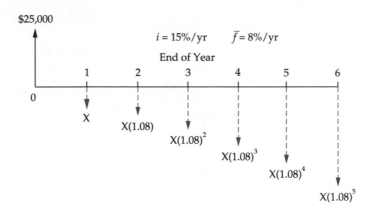

Figure P3-73
Figure for Problem P3-73

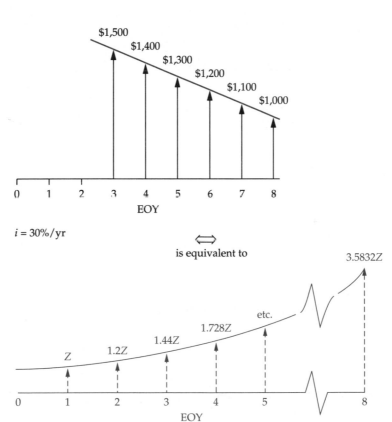

3-74. An individual makes six annual deposits of $2,000 in a savings account that pays interest at a rate of 4% compounded annually. Two years after making the last deposit, the inter- est rate changes to 7% compounded annually. Twelve years after the last deposit the accumu- lated money is withdrawn from the account. How much is withdrawn? (3.15)

3-75. Compute the effective annual interest rate in each of these situations: (3.16)

  a. 10% nominal interest, compounded semiannually.

  b. 10% nominal interest compounded quarterly.

  c. 10% nominal interest compounded weekly.

3-76. Sixty monthly deposits are made into an account paying 6% nominal interest compounded monthly. If the objective of these deposits is to accumulate $100,000 by the end of the fifth year, what is the amount of each deposit? (3.17)

  a. $1,930.   **b.** $1,478.   **c.** $1,667.
  d. $1,430.   **e.** $1,695.

3-77.

  a. What extra semiannual expenditure for five years would be justified for the maintenance of a machine in order to avoid an overhaul costing $3,000 at the end of five years? Assume nominal interest at 8%, compounded semiannually. (3.17)

  b. What is the annual equivalent value of $125,000 now when 12% nominal interest per year is compounded monthly? Let $N = 10$ years. (3.17)

3-78.

  a. What equal monthly payments will repay an original loan of $10,000 in six months at a nominal rate of 6% compounded monthly? What is the effective annual interest rate? (3.17)

  b. For part (a), what is the effective quarterly interest rate? (3.18)

3-79. Determine the current amount of money that must be invested at 12% nominal interest, compounded monthly, to provide an annuity of $10,000 (per year) for six years, starting 12 years from now. The interest rate remains constant over this entire period of time. (3.17)

3-80. Find the present equivalent value of the following series of payments: $100 at the end of each month for 72 months at a nominal rate of 15% compounded monthly. (3.17)

3-81. Determine the present equivalent value of $5,000 paid every three months over a period of seven years in each of the following situations: (3.18)

  a. The nominal interest rate is 12%, compounded annually.

  b. The nominal interest rate is 12%, compounded quarterly.

  c. The nominal interest rate is 12%, compounded weekly.

3-82. Suppose that you have just borrowed $7,500 at 12% nominal interest compounded quarterly. What is the total lump-sum, compounded amount to be paid by you at the end of a 10-year loan period? (3.17)

3-83. How many deposits of $100 each must you make at the end of each month if you desire to accumulate $3,350 for a new home entertainment center? Your savings account pays 9% nominal interest, compounded monthly. (3.17)

3-84. You have used your credit card to purchase automobile tires for $340. Unable to make payments for 11 months, you then write a letter of apology and enclose a check to pay your bill in full. The credit card company's nominal interest rate is 16.5% compounded monthly. For what amount should you write the check? (3.17)

3-85. How long does it take a given amount of money to double if the money is invested at a nominal rate of 12%, compounded monthly? (3.17)

3-86. What is the principal remaining after 20 monthly payments have been made on a $20,000 five-year loan? The annual interest rate is 12% nominal compounded monthly.

  a. $10,224.   **b.** $13,333.   **c.** $14,579.
  d. $16,073.   **e.** $17,094.

3-87.

  a. A certain savings-and-loan association advertises that it pays 8% nominal interest, compounded quarterly. What is the *effective* interest rate per annum? If you deposit $5,000 now and plan to withdraw it in three years, how much would your account be worth at that time? (3.17)

  b. If instead you decide to deposit $800 every year for three years, how much could be withdrawn at the end of the third year? Suppose that, instead, you deposit $400 every six months for three years. What would the accumulated amount be? (3.18)

3-88. The effective annual interest rate, i, has been determined to be 26.82% (based on monthly compounding). Calculate how much can be spent now to avoid future computer software maintenance expenses of $1,000 *per quarter* for the next five years. (3.18)

3-89. If the nominal interest rate is 8% and compounding is semiannual, what is the present equivalent value of the receipts in the following diagram? (3.13, 3.17)

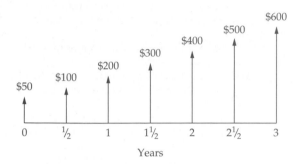

3-90. What is the monthly payment on a loan of $15,000 for five years at a nominal rate of interest of 9% compounded monthly? (3.17)

    a. $214.   **b.** $250.   **c.** $312.

    d. $324.   **e.** $381.

3-91. The effective annual interest rate, i, is stated to be 19.2%. What is the nominal interest rate

per year, r, if continuous compounding is being used? (3.19)

3-92. Find the value of A that is equivalent to the uniform gradient shown in Figure P3-92 if the nominal interest rate is 10% compounded monthly. (3.13, 3.18)

3-93. Suppose that you have a money market certificate earning an annual rate of interest, which varies over time as follows:

Year k	1	2	3	4	5
i_k	14%	12%	10%	10%	12%

If you invest $10,000 in this certificate at the beginning of year one and do not add or withdraw any money for five years, what is the value of the certificate at the end of the fifth year? (3.15)

3-94. Determine the present equivalent value of the cash-flow diagram of Figure P3-94 when the annual interest rate, i_k, varies as indicated. (3.15)

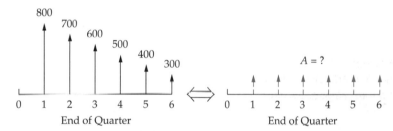

Figure P3-92 Figure for Problem P3-92

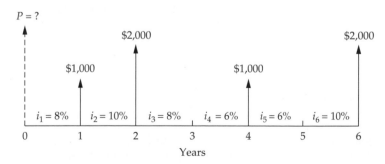

Figure P3-94 Figure for Problem P3-94

3-95. What is the value of F_4 in the following cash flow diagram? (3.15)

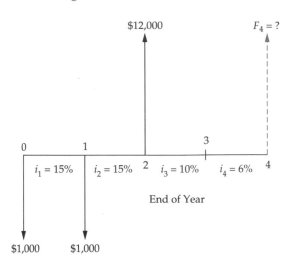

3-96. Indicate whether each of the following statements is true (T) or False (F). (all sections)

a. T F Interest is money paid for the use of equity capital.

b. T F $(A/F, i\%, N) = (A/P, i\%, N) + i$.

c. T F Simple interest ignores the time value of money principle.

d. T F Cash-flow diagrams are analogous to free-body diagrams for mechanics problems.

e. T F $1,791 10 years from now is equivalent to $900 now if the interest rate equals 8% per year.

f. T F It is always true that $i > r$ when $M \geq 2$.

g. T F Suppose that a lump-sum of $1,000 is invested at $\underline{r} = 10\%$ for eight years. The future equivalent is greater for daily compounding than it is for continuous compounding.

h. T F For a fixed amount, F dollars, that is received at EOY N, the "A equivalent" increases as the interest rate increases.

i. T F For a specified value of F at EOY N, P at time zero will be larger for $\underline{r} = 10\%$ per year than it will be for $r = 10\%$ per year, compounded monthly.

3-97. If a nominal interest rate of 8% is compounded continuously, determine the unknown quantity in each of the following situations: (3.19)

a. What uniform end-of-year amount for 10 years is equivalent to $8,000 at the end of year 10?

b. What is the present equivalent value of $1,000 per year for 12 years?

c. What is the future equivalent at the end of the sixth year of $243 payments made every six months during the six years? The first payment occurs six months from the present and the last occurs at the end of the sixth year.

d. Find the equivalent lump-sum amount at the end of year nine when $P_0 = \$1,000$ and a nominal interest rate of 8% is compounded continuously.

3-98. Find the value of the unknown quantity Z in the following diagram such that the equivalent cash outflow equals the equivalent cash inflows when $\underline{r} = 20\%$ compounded continuously: (3.19)

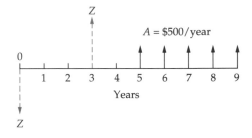

3-99. A man deposited $10,000 in a savings account when his son was born. The nominal interest rate was 8% per year, compounded continuously. On the son's 18th birthday, the accumulated sum is withdrawn from the account. How much will this accumulated amount be? (3.19)

3-100. Find the value of P in the cash-flow diagram in Figure P3-100. (3.19)

3-101. Your rich uncle has just offered to make you wealthy! For every dollar you save in an insured, continuously compounded, bank account during the next 10 years, he will give you a dollar to match it. Because your modest income permits you to save $3,000 per year for each of the next 10 years, your uncle will be willing to give you $30,000 at the end of the 10th year.

Figure P3-100
Figure for Problem
P3-100

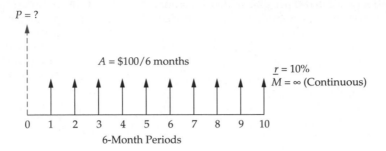

Figure P3-104
Figure for Problem
P3-104

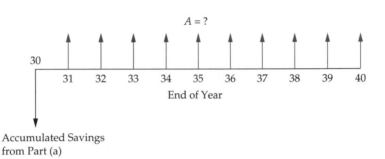

If you desire a *total* of $75,000 ten years from now, what annual interest rate would you have to earn on your insured bank account to make your goal possible? (3.19)

3-102. A person needs $18,000 immediately as a down payment on a new home. Suppose that she can borrow this money from her company credit union. She will be required to repay the loan in equal payments *made every six months* over the next 12 years. The annual interest rate being charged is 10% compounded continuously. What is the amount of each payment? (3.19)

3-103.

a. What is the present equivalent of a uniform series of annual payments of $3,500 each for five years if the interest rate, compounded continuously, is 10%?

b. The amount of $7,000 is invested in a Certificate of Deposit (CD) and will be worth $16,000 in nine years. What is the continuously compounded nominal (annual) interest rate for this CD? (3.19)

3-104.

a. Many persons prepare for retirement by making monthly contributions to a savings

program. Suppose that $2,000 is set aside each year and invested in a savings account that pays 10% interest per year, compounded continuously. Determine the accumulated savings in this account at the end of 30 years.

b. In part (a), suppose that an annuity will be withdrawn from savings that have been accumulated at the end of year 30. The annuity will extend from the end of year 31 to the end of year 40. What is the value of this annuity if the interest rate and compounding frequency in part (a) do not change? Refer to Figure P3-104. (3.19)

3-105.

a. What is the future equivalent of a continuous funds flow amounting to $10,500 per year when $r = 20\%$, $M = \infty$, and $N = 12$ years?

b. If the nominal interest rate is 10% per year, continuously compounded, what is the future equivalent of $10,000 per year flowing continuously for 8.5 years? See the cash-flow diagram in Figure P3-105.

c. Let $\bar{A} = \$7,859$ per year with $r = 20\%$, $M = \infty$. How many years will it take to have $1 million in this account? (3.20)

3-106. For how many years must an investment of $63,000 provide a continuous flow of funds at the rate of $16,000 per year so that a nominal interest rate of 10%, continuously compounded, will be earned? (3.20)

3-107. What is the present equivalent of the following continuous funds flow situations?

a. $1,000,000 per year for four years at 10% compounded continuously.

b. $6,000 per year for 10 years at 8% compounded annually.

c. $500 per quarter for 6.75 years at 20% compounded continuously. (3.20)

3-108. What is the difference in present equivalents for the cash-flow diagram in Figure P3-108 and the one from Problem 3-105? (3.20)

3-109. Mark each statement true (T) or false (F), and fill in the blanks in part **f**. (3.19, 3.20)

a. **T F** The nominal interest rate will always be less than the effective interest rate when $r = 10\%$ and $M = \infty$.

b. **T F** A certain loan involves monthly repayments of $185 over a 24-month period. If $\underline{r} = 10\%$ per year, more than half of the principal is still owed on this loan after the 10th monthly payment is made.

c. **T F** $1,791 10 years from now is equivalent to $900 now if the nominal interest rate is 8% compounded semiannually.

d. **T F** If i (expressed as a decimal) is added to the series capital-recovery factor, the series sinking-fund factor will be obtained.

e. **T F** The $(P/A, i\%, N)$ factor equals $N \cdot (P/F, i\%, 1)$.

f. Fill in the missing interest factor:

i. $(P/A, i\%, N)(\underline{\hspace{1.5cm}}) = (F/A, i\%, N)$.

ii. $(A/G, i\%, N)(P/A, i\%, N) = (\underline{\hspace{1.5cm}})$.

$\bar{A} = \$10,000/\text{yr}$ $F = ?$

End of Year

Figure P3-105 **Figure for Problem P3-105**

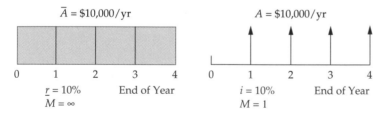

$\bar{A} = \$10,000/\text{yr}$ $A = \$10,000/\text{yr}$

$\underline{r} = 10\%$ End of Year $i = 10\%$ End of Year
$M = \infty$ $M = 1$

Figure P3-108 **Figure for Problem P3-108**

*B*asic Topics in Engineering Economy

The final test of any system is, does it pay?
—Frederick W. Taylor, testimony before the Special Committee of the U.S. House of
Representatives (January 25, 1912)

Applications of Money–Time Relationships

*T*he two primary objectives of this chapter are (1) to illustrate several basic methods for making engineering economy studies considering the time value of money and (2) to describe briefly the underlying assumptions and interrelationships among these methods.

The following topics are discussed in this chapter:

Determining the minimum attractive rate of return

The present worth method

The future worth method

The annual worth method

The internal rate of return method

The external rate of return method

The payback (payout) period method

Investment balance diagrams

4.1 Introduction

All engineering economy studies of capital projects should consider the return that a given project will or should produce. A basic question this book addresses is whether a proposed capital investment and its associated expenditures can be recovered by revenue (or savings) over time *in addition to* a return on the capital that is sufficiently attractive in view of the risks involved and the potential alternative uses. The interest and money–time relationships discussed in Chapter 3 emerge as essential ingredients in answering this question, and they are applied to many different types of problems in this chapter.

Because patterns of capital investment, revenue (or savings) cash flows, and disbursement cash flows can be quite different in various projects, there is no single method for performing engineering economic analyses that is ideal for all cases. Consequently, several methods are commonly used.

In this chapter, we concentrate on the correct use of five methods for evaluating the economic profitability of a single proposed problem solution (i.e., *alternative*).* Later, in Chapter 5, multiple alternatives are evaluated. The five methods described in Chapter 4 are Present Worth (PW), Future Worth (FW), Annual Worth (AW), Internal Rate of Return (IRR), and External Rate of Return (ERR). The first three methods convert cash flows resulting from a proposed problem solution into their equivalent worth at some point (or points) in time by using an interest rate known as the *Minimum Attractive Rate of Return (MARR)*. The concept of a MARR, as well as the determination of its value, is discussed in the next section. The IRR and ERR methods produce annual rates of profit, or returns, resulting from an investment, and are then compared to the MARR.

The payback period is also discussed briefly in this chapter. The payback period is a measure of the *speed* with which an investment is recovered by the cash inflows it produces. This measure, in its most common form, ignores time value of money principles. For this reason, the payback method is often used to supplement information produced by the five primary methods featured in this chapter. Another measure of liquidity is provided by an investment balance diagram. This measure is described in Section 4.9.

> Unless otherwise specified, the end-of-period cash flow convention and discrete compounding of interest are used throughout this and subsequent chapters. A planning horizon, or study (analysis) period, of N compounding periods (usually years) is used to evaluate prospective investments throughout the remainder of the book.

▼ 4.2 Determining the Minimum Attractive Rate of Return

The Minimum Attractive Rate of Return (MARR) is usually a policy issue resolved by the top management of an organization in view of numerous considerations. Among these considerations are the following:

1. The amount of money available for investment, and the source and cost of these funds (i.e., equity funds or borrowed funds).
2. The number of good projects available for investment and their purpose (i.e., whether they sustain present operations and are *essential* or whether they expand on present operations and are *elective*).
3. The amount of perceived risk associated with investment opportunities available to the firm and the estimated cost of administering projects over short planning horizons versus long planning horizons.

---

*The analysis of engineering projects using the benefit–cost ratio method is discussed in Chapter 11.

4. The type of organization involved (i.e., government, public utility, or competitive industry).

In theory the MARR, which is sometimes called the *hurdle rate,* should be chosen to maximize the economic well-being of an organization, subject to the types of considerations just listed. How an individual firm accomplishes this in practice is far from clear-cut and is frequently the subject of discussion. One popular approach to establishing a MARR involves the *opportunity cost* viewpoint described in Chapter 2, and it results from the phenomenon of *capital rationing.* For the purposes of this chapter, capital rationing exists when management decides to limit the total amount of capital invested. This situation may arise when the amount of available capital is insufficient to sponsor all worthy investment opportunities.

A simple example of capital rationing is given in Figure 4-1, where the cumulative investment requirements of seven acceptable projects are plotted against the prospective annual rate of profit of each. Figure 4-1 shows a limit of $6 million on available capital. In view of this limitation, the last funded project would be E, with a prospective rate of profit of 19% per year, and the best *rejected* project is F. In this case, the MARR by the opportunity cost principle would be 16% per year. By *not* being able to invest in project F, the firm would presumably be forfeiting the chance to realize a 16% annual return. As the amount of investment capital and opportunities available change over time, the firm's MARR will also change.

Superimposed on Figure 4-1 is the approximate cost of obtaining the $6 million, illustrating that project E is acceptable only as long as its annual rate of profit exceeds the cost of raising the last $1 million. As shown in Figure 4-1, the cost of capital will tend to increase gradually as larger sums of money are acquired through increased borrowing (debt) or new issuances of common stock (equity). One last observation in connection with Figure 4-1 is that the perceived risk associated

Figure 4-1
Determination of the
MARR Based on the
Opportunity Cost
Viewpoint. A popular
measure of annual rate
of profit is "internal rate
of return." (Discussed
later in this chapter.)

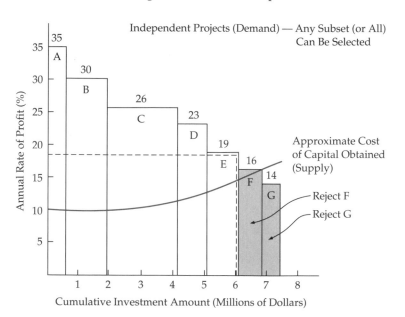

with financing and undertaking the seven projects has been determined by top management to be acceptable.

EXAMPLE 4-1

Consider the following schedule, which shows prospective annual rates of profit for a company's portfolio of capital investment projects (this is the *demand* for capital):

Expected Annual Rate of Profit	Investment Requirements (Thousands of Dollars)	Cumulative Investment
40% and over	$2,200	$ 2,200
30–39.9%	3,400	5,600
20–29.9%	6,800	12,400
10–19.9%	14,200	26,600
Below 10%	22,800	49,400

Note: All projects with a rate of profit of 10% or greater are acceptable.

If the supply of capital obtained from internal and external sources has a cost of 15% per year for the first $5,000,000 invested and then increases 1% for every $5,000,000 thereafter, what is this company's MARR when using an opportunity cost viewpoint?

SOLUTION

Cumulative capital demand versus supply can be plotted against prospective annual rate of profit, as shown in Figure 4-2. The point of intersection is approximately 18% per year, which represents a realistic estimate of this company's MARR when using the opportunity cost viewpoint.

Figure 4-2
Solution Graph for
Example 4-1

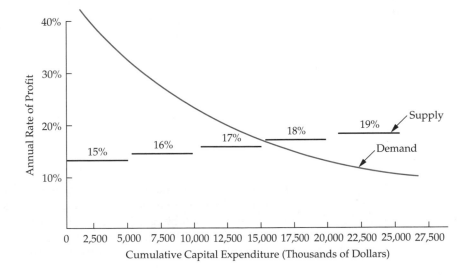

▼ 4.3 **The Present Worth Method**

The *Present Worth* (PW) method is based on the concept of equivalent worth of all cash flows relative to some base or beginning point in time called the present. That is, all cash inflows and outflows are discounted to the present point in time at an interest rate that is generally the MARR.

The PW of an investment alternative is a measure of how much money an individual or a firm could afford to pay for the investment in excess of its cost. Or, stated differently, a positive PW for an investment project is a dollar amount of profit over the minimum amount required by investors. It is assumed that cash generated by the alternative is available for other uses that earn interest at a rate equal to the MARR.

To find the PW as a function of $i\%$ (per interest period) of a series of cash inflows and outflows, it is necessary to discount future amounts to the present by using the interest rate over the appropriate study period (years, for example) in the following manner:

$$PW(i\%) = F_0(1+i)^0 + F_1(1+i)^{-1} + F_2(1+i)^{-2} + \cdots$$

$$+ F_k(1+i)^{-k} + \cdots + F_N(1+i)^{-N}$$

$$= \sum_{k=0}^{N} F_k(1+i)^{-k}. \tag{4-1}$$

Here

i = effective interest rate, or MARR, per compounding period,

k = index for each compounding period ($0 \le k \le N$),

F_k = future cash flow at the end of period k,

N = number of compounding periods in the planning horizon (i.e., study period).

The relationship given in Equation (4-1) is based on the assumption of a *constant interest rate* throughout the life of a particular project. If the interest rate is assumed to change, the PW must be computed in two or more steps, as was illustrated in Chapter 3.

The higher the interest rate and the further into the future a cash flow occurs, the lower its PW is. This is shown graphically in Figure 4-3. As long as the PW (i.e., present equivalent of cash inflows minus cash outflows) is greater than or equal to zero, the project is economically justified; otherwise, it is not acceptable.

EXAMPLE 4-2

An investment of $10,000 can be made in a project that will produce a uniform annual revenue of $5,310 for five years and then have a market (salvage) value of $2,000. Annual expenses will be $3,000 each year. The company is willing to accept any project that will earn 10% per year or more, on all invested capital. Show whether this is a desirable investment by using the PW method.

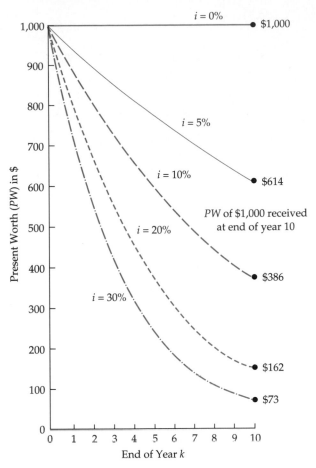

Figure 4-3 PW of $1,000 Received at the End of Year *k* at an Interest Rate of *i*% per year

SOLUTION

	PW	
	Cash Outflows	Cash Inflows
Annual revenue: $5,310(P/A, 10\%, 5)$		$20,129
Market (salvage) value: $2,000(P/F, 10\%, 5)$		1,242
Investment	$10,000	
Annual expenses: $3,000(P/A, 10\%, 5)$	11,372	
Total	$21,372	$21,371
Total PW		$0

Because total PW(10%) $\simeq$ $0, the project is shown to be marginally acceptable.

Figure 4-4
Cash Flow Diagram for
Example 4-3

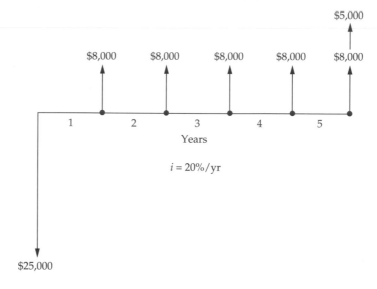

$i = 20\%/\mathrm{yr}$

EXAMPLE 4-3

A piece of new equipment has been proposed by engineers to increase the productivity of a certain manual welding operation. The investment cost is $25,000, and the equipment will have a market value of $5,000 at the end of a study period of five years. Increased productivity attributable to the equipment will amount to $8,000 per year after extra operating costs have been subtracted from the revenue generated by the additional production. A cash-flow diagram for this investment opportunity is given in Figure 4-4. If the firm's MARR is 20% per year, is this proposal a sound one? Use the PW method.

SOLUTION

$$PW = PW \text{ of cash inflows} - PW \text{ of cash outflows,}$$

or

$$PW(20\%) = \$8,000(P/A, 20\%, 5) + \$5,000(P/F, 20\%, 5) - \$25,000$$

$$= \$934.29.$$

Because $PW(20\%) > 0$, this equipment is economically justified.

Based on Example 4-3, Table 4-1 can be used to plot the cumulative PW of cash flows through year k. The graphs of cumulative PW shown in Figure 4-5 at $i = 20\%$ and $i = 0\%$ are plotted from Columns (C) and (D) of Table 4-1, respectively.

The MARR in Example 4-3 (and in other examples throughout this chapter) is to be interpreted as an effective interest rate (i). Here $i = 20\%$ per year. Cash flows are discrete, end-of-year amounts. If *continuous compounding* had been specified

TABLE 4-1 Cumulative PW Calculations for Example 4-3

End of Year k	(A) Net Cash Flow	(B) PW of Cash Flow at $i = 20\%$/yr	(C) Cumulative PW at $i = 20\%$/yr through Year k	(D) Cumulative PW at $i = 0\%$/yr through Year k
0	−$25,000	−$25,000	−$25,000	−$25,000
1	8,000	6,667	−18,333	−17,000
2	8,000	5,556	−12,777	−9,000
3	8,000	4,630	−8,147	−1,000
4	8,000	3,858	−4,289	7,000
5	13,000	5,223	+934	20,000

for a nominal interest rate (r) of 20% per year, the PW would have been calculated by using the interest factors presented in Appendix D:

$$\text{PW } (\underline{r} = 20\%) = -\$25,000 + \$8,000(P/A, \underline{r} = 20\%, 5)$$

$$+ \$5,000(P/F, \underline{r} = 20\%, 5)$$

$$= -\$25,000 + \$8,000(2.8551) + \$5,000(0.3679)$$

$$= -\$319.60.$$

Consequently, with continuous compounding, the equipment would not be economically justifiable. The reason is that the higher effective annual interest rate

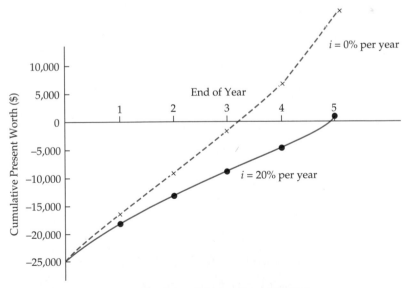

Figure 4-5 Graph of Cumulative PW for Example 4-3

$(e^{0.20} - 1 = 0.2214)$ reduces the PW of future positive cash flows, but does not affect the PW of the capital invested at the beginning of Year 1.

4.3.1 Bond Value

A bond provides an excellent example of commercial value being the PW of the future net cash flows that are expected to be received through ownership of an interest-bearing certificate. Thus, the value of a bond, at any time, is the PW of future cash receipts. For a bond, let

Z = face, or par, value,

C = redemption or disposal price (usually equal to Z),

r = bond rate (nominal interest) per interest period,

N = number of periods before redemption,

i = bond *yield* rate per period, and

V_N = value (price) of the bond N interest periods prior to redemption— this is a PW measure of merit.

The owner of a bond is paid two types of payments by the borrower. The first consists of the series of periodic interest payments he or she will receive until the bond is retired. There will be N such payments, each amounting to rZ. These constitute an annuity of N payments. In addition, when the bond is retired or sold, the bondholder will receive a single payment equal in amount to C. The PW of the bond is the sum of present worths of these two types of payments at the bond's yield rate ($i\%$):

$$V_N = C(P/F, i\%, N) + rZ(P/A, i\%, N). \tag{4-2}$$

EXAMPLE 4-4

Find the current price (PW) of a 10-year bond paying 6% per year (payable semi-annually) that is redeemable at par value, if bought by a purchaser to *yield 10% per year.* The face value of the bond is $1,000:

$$N = 10 \times 2 = 20 \text{ periods,}$$

$$r = 6\%/2 = 3\% \text{ per period,}$$

$$i = [(1.10)^{1/2} - 1]100 \simeq 4.9\% \text{ per semiannual period,}$$

$$C = Z = \$1,000.$$

SOLUTION

Using Equation (4-2), we obtain

$$V_N = \$1,000(P/F, 4.9\%, 20) + \$1,000(0.03)(P/A, 4.9\%, 20)$$

$$= \$384.10 + \$377.06 = \$761.16.$$

EXAMPLE 4-5

A bond with a face value of $5,000 pays interest of 8% per year. This bond will be redeemed at par value at the end of its 20-year life, and the first interest payment is due one year from now.

(a) How much should be paid now for this bond in order to receive a yield of 10% per year on the investment?

(b) If this bond is purchased now for $4,600, what annual yield would the buyer receive?

SOLUTION

(a) By using Equation (4-2), the value of V_N can be determined:

$$V_N = \$5,000(P/F, 10\%, 20) + \$5,000(0.08)(P/A, 10\%, 20)$$

$$= \$743.00 + \$3,405.44 = \$4,148.44.$$

(b) Here we are given $V_N = \$4,600$ and we must find the value of $i\%$ in Equation (4-2):

$$\$4,600 = \$5,000(P/F, i'\%, 20) + \$5,000(0.08)(P/A, i'\%, 20).$$

To solve for $i'\%$, we can resort to an iterative trial-and-error procedure (e.g., try 8.5%, 9.0%, etc.), to determine that $i'\% = 8.9\%$ per year.

EXAMPLE 4-6

A certain U.S. Treasury bond that matures in eight years has a face value of $10,000. This means that the bondholder will receive $10,000 cash when the bond's maturity date is reached. The bond stipulates a fixed nominal interest rate of 8% per year, but interest payments are made to the bondholder every three months; therefore each payment amounts to 2% of the face value.

A prospective buyer of this bond would like to earn 10% nominal interest (compounded quarterly) per year on his or her investment because interest rates in the economy have risen since the bond was issued. How much should this buyer be willing to pay for the bond?

SOLUTION

To establish the value of this bond in view of stated conditions, the PW of future cash flows during the next eight years (the study period) must be evaluated. Interest payments are quarterly. Because the prospective buyer desires to obtain 10% *nominal interest per year* on the investment, the PW is computed at $i = 10\%/4 = 2.5\%$ per quarter for the remaining $8(4) = 32$ quarters of the bond's life:

$$V_N = \$10,000(P/F, 2.5\%, 32) + \$10,000(0.02)(P/A, 2.5\%, 32)$$

$$= \$4,537.71 + \$4,369.84 = \$8,907.55.$$

Thus, the buyer should pay no more than $8,907.55 when 10% nominal interest per year is desired.

4.4 The Future-Worth Method

Because a primary objective of all time value of money methods is to maximize the future wealth of the owners of a firm, the economic information provided by the Future Worth (FW) method is very useful in capital investment decision situations. The future worth is based on the equivalent worth of all cash inflows and outflows at the end of the planning horizon (study period) at an interest rate that is generally the MARR. Also, the FW of a project is equivalent to its PW; that is, $FW = PW(F/P, i\%, N)$. If $FW \geq 0$ for a project, it would be economically justified.

Equation (4-3) summarizes the general calculations necessary to determine a project's future worth:

$$FW(i\%) = F_0(1+i)^N + F_1(1+i)^{N-1} + \cdots + F_N(1+i)^0$$

$$= \sum_{k=0}^{N} F_k(1+i)^{N-k}. \tag{4-3}$$

EXAMPLE 4-7

Evaluate the FW of the potential improvement project described in Example 4-3. Show the relationship between FW and PW for this example.

SOLUTION

$$FW\ (20\%) = -\$25{,}000(F/P, 20\%, 5)$$

$$+ \$8{,}000(F/A, 20\%, 5) + \$5{,}000$$

$$= \$2{,}324.80.$$

Again, the project is shown to be a good investment (FW $\geq$ 0). The FW is a multiple of the equivalent PW value:

$$PW\ (20\%) = \$2{,}324.80(P/F, 20\%, 5) = \$934.29.$$

To this point, the PW and FW methods have used a known and constant MARR over the study period. Each method produces a measure of merit expressed in dollars and is equivalent to the other. The difference in economic information provided is relative to the point in time used (i.e., the present for the PW vs. the future, or end of the study period, for the FW).

4.5 The Annual-Worth Method

The Annual Worth (AW) of a project is an equal annual series of dollar amounts, for a stated study period, that is *equivalent* to the cash inflows and outflows at an interest rate that is generally the MARR. Hence, the AW of a project is annual equivalent revenues or savings ($\underline{R}$) minus annual equivalent expenses ($\underline{E}$), less its annual equivalent Capital Recovery (CR) amount, which is defined in Equation (4-5). An annual equivalent value of $\underline{R}$, $\underline{E}$, and CR is computed for the study period, N, which is usually in years. In equation form, the AW, which is a function of $i\%$, is

$$AW(i\%) = \underline{R} - \underline{E} - CR(i\%). \tag{4-4}$$

Also, we need to notice that the AW of a project is equivalent to its PW and FW. That is, $AW = PW(A/P, i\%, N)$, and $AW = FW(A/F, i\%, N)$. Hence, it can be easily computed for a project from these other equivalent values.

As long as the AW is greater than or equal to zero, the project is economically attractive; otherwise, it is not. An AW of zero means that an annual return exactly equal to the MARR has been earned.

When revenues are *absent* in Equation (4-4), we designate this metric as EUAC ($i\%$) and call it "equivalent uniform annual cost." A low-valued EUAC($i\%$) is preferred to a high-valued EUAC($i\%$).

The Capital Recovery (CR) amount for a project is the equivalent uniform annual *cost* of the capital invested. It is an annual amount that covers the following two items:

1. Loss in value of the asset;
2. Interest on invested capital (i.e., at the MARR).

As an example, consider a machine or other asset that will cost \$10,000, last five years, and have a salvage (market) value of \$2,000. Thus, the loss in value of this asset over five years is \$8,000. Additionally, the MARR is 10% per year.

It can be shown that no matter which method of calculating an asset's loss in value over time is used, the equivalent annual CR amount is the same. For example, if uniform loss in value is assumed, the equivalent annual CR amount is calculated to be \$2,310, as shown in Table 4-2.

There are several convenient formulas by which the CR amount (cost) may be calculated to obtain the result in Table 4-2. Probably the easiest formula to understand involves finding the annual equivalent of the initial capital investment and then subtracting the annual equivalent of the salvage value. Thus,

$$CR(i\%) = I(A/P, i\%, N) - S(A/F, i\%, N), \tag{4-5}$$

where I = initial investment for the project,[*]

         S = salvage (market) value at the end of the study period,

         N = project study period.

---

[*]In some cases, the investment will be spread over several periods. In such situations, I is the PW of all investment amounts.

TABLE 4-2 Calculation of Equivalent Annual CR Amount

Year	Value of Investment at Beginning of Year[a]	Uniform Loss in Value	Interest on Beginning-of-Year Investment at $i = 10\%$	CR Amount for Year	PW of CR Amount at $i = 10\%$
1	$10,000	$1,600	$1,000	$2,600	$2,600(P/F, 10\%, 1) = \$2,364$
2	8,400	1,600	840	2,440	$2,440(P/F, 10\%, 2) = \$2,016$
3	6,800	1,600	680	2,280	$2,280(P/F, 10\%, 3) = \$1,713$
4	5,200	1,600	520	2,120	$2,120(P/F, 10\%, 4) = \$1,448$
5	3,600	1,600	360	1,960	$1,960(P/F, 10\%, 5) = \$1,217$
					$8,758

$CR = \$8,758(A/P, 10\%, 5) = \$2,310$

[a] This is also referred to later as the *beginning-of-year unrecovered investment.*

When Equation (4-5) is applied to the example in Table 4-2, the CR amount is

$$CR(10\%) = \$10,000(A/P, 10\%, 5) - \$2,000(A/F, 10\%, 5)$$

$$= \$10,000(0.2638) - 2,000(0.1638) = \$2,310.$$

Another way to calculate the CR amount is to add an annual sinking fund amount (or deposit) to the interest on the original investment. Thus,

$$CR(i\%) = (I - S)(A/F, i\%, N) + I(i\%). \tag{4-6}$$

When Equation (4-6) is applied to the example in Table 4-2, the CR amount is

$$CR(10\%) = (\$10,000 - \$2,000)(A/F, 10\%, 5) + \$10,000(10\%)$$

$$= \$8,000(0.1638) + \$10,000(0.10) = \$2,310.$$

Yet another way to calculate the CR amount is to add the equivalent annual cost of the uniform loss in value of the investment to the interest on the salvage value:

$$CR(i\%) = (I - S)(A/P, i\%, N) + S(i\%). \tag{4-7}$$

Applied to the example used previously,

$$CR(10\%) = (\$10,000 - \$2,000)(A/P, 10\%, 5) + \$2,000(10\%)$$

$$= \$8,000(0.2638) + \$2,000(0.10) = \$2,310.$$

EXAMPLE 4-8

By using the AW method and Equation (4-4), determine whether the equipment described in Example 4-3 should be recommended.

SOLUTION

The AW method applied to Example 4-3 yields the following:

$$\text{AW}(20\%) = \overbrace{\$8,000}^{R-E} - \overbrace{[\$25,000(A/P, 20\%, 5) - \$5,000(A/F, 20\%, 5)]}^{\text{CR amount (Equation 4-5)}}$$

$$= \$8,000 - (\$8,359.50 - \$671.90)$$

$$= \$312.40$$

Because its AW(20%) is positive, the equipment more than pays for itself over a period of five years while earning a 20% return per year on the unrecovered investment. In fact, the annual equivalent "surplus" is $312.40, which means that the equipment provided more than a 20% return on beginning-of-year unrecovered investment. This piece of equipment should be recommended as an attractive investment opportunity. Also, we can confirm that the AW(20%) in Example 4-8 is equivalent to PW(20%) = $934.29 in Example 4-3 and FW(20%) = $2,324.80 in Example 4-7. That is, AW(20%) = $934.29(A/P, 20%, 5) = $312.40 and also AW(20%) = $2,324.80(A/F, 20%, 5) = $312.40.

EXAMPLE 4-9

An investment company is considering building a 25-unit apartment complex in a growing town. Because of the long-term growth potential of the town, it is felt that the company could average 90% of full occupancy for the complex each year. If the following items are reasonably accurate estimates, what is the minimum monthly rent that should be charged if a 12% MARR (per year) is desired (use the AW method)?

Land investment cost	$50,000
Building investment cost	$225,000
Study period, N	20 years
Rent per unit per month	?
Upkeep expense per unit per month	$35
Property taxes and insurance per year	10% of *total* initial investment

SOLUTION

The procedure for solving this problem is first to determine the equivalent AW of all costs at the MARR of 12% per year. To earn exactly 12% on this project, the annual rental income, adjusted for 90% occupancy, must equal the AW of costs:

$$\text{Initial investment cost} = \$50{,}000 + \$225{,}000 = \$275{,}000,$$

$$\text{Taxes and insurance/year} = 0.1(\$275{,}000) = \$27{,}500,$$

$$\text{Upkeep expenses/year} = \$35(12 \times 25)(0.9) = \$9{,}450,$$

$$\text{CR cost/year [Equation (4-5)]} = \$275{,}000(A/P, 12\%, 20) - \$50{,}000(A/F, 12\%, 20)$$

$$= \$36{,}123.$$

(We assume that investment in land is recovered at the end of year 20 and that annual upkeep is directly proportional to the occupancy rate.) Thus,

$$\text{AW (of costs)} = \$27{,}500 + \$9{,}450 + \$36{,}123 = \$73{,}073.$$

Therefore, the minimum *annual* rental required equals $73,073 and with annual compounding $(M = 1)$ the monthly rental amount, $\hat{R}$, is

$$\hat{R} = \frac{\$73{,}073}{(12 \times 25)(0.9)} = \$270.64.$$

Many decision makers prefer the AW method because it is relatively easy to interpret when one is accustomed to working with annual income statements and cash-flow summaries.

Companion Web Site (http://www.prenhall.com/sullivan_engineering/):   Material costs comprise a significant portion of total construction costs. Visit the Web site to view an *annual worth (AW)* comparison of using either concrete or steel for beams. A spreadsheet cost calculator is included for you to try your own analysis.

4.6 The Internal-Rate-of-Return-Method

The Internal Rate of Return (IRR) method is the most widely used rate of return method for performing engineering economic analyses. It is sometimes called by several other names, such as the *investor's method*, the *discounted cash flow method*, and the *profitability index*.

This method solves for the interest rate that equates the equivalent worth of an alternative's cash inflows (receipts or savings) to the equivalent worth of cash outflows (expenditures, including investment costs). Equivalent worth may be computed with any of the three methods discussed earlier. The resultant interest rate is termed the *Internal Rate of Return (IRR)*.

For a single alternative, from the lender's viewpoint, the IRR is not positive unless (1) both receipts and expenses are present in the cash flow pattern and (2) the sum of receipts exceeds the sum of all cash outflows. Be sure to check both of these conditions in order to avoid the unnecessary work involved with finding that the

IRR is *negative*. (Visual inspection of the total net cash flow will determine whether the IRR is zero or less.)

By using a PW formulation, we see that the IRR is the $i'\%$* at which

$$\sum_{k=0}^{N} R_k(P/F, i'\%, k) = \sum_{k=0}^{N} E_k(P/F, i'\%, k),\tag{4-8}$$

where R_k = net revenues or savings for the kth year,

E_k = net expenditures including any investment costs for the kth year,

N = project life (or study period).

Once i' has been calculated, it is compared with the MARR to assess whether the alternative in question is acceptable. *If $i' \geq$ MARR, the alternative is acceptable; otherwise, it is not.*

A popular variation of Equation (4-8) for computing the IRR for an alternative is to determine the i' at which its *net* PW is zero. In equation form, the IRR is the value of i' at which

$$PW = \sum_{k=0}^{N} R_k(P/F, i'\%, k) - \sum_{k=0}^{N} E_k(P/F, i'\%, k) = 0.\tag{4-9}$$

For an alternative with a single investment cost at the present time ($k = 0$) followed by a series of positive cash inflows over N, a graph of PW versus the interest rate typically has the general convex form shown in Figure 4-6. The point at which PW = 0 in Figure 4-6 defines $i'\%$, which is the project's IRR.

The value of $i'\%$ can also be determined as the interest rate at which FW = 0 or AW = 0. For example, by setting net FW equal to zero, we find that

$$FW = \sum_{k=0}^{N} R_k(F/P, i'\%, N - k) - \sum_{k=0}^{N} E_k(F/P, i'\%, N - k) = 0.\tag{4-10}$$

Another way to interpret the IRR is through an *investment-balance diagram.* (See also Section 4.9.) Figure 4-7 shows how much of the original investment

Figure 4-6
Plot of PW Versus
Interest Rate

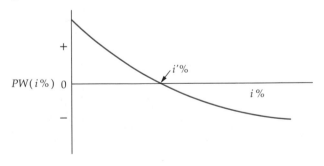

* i' is often used in place of i to mean the interest rate that is to be determined.

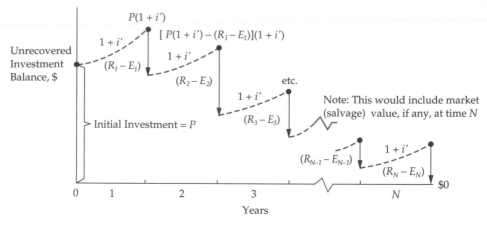

Figure 4-7 Investment Balance Diagram Showing IRR

in an alternative is still to be recovered as a function of time. The downward arrows in Figure 4-7 represent annual returns, $(R_k - E_k)$ for $1 \le k \le N$, against the unrecovered investment and the dashed lines indicate the opportunity cost of interest, or profit, on the beginning-of-year investment balance. *The IRR is the value of i' in Figure 4-7 that causes the unrecovered investment balance to exactly equal 0 at the end of the study period (year N)* and thus represents the *internal* earning rate of a project. It is important to notice that $i'\%$ is calculated on the *beginning-of-year unrecovered* investment through the life of a project rather than on the total initial investment. Additional examples of investment balance diagrams are given in Section 4.9.

The method of solving Equations (4-8) through (4-10) normally involves trial-and-error calculations until the $i'\%$ is converged upon or can be interpolated. Example 4-10 presents a typical solution.

EXAMPLE 4-10 (Restatement of Example 4-2)

A capital investment of $10,000 can be made in a project that will produce a uniform annual revenue of $5,310 for five years and then have a salvage (market) value of $2,000. Annual expenses will be $3,000. The company is willing to accept any project that will earn at least 10% per year on all invested capital. Determine whether it is acceptable by using the IRR method.

SOLUTION

In this example we immediately see that the sum of positive cash flows ($13,550) exceeds the sum of negative cash flows ($10,000). Thus, it is likely that a positive-valued $i'\%$ can be determined. By writing an equation for the PW of the project's total net cash flow and setting it equal to zero, we can compute the IRR:

$$\text{PW} = 0 = -\$10{,}000 + (\$5{,}310 - \$3{,}000)(P/A, i'\%, 5)$$

$$+ \$2{,}000(P/F, i'\%, 5); \quad i'\% = ?$$

Figure 4-8
Use of Linear
Interpolation to Find
the Approximation of
IRR for Example 4-10

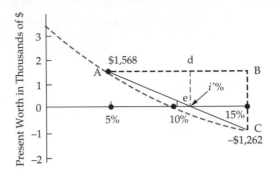

If we did not already know the answer from Example 4-2 ($i' = 10\%$), we would probably try a relatively low i', such as 5%, and a relatively high i', such as 15%. Linear interpolation will be used to solve for i', and the procedure illustrated in Figure 4-8 should not exceed a range of 10%. We have

$$\text{At } i' = \; 5\%: \; PW = -\$10{,}000 + \$2{,}310(4.3295)$$

$$+ \$2{,}000(0.7835) = +\$1{,}568;$$

$$\text{At } i' = 15\%: \; PW = -\$10{,}000 + \$2{,}310(3.3522)$$

$$+ \$2{,}000(0.4972) = -\$1{,}262.$$

Because we have both a positive and a negative PW, the answer is bracketed. The dashed curve in Figure 4-8 is what we are linearly approximating. The answer, $i'\%$, can be determined by using the similar triangles dashed in Figure 4-8:

$$\frac{\text{line } BA}{\text{line } BC} = \frac{\text{line } dA}{\text{line } de}.$$

Where BA is the line segment $B - A = 15\% - 5\%$. Thus,

$$\frac{15\% - 5\%}{\$1{,}568 - (-\$1{,}262)} = \frac{i'\% - 5\%}{\$1{,}568 - \$0},$$

or

$$i'\% = 5\% + \frac{\$1{,}568}{\$1{,}568 - (-\$1{,}262)}(15\% - 5\%)$$

$$= 5\% + 5.5\% = 10.5\%.$$

Because the IRR of the project (10.5%) is greater than the MARR, the project is acceptable. This approximate solution illustrates the trial-and-error process, together with linear interpolation. The error in this answer (actual $i' = 10\%$) is due to the nonlinearity of the PW function and would be less if the range of interest rates used in the interpolation had been smaller.

From the result of Example 4-2, we already know that the project is minimally acceptable and that $i' = \text{MARR} = 10\%$ per year. We can confirm this by substituting $i = 10\%$ in the PW equation as follows:

$$\text{PW}(10\%) = -\$10{,}000 + (\$5{,}310 - \$3{,}000)(P/A, 10\%, 5)$$

$$+ \$2{,}000(P/F, 10\%, 5) = 0.$$

EXAMPLE 4-11 (Restatement of Example 4-3)

A piece of new equipment has been proposed by engineers to increase the productivity of a certain manual welding operation. The investment cost is $25,000, and the equipment will have a market (salvage) value of $5,000 at the end of its expected life of five years. Increased productivity attributable to the equipment will amount to $8,000 per year after extra operating costs have been subtracted from the value of the additional production. A cash-flow diagram for this equipment was given in Figure 4-4. Evaluate the IRR of the proposed equipment. Is the investment a good one? Recall that the MARR is 20% per year.

SOLUTION

By using Equation (4-9), the following expression is obtained:

$$\text{PW}(i\%) = \$8{,}000(P/A, i'\%, 5) + \$5{,}000(P/F, i'\%, 5) - \$25{,}000 = 0; \quad i' = ?$$

To solve this equation by trial and error, use Table 4-3. The PW computations in Table 4-3 are illustrated in Figures 4-9 and 4-10.

By inspection, the value of $i'\%$ where PW = 0 is about 22%. For most applications an $i'\%$ value of 22% is accurate enough because our major concern is whether $i'\%$ equals or exceeds the MARR. A more precise value of i' can be determined by directly solving the above equation with repeated trial-and-error calculations ($i' = 21.577\%$). Clearly, this equipment is economically attractive because 21.577% > 20%.

TABLE 4-3 Computation of Selected PW(i) in Example 4-11

i' (as a decimal)	PW($i'\%$)
0.00	$\$8{,}000(5) + \$5{,}000(1) - \$25{,}000 = \$20{,}000$
0.10	$8{,}000(3.7908) + 5{,}000(0.6209) - 25{,}000 = 8{,}430.90$
0.20	$8{,}000(2.9906) + 5{,}000(0.4019) - 25{,}000 = 934.30$
0.25	$8{,}000(2.6893) + 5{,}000(0.3277) - 25{,}000 = -1{,}847.10$
0.30	$8{,}000(2.436) + 5{,}000(0.2693) - 25{,}000 = -4{,}165.50$

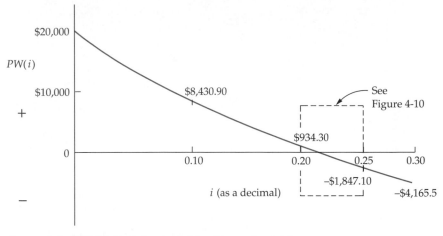

Figure 4-9 PW Plotted Against *i* for Example 4-11

A final point needs to be made for Example 4-11. The investment-balance diagram is provided in Figure 4-11, and the reader should notice that $i' = 21.577\%$ is a rate of return calculated on the beginning-of-year unrecovered investment. The IRR is *not* an average return each year based on the total investment of $25,000.

A rather common application of the IRR method is in so-called *installment financing* types of problems. These problems are associated with financing arrange-

Figure 4-10
Use of Linear
Interpolation to Find
the Approximate IRR
in Example 4-11 and
Figure 4-9

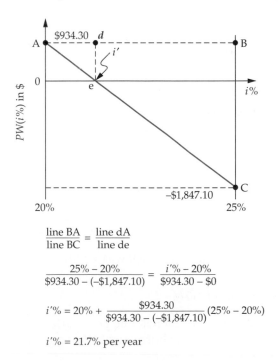

$$\frac{\text{line BA}}{\text{line BC}} = \frac{\text{line dA}}{\text{line de}}$$

$$\frac{25\% - 20\%}{\$934.30 - (-\$1,847.10)} = \frac{i'\% - 20\%}{\$934.30 - \$0}$$

$$i'\% = 20\% + \frac{\$934.30}{\$934.30 - (-\$1,847.10)}(25\% - 20\%)$$

$$i'\% = 21.7\% \text{ per year}$$

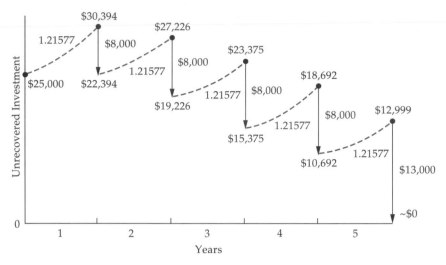

Figure 4-11 Investment Balance Diagram for Example 4-11

ments for purchasing merchandise "on time." The total interest, or finance, charge is often paid by the borrower based on what amount is owed at the beginning of the loan *instead* of on the unpaid loan balance, as illustrated by Figure 4-11. Usually, the average unpaid loan balance is about one-half of the initial amount borrowed. Clearly, a finance charge based solely on the *entire* amount of money borrowed involves payment of interest on money not actually borrowed for the full term. This practice leads to an actual interest rate that often greatly exceeds the stated interest rate. To determine the true interest rate being charged in such cases, the IRR method is frequently employed. Examples 4-12, 4-13, and 4-14 are representative installment financing problems.

EXAMPLE 4-12

In 1915, Albert Epstein allegedly borrowed $7,000 from a large New York bank on the condition that he would repay 7% of the loan every three months, until a total of 50 payments had been made. At the time of the 50th payment, the $7,000 loan would be completely repaid. Albert computed his annual interest rate to be $[0.07(\$7,000) \times 4]/\$7,000 = 0.28$ (28%).

(a) What true *effective* annual interest rate did Albert pay?

(b) What, if anything, was wrong with his calculation?

SOLUTION

(a) The true interest rate per quarter is found by equating the equivalent value of the amount borrowed to the equivalent value of the amounts repaid. Equating the AW amounts per quarter, we find

$$\$7{,}000(A/P, i'\%/\text{qtr.}, 50 \text{ qtrs.}) = 0.07(\$7{,}000) \text{ per quarter,}$$

$$(A/P, i'\%, 50) = 0.07.$$

Linearly interpolating to find $i'\%/\text{qtr.}$ by using similar triangles is the next step:

$$(A/P, 6\%, 50) = 0.0634,$$

$$(A/P, 7\%, 50) = 0.0725.$$

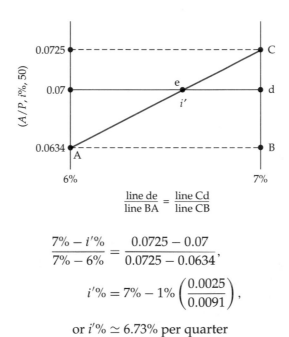

$$\frac{\text{line de}}{\text{line BA}} = \frac{\text{line Cd}}{\text{line CB}}$$

$$\frac{7\% - i'\%}{7\% - 6\%} = \frac{0.0725 - 0.07}{0.0725 - 0.0634},$$

$$i'\% = 7\% - 1\% \left(\frac{0.0025}{0.0091} \right),$$

or $i'\% \simeq 6.73\%$ per quarter

Now we can compute the effective $i'\%$ per year that Albert was paying:

$$i'\% = [(1.0673)^4 - 1]100\%$$

$$\simeq 30\% \text{ per year.}$$

(b) Even though Albert's answer of 28% is close to the true value of 30%, his calculation is insensitive to how long his payments were made. For instance, he would get 28% for an answer when 20, 50, or 70 quarterly payments of $490 were made! For 20 quarterly payments, the true effective interest rate is 14.5% per year, and for 70 quarterly payments, it is 31% per year. As more payments are made, the true annual effective interest rate being charged by the bank will increase, but Albert's method would not reveal by how much.

EXAMPLE 4-13

The Fly-by-Night finance company advertises a "bargain 6% plan" for financing the purchase of automobiles. To the amount of the loan being financed, 6% is added for each year money is owed. This total is then divided by the number of months over which the payments are to be made, and the result is the amount of the monthly payments. For example, a woman purchases a $10,000 automobile under this plan and makes an initial cash payment of $2,500. She wishes to pay the $7,500 balance in 24 monthly payments:

Purchase price	=	$10,000
− Initial payment	=	2,500
= Balance due, (P_0)	=	7,500
+ 6% finance charge = 0.06 × 2 years × $7,500	=	900
= Total to be paid	=	8,400
∴ Monthly payments (A) = $8,400/24	=	$350

What effective annual rate of interest does she actually pay?

SOLUTION

Because there are to be 24 payments of $350 each, made at the end of each month, these constitute an annuity (A) at some unknown rate of interest, $i'\%$, that should be computed only upon the unpaid balance instead of on the entire $7,500 borrowed. A cash-flow diagram of this situation is shown in Figure 4-12. In this example, the amount owed on the automobile (i.e., the initial unpaid balance) is $7,500, so the following equivalence expression is utilized to compute the unknown monthly interest rate:

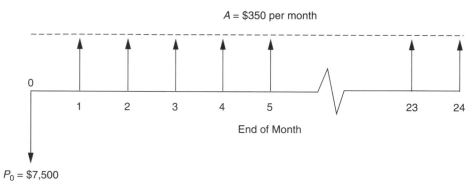

Figure 4-12 Cash-Flow Diagram of Example 4-13 from the Finance Company's Viewpoint

Figure 4-13
Use of Linear
Interpolation to Find
Approximate IRR in
Example 4-13

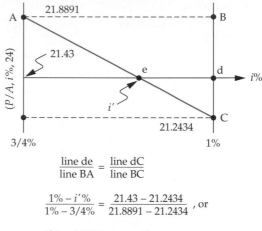

$$\frac{\text{line } de}{\text{line } BA} = \frac{\text{line } dC}{\text{line } BC}$$

$$\frac{1\% - i'\%}{1\% - 3/4\%} = \frac{21.43 - 21.2434}{21.8891 - 21.2434} \text{ , or}$$

$$i'\% = 0.93\% \text{ per month}$$

$$P_0 = A(P/A, i'\%, N),$$

$$\$7,500 = \$350/\text{mo } (P/A, i'\%, 24 \text{ months}),$$

$$(P/A, i'\%, 24) = \frac{\$7,500}{\$350} = 21.43.$$

Examining the interest tables for P/A factors at $N = 24$ that come closest to 21.43, one finds that $(P/A, 3/4\%, 24) = 21.8891$ and $(P/A, 1\%, 24) = 21.2434$.

Linear interpolation for the unknown IRR is demonstrated in Figure 4-13. Because payments are monthly, 0.93% is the interest rate being charged per month. The nominal rate paid on the borrowed money is 0.93%(12) = 11.16% compounded monthly. This corresponds to an effective annual interest rate of $[(1 + 0.0093)^{12} - 1] \times 100\% \cong 12\%$. What appeared at first to be a real bargain turns out to involve effective annual interest at twice the stated rate. The reason is that on the average only \$3,750 is borrowed over the two-year period, but interest on \$7,500 over 24 months was charged by the finance company.

EXAMPLE 4-14

A small company needs to borrow \$160,000. The local (and only) banker makes this statement: "We can loan you \$160,000 at a very favorable rate of 12% per year for a five-year loan. However, to secure this loan, you must agree to establish a checking account (with no interest) in which the *minimum* average balance is \$32,000. In addition, your interest payments are due at the end of each year, and the principal will be repaid in a lump-sum amount at the end of year five." What is the true effective annual interest rate being charged?

SOLUTION

The cash-flow diagram from the banker's viewpoint appears in Figure 4-14. When solving for an unknown interest rate, it is good practice to draw a *cash-flow diagram*

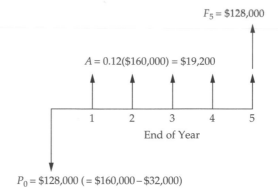

Figure 4-14
Cash-Flow Diagram for
Example 4-14

$F_5 = \$128,000$

$A = 0.12(\$160,000) = \$19,200$

End of Year

$P_0 = \$128,000 \,(= \$160,000 - \$32,000)$

prior to writing an equivalence relationship. The interest rate (IRR) that establishes equivalence between positive and negative cash flows can now easily be computed:

$$P_0 = F_5(P/F, i'\%, 5) + A(P/A, i'\%, 5),$$

$$\$128,000 = \$128,000(P/F, i'\%, 5) + \$19,200(P/A, i'\%, 5).$$

If we try $i' = 15\%$, we discover that $\$128,000 = \$128,000$. Therefore, the true effective interest rate is 15% per year.

4.6.1 Difficulties Associated with the IRR Method

The PW, AW, and FW methods assume that net receipts less expenses (positive recovered funds) each period are reinvested at the MARR during the study period, N. However, the IRR method is not limited by this assumption and measures the internal earning rate of an investment.*

Other difficulties with the IRR method include its computational difficulty and the occurrence of multiple IRRs in some types of problems. A procedure for dealing with seldom-experienced multiple rates of return is discussed and demonstrated in Appendix 4-A. Generally speaking, multiple rates are not meaningful for decision-making purposes, and another method of evaluation (e.g., PW) should be utilized.

Another possible drawback to the IRR method is that it must be carefully applied and interpreted in the analysis of two or more alternatives when only one of them is to be selected (i.e., mutually exclusive alternatives). This is discussed further in Chapter 5. The key advantage of the method is its widespread acceptance by industry, where various types of rates of return and ratios are routinely used in making project selections. The difference between a project's IRR and the required return (i.e., MARR) is viewed by management as a measure of investment safety. A large difference signals a wider margin of safety (or less relative risk).

---

*See H. Bierman and S. Smidt, *The Capital Budgeting Decision: Economic Analysis of Investment Projects* (New York: Macmillan Publishing Company, 1984). The term *internal* rate of return means that the value of this measure depends only on the cash flows from an investment and not on any assumptions about reinvestment rates: "One does not need to know the reinvestment rates to compute the internal rate of return. However, one may need to know the reinvestment rates to compare alternatives" (p. 60).

4.7 The External Rate of Return Method*

The reinvestment assumption of the IRR method noted previously may not be valid in an engineering economy study. For instance, if a firm's MARR is 20% per year and the IRR for a project is 42.4%, it may not be possible for the firm to reinvest net cash proceeds from the project at much more than 20%. This situation, coupled with the computational demands and possible multiple interest rates associated with the IRR method, has given rise to other rate of return methods that can remedy some of these weaknesses.

One such method is the external-rate-of-return (ERR) method. It directly takes into account the interest rate (ϵ) external to a project at which net cash flows generated (or required) by the project over its life can be reinvested (or borrowed). If this external reinvestment rate, which is usually the firm's MARR, happens to equal the project's IRR, then the ERR method produces results identical to those of the IRR method.

In general, *three* steps are used in the calculating procedure. First, all net cash *outflows* are discounted to time 0 (the present) at $\epsilon\%$ per compounding period. Second, all net cash *inflows* are compounded to period N at $\epsilon\%$. Third, the external rate of return, which is the interest rate that establishes equivalence between the two quantities, is determined. The *absolute value* of the present equivalent worth of the net cash outflows at $\epsilon\%$ (first step) is used in this last step. In equation form, the ERR is the $i'\%$ at which

$$\sum_{k=0}^{N} E_k(P/F, \epsilon\%, k)(F/P, i'\%, N) = \sum_{k=0}^{N} R_k(F/P, \epsilon\%, N - k), \qquad (4\text{-}11)$$

where
R_k = excess of receipts over expenses in period k,

E_k = excess of expenditures over receipts in period k,

N = project life or number of periods for the study,

ϵ = external reinvestment rate per period.

Graphically, we have the following (the numbers relate to the three steps):

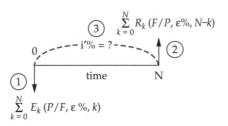

A project is acceptable when $i'\%$ of the ERR method is greater than or equal to the firm's MARR.

---

*This method is also known as the "modified internal rate of return" (MIRR) method. For example, see C. S. Park, and G. P. Sharp-Bette, *Advanced Engineering Economy*. New York: John Wiley & Sons, 1990, pp. 223–226.

The ERR method has two basic advantages over the IRR method:

1. It can usually be solved for directly, without needing to resort to trial and error.
2. It is not subject to the possibility of multiple rates of return. (*Note:* The multiple-rate-of-return problem with the IRR method is discussed in Appendix 4-A.)

EXAMPLE 4-15

Referring to Example 4-11, suppose that $\epsilon = \text{MARR} = 20\%$ per year. What is the project's ERR, and is the project acceptable?

SOLUTION

By utilizing Equation (4-11), we have the following relationship to solve for i':

$$\$25,000(F/P, i'\%, 5) = \$8,000(F/A, 20\%, 5) + \$5,000,$$

$$(F/P, i'\%, 5) = \frac{\$64,532.80}{\$25,000} = 2.5813 = (1 + i')^5,$$

$$i' = 20.88\%.$$

Because $i' > \text{MARR}$, the project is justified, but just barely.

EXAMPLE 4-16

When $\epsilon = 15\%$ and MARR $= 20\%$ per year, determine whether the project whose total cash flow diagram appears next is acceptable. Notice in this example that the use of an $\epsilon\%$ different from the MARR is illustrated. This might occur if, for some reason, part or all of the funds related to a project are "handled" outside the firm's normal capital structure.

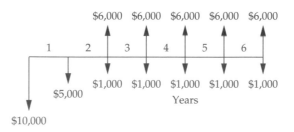

SOLUTION

$$E_0 = \$10,000 \ (k = 0),$$

$$E_1 = \$5,000 \ (k = 1),$$

$$R_k = \$5,000 \text{ for } k = 2, 3, \ldots, 6,$$

$$[\$10,000 + \$5,000(P/F, 15\%, 1)](F/P, i'\%, 6) = \$5,000(F/A, 15\%, 5); \ i'\% = 14.2\%.$$

The $i'\%$ is less than the MARR $= 20\%$; therefore, this project would be unacceptable according to the ERR method.

4.8 The Payback (Payout) Period Method

All methods presented thus far reflect the *profitability* of a proposed alternative for a study period of N. The payback method, which is often called the *simple payout method*, mainly indicates a project's *liquidity* rather than its profitability. Historically, the payback method has been used as a measure of a project's riskiness, since liquidity deals with how fast an investment can be recovered. A low-valued payback period is considered desirable. Quite simply, the payback method calculates the number of years required for cash inflows to just equal cash outflows. Hence the simple payback period is the *smallest* value of θ $(\theta \leq N)$ for which this relationship is satisfied under our normal end-of-year cash-flow convention. For a project where all capital investment occurs at time 0, we have

$$\sum_{k=1}^{\theta}(R_k - E_k) - I \geq 0. \tag{4-12}$$

The *simple* payback period, θ, ignores the time value of money and all cash flows that occur after θ. If this method is applied to the investment project in Example 4-3, the number of years required for the undiscounted sum of cash inflows to exceed the initial investment is four years. This calculation is shown in Column 3 of Table 4-4. Only when $\theta = N$ (the last time period in the planning horizon), is the market (salvage) value included in the determination of a payback period. As can be seen from Equation (4-12), the payback period does not indicate anything about project desirability except the speed with which the investment will be recovered. The payback period can produce misleading results, and it is recommended as supplemental information only in conjunction with one or more of the five methods previously discussed.

Sometimes, the *discounted* payback period, $\theta'(\theta' \leq N)$, is calculated so that the time value of money is considered. In this case,

$$\sum_{k=1}^{\theta'}(R_k - E_k)(P/F, i\%, k) - I \geq 0, \tag{4-13}$$

where $i\%$ is the minimum attractive rate of return, I is the capital investment usually made at the present time $(k = 0)$, and θ' is the smallest value that satisfies Equation (4-13). Table 4-4 (Columns 4 and 5) also illustrates the determination of θ' for Example 4-3. Notice that θ' is the first year in which the cumulative discounted cash inflows exceed the $25,000 capital investment. Payback periods of three years or less are often desired in U.S. industry, so the project in Example 4-3 could be *rejected* even though it is profitable (PW at 20% = $934.29).

TABLE 4-4 Calculation of the Simple Payback Period (θ) and the Discounted Payback Period (θ') at MARR = 20% for Example 4-3[a]

Column 1 End of Year k	Column 2 Net Cash Flow	Column 3 Cumulative PW at $i = 0\%$/yr. through Year k	Column 4 PW of Cash Flow at $i = 20\%$/yr.	Column 5 Cumulative PW at $i = 20\%$/yr. through Year k
0	−$25,000	−$25,000	−$25,000	−$25,000
1	8,000	− 17,000	6,667	−18,333
2	8,000	− 9,000	5,556	−12,777
3	8,000	− 1,000	4,630	−8,147
4	8,000	+7,000	3,858	−4,289
5	13,000		5,223	+934
		↑ $\theta = 4$ years be- cause the cumula- tive balance turns positive at EOY 4		↑ $\theta' = 5$ years be- cause the cumu- lative discounted balance turns pos- itive at EOY 5

[a] Notice that $\theta' \geq \theta$ for MARR $\geq 0\%$.

This variation (θ') of the simple payback period produces the *breakeven life* of a project in view of the time value of money. However, neither payback period calculation includes cash flows occurring after θ (or θ'). This means that θ (or θ') may not take into consideration the entire useful life of physical assets. Thus, these methods will be misleading if one alternative that has a longer (less desirable) payback period than another but produces a higher rate of return (or PW) on the invested capital.

Using the payback period to make investment decisions should generally be avoided except as a measure of how quickly invested capital will be recovered, which is an indicator of project risk. The simple payback and discounted payback period methods tell us how long it takes cash inflows from a project to accumulate to equal (or exceed) the project's cash outflows. The longer it takes to recover invested monies, the greater is the perceived riskiness of a project.

4.9 Investment Balance Diagrams

Another useful method for describing how much money is tied up in a project and how the recovery of funds behaves over its estimated life is the *investment balance diagram*. The mechanics of this method were illustrated for a particular project in Figure 4-7 (where i' was specified to be the IRR and negative amounts were drawn above the line).

Figure 4-15
Investment Balance
Diagram for
Example 4-10

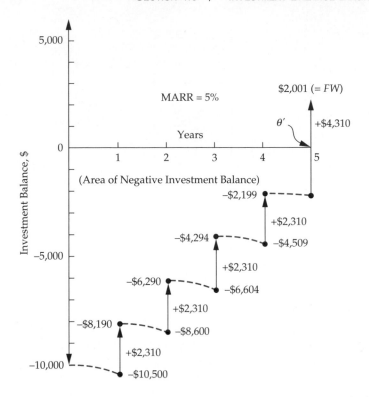

Suppose we return to Example 4-10 and develop an investment balance diagram for this project when the MARR = 5% per year. This diagram, with positive amounts above the time axis, is shown in Figure 4-15. It provides us with several pieces of information: the discounted payback period (θ') is five years, the FW is $2,001, and the project has a negative investment balance until the end of the fifth year. An investor in this venture is "at risk" until the last year of the study period. This is not a comfortable situation when one fears losing money on capital investments that have an uncertain future. In summary, the investment balance diagram provides additional insight into the "worthiness" of a proposed capital investment opportunity and helps to communicate important economic information.

EXAMPLE 4-17

Construct an investment balance diagram for the following cash-flow table (the MARR is 10% per year):

End of Year	Net Cash Flow	Three Sign Changes
0	−$5,000	—
1	6,000	Negative to Positive
2	−1,000	Positive to Negative
3	4,000	Negative to Positive

Figure 4-16
Example 4-17
Investment Balance
Diagram

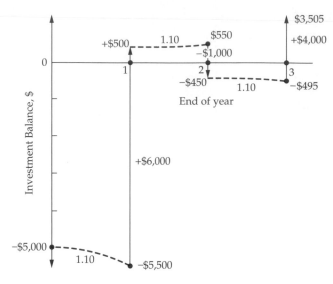

SOLUTION

The investment balance diagram is shown in Figure 4-16. You can see that money is committed to the project in years one and three and that the initial investment cost is completely recovered by the end of year one. The exposure to loss is much less in Figure 4-16 than it is in Figure 4-15. In fact, FW(10%) = $3,505 and IRR = 45%, which reinforces our good feeling about this capital investment. Furthermore, *the IRR is unique,* as demonstrated by the plot of PW(*i*%) versus *i*% in Figure 4-17 for this example. Unique in this case means that the PW(*i*%) curve only intersects with the *i*% axis at one point. Thus, the IRR is unique even though there are three sign changes in the project's cash-flow profile.*

Figure 4-17
Example 4-17 Plot of
PW versus *i*%

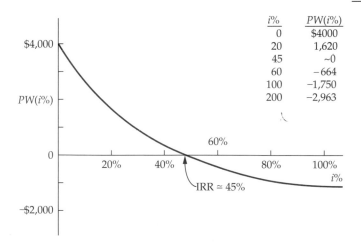

i%	PW(*i*%)
0	$4000
20	1,620
45	~0
60	−664
100	−1,750
200	−2,963

---

*More than one sign change in a project's cash-flow profile signals the possible existence of multiple IRRs. (This topic is discussed in Appendix 4-A.)

▼ 4.10 An Example of a Proposed Capital Investment to Improve Process Yield

Many engineering projects are aimed at improving facility utilization and process yields. Example 4-18 illustrates an engineering economy analysis related to the redesign of a major component in the manufacture of semiconductors.

EXAMPLE 4-18

Semiconductor manufacturing involves taking a flat disc of silicon, called a wafer, and depositing many layers of material on top of it. Each layer has a pattern on it that, upon completion, defines the electrical circuits of the finished microprocessor. Each eight-inch wafer has up to 100 microprocessors on it. However, the typical average yield of the production line is 75 percent good microprocessors per wafer.

The process engineers responsible for the chemical-vapor-deposition (CVD) tool (i.e., process equipment) that deposits *one* of the many layers have an idea for improving overall yield. They propose to improve this tool's vacuum with a re-design of one of its major components. The engineers believe the project will result in a 2% increase in the average production yield of nondefective microprocessors per wafer.

This company has only one CVD tool, and it can process 10 wafers per hour. The CVD tool has an average utilization rate (i.e., "time running") of 80%. A wafer costs $5,000 to manufacture, and a good microprocessor can be sold for $100. These semiconductor fabrication plants ("fabs") operate 168 hours per week, and all good microprocessors produced can be sold.

The capital investment required for the project is $250,000, and maintenance and support expenses are expected to be $25,000 per month. The lifetime of the modified tool will be five years, and the company uses a 12% MARR per year (compounded monthly) as its "hurdle rate."

(a) Should the project be approved? Use the PW method.

(b) If process engineers tend to overestimate the achievable improvement in production yield, at what percent yield improvement would the project break even?

SOLUTION

(a) The average number of wafers per week is (10 wafers/hour) × (168 hr/week) × (0.80) = 1,344. Because the cost per wafer is $5,000 and good microprocessors can be sold for $100 each, profit is earned on each microprocessor produced and sold after the 50th microprocessor on each wafer. Thus, the 2% increase in production yield is all profit (i.e., from 75 good microprocessors per wafer on the average to 76.5). The corresponding additional profit per wafer is $150. The added profit per month, assuming a month is 52 weeks/year ÷ 12 months per year = 4.333 weeks, is (1,344 wafers/wk.) (4.333 wks./month) ($150/wafer) = $873,533.

Therefore, the present worth of the project is

$$PW(1\%) = -\$250,000 - \$25,000(P/A, 1\%/\text{month}, 60 \text{ months})$$

$$+ \$873,533(P/A, 1\%, 60)$$

$$= \$37,898,813.$$

The project *should* be undertaken.

(b) At the breakeven point, profit equals zero. That is, the PW of the project is equal to zero, or PW of costs = PW of revenues. In other words,

$$\$1,373,875 = (1,344 \text{ wafers/wk}) \times (4.333 \text{ wk/mo}) \times (\$X/\text{wafer}) \times (P/A, 1\%, 60),$$

where $X = \$100$ times the number of extra microprocessors per wafer:

$$\frac{\$1,373,875}{44.955(1,344)(4.333)} = X, \text{ or } X \simeq \$5.25 \text{ per wafer.}$$

Thus, $\$5.25/\$100 = 0.0525$ *extra* microprocessors per wafer (total of 75.0525) equates PW of costs to PW of revenues. This corresponds to a breakeven increase in yield of

$$\frac{1.5 \text{ die per wafer}}{0.0525 \text{ die per wafer}} = \frac{2.0\% \text{ increase}}{\text{BE increase}},$$

or BE increase in yield = 0.07%.

▼ 4.11 Spreadsheet Applications

In this chapter, several measures of merit were introduced for evaluating engineering projects. Most spreadsheet packages include financial functions that can be used to simplify the computation of these measures. Microsoft Excel functions and their parameters are described in the following table:

Function	Description
NPV(i, *range*)	Returns the net present worth of the cash flows in *range*, using i as the interest rate one period prior to the first cash flow in *range*.
PMT(i, n, P, F, |type|)	Returns the value of the uniform end-of-period payments made on a loan with interest rate i, repayment period n, and principal amount P or, when the P parameter is set to 0, it returns the value of the n uniform end-of-period payments required to accumulate a future amount F when the interest rate is i.
FV(i, n, A, P, |type|)	Returns the future worth (at end of period n) of n uniform payments of A dollars when the interest rate is i or, when the A parameter is set to 0, it returns the future value of P, n interest periods later.
IRR(*range*, *guess*)	Returns the internal rate of return of the cash flows in *range*, where *guess* is an initial estimate of the IRR. The MARR is usually a good guess.
MIRR(*range*, i, ϵ)	Returns the external rate of return of the cash flows in *range* where i is the interest rate charged on cash outflows and ϵ is the reinvestment rate for cash inflows.
type $= 0$	Default, end-of-period cash flow
type $= 1$	Beginning-of-period cash flow

The financial functions are based on the following assumptions, which agree with those presented in this book:

1. The per period interest rate, i, remains constant.
2. There is exactly one period between cash flows.
3. The period length remains constant.
4. The end-of-period cash-flow convention is utilized.
5. The first cash flow in the NPV() function is at the end of the first period.

The NPV() function is the most useful of the equivalent value financial functions; however, care should be taken to observe the assumptions of this function. The function is designed to compute the net present worth of a series of cash flows. According to assumption 5, the timing of net present worth returned is one interest period before the first cash flow. Thus, if you include the investment amount at $t = 0$ in the range of cash flows, the net present worth returned by the NPV() function is related to $t = -1$. One way to address this issue is to include cash flows for periods 1 through N in P_range and then add to this value the capital investment amount. This is the approach taken in this book.

The equivalent worth and rate of return measures of merit are obtained with the following combinations of functions:

$$PW = NPV(MARR, P\_range) + Capital\ Investment,$$

$$AW = -PMT(MARR, n, NPV(MARR, P\_range) + Capital\ Investment),$$

$$FW = FV(MARR, n, PMT(MARR, n, NPV(MARR, P\_range) + Capital\ Investment)),$$

$$IRR = IRR(range, MARR),$$

$$ERR = MIRR(range, \epsilon, \epsilon).$$

The payback period of a project can also be easily computed using a spreadsheet. By computing the cumulative present worth with $i = 0\%$ and $i = MARR$, it is easy to identify the simple and discounted payback periods, respectively.

Figure 4-18 shows a spreadsheet that calculates all of the economic measures of merit discussed in this chapter for the proposed project discussed in Example 4-11. The formulas in the highlighted cells are given in the following table:

Cell	Contents
C13	= B13 + C12
D13	= IF(AND(C13 >= 0, C12 < 0), "*", "")
E13	= $B$9 + NPV($C$3, B$10 : B13)
F13	= IF(AND(E13 >= 0, E12 < 0), " * *", "")
C16	= NPV($C$3, B10 : B14) + B9
C17	= PMT($C$3, 5, −(NPV($C$3, B10 : B14) + B9))
C18	= FV($C$3, 5, PMT($C$3, 5, (NPV($C$3, B10 : B14) + B9)))
C20	= IRR(B9 : B14, $C$3)
C21	= MIRR(B9 : B14, C4, C4)

▼ 4.12 Summary

Throughout this chapter, we have examined five basic methods for evaluating the financial *profitability* of a single project: present worth, annual worth, future worth, internal rate of return, and external rate of return. Three supplemental methods for assessing a project's *liquidity* were also presented: the simple payback period, the discounted payback period, and the investment balance diagram. Computational procedures, assumptions, and acceptance criteria for all methods were discussed and illustrated with examples. Appendix B provides a listing of new abbreviations and notations that have been introduced in this chapter.

Figure 4-18
Spreadsheet for
Computing Economic
Measures of Merit for
Example 4-11

	A	B	C	D	E	F
1	Economic Measures of Merit:					
2						
3	MARR		20%			
4	Reinvestment Rate (ε)		20%			
5						
6	End of	Net	Cumulative		Cumulative	
7	Period	Cash Flow	PW (0%)		PW(MARR)	
8						
9	0	$ (25,000)	$ (25,000)		$ (25,000)	
10	1	$ 8,000	$ (17,000)		$ (18,333)	
11	2	$ 8,000	$ (9,000)		$ (12,778)	
12	3	$ 8,000	$ (1,000)		$ (8,148)	
13	4	$ 8,000	$ 7,000	*	$ (4,290)	
14	5	$ 13,000	$ 20,000		$ 934	**
15						
16	Present Worth		$ 934.28			
17	Annual Worth		$ 312.41			
18	Future Worth		$ 2,324.80			
19						
20	Internal Rate of Return		21.58%			
21	External Rate of Return		20.88%			
22						
23	Note:	* denotes the simple payback period, and				
24		** denotes the discounted payback period				

4.13 References

CANADA, J. R., W. G. SULLIVAN, and J. A. WHITE. *Capital Investment Decision Analysis for Engineering and Management.* 2nd ed. (Englewood Cliffs, NJ: Prentice Hall, Inc., 1996).

GRANT, E. L., W. G. IRESON, and R. S. LEAVENWORTH. *Principles of Engineering Economy,* 8th ed. (New York: John Wiley & Sons, 1989).

MORRIS, W. T. *Engineering Economic Analysis.* (Reston, VA: Reston Publishing Co., 1976).

THUESEN, G. J., and W. J. FABRYCKY. *Engineering Economy,* 9th ed. (Upper Saddle River, NJ: Prentice Hall, Inc., 2001).

▼ 4.14 Problems

Unless stated otherwise, discrete compounding of interest and end-of-period cash flows should be assumed in all problem exercises for the remainder of the book. All MARRs are "per year." The number in parentheses at the end of a problem refers to the chapter section(s) most closely related to the problem.

4-1. "The higher the MARR, the higher the price that a company should be willing to pay for equipment that reduces annual operating expenses." Do you agree with this statement? Explain your answer. (4.2)

4-2. Your are faced with making a decision on a large capital investment proposal. The capital investment amount is $640,000. Estimated annual revenue at the end of each year in the eight-year study period is $180,000. The estimated annual year-end expenses are $42,000 starting in year one. These expenses begin *decreasing* by $4,000 per year at the end of year four and continue decreasing through the end of year eight. Assuming a $20,000 market value at the end of year eight and a MARR = 12% per year, answer the following questions. (4,3, 4.6)

a. What is the PW of this proposal?
b. What is the IRR of this proposal?
c. What is the simple payback period for this proposal?
d. What is your conclusion about the acceptability of this proposal?

4-3.

a. Evaluate machine XYZ on the basis of the PW method when the MARR is 12% per year. Pertinent cost data are as follows: (4.3)

	Machine *XYZ*
Investment cost	$13,000
Useful life	15 years
Market value	$3,000
Annual operating expenses	$100
Overhaul cost—end of fifth year	$200
Overhaul cost—end of tenth year	$550

b. Determine the capital recovery amount of machine XYZ by all three formulas presented in the text. (4.5)

4-4.

a. Determine the PW, FW, and AW of the following engineering project when the MARR is 15% per year. Is the project acceptable? (4.3, 4.5)

	Proposal *A*
Investment cost	$10,000
Expected life	5 years
Market (salvage) value[a]	−$1,000
Annual receipts	$8,000
Annual expenses	$4,000

[a] A negative market value means that there is a net cost to dispose of an asset.

b. Determine the IRR of the project. Is it acceptable? (4.6)
c. What is the ERR for this project? Assume that $\epsilon = 15\%$ per year. (4.7)

4-5. Uncle Wilbur's trout ranch is now for sale for $30,000. Annual property taxes, maintenance, supplies, and so on are estimated to continue to be $3,000 per year. Revenues from the ranch are expected to be $10,000 next year and then to decline by $400 per year thereafter through the 10th year. If you bought the ranch, you would plan to keep it for only five years and at that time to sell it for the value of the land, which is $15,000. If your desired MARR is 12% per year, should you become a trout rancher? Use the PW method. (4.3)

4-6. A company is considering constructing a plant to manufacture a proposed new product. The land costs $300,000, the building costs $600,000, the equipment costs $250,000, and $100,000 additional working capital is required. It is expected that the product will result in sales of $750,000 per year for 10 years, at which time the land can be sold for $400,000, the building for $350,000, and the equipment for $50,000. All of the working capital would be recovered at the end of year 10. The annual expenses for labor, materials, and all other items are estimated to total $475,000. If the company requires a MARR of 15% per year on projects of comparable risk, determine if it should invest in the new product line. Use the PW method. (4.3)

4-7.

a. Draw a cash-flow diagram for the bond that was described in Example 4-4.

b. If the bond in Example 4-4 is purchased to yield 5% per six-month period (rather than $i = 10\%$ per year), the current purchase price would be how much? (4.3)

4-8. How much can be paid for a $5,000, 10% bond, with interest paid semiannually, if the bond matures 12 years hence? Assume that the purchaser will be satisfied with 6% nominal interest compounded semiannually. (4.3)

4-9. A 20-year bond with a face value of $5,000 is offered for sale at $3,800. The nominal rate of interest on the bond is 7%, paid semiannually. This bond is now 8 years old. (That is, the owner has received 16 semiannual interest payments.) If the bond is purchased for $3,800, what effective annual rate of interest would be realized on this investment opportunity? (4.3)

4-10.

a. A company has issued 10-year bonds, with a face value of $1,000,000, in $1,000 units. Interest at 8% is paid quarterly. If an investor desires to earn 12% nominal interest (compounded quarterly) on $10,000 worth of these bonds, what would the purchase price have to be?

b. If the company plans to redeem these bonds in total at the end of 10 years and establishes a sinking fund that earns 8%, compounded semiannually, for this purpose, what is the *annual* cost of interest and redemption? (4.3)

4-11. You bought a $1,000 bond at par (face value) that paid nominal interest at the rate of 10%, payable semiannually, and held it for 10 years. You then sold it at a price that resulted in a yield of 8% nominal interest compounded semiannually on your capital. What was the selling price? (4.3)

4-12. A small company bought a BMI bond at its face value on January 1, 1991. This bond pays interest of 7.25% every six months (14.5% per year). The face value of the bond is $100,000, and it matures on December 31, 2006. On January 1, 2001, this bond was sold for $110,000. What interest rate (per six months) was earned by the company on the BMI bond? (4.3)

4-13. Susie Queue has a $100,000 mortgage on her deluxe townhouse in urban Philadelphia. She makes monthly payments on a 10% nominal interest rate (compounded monthly) loan and has a 30-year mortgage. Home mortgages are presently available at a 7% nominal interest rate on a 30-year loan. Susie has lived in the townhouse for only two years, and she is considering refinancing her mortgage at a 7% nominal interest rate. The mortgage company informs her that the one-time cost to refinance the present mortgage is $4,500.

How many months must Susie continue to live in her townhouse to make the decision to refinance a good one? Her MARR is the return she can earn on a 30-month certificate of deposit that pays 1/2% per month (6% nominal interest). (4.3, 4.5)

4-14. On January 1, 1997, your brother bought a used car for $8,200, and he agreed to make a down payment of $1,500 and repay the balance in 36 equal payments, with the first payment due February 1. The nominal interest rate is 13.8% per year compounded monthly. During the summer, your brother made enough money so that he decided to repay the entire balance due on the car as of September 1. How much did he repay on September 1? (4.3)

4-15. An apartment complex wishes to establish a fund at the end of 2002 that by the end of year 2019 will grow to an amount large enough to place new roofs on its 39 apartment units. Each new roof is estimated to cost $2,500 in 2017, at which time 13 apartments will be reroofed. In 2018, another 13 apartments will be reroofed, but the unit cost will be $2,625. The last 13 apartments will be reroofed in 2019 at a unit cost of $2,750.

The annual effective interest rate that can be earned on this fund is 4%. How much money *each year* must be put aside (saved) starting at the end of 2003 to pay for all 39 new roofs? State any assumptions you make.

4-16. The Anirup Food Processing Company is presently using an outdated method for filling 25-pound sacks of dry dog food. To compensate for weighing inaccuracies inherent to this packaging method, the process engineer at the plant has estimated that each sack is overfilled by 1/8 pound on the average. A better method of packaging is now available that would eliminate overfilling (and underfilling). The production quota for the plant is 300,000 sacks per year for the next six years, and a pound of dog food costs

TABLE P4-17 Table for Problem 4-17

Year	Investment Beginning of Year of Year	Opportunity Cost of Interest ($i = 15\%$)	Loss in Value of Asset During Year	Capital Recovery Amount for Year
1	$10,000		$3,000	
2			$2,000	
3			$2,000	
4				

this plant $0.15 to produce. The present system has no market value and will last another four years, and the new method has an estimated life of four years with a market value equal to 10% of its investment cost, I. The present packaging operation expense is $2,100 per year more to maintain than the new method. If the MARR is 12% per year for this company, what amount, I, could be justified for the purchase of the new packaging method? (4.3)

4-17. Fill in Table P4-17 when P = $10,000, S = $2,000 (at the end of four years), and i = 15%/yr. Complete the accompanying table and show that the equivalent uniform CR amount equals $3,102.12. (4.5)

4-18. A certain service can be performed satisfactorily by process R, which has a capital investment cost of $8,000, an estimated life of 10 years, no market value, and annual net receipts (revenues – expenses) of $2,400. Assuming a MARR of 18% before income taxes, find the FW and AW of this process and specify whether you would recommend it. (4.4, 4.5)

4-19. You purchased a building five years ago for $100,000. Its annual maintenance expense has been $5,000 per year. At the end of three years, you spent $9,000 on roof repairs. At the end of five years (now), you sell the building for $120,000. During the period of ownership, you rented out the building for $10,000 per year paid at the *beginning* of each year. Use the AW method to evaluate this investment when your MARR is 8% per year. (4.5)

4-20. Given that the purchase price of a machine is $1,000 and its market value at the end of year four is $300, complete Table P4-20 (values (a) through (f)) using an opportunity cost of 5% per year. Compute the equivalent uniform capital recovery amount, based on information from the completed table. (4.5)

4-21. Based on the accompanying cash-flow diagram on the top of the next page answer the following questions (4.3, 4.5, 4.8):

  a. As $i \rightarrow \infty$, the PW equals _____.
  b. The discounted payback period (θ') is _____ years. Let MARR = 12% per year.
  c. If the cash flow at the end of year six had been $-$2,000 instead of $+$2,000, AW (0%) = _____.

TABLE P4-20 Table for Problem 4-20

Year	Investment at Beginning of Yr	Opportunity Cost (5% per Yr)	Loss in Value of Asset During Yr	Capital Recovery Amount for Year
1	$1,000	$50	$(a)	$250
2	(b)	(c)	200	240
3	600	30	200	230
4	(d)	20	(e)	(f)

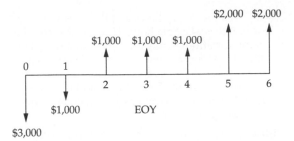

$2,000 $2,000

$1,000 $1,000 $1,000

0   1

      2   3   4   5   6

      $1,000   EOY

$3,000

4-22. A manufacturing firm has considerable excess capacity in its plant and is seeking ways to utilize it. The firm has been invited to submit a bid to become a subcontractor on a product that is not competitive with the one it produces but that, with the addition of $75,000 in new equipment, could readily be produced in its plant. The contract would be for five years at an annual output of 20,000 units.

  In analyzing probable costs, direct labor is estimated at $1.00 per unit and new materials at $0.75 per unit. In addition, it is discovered that in each new unit, one pound of scrap material can be used from the present operation, which is now selling for $0.30 per pound of scrap. The firm has been charging overhead at 150% of prime cost, but it is believed that for this new operation the incremental overhead, above maintenance, taxes, and insurance on the new equipment, would not exceed 60% of the direct labor cost. The firm estimates that the maintenance expenses on this equipment would not exceed $2,000 per year, and annual taxes and insurance would average 5% of the investment cost. (*Note:* Prime cost = direct labor + direct materials cost.)

  While the firm can see no clear use for the equipment beyond the five years of the proposed contract, the owner believes it could be sold for $3,000 at that time. He estimates that the project will require $15,000 in working capital (which would be fully recovered at the end of the fifth year), and he wants to earn at least a 20% before-tax annual rate of return on all capital utilized. (4.3, 4.5)

  a. What unit price should be bid?
  b. Suppose that the purchaser of the product wants to sell it at a price that will result in a profit of 20% of the selling price. What should be the selling price?

4-23. To purchase a used automobile, you borrow $8,000 from Loan Shark Enterprises. They tell

you the interest rate being charged is 1% per month for 35 months. They also charge you $200 for a credit investigation, so you leave with $7,800 in your pocket. The monthly payment they calculated for you is

$$\frac{8,000(0.01)(35) + \$8,000}{35} = \$308.57/\text{month}$$

If you agree to these terms and sign their contract, what is the actual APR (annual percentage rate) that you are paying? (4.6)

4-24. Suppose that you borrow $1,000 from the Easy Credit Company with the agreement to repay it over a 5-year period. Their stated interest rate is 9% per year. They show you the following items in determining the monthly payment: (4.6)

Principal	$1,000
Total interest: 0.09 (5 years) ($1,000)	$450

They ask you to pay 20% of the *interest* immediately, so you leave with $1,000 − $90 = $910 in your pocket. Your monthly payment is calculated as follows:

$$\frac{\$1,000 + \$450}{60} - \$24.17/\text{month}.$$

  a. Draw a cash-flow diagram of this transaction.
  b. Determine the effective *annual* interest rate.

4-25. An individual approaches the Ajax Loan Company for $1,000 to be repaid in 24 monthly installments. The agency advertises an interest rate of 1.5% per month. They proceed to calculate a monthly payment in the following manner:

Amount requested	$1,000
Credit investigation	25
Credit risk insurance	5
Total	$1,030

Interest: ($1,030)(24)(0.015) = $371
Total owed: $1,030 + $371 = $1,401
Payment: $\dfrac{\$1,401}{24} = \$58.50$

What effective annual interest rate is the individual paying if this individual leaves Ajax with $1,000 cash?

4-26. Refer to Problem 4-25 and the "deal" presented next that was actually offered to an engineering student. Your job is to give advice to the student concerning the true effective annual interest rate being charged in the situation below.

An agent of the Ajax Loan Agency offers the individual who agreed to the terms in Problem 4-25 a special deal: "If you're interested in prepaying the loan, I can let you do this. For each prepayment of $58.50, a month and its corresponding payment will be dropped from the original 24-month loan repayment schedule."

If the individual has the money to make two payments of $117 in months one and two, then $58.50 will still be owed in months 3–22. What is the effective annual interest rate in this situation? (4.6)

4-27. Suppose you are now 20 years old. You decide to save $A per year starting on your 21st birthday and continuing through your 60th birthday. At age 60 you will have saved an accumulated (compounded) amount of $F.

A friend of yours waits five years to start her savings plan. Starting on her 26th birthday, it takes annual payments of $2A for her to accumulate $F when she becomes 60 years old.

Still another friend delays his savings plan until 10 years after you started yours. He finds that it takes $4A each year from his 31st until his 60th birthday to accumulate $F.

What effective annual interest rate (i') makes the preceding three savings plans *equivalent*? What can you generalize from this problem?

4-28. Your roommate borrowed money from a banker on the condition that the roommate would pay 7% of the loan every three months, until a total of 35 payments were made. Then the loan would be considered repaid. What effective *annual* interest rate did your roommate pay? Solve for the interest rate to the nearest 1/10 percent. (Use linear interpolation.) (4.6)

4-29. A machine that is not equipped with a brake "coasts" 30 seconds after the power is turned off upon completion of each workpiece, thus preventing removal of the work from the machine. The time per piece, exclusive of this stopping time, is two minutes. The machine is used to produce 40,000 pieces per year. The operator receives $16.50 per hour, and the machine overhead rate is $4.00 per hour. How much could the company afford to pay for a brake that would reduce the stopping time to three seconds if it had a life span of five years? Assume zero market value, a MARR of 15% per year, and repairs and maintenance on the brake totaling no more than $250 per year. (4.3)

4-30. Your boss has just presented you with the summary in the accompanying table of projected costs and annual receipts for a new product line. He asks you to calculate the IRR for this investment opportunity. What would you present to your boss, and how would you explain the results of your analysis? (It is widely known that the boss likes to see graphs of present worth versus interest rate for this type of problem.) The company's MARR is 10% per year. (4.6)

End of Year	Net Cash Flow
0	−$450,000
1	−42,500
2	+92,800
3	+386,000
4	+614,600
5	−$202,200

4-31. Determine the single (and unique) IRR in each of these situations: (4.6)

a.

EOY	Cash Flow	
0–3	0	
4	−$1,000	
5	300	
6	300	
7	300	(Ans. = 15.2%)
8	300	
9	300	

b.

EOY	Cash Flow	
0	−$1,800	
1	−700	
2	1,830	(Ans. = 18.8%)
3	1,830	

c.

EOY	Cash Flow	
0	−$450	
1	−42.5	
2	92.8	
3	386.0	(Ans. = 21.5%)
4	614.6	
5	−202.2	

d.

EOY	Cash Flow	
0	0	
1	−$3,000	
2	1,000	
3	1,900	(Ans. = 20%)
4	−800	
5	2,720	

4-32. Find the IRR in each of these situations: (4.6)

a.

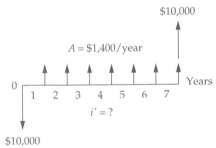

b.

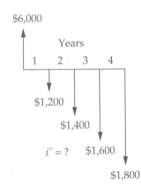

c. You purchased a used car for $4,200. After you make a $1,000 down payment on the car, the salesperson looks in her *Interest Calculations Made Simple* handbook and announces: "The monthly payments will be $160 for the next 24 months and the first payment is due one month from now." (Draw a cash-flow diagram.)

4-33. Rework Part (a) of Problem 4-32 by using the ERR method when $\epsilon = 8\%$ per year. (4.7)

4-34. Plot the PW of Part (a) of Problem 4-32 as a function of the interest rate. The MARR is equal to 8% per year. (4.3)

4-35. Draw an investment balance diagram for Part (a) of Problem 4-32 using $i = $ IRR (determined in that problem). (4.9)

4-36. A zero-coupon certificate involves payment of a fixed sum of money now with a future lump-sum withdrawal of an accumulated amount. Earned interest is not paid out periodically, but instead compounds to become the major component of the accumulated amount paid when the zero-coupon certificate matures. Consider a certain zero-coupon certificate that was issued on March 25, 1993, and matures on January 30, 2010. A person who purchases a certificate for $13,500 will receive a check for $54,000 when the certificate matures. What is the annual interest rate (yield) that will be earned on this certificate? Assume monthly compounding. (4.3)

4-37. A small company purchased now for $23,000 will lose $1,200 each year the first four years. An additional $8,000 invested in the company during the fourth year will result in a profit of $5,500 each year from the fifth year through the fifteenth year. At the end of 15 years, the company can be sold for $33,000.

a. Determine the IRR. (4.6)
b. Calculate the FW if MARR = 12%. (4.4)
c. Calculate the ERR when $\epsilon = 12\%$. (4.7)

4-38. Construct the investment balance diagram for Problem 4-30. What additional insights into the profitability and liquidity of this new product line do you gain? (4.9)

4-39. A $20,000 ordinary life insurance policy for a 22-year-old female can be obtained for annual premiums of approximately $250. This type of policy (ordinary life) would pay a death benefit of $20,000 in exchange for annual premiums of $250 that are paid during the lifetime of the insured person. If the average life expectancy of a 22-year-old female is 77 years, what interest rate establishes equivalence between cash outflows and inflows for this type of insurance policy? Assume that all premiums are paid on a beginning-of-year basis and that the last premium is paid on the female's 76th birthday. (4.6)

4-40. Evaluate the acceptability of the following project with all methods discussed in Chapter 4. Let MARR = ϵ = 15% per year, maximum acceptable θ = 5 years, and maximum acceptable θ' = 6 years.

> *Project:* R137-A
> *Title:* Syn-Tree Fabrication
> *Description:* Establish a production facility to manufacture synthetic palm trees for sale to resort areas in Alaska.

Cash Flow Estimates:

Year	Amount (thousands)
0	−$1,500
1	200
2	400
3	450
4	450
5	600
6	900
7	1,100

4-41. Refer to the following cash-flow diagram:

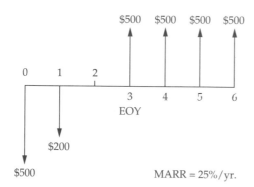

MARR = 25%/yr.

a. What is the breakeven life [θ'] of this project? (4.8)
b. What is the breakeven interest rate (i')? (4.6)
c. Draw the investment balance diagram. (4.9)

4-42. The Going Aircraft Corporation is manually producing a certain subassembly at a direct labor cost of $100,000 per year. This manual work can be totally automated so that $80,000 in direct labor and $20,000 in indirect labor and overhead

costs will be saved each year. Annual maintenance for the automated system will be $10,000, and its market (salvage) value will be $7,000 at any time in the future. The system's useful life is 5 to 10 years, inclusive.

a. If the firm's MARR is 12% per year, develop a graph that shows how much money can be spent on the automated equipment. (*Hint:* Plot the PW of positive cash flows versus the useful life.) (4.3)
b. When N = 6 years and P = $344,000, what is the simple payback period? (4.8)

4-43. Consider the following cash-flow diagram:

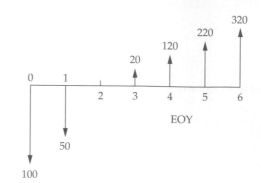

a. If the MARR is 15% per year, is this project financially profitable? (4.3)
b. Calculate the simple payback period, θ. (4.8)
c. Calculate the discounted payback period, θ'. (4.8)

4-44. Advanced Manufacturing Technology (AMT) typically exhibits net annual revenues that increase over a fairly long period. In the long run, an AMT project may be profitable as measured by IRR, but its simple payback period may be unacceptable. Evaluate this AMT project when the company MARR is 15% per year and its maximum allowable payback period is three years: (4.6, 4.8)

Capital investment at time 0	$100,000
Net revenues in year k	$20,000+ $10,000 \cdot (k-1)$
Market (salvage) value	$10,000
Life	5 years

a. The IRR equals _____. Use linear interpolation to determine the IRR.

b. The simple payback period equals _____.

c. What is your recommendation?

4-45. A company has the opportunity to take over a redevelopment project in an industrial area of a city. No immediate investment is required, but it must raze the existing buildings over a four-year period and at the end of the fourth year invest $2,400,000 for new construction. It will collect all revenues and pay all costs for a period of 10 years, at which time the entire project, and properties thereon, will revert to the city. The net cash flows are estimated to be as follows:

Year End	Net Cash Flow
1	$500,000
2	300,000
3	100,000
4	−2,400,000
5	150,000
6	200,000
7	250,000
8	300,000
9	350,000
10	400,000

Tabulate the PW versus the interest rate and determine whether multiple IRRs exist. If so, use the ERR method when $\epsilon = 8\%$ per year to determine a rate of return. (4.7)

4-46. A certain project has net receipts equaling $1,000 now, has costs of $5,000 at the end of the first year, and earns $6,000 at the end of the second year.

a. Show that multiple rates of return exist for this problem when using the IRR method ($i' = 100\%, 200\%$). (Appendix 4-A)

b. If an external reinvestment rate of 10% is available, what is the rate of return for this project using the ERR method? (4.7)

4-47. The prospective exploration for oil in the outer continental shelf by a small, independent drilling company has produced a rather curious pattern of cash flows, as follows:

End of Year	Net Cash Flow
0	−$520,000
1–10	+200,000
10	−1,500,000

The $1,500,000 expense at the end of year 10 will be incurred by the company in dismantling the drilling rig.

a. Over the 10-year period, plot PW versus the interest rate (i) in an attempt to discover whether multiple rates of return exist. (4.6)

b. Based on the projected net cash flows and results in Part (a), what would you recommend regarding the pursuit of this project? Customarily, the company expects to earn at least 20% per year on invested capital before taxes. Use the ERR method ($\epsilon = 20\%$). (4.7)

4-48.

a. Calculate the IRR for each of the three cash-flow diagrams presented below. Use end-of-year 0 for i and end-of-year 4 for ii and iii as the reference points in time. What can you conclude about "reference year shift" and "proportionality" issues of the IRR method?

b. Calculate the PW at MARR = 10% per year at the end-of-year 0 for i and ii and end-of-year 4 for ii and iii. How do the IRR and PW methods compare?

i.

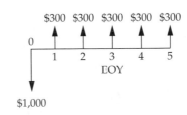

ii.

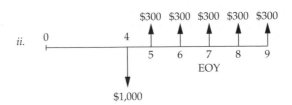

iii.

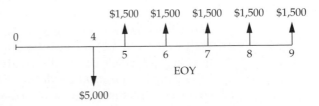

4-49. At a July 4 family reunion last summer, your uncle Sidney learned that you took a course in engineering economy. Uncle Sidney has been working as a skilled machinist for Ford Motor Company since 1965. At the picnic he was curious about a couple of things, so he asked you these questions pertaining to the engineering economy course:

a. Your uncle is thinking about "early" retirement at age 62 (he's now 54 years old), at which time he will receive a *monthly* Social Security check for $800. Alternatively, he can wait until age 65 to start receiving monthly Social Security checks for $1,000. If you conclude that his personal MARR is about 1/2% per month (conservative), how old will your uncle be when both Social Security plans are equally desirable to him? What advice would you give him?

b. Referring to Part (a), what's the answer if your uncle's MARR is 1.5% per month? (In this case he's a fairly aggressive investor!) What can you generalize from your answer to Parts (b) and (c)?

c. Suppose your uncle's MARR is 0%, what should he do then?

4-50. A company is producing a high-volume item that sells for $0.75 per unit. The variable production cost is $0.30 per unit. The company is able to produce and sell 10,000,000 items per year when operating at full capacity.

The critical attribute for this product is weight. The target value for weight is 1,000 grams, and the specification limits are set at ±50 grams. The filling machine used to dispense the product is capable of weights following a normal distribution with an average (μ) of 1,000 grams

and a standard deviation (σ) of 40 grams. Because of the large standard deviation (with respect to the specification limits), 21.12% of all units produced are not within the specification limits. (They either weigh less than 950 grams or more than 1050 grams.) This means that 2,112,000 out of 10,000,000 units produced are nonconforming and cannot be sold without being reworked.

Assume that nonconforming units can be reworded to specification at an additional fixed cost of $0.10 per unit. Reworked units can be sold for $0.75 per unit. It has been estimated that the demand for this product will remain at 10,000,000 units per year for the next five years.

To improve the quality of this product, the company is considering the purchase of a new filling machine. The new machine will be capable of dispensing the product with weights following a normal distribution with $\mu = 1,000$ grams and $\sigma = 20$ grams. As a result, the percent of nonconforming units will be reduced to 1.24% of production. The new machine will cost $710,000 and will last for at least five years. At the end of five years, this machine can be sold for $100,000.

a. If the company's MARR is 15% per year, is the purchase of the new machine to improve quality (reduce variability) economically attractive? Use the AW method to make your recommendation.

b. Compute the IRR, simple payback period, and discounted payback period of the proposed investment.

c. What other factors, in addition to reduced total rework costs, may influence the company's decision about quality improvement?

Appendix 4-A   The Multiple Rate of Return Problem with the IRR Method

Whenever the IRR method is used and the cash flows reverse sign (from net cash outflow to net cash inflow or the opposite) more than once over the study period, one should be aware of the rather remote possibility that either no interest rate or multiple interest rates may exist. Actually, the *maximum* number of possible IRRs in the $(-1, \infty)$ interval for any given project is equal to the number of cash flow sign reversals during the study period. *The simplest way to check for multiple IRRs is to plot equivalent worth (e.g., PW) against the interest rate.* If the resulting plot crosses the interest rate axis more than once, multiple IRRs are present and another equivalence method is recommended for determining project acceptability.

As an example, consider the following project for which the IRR is desired:

EXAMPLE 4-A-1

Plot the present worth versus interest rate for the following cash flows. Are there multiple IRRs? If so, what do they mean?

Year, k	Net Cash Flow		$i\%$	PW($i\%$)
0	$500		0	$250
1	−1,000		10	150
2	0		20	32
3	250		30	~0
4	250		40	−11
5	250		62	~0
			80	24

Thus, the PW of the net cash flows equals zero at interest rates of about 30% and 62%, so multiple IRRs do exist. Whenever there are multiple IRRs, which is seldom, it is likely that none are correct.

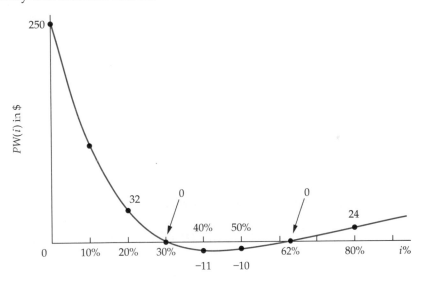

In this situation, the ERR method (see Section 4.7) could be used to decide whether the project is worthwhile. Or, we usually have the option of using an equivalent worth method. In Example 4-A-1, if the external reinvestment rate (ϵ) is 10% per year, we see that the ERR is 12.4%:

$$\$1,000(P/F, 10\%, 1)(F/P, i'\%, 5) = \$500(F/P, 10\%, 5) + \$250(F/A, 10\%, 3)$$

$$(P/F, 10\%, 1)(F/P, i', 5) = 1.632$$

$$i' = 0.124 \ (12.4\%).$$

In addition PW(10%) = $105, so both the ERR and PW methods indicate that this project is acceptable when the MARR is 10% per year..

EXAMPLE 4-A-2

Use the ERR method to analyze the cash flow pattern shown in the accompanying table. The IRR is indeterminant (none exists), so the IRR is not a workable procedure. The external reinvestment rate (ϵ) is 12% per year, and the MARR equals 15%.

Year	Cash Flows
0	$5,000
1	−7,000
2	2,000
3	2,000

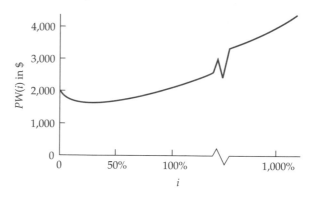

SOLUTION

The ERR method provides this result:

$$\$7{,}000(P/F, 12\%, 1)(F/P, i'\%, 3) = \$5{,}000(F/P, 12\%, 3)$$
$$+ \$2{,}000(F/P, 12\%, 1) + \$2{,}000$$
$$(F/P, i', 3) = 1.802$$
$$i' = 21.7\%.$$

Thus, the ERR is greater than the MARR. Hence, the project having this cash flow pattern would be acceptable. The PW at 15% is equal to $1,740.36, which confirms the acceptability of this project.

Comparing Alternatives

The primary objective of Chapter 5 is to develop and demonstrate the economic analysis and comparison of mutually exclusive design alternatives for an engineering project.

The following topics are discussed in this chapter:

Basic concepts for comparing alternatives

The study (analysis) period

Useful lives equal to the study period

Useful lives are different among the alternatives

The capitalized worth method

Mutually exclusive combinations of projects

5.1 Introduction

Most engineering projects can be accomplished by more than one feasible design alternative. When the selection of one of these alternatives excludes the choice of any of the others, the alternatives are called *mutually exclusive.* Typically, the alternatives being considered require the investment of different amounts of capital, and their annual revenues and costs may vary. Sometimes the alternatives may have different useful lives. Because different levels of investment normally produce varying economic outcomes, we must perform an analysis to determine which one of the mutually exclusive alternatives is preferred and, consequently, how much capital should be invested.

A seven-step procedure for accomplishing engineering economy studies was discussed in Chapter 1. In this chapter, we address Step 5 (analysis and comparison of the feasible alternatives) and Step 6 (selection of the preferred alternative) of this procedure, and we compare mutually exclusive alternatives on the basis of economic considerations alone.

Five of the basic methods discussed in Chapter 4 for analyzing cash flows are used in the analyses in this chapter (PW, AW, FW, IRR, and ERR). These methods provide a *basis for economic comparison* of alternatives for an engineering project. When correctly applied, these methods result in the correct selection of a preferred alternative from a set of mutually exclusive alternatives. The comparison of mutually exclusive alternatives using the benefit–cost-ratio method is discussed in Chapter 11.

5.2 Basic Concepts for Comparing Alternatives

Principle 1 (Chapter 1) emphasized that a choice (decision) is among alternatives. Such choices must incorporate the fundamental purpose of capital investment; namely, to obtain at least the MARR for each dollar invested. In practice, there are usually a limited number of feasible alternatives to consider for an engineering project. The problem of deciding which mutually exclusive alternative should be selected is made easier if we adopt this rule based on Principle 2 in Chapter 1: *The alternative that requires the minimum investment of capital and produces satisfactory functional results will be chosen unless the incremental capital associated with an alternative having a larger investment can be justified with respect to its incremental benefits.*

Under this rule, we consider the acceptable alternative that requires the least investment of capital to be the *base alternative*. The investment of additional capital over that required by the base alternative usually results in increased capacity, increased quality, increased revenues, decreased operating expenses, or increased life. Therefore, before additional money is invested, it must be shown that each avoidable increment of capital can pay its own way relative to other available investment opportunities.

> In summary, *if* the extra benefits obtained by investing additional capital are better than those that could be obtained from investment of the same capital elsewhere in the company at the MARR, the investment should be made. If this is not the case, we obviously would not invest more than the minimum amount of capital required, including the possibility of doing nothing at all. Stated simply, our rule will keep as much capital as possible invested at a rate of return equal to or greater than the MARR.

5.2.1 Investment and Cost Projects and Alternatives

This basic policy for the comparison of mutually exclusive alternatives can be demonstrated with two examples. The *first example* involves an investment project situation. Alternatives *A* and *B* are two mutually exclusive *investment alternatives* with estimated net cash flows,* as shown. *Investment alternatives are those with initial (or front-end) capital investment(s) that produce positive cash flows from increased revenue, savings through reduced costs, or both.* The useful life of each alternative in this example is four years.

	Alternative	
	A	B
Capital investment	− $60,000	− $73,000
Annual revenues less expenses	22,000	26,225

The cash-flow diagrams for Alternatives A and B, and for the year-by-year differences between them (i.e., B minus A), are shown in Figure 5-1. These diagrams typify those for investment project alternatives. In this first example, at MARR = 10% per year, the PW values are

$$PW(10\%)_A = -\$60,000 + \$22,000(P/A, 10\%, 4) = \$9,738,$$

$$PW(10\%)_B = -\$73,000 + \$26,225(P/A, 10\%, 4) = \$10,131.$$

Since the PW_A is greater than 0 at $i =$ MARR, it is the base alternative and would be selected *unless* the additional (incremental) capital associated with Alternative B ($13,000) is justified. In this case, Alternative B is preferred to A, because it has a greater PW value. Hence, *the extra benefits obtained by investing the additional $13,000*

Figure 5-1
Cash-Flow Diagrams for Alternatives A and B and Their Difference

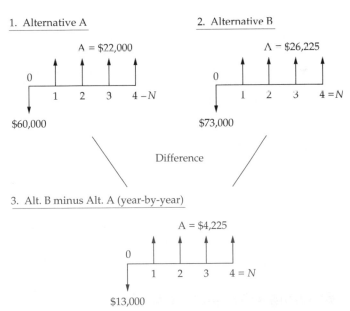

*In this book, the terms *net cash flow* and *cash flow* will be used interchangeably when referring to periodic cash inflows and cash outflows for an alternative.

of capital in B (diagram 3, Figure 5-1), have a present worth of $10,131 − $9,738 = $393. That is,

$$PW(10\%)_{Diff} = -\$13,000 + \$4,225(P/A, 10\%, 4) = \$393,$$

and the additional capital invested in *B* is justified.

The *second example* involves a cost project situation. Alternatives *C* and *D* are two mutually exclusive *cost alternatives* with estimated net cash flows as shown over a three-year life. *Cost alternatives are those with all negative cash flows, except for a possible positive cash flow element from disposal of assets at the end of the project's useful life.* This situation occurs when the organization must take some action, and the decision involves the most economical way of doing it (e.g., the addition of environmental control capability to meet new regulatory requirements).

End of Year	Alternative	
	C	*D*
0	−$380,000	−$415,000
1	−38,100	−27,400
2	−39,100	−27,400
3	−40,100	−27,400
3[a]	0	26,000

[a] Market value.

The cash-flow diagrams for Alternatives *C* and *D*, and for the year-by-year differences between them (i.e., *D* minus *C*), are shown in Figure 5-2. These diagrams typify those for cost project alternatives. In this "must take action" situation, Alternative *C*, which has the lesser capital investment, is automatically the base alternative and would be selected *unless* the additional (incremental) capital associated with Alternative *D* ($35,000) is justified. With the greater capital investment, Alternative *D* in this illustration has smaller annual expenses. Otherwise, it would not be a feasible alternative. (It would not be logical to invest more capital in an alternative without obtaining additional revenues or savings.) Note in diagram 3, Figure 5-2, that the difference between two feasible cost alternatives is an investment alternative.

In this second example, at MARR = 10% per year, the $PW(10\%)_C = -\$477,077$ and the $PW(10\%)_D = -\$463,607$. Alternative *D* is preferred to *C* because it has the less negative PW (minimizes costs). Hence, *the lower annual expenses obtained by investing the additional $35,000 of capital in Alternative D* have a present worth of −$463,607 − (−$477,077) = $13,470. That is, the $PW(10\%)_{Diff} = \$13,470$ and the additional capital invested in Alternative *D* is justified.

5.2.2 Ensuring a Comparable Basis

Each feasible mutually exclusive alternative selected for detailed analysis meets the functional requirements established for the engineering project (Section 1.4.2). However, differences among the alternatives may occur in many forms. Ensuring

Figure 5-2
Cash-Flow Diagrams
for Alternatives *C* and
D and Their Difference

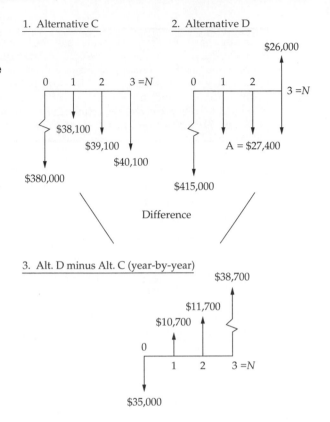

a comparable basis for their analysis requires that any economic impacts of these differences be included in the estimated cash flows for the alternatives (as well as comparing them over the same analysis period—see Section 5.3). Otherwise, the wrong design alternative may be selected for implementing the project. The following are examples of the types of differences that may occur:

1. Operational performance factors such as output capacity, speed, thrust, heat dissipation rate, reliability, fuel efficiency, setup time, and so on.

2. Quality factors such as the number of defect-free (nondefective) units produced per period or the percent of defective units (reject rate).

3. Useful life, capital investment required, revenue changes, various annual expenses or cost savings, and so on.

This list of examples could be expanded. The specific differences, however, for each engineering project and its design alternatives must be identified. Then, because the economic analysis of mutually exclusive alternatives focuses on the differences among them (Principle 2 in Chapter 1), the cash-flow estimates for the alternatives must include the economic costs involved.

> In summary, the economic analysis of the mutually exclusive alternatives for an engineering project must be done on a comparable basis. Since each alternative meets the same functional requirements established for the project, and some differences in performance capabilities, useful lives, output quality, or other factors still exist among them, the economic impacts of these differences (from the perspective of the firm) must be included in the cash-flow estimates and the analysis method. *This is a fundamental premise for comparing alternatives in Chapter 5, and in the chapters that follow.*

Two rules were given in Section 2.5 for facilitating the correct analysis and comparison of mutually exclusive alternatives when the time value of money is not a factor (present economy studies). For convenience, these rules are repeated here and *extended to account for the time value of money:*

Rule 1: When revenues and other economic benefits are present and vary among the alternatives, choose the alternative that *maximizes* overall profitability. That is, select the alternative that has the greatest positive equivalent worth at i = MARR and satisfies all project requirements.

Rule 2: When revenues and other economic benefits are *not* present *or* are constant among the alternatives, consider only the costs and select the alternative that *minimizes* total cost. That is, select the alternative that has the least negative equivalent worth at i = MARR and satisfies all project requirements.

In the remainder of this chapter, we will highlight these considerations in several of the example problems.

5.3 The Study (Analysis) Period

The study (analysis) period, sometimes called the *planning horizon,* is the selected time period over which mutually exclusive alternatives are compared. The determination of the study period for a decision situation may be influenced by several factors—for example, the service period required, the useful life* of the shorter-lived alternative, the useful life of the longer-lived alternative, company policy, and so on. *The key point is that the selected study period must be appropriate for the decision situation under investigation.*

The useful lives of alternatives being compared, relative to the selected study period, can involve two situations:

1. Case 1: Useful lives are the same for all alternatives and equal to the study period.
2. Case 2: Useful lives are different among the alternatives and at least one does not match the study period.

---

*The useful life of an asset is the period during which it is kept in productive use in a trade or business.

Unequal lives among alternatives somewhat complicate their analysis and comparison. To make engineering economy studies in such cases, we adopt the rule of comparing mutually exclusive alternatives over the same period of time. The repeatability assumption and the coterminated assumption are the two types of assumptions used for these comparisons.

The *repeatability assumption* involves two main conditions:

1. The study period over which the alternatives are being compared is either indefinitely long or equal to a common multiple of the lives of the alternatives.
2. The economic consequences that are estimated to happen in an alternative's initial useful life span will also happen in all succeeding life spans (replacements).

Actual situations in engineering practice seldom meet both conditions. This has tended to limit the use of the repeatability assumption, except in those situations where the difference between the annual worth of the first life cycle and the annual worth over more than one life cycle of the assets involved is quite small.*

The *coterminated assumption* uses a finite and identical study period for all alternatives. This planning horizon, combined with appropriate adjustments to the estimated cash flows, puts the alternatives on a common and comparable basis. For example, if the situation involves providing a service, the same time period requirement applies to each alternative in the comparison. To force a match of cash-flow durations to the cotermination time, adjustments (based on additional assumptions) are made to cash-flow estimates of project alternatives having useful lives different from the study period. For example, if an alternative has a useful life shorter than the study period, the estimated annual cost of contracting for the activities involved might be assumed and used during the remaining years. Similarly, if the useful life of an alternative is longer than the study period, a reestimated market value is normally used as a terminal cash flow at the end of a project's coterminated life.

5.4 Case 1: Useful Lives Are Equal to the Study Period

When the useful life of an alternative is equal to the selected study period, adjustments to the cash flows are not required. In this section, we discuss the comparison of mutually exclusive alternatives using equivalent-worth methods and rate-of-return methods when the useful lives of all alternatives are equal to the study period.

---

*T. G. Eschenbach and A. E. Smith, "Violating the Identical Repetition Assumption of EAC," *Proceedings, International Industrial Engineering Conference* (May 1990), The Institute of Industrial Engineers, Norcross, GA, pp. 99–104.

5.4.1 Equivalent-Worth Methods

In Chapter 4, we learned that the equivalent-worth methods convert all relevant cash flows into equivalent present, annual, or future amounts. When these methods are used, consistency of alternative selection results from this equivalency relationship. Also, the economic ranking of mutually exclusive alternatives will be the same using the three methods. Consider the general case of two alternatives, A and B. If

$$PW(i\%)_A < PW(i\%)_B,$$

then

$$PW(i\%)_A (A/P, i\%, N) < PW(i\%)_B (A/P, i\%, N)$$

and

$$AW(i\%)_A < AW(i\%)_B.$$

Similarly,

$$PW(i\%)_A (F/P, i\%, N) < PW(i\%)_B (F/P, i\%, N)$$

and

$$FW(i\%)_A < FW(i\%)_B.$$

> The most straightforward technique for comparing mutually exclusive alternatives when all useful lives are equal to the study period, is to determine the equivalent worth of each alternative based on total investment at $i = $ MARR. Then, for investment alternatives, the one with the greatest positive equivalent worth is selected. And, in the case of cost alternatives, the one with the least negative equivalent worth is selected.

EXAMPLE 5-1

Three mutually exclusive investment alternatives for implementing an office automation plan in an engineering design firm are being considered. Each alternative meets the same service (support) requirements, but differences in capital investment amounts and benefits (cost savings) exist among them. The study period is 10 years, and the useful lives of all three alternatives are also 10 years. Market values of all alternatives are assumed to be zero at the end of their useful lives. If the firm's MARR is 10% per year, which alternative should be selected in view of the following estimates?

	Alternative		
	A	B	C
Capital investment	$390,000	$920,000	$660,000
Annual cost savings	69,000	167,000	133,500

S O L U T I O N: *Solution of Example 5-1 by the PW Method*

$$PW(10\%)_A = -\$390,000 + \$69,000(P/A, 10\%, 10) = \$33,977,$$

$$PW(10\%)_B = -\$920,000 + \$167,000(P/A, 10\%, 10) = \$106,148,$$

$$PW(10\%)_C = -\$660,000 + \$133,500(P/A, 10\%, 10) = \$160,304.$$

Based on the PW method, Alternative C would be selected because it has the largest PW value ($160,304). The order of preference is $C > B > A$, where $C > B$ means C is preferred to B.

S O L U T I O N: *Solution of Example 5-1 by the AW method*

$$AW(10\%)_A = -\$390,000(A/P, 10\%, 10) + \$69,000 = \$5,547,$$

$$AW(10\%)_B = -\$920,000(A/P, 10\%, 10) + \$167,000 = \$17,316,$$

$$AW(10\%)_C = -\$660,000(A/P, 10\%, 10) + \$133,500 = \$26,118.$$

Alternative C is again chosen because it has the largest AW value ($26,118).

S O L U T I O N: *Solution of Example 5-1 by the FW method*

$$FW(10\%)_A = -\$390,000(F/P, 10\%, 10) + \$69,000(F/A, 10\%, 10) = \$88,138,$$

$$FW(10\%)_B = -\$920,000(F/P, 10\%, 10) + \$167,000(F/A, 10\%, 10) = \$275,342,$$

$$FW(10\%)_C = -\$660,000(F/P, 10\%, 10) + \$133,500(F/A, 10\%, 10) = \$415,801.$$

Based on the FW method, the choice is again Alternative C because it has the largest FW value ($415,801). For all three methods (PW, AW, and FW) in this example, notice that $C > B > A$ because of the equivalency relationship among the methods. Also, notice that Rule 1 (Section 5.2.2) applies in this example, since the economic benefits (cost savings) vary among the alternatives.

The two parts of Example 5-2 illustrate the impact that estimated differences in the capability of alternatives to produce defect-free products have on the economic analysis. In the first part, each of the plastic-molding presses produces the same total amount of output units, all of which are defect-free. Then, in the second part of the example, each press still produces the same total amount of output units, but the percent of defective units (reject rate) varies among the presses.

Companion Web Site (http://www.prenhall.com/sullivan_engineering/): A number of European governments are limiting the amount of plastic waste products from cars that can be land filled. Economically environmental or "ecological friendly" solutions that are cost effective are desired. Visit the Web site to view a *cost comparison* of several "eco-friendly" alternatives.

EXAMPLE 5-2

A company is planning to install a new automated plastic-molding press. Four different presses are available. The initial capital investments and annual expenses for these four mutually exclusive alternatives are as follows:

	Press			
	P1	P2	P3	P4
Capital investment	$24,000	$30,400	$49,600	$52,000
Useful life (years)	5	5	5	5
Annual expenses				
Power	2,720	2,720	4,800	5,040
Labor	26,400	24,000	16,800	14,800
Maintenance	1,600	1,800	2,600	2,000
Property taxes and insurance	480	608	992	1,040
Total annual expenses	$31,200	$29,128	$25,192	$22,880

Assume that each press has the same output capacity (120,000 units per year) and has no market value at the end of its useful life; the selected analysis period is five years; and any additional capital invested is expected to earn at least 10% per year. Which press should be chosen if (a) 120,000 nondefective units per year are produced by each press and all units can be sold, and (b) each press still produces 120,000 units per year but the estimated reject rate is 8.4% for P1, 0.3% for P2, 2.6% for P3, and 5.6% for P4 (all nondefective units can be sold). The selling price is $0.375 per unit.

SOLUTION

(a) Since the same number of nondefective units per year will be produced and sold using each press, revenue can be disregarded and the preferred alternative will minimize the equivalent worth of total costs over the five-year analysis period (Rule 2, Section 5.2.2). That is, the four alternatives can be compared as cost alternatives. The PW, AW, and FW calculations for Alternative P1 are

$$PW(10\%)_{P1} = -\$24,000 - \$31,200(P/A, 10\%, 5) = -\$142,273,$$

$$AW(10\%)_{P1} = -\$24,000(A/P, 10\%, 5) - \$31,200 = -\$37,531,$$

$$FW(10\%)_{P1} = -\$24,000(F/P, 10\%, 5) - \$31,200(F/A, 10\%, 5) = -\$229,131.$$

The PW AW, and FW values for Alternatives P2, P3, and P4 are determined with similar calculations and shown for all four presses in Table 5-1. Alternative P4 minimizes all three equivalent worth values of total costs and is the preferred alternative. The preference ranking (P4 > P2 > P1 > P3) resulting from the analysis is the same for all three methods.

TABLE 5-1	Comparison of Four Molding Presses Using the PW, AW, and FW Methods to Minimize Total Costs [Part (a), Example 5-2]			
	Press (Equivalent-Worth Values)			
Method	P1	P2	P3	P4
Present worth	−$142,273	−$140,818	−$145,098	−$138,734
Annual worth	−37,531	−37,148	−38,276	−36,598
Future worth	−229,131	−226,788	−233,689	−223,431

(b) In this part, each of the four alternative presses produces 120,000 units per year, but they have different estimated reject rates. Therefore, the number of nondefective output units produced and sold per year, as well as the annual revenues received by the company, vary among the alternatives. But the annual expenses are assumed to be unaffected by the reject rates. In this situation, the preferred alternative will maximize overall profitability (Rule 1, Section 5.2.2). That is, the four presses need to be compared as investment alternatives. The PW, AW, and FW calculations for Alternative P4 are

$$PW(10\%)_{P4} = -\$52,000 + [(1 - 0.056)(120,000)(\$0.375) - \$22,880](P/A, 10\%, 5)$$

$$= \$22,300,$$

$$AW(10\%)_{P4} = -\$52,000(A/P, 10\%, 5) + [(1 - 0.056)(120,000)(\$0.375) - \$22,880]$$

$$= \$5,882,$$

$$FW(10\%)_{P4} = -\$52,000(F/P, 10\%, 5)$$

$$+ [(1 - 0.056)(120,000)(\$0.375) - \$22,800](F/A, 10\%, 5)$$

$$= \$35,914.$$

The PW, AW, and FW values for Alternatives P1, P2, and P3 are determined with similar calculations and shown for all four alternatives in Table 5-2. Alternative P2 maximizes all three equivalent worth measures of overall profitability and is preferred [versus P4 in Part (a)]. The preference ranking (P2 > P4 >

TABLE 5-2	Comparison of Four Molding Presses Using the PW, AW, and FW Methods to Maximize Overall Profitability [(Part (b), Example 5-2)]			
	Press (Equivalent-Worth Values)			
Method	P1	P2	P3	P4
Present worth	$13,984	$29,256	$21,053	$22,300
Annual worth	3,689	7,718	5,554	5,882
Future worth	22,521	47,117	33,906	35,914

P3 > P1) is the same for the three methods, but is different from the ranking in Part (a). The different preferred alternative and preference ranking in Part (b) are the result of the varying capability among the presses to produce nondefective output units.

5.4.2 Rate-of-Return Methods

Annual return on investment is a popular metric of profitability in the United States. When using rate of return methods to evaluate mutually exclusive alternatives, the best alternative produces satisfactory functional results and requires the minimum investment of capital. This is true unless a larger investment can be justified in terms of its incremental benefits and costs. Accordingly, these three guidelines are applicable to rate-of-return methods:

1. Each increment of capital must justify itself by producing a sufficient rate of return (greater than or equal to the MARR) on that increment.
2. Compare a higher investment alternative against a lower investment alternative only when the latter is acceptable. The difference between the two alternatives is usually an *investment alternative* and permits the better one to be determined.
3. Select the alternative that requires the largest investment of capital as long as the incremental investment is justified by benefits that earn at least the MARR. This maximizes equivalent worth on total investment at $i = MARR$.

> Do not compare the IRRs of mutually exclusive alternatives (or IRRs of the differences between mutually exclusive alternatives) against those of other alternatives. Compare an IRR only against the MARR (IRR ≥ MARR) in determining the acceptability of an alternative.

These guidelines can be implemented using the *incremental investment analysis technique* with rate of return methods.* First, however, we will discuss the *inconsistent ranking problem* that can occur with incorrect use of rate-of-return methods in the comparison of alternatives.

5.4.2.1 The Inconsistent Ranking Problem
In Section 5.2, we discussed a small investment project involving two alternatives, A and B. The cash flow for each alternative is restated here, as well as the cash flow (incremental) difference.

| | Alternative | | Difference |
	A	B	$\Delta(B - A)$
Capital investment	$60,000	$73,000	$13,000
Annual revenues less expenses	22,000	26,225	4,225

---

*The IRR method is the most celebrated time-value-of-money based profitability metric in the United States. The incremental analysis technique must be learned so that the IRR method can be correctly applied in the comparison of mutually exclusive alternatives.

The useful life of each alternative and the study period is four years. Also, assume that the MARR = 10% per year. First, check to see if the sum of positive cash flows exceeds the sum of negative cash flows. This is the case here, so the IRR and PW(10%) of each alternative are calculated and shown below:

Alternative	IRR	PW(10%)
A	17.3%	$ 9,738
B	16.3	10,131

If, at this point, a choice were made based on maximizing the IRR of the total cash flow, Alternative A would be selected. But, based on maximizing the PW of the total investment at i = MARR, Alternative B is preferred. Obviously, here we have an inconsistent ranking of the two mutually exclusive investment alternatives.

Now that we know Alternative A is acceptable (IRR > MARR; PW at MARR > 0), we will analyze the incremental cash flow between the two alternatives, which we shall refer to as $\Delta(B - A)$. The IRR of this increment, IRR$_\Delta$, is 11.4%. This is greater than the MARR of 10%, and the incremental investment of $13,000 is justified. This outcome is confirmed by the PW of the increment, PW$_\Delta$(10%), which is equal to $393. Thus, when the IRR of the incremental cash flow is used, versus the IRR of the total cash flow for each alternative, the rankings of A and B are consistent with that based on the PW on total investment.

The fundamental role that the incremental net cash flow, $\Delta(B - A)$, plays in the comparison of two alternatives (where B has the greater capital investment) is based on the relationship:

Cash flow of B = cash flow of A + cash flow of the difference.

Clearly, the cash flow of B is made up of two parts. The first part is equal to the cash flow of Alternative A, and the second part is the incremental cash flow between A and B, $\Delta(B - A)$. Obviously, if the equivalent worth of the difference is greater than or equal to zero at i = MARR, then Alternative B is preferred. Otherwise, given that Alternative A is justified (an acceptable *base alternative*), Alternative A is preferred. It is always true that if PW$_\Delta \geq 0$, then IRR$_\Delta \geq$ MARR. Therefore, in this example Alternative B is preferred to A.

Figure 5-3 illustrates how ranking errors can occur when a selection among mutually exclusive alternatives is based wrongly on maximization of IRR on the total cash flow. When the MARR lies to the left of IRR$_\Delta$ (11.4% in this case), an incorrect choice will be made by selecting an alternative that maximizes internal rate of return. This is because the IRR method assumes reinvestment of cash flows at the calculated rate of return (17.3% and 16.3%, respectively, for Alternatives A and B in this case), whereas the PW method assumes reinvestment at the MARR (10%).

Figure 5-3 shows our previous results with PW$_B$ > PW$_A$ at MARR = 10%, even though IRR$_A$ > IRR$_B$. Also, the figure shows how to avoid this ranking

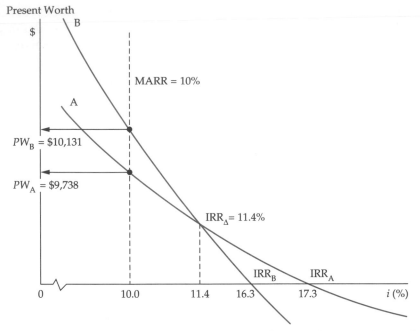

Figure 5-3 Illustration of the Ranking Error in Studies Using the Internal-Rate-of-Return Method

inconsistency by examining the IRR of the increment, IRR_Δ, which correctly leads to the selection of Alternative B, the same as with the PW method.

5.4.2.2 The Incremental Investment Analysis Procedure

> We recommend the incremental investment analysis procedure to avoid incorrect ranking of mutually exclusive alternatives when using rate of return methods. We will use this procedure in the remainder of the book.

The incremental analysis procedure for the comparison of mutually exclusive alternatives is summarized in three basic steps (and illustrated in Figure 5-4):

1. Arrange (rank order) the feasible alternatives based on increasing capital investment.*

---

*This ranking rule assumes a logical set of mutually exclusive alternatives. That is to say, for investment or cost alternatives, increased initial investment results in additional economic benefits, whether from added revenues, reduced costs, or a combination of both. Also, this rule assumes that for any nonconventional investment cash flow, the PW, AW, or FW or ERR analysis method would be used instead of IRR. Simply stated, a nonconventional investment cash flow involves multiple sign changes or positive cash flow at time 0, or both. For a more detailed discussion of ranking rules, see C. S. Park and G. P. Sharp-Bette, *Advanced Engineering Economy* (New York: John Wiley & Sons, 1990).

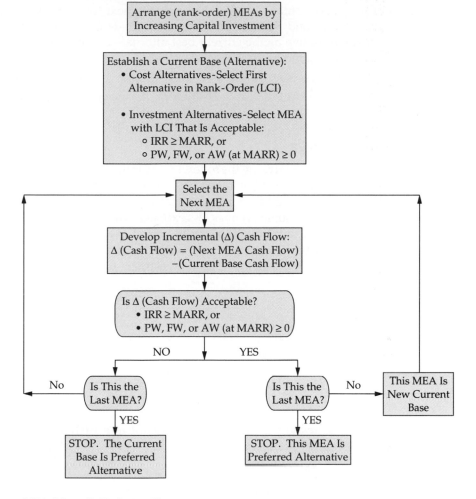

MEA: Mutually Exclusive Alternative
LCI: Least Capital Investment

Figure 5-4 **Incremental Investment Analysis Procedure**

2. Establish a base alternative.

   (a) Cost alternatives—The first alternative (least capital investment) is the base.

   (b) Investment alternatives—If the first alternative is acceptable (IRR $\geq$ MARR; PW, FW, or AW at MARR ≥ 0), select it as the base. If the first alternative is not acceptable, choose the next alternative in order of increasing capital investment and check the profitability criterion (PW, etc.) values. Continue until an acceptable alternative is obtained. If none is obtained, the do-nothing alternative is selected.

3. Use iteration to evaluate differences (incremental cash flows) between alternatives until all alternatives have been considered.

(a) If the incremental cash flow between the next (higher capital investment) alternative and the current selected alternative is acceptable, choose the next alternative as the current best alternative. Otherwise, retain the last acceptable alternative as the current best.

(b) Repeat, and select as the preferred alternative the last one for which the incremental cash flow was acceptable.

EXAMPLE 5-3

Suppose that we are analyzing the following six mutually exclusive alternatives for a small investment project (arranged by increasing capital investment) using the IRR method. The useful life of each alternative is 10 years, and the MARR is 10% per year. Also, net annual revenues less expenses vary among all alternatives, and Rule 1, Section 5.2.2, applies. If the study period is 10 years, and the market (salvage) values are 0, which alternative should be chosen? Notice that the alternatives have been *rank ordered* from low capital investment to high capital investment.

	Alternative					
	A	B	C	D	E	F
Capital investment	$900	$1,500	$2,500	$4,000	$5,000	$7,000
Annual revenues less expenses	150	276	400	925	1,125	1,425

SOLUTION

For each of the feasible alternatives, the IRR on the total cash flow can be computed by determining the interest rate at which the PW, FW, or AW equals zero (use of AW is illustrated for Alternative A):*

$$0 = -\$900(A/P, i'_A\%, 10) + \$150; \quad i'\% = ?$$

By trial and error, we determine that $i'_A\% = 10.6\%$. In the same manner, the IRRs of all the alternatives are computed and summarized:

	A	B	C	D	E	F
IRR on total cash flow	10.6%	13.0%	9.6%	19.1%	18.3%	15.6%

At this point, *only Alternative C is unacceptable* and can be eliminated from the comparison because its IRR is less than the MARR of 10% per year. Also, *A is the base alternative from which to begin the incremental investment analysis procedure,*

---

*The three steps of the incremental analysis procedure previously discussed (and illustrated in Figure 5-4) do not require the calculation of the IRR value for each alternative. In this example, the IRR of each alternative is used for teaching purposes.

TABLE 5-3 Comparison of Five Acceptable Investment Alternatives Using the IRR Method (Example 5-3)

Increment Considered	A	$\Delta(B-A)$	$\Delta(D-B)$	$\Delta(E-D)$	$\Delta(F-E)$
Δ Capital investment	$900	$600	$2,500	$1,000	$2,000
Δ Annual revenues less expenses	$150	$126	$649	$200	$300
IRR_Δ	10.6%	16.4%	22.6%	15.1%	8.1%
Is increment justified?	Yes	Yes	Yes	Yes	No

because it is the mutually exclusive alternative with the lowest capital investment whose IRR (10.6%) is equal to or greater than the MARR (10%). This pre-analysis of the feasibility of each alternative using the IRR, PW, FW, or AW method is not required by the incremental analysis procedure. However, it is useful when analyzing a larger set of mutually exclusive alternatives. You can immediately eliminate nonfeasible (nonprofitable) alternatives, as well as easily identify the base alternative.

As discussed in Section 5.4.2.1, it is not necessarily correct to select the alternative that maximizes the IRR on total cash flow. That is to say, Alternative D may not be the best choice, since maximization of IRR does not guarantee maximization of equivalent worth on total investment at the MARR, and resultant maximization of an organization's future wealth to its owners. Therefore, to make the correct choice, we must examine each increment of capital investment to see if it will pay its own way. Table 5-3 provides the analysis of the five remaining alternatives, and the IRRs on incremental cash flows are again computed by setting $\text{AW}_\Delta(i') = 0$ for cash-flow differences between alternatives.

From Table 5-3, it is apparent that Alternative E will be chosen (not D) because it requires the largest investment for which the last increment of capital investment is justified. That is, we desire to invest additional increments of the $7,000 presumably available for this project as long as each avoidable increment of investment can earn 10% per year or better.

It was assumed in Example 5-3 (and in all other examples involving mutually exclusive alternatives, unless noted to the contrary) that available capital for a project *not* committed to one of the feasible alternatives is invested in some other project where it will earn a return equal to the MARR. Therefore, in this case, the $2,000 left over by selecting Alternative E instead of F is assumed to earn 10% per year elsewhere, which is more than we could obtain by investing it in F.

In summary, three errors commonly made in this type of analysis are to choose the mutually exclusive alternative (1) with the highest overall IRR on total cash flow, or (2) with the highest IRR on an incremental capital investment, or (3) with the largest capital investment that has an IRR greater than or equal to the MARR. None of these criteria are generally correct. For instance, in Example 5-3, one might erroneously choose Alternative D rather than E because the IRR for the increment from B to D is 22.6% and that from D to E is only 15.1% (error 2). A more obvious error, as previously discussed, is the temptation to maximize the IRR on total cash flow and select Alternative D (error 1). The third error would be committed by selecting Alternative F for the reason that it has the largest total investment with an IRR greater than the MARR (15.6% > 10%).

The equivalent worth methods may also be applied using the incremental analysis procedure to compare mutually exclusive alternatives. The ranking of the alternatives will be consistent with that of the equivalent worth values based on total investment for each alternative. The ranking will also be consistent with that of the rate of return methods when using an incremental analysis. When the equivalent worth of an investment cash flow is greater than zero at i = MARR, its IRR is greater than the MARR. Therefore, the equivalent worth methods, using incremental investment analysis, can be used as a *screening method* for the IRR method. That is, the same decisions are made concerning additional increments of capital investment. These points are included in Example 5-4.

EXAMPLE 5-4

The estimated capital investment and the annual expenses (based on 1,500 hours of operation per year) for four alternative designs of a diesel-powered air compressor are shown, as well as the estimated market value for each design at the end of the common five-year useful life. The perspective (Principle 3, Chapter 1) of these cost estimates is that of the typical user (construction company, plant facilities department, government highway department, and so on). The study period is five years, and the MARR is 20% per year. One of the designs must be selected for the compressor, and each design provides the same level of service. Based on this information, (1) determine the preferred design alternative using the IRR method, and (2) show that the PW method (i = MARR), using the incremental analysis procedure, results in the same decision. Observe that this example is a *cost-type situation with four mutually exclusive cost alternatives*. The following solution demonstrates the use of the incremental analysis procedure to compare cost alternatives and applies Rule 2 in Section 5.2.2:

	Design Alternative			
	D1	D2	D3	D4
Capital investment	$100,000	$140,600	$148,200	$122,000
Annual expenses	29,000	16,900	14,800	22,100
Useful life (years)	5	5	5	5
Market value	10,000	14,000	25,600	14,000

SOLUTION

The first step is to arrange (rank order) the four mutually exclusive cost alternatives based on their increasing capital investment costs. Therefore, the order of the alternatives for incremental analysis is D1, D4, D2, and D3.

TABLE 5-4 Comparison of Four Cost (Design) Alternatives Using the IRR and PW Methods with Incremental Analysis (Example 5-4)

Increment Considered	Δ(D4 − D1)	Δ(D2 − D4)	Δ(D3 − D4)
Δ Capital investment	$22,000	$18,600	$26,200
Δ Annual expense (savings)	6,900	5,200	7,300
Δ Market value	4,000	0	11,600
Useful life (years)	5	5	5
IRR$_\Delta$	20.5%	12.3%	20.4%
Is increment justified?	Yes	No	Yes
PW_Δ(20%)	$243	−$3,049	$293
Is increment justified?	Yes	No	Yes

Since these are cost alternatives, the one with the least capital investment, D1, is the base alternative. Therefore, the base alternative will be preferred unless additional increments of capital investment can produce cost savings (benefits) that lead to a return equal to or greater than the MARR.

The first incremental cash flow to be analyzed is that between designs D1 and D4, Δ(D4−D1). The results of this analysis, and of subsequent differences between the cost alternatives, are summarized in Table 5-4, and the incremental investment analysis for the IRR method is illustrated in Figure 5-5. These results show the following:

1. The incremental cash flows between the cost alternatives are, in fact, investment alternatives.
2. The first increment, Δ(D4 − D1), is justified (IRR$_\Delta$ = 20.5% is greater than MARR = 20%, and PW$_\Delta$(20%) = $243 > 0); the increment Δ(D2 − D4) is not

Figure 5-5
Representation of Capital Investment Increments and IRR on Increments Considered in Selecting Design 3 (D3) in Example 5-4

Incremental Investment Analysis				Selection	
Increment of Investment	Capital Investment	IRR$_\Delta$		Design	Capital Investment
Δ (D3 − D4)	$26,200	20.4% (Accept)			
Δ (D2 − D4)	$18,600	12.3% (Reject)			
Δ (D4 − D1)	$22,000	20.5% (Accept)			
				D3*	$148,200
D1	$100,000	Base Alternative*			

*Since these are cost alternatives, the IRR cannot be determined.

justified; and the last increment, $\Delta(D3 - D4)$—not $\Delta(D3 - D2)$ because Design D2 has already been shown to be unacceptable—is justified, resulting in the selection of Design D3 for the air compressor. It is the highest investment for which each increment of investment capital is justified from the user's perspective.

3. The same capital investment decision results from the IRR method and the PW method using the incremental analysis procedure, because *when the equivalent worth of an investment at i = MARR is greater than zero, its IRR is greater than the MARR* (from the definition of the IRR; Chapter 4).

The external-rate-of-return (ERR) method was explained in Chapter 4. Also, in Appendix 4-A, the ERR method was illustrated as a substitute for the IRR method when analyzing a nonconventional investment type of cash flow. In Example 5-5, the ERR method is applied using the incremental investment analysis procedure to compare the mutually exclusive alternatives for an engineering improvement project.

EXAMPLE 5-5

In an automotive parts plant, an engineering team is analyzing an improvement project to increase the productivity of a flexible manufacturing center. The estimated net cash flows for the three feasible alternatives being compared are shown in Table 5-5. The analysis period is six years, and the MARR for capital investments at the plant is 20% per year. Using the ERR method, which alternative should be selected? ($\epsilon = $ MARR.)

TABLE 5-5 Comparison of Three Mutually Exclusive Alternatives Using the ERR Method (Example 5-5)

	Alternative Cash Flows			Incremental Analysis of Alternatives		
End of Period	A	B	C	A^a	$\Delta(B - A)$	$\Delta(C - A)$
0	−$640,000	−$680,000	−$755,000	−$640,000	−$40,000	−$115,000
1	262,000	−40,000	205,000	262,000	−302,000	−57,000
2	290,000	392,000	406,000	290,000	102,000	116,000
3	302,000	380,000	400,000	302,000	78,000	98,000
4	310,000	380,000	390,000	310,000	70,000	80,000
5	310,000	380,000	390,000	310,000	70,000	80,000
6	260,000	380,000	324,000	260,000	120,000	64,000
Incremental analysis:						
Δ PW of negative cash-flow amounts				640,000	291,657	162,498
Δ FW of positive cash-flow amounts				2,853,535	651,091	685,082
ERR				28.3%	14.3%	27.1%
Is increment justified?				Yes	No	Yes

[a] The net cash flow for Alternative A, which is the incremental cash flow between making no change ($0) and implementing Alternative A.

SOLUTION

The procedure for using the ERR method to compare mutually exclusive alternatives is the same as for the IRR method. The only difference is in the calculation methodology.

Table 5-5 provides a tabulation of the calculation and acceptability of each increment of capital investment considered. Since these three feasible alternatives are a mutually exclusive set of investment alternatives, the base alternative is the one with the least capital investment cost that is economically justified. For Alternative A, the PW of the negative cash-flow amounts (at $i = \epsilon\%$) is just the $640,000 investment cost. Therefore, the ERR for Alternative A is the following:

$$\$640,000(F/P, i'\%, 6) = \$262,000(F/P, 20\%, 5) + \cdots + \$260,000$$

$$= \$2,853,535$$

$$(F/P, i'\%, 6) = (1 + i')^6 = \$2,853,535/\$640,000 = 4.4586$$

$$(1 + i') = (4.4586)^{1/6} = 1.2829$$

$$i' = 0.2829, \text{ or ERR} = 28.3\%.$$

Using a MARR $= 20\%$ per year, this capital investment is justified and Alternative A is an acceptable base alternative. By using similar calculations, the increment $\Delta(B-A)$, earning 14.3%, is not justified and the increment $\Delta(C-A)$, earning 27.1%, is justified. Therefore, Alternative C is the preferred alternative for the improvement project. Note in this example that revenues varied among the alternatives and that Rule 1, Section 5.2.2, was applied.

By this point in the chapter, three key observations are clear concerning the comparison of mutually exclusive alternatives: (1) Equivalent-worth methods are computationally less cumbersome to use, (2) both the equivalent-worth and rate-of-return methods, if used properly, will consistently recommend the best alternative, but (3) rate-of-return methods may not produce correct choices if the analyst or the manager insists on maximizing the rate of return on the total cash flow. That is, incremental investment analysis must be used with rate of return methods to ensure that the best alternative is selected.

To reinforce these points further, consider the assignment given to Cynthia Jones in Example 5-6.

EXAMPLE 5-6

The owner of a downtown parking lot has retained an architectural engineering firm to determine whether it would be financially attractive to construct an office building on the site now being used for parking. If the site is retained for parking, improvements are required for its continued use. Cynthia Jones, a newly hired civil engineer and member of the project team, has been requested to perform the analysis and offer a recommendation. The data she has assembled on the four feasible, mutually exclusive alternatives developed by the project team are summarized as follows:

Alternative	Capital Investment (Including Land)	Net Annual Income
P. Keep existing parking lot, but improve	$200,000	$22,000
B1. Construct one-story building	4,000,000	600,000
B2. Construct two-story building	5,550,000	720,000
B3. Construct three-story building	7,500,000	960,000

(a) The study period selected is 15 years. For each alternative, the property has an estimated residual value at the end of 15 years that is *equal* to 50% of the capital investment shown. The owner of the parking lot prefers the information from the IRR method, but the firm's manager has always insisted on a PW analysis. Therefore, she decides to perform the analysis using both methods. If the MARR equals 10% per year, which alternative should Cynthia recommend?

(b) Which rule (Section 5.2.2) was applied in the solution to Part (a)? Why?

SOLUTION

(a) The PW of the parking lot alternative (P) is computed as follows:

$$PW_P(10\%) = -\$200,000 + \$22,000(P/A,\ 10\%, 15) + \$100,000\ (P/F,\ 10\%, 15)$$

$$= -\$8,726.$$

By a similar calculation, the PW values for Alternatives B1, B2, and B3 are

$$PW(10\%)_{B1} = \$1,042,460,$$

$$PW(10\%)_{B2} = \$590,727,$$

$$PW(10\%)_{B3} = \$699,606.$$

Based on the PW method, the one-story building (Alternative B1) would be recommended. (Alternative P is unacceptable, and the preference ranking of the remaining alternatives is B1 > B3 > B2).

The IRR method of analysis requires more time and computation effort:

	Mutually Exclusive Alternatives			
	P	B1	B2	B3
Capital investment	$200,000	$4,000,000	$5,550,000	$7,500,000
Net annual income	22,000	600,000	720,000	960,000
Residual value	100,000	2,000,000	2,775,000	3,750,000
IRR[a]	9.3%	13.8%	11.6%	11.4%

[a] For example, the IRR of Alternative P is computed as follows: $0 = -\$200,000 + \$22,000$ $(P/A, i'\%, 15) + \$100,000\ (P/F, i'\%, 15)$; $i'\% = ?$ By trial and error, $i' = 9.3\%$.

TABLE 5-6 Example 5-6 (IRR Method)

	Incremental Analysis of Alternatives		
	B1[b]	Δ(B2 − B1)	Δ(B3 − B1)
Δ Capital investment	$4,000,000	$1,550,000	$3,500,000
Δ Annual income	600,000	120,000	360,000
Δ Residual value	2,000,000	755,000	1,750,000
IRR_{Δ}^{a}	13.8%	5.5%	8.5%
Decision	Accept one-story bldg.	Keep one-story bldg., reject two-story bldg.	Keep one-story bldg., reject three-story bldg.

[a] For instance, the IRR of Δ(B2-B1) is determined as follows: $0 = -\$1,550,000 + \$120,000(P/A, i'\%, 15) + \$775,000(P/F, i'\%, 15); \ i' = 5.5\%$.
[b] The net cash flow for Alternative B1, which is the incremental cash flow between making no change ($0) and implementing Alternative B1.

Alternative P is unacceptable (9.3% < 10%), which confirms our result in Part (a), and cannot serve as the base alternative from which to proceed with the incremental analysis procedure. However, Alternative B1 is acceptable and has the least capital investment of the remaining three feasible alternatives, so the incremental analysis would proceed as shown in Table 5-6.

Finally, Cynthia concludes that the one-story building is also the best alternative by using the IRR method. At this point, she tells her manager: "If I ever have to repeat this sort of analysis involving mutually exclusive alternatives, I'm going to insist on using an equivalent-worth method such as PW or having a better computer program."

(b) Rule 1 was used in the solution to Part (a) because the net annual income amounts varied among the alternatives.

5.5 Case 2: Useful Lives Are Different Among the Alternatives

When the useful lives of mutually exclusive alternatives are different, the *repeatability assumption* may be used in their comparison if the study period can be infinite in length or a common multiple of the useful lives. This assumes that the economic estimates for an alternative's initial useful life cycle will be repeated in all subsequent replacement cycles. As we discussed in Section 5.3, this condition is more robust for practical application than it may appear. Another viewpoint is to consider the repeatability assumption as a modeling convenience for the purpose of making a current decision. *When this assumption is applicable to a decision situation, it simplifies comparison of the mutually exclusive alternatives.* One solution method often used is to compute the AW of each alternative over its useful life and select the one with the best value (i.e., the alternative with the largest positive AW value for investment alternatives, and the one with the least negative AW value for cost alternatives).

If the repeatability assumption is not applicable to a decision situation, then an appropriate study period needs to be selected (*coterminated assumption*). This is the approach most frequently used in engineering practice because product life cycles are becoming shorter. Often, one or more of the useful lives will be shorter or longer than the selected study period. When this is the case, cash-flow adjustments based on additional assumptions need to be used *so all the alternatives are compared over the same study period.* The following guidelines apply to this situation:

1. (Useful life) < (Study period)
   (a) Cost alternatives: Because each cost alternative has to provide the same level of service over the study period, contracting for the service or leasing the needed equipment for the remaining years may be appropriate. Another potential course of action is to repeat part of the useful life of the original alternative, and then use an estimated market value to truncate it at the end of the study period.
   (b) Investment alternatives: The first assumption is that all cash flows will be reinvested in other opportunities available to the firm at the MARR to the end of the study period. A second assumption involves replacing the initial investment with another asset having possibly different cash flows over the remaining life. A convenient solution method is to calculate the FW of each mutually exclusive alternative at the end of the study period. The PW can also be used for investment alternatives since the FW at the end of the study period, say N, of each alternative is its PW times a common constant $(F/P, i\%, N)$, where $i\% = $ MARR.
2. (Useful life) > (Study period): The most common technique is to truncate the alternative at the end of the study period using an estimated market value. This assumes that the disposable assets will be sold at the end of the study period at that value.

> The underlying principle, as discussed in Section 5.3, is to compare the mutually exclusive alternatives being considered in a decision situation over the same study (analysis) period.

EXAMPLE 5-7

The following data on page 221 have been estimated for two mutually exclusive investment alternatives, A and B, associated with a small engineering project for which revenues as well as expenses are involved. They have useful lives of four and six years, respectively. If the MARR = 10% per year, show which alternative is more desirable by using equivalent worth methods. Use the repeatability assumption.

	A	B
Capital investment	$3,500	$5,000
Annual cash flow	1,255	1,480
Useful life (years)	4	6
Market value at end of useful life	0	0

SOLUTION

The least common multiple of the useful lives of Alternatives A and B is 12 years. Using the repeatability assumption and a 12-year study period, the first like (identical) replacement of Alternative A would occur at the end of year four, and the second would be at the end of year eight. For Alternative B, one like replacement would occur at the end of year six. This is illustrated in Part 1 of Figure 5-6.

SOLUTION: Solution of Example 5-7 by the PW Method

The PW (or FW) solution must be based on the total study period (12 years). The PW of the initial useful life cycle will be different than the PW of subsequent replacement cycles:

$$\mathrm{PW}(10\%)_A = -\$3,500 - \$3,500[(P/F, 10\%, 4) + (P/F, 10\%, 8)]$$

$$+ (\$1,255)(P/A, 10\%, 12)$$

$$= \$1,028,$$

Figure 5-6
Illustration of Repeatability Assumption (Example 5-7), and Coterminated Assumption (Example 5-8)

Part 1: Repeatability Assumption, Example 5-7, Least Common Multiple of Useful Lives Is 12 years

Three cycles of alternative A:

A1	A2	A3

0   4   8   12   years

Two cycles of alternative B:

B1	B2

0   6   12   years

Part 2: Coterminated Assumption, Example 5-8, 6-Year Analysis Period.

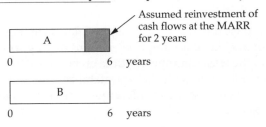

Assumed reinvestment of cash flows at the MARR for 2 years

A

0   6   years

B

0   6   years

$$PW(10\%)_B = -\$5,000 - \$5,000(P/F, 10\%, 6)$$

$$+ (\$1,480)(P/A, 10\%, 12)$$

$$= \$2,262.$$

Based on the PW method, we would select Alternative B because it has the larger value ($2,262).

SOLUTION: Solution of Example 5-7 by the AW method
The like replacement of assets assumes that the economic estimates for the initial useful life cycle will be repeated in each subsequent replacement cycle. Consequently, the AW will have the same value for each cycle and for the study period (12 years). This is demonstrated in the next AW solution by calculating (1) the AW of each alternative over the 12-year analysis period based on the previous PW values, and (2) determining the AW of each alternative over one useful life cycle. Based on the previously calculated PW values, the AW values are

$$AW(10\%)_A = \$1,028(A/P, 10\%, 12) = \$151,$$

$$AW(10\%)_B = \$2,262(A/P, 10\%, 12) = \$332.$$

Next, the AW of each alternative is calculated over one useful life cycle:

$$AW(10\%)_A = -\$3,500(A/P, 10\%, 4) + (\$1,255) = \$151,$$

$$AW(10\%)_B = -\$5,000(A/P, 10\%, 6) + (\$1,480) = \$332.$$

This confirms that both calculations for each alternative result in the same AW value, and we would again select Alternative B because it has the larger value ($332).

Companion Web Site (http://www.prenhall.com/sullivan_engineering/): Suppliers of aerospace products, such as inflatable evacuation slides, life rafts, and helicopter floats, use large industrial cutting devices. However, normal wear and tear and advances in new technology necessitate periodic replacement of these machines. Visit the Web site to view a *replacement analysis* from an industrial company that makes use of the *repeatability assumption.*

EXAMPLE 5-8

Suppose that Example 5-7 is modified such that an analysis period of 6 years is used (coterminated assumption) instead of 12 years, which was based on repeatability and the least common multiple of the useful lives. Perhaps the responsible manager did not agree with the repeatability assumption and wanted a six-year analysis period because it is the planning horizon used in the company for small investment projects.

SOLUTION

An assumption used for an investment alternative (when useful life is less than the study period) is that all cash flows will be reinvested by the firm at the MARR until the end of the study period. This assumption applies to Alternative A, which has a four-year useful life (two years less than the study period), and it is illustrated in Part 2 of Figure 5-6. We use the FW method to analyze this situation:

$$FW(10\%)_A = [-\$3,500(F/P, 10\%, 4) + (\$1,255)(F/A, 10\%, 4)](F/P, 10\%, 2)$$

$$= \$847,$$

$$FW(10\%)_B = -\$5,000(F/P, 10\%, 6) + (\$1,480)(F/A, 10\%, 6)$$

$$= \$2,561.$$

Based on the FW of each alternative at the end of the six-year study period, we would select Alternative B because it has the larger value ($2,561).

EXAMPLE 5-9

You are a member of an engineering project team that is designing a new processing facility. Your present design task involves the portion of the catalytic system that requires pumping a hydrocarbon slurry that is corrosive and contains abrasive particles. For final analysis and comparison, you have selected two fully lined slurry pump units, of equal output capacity, from different manufacturers. Each unit has the larger diameter impeller required and an integrated electric motor with solid state controls. Both units will provide the same level of service (support) to the catalytic system but have different useful lives and costs.

	Pump Model	
	SP240	HEPS9
Capital investment	$33,200	$47,600
Annual expenses:		
Electrical energy	$2,165	$1,720
Maintenance	$1,100 in year 1, and increasing $500/yr there-after	$500 in year 4, and increasing $100/yr there-after
Useful life (years)	5	9
Market value (end of useful life)	0	5,000

The new processing facility is needed by your firm at least as far into the future as the strategic plan forecasts operating requirements. The MARR is 20% per year. Based on this information, which model slurry pump should you select?

SOLUTION

The repeatability assumption is a logical choice for this analysis, and a study period, either infinite or 45 years (least common multiple of the useful lives) in length can be used. With repeatability, the AW over the initial *useful life* of each alternative is the same as its AW over the length of either study period:

$$AW(20\%)_{SP240} = -\$33,200(A/P, 20\%, 5) - \$2,165 - [\$1,100 + \$500(A/G, 20\%, 5)]$$

$$= -\$15,187,$$

$$AW(20\%)_{HEPS9} = -\$47,600(A/P, 20\%, 9) + \$5,000(A/F, 20\%, 9)$$

$$-\$1,720 - [\$500(P/A, 20\%, 6)$$

$$+\$100(P/G, 20\%, 6)] \times (P/F, 20\%, 3) \times (A/P, 20\%, 9)$$

$$= -\$13,622.$$

Based on Rule 2 (Section 5.2.2), you should select pump model HEPS9, since the AW over its useful life (nine years) has the lesser negative value (−$13,622).

As additional information, the following two points support choosing the repeatability assumption in Example 5-9:

1. The repeatability assumption is commensurate with the long planning horizon for the new processing facility, and with the design and operating requirements of the catalytic system.
2. If the initial estimated costs change for future pump replacement cycles, a logical assumption is that the ratio of the AW values for the two alternatives will remain approximately the same. Competition between the two manufacturers should cause this to happen. Hence, the pump selected (model HEPS9) should continue to be the preferred alternative.

If redesigned or new models of slurry pumps become available, however, another study to analyze and compare all feasible alternatives is required before a replacement of the selected pump occurs.

The next example illustrates the comparison of two mutually exclusive alternatives to increase the production capacity of a critical system in a plant by improving its operational availability.

EXAMPLE 5-10

The reliability engineer for an electronic manufacturing plant is trying to decrease the downtime of critical production systems. It is desired to improve the operational availability of these systems so the potential capacity of the plant will be increased. One critical system under review is used to manufacture an electronic control unit used in major home appliances. A reliability improvement team has developed two mutually exclusive alternatives for improving the operational availability of this system. The alternatives have differences in real-time monitoring technologies (predictive maintenance), preplanned preventive maintenance actions, computer information system support, and personnel training. Also, there are differences in annual maintenance expenses and the amount of increase in system availability. Relative to the current operation of the system, the following estimates have been developed:

	Alternative	
Factor	A1	A2
Capital investment	$260,000	$505,000
Annual maintenance expenses:		
Increase	$9,400	0
Decrease (savings)	0	$6,200
Increased system availability	4%	6.5%

Assume that the MARR = 18% per year, the analysis period is five years, any additional units produced can be immediately sold, the current average availability of the system (80.3%) results in 7,400 units being produced and sold per month, each 1% increase in the average availability of the system results in a 0.7% increase in plant capacity for production of the product, and each additional unit sold increases revenues by $48.20. (a) Select the preferred alternative using the IRR method, and (b) which rule (Section 5.2.2) was used in the selection? Why?

SOLUTION

(a) Rate of return methods require the use of the incremental investment analysis procedure. The rank order of the alternatives for incremental analysis, based on capital investment, is do nothing, A1, A2. Since these are investment alternatives, the next step is to check whether A1 is an acceptable base alternative:

$$PW(18\%)_{A1} = -\$260,000 - \$9,400(P/A, 18\%, 5)$$

$$+ 4(0.007)(7,400)(12)(\$48.20)(P/A, 18\%, 5)$$

$$= \$85,382.$$

Because the PW(MARR = 18%)$_{A1}$ > 0, we know that the IRR$_{A1}$ > MARR, and the alternative (A1) is an acceptable base alternative. Next, we need to find the IRR of the incremental cash flow between Alternatives A1 and A2:

$$0 = [-\$505,000 - (-\$260,000)] + [\$6,200 - (-\$9,400)](P/A, i'\%, 5)$$

$$+ (6.5 - 4.0)(0.007)(7,400)(12)(\$48.20)(P/A, i'\%, 5).$$

By linear interpolation (Section 4.6), we find that $i'\% = 24.7\%$ per year, which is greater than the MARR of 18% per year. Therefore, the additional capital investment in A2 over A1 is economically justified, and Alternative A2 would be selected.

(b) Rule 1, Section 5.2.2, applies in this example because the economic benefits vary among the alternatives and we want to maximize overall profitability.

Example 5-11 demonstrates how to deal with situations in which multiple machines are required to satisfy a fixed annual demand for a product or service. Such problems can be solved by using Rule 2 and the repeatability assumption.

EXAMPLE 5-11

Three products will be manufactured in a new facility at the Apex Manufacturing Company. They each require an identical manufacturing operation, but different production times, on a broaching machine. Two alternative types of broaching machines (M1 and M2) are being considered for purchase. One machine type must be selected.

For the same level of annual demand for the three products, *annual* production requirements (machine hours) and annual operating expenses (per machine) are listed next. Which machine should be selected if the MARR is 20% per year? Show all work to support your recommendation (use Rule 2 on page 202 to make your recommendation).

Product	Machine M1	Machine M2
ABC	1,500 hr	900 hr
MNQ	1,750 hr	1,000 hr
STV	2,600 hr	2,300 hr
	5,850 hr	4,200 hr
Capital Investment	$15,000 per machine	$22,000 per machine
Expected Life	5 years	8 years
Annual Expenses	$4,000 per machine	$6,000 per machine

Assumptions: The facility will operate 2,000 hours per year. Machine availability is 90% for Machine M1 and 80% for Machine M2. The yield of Machine M1 is 95%, and the yield of Machine M2 is 90%. Annual operating expenses are based on an assumed operation of 2,000 hours per year, and workers are paid during any idle time of Machine M1 or Machine M2. Salvage (market) values of both machines are negligible.

SOLUTION

The company will need 5,850 hr/[2,000 hr (0.90)(0.95)] = 3.42 (4 machines of type M1) or 4,200 hr/[2,000 hr (0.80)(0.90)] = 2.92 (3 machines of type M2). The maximum operation time of 2,000 hours per year in the denominator must be multiplied by the availability of each machine and the yield of each machine as indicated.

The annual cost of ownership, assuming a MARR $= 20\%$ per year, is $15,000(4) \cdot (A/P, 20\%, 5) = \$20,064$ for Machine M1 and $22,000(3)(A/P, 20\%, 8) = \$17,200$ for Machine M2.

There is excess capacity when four Machine M1s and three Machine M2s are used to provide the machine-hours (5,850 and 4,200, respectively) just given. If we assume that the operator is paid for idle time he or she may experience on M1 or M2, the annual expense for operation four M1s is 4 machines $\times$ $4,000 per machine $=$ $16,000. For three M2s, the annual expense is 3 machines $\times$ $6,000 per machine $=$ $18,000.

The total equivalent annual cost for four Machine M1s is $20,064 + \$16,000 = \$36,064$. Similarly, the total equivalent annual expense for three Machine M2s is $17,200 + \$18,000 = \$35,200$. By a slim margin, Machine M2 is the preferred choice to minimize equivalent annual costs with the repeatability assumption.

5.5.1 The Imputed Market Value Technique

Obtaining a current estimate from the marketplace for a piece of equipment or another type of asset is the preferred procedure in engineering practice when a market value at time $T <$ (useful life) is required. This approach, however, may not be feasible in some cases. For example, a type of asset may have low turnover in the marketplace and information for recent transactions is not available. Hence, it is sometimes necessary to estimate the market value for an asset without current and representative historical data.

The *imputed market value* technique, which is sometimes called the *implied* market value, can be used for this purpose as well as for comparison with marketplace values when current data are available. The estimating procedure used in the technique is based on logical assumptions about the value of the remaining useful life for an asset. If an imputed market value is needed for a piece of equipment, say, at the end of year $T <$ (useful life), the estimate is calculated based on the sum of two parts as follows:

$$MV_T = [\text{PW at end of year } T \text{ of remaining capital recovery amounts}]$$
$$+ [\text{PW at end of year } T \text{ of original market value at end of useful life}],$$

where PW means present worth at $i =$ MARR.

The next example uses information from Example 5-9 to illustrate the technique.

EXAMPLE 5-12

Use the imputed market value technique to develop an estimated market value for pump model HEPS9 (Example 5-9) at the end of year five. The MARR remains 20% per year.

SOLUTION

The original information from Example 5-9 will be used in the solution: capital investment = $47,600, useful life = nine years, and market value = $5,000 at the end of useful life.

Compute the PW at end of year five of the remaining CR amounts [Equation (4-5)]:

$$PW(20\%)_{CR} = [\$47,600(A/P, 20\%, 9) - \$5,000(A/F, 20\%, 9)] \times (P/A, 20\%, 4)$$

$$= \$29,949.$$

Compute the PW at end of year five, based on the original MV at end of useful life (9 years):

$$PW(20\%)_{MV} = \$5,000(P/F, 20\%, 4) = \$2,412.$$

Then, the estimated market value at the end of year five ($T = 5$) is as follows:

$$MV_5 = PW_{CR} + PW_{MV}$$

$$= \$29,949 + \$2,412 = \$32,361.$$

As additional information, if we use the estimated $MV_5 = \$32,361$ that we calculated for Pump HEPS9 in Example 5-12 to determine the AW of the pump over a five-year analysis period, the result is AW = −$13,449 (calculation not shown). This result is very close to AW = −$13,622 for the same pump over its useful life (nine years) in Example 5-9 when the repeatability assumption is utilized. The difference (−$172) is due to the maintenance expenses being a deferred uniform *gradient* series of cash flows. If the maintenance expenses had been equal annual amounts over both study periods, the two AW values for the pump would have been exactly the same. That is, when the annual cash flows (e.g., energy, maintenance, and so on) over the useful life of an asset are the same as for a truncated study period that is less than the useful life, the AW values over both periods will be equal. In this case, the repeatability assumption, or the use of an imputed market value to truncate useful life at the end of a shorter study period, provide the same AW results.

In summary, utilizing the repeatability assumption for Case 2 reduces to the simple rule of "comparing alternatives over their useful lives using the AW method, at i = MARR." However, this simplification may not apply when a study period, selected to be shorter or longer than the common multiple of lives (coterminated assumption), is more appropriate for the decision situation. When utilizing the coterminated assumption, cash flows of alternatives need to be adjusted to terminate at the end of the study period. Adjusting these cash flows usually requires estimating the market value of assets at the end of the study period or extending service to the end of the study period through leasing or some other assumption.

5.6 Comparison of Alternatives Using the Capitalized Worth Method

One special variation of the PW method discussed in Chapter 4 involves determining the present worth of all revenues or expenses over an infinite length of time. This is known as the *Capitalized Worth* (CW) method. If expenses only are considered, results obtained by this method are sometimes referred to as *capitalized cost*. This is a convenient basis for comparing mutually exclusive alternatives when the period of needed service is indefinitely long and the repeatability assumption is applicable.

The CW of a perpetual series of end-of-period uniform payments A, with interest at $i\%$ per period, is $A(P/A, i\%, \infty)$. From the interest formulas, it can be seen that $(P/A, i\%, N) \to 1/i$ as N becomes very large. Thus, $CW = A/i$ for such a series, as can also be seen from the relation

$$CW(i\%) = PW_{N \to \infty} = A(P/A, i\%, \infty) = A\left[\lim_{N \to \infty} \frac{(1+i)^N - 1}{i(1+i)^N}\right] = A\left(\frac{1}{i}\right).$$

Hence, the CW of a project with interest at $i\%$ per year is the annual equivalent of the project over its useful life divided by i.

The AW of a series of payments of amount $\$X$ at the end of each kth period with interest at $i\%$ per period is $\$X(A/F, i\%, k)$. The CW of such a series can thus be calculated as $\$X(A/F, i\%, k)/i$.

EXAMPLE 5-13

Suppose that a firm wishes to endow an advanced manufacturing processes laboratory at a university. The endowment principal will earn interest that averages 8% per year, which will be sufficient to cover all expenditures incurred in the establishment and maintenance of the laboratory for an indefinitely long period of time (forever). Cash requirements of the laboratory are estimated to be $100,000 now (to establish it), $30,000 per year indefinitely, and $20,000 at the end of every fourth year (forever) for equipment replacement.

(a) For this type of problem, what study period (N) is, practically speaking, defined to be "forever"?

(b) What amount of endowment principal is required to establish the laboratory and then earn enough interest to support the remaining cash requirements of this laboratory forever?

SOLUTION

(a) A practical approximation of "forever" (infinity) is dependent on the interest rate. By examining the $(P/A, i\%, N)$ factor as N increases, we observe that this factor approaches a value of $1/i$. For $i = 8\%$ ($1/i = 12.5000$), note that the $(P/A, 8\%, N)$ factor equals 12.4943 when $N = 100$. Therefore, $N = 100$ is essentially forever (∞) when $i = 8\%$. As the interest rate gets larger, the approximation of forever drops dramatically. For instance, when $i = 20\%$ ($1/i = 5.0000$), forever can be approximated using about 40 years; the $(P/A, 20\%, N)$ factor equals 4.9966 when $N = 40$.

(b) The CW of the cash requirements is synonymous with the endowment principal initially needed to establish and then support the laboratory forever. By using the relationship that $CW = A/i = $ (equivalent annual cost)$/i$, we can compute the amount of the endowment as

$$CW(8\%) = \frac{-\$100,000(A/P, 8\%, \infty) - \$30,000 - \$20,000(A/F, 8\%, 4)}{0.08}$$

$$= \frac{-\$8,000 - \$30,000 - \$4,438}{0.08}$$

$$= -\$530,475,$$

where the factor value for $(A/P, 8\%, \infty)$ is given in Table C–11 (Appendix C) as equal to 0.08000.

Another way of considering the amount of endowment principal needed in this example is to have enough to establish the facility ($100,000) and then have enough principal left in the fund to earn a return that will meet the annual maintenance costs ($30,000) and the periodic replacement of equipment needs ($20,000 at the end of each fourth year). Using this logic, we have

$$CW(8\%) = -\$100,000 - \left[\frac{\$30,000 + \$20,000(A/F, 8\%, 4)}{0.08}\right]$$

$$= -\$100,000 - \left[\frac{(\$30,000 + \$4,438)}{0.08}\right]$$

$$= -\$530,475,$$

which, of course, is the same CW amount as our previous calculation.

EXAMPLE 5-14

A selection is to be made between two structural designs. Because revenues do not exist (or can be assumed to be equal), only negative cash-flow amounts (costs) and the market value at the end of useful life are estimated, as follows:

	Structure *M*	Structure *N*
Capital investment	$12,000	$40,000
Market value	0	$10,000
Annual expenses	$2,200	$1,000
Useful life (years)	10	25

Using the repeatability assumption and the CW method of analysis, determine which structure is better if the MARR is 15% per year.

SOLUTION
The annual equivalent value (AW) over the useful life of each alternative structure, at MARR = 15% per year, is calculated as follows:

$$AW(15\%)_M = -\$12,000(A/P, 15\%, 10) - \$2,200 = -\$4,592$$

$$AW(15\%)_N = -\$40,000(A/P, 15\%, 25) + \$10,000(A/F, 15\%, 25) - \$1,000$$

$$= -\$7,141.$$

Then, the CWs of structures M and N are as follows:

$$CW(15\%)_M = \frac{AW_M}{i} = \frac{-\$4,592}{0.15} = -\$30,613,$$

$$CW(15\%)_N = \frac{AW_N}{i} = \frac{-\$7,141}{0.15} = -\$47,607.$$

Based on the CW of each structural design, alternative M should be selected be-cause it has the lesser negative value ($-\$30,613$).

5.7 Defining Mutually Exclusive Investment Alternatives in Terms of Combinations of Projects

It is helpful to categorize investment opportunities (projects) into three major groups as follows:

1. *Mutually exclusive:* At most one project out of the group can be chosen.
2. *Independent:* The choice of a project is independent of the choice of any other project in the group, so that all or none of the projects may be selected or some number in between.
3. *Contingent:* The choice of a project is conditional on the choice of one or more other projects.

It is common for decision makers to be faced with sets of mutually exclusive, inde-pendent, or contingent investment projects. For example, a construction contractor might be considering investing in a dump truck, in a backhoe, or in the expansion of the headquarters office building. For each of these investment projects, there may be two or more mutually exclusive alternatives (i.e., brands of dump trucks, types of backhoes, and designs for expansion of the office building). While the choice of a design for the office building is probably independent of that of either dump trucks or backhoes, the choice of any type of backhoe may be contingent (conditional) on the decision to purchase a dump truck.

TABLE 5-7 Combinations of Three Mutually Exclusive Projects[a]

Mutually Exclusive Combination	Project			Explanation
	X_A	X_B	X_C	
1	0	0	0	Accept none
2	1	0	0	Accept A
3	0	1	0	Accept B
4	0	0	1	Accept C

[a] For each investment project there is a binary variable X_j that will have the value 0 or 1 indicating that project j is rejected (0), or accepted (1). Each row of binary numbers represents an investment alternative in terms of a combination of projects (mutually exclusive combination). This convention is used, as required, in the remainder of this book.

A general approach, then, requires that all projects be listed and that all the feasible combinations of projects be enumerated. *Such combinations of projects will then be mutually exclusive.* Each combination of projects is mutually exclusive because each is unique and the acceptance of one combination of investment projects precludes the acceptance of any of the other combinations. The total net cash flow of each combination is determined simply by adding, period by period, the cash flows of each project included in the mutually exclusive combination being considered.

For example, suppose that we have three projects: A, B, and C. Each project can be selected once or not at all. (That is, multiple project A's are not possible.) If the projects themselves are all mutually exclusive, then the four possible mutually exclusive combinations are shown in binary form in Table 5-7. If, by chance, the firm feels that one of the projects must be chosen (i.e., it is not permissible to turn down all projects), then mutually exclusive combination one would be eliminated from consideration.

If the three projects are independent, there are eight mutually exclusive combinations, as shown in Table 5-8.

TABLE 5-8 Mutually Exclusive Combinations of Three Independent Projects

Mutually Exclusive Combination	Project			Explanation
	X_A	X_B	X_C	
1	0	0	0	Accept none
2	1	0	0	Accept A
3	0	1	0	Accept B
4	0	0	1	Accept C
5	1	1	0	Accept A and B
6	1	0	1	Accept A and C
7	0	1	1	Accept B and C
8	1	1	1	Accept A, B, and C

TABLE 5-9 Mutually Exclusive Combinations for Two Independent Sets of Mutually Exclusive Projects

Mutually Exclusive Combination	Project				Explanation
	X_{A1}	X_{A2}	X_{B1}	X_{B2}	
1	0	0	0	0	Accept none
2	1	0	0	0	Accept A1
3	0	1	0	0	Accept A2
4	0	0	1	0	Accept B1
5	0	0	0	1	Accept B2
6	1	0	1	0	Accept A1 and B1
7	1	0	0	1	Accept A1 and B2
8	0	1	1	0	Accept A2 and B1
9	0	1	0	1	Accept A2 and B2

To illustrate one of the many possible instances of contingent projects, suppose that A is contingent on the acceptance of both B and C and that C is contingent on the acceptance of B. Now there are four mutually exclusive combinations: (1) do nothing, (2) B only, (3) B and C, and (4) A, B, and C.

Suppose that a company is considering two independent sets of mutually exclusive projects. That is, projects A1 and A2 are mutually exclusive, as are B1 and B2. However, the selection of any project from the set of projects A1 and A2 is independent of the selection of any project from the set of projects B1 and B2. *Independent* means that the choice of a project from the A set does not affect the choice from the B set. Table 5-9 shows all mutually exclusive combinations for this situation.

EXAMPLE 5-15

Given the three independent engineering projects for improving energy efficiency shown here, determine which should be chosen using the AW method. The MARR = 10% per year, and there is no budget limitation on total investment funds available for this type of project.

Project	Capital Investment, I	Net Annual Cash Flow	Useful Life (yrs)	Market Value (at end of life)
E1	$10,000	$2,300	5	$10,000
E2	12,000	2,800	5	0
E3	15,000	4,067	5	0

	TABLE 5-10 Example 5-15 (AW Method)		
	(1)	(2)	(3) = (1) − (2)
Project	Net Annual Cash Flow	Capital Recovery Amount (Cost)	AW
E1	$2,300	$1,000	$1,300
E2	2,800	3,166	−366
E3	4,067	3,957	110

SOLUTION

As shown in Table 5-10 projects E1 and E3, having positive AWs, would be satisfactory for investment, but project E2 would not. The same indication of satisfactory projects and the unsatisfactory project would be obtained using other equivalent worth methods or the rate of return methods. Because there is no budget limitation on total investment funds available, both projects E1 and E3 would be recommended for implementation.

Example 5-16 illustrates how to enumerate the mutually exclusive project combinations (investment alternatives) from sets of projects that have all three basic relationships among them (mutually exclusive, independent, and contingent), and then to select an optimal set (*portfolio*) of projects under a capital investment budget constraint.

EXAMPLE 5-16

The following are five proposed projects being considered by an engineer in an integrated transportation company for upgrading an intermodal shipment transfer facility for less than truckload lots of consumer goods. The interrelationships among the projects, and their respective cash flows for the coming budgeting period, are as shown. Some of the projects are mutually exclusive, as noted, and B1 and B2 are independent of C1 and C2. Also, certain projects are dependent on others that may be included in the final portfolio. Using the PW method and MARR = 10% per year, determine what combination of projects is best if the capital to be invested is (a) unlimited, and (b) limited to $48,000.

Project B1 ⎱
Project B2 ⎰ mutually exclusive and independent of C set

Project C1 ⎱ mutually exclusive and dependent (contingent) on the acceptance
Project C2 ⎰ of B2

Project *D*   contingent on the acceptance of C1

TABLE 5-11	Project Cash Flows and PWs (Example 5-16)					
	Cash Flow ($000s) for End of Year					PW ($000s) at
Project	0	1	2	3	4	MARR = 10%/yr
B1	−$50	$20	$20	$20	$20	$13.4
B2	−30	12	12	12	12	8.0
C1	−14	4	4	4	4	−1.3
C2	−15	5	5	5	5	0.8
D	−10	6	6	6	6	9.0

SOLUTION

The PW for each project by itself is shown in the right-hand column of Table 5-11. As a sample calculation, the PW for project B1 is

$$PW(10\%)_{B1} = -\$50,000 + \$20,000(P/A, 10\%, 4) = \$13,400.$$

The mutually exclusive project combinations are shown in Table 5-12. Project C1 (which has a PW < 0) has not been eliminated from further consideration because project D is contingent on it.

The combined cash flows and the PW for each mutually exclusive combination are shown in Table 5-13. Examination of the right-hand column reveals that mutually exclusive combination 6 has the highest PW if capital available (in year 0) is unlimited, as specified in part (a). If, however, capital available is limited to $48,000, as specified in part (b), mutually exclusive combinations 2 and 6 are not feasible. Of the remaining mutually exclusive combinations, 5 is best, which means that a portfolio consisting of projects B2 and C2 would be selected with a PW = $8,888.

TABLE 5-12	Mutually Exclusive Project Combinations (Example 5-16)				
Mutually Exclusive Combination	Project				
	B1	B2	C1	C2	D
1	0	0	0	0	0
2	1	0	0	0	0
3	0	1	0	0	0
4	0	1	1	0	0
5	0	1	0	1	0
6	0	1	1	0	1

TABLE 5-13 Combined Project Cash Flows and PWs (Example 5-16)

Mutually Exclusive Combination	Cash Flow ($000s) for End of Year					Invested Capital ($000s)	PW ($000s) at MARR = 10%/yr
	0	1	2	3	4		
1	$0	$0	$0	$0	$0	$0	$0
2	−50	20	20	20	20	50	13.4
3	−30	12	12	12	12	30	8.0
4	−44	16	16	16	16	44	6.7
5	−45	17	17	17	17	45	8.9
6	−54	22	22	22	22	54	15.7

For problems that involve a relatively small number of projects, the general technique just presented for arranging various types of projects into mutually exclusive combinations is computationally practical. However, for larger numbers of projects, the number of mutually exclusive combinations becomes quite large, and a computer program should be used to perform the calculations.

In many problems involving selections among independent projects, different revenues (or savings) and useful lives are present. Because these projects are typically nonrepeating, it is assumed that cash flows of shorter lived projects are reinvested at the MARR over a period corresponding to the life of the longest-lived project (Section 5.5). The next example illustrates this assumption.

EXAMPLE 5-17

A large corporation is considering the funding of three independent, nonrepeating projects for enlarging freshwater harbors supporting its operations in three areas of the country. Its available capital investment budget this year for such projects is $200 million, and the firm's MARR is 10% per year. In view of the following data, which project(s), if any, should be funded?

Project	Capital Investment, I	Net Annual Benefits, A	Useful Life, N	$PW(10\%) = -I + A(P/A, 10\%, N)$
H1	$93,000,000	$13,000,000	15 years	$5,879,300
H2	55,000,000	9,500,000	10 years	3,373,700
H3	71,000,000	10,400,000	30 years	27,039,760

SOLUTION

Based on the PW values, each project is economically justified. Hence, the eight feasible mutually exclusive combinations of the three independent projects (reference the general case that was enumerated in Table 5-8) need to be evaluated. The total PW of each combination can be used for this purpose. The total FW of each combination of projects, at the end of the longest-lived project (30 years), is its PW times a common constant, $(F/P, 10\%, 30)$, and will result in the same selection.

A review of the investment cost and PW values for each of the three projects indicates that only the three mutually exclusive combinations involving two of the projects need to be considered. The capital investment budget constraint will not allow all three projects to be implemented, and the do nothing alternative is not preferred because each project adds wealth to the firm. Furthermore, each of the three combinations of two projects are within the budget constraint and will add more wealth to the firm than a single project alone. Since projects H1 and H3 have the greatest positive PW values, this combination should be selected. The total PW of this combination is $32,919,060, and its total FW at the end of 30 years is $32,919,060 $(F/P, 10\%, 30) = \$574,417,850$. It is assumed that the remaining $36,000,000 of the $200,000,000 capital investment budget will be invested by the firm in other projects earning at least the MARR $= 10\%$ per year.

5.8 Spreadsheet Applications

Because of the repetitive nature of the previous calculations, spreadsheets can be very useful when comparing mutually exclusive alternatives. Given the net-cash-flow profile for each alternative being considered, we can use the spreadsheet's financial functions (as described in Section 4.11) to compute the equivalent-worth measures of merit for each alternative. We can also use a spreadsheet to analyze alternatives with the incremental IRR and ERR procedures.

The analysis of five alternatives (Alpha, Beta, Gamma, Delta, and Theta) using the equivalent worth methods is shown in Figure 5-7. The equivalent-worth measures are calculated based on the net-cash-flow profile presented. The alternative having the largest equivalent worth (Beta) is identified as the recommended alternative. The formulas for the highlighted cells are shown as follows:

Cell	Contents
C11	=NPV($B$1, C5:C9) + C4
C12	=PMT($B$1, 5, −(NPV($B$1, C5:C9) + C4))
C13	=FV($B$1, 5, PMT($B$1, 5, (NPV($B$1, C5:C9) + C4)))
C14	=IF(C11 = MAX(B11:F11), "Recommend", "")

To analyze alternatives using the rate-of-return methods, we need to perform an incremental analysis. Although there is no financial function for computing an incremental rate of return, we can modify the cash flows and use the IRR( ) financial function. The approach is straightforward:

1. Arrange the alternatives in order of increasing capital investment.

	A	B	C	D	E	F
1	MARR	10%				
2						
3	EOY	Alpha	Beta	Gamma	Delta	Theta
4	0	$ (8,000)	$ (16,000)	$ (10,000)	$ (13,000)	$ (9,500)
5	1	$ 2,500	$ 5,000	$ 2,800	$ 3,800	$ 2,000
6	2	$ 2,500	$ 5,000	$ 3,200	$ 3,800	$ 2,200
7	3	$ 2,500	$ 5,000	$ 3,400	$ 3,800	$ 2,600
8	4	$ 2,500	$ 5,000	$ 3,700	$ 3,800	$ 2,800
9	5	$ 2,500	$ 6,000	$ 3,800	$ 3,800	$ 3,000
10						
11	PW	$ 1,476.97	$ 3,574.86	$ 2,631.20	$ 1,404.99	$ (135.02)
12	AW	$ 389.62	$ 943.04	$ 694.10	$ 370.63	$ (35.62)
13	FW	$ 2,378.67	$ 5,757.34	$ 4,237.58	$ 2,262.75	$ (217.45)
14			Recommend			

Figure 5-7 Spreadsheet for Comparing MEAs Using Equivalent-Worth Methods

2. Determine the IRR for each alternative to decide if it is greater than or equal to the MARR. Eliminate any unacceptable alternatives from further consideration.*

3. Create a column that determines the difference between the alternative having the least capital investment (the base alternative) and the next most expensive alternative. Remember that the difference is computed by subtracting the lower investment column from the higher investment column, so that the difference column will have a negative cash flow at time 0.

4. Compute the IRR of the difference column. This is the IRR_Δ. Accept the more expensive alternative only if the $IRR_\Delta \geq MARR$.

5. Repeat the procedure, forming a new difference column for each comparison, until all alternatives have been compared.

The incremental IRR analysis of the five alternatives considered earlier is shown in Figure 5-8. The alternatives have been reordered according to increasing capital investment. Alternative Theta is eliminated from further consideration because it has an IRR < MARR. Alpha is the base alternative since it requires the smallest capital investment and has an IRR > MARR. The next least expensive alternative is Gamma. Comparing Alpha with Gamma, we see that the incremental investment is justified because $IRR_\Delta \geq MARR$.

---

*This step applies only when we are comparing investment alternatives. Recall that in the case of cost alternatives, the rate of return is typically less than zero.

	A	B	C	D	E	F
1	MARR	10%				
2	ε	8%				
3						
4	EOY	Alpha	Theta	Gamma	Delta	Beta
5	0	$ (8,000)	$ (9,500)	$ (10,000)	$ (13,000)	$ (16,000)
6	1	$ 2,500	$ 2,000	$ 2,800	$ 3,800	$ 5,000
7	2	$ 2,500	$ 2,200	$ 3,200	$ 3,800	$ 5,000
8	3	$ 2,500	$ 2,600	$ 3,400	$ 3,800	$ 5,000
9	4	$ 2,500	$ 2,800	$ 3,700	$ 3,800	$ 5,000
10	5	$ 2,500	$ 3,000	$ 3,800	$ 3,800	$ 6,000
11						
12	IRR	16.99%	9.48%	19.29%	14.15%	18.20%
13	ERR	12.89%	8.90%	14.41%	11.39%	13.65%
14						
15	Incremental Analysis					
16						
17	EOY	Δ(Gamma-Alpha)	Δ(Delta-Gamma)	Δ(Beta-Gamma)		
18	0	$ (2,000)	$ (3,000)	$ (6,000)		
19	1	$ 300	$ 1,000	$ 2,200		
20	2	$ 700	$ 600	$ 1,800		
21	3	$ 900	$ 400	$ 1,600		
22	4	$ 1,200	$ 100	$ 1,300		
23	5	$ 1,300	$ -	$ 2,200		
24						
25	IRR Δ	26.28%	-17.20%	16.18%		
26	Decision	Accept	Reject	Accept		
27						
28	ERR Δ	19.80%	-2.15%	12.33%		
29	Decision	Accept	Reject	Accept		

Figure 5-8 Spreadsheet for Comparing MEAs Using Rate-of-Return Methods

The next comparison made is between Gamma and Delta. Comparing the IRR$_\Delta$ to the MARR, we find that the incremental investment is not justified. The same conclusion can be reached by noting that the sum of the nondiscounted positive cash flows is less than the required incremental investment. Finally, we compare Gamma to Beta. Since IRR$_\Delta \geq$ MARR and there are no more alternatives to be considered, alternative Beta is recommended. This recommendation is consistent with the recommendation using the equivalent worth methods. (See Figure 5-7.) Note that Gamma, which has the highest overall IRR, is not selected as the recommended alternative.

This same procedure applies to an ERR analysis of alternatives. We simply specify the reinvestment rate and substitute the MIRR( ) financial function for the IRR( ) function. The results of an incremental ERR analysis (when $\epsilon = 8\%$) are shown at the bottom of Figure 5-8. The formulas for the highlighted cells are shown in the following table.

Cell	Contents
B12	= IRR(B5:B10, $B$1)
B13	= MIRR(B5:B10, $B$1, $B$2)
B18	= D5 − B5
C18	= E5 − D5
D18	= F5 − D5
B25	= IRR(B18:B23, $B$1)
B26	= IF(B25>=$B$1, "Accept", "Reject")
C28	= MIRR(C18:C23, $B$1, $B$2)
C29	= IF(C28>=$B$1, "Accept", "Reject")

5.9 Summary

Chapter 5 has built on the previous chapters, in which the principles and applications of money–time relationships were developed. Specifically, Chapter 5 has (1) introduced several difficulties associated with selecting the best alternative from a mutually exclusive set of feasible candidates when using time value of money concepts, and (2) demonstrated the application of the profitability analysis methods discussed in Chapter 4 to select the preferred alternative. Moreover, alternatives with unequal lives, various types of dependencies, cost-only versus different revenues and costs, and funding constraints were considered in deciding how to maximize the productivity of invested capital based on the MARR. In summary, we learned that choosing the alternative with the largest equivalent worth (or least negative in the case of cost alternatives) using the MARR would produce this desired result.

If a rate-of-return method is being used to analyze mutually exclusive alternatives, each avoidable increment of additional capital must earn at least the MARR to ensure that the best alternative is chosen. Therefore, alternatives are rank ordered from the least capital investment to the greatest capital investment. Examples were provided to illustrate correct computational procedures for avoiding the ranking inconsistency that sometimes occurs when equivalent worth and rate-of-return methods are applied to the same set of mutually exclusive alternatives. We also considered projects with perpetual lives, applying the capitalized worth method of economic evaluation. The chapter concluded by demonstrating the evaluation of combinations of mutually exclusive, independent, or contingent projects using these same methods.

5.10 References

BUSSEY, L. E. and T. G. ESCHENBACH, *The Economic Analysis of Industrial Projects*. 2nd ed. (Englewood Cliffs, NJ: Prentice Hall, 1992).

FLEISCHER, GERALD A. "Two Major Issues Associated with The Rate of Return Method for Capital Allocation: The 'Ranking Error' and 'Preliminary Selection.'" *The Journal of Industrial Engineering*, vol. 17, no. 4, April 1966, pp. 202–208.

GRANT, E. L., W. G. IRESON, and R. S. LEAVENWORTH. *Principles of Engineering Economy*, 8th ed. (New York: John Wiley & Sons, 1989).

PARK, C. S., and G. P. SHARP-BETTE. *Advanced Engineering Economics* (New York: John Wiley & Sons, 1990).

5.11 Problems

The number in parentheses () that follows each problem refers to the section from which the problem is taken.

5-1. Four mutually exclusive alternatives are being evaluated, and their costs and revenues are itemized in Table P5-1. (5.4)

a. If the MARR is 15% per year, and the analysis period is 10 years, use the PW method to determine which alternatives are economically acceptable and which one should be selected.

b. If the total capital investment budget available is $200,000, which alternative should be selected?

c. Which rule (Section 5.2.2) applies? Why?

5-2. In the design of a new facility, the mutually exclusive alternatives in Table P5-2 are under consideration. Assume that the interest rate (MARR) is 15% per year and the analysis period is 10 years. Use the following methods to choose the best of these three design alternatives: (5.4)

a. AW method.

b. FW method.

5-3. The Consolidated Oil Company must install antipollution equipment in a new refinery to meet federal clean-air standards. Four design

TABLE P5-1 Table for Problem P5-1

	Mutually Exclusive Alternative			
	I	II	III	IV
Capital investment	$100,000	$152,000	$184,000	$220,000
Annual revenues less expenses	15,200	31,900	35,900	41,500
Market value (end of useful life)	10,000	0	15,000	20,000
Useful life (years)	10	10	10	10

TABLE P5-2 Table for Problem P5-2

	Design 1	Design 2	Design 3
Capital investment	$28,000	$16,000	$23,500
Annual revenues less expenses	5,500	3,300	4,800
Market value	1,500	0	500
Useful life (years)	10	10	10

TABLE P5-3 Table for Problem P5-3

	Alternative Design			
	D1	D2	D3	D4
Capital investment	$600,000	$760,000	$1,240,000	$1,600,000
Annual expenses:				
Power	68,000	68,000	120,000	126,000
Labor	40,000	45,000	65,000	50,000
Maintenance	660,000	600,000	420,000	370,000
Taxes and insurance	12,000	15,000	25,000	28,000

alternatives are being considered, which will have capital investment and annual operating expenses as shown in Table P5-3. Assuming a useful life of 10 years for each design, no market value, a desired MARR of 10% per year, and an analysis period of 10 years, determine which design should be selected based on the PW method. Confirm your selection by using the IRR method. Which rule (Section 5.2.2) applies? Why? (5.4)

5-4. The 21st-Century Development Corporation has a 30-year lease on a plot of land. Estimates of the annual expenses and revenues of various types of structures on the property are as shown in the accompanying table.

Each structure is expected to have a market value equal to 20% of its capital investment at the end of a 30-year analysis period. If the investor requires an MARR of at least 12% per year on all investments, which structure (if any) should be selected? Use the AW method. (5.4)

	Capital Investment	Annual Revenues Less Expenses
Apartment house	$300,000	$69,000
Theater	200,000	40,000
Department store	250,000	55,000
Office building	400,000	76,000

5-5. The following cash-flow estimates have been developed for two small, mutually exclusive investment alternatives:

End of Year	Alternative 1	Alternative 2
0	−$2,500	−$4,000
1	750	1,200
2	750	1,200
3	750	1,200
4	750	1,200
5	2,750	3,200

The MARR = 12% per year. For parts (a) through (d), select the closest answer. (5.4)

a. What is the AW of Alternative 1?
   1. $371   **2.** −$162   **3.** $135
   4. $1,338   **5.** $1,590

b. What is the IRR of Alternative 1?
   1. 12%   **2.** 31%   **3.** 16%
   4. 28%   **5.** 25%

c. What is the IRR of the incremental net cash flow?
   1. 18%   **2.** 21%   **3.** 12%
   4. 24%   **5.** 15%

d. Given your answers for parts (a) through (c), which alternative should be selected?
   1. Alternative 1
   2. Alternative 2
   3. Neither
   4. Both Alternatives 1 and 2

5-6. An electronics company is trying to determine to which new product they should commit their limited capital resources. (There is not enough investment capital for both products.) The information in the following table shows the estimated net cash flow for each of the two proposed products:

TABLE P5-7 Table for Problem P5-7

| | Packaging Equipment | | | | |
	A	B	C	D	E
Capital investment	$38,000	$50,000	$55,000	$60,000	$70,000
Annual revenues less expenses	11,000	14,100	16,300	16,800	19,200
Rate of return (IRR)	26.1%	25.2%	26.9%	25.0%	24.3%

End of Year, k	Product 1	Product 2
0	−$150,000	−$520,000
1	50,000	30,000
2	50,000	130,000
3	50,000	230,000
4	50,000	330,000
IRR	12.6%	11.0%

If the MARR = 10% per year, show that the same project selection would be made with proper application of (a) the PW method, and (b) the IRR method. (5.4)

5-7. In the Rawhide Company (a leather products manufacturer), decisions regarding approval of proposals for capital investment are based upon a stipulated MARR of 18% per year. The five packaging devices listed in Table P5-7 were compared assuming a 10-year life and zero market value for each at that time. Which one (if any) should be selected? Make any additional calculations you think are needed to make a comparison using the IRR method. (5.4)

5-8. Two mutually exclusive alternatives are available. If MARR = 15%, select the *one best alternative* using the IRR method. In this problem, "do nothing" is an option. The cash-flow profiles for the alternatives are as follows:

	A	B
Initial Investment	$9,000	$6,000
Net Annual Cash Flow	$2,400	$1,600
Salvage Value	$0	$300
Useful Life (Years)	6	6
IRR	15.3%	16.1%

5-9. Rework Problem 5-2 using the IRR method. (5.4)

5-10. Three alternative designs are being considered for a potential improvement project related to the operation of your engineering department. The prospective net cash flows for these alternatives are shown in the following table, and the MARR is 15% per year:

| End of Year, k | Alternative Net Cash Flows | | |
	A	B	C
0	−$200,000	−$230,000	−$212,500
1	90,000	108,000	−15,000
2			122,500
3			
4	a	a	a
5			
6	90,000	108,000	122,500

a Continuing uniform cash flow.

Show that the same capital investment decision results from the IRR method and the PW method applied using the incremental investment analysis procedure. (5.4)

5-11. Refer to Example 5-10. Assume the company's marketing department estimates that 91,000 units of the electronic control unit is the maximum that can be sold in any year. Based on this assumption, should the project still be implemented? If yes, which alternative (A1, A2) should be selected? Why? (*Note*: Use the AW method of analysis in your solution.) (5.4)

5-12. The net cash flows are shown in the following table for three preliminary design alternatives for a heavy-duty industrial compressor:

End of	Net Cash Flows		
Year, k	A	B	C
0	−$85,600	−$63,200	−$71,800
1	−7,400	−12,100	−10,050
.			
.	a	a	a
.	↓	↓	↓
7	−7,400	−12,100	−10,050

^{a} Continuing uniform cash flow.

The perspective of the cash flows is that of the typical user. The MARR = 12% per year, and the study period is seven years. Which preliminary design is economically preferred based on (a) the AW method, and (b) the ERR method (ϵ = MARR = 12% per year)? (5.4)

5-13. A new highway is to be constructed. Design A calls for a *concrete* pavement costing $90 per foot with a 20-year life; two paved ditches costing $3 per foot each; and three box culverts every mile, each costing $9,000 and having a 20-year life. Annual maintenance will cost $1,800 per mile; the culverts must be cleaned every five years at a cost of $450 each per mile.

Design B calls for a *bituminous* pavement costing $45 per foot with a 10-year life; two sodded ditches costing $1.50 per foot each; and three pipe culverts every mile, each costing $2,250 and having a 10-year life. The replacement culverts will cost $2,400 each. Annual maintenance will cost $2,700 per mile; the culverts must be cleaned yearly at a cost of $225 each per mile; and the annual ditch maintenance will cost $1.50 per foot per ditch.

Compare the two designs on the basis of equivalent worth per mile for a 20-year period. Find the most economical design on the basis of equivalent annual worth and present worth if the MARR is 6% per year. (5.3)

5-14. A designer is evaluating two electric motors for an automated paint booth application. Each motor's output must be 10 horsepower (hp). She estimates the typical user will operate the booth an average of six hours per day for 250 days per year. Past experience indicates that (a) the annual expense for taxes and insurance averages 2.5% of the capital investment, (b) the MARR is 10% per year, and (c) the capital invested in machinery must be recovered within five years. Motor A costs $850 and has a guaranteed efficiency of 85% at the indicated operating load.

Motor B costs $700 and has a guaranteed efficiency of 80% at the same operating load. Electric energy costs the typical user 5.1 cents per kilowatt-hour (kWh), and 1 hp = 0.746 kW. Recall that electrical input to a motor equals output ÷ efficiency.

Use the IRR method to choose the better electric motor for the design application. Confirm your selection using the PW method. (5.4)

5-15. Two electric motors (A and B) are being considered to drive a centrifugal pump. Each motor is capable of delivering 50 horsepower (output) to the pumping operation. It is expected that the motors will be in use 1,000 hours per year. If electricity costs $0.07 per kilowatt-hour, *which motor should be selected if MARR = 8% per year?* Refer to the following data. Recall that 1 hp = 0.76 kW. (5.5)

	Motor A	Motor B
Initial Cost	$1,200	$1,000
Electrical Efficiency	0.82	0.77
Annual Maintenance	$60	$100
Life	4 years	5 years

5-16. Consider the three small mutually exclusive investment alternatives in the table below. The feasible alternative chosen must provide service for a 10-year period. The MARR is 12% per year, and the market value of each is 0 at the end of useful life. State all assumptions you make in your analysis. Which alternative should be chosen? (5.4, 5.5)

	A	B	C
Capital investment	$2,000	$8,000	$20,000
Annual revenues less expenses	600	2,200	3,600
Useful life (years)	5	5	10

5-17. A certain service can be performed satisfactorily by either process R or process S. Process R has a first cost of $8,000, an estimated service life of 10 years, no market value, and annual revenues less expenses of $2,400. The corresponding figures for process S are $18,000, 20 years, market value equal to 20% of the first cost, and $4,000. Assuming a MARR of 12% per year, find the AW of each process and specify which you would recommend. Use the repeatability assumption. (5.4)

5-18. A new manufacturing facility will produce two products, each of which requires a drilling operation during processing. Two alternative types of drilling machines (D1 and D2) are being considered for purchase. One of these machines must be selected. For the same annual demand, the *annual* production requirements (machine hours) and the annual operating expenses (per machine) are listed in Table P5-18. *Which machine should be selected if the MARR is 15% per year?* Show all your work to support your recommendation. (5.5)

   Assumptions: The facility will operate 2,000 hours per year. Machine availability is 80% for Machine D1 and 75% for Machine D2. The yield of D1 is 90% and the yield of D2 is 80%. Annual operating expenses are based on an assumed operation of 2,000 hours per year, and workers are paid during any idle time of Machine D1 or Machine D2. State any other assumptions needed to solve the problem.

5-19. As the supervisor of a facilities engineering department, you consider mobile cranes to be critical equipment. The purchase of a new medium-sized, truck-mounted crane is being evaluated. The economic estimates for the two best alternatives are shown in the accompanying table. You have selected the longest useful life (nine years) for the study period and would lease a crane for the final three years under Alternative A. Based on previous experience, the estimated annual leasing cost at that time will be $66,000 per year (plus the annual expenses of

$28,800 per year). The MARR is 15% per year. Show that the same selection is made based on (a) the PW method, (b) the IRR method, and (c) the ERR method. Also, (d) would leasing crane A for nine years, assuming the same costs per year as for three years, be preferred to your present selection? (ϵ = MARR = 15%). (5.4, 5.5)

	Alternatives	
	A	B
Capital investment	$272,000	$346,000
Annual expenses[a]	28,800	19,300
Useful life (years)	6	9
Market value (at end of life)	$25,000	$40,000

[a] Excludes the cost of an operator, which is the same for both alternatives.

5-20. A set of six long-life light bulbs costs $15.95. Each bulb is rated for 20,000 hours of service and 60 watts of output. The electrical efficiency of each bulb is 85%. The alternative to these long-life bulbs is a standard 60-watt bulb that costs 60 cents ($0.60), is rated for 1,000 hours of service and is 95% efficient. (5.4)

a. If the cost of electricity is 10 cents ($0.10) per kilowatt-hour, which type of bulb is better when lighting is required for 5,000 hours per year? The MARR is 12% per year. Assume end-of-year cash-flow convention.

b. What factors besides cost may dictate the choice of the better light bulb?

TABLE P5-18 Table for Problem P5-18		
Product	Machine D1	Machine D2
R-43	1,200 hours	800 hours
T-22	2,250 hours	1,550 hours
	3,450 hours	2,350 hours
Capital Investment	$16,000/machine	$24,000/machine
Useful Life	6 years	8 years
Annual Expenses	$5,000/machine	$7,500/machine
Salvage Value	$3,000/machine	$4,000/machine

5-21. Consider the following two mutually exclusive alternatives related to an improvement project, and recommend which one (if either) should be implemented using (a) the AW method and (b) the PW method. Also, (c) confirm your selection in Parts (a) and (b) using the IRR method. The MARR = 15% per year, and the study period is 10 years. Assume repeatability is applicable. (5.5)

	Machine	
	A	B
Capital investment	$20,000	$30,000
Annual cash flow	5,600	5,400
Market value	$4,000	$0
Useful life (years)	5	10

5-22. Select the preferred investment alternative from the mutually exclusive pair shown in the following table based on (a) the repeatability assumption, (b) the coterminated assumption with a four-year study period and the market value of Alternative 2 (at the end of year four) determined using the imputed market value technique, and (c) the coterminated assumption with an eight-year study period (Alternative 1 would not be repeated). The MARR is 10% per year. (5.5)

End of Year	Alternative 1	Alternative 2
0	−$40,000	−$50,000
1	12,000	10,000
2	12,000	10,000
3	12,000	10,000
4	36,000	10,000
5		10,000
6		10,000
7		10,000
8		10,000
8 (MV)		40,000

5-23. Three models of baseball bats will be manufactured in a new plant in Pulaski. Each bat requires some manufacturing time at either Lathe 1 or Lathe 2 according to the following table. Your task is to help decide on which type of lathe to install. Show and explain all work to support your recommendation. (5.5)

Machining Hours for the Production of Baseball Bats

Product	Lathe 1 (L1)	Lathe 2 (L2)
Wood Bat	1,600 hr	950 hr
Aluminum Bat	1,800 hr	1,100 hr
Kevlar Bat	2,750 hr	2,350 hr
Total Machining Hours	6,150 hr	4,400 hr

The plant will operate 3,000 hours per year. Machine availability is 85% for Lathe 1 and 90% for Lathe 2. Scrap rates for the two lathes are 5% vs. 10% for L1 and L2, respectively. Cash flows and expected lives for the two lathes are as follows:

Cash-Flows and Expected Lives for L1 and L2

	Lathe 1 (L1)	Lathe 2 (L2)
Capital Investment	$18,000 per lathe	$25,000 per lathe
Expected Life	7 years	11 years
Annual Expenses	$5,000 per lathe	$9,500 per lathe

Annual operating expenses are based on an assumed operation of 3,000 hours per year, and workers are paid during any idle time of L1 and L2. Upper management has decided that MARR = 18% per year.

a. How many type L1 lathes will be required to meet the machine-hour requirement?
(i) 2 lathes   (ii) 3 lathes
(iii) 4 lathes   (iv) 1 lathe

b. What is the capital recovery cost of the required type L2 lathes (most close answer)?
(i) $9,555   (ii) $14,168
(iii) $10,740   (iv) $5,370

c. What is the annual operating expense of the type L2 lathes (most close answer)?
(i) $5,375   (ii) $9,500
(iii) $21,000   (iv) $19,000

d. What type of lathe has the lowest *total* equivalent annual cost?
(i) Lathe L1   (ii) Lathe L2

5-24. A piece of production equipment is to be replaced immediately because it no longer meets quality requirements for the end product. The two best alternatives are a used piece of equipment (E1) and a new automated model (E2). The economic estimates for each are shown in the accompanying table.

	Alternative	
	E1	E2
Capital investment	$14,000	$65,000
Annual expenses	$14,000	$9,000
Useful life (years)	5	20
Market value		
(at end of useful life)	$8,000	$13,000

The MARR is 15% per year.

a. Which alternative is preferred, based on the repeatability assumption? (5.5)

b. Show, for the coterminated assumption with a five-year study period and an imputed market value for Alternative B, that the AW of B remains the same as it was in part (a) [and obviously the selection is the same as in part (a)]. Explain why that occurs in this problem. (5.5)

5-25. Estimates for a proposed small public facility are as follows. Plan A has a first cost of $50,000, a life of 25 years, a $5,000 market value, and annual maintenance expenses of $1,200. Plan B has a first cost of $90,000, a life of 50 years, no market value, and annual maintenance expenses of $6,000 for the first 15 years and $1,000 per year for years 16 through 50. Assuming interest at 10% per year, compare the two plans using the CW method. (5.6)

5-26. In the design of a special-use structure, two mutually exclusive alternatives are under consideration. These design alternatives are as follows:

	D1	D2
Capital investment	$50,000	$120,000
Annual expenses	$9,000	$5,000
Useful life (years)	20	50
Market value (at end of useful life)	$10,000	$20,000

If *perpetual service* from the structure is assumed, which design alternative do you recommend? The MARR is 10% per year. (5.6)

5-27. Use the CW method to determine which mutually exclusive bridge design (L or H) to recommend based on the data provided in the table below. The MARR is 15% per year. (5.6)

	Bridge Design L	Bridge Design H
Capital Investment	$274,000	$326,000
Annual Expenses	$10,000	$8,000
Periodic upgrade cost	$50,000 (every sixth year)	$42,000 (every seventh year)
Market value	0	0
Useful life (years)	83	92

5-28.

a. What is the capitalized worth, when $i = 10\%$ per year, of $1,500 per year, starting in year one and continuing forever and $10,000 in year five, repeating every four years thereafter, and continuing forever? (5.6)

b. When $i = 10\%$ per year in this type of problem, what value of N, practically speaking, defines "forever"? (5.6)

5-29. You are involved with an equipment selection task. The operating requirement to be met by the selected piece of equipment is for the next six years *only*. The economic decision criterion presently being used by your company is 15% per year. Your choice has been reduced to two alternatives (E1 or E2; see Table P5-29).

Which alternative ought to be selected? Use the PW method of analysis. Which rule, Section 5.2.2, did you use? (5.2, 5.5)

5-30. The estimated information for two design alternatives for an engineering project is given below. Assume the MARR = 12% per year, and a ten-year analysis period is being used. Also, the useful life of each design is ten years. (5.4)

a. Select the preferred alternative using the FW method.

b. What is the IRR of the incremental cash flow? Does it confirm your answer to Part (a)? Why?

c. Given: $IRR_{D1} = 16.43\%$, and $IRR_{D2} = 15.27\%$. Why does inconsistent ranking of the two alternatives *not* occur in this problem?

Factor	D1	D2
Capital investment	$152,000	$184,000
Annual net cash flow	$31,900	$35,900
Market value (end of useful life)	0	$15,000

TABLE P5-29 Table for Problem P5-29		
Factor	E1	E2
Capital Investment	$210,000	$264,000
Useful life (years)	6	10
Annual Expenses	$31,000 in the first year and increasing $2,000 per year thereafter	$19,000 in the first year and increasing 5.7% per year thereafter
Market value (end of useful life)	$21,000	$38,000

5-31. The alternatives for an engineering project to recover most of the energy presently being lost in the primary cooling stage of a chemical processing system have been reduced to three designs. The estimated capital investment amounts and annual expense *savings* are as follows:

	Design		
EOY	ER1	ER2	ER3
0	−$98,600	−$115,000	−$81,200
1	25,800	29,000	19,750
2			
3			
	$f = 6\%^a$	$G = \$150^b$	c
4			
5			
6	34,526	29,750	19,750

[a] After year one, the annual savings are estimated to increase at the rate of 6% per year.
[b] After year one, the annual savings are estimated to increase $150 per year.
[c] Uniform sequence of annual savings.

Assume that the MARR is 12% per year, the study period is six years, and the market value is zero for all three designs. Apply an analysis method *using the incremental analysis procedure* to determine the preferred alternative. (5.4)

5-32. A small company has $20,000 in surplus capital that it wishes to invest in new revenue-producing projects. Three independent sets of mutually exclusive projects have been developed. The useful life of each is five years, and all market values are zero. You have been asked to perform an IRR analysis to select the best combination of projects. If the MARR is 12% per

year, which combination of projects would you recommend? (See the table below.) (5.7)

	Project	Capital Investment	Net Annual Benefits
Mutually exclusive	A1	$5,000	$1,500
	A2	7,000	1,800
Mutually exclusive	B1	12,000	2,000
	B2	18,000	4,000
Mutually exclusive	C1	14,000	4,000
	C2	18,000	4,500

5-33. A firm is considering the development of several new products. The products under consideration are listed here; the products in each project group are mutually exclusive.

Project Group	Products	Development Cost	Annual Net Cash Inflow
A	A1	$500,000	$90,000
	A2	650,000	110,000
	A3	700,000	115,000
B	B1	600,000	105,000
	B2	675,000	112,000
C	C1	800,000	150,000
	C2	1,000,000	175,000

At most one product from each group will be selected. The firm has a MARR of 10% per year and a capital investment budget limitation on development costs of $2,100,000. The life of all products is assumed to be ten years. Assume no market values at the end of ten years. (5.7)

a. List all mutually exclusive combinations (investment alternatives).

b. Use the PW method to determine which combination of products should be selected.

5-34. Three independent investment projects are being considered:

	Project		
	X	Y	Z
Capital investment*	$100	$150	$200
Annual savings*	16.28	22.02	40.26
Useful life (years)	10	15	8
IRR over the useful life	10%	12%	12%

*In thousands of dollars.

The MARR is 10% per year, so all projects appear to be acceptable. Assume a study period of 15 years. Which project(s) should be chosen if investment funds are limited to $250,000? State any assumptions. (5.7)

5-35. Engineering projects A, B_1, B_2, and C are being considered with cash flows estimated over 10 years as shown in the accompanying table. Projects B_1 and B_2 are mutually exclusive, Project C depends upon B_2, and Project A depends upon B_1. The capital investment budget limit is $100,000, and the MARR is 12% per year. (5.7)

a. List all possible alternatives.

b. Develop the net cash flows for all feasible alternatives.

c. Which investment alternative (combination of projects) should be selected? Use the PW method.

	A	B_1	B_2	C
Capital investment	$30,000	$22,000	$70,000	$82,000
Annual revenues less expenses	8,000	6,000	14,000	18,000
Market value	3,000	2,000	5,000	7,000

5-36. There is a continuing requirement for standby electrical power at a public utility service facility. Equipment alternative S1 involves an initial cost of $72,000, a 9-year useful life, annual expenses of $2,200 the first year and increasing $300 per

year thereafter, and a net market value of $8,400 at the end of the useful life. Alternative S2 has an initial cost of $90,000, a 12-year-useful life, annual expenses of $2,100 the first year and increasing at the rate of 5% per year thereafter, and a net market value of $13,000 at the end of the useful life. The current interest rate is 10% per year. Which alternative is preferred using the capitalized worth method of analysis? (5.6)

5-37. A single-stage centrifugal blower is to be selected for an engineering design application. Suppliers have been consulted, and the choice has been narrowed down to two new models, both made by the same company and both having the same rated capacity and pressure. Both are driven at 3,600 rpm by identical 90-hp electric motors (output).

One blower has a guaranteed efficiency of 72% at full load and is offered installed for $42,000. The other is more expensive because of aerodynamic refinement, which gives it a guaranteed efficiency of 81% at full load.

Except for these differences in efficiency and installed price, the units are equally desirable in other operating characteristics such as durability, maintenance, ease of operation, and quietness. In both cases, plots of efficiency versus amount of air handled are flat in the vicinity of full rated load. The application is such that whenever the blower is running, it will be at full load.

Assume that both blowers have negligible market values at the end of the useful life, and the firm's MARR is 20% per year. Develop a formula for calculating how much the user could afford to pay for the more efficient unit. (*Hint:* You need to specify important parameters and use them in your formula, and remember 1 hp = 0.746 kW.) (5.4)

5-38. A study has been made of the most economical height of skyscrapers. This study grew out of experience related to the Empire State Building, whose height was uneconomical at the time it was constructed. Data are summarized in the graph of Figure P5-38 for a theoretical office building of different heights and corresponding investments. The heights for the building were considered to be 8, 15, 22, 30, 37, 50, 63, and 75 stories. If the owners of this building expect at least a 15% per year return on their capital investment, how many stories should be constructed?

Figure P5-38
Bar Chart for Problem
P5-38

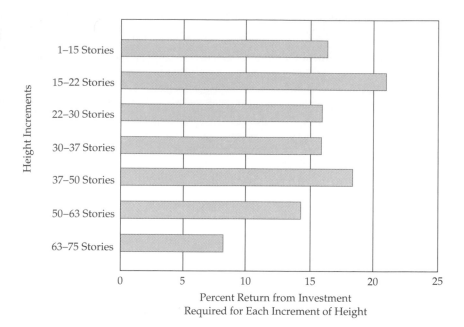

5-39. The annual performance report for Ned and Larry's Ice Cream Company praised the firm for its progressive policies but noted that environmental issues like packaging disposal were a concern. In an effort to reduce the effects of consumer disposal of product packaging, the report stated that Ned and Larry's should consider the following proposals:

a. Package all ice cream and frozen yogurt in quarts;

b. Package all ice cream and frozen yogurt in half-gallons.

By packaging the product in containers larger than the current pints, the plastic-coated bleached sulfate board containers will hold more ounces of product per square inch of surface area. The net result is less discarded packaging per ounce of product consumed. Changing to a larger container requires redesign of the packaging and modification to the filling production line. The existing material handling equipment can handle the pints and quarts, but additional equipment will be required to handle half-gallons. Any new equipment purchased for proposals (a) and (b) has an expected useful life of six years. The total capital investment for each proposal is shown in Table P5-39. Additional advantages

TABLE P5-39 Table for Problem P5-39			
	(Current) Pints	(A) Quarts	(B) Half-Gallons
Capital investment	$0	$1,200,000	$1,900,000
Packaging cost per gallon	$0.256	$0.225	$0.210
Handling labor cost per gallon	$0.128	$0.120	$0.119
Postconsumer landfill contribution from discarded packaging (yd^3/yr)	6,500	5,200	4,050

to using larger containers include lower packaging costs per ounce and less handling labor per ounce. The table summarizes the details of these proposals as well as the current production of pints.

Because Ned and Larry's promotes partnering with suppliers, customers, and the community, they wish to include a portion of the cost to society when evaluating these alternatives. They will consider 50% of the postconsumer landfill cost as part of the costs for each alternative. They have estimated landfill costs to average $20 per cubic yard nationwide.

Assume a MARR of 15% per year, a study period of six years, and that production will remain constant at 10,625,000 gallons per year. Use the IRR method to determine whether Ned and Larry's should package ice cream and frozen yogurt in pints, quarts, or half-gallons.

5-40. Refer to column 3 in Table 5-8. Show that multiple IRRs do *not* exist for this incremental net cash flow.

5-41. A foundation was endowed with $10,000,000 in July 2000. In July 2004, $3,000,000 was expended for facilities, and it was decided to provide $250,000 at the end of each year forever to cover operating expenses. The first operating expense is in July 2005, and the first replacement expense is in July 2009. If all money earns interest at 5% after the time of endowment, what amount would be available for capital replacements at the end of every fifth year forever? (*Hint:* Draw a cash-flow diagram first.)

5-42. Write a computer program that will calculate the AWs of three mutually exclusive electric motor alternatives and select the best alternative based on the assumptions of *repeatability* and *cotermination.* For cotermination, the user must enter estimated market values for the year corresponding to the *shortest life* of the three alternatives. The program will also calculate the incremental internal rate of return to the nearest 0.1% between any two alternatives under the *cotermination* assumption. The user will select the initial two alternatives for calculating the IRR of the incremental cash flow.

Problem Specifics:

a. Name the program MOTORS and write it in FORTRAN, PASCAL, or C. Provide an executable program MOTORS.EXE on a high-density 3.5-inch diskette.

b. The program will first prompt the user for the following information:
  - The MARR (10% per year will be entered as 10)
  - The output horsepower of the motors (same for all three)
  - The number of hours per day the motor will be used (not to exceed 24)
  - The number of days per year the motor will be used (not to exceed 365)
  - The cost per kilowatt-hour in dollars (remember that 1 hp = 0.746 kW)

c. For each alternative (call them 1, 2, and 3), the user will enter the following:
  - Useful life of the motor
  - Capital investment
  - Market value (if any)
  - Efficiency

d. After the user inputs this information, the computer will present a table of economic data that resembles the one shown in Problem 5-16. The computer will display the resultant equivalent uniform annual costs for each alternative, along with a brief statement indicating the best choice. Use the repeatability assumption in this part.

e. Next, the computer will determine the shortest of the three lives and prompt the user for estimated market values at that year for all but the shortest alternative. Again, the computer will display a table of economic data, the AWs, and a statement indicating the best choice. This part utilizes the cotermination assumption.

f. In addition to Parts (a)–(e), prompt the user to select two alternatives for calculating an incremental rate of return. Display the incremental internal rate of return and allow the user to calculate another incremental rate of return if desired.

5-43. Winfield County is attempting to develop a site for a convention center. Under the previous county government, a site was acquired and preliminary grading and land development were done, at a total cost of $284,000. The present government is faced with the problem of what to do. The development could be continued, in which case it is anticipated that $320,000 more will be spent during the course of one year (treat this as if it were paid at the end of the year). At the end of that year, the site could be rented, and the county believes that it would produce an income stream of $24,000 per year (payable at the beginning of each year) for the foreseeable future. The county thinks that this estimate is very close to the actual amounts that would be produced. Alternatively, the land could be sold as is, now. Because of the extensive grading work, the sale price would be only $20,000. The county's MARR for decision-making is 7% per year.

One of the county commissioners has argued for abandonment, saying, "The total cost of the development will be $584,000, and at 7% this means we should be bringing in nearly $41,000 per year to justify the investment. We're not even close. Let's cut our losses and at least recover $20,000 out of this mess."

The county has asked you to advise them. Explain what you recommend and justify your answer.

6

Depreciation and Income Taxes

The objectives of this chapter are to illustrate some of the concepts and mechanics of depreciation and depletion and to describe their role in after-tax analysis. Additionally, we illustrate the differences between before-tax and after-tax analysis in engineering economy studies.

The following topics are discussed in this chapter:

The nature of depreciation

Classical (historical) depreciation methods

The Modified Accelerated Cost Recovery System (MACRS)

Depletion

Distinctions between different types of taxes

The after-tax MARR

Corporate taxable income

The effective (marginal) income tax rate

Gain or loss on the disposal of an asset

General procedure for making after-tax analyses

Illustrations of after-tax analyses

The economic value-added criterion

The after-tax effect of depletion allowances

6.1 Introduction

Taxes have been collected since the dawn of civilization. Interestingly, in the United States, a federal income tax did not exist until March 13, 1913, when Congress enacted the Sixteenth Amendment to the Constitution.* Most organizations consider the effect of income taxes on the financial results of a proposed engineering project because income taxes usually represent a significant cash outflow that cannot be ignored in decision making. Based on the previous chapters, we can now describe how income tax liabilities (or credits) and after-tax cash flows are determined in engineering practice. In this chapter, an *After-Tax Cash Flow (ATCF) procedure* will be used in capital investment analysis because it avoids innumerable problems associated with measuring corporate net income. This theoretically sound procedure provides a quick and relatively easy way to define a project's profitability.

Because the amount of material included in the Internal Revenue Code (and in state and municipal laws, where they exist) is very extensive, only selected parts of the subject can be discussed within the scope of this textbook. Our focus in this chapter is *federal corporate income taxes* and their general effect on the financial results of proposed engineering projects. The material presented is for educational purposes. In practice, you should seek expert counsel when analyzing a specific project.

Due to the effect of *depreciation* on the after-tax cash flows of a project, this topic is discussed first. The selected material on depreciation will then be used in the remainder of the chapter for accomplishing after-tax analyses of engineering projects.

6.2 Depreciation Concepts and Terminology

> Depreciation is the decrease in value of physical properties with the passage of time and use. More specifically, depreciation is an *accounting concept* that establishes an annual deduction against before-tax income such that the effect of time and use on an asset's value can be reflected in a firm's financial statements. Annual depreciation deductions are intended to "match" the yearly fraction of value used by an asset in the production of income over the asset's actual economic life. The actual amount of depreciation can never be established until the asset is retired from service. Because depreciation is a *noncash cost* that affects income taxes, we must consider it properly when making after-tax engineering economy studies.

Depreciable property is property for which depreciation is allowed under federal, state, or municipal income tax laws and regulations. To determine if depreciation deductions can be taken, the classification of various types of property must

---

*During the Civil War, a federal income tax rate of 3% was initially imposed in 1862 by the Internal Revenue Service to help pay for war expenditures. The federal rate was later raised to 10%, but eventually eliminated in 1872.

be understood. In general, property is depreciable if it meets the following basic requirements:

1. It must be used in business or held to produce income.
2. It must have a determinable useful life (defined in Section 6.2.2), and the life must be longer than one year.
3. It must be something that wears out, decays, gets used up, becomes obsolete, or loses value from natural causes.
4. It is not inventory, stock in trade, or investment property.

Depreciable property is classified as either *tangible* or *intangible*. Tangible property can be seen or touched, and it includes two main types called *personal property* and *real property*. Personal property includes assets such as machinery, vehicles, equipment, furniture, and similar items. In contrast, real property is land and generally anything that is erected on, growing on, or attached to land. Land itself, however, is not depreciable, because it does not have a determinable life.

Intangible property is personal property such as a copyright, patent, or franchise. We will not discuss the depreciation of intangible assets in this chapter because engineering projects rarely include this class of property.

A company can begin to depreciate property it owns when the property is *placed in service* for use in the business and for the production of income. Property is considered to be placed in service when it is ready and available for a specific use, even if it is not actually used yet. Depreciation stops when the cost of placing an asset in service has been recovered.

6.2.1 Depreciation Methods and Related Time Periods

The depreciation methods permitted under the Internal Revenue Code have changed with time. In general, the following summary indicates the *primary methods used for property placed in service during three distinct time periods:*

Before 1981 Several methods could be elected for depreciating property placed in service before 1981. The primary methods used were Straight Line (SL), Declining Balance (DB), and Sum of the Years Digits (SYD). We will refer to these methods, collectively, as the *classical, or historical, methods* of depreciation.

After 1980 and before 1987 For federal income taxes, tangible property placed in service during this period must be depreciated using the Accelerated Cost Recovery System (ACRS). This system was implemented by the Economic Recovery Tax Act of 1981 (ERTA).

After 1986 The Tax Reform Act of 1986 (TRA 86) was one of the most extensive income tax reforms in the history of the United States. This act modified the previous ACRS implemented under ERTA and requires the use of the Modified Accelerated Cost–Recovery System (MACRS) for the depreciation of tangible property placed in service after 1986. A description of the classical (historical) methods of depreciation is included in the chapter for several important reasons. They apply directly to property placed in service prior to 1981, as well as for property, such as intangible property (which requires the SL method), that does not qualify for ACRS or MACRS. Also, these methods are often specified by the tax laws and regulations of state and municipal governments in the United States and are used for depreciation purposes in other countries. In addition, as we shall see in Section 6.4, the DB and the SL methods are used in determining the annual recovery rates under MACRS.

We do not discuss the application of ACRS in the chapter, but readily available Internal Revenue Service (IRS) publications describe its use.* Selected parts of the MACRS, however, are described and illustrated because this system applies to depreciable property in present and future engineering projects.

6.2.2   Additional Definitions

Because this chapter uses many terms that are not generally included in the vocabulary of engineering education and practice, an abbreviated set of definitions is presented here. This list is intended to supplement the previous definitions provided in this section:

Adjusted (cost) basis   The original cost basis of the asset, adjusted by allowable increases or decreases, is used to compute depreciation and depletion deductions. For example, the cost of any improvement to a capital asset with a useful life *greater* than one year increases the original cost basis, and a casualty or theft loss decreases it. If the basis is altered, the depreciation deduction may need to be adjusted.

Basis, or cost basis   The initial cost of acquiring an asset (purchase price plus any sales taxes), including transportation expenses and other normal costs of making the asset serviceable for its intended use. This amount is also called the *unadjusted cost basis.*

Book Value (BV)   The worth of a depreciable property as shown on the accounting records of a company. It is the original cost basis of the property, including any adjustments, less all allowable depreciation or depletion deductions. It thus represents the amount of capital that remains invested in the property and must be recovered in the future through the accounting process. The BV of a property may not be a useful measure of its market value. In general, the BV of a property at the end of year k is

$$(\text{Book value})_k = \text{adjusted cost basis} - \sum_{j=1}^{k} (\text{depreciation deduction})_j. \qquad (6\text{-}1)$$

---

*Useful references on material in this chapter, available from the Internal Revenue Service in an annually updated version, are Publication 534 (*Depreciation*), Publication 334 (*Tax Guide for Small Business*), Publication 542 (*Tax Information on Corporations*), and Publication 544 (*Sales and Other Dispositions of Assets*).

Market Value (MV)   The amount that will be paid by a willing buyer to a willing seller for a property where each has equal advantage and is under no compulsion to buy or sell. The MV approximates the present value of what will be received through ownership of the property, including the time value of money (or profit).

Recovery period   The number of years over which the basis of a property is recovered through the accounting process. For the classical methods of depreciation, this period is normally the useful life. Under MACRS, this period is the *property class* for the General Depreciation System (GDS), and it is the *class life* for the Alternative Depreciation System (ADS). (See Section 6.4.)

Recovery rate   A percentage (expressed in decimal form) for each year of the MACRS recovery period that is utilized to compute an annual depreciation deduction.

Salvage Value (SV)   The estimated value of a property at the end of its useful life.[*] It is the expected selling price of a property when the asset can no longer be used productively by its owner. The term *net salvage value* is used when the owner will incur expenses in disposing of the property, and these cash outflows must be deducted from the cash inflows to obtain a final net SV. When the classical methods of depreciation are applied, an estimated salvage value is initially established and used in the depreciation calculations. Under MACRS, the SV of depreciable property is defined to be zero.

Useful life   The expected (estimated) period that a property will be used in a trade or business to produce income. It is not how long the property will last, but how long the owner expects to productively use it.

6.3 The Classical (Historical) Depreciation Methods

This section describes and illustrates the SL, DB, and SYD methods of calculating depreciation deductions. As mentioned in Section 6.2, these historical methods continue to apply, directly and indirectly, to the depreciation of property. Also included is a discussion of the units of production method.

6.3.1 Straight-Line (SL) Method

SL depreciation is the simplest depreciation method. It assumes that a constant amount is depreciated each year over the depreciable (useful) life of the asset. If we define

---

[*] We often use the term Market Value (MV) in place of Salvage Value (SV).

N = depreciable life of the asset in years,
B = cost basis, including allowable adjustments,
d_k = annual depreciation deduction in year k $(1 \leq k \leq N)$,
BV_k = book value at end of year k,
SV_N = estimated salvage value at end of year N, and
d_k^* = cumulative depreciation through year k,

then

$$d_k = (B - SV_N)/N, \qquad (6\text{-}2)$$

$$d_k^* = kd_k \text{ for } 1 \leq k \leq N, \qquad (6\text{-}3)$$

$$BV_k = B - d_k^*. \qquad (6\text{-}4)$$

Note that, for this method, you must have an estimate of the final SV, which will also be the final book value at the end of year N. In some cases, the estimated SV_N may not equal an asset's actual terminal MV.

EXAMPLE 6-1

A new electric saw for cutting small pieces of lumber in a furniture manufacturing plant has a cost basis of $4,000 and a 10-year depreciable life. The estimated SV of the saw is zero at the end of 10 years. Determine the annual depreciation amounts using the straight-line method. Tabulate the annual depreciation amounts and the book value of the saw at the end of each year.

SOLUTION

The depreciation amount, cumulative depreciation, and book value for each year are obtained by applying Equations (6-2), (6-3), and (6-4). Sample calculations for year five are as follows:

$$d_5 = \frac{\$4,000 - 0}{10} = \$400,$$

$$d_5^* = \frac{5(\$4,000 - 0)}{10} = \$2,000,$$

$$BV_5 = \$4,000 - \frac{5(\$4,000 - 0)}{10} = \$2,000.$$

The depreciation and book value amounts for each year are shown in the following table:

EOY, k	d_k	BV_k
0	—	$4,000
1	$400	3,600
2	400	3,200
3	400	2,800
4	400	2,400
5	400	2,000
6	400	1,600
7	400	1,200
8	400	800
9	400	400
10	400	0

6.3.2   Declining-Balance (DB) Method

In the DB method, sometimes called the *constant-percentage method* or the *Matheson formula*, it is assumed that the annual cost of depreciation is a fixed percentage of the BV at the *beginning* of the year. The ratio of the depreciation in any one year to the BV at the beginning of the year is constant throughout the life of the asset and is designated by R $(0 \leq R \leq 1)$. In this method, $R = 2/N$ when a 200% declining balance is being used (i.e., twice the SL rate of $1/N$), and N equals the depreciable (useful) life of an asset. If the 150% DB method is specified, then $R = 1.5/N$. The following relationships hold true for the DB method:

$$d_1 = B(R), \qquad (6\text{-}5)$$

$$d_k = B(1 - R)^{k-1}(R), \qquad (6\text{-}6)$$

$$d_k^* = B[1 - (1 - R)^k], \qquad (6\text{-}7)$$

$$BV_k = B(1 - R)^k, \qquad (6\text{-}8)$$

$$BV_N = B(1 - R)^N. \qquad (6\text{-}9)$$

Notice that Equations (6-5) through (6-9) do not contain a term for SV_N.

EXAMPLE 6-2

Rework Example 6-1 with the DB method when (a) $R = 2/N$ (200% DB method) and (b) $R = 1.5/N$ (150% DB method). Again, tabulate the annual depreciation amount and book value for each year.

SOLUTION

Annual depreciation, cumulative depreciation, and book value are determined by using Equations (6-6), (6-7), and (6-8), respectively. Sample calculations for year six are as follows:

(a)

$$R = 2/10 = 0.2,$$

$$d_6 = \$4,000(1 - 0.2)^5(0.2) = \$262.14,$$

$$d_6^* = \$4,000[1 - (1 - 0.2)^6] = \$2,951.42,$$

$$BV_6 = \$4,000(1 - 0.2)^6 = \$1,048.58.$$

(b)

$$R = 1.5/10 = 0.15,$$

$$d_6 = \$4,000(1 - 0.15)^5(0.15) = \$266.22,$$

$$d_6^* = \$4,000[1 - (1 - 0.15)^6] = \$2,491.40,$$

$$BV_6 = \$4,000(1 - 0.15)^6 = \$1,508.60.$$

The depreciation and BV amounts for each year, when $R = 2/N = 0.2$, are shown in the following table:

	200% DB Method Only	
EOY, k	d_k	BV_k
0	—	$4,000
1	$800	3,200
2	640	2,560
3	512	2,048
4	409.60	1,638.40
5	327.68	1,310.72
6	262.14	1,048.58
7	209.72	838.86
8	167.77	671.09
9	134.22	536.87
10	107.37	429.50

6.3.3 Sum-of-the-Years-Digits (SYD) Method

To compute the depreciation deduction by the SYD method, the digits corresponding to the number for each permissible year of depreciable life are first listed in reverse order. The sum of these digits is then determined. The depreciation factor for any year is the number from the reverse-ordered listing for that year divided by the sum of the digits. For example, for a property having a depreciable (useful) life of five years, SYD depreciation factors are as follows:

Year	Number of the Year in Reverse Order (digits)	SYD Depreciation Factor
1	5	5/15
2	4	4/15
3	3	3/15
4	2	2/15
5	1	1/15
Sum of the digits	$\overline{15}$	

The depreciation for any year is the product of the SYD depreciation factor for that year and the difference between the cost basis (B) and the estimated final SV. The general expression for the annual cost of depreciation for any year k, when N equals the depreciable life of an asset, is

$$d_k = (B - SV_N) \cdot \left[\frac{2(N - k + 1)}{N(N + 1)} \right]. \tag{6-10}$$

The BV at the end of the year k is

$$BV_k = B - \left[\frac{2(B - SV_N)}{N} \right] k + \left[\frac{(B - SV_N)}{N(N + 1)} \right] k(k + 1), \tag{6-11}$$

and the cumulative depreciation through the kth year is simply

$$d_k^* - B - BV_k. \tag{6-12}$$

EXAMPLE 6-3

Rework Example 6-1 using the SYD method. Tabulate the annual depreciation amount and book value for each year.

SOLUTION
With Equations (6-10), (6-11), and (6-12), respectively, the annual depreciation, BV, and cumulative depreciation amounts are obtained. Sample calculations for year four are

$$d_4 = \$4,000 \left[\frac{2(10 - 4 + 1)}{10(11)} \right] = \$509.09,$$

$$BV_4 = \$4,000 - \left[\frac{2(\$4,000)}{10} \right] \cdot 4 + \left[\frac{\$4,000}{10(11)} \right] \cdot 4 \cdot 5 = \$1,527.27,$$

$$d_4^* = \$4,000 - \$1,527.27 = \$2,472.73.$$

Depreciation and BV amounts for each year are shown in the following table:

EOY, k	d_k	BV_k
0	—	$4,000
1	$727.27	3,272.73
2	654.55	2,618.18
3	581.82	2,036.36
4	509.09	1,527.27
5	436.36	1,090.91
6	363.64	727.27
7	290.91	436.36
8	218.18	218.18
9	145.45	72.73
10	72.73	0

6.3.4 DB with Switchover to SL

Because the DB method never reaches a BV of zero, it is permissible to switch from this method to the SL method so that an asset's BV_N will be zero (or some other determined amount, such as SV_N). Also, this method is used in calculating the MACRS recovery rates in Table 6-3.

Table 6-1 illustrates a switchover from double DB depreciation to SL depreciation for Example 6-1. The switchover occurs in the year in which a larger depreciation amount is obtained from the SL method. From Table 6-1, it is apparent

TABLE 6-1 The 200% DB Method with Switchover to the SL Method (Example 6-1)

		Depreciation Method		
	(1)	(2)	(3)	(4)
	Beginning-	(2) 200% DB	SL	Depreciation
Year, k	of-Year BV[a]	Method[b]	Method[c]	Amount Selected[d]
1	$4,000.00	$800.00	> $400.00	$800.00
2	3,200.00	640.00	> 355.56	640.00
3	2,560.00	512.00	> 320.00	512.00
4	2,048.00	409.60	> 292.57	409.60
5	1,638.40	327.68	> 273.07	327.68
6	1,310.72	262.14	= 262.14	262.14 (switch)
7	1,048.58	209.72	< 262.14	262.14
8	786.44	167.77	< 262.14	262.14
9	524.30	134.22	< 262.14	262.14
10	262.16	107.37	< 262.14	262.14
		$3,570.50		$4,000.00

[a] Column 1 for year k less column 4 for year k equals the entry in column 1 for year $k+1$.
[b] 200% (= 2/10) of column 1.
[c] Column 1 minus estimated SV_N divided by the remaining years from the beginning of the year through the 10th year.
[d] Select the larger amount in column 2 or column 3.

that $d_6 = \$262.14$. The BV at the end of year six (BV$_6$) is \$1,048.58. Additionally, observe that BV$_{10}$ is \$4,000 − \$3,570.50 = \$429.50 without switchover to the SL method in Table 6-1. With switchover, BV$_{10}$ equals 0. It is clear that this asset's d_k, d_k^*, and BV$_k$ in years 7 through 10 are established from the SL method, which permits the full cost basis to be depreciated over the 10-year recovery period.

6.3.5 Units-of-Production Method

All the depreciation methods discussed to this point are based on elapsed time (years) on the theory that the decrease in value of property is mainly a function of time. When the decrease in value is mostly a function of use, depreciation may be based on a method not expressed in terms of years. The units-of-production method is normally used in this case.

This method results in the cost basis (minus final SV) being allocated equally over the estimated number of units produced during the useful life of the asset. The depreciation rate is calculated as

$$\text{Depreciation per unit of production} = \frac{B - SV_N}{(\text{Estimated lifetime production in units})}.$$
(6-13)

EXAMPLE 6-4

A piece of equipment used in a business has a basis of \$50,000 and is expected to have a \$10,000 salvage value when replaced after 30,000 hours of use. Find its depreciation rate per hour of use, and find its BV after 10,000 hours of operation.

SOLUTION

$$\text{Depreciation per unit of production} = \frac{\$50,000 - \$10,000}{30,000 \text{ hours}} = \$1.33 \text{ per hour.}$$

$$\text{After 10,000 hours, BV} = \$50,000 - \frac{\$1.33}{\text{hour}}(10,000 \text{ hours}), \text{ or BV} = \$36,700.$$

▼ ## 6.4 The Modified Accelerated Cost Recovery System

As we discussed in Section 6.2.1, the MACRS was created by TRA 86 and is now the principal method for computing depreciation deductions for property in engineering projects. MACRS applies to most tangible depreciable property placed in service after December 31, 1986. Examples of assets that cannot be depreciated under MACRS are property you elect to exclude because it is to be depreciated under a method that is not based on a term of years (units-of-production method) and intangible property. Previous depreciation methods have required estimates of useful life (N) and SV at the end of useful life (SV$_N$). Under MACRS, however, SV$_N$ is defined to be 0, and useful life estimates are not used directly in calculating depreciation amounts.

> MACRS consists of two systems for computing depreciation deductions. The main system is called the *General Depreciation System (GDS)* and the second system is called the *Alternative Depreciation System (ADS)*. In general, ADS provides a longer recovery period and uses only the straight-line method of depreciation. Property that is placed in any tax-exempt use and property used predominantly outside the United States are examples of assets that must be depreciated under ADS. Any property that qualifies under GDS, however, can be depreciated under ADS, if elected.

When an asset is depreciated under MACRS, the following information is needed before depreciation deductions can be calculated:

1. The cost basis (B)
2. The date the property was placed in service
3. The property class and recovery period
4. The MACRS depreciation method to be used (GDS or ADS)
5. The time convention that applies (half year)

The first two items were discussed in Section 6.2. Items 3 through 5 are discussed in the following sections.

6.4.1 Property Class and Recovery Period

Under MACRS, tangible depreciable property is categorized (organized) into asset classes. The property in each asset class is then assigned a *class life, GDS recovery period (and property class), and ADS recovery period.* For our use, a partial listing of depreciable assets used in business is provided in Table 6-2. The types of depreciable property grouped together are identified in the second column. Then the class life, GDS recovery period (and property class), and ADS recovery period (all in years) for these assets are listed in the remaining three columns.

Under the General Depreciation System (GDS), the basic information about property classes and recovery periods is as follows:

1. Most tangible personal property is assigned to one of six *personal property classes* (3-, 5-, 7-, 10-, 15-, and 20-year property). The personal property class (in years) is the same as the *GDS recovery period.* Any depreciable personal property that does not "fit" into one of the defined asset classes is depreciated as being in the seven-year property class.
2. Real property is assigned to two *real property classes*: Nonresidential real property and residential rental property.
3. The *GDS recovery period* is 39 years for nonresidential real property (31.5 years if placed in service before May 13, 1993) and 27.5 years for residential real property.

The following is a summary of basic information for the ADS:

1. For tangible personal property, the *ADS recovery period* is shown in the last column on the right of Table 6-2 (and is normally the same as the class life of the property; there are exceptions such as those shown in asset classes 00.12 and 00.22).

TABLE 6-2 MACRS Class Lives and Recovery Periods[a]

Asset Class	Description of Assets or Depreciable Assets Used in Business	Class Life	Recovery Period GDS[b]	Recovery Period ADS
00.11	Office furniture and equipment	10	7	10
00.12	Information systems, including computers	6	5	5
00.22	Automobiles, taxis	3	5	5
00.23	Buses	9	5	9
00.241	Light general purpose trucks	4	5	5
00.242	Heavy general purpose trucks	6	5	6
00.26	Tractor units for use over the road	4	3	4
10.0	Mining	10	7	10
13.2	Production of petroleum and natural gas	14	7	14
13.3	Petroleum refining	16	10	16
15.0	Construction	6	5	6
22.3	Manufacture of carpets	9	5	9
24.4	Manufacture of wood products	10	7	10
28.0	Manufacture of chemicals and allied products	9.5	5	9.5
30.1	Manufacture of rubber products	14	7	14
32.2	Manufacture of cement	20	15	20
34.0	Manufacture of fabricated metal products	12	7	12
36.0	Manufacture of electronic components, products, and systems	6	5	6
37.11	Manufacture of motor vehicles	12	7	12
37.2	Manufacture of aerospace products	10	7	10
48.12	Telephone central office equipment	18	10	18
49.13	Electric utility steam production plant	28	20	28
49.21	Gas utility distribution facilities	35	20	35

[a] Partial listing abstracted from *How to Depreciate Property*, IRS Publication 946, Tables B-1 and B-2, 1998.
[b] Also the *GDS property class*.

2. Any tangible personal property that does not fit into one of the asset classes is depreciated using a 12-year ADS recovery period.
3. The *ADS recovery period* for nonresidential real property is 40 years.

The use of these rules under the MACRS is discussed further in the next section.

6.4.2 Depreciation Methods, Time Convention, and Recovery Rates

The primary methods used under MACRS for calculating the depreciation deductions over the recovery period of an asset are summarized as follows:

1. GDS 3-, 5-, 7-, and 10-year personal property classes: The 200% DB method, which switches to the SL method when that method provides a greater deduction. The DB method with switchover to SL was illustrated in Section 6.3.4.
2. GDS 15- and 20-year personal property classes: The 150% DB method, which switches to the SL method when that method provides a greater deduction.

3. GDS nonresidential real and residential rental property classes: The SL method over the fixed GDS recovery periods.
4. ADS: The SL method for both personal and real property over the fixed ADS recovery periods.

A *half-year time convention* is used in MACRS depreciation calculations for tangible personal property. This means that all assets placed in service during the year are treated as if use began in the middle of the year, and one-half year of depreciation is allowed. When an asset is disposed of, the half-year convention is also allowed. *If the asset is disposed of before the full recovery period is used, then only half of the normal depreciation deduction can be taken for that year.*

The GDS *recovery rates* (r_k) for the six personal property classes that we will use in our depreciation calculations are listed in Table 6-3. The GDS personal property rates in Table 6-3 include the half-year convention as well as switchover from the DB method to the SL method when that method provides a greater deduction.

TABLE 6-3   GDS Recovery Rates (r_k) for the Six Personal Property Classes

Year	Recovery Period (and Property Class)					
	3-year[a]	5-year[a]	7-year[a]	10-year[a]	15-year[b]	20-year[b]
1	0.3333	0.2000	0.1429	0.1000	0.0500	0.0375
2	0.4445	0.3200	0.2449	0.1800	0.0950	0.0722
3	0.1481	0.1920	0.1749	0.1440	0.0855	0.0668
4	0.0741	0.1152	0.1249	0.1152	0.0770	0.0618
5		0.1152	0.0893	0.0922	0.0693	0.0571
6		0.0576	0.0892	0.0737	0.0623	0.0528
7			0.0893	0.0655	0.0590	0.0489
8			0.0446	0.0655	0.0590	0.0452
9				0.0656	0.0591	0.0447
10				0.0655	0.0590	0.0447
11				0.0328	0.0591	0.0446
12					0.0590	0.0446
13					0.0591	0.0446
14					0.0590	0.0446
15					0.0591	0.0446
16					0.0295	0.0446
17						0.0446
18						0.0446
19						0.0446
20						0.0446
21						0.0223

Source: *Depreciation.* IRS Publication 534. Washington, D.C.: U.S. Government Printing Office, for 1998 tax returns.
[a] These rates are determined by applying the 200% DB method (with switchover to the SL method) to the recovery period with the half-year convention applied to the first and last years. Rates for each period must sum to 1.0000.
[b] These rates are determined with the 150% DB method instead of the 200% DB method (with switchover to the SL method) and are rounded off to four decimal places.

Note that if an asset is disposed of in year $N + 1$ the final BV of the asset will be zero. Furthermore, there are $N + 1$ recovery rates shown for each GDS property class for a recovery period of N years.

The information in Table 6-4 provides a summary of the principal features of the main General Depreciation System under MACRS. Included are some selected special rules about depreciable assets. A flow diagram for computing depreciation deductions under MACRS is shown in Figure 6-1. As indicated in the figure, an important choice is whether the main GDS is to be used for an asset or whether ADS is elected instead (or required). Normally, the choice would be to use the GDS for calculating the depreciation deductions.

TABLE 6-4 MACRS (GDS) Property Classes and Primary Methods for Calculating Depreciation Deductions

GDS Property Class and Depreciation Method	Class Life	Special Rules
3-year, 200% DB with Switchover to SL	Four years or less	Includes some race horses. Excludes cars and light trucks.
5-year, 200% DB with Switchover to SL	More than four years to less than 10	Includes cars and light trucks, semiconductor manufacturing equipment, qualified technological equipment, computer-based central office switching equipment, some renewable and biomass power facilities, and research and development property.
7-year, 200% DB with Switchover to SL	10 years to less than 16	Includes single-purpose agricultural and horticultural structures and railroad track. Includes property not assigned to a property class.
10-year, 200% DB with Switchover to SL	16 years to less than 20	None.
15-year, 150% DB with Switchover to SL	20 years to less than 25	Includes sewage treatment plants, telephone distribution plants, and equipment for two-way voice and data communication.
20-year, 150% DB with Switchover to SL	25 years or more	Excludes real property of 27.5 years or more. Includes municipal sewers.
27.5 year, SL	N/A	Residential rental property.
39-year, SL	N/A	Nonresidential real property.

Source: Arthur Andersen & Company, *Tax Reform 1986: Analysis and Planning,* Chicago, 1986, p. 112. Reproduced with permission of Arthur Andersen & Co.

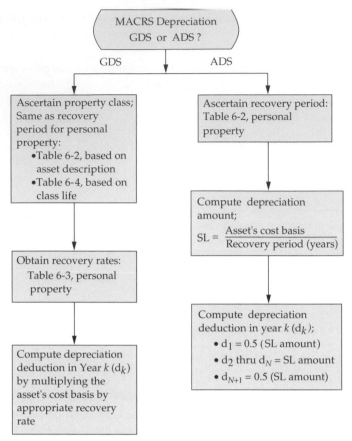

Figure 6-1 Flow Diagram for Computing Depreciation Deductions under MACRS

EXAMPLE 6-5

A firm purchased and placed in service a new piece of semiconductor manufacturing equipment. The cost basis for the equipment is $100,000. Determine (a) the depreciation charge permissible in the fourth year, (b) the BV at the end of the fourth year, (c) the cumulative depreciation through the third year, and (d) the BV at the end of the fifth year if the equipment is disposed of at that time.

SOLUTION

From Table 6-2, it may be seen that the semiconductor (electronic) manufacturing equipment has a class life of six years and a GDS recovery period of five years. The recovery rates that apply are given in Table 6-3.

(a) The depreciation deduction, or cost-recovery allowance, that is allowable in year four (d_4) is 0.1152 ($100,000) = $11,520.

(b) The BV at the end of year four (BV_4) is the cost basis less depreciation charges in years one through four:

$$BV_4 = \$100,000 - \$100,000(0.20 + 0.32 + 0.192 + 0.1152)$$

$$= \$17,280.$$

(c) Accumulated depreciation through year three, d_3^*, is the sum of depreciation amounts in years one through three:

$$d_3^* = d_1 + d_2 + d_3$$

$$= \$100,000(0.20 + 0.32 + 0.192)$$

$$= \$71,200.$$

(d) The depreciation deduction in year five can only be $(0.5)(0.1152) \cdot (\$100,000) = \$5,760$ when the equipment is disposed of prior to year six. Thus, the BV at the end of year five is $BV_4 - \$5,760 = \$11,520$.

From Example 6-5, we can conclude that Equation (6-14) is true from the buyer's viewpoint when property of the same type and class is exchanged:

$$\text{Basis} = \text{Actual Cash Cost} + \text{Book Value of the Trade-in} \qquad (6\text{-}14)$$

To illustrate Equation (6-14), suppose your company has operated an optical character recognition (OCR) scanner for two years. Its BV now is $35,000, and its fair market value is $45,000. The company is considering a new OCR scanner that costs $105,000. Ordinarily, you would trade the old scanner for the new one and pay the dealer $60,000. The basis ($B$) for depreciation is then $60,000+$35,000 = $95,000.*

EXAMPLE 6-6

In May 1999, your company traded in a computer and peripheral equipment, used in its business, that had a BV at that time of $25,000. A new, faster computer system having a fair MV of $400,000 was acquired. Because the vendor accepted the older computer as a trade-in, a deal was agreed to whereby your company would pay $325,000 cash for the new computer system.

(a) What is the GDS property class of the new computer system?

(b) How much depreciation can be deducted each year based on this class life? (Refer to Figure 6-1.)

---

*The *exchange price* of the OCR scanner is $105,000 − $45,000 (= actual cash cost). Equation (6-15) prevents exaggerated cost bases to be claimed for new assets having large "sticker prices" compared with their exchange price.

SOLUTION

(a) The new computer is in Asset Class 00.12 and has a class life of six years. (See Table 6-2.) Hence, its GDS property class and recovery period is five years.

(b) The cost basis for this property is $350,000, which is the sum of the $325,000 cash price of the computer and the $25,000 book value remaining on the trade-in. (In this case, the trade-in was treated as a nontaxable transaction.)

MACRS (GDS) rates that apply to the $350,000 cost basis are found in Table 6-3. An allowance (half-year) is built into the year-one rate, so it does not matter that the computer was purchased in May 1999 instead of, say, November 1999. The depreciation deductions (d_k) for 1999 through 2004 can be computed with

$$d_k = r_k \cdot B; \quad 1 \le k \le N + 1, \tag{6-15}$$

where r_k = recovery rate for year k (from Table 6-3):

Property	Data Placed in Service	Cost Basis	Class Life	MACRS (GDS) Recovery Period
Computer System	May 1999	$350,000	6 years	5 years

Year	Depreciation Deductions
1999	$0.20 \times \$350,000 = \$\ 70,000$
2000	$0.32 \ \times\ \ 350,000 = \ \ 112,000$
2001	$0.192 \times\ \ 350,000 = \ \ \ 67,200$
2002	$0.1152 \times\ \ 350,000 = \ \ \ 40,320$
2003	$0.1152 \times\ \ 350,000 = \ \ \ 40,320$
2004	$0.0576 \times\ \ 350,000 = \ \ \ 20,160$
	Total $\$350,000$

EXAMPLE 6-7

A large manufacturer of sheet metal products in the Midwest purchased and placed in service a new, modern, computer-controlled flexible manufacturing system for $3.0 million. Because this company would not be profitable until the new technology had been in place for several years, it elected to utilize the ADS under MACRS in computing its depreciation deductions. Thus, the company could slow down its depreciation allowances in hopes of postponing its income tax advantages until it became a profitable concern. What depreciation deductions can be claimed for the new system?

SOLUTION

From Table 6-2, the ADS recovery period for a manufacturer of fabricated metal products is 12 years. Under ADS, the SL method with no SV is applied to the 12-year recovery period using the half-year convention. Consequently, the depreciation in year one would be

$$\frac{1}{2}\left(\frac{\$3,000,000}{12}\right) = \$125,000.$$

Depreciation deductions in years 2 through 12 would be $250,000 each year, and depreciation in year 13 would be $125,000. Notice how the half-year convention extends depreciation deductions over 13 years $(N + 1)$.

6.5 A Comprehensive Depreciation Example

We now consider an asset for which depreciation is computed by the historical and MACRS (GDS) methods previously discussed. Be careful to observe the differences in the mechanics of each method, as well as the differences in the annual depreciation amounts themselves. Also, we compare the present worths at $k = 0$ of selected depreciation methods when MARR $= 10\%$ per year. As we shall see later in this chapter, depreciation methods that result in larger present worths (of the depreciation amounts) are preferred by a firm that wants to reduce the present worth of its *income taxes paid* to the government.

EXAMPLE 6-8

The La Salle Bus Company has decided to purchase a new bus for $85,000 with a trade-in of their old bus. The old bus has a BV of $10,000 at the time of the trade-in. The new bus will be kept for 10 years before being sold. Its estimated salvage value at that time is expected to be $5,000.

First, we must calculate the cost basis. The basis is the original purchase price of the bus plus the book value of the old bus that was traded in [Equation (6-15)]. Thus, the basis is $85,000 + $10,000, or $95,000. We need to look at Table 6-2 and find buses, which are asset class 00.23. Hence, we find that buses have a nine-year class life, over which we depreciate the bus with historical methods discussed in Section 6.3, and a five-year GDS recovery period.

SOLUTION: SL Method

For the straight-line method, we use the class life of nine years, even though the bus will be kept for 10 years. By using Equations (6-2) and (6-4), we obtain the following information:

$$d_k = \frac{\$95,000 - \$5,000}{9 \text{ years}} = \$10,000, \qquad \text{for } k = 1 \text{ to } 9.$$

	SL Method	
EOY k	d_k	BV_k
0	—	$95,000
1	$10,000	85,000
2	10,000	75,000
3	10,000	65,000
4	10,000	55,000
5	10,000	45,000
6	10,000	35,000
7	10,000	25,000
8	10,000	15,000
9	10,000	5,000

Notice that no depreciation was taken after year nine because the class life was only nine years. Also, the final BV was the estimated SV, and the BV will remain at $5,000 until the bus is sold.

SOLUTION: *DB Method*
To demonstrate this method, we will use the 200% DB equations. With Equations (6-6) and (6-8), we calculate the following:

$$R = 2/9 = 0.2222;$$

$$d_1 = \$95,000(0.2222) = \$21,111;$$

$$d_5 = \$95,000(1 - 0.2222)^{5-1}(0.2222) = \$7,726;$$

$$BV_5 = \$95,000(1 - 0.2222)^5 = \$27,040.$$

	200% DB Method	
EOY k	d_k	BV_k
0	—	$95,000
1	$21,111	73,889
2	16,420	57,469
3	12,771	44,698
4	9,932	34,765
5	7,726	27,040
6	6,009	21,031
7	4,674	16,357
8	3,635	12,722
9	2,827	9,895

SOLUTION: *SYD Method*
Once again, we will use the class life of nine years. The SYD depreciation amounts are as follows:

EOY k	Number of the Year in Reverse Order	SYD Depreciation Factor	$d_k =$ $(B - SV_N) * $ Factor	BV_k
0	—	—	—	$95,000
1	9	9/45	$18,000.00	77,000
2	8	8/45	16,000.00	61,000
3	7	7/45	14,000.00	47,000
4	6	6/45	12,000.00	35,000
5	5	5/45	10,000.00	25,000
6	4	4/45	8,000.00	17,000
7	3	3/45	6,000.00	11,000
8	2	2/45	4,000.00	7,000
9	1	1/45	2,000.00	5,000
	Sum = 45			

Using Equations (6-10) and (6-11), we compute the following:

$$d_5 = (\$95,000 - \$5,000) \left[\frac{2(9 - 5 + 1)}{9(9 + 1)} \right] = \$10,000;$$

$$BV_5 = \$95,000 - \frac{2(\$95,000 - \$5,000)}{9}(5) + \frac{(\$95,000 - \$5,000)5(5 + 1)}{9(9 + 1)} = \$25,000.$$

SOLUTION: *DB with Switchover to SL Depreciation*

To illustrate the mechanics of Table 6-1 for this example, we first specify that the bus will be depreciated by the 200% DB method ($R = 2/N$). Because DB methods never reach a zero BV, suppose that we further specify that a switchover to SL depreciation will be made to ensure a BV of $5,000 at the end of the vehicle's nine-year class life.

EOY k	Beginning-of-Year BV	200% DB Method	SL Method $(BV_9 = \$5,000)$	Depreciation Amount Selected
1	$95,000	$21,111	$10,000	$21,111
2	73,889	16,420	8,611	16,420
3	57,469	12,771	7,496	12,771
4	44,698	9,933	6,616	9,933
5	34,765	7,726	5,953	7,726
6	27,040	6,009	5,510	6,009
7	21,031	4,674	5,344	5,344[a]
8	15,687	3,635	5,344	5,344
9	10,344	2,827	5,344	5,344

[a] Switchover occurs in year seven.

SOLUTION: *MACRS (GDS) with Half-Year Convention*

To demonstrate the GDS with the half-year convention, we will change the La Salle bus problem so that the bus is now sold in year five in part (a) and in year six for part (b).

(a) Selling Bus in Year Five:

EOY k	Factor	d_k	BV_k
0	—	—	$95,000
1	0.2000	$19,000	76,000
2	0.3200	30,400	45,600
3	0.1920	18,240	27,360
4	0.1152	10,944	16,416
5	0.0576	5,472	10,944

(b) Selling Bus in Year Six:

EOY k	Factor	d_k	BV_k
0	—	—	$95,000
1	0.2000	$19,000	76,000
2	0.3200	30,400	45,600
3	0.1920	18,240	27,360
4	0.1152	10,944	16,416
5	0.1152	10,944	5,472
6	0.0576	5,472	0

Notice that when we sold the bus in year five before the recovery period had ended, we took only half of the normal depreciation. The other years (years one through four) were not changed. When the bus was sold in year six, at the end of the recovery period, we did not divide the last year amount by two.

Selected methods of depreciation, illustrated in Example 6-8, are compared in Figure 6-2. In addition, the PW (10%) of each method is shown in Figure 6-2. Because large present worths of depreciation deductions are generally viewed as desirable, it is clear that the MACRS method is very attractive to profitable companies.

6.6 Depletion

When natural resources are being consumed in producing products or services, the term *depletion* is used to indicate the decrease in value of the resource base that has occurred. The term is commonly used in connection with mining properties, oil and gas wells, timberlands, and so on. In any given parcel of mineral property, for example, there is a definite quantity of ore, oil, or gas available. As some of the

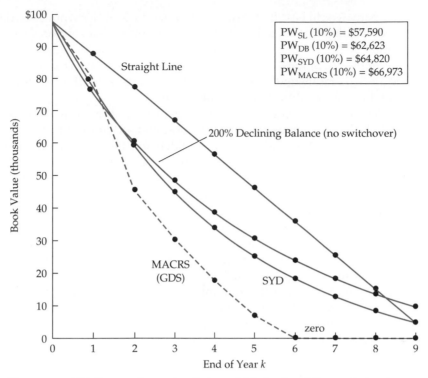

$PW_{SL} (10\%) = \$57,590$
$PW_{DB} (10\%) = \$62,623$
$PW_{SYD} (10\%) = \$64,820$
$PW_{MACRS} (10\%) = \$66,973$

Figure 6-2 BV Comparisons for Selected Methods of Depreciation in Example 6-8 (*Note*: The bus is assumed to be sold in year 6 for the MACRS-GDS method.)

resource is extracted and sold, the reserve decreases, and the value of the property normally diminishes.

However, there is a difference in the manner in which the amounts recovered through depletion and depreciation must be handled. In the case of depreciation, the property involved usually may be replaced with similar property when it has become fully depreciated. In the case of depletion of mineral or other natural resources, such replacement usually is not possible. Once the gold has been removed from a mine, or the oil from an oil well, it cannot be replaced. Thus, in a manufacturing or other business in which depreciation occurs, the principle of maintenance of capital is practiced, and the amounts charged for depreciation expense are reinvested in new equipment so that the business may continue in operation indefinitely. On the other hand, in the case of a mining or other mineral industry, the amounts charged as depletion cannot be used to replace the sold natural resource, and the company, in effect, may sell itself out of business, bit by bit, as it carries out its normal operations. Such companies frequently pay out to the owners each year the amounts recovered as depletion. Thus, the annual payment to the owners is made up of two parts: (1) the profit that has been earned and (2) a portion of the owner's capital that is being returned, marked as depletion. In such cases, if the natural resource were eventually completely consumed, the company would be out of business, and the stockholder would hold stock that was theoretically worthless, but would have received back all of his or her invested capital.

In the operation of many natural resource businesses, the depletion funds may be used to acquire new properties, such as new mines and oil-producing properties, and thus give continuity to the enterprise.

There are two ways to compute depletion allowances: (1) the cost method and (2) the percentage method. The cost method applies to all types of property subject to depletion and is the more widely used method. Under the cost method, a *depletion unit* is determined by dividing the adjusted cost basis of a property by the number of units remaining to be mined or harvested. (Units may be feet of timber, tons of ore, etc.) The deduction (depletion allowance) for a given tax year is then calculated as the product of number of units *sold* during the year and the depletion unit, in dollars.

In practice, depletion may also be based on a percentage of the year's income in accordance with IRS regulations. Depletion allowances on mines and other natural deposits, including geothermal deposits, may be computed as a percentage of gross income, provided that the amount charged for depletion does not exceed 50% of the *net income* (100% for oil and gas property) before deduction of the depletion allowance. The percentage method can be used for most types of metal mines, geothermal deposits, and coal mines, but not for timber. Generally, the use of percentage depletion for oil and gas is not allowed except for certain domestic oil and gas production. Some examples of depletion allowances* are as follows:

Sulfur and uranium; domestically mined lead, zinc, nickel, and asbestos	22%
Gold, silver, copper, iron ore, and oil shale from U.S. deposits; geothermal wells in the United States	15%
Coal, lignite, and sodium chloride	10%
Clay, gravel, sand, and stone	5%

It is possible that the total amount charged for depletion over the life of a property under this procedure may be far more than the original cost. When the percentage method applies to a property, depletion allowances must be calculated by both the cost method and the percentage method. The larger allowance may be taken and used to reduce the basis of the property for purposes of refiguring the depletion unit as necessary. Figure 6-3 provides the logic for determining whether percentage or cost depletion is allowable in a given tax year.

Example 6-9 illustrates the cost method of determining a depletion allowance.

---

*Depletion allowances are established by the IRS and may be revised with new federal income tax legislation.

Figure 6-3

Logic for Determining
Whether Percentage
or Cost Depletion Is
Allowed

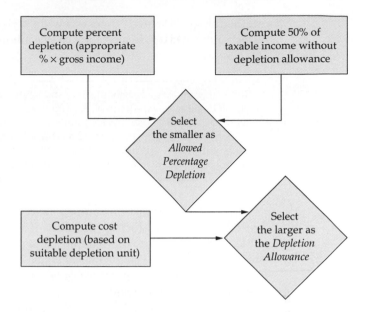

EXAMPLE 6-9

The WGS Zinc Company recently bought an ore-bearing parcel of land for $2,000,000. The recoverable reserves in the mine were estimated to be 500,000 tons.

(a) If 75,000 tons of ore were mined during the first year and 50,000 tons were sold, what was the depletion allowance for year one?

(b) Suppose at the end of year one reserves were reevaluated and found to be only 400,000 tons. If 50,000 additional tons are sold in the second year, what is the depletion allowance for year two?

SOLUTION

(a) The depletion unit is $2,000,000/500,000 tons = $4.00 per ton. A depletion allowance, based on units sold, for year one is 50,000 tons ($4.00/ton) = $200,000.

(b) The adjusted cost basis at the beginning of the second year would be $2,000,000−$200,000 = $1,800,000. The depletion unit would be $1,800,000/400,000 tons = $4.50/ton, and a depletion allowance of 50,000 tons ($4.50/ton) = $225,000 is permissible in year two.

6.7 Introduction to Income Taxes

Up to this point, there has been no consideration of income taxes in our discussion of engineering economy, except for the influence of depreciation and other types of deductions. By not complicating our studies with income-tax effects, we have placed primary emphasis on basic engineering economy principles and methodology. However, there is a wide variety of capital investment problems in which income taxes do affect the choice among alternatives, and after-tax studies are essential.

> In the remainder of this chapter, we shall be concerned with how income taxes affect a project's estimated cash flows. Income taxes resulting from the profitable operation of a firm are usually taken into account in evaluating engineering projects. The reason is quite simple: Income taxes associated with a proposed project may represent a major cash outflow that should be considered together with other cash inflows and outflows in assessing the overall economic profitability of that project.

There are other taxes discussed in Section 6.7.1 that are not directly associated with the income-producing capability of a new project, but they are usually negligible when compared with federal and state income taxes. *When other types of taxes are included in engineering economy studies, they are normally deducted from revenue, as any other operating expense would be, in determining the before-tax cash flow that we considered in Chapters 4 and 5.*

The mystery behind the sometimes complex computation of income taxes is reduced when we recognize that income taxes paid are just another type of expense, but income taxes saved (through depreciation, expenses, and direct tax credits) are identical to other kinds of reduced expenses (e.g., savings in maintenance expenses).

The basic concepts underlying federal and state income tax laws and regulations that apply to most economic analyses of capital investments generally can be understood and applied without difficulty. This discussion of income taxes is not intended to be a comprehensive treatment of the subject. Rather, we utilize some of the more important provisions of the federal Tax Reform Act of 1986 (TRA 86), followed by illustrations of a general procedure for computing the net after-tax cash flow (ATCF) for an engineering project and conducting after-tax economic analyses. Where appropriate, important changes to TRA 86 enacted by the Omnibus Budget Reconciliation Act of 1993 (OBRA 93) and the Taxpayer Relief Act of 1997 are also included in this chapter.

6.7.1 Distinctions between Different Types of Taxes

Before discussing the consequences of income taxes in engineering economy studies, we need to distinguish between income taxes and several other types of taxes:

1. *Income taxes* are assessed as a function of gross revenues minus allowable deductions. They are levied by the federal, most state, and occasionally municipal governments.
2. *Property taxes* are assessed as a function of the value of property owned, such as land, buildings, equipment, and so on, and the applicable tax rates. Hence, they are independent of the income or profit of a firm. They are levied by municipal, county, or state governments.

3. *Sales taxes* are assessed on the basis of purchases of goods or services and are thus independent of gross income or profits. They are normally levied by state, municipal, or county governments. Sales taxes are relevant in engineering economy studies only to the extent that they add to the cost of items purchased.
4. *Excise taxes* are federal taxes assessed as a function of the sale of certain goods or services often considered nonnecessities, and are hence independent of the income or profit of a business. Although they are usually charged to the manufacturer or original provider of the goods or services, a portion of the cost is passed on to the purchaser.

Income taxes are usually the most significant type of tax encountered in engineering economic analysis.

6.7.2 The Before-Tax and After-Tax Minimum Attractive Rates of Return

In the preceding chapters, we have treated income taxes as if they are not applicable, or we have taken them into account, in general, by using a before-tax MARR, which is larger than the after-tax MARR. An approximation of the before-tax MARR requirement, which includes the effect of income taxes, for studies involving only before-tax cash flows can be obtained from the following relationship:

(Before-tax MARR)[(1 − effective income tax rate)] $\cong$ after-tax MARR.

Thus,

$$\text{Before-tax MARR} \approx \frac{\text{After-tax MARR}}{(1 - \text{effective income tax rate})}. \tag{6-16}$$

Determining the effective income tax rate for a firm is discussed in Section 6.8.

This approximation is exact if the asset is nondepreciable and there are no gains or losses on disposal, tax credits, or other types of deductions involved. Otherwise, these factors affect the amount and timing of income tax payments, and some degree of error is introduced into the relationship in Equation (6-16).

6.7.3 Taxable Income of Corporations (Business Firms)

At the end of each tax year, a corporation must calculate its net (i.e., taxable) before-tax income or loss. Several steps are involved in this process, beginning with the calculation of *gross income.* The corporation may deduct from gross income all ordinary and necessary operating expenses, including interest, to conduct the business *except capital investments.* Deductions for depreciation are permitted each tax period as a means of consistently and systematically recovering capital investment. Consequently, allowable expenses and deductions may be used to determine taxable income:

Taxable income = gross income − all expenses except capital investments

− depreciation (depletion) deductions.

$$\tag{6-17}$$

This taxable income is also referred to as *net income before taxes (NIBT)*. When income taxes are subtracted, the remainder is called the *net income after taxes (NIAT)*. In summary,

$$\text{Net income after taxes} = \left\{ \begin{array}{c} \text{taxable income} \\ \text{(i.e., NIBT)} \end{array} \right\} - \text{income taxes.} \qquad (6\text{-}18)$$

EXAMPLE 6-10

A company generates $1,500,000 of gross income during its tax year and incurs operating expenses of $800,000. Interest payments on borrowed capital amount to $48,000. The total depreciation deductions for the tax year equal $114,000. What is the taxable income (NIBT) of this firm?

SOLUTION

Based on Equation (6-17), this company's taxable income for the tax year would be

$$\$1,500,000 - \$800,000 - \$48,000 - \$114,000 = \$538,000$$

6.8 The Effective (Marginal) Corporate Income Tax Rate

The federal corporate income tax rate structure in 2001 is shown in Table 6-5. Depending on the taxable income bracket that a firm is in for a tax year, the marginal federal rate can vary from 15% to a maximum of 39%. However, note that the weighted average tax rate at taxable income = $335,000 is 34%, and the weighted average tax rate at taxable income = $18,333,333 is 35%. Therefore, if a corporation has a taxable income for a tax year *greater than* $18,333,333, federal taxes are computed using a flat rate of 35%.

TABLE 6-5 Corporate Federal Income Tax Rates (2001)

If taxable income is:		The tax is:	
Over	but not over		of the amount over
0	$50,000	15%	0
$50,000	75,000	$7,500 + 25%	$50,000
75,000	100,000	13,750 + 34%	75,000
100,000	335,000	22,250 + 39%	100,000
335,000	10,000,000	113,900 + 34%	335,000
10,000,000	15,000,000	3,400,000 + 35%	10,000,000
15,000,000	18,333,333	5,150,000 + 38%	15,000,000
18,333,333	....	6,416,667 + 35%	18,333,333

Source: *Tax Information on Corporations*, IRS Publication 542, 1994.

EXAMPLE 6-11

Suppose that a firm for a tax year has a gross income of $5,270,000, expenses (excluding capital) of $2,927,500, and depreciation deductions of $1,874,300. What would be its taxable income and federal income tax for the tax year based on Equation (6-17) and Table 6-5?

SOLUTION

Taxable income = gross income − expenses − depreciation deductions

$$= \$5,270,000 - \$2,927,500 - \$1,874,300$$

$$= \$468,200$$

Income tax = 15% of first $50,000	$7,500
+ 25% of the next $25,000	6,250
+ 34% of the next $25,000	8,500
+ 39% of the next $235,000	91,650
+ 34% of the next $133,200	45,288
Total	$159,188

The total tax liability in this case is $159,188. As an added note, we could have used a flat rate of 34% in this example because the federal weighted average tax rate at taxable income = $335,000 is 34%. The remaining $133,200 of taxable income above this amount is in a 34% tax bracket (Table 6-5). So we have 0.34($468,200) = $159,188.

Although the tax laws and regulations of most of the states (and some municipalities) with income taxes have the same basic features as the federal laws and regulations, there is significant variation in income tax rates. State income taxes are in most cases much less than federal taxes and often can be closely approximated as ranging from 6% to 12% of taxable income. No attempt will be made here to discuss the details of state income taxes. However, to illustrate the calculation of an effective income tax rate (t) for a large corporation based on the consideration of both federal and state income taxes, assume that the applicable federal income tax rate is 35% and the state income tax rate is 8%. Further assume the common case in which taxable income is computed the same way for both types of taxes, except that state income taxes are deductible from taxable income for federal tax purposes but federal income taxes are not deductible from taxable income for state tax purposes. Based on these assumptions, the general expression for the effective income tax rate is

$$t = \text{state rate} + \text{federal rate}(1 - \text{state rate}), \qquad (6\text{-}19)$$

and, in this example, the effective income tax rate for the corporation would be

$$t = 0.08 + 0.35(1 - 0.08) = 0.402, \text{ or approximately } 40\%.$$

In this chapter, we will often use an effective corporate income tax rate of *approximately* 40% as a representative value that includes state income taxes.

The effective income tax rate on *increments* of taxable income is of importance in engineering economy studies. This concept is illustrated in Figure 6-4, which plots the federal income tax rates and brackets listed in Table 6-5 and shows the added (incremental) taxable income and federal income taxes that would result from a proposed engineering project. In this case, the corporation is assumed to have a taxable income for their tax year greater than $18,333,333. However, the same concept applies to a smaller firm with less taxable income for its tax year, which is illustrated in Example 6-12.

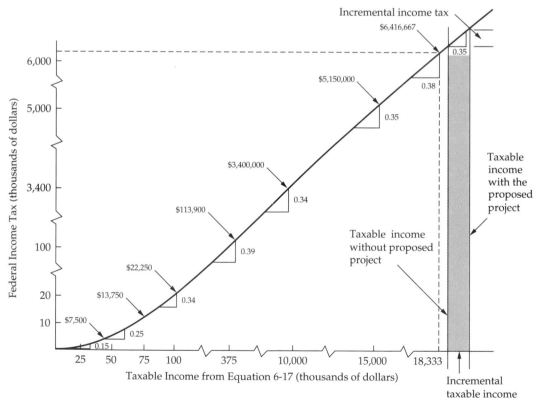

Figure 6-4 The Federal Income Tax Rates for Corporations (Table 6-5) with Incremental Income Tax for a Proposed Project (Assumes, in this case, corporate taxable income without project > $18,333,333)

EXAMPLE 6-12

A small corporation is expecting an annual taxable income of $45,000 for its tax year. It is considering an additional capital investment of $100,000 in an engineering project, which is expected to create an added annual net cash flow (revenues minus expenses) of $35,000 and an added annual depreciation deduction of $20,000. What is the corporation's federal income tax liability (a) without the added capital investment and (b) with the added capital investment?

SOLUTION

(a)

Income Taxes	Rate	Amount
On first $45,000	15%	$6,750
	Total	$6,750

(b)

Taxable Income		
Before added investment		$45,000
+ added net cash flow		+35,000
− depreciation deduction		−20,000
	Net total	$60,000

Income Taxes on $60,000	Rate	Amount
On first $50,000	15%	$7,500
On next $10,000	25%	2,500
	Total	$10,000

The increased income tax liability from the investment is $10,000 − $6,750 = $3,250.

As an added note, the change in tax liability can usually be determined more readily by an incremental approach. For instance, this example involved changing the taxable income from $45,000 to $60,000 as a result of the new investment. Thus, the change in income taxes for the tax year could be calculated as follows:

$$\text{First } \$50,000 - \$45,000 = \$5,000 \text{ at } 15\% = \$ \quad 750$$

$$\text{Next } \$60,000 - \$50,000 = \$10,000 \text{ at } 25\% = \quad 2,500$$

$$\text{Total} \quad \$3,250$$

The average federal income tax rate on the additional $35,000 − $20,000 = $15,000 of taxable income is calculated as ($3,250/$15,000) = 0.2167, or 21.67%.

In addition to lowering the maximum rate on corporate taxable income from 46% to 35%, TRA 86 created a new *alternative minimum tax (AMT)* system that is intended to ensure that any corporation with economic income pays a minimum amount of federal income tax. Now corporations must compute their income tax liability as illustrated in this section, and many must also compute their AMT according to a rather complex set of rules that is beyond the scope of our discussion.

Corporations now pay the *maximum* income tax resulting from the system using rates in Table 6-5 or from the AMT system. It is generally acknowledged that the AMT is the most far-reaching and complex business tax provision in TRA 86.

6.9 Gain (Loss) on the Disposal of an Asset

When a *depreciable asset* (tangible personal or real property, Section 6.2) is sold, the market value is seldom equal to its BV [Equation (6-1)]. In general, the gain (loss) on sale of depreciable property is the fair market value minus its book value at that time. That is,

$$[\text{gain (loss) on disposal}]_N = MV_N - BV_N. \tag{6-20}$$

When the sale results in a gain, it is often referred to as *depreciation recapture*. The tax rate for the gain (loss) on disposal of depreciable personal property is usually the same as for ordinary income or loss, which is the effective income tax rate, t.

When a *capital asset* is sold or exchanged, the gain (loss) is referred to as a *capital gain (loss)*. Examples of capital assets are stocks, bonds, gold, silver, and other metals, as well as real property such as a home. Because engineering economic analysis seldom involves an actual capital gain (loss), the more complex details of this situation are not discussed further.

EXAMPLE 6-13

A corporation sold a piece of equipment during the current tax year for $78,600. The accounting records show that its cost basis, B, is $190,000 and the accumulated depreciation is $139,200. Assume that the effective income tax rate is 40%. Based on this information, what is (a) the gain (loss) on disposal, (b) the tax liability (or credit) resulting from this sale, and (c) the tax liability (or credit) if the accumulated depreciation was $92,400 instead of $139,200?

SOLUTION

(a) The BV at time of sale is $190,000 − $139,200 = $50,800. Therefore, the gain on disposal is $78,600 − $50,800 = $27,800.

(b) The tax owed on this gain is −0.40($27,800) = −$11,120.

(c) With $d_k^* = \$92,400$, the BV at the time of sale is $190,000 − $92,400 = $97,600. The gain is $78,600 − $97,600 = −$19,000. The tax credit resulting from this loss on disposal is −0.40(−$19,000) = $7,600.

6.10 General Procedure for Making After-Tax Economic Analyses

After-tax economic analyses usually utilize the same profitability measures as do before-tax analyses. The only difference is that after-tax cash flows (ATCFs) are used in place of Before-Tax Cash Flows (BTCFs) by including expenses (or savings) due to income taxes and then making equivalent worth calculations *using an after-tax MARR*. The tax rates and governing regulations may be complex and subject to changes, but once those rates and regulations have been translated into their effect on ATCFs, the remainder of the after-tax analysis is relatively straightforward. To formalize the procedure, let

R_k = revenues (and savings) from the project; this is the cash inflow from the project during period k,

E_k = cash outflows during year k for deductible expenses and interest,

d_k = sum of all noncash, or book, costs during year k, such as depreciation and depletion,

t = effective income tax rate on *ordinary* income (federal, state, and other), t is assumed to remain constant during the study period,

T_k = income tax consequences during year k, and

$ATCF_k$ = ATCF from the project during year k.

Because the NIBT (i.e., taxable income) is $(R_k - E_k - d_k)$, the *ordinary income tax consequences* during year k are computed with Equation (6-21):

$$T_k = -t(R_k - E_k - d_k). \tag{6-21}$$

Therefore, when $R_k > (E_k + d_k)$, a tax liability (i.e., negative cash flow) occurs. When $R_k < (E_k + d_k)$, a decrease in the tax amount occurs. The NIAT [Equation (6-18)] is then simply taxable income (i.e., net income before taxes) algebraically added to the tax amount determined by Equation (6-21), so

$$NIAT_k = \underbrace{(R_k - E_k - d_k)}_{\text{taxable income}} - \underbrace{t(R_k - E_k - d_k)}_{\text{income taxes}};$$

or

$$NIAT_k = (R_k - E_k - d_k)(1 - t). \tag{6-22}$$

The ATCF associated with a project equals the NIAT plus noncash items such as depreciation, so

$$ATCF_k = NIAT_k + d_k \tag{6-23}$$

$$= (R_k - E_k - d_k)(1 - t) + d_k, \tag{6-24}$$

or $\qquad ATCF_k = (1 - t)(R_k - E_k) + td_k. \tag{6-25}$

In many economic analyses of engineering projects, ATCFs in year k are computed in terms of BTCF_k (i.e., year k before-tax cash flows):

$$\text{BTCF}_k = R_k - E_k. \tag{6-26}$$

Thus,*

$$\text{ATCF}_k = \text{BTCF}_k + T_k \tag{6-27}$$

$$= (R_k - E_k) - t(R_k - E_k - d_k)$$

$$= (1 - t)(R_k - E_k) + td_k \tag{6-28}$$

Equations (6-25) and (6-28) are obviously identical.

Tabular headings to facilitate the computation of after-tax cash flows with Equations (6-21) and (6-28) are as follows:

Year	(A) BTCF	(B) Depreciation	$(C) = (A) - (B)$ Taxable Income	$(D) = -t(C)$ Cash Flow for Income Taxes	$(E) = (A) + (D)$ ATCF
k	$R_k - E_k$	d_k	$R_k - E_k - d_k$	$-t(R_k - E_k - d_k)$	$(1 - t)(R_k - E_k) + td_k$

Column A consists of the same information used in before-tax analyses, namely, the cash revenues (or savings) less the deductible expenses. Column B contains depreciation that can be claimed for tax purposes. Column C is the taxable income, or amount subject to income taxes. Column D contains the income taxes paid (or saved). Finally, column E shows the ATCFs to be used directly in after-tax economic analyses.

A summary of the process of determining NIAT and ATCF during each year of an N-year study period is provided in Figure 6-5. NIAT is well understood in most companies, and it can be easily obtained from Figure 6-5 for making presentations to upper-level management. The format of Figure 6-5 is used extensively throughout the remainder of this chapter, and it provides a convenient way to organize data in after-tax studies.

The column headings of Figure 6-5 indicate the arithmetic operations for computing columns C, D, and E when $k = 1, 2, \ldots, N$. When $k = 0$, capital investments are usually involved, and their tax treatment (if any) is illustrated in the examples that follow. The table should be used with the conventions of $+$ for cash inflow or savings and $-$ for cash outflow or opportunity forgone.

---

* In Figure 6-5, we use $-t$ in column D, so algebraic subtraction of income taxes in Equation (6-27) is accomplished.

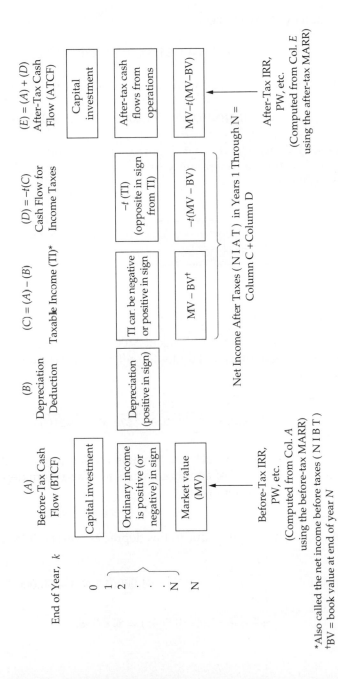

Figure 6-5 General Format (Worksheet) for After-Tax Analysis; Determining the ATCF (and NIAT)

EXAMPLE 6-14

If the revenue from a project is $10,000 during a tax year, out-of-pocket expenses are $4,000, and depreciation deductions for income tax purposes are $2,000, what is the ATCF when $t = 0.40$? What is the NIAT?

SOLUTION
From Equation (6-24), we have

$$\text{ATCF} = (1 - 0.4)(\$10,000 - \$4,000 - \$2,000) + \$2,000 = \$4,400.$$

The same result can be obtained with Equation (6-25) or Equation (6-28):

$$\text{ATCF} = (1 - 0.4)(\$10,000 - \$4,000) + 0.4(\$2,000) = \$4,400.$$

Equation (6-25) shows clearly that depreciation contributes a credit of $t \cdot d_k$ to the after-tax cash flow in operating year k. The NIAT, from Equation (6-23), is $4,400 − $2,000 = $2,400.

The ATCF attributable to depreciation (a tax savings) is td_k in year k. After income taxes, an expense becomes $(1 - t)E_k$.

EXAMPLE 6-15

Suppose that an asset with a cost basis of $100,000 and an ADS recovery period of five years is being depreciated under the *Alternate Depreciation System (ADS) of MACRS* as follows:

Year	1	2	3	4	5	6
Depreciation deduction	$10,000	$20,000	$20,000	$20,000	$20,000	$10,000

If the firm's effective income tax rate remains constant at 40% during this six-year period, what is the PW of after-tax savings resulting from depreciation when MARR = 10% per year (after taxes)?

SOLUTION
The PW of tax credits (savings) due to this depreciation schedule is

$$\text{PW}(10\%) = \sum_{k=1}^{6} 0.4d_k(1.10)^{-k} = \$4,000(0.9091) + \$8,000(0.8264)$$

$$+ \cdots + \$4,000(0.5645) = \$28,948.$$

EXAMPLE 6-16

The asset in Example 6-15 is expected to produce net cash inflows (net revenues) of $30,000 per year during the six-year period, and its terminal market value is negligible. If the effective income tax rate is 40%, how much can a firm afford to spend for this asset and still earn the MARR? What is the meaning of any excess in affordable amount over the $100,000 cost basis given in Example 6-15?

SOLUTION

After income taxes, the PW of net revenues is $(1 - 0.4)(\$30,000) \cdot (P/A, 10\%, 6) =$ $18,000(4.3553) = \$78,395$. After adding to this the PW of tax savings computed in Example 6-15, the affordable amount is $107,343. Because the capital investment is $100,000, the net PW equals $7,343. This same result can be obtained by using the general format (worksheet) of Figure 6-5:

EOY	(A) BTCF	(B) Depreciation Deduction	(C) = (A) − (B) Taxable Income	(D) = −0.4(C) Income Taxes	(E) = (A) + (D) ATCF
0	−$100,000				−$100,000
1	30,000	$10,000	$20,000	−$8,000	22,000
2	30,000	20,000	10,000	−4,000	26,000
3	30,000	20,000	10,000	−4,000	26,000
4	30,000	20,000	10,000	−4,000	26,000
5	30,000	20,000	10,000	−4,000	26,000
6	30,000	10,000	20,000	−8,000	22,000
Total	$80,000		Total $80,000		PW(10%) of ATCF = $7,343

6.11 Illustration of Computations of ATCFs

The following problems (Examples 6-17, 6-18, 6-19, and 6-20) illustrate the computation of ATCFs, as well as many common situations that affect income taxes. All problems include the assumption that income tax expenses (or savings) occur at the same time (year) as the revenue or expense that gives rise to the taxes. For purposes of comparing the effects of various situations, the after-tax IRR or PW is computed for each example. We can observe from the results of Examples 6-17 and 6-19 that the faster (i.e., earlier) the depreciation deduction is, the more favorable the after-tax IRR and PW will become.

EXAMPLE 6-17

Certain new machinery when placed in service is estimated to cost $180,000. It is expected to *reduce* net annual operating expenses by $36,000 per year for 10 years and to have a $30,000 MV at the end of the 10th year. (a) Develop the before-tax and after-tax cash flows, and (b) calculate the before-tax and after-tax IRR. Assume that the firm is in the federal taxable income bracket of $335,000 to $10,000,000 and that the state income tax rate is 6%. State income taxes are deductible from federal taxable income. This machinery is in the MACRS (GDS) five-year property class. (c) Calculate the after-tax PW when the *after-tax* MARR = 10% per year. In this example, the study period is 10 years, but the property class of the machinery is 5 years.

SOLUTION

(a) Table 6-6 applies the format illustrated in Figure 6-5 to calculate the BTCF and ATCF for this example. In column D the effective income tax rate is very close to 0.38 [from Equation (6-19)] based on the information just provided.

(b) The before-tax IRR is computed from column A:

$$0 = -\$180{,}000 + \$36{,}000(P/A, i'\%, 10) + \$30{,}000(P/F, i'\%, 10).$$

By trial and error, we find that $i' = 16.1\%$.

   The entry in the last year is shown to be $30,000 because the machinery will have this estimated MV. However, the asset was depreciated to zero with the GDS method. Therefore, when the machine is sold at the end of year ten, there will be $30,000 of *recaptured depreciation*, or gain on disposal [Equation (6-20)], which is taxed at the effective income tax rate of 38%. This tax entry is shown in column D (EOY 10).

   By trial and error, the after-tax IRR for Example 6-17 is found to be 12.4%.

(c) When MARR = 10% per year is inserted into the PW equation at the bottom of Table 6-6, it can be determined that the after-tax PW of this investment is $17,209.

If the machinery in Example 6-17 had been classified in the 10-year MACRS (GDS) property class instead of 5-year property class, depreciation deductions would be slowed down in the early years of the study period and shifted into later years, as shown in Table 6-7. Compared with entries in Table 6-6, entries in columns C, D, and E of Table 6-7 are less favorable in the sense that a fair amount of ATCF is deferred until later years, producing a lower after-tax IRR and PW. For instance, the PW is reduced from $17,208 in Table 6-6 to $9,136 in Table 6-7. The basic difference between Table 6-6 and Table 6-7 is the *timing of the ATCF*, which is a function of the timing and magnitude of the depreciation deductions. In fact, the curious reader can confirm that the sums of entries in columns A through E of Tables 6-6 and 6-7 are nearly the same (except for the half-year of depreciation only in year ten of Table 6-7). The timing of cash flows does, of course, make a difference!

TABLE 6-6 ATCF Analysis of Example 6-17

End of Year, k	(A) BTCF	(B) Cost Basis	×	Depreciation Deduction GDS Recovery Rate	=	Deduction	(C) = (A) − (B) Taxable Income	(D) = −0.38(C) Cash Flow for Income Taxes	(E) = (A) + (D) ATCF
0	−$180,000	—		—		—			−$180,000
1	36,000	$180,000	×	0.2000	=	$36,000	0	0	36,000
2	36,000	180,000	×	0.3200	=	57,600	−21,600	+8,208	44,208
3	36,000	180,000	×	0.1920	=	34,560	1,440	−547	35,453
4	36,000	180,000	×	0.1152	=	20,736	15,264	−5,800	30,200
5	36,000	180,000	×	0.1152	=	20,736	15,264	−5,800	30,200
6	36,000	180,000	×	0.0576	=	10,368	25,632	−9,740	26,260
7–10	36,000	0				0	36,000	−13,680	22,320
10	30,000						30,000 [a]	−11,400 [b]	18,600
Total $	210,000							Total	$130,201
								PW (10%) =	$17,208

[a] Depreciation recapture = $MV_{10} − BV_{10} = \$30,000 − 0 = \$30,000$ (gain on disposal).

[b] Tax on depreciation recapture = $\$30,000(0.38) = \$11,400$.

After-tax IRR: Set PW of column E = 0 and solve for i' in the following equation:

$$0 = -\$180,000 + \$36,000(P/F, i', 1) + \$44,208(P/F, i', 2) + \$35,453(P/F, i', 3) + \$30,200(P/F, i', 4) + \$30,200(P/F, i', 5) + \$26,260(P/F, i', 6)$$

$$+ \$22,320(P/A, i', 4)(P/F, i', 6) + \$18,600(P/F, i', 10); \quad IRR = 12.4\%$$

TABLE 6-7 Reworked Example 6-17 with Machinery in the 10-Year MACRS (GDS) Property Class

End of Year, k	(A) BTCF	(B) Cost Basis	Depreciation × GDS Recovery Rate	=	Deduction Deduction	(C) = (A) − (B) Taxable Income	(D) = −0.38(C) Cash Flow for Income Taxes	(E) = (A) + (D) ATCF
0	−$180,000	—	—		—			−$180,000
1	36,000	$180,000 ×	0.1000	=	$18,000	$18,000	−$6,840	29,160
2	36,000	180,000 ×	0.1800	=	32,400	3,600	−1,368	34,632
3	36,000	180,000 ×	0.1440	=	25,920	10,080	−3,830	32,170
4	36,000	180,000 ×	0.1152	=	20,736	15,264	−5,800	30,200
5	36,000	180,000 ×	0.0922	=	16,596	19,404	−7,374	28,626
6	36,000	180,000 ×	0.0737	=	13,266	22,734	−8,639	27,361
7	36,000	180,000 ×	0.0655	=	11,790	24,210	−9,200	26,800
8	36,000	180,000 ×	0.0655	=	11,790	24,210	−9,200	26,800
9	36,000	180,000 ×	0.0656	=	11,808	24,192	−9,193	26,807
10	36,000	180,000 ×	0.0655/2	=	5,895	30,105	−11,440	24,560
10	30,000					18,201 [a]	−6,916	23,084

Total $130,196

PW (10%) ≅ $9,136

IRR = 11.2%

[a] Gain on disposal = $MV_{10} - BV_{10} = \$30.000 - \left(\dfrac{0.0655}{2} + 0.0328\right)(\$180,000) = \$18,201$.

> Depreciation does not affect BTCF. Fast (accelerated) depreciation produces a larger PW of tax savings than does the same amount of depreciation claimed later in an asset's life.

A minor complication is introduced in ATCF analyses when the study period is shorter than an asset's MACRS recovery period (e.g., for a five-year recovery period, the study period is five years or less). In such cases, we shall assume throughout this book that the asset is sold for its MV in the last year of the study period. Due to the half-year convention, only one-half of the normal MACRS depreciation can be claimed in the year of disposal or end of the study period, so there will usually be a difference between an asset's BV and its MV. Resulting income tax adjustments will be made at the time of the sale (see the last row in Table 6-7) unless the asset in question is not sold, but instead kept for standby service. In such a case, depreciation deductions usually continue through the end of the asset's MACRS recovery period. Our assumption of project termination at the end of the study period makes good economic sense, as illustrated in Example 6-18.

EXAMPLE 6-18

A highly specialized piece of OCR equipment has a first cost of $50,000. If this equipment is purchased, it will be used to produce income (through rental) of $20,000 per year for only four years. At the end of year four, the equipment will be sold for a negligible amount. Estimated annual expenses for upkeep are $3,000 during each of the four years. The MACRS (GDS) recovery period for the equipment is seven years, and the firm's effective income-tax rate is 40%.

(a) If the after-tax MARR is 7% per year, should the equipment be purchased?

(b) Rework the problem, assuming that the equipment is placed on standby status such that depreciation is taken over the full MACRS recovery period.

SOLUTION

(a)

End of Year, k	(A) BTCF	(B) Depreciation Deduction	(C) = (A) − (B) Taxable Income	(D) = −0.4(C) Cash Flow for Income Taxes	(E) = (A) + (D) ATCF
0	−$50,000				−$50,000
1	17,000	$7,145	$9,855	−$3,942	13,058
2	17,000	12,245	4,755	−1,902	15,098
3	17,000	8,745	8,255	−3,302	13,698
4	17,000	3,123[a]	13,877	−5,551	11,449
4	0		−18,742[b]	7,497	7,497

[a] Half-year convention applies with disposal in year four.
[b] Remaining BV.

PW(7%) = $1,026. Because the PW > 0, the equipment should be purchased.

(b)

End of Year, k	(A) BTCF	(B) Depreciation Deduction	(C) Taxable Income	(D) Cash Flow for Income Taxes	(E) ATCF
0	−$50,000				−$50,000
1	17,000	$7,145	$9,855	−$3,942	13,058
2	17,000	12,245	4,755	−1,902	15,098
3	17,000	8,745	8,255	−3,302	13,698
4	17,000	6,245	10,755	−4,302	12,698
5	0	4,465	−4,465	1,786	1,786
6	0	4,460	−4,460	1,784	1,784
7	0	4,465	−4,465	1,786	1,786
8	0	2,230	−2,230	892	892
8	0				0

PW(7%) = $353, so the equipment should be purchased.

The present worth is $673 higher in part (a), which equals the PW of deferred depreciation deductions in part (b). A firm would select the situation in part (a) if it had a choice.

An illustration of determining ATCFs for a somewhat more complex, though realistic, capital investment opportunity is provided in Example 6-19.

EXAMPLE 6-19

The Ajax Semiconductor Company is attempting to evaluate the profitability of adding another integrated circuit production line to its present operations. The company would need to purchase two or more acres of land for $275,000 (total). The facility would cost $60,000,000 and have no net MV at the end of five years. The facility could be depreciated using a GDS recovery period of five years. An increment of working capital would be required, and its estimated amount is $10,000,000. Gross income is expected to increase by $30,000,000 per year for five years, and operating expenses are estimated to be $8,000,000 per year for five years. The firm's effective income tax rate is 40%. (a) Set up a table and determine the ATCF for this project. (b) What is the NIAT in year three? (c) Is the investment worthwhile when the after-tax MARR is 12% per year?

SOLUTION

(a) The format recommended in Figure 6-5 is followed in Table 6-8 to obtain ATCFs in years zero through five. Acquisitions of land, as well as additional working capital, are treated as nondepreciable capital investments whose MVs at the

TABLE 6-8 After-Tax Analysis of Example 6-19

End of Year, k	(A) BTCF	(B) Depreciation Deduction	$(C) = (A) - (B)$ Taxable Income	$(D) = -0.4(C)$ Cash Flow for Income Taxes	$(E) = (A) + (D)$ ATCF
0	$\begin{cases} -\$60,000,000 \\ -10,000,000 \\ -275,000 \end{cases}$				$-\$70,275,000$
1	22,000,000	\$12,000,000	\$10,000,000	$-\$4,000,000$	18,000,000
2	22,000,000	19,200,000	2,800,000	$-1,120,000$	20,880,000
3	22,000,000	11,520,000	10,480,000	$-4,192,000$	17,808,000
4	22,000,000	6,912,000	15,088,000	$-6,035,200$	15,964,800
5	22,000,000	3,456,000	18,544,000	$-7,417,600$	14,582,400
5	10,275,000[a]		$-6,912,000$[b]	2,764,800[b]	13,039,800

[a] MV of working capital and land.
[b] Because BV_5 of the production facility is \$6,912,000 and net $MV_5 = 0$, a loss on disposal would be taken at EOY 5.

end of year five are estimated to equal their first costs. (In economic evaluations, it is customary to assume that land and working capital do not inflate in value during the study period because they are "nonwasting" assets.) By using Equation (6-24), we are able to compute ATCF in year three (for example) to be

$$ATCF_3 = (\$30,000,000 - \$8,000,000 - \$11,520,000)(1 - 0.40) + \$11,520,000$$

$$= \$17,808,000.$$

(b) The NIAT in year three can be determined with Equation (6-22):

$$NIAT_3 = (\$30,000,000 - \$8,000,000 - \$11,520,000)(1 - 0.40) = \$6,288,000.$$

This can also be obtained directly from Table 6-8 by adding the year three entries from columns C and D: \$10,480,000 - \$4,192,000 = \$6,288,000.

(c) The depreciable property in Example 6-19 (\$60,000,000) will be disposed of for \$0 at the end of year five, and a loss on disposal of \$6,912,000 will be claimed at the end of year five. Only a half-year of depreciation (\$3,456,000) can be claimed as a deduction in year five, and the BV is \$6,912,900 at the end of year five. Because the selling price (MV) is zero, the loss on disposal equals our BV of \$6,912,000. As seen from Figure 6-5, a tax credit of 0.40(\$6,912,000) = \$2,764,800 is created at the end of year five. The after-tax IRR is obtained from entries in column E of Table 6-8 and is found to be 12.5%. The after-tax PW equals \$936,715 at MARR = 12% per year. Based on economic considerations, this integrated circuit production line should be recommended because it appears to be quite attractive.

In the next example, the after-tax comparison of mutually exclusive alternatives involving costs only is illustrated.

EXAMPLE 6-20

An engineering consulting firm can purchase a fully configured Computer-Aided Design (CAD) workstation for $20,000. It is estimated that the useful life of the workstation is seven years, and its MV in seven years should be $2,000. Operating expenses are estimated to be $40 per eight-hour workday, and maintenance will be performed under contract for $8,000 per year. The MACRS (GDS) property class is five years, and the effective income tax rate is 40%.

As an alternative, sufficient computer time can be leased from a service company at an annual cost of $20,000. If the after-tax MARR is 10% per year, how many workdays per year must the workstation be needed in order to justify *leasing* it?

SOLUTION

This example involves an after-tax evaluation of purchasing depreciable property versus leasing it. We are to determine how much the workstation must be utilized so that the lease option is a good economic choice. A *key* assumption is that the cost of engineering design time (i.e., operator time) is unaffected by whether the workstation is purchased or leased. Variable operations expenses associated with ownership result from the purchase of supplies, utilities, and so on. Hardware and software maintenance cost is contractually fixed at $8,000 per year. It is further assumed that the maximum number of working days per year is 250.

Lease fees are treated as an annual expense, and the consulting firm (the lessee) may *not* claim depreciation of the equipment to be an additional expense. (The leasing company presumably has included the cost of depreciation in its fee.) Determination of ATCF for the lease option is relatively straightforward and is not affected by how much the workstation is utilized:

$$(\text{After-tax expense of the lease})_k = -\$20,000(1 - 0.40) = -\$12,000; \ k = 1, \ldots, 7.$$

ATCFs for the purchase option involve expenses that are fixed (not a function of equipment utilization) in addition to expenses that vary with equipment usage. If we let X equal the number of working days per year that the equipment is utilized, the variable cost per year of operating the workstation is $40X$. The after-tax analysis of the purchase alternative is shown in Table 6-9.

The after-tax annual worth of purchasing the workstation is

$$\text{AW}(10\%) = -\$20,000(A/P, 10\%, 7) - \$24X - [\$3,200(P/F, 10\%, 1) + \cdots$$

$$+ \$4,800(P/F, 10\%, 7)](A/P, 10\%, 7) + \$1,200(A/F, 10\%, 7)$$

$$= -\$24X + -\$7,511.$$

To solve for X, we equate the after-tax annual worth of both alternatives:

$$-\$12,000 = -\$24X - \$7,511.$$

TABLE 6-9 After-Tax Analysis of Purchase Alternative (Example 6-20)

End of Year, k	(A) BTCF	(B) Depreciation Deduction[a]	$(C) = (A) - (B)$ Taxable Income	$(D) = -t(C)$ Cash Flow for Income Taxes	$(E) = (A) + (D)$ ATCF
0	−$20,000				−$20,000
1	−40X − 8,000	$4,000	−$40X − $12,000	$16X + $4,800	−24X − 3,200
2	−40X − 8,000	6,400	− 40X − 14,400	16X + 5,760	−24X − 2,240
3	−40X − 8,000	3,840	− 40X − 11,840	16X + 4,736	−24X − 3,264
4	−40X − 8,000	2,304	− 40X − 10,304	16X + 4,122	−24X − 3,878
5	−40X − 8,000	2,304	− 40X − 10,304	16X + 4,122	−24X − 3,878
6	−40X − 8,000	1,152	− 40X − 9,152	16X + 3,661	−24X − 4,339
7	−40X − 8,000	0	− 40X − 8,000	16X + 3,200	−24X − 4,800
7	2,000		2,000	−800	1,200

[a]Depreciation deduction$_k$ = $20,000 × (GDS recovery rate).

Thus, $X = 187$ days per year. Therefore, if the firm expects to utilize the CAD workstation in its business *more than* 187 days per year, the equipment should be leased. The graphic summary of Example 6-20 shown in Figure 6-6 provides the rationale for this recommendation. The importance of the workstation's estimated utilization, in workdays per year, is now quite apparent.

Companion Web Site (http://www.prenhall.com/sullivan_engineering/): For many individuals or companies, deciding which desktop computer to purchase may present a challenge. Visit the Web site to view a comparison of three alternative PC systems that uses an *After-Tax Cash-Flow (ATCF)* analysis.

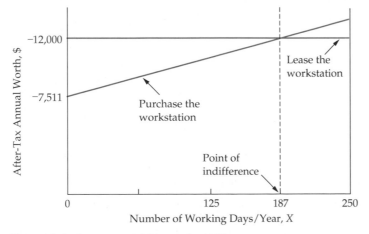

Figure 6-6 Summary of Example 6-20

▼ 6.12 Economic Value Added

This section discusses an economic measure for estimating the wealth creation potential of capital investments that is experiencing increased attention and use. The measure, called economic value added (EVA),[*] can be determined from some of the data available in an after-tax analysis of cash flows generated by a capital investment. Through retroactive analysis of a firm's common stock valuation, it has been established that some companies experience a statistically significant relationship between the EVA metric and the historical value of their common stock.[†] For our purposes, EVA can also be used to estimate the profit-earning *potential* of proposed capital investments in engineering projects.

Simply stated, EVA is the difference between the company's adjusted net operating profit after taxes (NOPAT) in a particular year and its after-tax cost of capital during that year. Another way to characterize EVA is "the spread between the return on the capital and the cost of the capital."[‡] On a project-by-project basis (i.e., for discrete investments), the EVA metric can be used to gauge the wealth creation opportunity of proposed capital expenditures. We now define annual EVA as

$$\text{EVA}_k = (\text{Net Operating Profit After Taxes})_k -$$

$$(\text{Cost of Capital Used to Produce Profit})_k \qquad (6\text{-}29)$$

$$= \text{NOPAT}_k - i \cdot \text{BV}_{k-1},$$

where $k =$ an index for the year in question $(1 \leq k \leq N)$,

$i =$ after-tax MARR based on a firm's cost of capital,

$\text{BV}_{k-1} =$ beginning-of-year book value,

$N =$ the study (analysis) period in years.

Figure 6-5 presented earlier can be used to relate EVA amounts to after-tax cash flow (ATCF) amounts for a proposed capital investment. The annual EVA amount for year k can be obtained from Figure 6-5 by (1) algebraically *adding* the entry in Column C for year $k(1 \leq k \leq N)$ to the corresponding entry in Column D to yield net income after taxes (NIAT), which is the same as NOPAT, and then (2) *subtracting* the product of the project's after-tax MARR (based on the cost of capital) and its

---

[*] EVA is a registered trademark of Stern Stewart & Company, New York City, NY.

[†] See J. L. Dodd and S. Chen, "EVA: A New Panacea?" *B & E Review*, July–September 1996, pp 26–28, and W. Freedman "How Do You Add Up?" *Chemical Week*, October 9, 1996, pp. 31–34.

[‡] S. Tully, "The Real Key To Creating Wealth," *Fortune*, September 30, 1993, p. 38ff.

beginning-of-year book value. This calculation is obvious from Equation (6-29). Clearly, accurate BTCF estimates (forecasts) in Figure 6-5 are needed for acceptable predictions of annual ATCF and EVA amounts.

Using the notation on page 285, we find that $\text{NOPAT}_k = (1 - t)(R_k - E_k - d_k)$ and $\text{EVA}_k = (1-t)(R_k - E_k - d_k) - i \cdot \text{BV}_{k-1}$. The ATCF is defined as follows: When $k > 0 \text{ATCF}_k = (1 - t)(R_k - E_k - d_k) + d_k$, and $\text{ATCF}_0 = \text{BV}_0$ when $k = 0$. Now we see the relationship between ATCF_k and EVA_k to be $\text{ATCF}_k = \text{EVA}_k + i \cdot \text{BV}_{k-1} + d_k$.

Equation (6-29) and Figure 6-5 are demonstrated in Example 6-21 to determine the ATCF amounts, after-tax AW, and the EVA amounts related to a capital investment.

EXAMPLE 6-21

Consider the following proposed capital investment in a engineering project and determine its (a) year-by-year ATCF, (b) after-tax AW, and (c) annual equivalent EVA.

Proposed Capital Investment	= \$84,000
Salvage Value (end of year 4)	= \$0
Annual Expenses/Year	= \$30,000
Gross Revenues/Year	= \$70,000
Depreciation Method	= Straight Line
Useful Life	= 4 years
Effective Income Tax Rate (t)	= 50%
After-Tax MARR (i)	= 12% per year

SOLUTION

(a) Year-by-year ATCF amounts are shown in the following table:

EOY	BTCF	Deprec.	Taxable Income	Income Taxes	ATCF
0	−\$84,000	—	—	—	−\$84,000
1	70,000 − 30,000	\$21,000	\$19,000	−\$9,500	30,500
2	70,000 − 30,000	21,000	19,000	−9,500	30,500
3	70,000 − 30,000	21,000	19,000	−9,500	30,500
4	70,000 − 30,000	21,000	19,000	−9,500	30,500

(b) The annual equivalent worth of the ATCFs equals $-\$84,000(A/P, 12\%, 4) + \$30,500 = \$2,844$.

(c) The EVA in year k equals NOPAT_k (as defined on page 298) -0.12BV_{k-1} [Equation (6-29)]. The year-by-year EVA amounts and the annual equivalent worth of EVA ($2,844) are shown in the next table. Hence, the after-tax annual worth and the annual equivalent worth of EVA of the project are *identical*.

EOY_k	NOPAT	$\text{EVA} = \text{NOPAT} - i \cdot \text{BV}_{k-1}$
1	$19,000 - \$9,500 = \$9,500$	$9,500 - 0.12(\$84,000) = -\$580$
2	$= \$9,500$	$9,500 - 0.12(\$63,000) = \$1,940$
3	$= \$9,500$	$9,500 - 0.12(\$42,000) = \$4,460$
4	$= \$9,500$	$9,500 - 0.12(\$21,000) = \$6,980$

Annual equivalent EVA $= [-\$580(P/F, 12\%, 1) + \$1,940(P/F, 12\%, 2) + \$4,460(P/F, 12\%, 3) + \$6,980(P/F, 12\%, 4)](A/P, 12\%, 4) = \$2,844$.

In Example 6-21, it was shown that the after-tax AW (12%) of the proposed engineering project is identical to the annual equivalent EVA at the same interest rate. Therefore, the annual equivalent EVA is simply the annual worth, at the after-tax MARR, of a project's after-tax cash flows. This straightforward relationship is also valid when accelerated depreciation methods (such as MACRS) are used in the analysis of a proposed project. The reader is referred to Problems 6-40, 6-41, and 6-42 at the end of the chapter for EVA exercises.

6.13 The After-Tax Effect of Depletion Allowances

Income from investment in certain natural resources is subject to depletion allowances before income taxes are computed (Section 6.6). Under certain conditions, notably where the taxpayer is in a relatively high income tax bracket, depletion provisions in the tax law can provide considerable economic advantages.

As an example, consider the case of a profitable corporation that has a net taxable income of $600,000. During the tax year, the firm spends $400,000 to drill and develop a geothermal well that has an estimated reservoir of 10,000,000 gallons of water. Hot water is produced and sold at $0.20 per gallon in accordance with the schedule shown in column 2 of Table 6-10 to produce the gross income shown in column 3. Column 4 shows the net cash flow after production costs have been deducted.

The depletion allowance that can be deducted in a given year may be based on a fixed *percentage* of the gross income (15% for geothermal wells), provided that the deduction *does not exceed 50% (100% for oil and gas property) of the net income before such a deduction* (column 5). Depletion, computed in this manner, is shown in column 7 of Table 6-10. Another method is to base depletion on the estimated investment cost of the product. In this case, the estimated 10,000,000 gallons of water in the well cost $400,000. Such *cost depletion* may, if desired, be charged at the rate of $0.04 per gallon, shown in column 6 of the same table.

TABLE 6-10 Capital Recovery Provided by a Geothermal Well, with Cost and Percentage Depletion Allowances Used in Computation of Income Taxes

(1)	(2)	(3)	(4)	(5)	(6)	(7)	(8)	(9)	(10)
End of Year, k	Gallons of Water Sold	Gross Income (Cash Flow)	Net Income	50% of Net Income	Cost Depletion at $0.04 per Gallon[a]	Depletion Allowance at 15% of Gross Income	$[= (4) -$ either (6) or (7)] Taxable Income[b]	$[= -0.40(8)]$ Income Taxes	$[= (4) + (9)]$ ATCF
1	700,000	$140,000	$80,000	$40,000	$28,000	$21,000	$52,000	−$20,800	$59,200
2	600,000	120,000	70,000	35,000	24,000	18,000	46,000	−18,400	51,600
3	450,000	90,000	48,000	24,000	18,000	13,500	30,000	−12,000	36,000
4	200,000	40,000	24,000	12,000	8,000	6,000	16,000	−6,400	17,600
5	50,000	10,000	2,500	1,250	2,000	1,500	500	−200	2,300

[a]Cost depletion summary: Yr. 1 $\dfrac{700,000}{10,000,000}($400,000) = $28,000$;

Yr. 2 $\dfrac{600,000($400,000 - $28,000)}{10,000,000 - 700,000} = $24,000$;

Yr. 3 $\dfrac{450,000($372,000 - $24,000)}{9,300,000 - 600,000} = $18,000$;

Yr. 4 $\dfrac{200,000($348,000 - $18,000)}{8,700,000 - 450,000} = $8,000$;

Yr. 5 $\dfrac{50,000($330,000 - $8,000)}{8,250,000 - 200,000} = $2,000$.

[b]In computing taxable income, the larger allowance in column 6 or column 7 is chosen as long as percentage depletion does not exceed 50% of column 4. If cost depletion exceeds percentage depletion, then cost depletion must be used during that particular year. (*Note:* Percentage depletion is generally not allowed for oil and gas properties.)

The taxable income resulting from the most favorable application of depletion allowances (by the cost method or the percentage method) is shown in column 8. The advantage of percentage depletion allowances stems from the fact that the total depletion that can be claimed often exceeds the depreciable capital investment. However, this advantage is not shown in this particular situation because of the relatively small fraction of reservoir capacity that is sold in years one through five. In fact, the cost depletion allowance in column 6 is consistently better (higher) than the fixed percentage allowance in column 7. When cost depletion exceeds percentage depletion, the cost depletion allowance must be used in computing taxable income, given that the basis of the property has not been exhausted. It is also noteworthy that cost depletion is not limited to 50% of net income shown in column 5, which is determined before depletion deductions are considered. (Recall that Figure 6-3 summarized the procedure for determining allowable depletion.)

As the firm has a net taxable income of $600,000 even before returns from the geothermal well are considered, we shall assume that the total incremental tax rate (t) is 40%, thus giving the income tax shown in column 9 of Table 6-10. Column 10 shows the net ATCF provided to the investor for years one through five of the well's operation. The remaining 8,000,000 gallons of hot water would presumably be sold over the subsequent 10 to 15 years of the well's operation.

6.14 Spreadsheet Applications

This template is the workhorse for after-tax evaluation of engineering projects. It uses the form given in Figure 6-5 to convert BTCFs to ATCFs.

Cell B8 contains the cost basis, Cells B9:B15 contain the BTCFs, and the market value is in B16. The VDB function is used to determine the MACRS depreciation amounts in Column C. Note the negative sign on the first argument in the VDB function. Cell D3, which is hidden, is needed if MACRS depreciation is used. If the class life is greater than or equal to 15, it specifies that 150% declining balance is used in determining r_k, otherwise, 200% declining balance is used.

Any other depreciation method (SL, SYD, etc.) can be used by placing the appropriate formula in Column C. For simplicity, the formula in Column C is copied to the end of the class life, which is year 6 for this example. Once this point is reached, no further depreciation is allowed. The remainder of the table uses the same approach as illustrated in the text.

The adjusted ATCF column is necessary because period N (in this example $N = 7$) appears twice. The first appearance represents the ordinary BTCFs for that year, while the second represents the consequence of the sale of the asset. This column (G) simply carries forward all ATCFs for years 0 through $N - 1$ and combines the two rows representing year N. This column is used to calculate all financial measures of merit for comparing alternatives. The most common measures of merit appear in cells G18:G20.

	A	B	C	D	E	F	G	H
1	After Tax Analysis, MACRS Depreciation							
2								
3	Tax Rate =	40%		2,00				
4	MARR =	10%						
5	Class Life:	5						
6								
7	Year	BTCF	d(k)	Taxable Income	Income Tax Due	ATCF	Adjusted ATCF	
8	0	($100,000)				($100,000)	($100,000)	
9	1	$20,000	$20,000	$0	$0	$20,000	$20,000	
10	2	$20,000	$32,000	($12,000)	$4,800	$24,800	$24,800	
11	3	$20,000	$19,200	$800	($320)	$19,680	$19,680	
12	4	$20,000	$11,520	$8,480	($3,392)	$16,608	$16,608	
13	5	$20,000	$11,520	$8,480	($3,392)	$16,608	$16,608	
14	6	$20,000	$5,760	$14,240	($5,696)	$14,304	$14,304	
15	7	$20,000	*note*	$20,000	($8,000)	$12,000	$30,000	
16	7	$30,000	*note*	$30,000	($12,000)	$18,000		
17								
18						NPV =	($1,412)	
19						AW =	($290)	
20						IRR =	9.57%	
21	KEY:							
22		= User input		*note*				
23		= Unique formula		copy depreciation formula only to class life +1				
24								
25	Unique Formulas							
26	D3	=IF(B5>=15,1.5,2)						
27	F8	=B8						
28	G8	=F8						
29	C9	=IF(A9=$B$5+1,0.5*C8,VDB(-$B$8,0,$B$5,MAXA(0,A9-1.5),A9-0.5,$D$3,FALSE))						
30	D9	=B9-C9						
31	E9	=$B$3*D9						
32	F9	=B9-E9						
33	G15	=F15+F16						
34	G18	=NPV(B4,G9:G15)+G8						
35	G19	=PMT(B4,7,-G18)						
36	G20	=IRR(G8:G15,B4)						
37								

Spreadsheet template workhorse for after-tax evaluation of engineering projects

▼ 6.15 Summary

In this chapter, we have presented important aspects of federal legislation relating to depreciation, depletion, and income taxes. It is essential to understand these topics so that correct after-tax engineering economy evaluations of proposed projects may be conducted. Depreciation and income taxes are also integral parts of subsequent chapters in this book.

In this chapter, many concepts regarding current federal income tax laws were described. For example, topics such as taxable income, effective income tax rates, taxation of ordinary income, and gains and losses on disposal of assets were explained. A general format for pulling together and organizing all these apparently diverse subjects was presented in Figure 6-5. This format offers the student or practicing engineer a means of collecting on one worksheet information that is required for determining ATCFs and properly evaluating the after-tax financial results of a proposed capital investment. Figure 6-5 was then employed in numerous examples. The student's challenge is now to use this worksheet in organizing data presented in problem exercises at the end of this and subsequent chapters and to answer questions regarding the after-tax profitability of the proposed undertaking(s).

▼ 6.16 References

AMERICAN TELEPHONE AND TELEGRAPH COMPANY, Engineering Department. *Engineering Economy*, 3rd ed. (New York: American Telephone and Telegraph Company, 1977).

ARTHUR ANDERSEN & CO. *Tax Reform 1986: Analysis and Planning*, Subject File AA3010, Item 27. St. Louis, Mo., 1986.

COMMERCE CLEARING HOUSE, INC. *Explanation of Tax Reform Act of 1986.* Chicago, 1987.

LASSER, J. K. *Your Income Tax* [New York: Simon & Schuster (see the latest edition)].

U.S. DEPARTMENT OF THE TREASURY. *Tax Guide for Small Business,* IRS Publication 334, Washington, DC: U.S. Government Printing Office (revised annually).

—. *Depreciating Property Placed in Service Before 1987*, IRS Publication 534, Washington, DC: U.S. Government Printing Office (revised annually).

—. *Sales and Other Dispositions of Assets*, IRS Publication 544, Washington, DC: U.S. Government Printing Office (revised annually).

—. *Investment Income and Expenses*, IRS Publication 550, Washington, DC: U.S. Government Printing Office (revised annually).

—. *Basis of Assets*, IRS Publication 551, Washington, DC: U.S. Government Printing Office (revised annually).

—. *Tax Information on Corporations*, IRS Publication 542, Washington, DC: U.S. Government Printing Office (revised annually).

—. *How to Depreciate Property,* IRS Publication 946, Washington, DC: U.S. Government Printing Office (revised annually).

6.17 Problems

The number in parentheses ( ) that follows each problem refers to the section from which the problem is taken.

6-1. How are depreciation deductions different from other production or service expenses such as labor, material, and electricity? (6.2)

6-2. What conditions must a property satisfy to be considered depreciable? (6.2)

6-3. Explain the difference between real and personal property. (6.2)

6-4. Explain the difference between tangible and intangible property. (6.2)

6-5. Explain how the cost basis of depreciable property is determined. (6.2)

6-6. What is the depreciation deduction, using each of the following methods, for the second year for an asset that costs $35,000 and has an estimated MV of $7,000 at the end of its seven-year useful life? Assume its MACRS class life is also 7 years. (a) SYD, (b) 200% declining-balance, (c) GDS (MACRS), and (d) ADS (MACRS). (6.3, 6.4)

6-7. Your company has purchased a large new truck-tractor for over-the-road use (asset class 00.26). It has a basic cost of $180,000. With additional options costing $15,000, the cost basis for depreciation purposes is $195,000. Its market value at the end of five years is estimated as $40,000. Assume it will be depreciated under the GDS: (6.4)

a. The cumulative depreciation through the end of year three is closest to

  1. $195,000   **2.** $187,775   **3.** $180,000
  4. $151,671   **5.** $180,551

b. The MACRS depreciation in year four is most nearly

  1. 0   **2.** $13,350   3.$14,450
  4. $31,150   **5.** $45,400

c. The book value at the end of year two is most nearly

  1. $33,000   **2.** $36,000   **3.** $42,000
  4. $43,000   **5.** $157,000

6-8. Why would a business elect, under MACRS, to use the ADS rather than the GDS? (6.4)

6-9. The "Big-Deal" Company has purchased new furniture for their offices at a retail price of $100,000. An additional $20,000 has been charged for insurance, shipping, and handling. The company expects to use the furniture for 10 years (useful life = 10 years) and then sell it at a salvage (market) value of $10,000. (6.3, 6.4)

Using the 200% DB method for depreciation (with the above data), answer Questions a–c.

a. What is the depreciation during the second year?

  (i) $16,000   **(ii)** $17,600
  (iii) $24,000   **(iv)** $19,000

b. What is the *book value* of the asset the end of the first year?

  (i) $96,000   **(ii)** $86,000
  (iii) $88,000   **(iv)** $104,000

c. What is the book value of the asset after 10 years?

  (i) $10,000   **(ii)** Unknown
  (iii) $12,885   **(iv)** $16,106

Using the *MACRS* method (with the above data), answer Questions d–f.

d. What is the recovery period (property class) of the asset?

  (i) 10 years   **(ii)** 7 years
  (iii) 5 years   **(iv)** 15 years

e. What is the depreciation of the asset for the first year?

  (i) $17,148   **(ii)** $14,290
  (iii) $12,000   **(iv)** $24,000

f. What is the book value of the asset at the end of the third year?

  (i) $69,120   **(ii)** $52,476
  (iii) $73,464   **(iv)** $57,600

g. If the asset were to be sold at the end of the fourth year, what would be the depreciation during the fourth year?

  (i) $5,352   **(ii)** $7,494
  (iii) $13,842   **(iv)** $14,988

6-10. A company purchased a machine for $15,000. It paid sales taxes and shipping costs of $1,000 and nonrecurring installation costs amounting to $1,200. At the end of three years, the company had no further use for the machine, so it spent $500 to have the machine dismantled and was able to sell the machine for $1,500. (6.3)

a. What is the cost basis for this machine?

b. The company had depreciated the machine on a SL basis, using an estimated useful life of five years and $1,000 SV. By what amount did the depreciation deductions fail to cover the actual depreciation?

6-11. An asset for drilling was purchased and placed in service by a petroleum production company. Its cost basis is $60,000 and it has an estimated MV of $12,000 at the end of an estimated useful life of 14 years. Compute the depreciation amount in the third year and the BV at the end of the fifth year of life by each of these methods: (6.3, 6.4)

a. The SL method.

b. The SYD method.

c. The 200% DB method with switchover to straight line.

d. The GDS.

e. The ADS.

6-12. An optical scanning machine was purchased for $150,000 in the current tax year (year one). It is to be used for reproducing blueprints of engineering drawings, and its MACRS class life is nine years. The estimated MV of this machine at the end of 10 years is $30,000. (6.4)

a. What is the GDS recovery period of the machine?

b. Based on your answer to part (a), what is the depreciation deduction in year four?

c. What is the BV at the beginning of year five?

6-13. A piece of construction equipment (asset class 15.0) was purchased by the Jones Construction Company. The cost basis was $300,000.

a. Determine the GDS and ADS depreciation deductions for this property. (6.4)

b. Compute the difference in present worth of the two sets of depreciation deductions in (a) if $i = 12\%$ per year. (6.5)

6-14. A bowling alley costs $500,000 and has a useful life of 10 years. Its estimated market value at the end of year 10 is $20,000.

a. Determine the depreciation for years 1–10 using: (i) the SL method, (ii) the 200% DB method, and (iii) the MACRS method (GDS class life = 10 years). A table containing some of the depreciation values is provided below. Complete the accompanying table. (6.5)

b. Compute the present worth of the depreciation deductions (at EOY 0) for each of the three methods. The MARR is 10% per year.

c. If a large present worth in part (b) is desirable, what do you conclude regarding which method is preferred?

EOY	SL Method	DB Method	MACRS Method
1		$100,000	$71,450
2		$80,000	
3			
4			
5			
6		$32,768	$44,600
7		$26,214	$44,650
8			
9			
10	$48,000		

6-15. During its current tax year (year one) a pharmaceutical company purchased a mixing tank that had a fair market price of $120,000. It replaced an older, smaller mixing tank that had a BV of $15,000. Because a special promotion was underway, the old tank was used as a trade-in for the new one, and the cash price (including delivery and installation) was set at $99,500. The MACRS class life for the new mixing tank is 9.5 years. (6.4, 6.3)

a. Under the GDS, what is the depreciation deduction in year three?

b. Under the GDS, what is the BV at the end of year four?

c. If 200% DB depreciation had been applied to this problem, what would be the cumulative depreciation through the end of year four?

6-16. A special-purpose machine is to be depreciated as a linear function of use (units-of-production method). It costs $25,000 and is expected to produce 100,000 units and then be sold for $5,000. Up to the end of the third year it had produced 60,000 units, and during the fourth year it produced 10,000 units. What is the depreciation deduction for the fourth year and the BV at the end of the fourth year? (6.3)

6-17. A gold mine that is expected to produce 30,000 ounces of gold is purchased for $2,400,000. The gold can be sold for $450 per ounce; however, it costs $265 per ounce for mining and processing costs. If 3,500 ounces are produced this year, what will be the depletion allowance for (a) unit depletion, and (b) percentage depletion? (6.6)

6-18. A marble quarry is estimated to contain 900,000 tons of stone, and the ZARD Mining Company just purchased this quarry for $1,800,000. If 100,000 tons of marble can be sold each year and the average selling price per ton is $8.60, calculate the first year's depletion allowance for (a) the cost depletion method and (b) percentage depletion at 5% per year. ZARD's net income before deduction of a depletion allowance is $350,000. (6.6)

6-19. A gas well in Oklahoma has reserves of 2,000,000 mcf in the ground. The initial cost basis was $800,000, and during the first year of operation a depletion allowance of $280,000 was taken. At the beginning of the second year of operation, the reserves were reestimated to be 1,400,000 mcf. What is the new value of the depletion unit with the cost method? (6.6)

6-20. Consider a firm that had a taxable income of $90,000 in the current tax year, and total gross revenues of $220,000. Based on this information, answer these questions: (6.7, 6.10)

   a. How much federal income tax was paid for the tax year?
   b. What was the net income after taxes (NIAT)?
   c. What was the total amount of deductible expenses (e.g., materials, labor, fuel, interest) and depreciation deductions claimed in the tax year?

6-21. For the Surefire Automatic Casting Company, gross revenues amounted to $7,800,000. Operating expenses were $4,900,000 and depreciation deductions were $1,200,000. There was no interest on borrowed money. (6.7, 6.11)

   a. How much federal income tax was paid for the tax year?
   b. What was the NIAT?
   c. What was the firm's ATCF?

6-22. Your company is contemplating the purchase of a large stamping machine. The machine will cost $180,000. With additional transportation and installation costs of $5,000 and $10,000, the cost basis for depreciation purposes is $195,000. Its MV at the end of five years is estimated as $40,000. For simplicity, assume that this machine is in the three-year MACRS (GDS) property class. The justifications for this machine include $40,000 savings per year in labor and $30,000 per year in reduced materials. The before-tax MARR is 20% per year, and the effective income tax rate is 40%. (6.4, 6.7, 6.10)

   a. The NIAT at the end of year one is most nearly
      (i) −$13,000   (ii) $3,000   (iii) $23,000
      (iv) $68,000   (v) $130,000
   b. The GDS depreciation in year four is most nearly
      (i) 0   (ii) $13,350   (iii) $14,450
      (iv) $31,150   (v) $45,400
   c. The BV at the end of year two is most nearly
      (i) $33,000   (ii) $36,000   (iii) $42,000
      (iv) $43,000   (v) $157,000
   d. The *total* BTCF in year five is most nearly (assuming that you sell the machine at the end of year five)
      (i) $9,000   (ii) $40,000   (iii) $70,000
      (iv) $80,000   (v) $110,000
   e. The taxable income for year three is most nearly
      (i) $5,010   (ii) $16,450   (iii) $28,880
      (iv) $41,120   (v) $70,000
   f. The PW of the *after-tax savings* from the machine, *in labor and materials only* (neglecting the first cost, depreciation, and the MV), is most nearly (using the after-tax MARR)
      (i) $12,000   (ii) $95,000   (iii) $151,000
      (iv) $184,000   (v) $193,000
   g. Assume that the stamping machine will now be used for only three years due to the loss of several government contracts. The MV at the end of year three is $50,000. What is the income tax owed at the end of year three due to depreciation recapture (gain on disposal)?
      (i) $8,444   (ii) $14,220   (iii) $21,111
      (iv) $35,550   (v) $20,000

6-23. If the incremental federal income tax rate is 34% and the incremental state income tax rate is 6%, what is the effective combined income tax rate (t)? If state income taxes are 12% of taxable income, what now is the value of t? (6.8)

6-24. A corporation has estimated that its taxable income will be $57,000 in the current tax year. It has the opportunity to invest in a project that is expected to add $8,000 to this taxable income. How much federal tax will be owed *with* and *without* the proposed venture? (6.8)

6-25. Determine the after-tax yield (i.e., IRR on the ATCF) obtained by an individual who purchases a $10,000, 10-year, 10% nominal interest rate bond. The following information is given:

- Interest is paid semi-annually, and the bond was bought after the fifth payment had just been received by the previous owner.
- The purchase price for the bond was $9,000.
- All revenues (including capital gains) are taxed at an income rate of 28%.
- The bond is held to maturity.

6-26. In a chlorine fluxing installation in a large aluminum company, engineers are considering the replacement of existing plastic pipe fittings with more expensive, but longer lived, copper fittings. The following table gives a comparison of the capital investments, lives, salvage values, etc., of the two mutually exclusive alternatives under consideration:

	(A) Plastic	(B) Copper
Capital Investment	$5,000	$10,000
Useful (Class) Life	5 year	10 year
Salvage Value for Depreciation Purposes	$1,000(=$SV_5$)	$5,000(=$SV_{10}$)
Annual Expenses	$300	$100
Market Value at End of Useful life	$0	$0

Depreciation amounts are calculated with the SL method. Assume an income-tax rate of 40% and a minimum attractive rate of return after-taxes of 12% per year. Which pipe fitting would you select and why? Carefully list all assumptions that you make in performing the analysis. (6.10, 6.11)

6-27. Storage tanks to hold a highly corrosive chemical are currently made of material Z26. The capital investment in a tank is $30,000, and its useful life is eight years. Your company manufactures electronic components, and uses the alternative depreciation system (ADS) under MACRS to calculate depreciation deductions for these tanks. The net MV of the tanks at the end of their useful life is zero. When a tank is four years old, it must be relined at a cost of $10,000. This cost is not depreciated and can be claimed as an expense during year four.

Instead of purchasing the tanks, they can be leased. A contract for up to 20 years of storage tank service can be written with the Rent-All Company. If your firm's after-tax MARR is 12% per year, what is the greatest annual amount that you can afford to pay for tank leasing without causing purchasing to be the more economical alternative? Your firm's effective income tax rate is 40%. State any assumptions you make. (6.4, 6.10)

6-28. Two fixture are being considered for a particular job in the manufacturing firm. The pertinent data for their comparison are summarized in Table P6-28.

The effective federal and state income-tax rate is 50%. Depreciation recapture is also taxed at 50%. If the after-tax MARR is 8% per year, which of the two fixtures should be recommended? State any important assumptions you make in your analysis. (6.10)

6-29. A firm expects for the next several years to have annual taxable income in the $100,000-to-$335,000 tax rate bracket. A new project is proposed that will raise revenues by $30,000 per year and increase the cost of sales by $10,000 per year. If this new project necessitates a total

TABLE P6-28 Table for Problem P6-28

	Fixture X	Fixture Y
Capital Investment	$30,000	$40,000
Annual Operating Expenses	$3,000	$2,500
Useful Life	6 years	8 years
Market Value	$6,000	$4,000
Depreciation Method	SL to zero book value over 5 years	MACRS (GDS) with a 5-year recovery period

capital investment of $50,000, and has zero MV at the end of its six-year life, what is the IRR after federal income taxes are paid? Assume no state taxes and that MACRS depreciation is used (GDS with a recovery period of five years). (6.4, 6.8, 6.10)

6-30. Two alternative machines will produce the same product, but one is capable of higher-quality work, which can be expected to return greater revenue. The following are relevant data:

	Machine *A*	Machine *B*
First Cost	$20,000	$30,000
Life	$12 years	8 years
Terminal BV (and MV)	$4,000	$0
Annual Receipts	$150,000	$188,000
Annual Expenses	$138,000	$170,000

Determine which is the better alternative assuming "repeatability" and based on using SL depreciation, an income-tax rate of 40%, and an after-tax minimum attractive rate of return of 10% using the following methods: (6.10)

a. Annual Worth
b. Present Worth
c. Internal Rate of Return

6-31. A firm must decide between two system designs, S1 and S2, shown in the accompanying table. Their effective income tax rate is 40%, and MACRS (GDS) depreciation is used. If the after-tax desired return on investment is 10% per year, which design should be chosen? State your assumptions. (6.10)

	Design	
	S1	S2
Capital investment	$100,000	$200,000
GDS recovery period (years)	5	5
Useful life (years)	7	6
Market value at end of useful life	$30,000	$50,000
Annual revenues less expenses over useful life	$20,000	$40,000

6-32. Alternative Methods I and II are proposed for a plant operation. The following is comparative information:

	Method I	Method II
Initial Investment	$10,000	$40,000
Useful (ADR) Life	5 year	10 year
Terminal Market Value	$1,000	$5,000
Annual Expenses		
Labor	$12,000	$4,000
Power	$250	$300
Rent	$1,000	$500
Maintenance	$500	$200
Property Taxes and Insurance	$400	$2,000
Total Annual Expenses	$14,150	$7,000

Determine which is the better alternative based on an after-tax annual cost analysis with an effective income tax of 40% and an after-tax MARR of 12% assuming the following methods of depreciation (6.10)

a. SL
b. MACRS

6-33. Your firm can purchase a machine for $12,000 to replace a rented machine. The rented machine costs $4,000 per year. The machine that you are considering would have a useful life of eight years and a $5,000 MV at the end of its useful life. By how much could annual operating expenses increase and still provide a return of 10% per year after taxes? The firm is in the 40% income tax bracket, and revenues produced with either machine are identical. Assume that alternate MACRS (ADS) depreciation is utilized to recover the investment in the machine, and that the ADS recovery period is five years. (6.4, 6.10)

6-34. An injection molding machine can be purchased and installed for $90,000. It is in the seven-year GDS property class, and is expected to be kept in service for eight years. It is believed that $10,000 can be obtained when the machine is disposed of at the end of year eight. The net annual *value added* (i.e., revenues less expenses) that can be attributed to this machine is constant over eight years and amounts to $15,000. An effective income tax rate of 40% is used by the company, and the after-tax MARR equals 15% per year. (6.4, 6.10)

a. What is the approximate value of the company's before-tax MARR?

b. Determine the GDS depreciation amounts in years one through eight.

c. What is the taxable income at the end of year eight that is related to capital investment?

d. Set up a table and calculate the ATCF for this machine.

e. Should a recommendation be made to purchase the machine?

6-35. Your company has purchased equipment (for $50,000) that will reduce materials and labor costs by $14,000 each year for N years. After N years, there will be no further need for the machine, and because the machine is specially designed, it will have no MV at any time. However, the IRS has ruled that you must depreciate the equipment on an SL basis with a tax life of five years. If the effective income-tax rate is 40%, what is the minimum number of years your firm must operate the equipment to earn 10% per year after taxes on its investment? (6.10)

6-36. A manufacturing process can be designed for varying degrees of automation. The following is relevant cost information:

Degree	First Cost	Annual Labor Expense	Annual Power and Maintenance Expense
A	$10,000	$9,000	$ 500
B	14,000	7,500	800
C	20,000	5,000	1,000
D	30,000	3,000	1,500

Determine which is best by after-tax analysis using an income-tax rate of 40%, and after-tax MARR of 15%, and a straight-line depreciation. Assume that each has a life of five years and no BV or MV. Use each of the following methods: (6.10)

a. Annual Worth

b. Present Worth

c. Internal Rate of Return

6-37. The following information is for a proposed project that will provide the capability to produce a specialized product estimated to have a short market (sales) life:

- Capital investment is $1,000,000 (this includes land and working capital).
- The cost of depreciable property, *which is part* of the $1,000,000 total estimated project cost, is $420,000.
- Assume, for simplicity, that the depreciable property is in the MACRS (GDS) three-year property class.
- The analysis period is three years.
- Annual operating and maintenance expenses are $636,000 in the first year and they increase at the rate of 6% per year (i.e., $\bar{f} = 6\%$) thereafter (see geometric gradient, Chapter 3).
- Estimated MV of depreciable property from the project at the end of three years is $280,000.
- Federal income tax rate = 34%; state income tax rate = 4%.
- MARR (after taxes) is 10% per year.

Based on an after-tax analysis using the PW method, what minimum amount of equivalent uniform annual revenue is required to justify the project economically? (6.10, 6.11)

6-38. Your company has to obtain some new production equipment for the next six years, and leasing is being considered. You have been directed to perform an after-tax study of the leasing approach. The pertinent information for the study is as follows:

Lease costs: First year, $80,000; second year, $60,000; third through sixth years, $50,000 per year. Assume that a six-year contract has been offered by the lessor that fixes these costs over the six-year period. Other costs (not covered in the contract) are $4,000 per year, and the effective income tax rate is 40%.

a. Develop the annual ATCFs for the leasing alternative.

b. If the MARR after taxes is 8% per year, what is the AW for the leasing alternative? (6.10, 6.11)

6-39. Individual industries will use energy as efficiently as economically possible, and there are several incentives to improve the efficiency of energy consumption. One incentive for the purchase of more energy-efficient equipment is to reduce the time allowed to write off the initial cost. Another "incentive" might be to raise the price of energy in the form of an energy tax.

To illustrate these two incentives, consider the selection of a new motor-driven centrifugal pump for a refinery. The pump is to operate

8,000 hours per year. Pump *A* costs $1,600, consumes 10 hp, and has an overall efficiency of 65% (it delivers 6.5 hp). The other available alternative, Pump *B*, costs $1,000, consumes 13 hp, and has an overall efficiency of 50% (it delivers 6.5 hp). *Note*: 1 hp = 0.746 kW.

Compute the after-tax internal rate of return on extra investment in Pump *A*, assuming an effective income-tax rate of 40%, an ADR useful life of 10 years [parts (a) and (c) only], zero market values, and SL depreciation for each of these situations: (6.10)

a. The cost of electricity is $0.04/kWh.
b. A 5-year depreciation write-off period is allowed, the expected life of both pumps is still 10 years, and the cost of electricity is $0.04/kWh.
c. Repeat part (a), but now electricity costs $0.07/kWh.

6-40. AMT, Inc. is considering the purchase of a digital camera for maintenance of design specifications by feeding digital pictures directly into an engineering workstation where computer-aided design files can be superimposed over the digital pictures. Differences between the two images can be noted and corrections, as appropriate, can then be made by design engineers. (6.12)

a. You have been asked by management to *determine the present worth of the economic value added (EVA)* of this equipment assuming the following estimates: capital investment = $345,000; market value at end of year six = $120,000; annual revenues = $120,000; annual expenses = $8,000; equipment life = 6 years; effective income tax rate = 50% and after-tax MARR = 10% per year. MACRS depreciation will be used with a 5-year recovery period.
b. Compute the PW of the equipment's after-tax cash flows. Is your answer in part (a) the same as your answer in part (b)?

6-41. Refer to Example 6-17. Show that the present worth of the annual *economic value added (EVA)* amounts by the new machinery is the same as the PW of the ATCF amounts ($17,208) given in Table 6-6. (6.11, 6.12)

6-42. Rework Example 6-21 using the MACRS depreciation method (assume three-year property class) instead of the SL depreciation method. (6.12)

6-43. The Greentree Lumber Company is attempting to evaluate the profitability of adding another cutting line to its present sawmill operations. They would need to purchase 2 more acres of land for $30,000 (total). The equipment would cost $130,000 and could be depreciated over a five-year recovery period with the MACRS method. Gross revenue is expected to increase by $50,000 per year for five years and operating expenses will be $15,000 annually for five years. It is expected that this cutting line will be closed down after five years. The firm's effective income-tax rate is 50%. If the company's after-tax MARR is 5% per year, is this a profitable investment? (6.10)

6-44. Refer to Figure 6-3 and the depletion example in Table 6-10. Suppose that in years 6 through 10 of this well's operation, hot water can be sold for $0.22 per gallon, a constant 1,000,000 gallons per year can be sold, and the depletion allowance is 22%. The firm's expected net income, before any depletion allowance has been deducted, is $80,000 per year (in column 4 of Table 6-10). If the effective income tax rate remains at 40%, what is the net cash flow after taxes in years 6 through 10? (6.13)

6-45. A large mineral deposit in Wyoming is estimated to contain 1,000,000 tons of a mineral whose percentage depletion allowance is 22%. A mining company has made an initial investment of $40,000,000 to recover this ore, and the market price for the ore is $175 per ton. The company's after-tax MARR is 12% per year, and its effective income tax rate is 40%. It is anticipated that the ore will be sold at the rate of 100,000 tons per year and that operating expenses, exclusive of depletion deductions, will be approximately $9,000,000 per year. (6.13)

a. Determine the ATCF for this mining venture when percentage depletion (or cost depletion, if appropriate) is used.
b. Determine the PW of ATCF in part (a).

6-46. Allen International, Inc. manufactures chemicals. It needs to acquire a new piece of production equipment to work on production for a large order that Allen has received. The order is for a period of three years, and at the end of that time the machine would be sold.

Allen has received two supplier quotations, both of which would provide the required service. Quotation I has a first cost of $180,000 and an estimated salvage value of $50,000 at the end of three years. Its cost for operation and maintenance is estimated at $28,000 per year. Quotation II has a first cost of $200,000 and an estimated salvage value of $60,000 at the end of three years. Its cost for operation and maintenance is estimated at $17,000 per year. The company pays income tax at a rate of 40% on ordinary income and 28% on depreciation recovery The machine will be depreciated using MACRS-GDS (asset class 28.0). Allen uses an after-tax MARR of 12% for economic analysis and it plans to accept whichever of these two quotations costs less. (6.10)

To perform an after-tax analysis to determine which of these machines should be acquired, you must

a. State the study period you are using.
b. Show all numbers necessary to support your conclusions.
c. State what the company should do.

CHAPTER 7

Cost Estimation Techniques

The objectives of this chapter are (1) to discuss an integrated approach used to develop cash flows for the alternatives being analyzed in a study, and (2) to delineate and illustrate selected techniques that will be useful in making such estimates.

The following topics are discussed in this chapter:

An integrated approach for developing cash flows

Definition of a work breakdown structure

The cost and revenue structure

Estimating techniques (models)

Parametric cost estimating

Description of the learning curve effect

Cost estimation during the design process

Estimating cash flows for a typical small project

7.1 Introduction

In Chapter 1, we discussed the engineering economic analysis procedure in terms of seven steps, which are listed here:

1. Recognition and formulation of the problem.
2. Development of the feasible alternatives.
3. Development of the net cash flow (and other prospective outcomes) for each alternative.
4. Selection of a criterion (or criteria) for determining the preferred alternative.
5. Analysis and comparison of the alternatives.

6. Selection of the preferred alternative.
7. Performance monitoring and postevaluation of results.

In Chapters 3 through 6, the methodology needed to accomplish Steps 4, 5, and 6 was developed and demonstrated. In this chapter, we return to Step 3.

> Because engineering economy studies deal with outcomes that extend into the future, estimating the future cash flows for feasible alternatives is a critical step in the analysis procedure. A decision based on the analysis is economically sound only to the extent that these cost and revenue estimates are representative of what subsequently will occur.

In Step 1 of the procedure, the need for doing an analysis was identified; the specific situation (improvement opportunity, design project, new venture, etc.) was explicitly defined; the desired outcomes in terms of goals, objectives, and other results were developed; and any special conditions and constraints that needed to be met were delineated. Then, in Step 2, the feasible alternatives to be analyzed in the engineering economy study were selected and described using the systems approach.

Thus, in Step 3 the *alternatives* to be analyzed have *already been selected* and the differences between them *already highlighted.* Other important information (results to be achieved and requirements to be met) that is needed in the analysis is available from the first two steps.

Applying the concepts and methodology discussed in this chapter is an important part of engineering practice. A commercial building project is used as the basis for some of the examples in Chapter 7. Any other engineering project, such as the expansion of a chemical processing plant or the design of an electrical distribution system switching center, could have been chosen.

7.2 An Integrated Approach

An integrated approach to developing the net cash flows for the feasible project alternatives (Step 3) is shown in Figure 7-1. We will use the term *project* to refer to the undertaking that is the subject of the analysis. This integrated approach includes three basic components:

1. *Work Breakdown Structure (WBS).* This is a technique for explicitly defining, at successive levels of detail, the work elements of a project and their interrelationships (sometimes called a *work element structure*).
2. *Cost and revenue structure (classification).* Delineation of the cost and revenue categories and elements is made for estimates of cash flows at each level of the WBS.
3. *Estimating techniques (models).* Selected mathematical models are used to estimate the future costs and revenues during the analysis period.

These three basic components, together with integrating procedural steps, provide an organized approach for developing the cash flows for the alternatives.

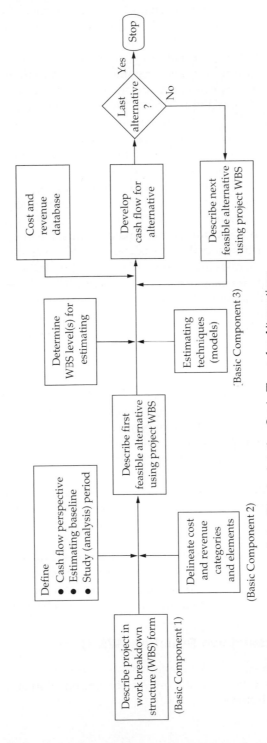

Figure 7-1 Integrated Approach for Developing the Cash Flows for Alternatives

As shown in Figure 7-1, the integrated approach begins with a description of the project in terms of a WBS. This project WBS is used to describe the project and each alternative's unique characteristics in terms of design, labor, material requirements, and so on. Then these variations in design, resource requirements, and other characteristics are reflected in the estimated future costs and revenues (net cash flow) for that alternative.

To estimate future costs and revenues for an alternative, the perspective (viewpoint) of the cash flow must be established and an estimating baseline and analysis period defined. Normally, cash flows are developed from the owner's viewpoint.

The net cash flow for an alternative represents what is estimated to happen to future revenues and costs from the perspective being used. Therefore, the estimated changes in revenues and costs associated with an alternative have to be relative to a baseline that is consistently used for all the alternatives being compared. This baseline is defined and applied in either of two ways.

The first method is the *total revenue and cost approach.* That is, the no-change (do nothing) alternative is explicitly included in the set of alternatives, and the total revenues and costs for it are estimated. Thus, when the total cost and revenue baseline approach is used, the net cash flow for the no-change alternative represents the projected revenues and costs of the current operation or situation. Similarly, the net cash flow for each of the other feasible alternatives is estimated.

The second method often used is the *differential approach.* Using this approach, the cash flow for the no-change alternative is defined as zero whether or not it is one of the feasible alternatives. The cash flow for each of the other feasible alternatives then represents the estimated differences (changes) in revenues and costs relative to the current situation (no-change alternative).

Whichever estimating baseline approach is used in a study, it must be consistently applied for all feasible alternatives. *A common error is to inadvertently use both baseline definitions when developing the individual cash flows.* For example, the total revenue and cost approach might be used in estimating maintenance costs for the no-change alternative, but in the other alternatives these costs might be estimated by using differences from current operations.

Before developing the cash flows, other procedural steps need to be accomplished. First, decide what level(s) of the WBS to use for developing the cost and revenue estimates. The purpose of the study will be a primary factor in this decision. If the study is a project feasibility analysis, cost and revenue estimating will be less accurate than in the detailed economic analysis that will be used to make the final decision about a project. (This is discussed further in Section 7.2.3.)

Next, organize cost and revenue information from sources internal and external to the organization and assemble the relevant data for the study. Use these data, together with selected estimating techniques (models), to develop the estimates.

7.2.1 The Work Breakdown Structure (WBS)

We briefly defined a work breakdown structure (work element structure) in Section 7.2 and identified it as the first basic component in an integrated approach to developing cash flows.

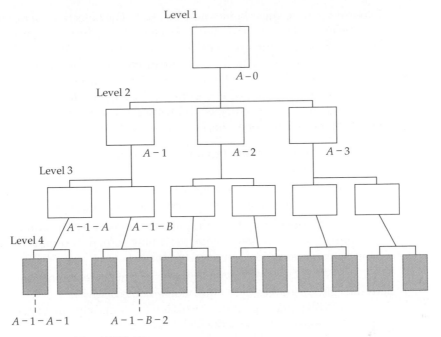

Figure 7-2 The WBS Diagram

The WBS technique is a basic tool in project management and is a vital aid in an engineering economy study. The WBS serves as a framework for defining all project work elements and their interrelationships, collecting and organizing information, developing relevant cost and revenue data, and integrating project management activities. If a WBS does not exist and the project is of reasonable size, the first step in preparing cash flows for the alternatives should be to develop one.

The WBS is essential in ensuring the inclusion of all work elements, eliminating duplications and overlaps between work elements, avoiding nonrelated activities, and preventing other errors that could be introduced into the study. A WBS description dictionary is often prepared for large projects to ensure that each work element in the hierarchy is uniquely defined.

Figure 7-2 shows a diagram of a typical four-level work breakdown structure. It is developed from the top (project level) down in successive levels of detail. The project is divided into its major work elements (Level 2). These major elements are then divided to develop Level 3, and so on. For example, an automobile (first level of the WBS) can be divided into second-level components (or work elements) such as the chassis, drive train, and electrical system. Then each second-level component of the WBS can be subdivided further into third-level elements. The drive train, for example, can be subdivided into third-level components such as the engine, differential, and transmission. This process is continued until the desired detail in the definition and description of the project or system is achieved.

Different numbering schemes may be used. The objectives of numbering are to indicate the interrelationships of the work elements in the hierarchy and to facilitate the manipulation and integration of data. The scheme illustrated in Figure 7-2 is an alphanumeric format. Another scheme often used is all numeric—Level 1: 1-0; Level 2: 1-1, 1-2, 1-3; Level 3: 1-1-1, 1-1-2, 1-2-1, 1-2-2, 1-3-1, 1-3-2; and so on (i.e., similar to the organization of this book). Usually, the level is equal (except for Level 1) to the number of characters indicating the work element.

Other characteristics of a project WBS are as follows:

1. Both functional (e.g., planning) and physical (e.g., foundation) work elements are included in it:
   (a) Typical functional work elements are logistical support, project management, marketing, engineering, and systems integration.
   (b) Physical work elements are the parts that make up a structure, product, piece of equipment, weapon system, or similar item; they require labor, materials, and other resources to produce or construct.
2. The content and resource requirements for a work element are the sum of the activities and resources of related subelements below it.
3. A project WBS usually includes recurring (e.g., maintenance) and nonrecurring (e.g., initial construction) work elements.

EXAMPLE 7-1

You have been appointed by your company to manage a project involving construction of a small commercial building with two floors of 15,000 gross square feet each. The ground floor is planned for small retail shops, and the second floor is planned for offices. Develop the first three levels of a representative WBS adequate for all project efforts from the time the decision was made to proceed with the design and construction of the building until initial occupancy is completed.

SOLUTION
There would be variations in the WBSs developed by different individuals for a commercial building. However, a representative three-level WBS is shown in Figure 7-3. Level 1 is the total project. At Level 2, the project is divided into seven major physical work elements and three major functional work elements. Then each of these major elements is divided into subelements as required (Level 3). The numbering scheme used in this example is all numeric.

7.2.2 The Cost and Revenue Structure

The second basic component of the integrated approach for developing cash flows (Figure 7-1) is the cost and revenue structure. This structure is used to identify and categorize the costs and revenues that need to be included in the analysis. Detailed data are developed and organized within this structure for use with the estimating techniques of Section 7.3 to prepare the cash-flow estimates.

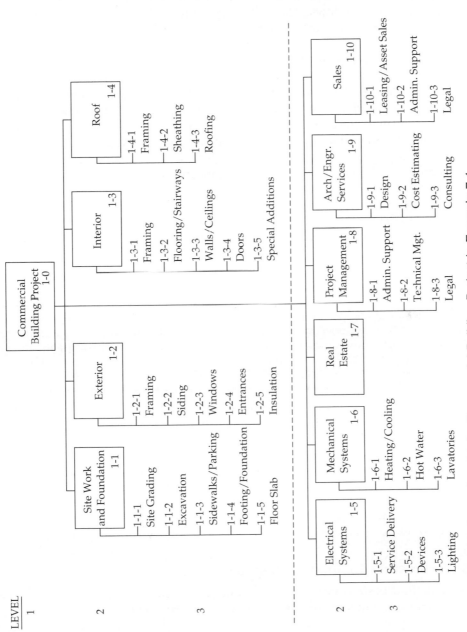

LEVEL
1

2

3

2

3

Commercial
Building Project
1-0

Site Work
and Foundation
1-1

1-1-1
Site Grading

1-1-2
Excavation

1-1-3
Sidewalks/Parking

1-1-4
Footing/Foundation

1-1-5
Floor Slab

Exterior
1-2

1-2-1
Framing

1-2-2
Siding

1-2-3
Windows

1-2-4
Entrances

1-2-5
Insulation

Interior
1-3

1-3-1
Framing

1-3-2
Flooring/Stairways

1-3-3
Walls/Ceilings

1-3-4
Doors

1-3-5
Special Additions

Roof
1-4

1-4-1
Framing

1-4-2
Sheathing

1-4-3
Roofing

Electrical
Systems
1-5

1-5-1
Service Delivery

1-5-2
Devices

1-5-3
Lighting

Mechanical
Systems
1-6

1-6-1
Heating/Cooling

1-6-2
Hot Water

1-6-3
Lavatories

Real
Estate
1-7

Project
Management
1-8

1-8-1
Admin. Support

1-8-2
Technical Mgt.

1-8-3
Legal

Arch/Engr.
Services
1-9

1-9-1
Design

1-9-2
Cost Estimating

1-9-3
Consulting

Sales
1-10

1-10-1
Leasing/Asset Sales

1-10-2
Admin. Support

1-10-3
Legal

Figure 7-3 WBS (Three Levels) for Commercial Building Project in Example 7-1

The life-cycle concept was discussed and illustrated in Chapter 2. The life cycle is divided into two general time periods: the acquisition phase and the operation phase. It begins with initial identification of the economic need or want (the requirement) and ends with retirement or disposal. Thus, it is intended to encompass all present and future costs and revenues.

The life-cycle concept and the WBS are important aids in developing the cost and revenue structure for a project. The life cycle defines a maximum time period and establishes a range of cost and revenue elements that need to be considered in developing cash flows. The WBS focuses the analyst's effort on the specific functional and physical work elements of a project, and on its related costs and revenues.

Ideally, the study period for a project is the life cycle of the product, structure, system, or service involved. This permits all relevant costs and revenues, both present and future, to be fully considered in decision making. Also, the study period directs attention to *explicit trade-offs between initial costs during the acquisition phase and all subsequent costs and revenues during the operation phase.*

However, the accuracy of cost and revenue estimates decreases with increases in the length of the study period. Also, the effort required to develop cash flows increases with the length of the study period. Thus, a time horizon for the study period is selected to balance these factors and provide a sound basis for decision making.

As previously discussed, judgment is required, based on the decision situation, to determine the study period, and thus how far into the future to estimate costs and revenues in an engineering economy study. This judgment should also weigh which cost and revenue elements are the most important and deserve more detailed study and which elements, even if drastically misjudged, will not produce significant changes in the estimated cash flows.

> Perhaps the most serious source of errors in developing cash flows is overlooking important categories of costs and revenues. The cost and revenue structure, prepared in tabular or checklist form, is a good means of preventing such oversights. Technical familiarity with the project is essential in ensuring the completeness of the structure, as are using the life-cycle concept and the WBS in its preparation.

The following is a brief listing of some categories of costs and revenues that are typically needed in an engineering economy study (some of these terms were discussed in Chapter 2):

1. Capital investment (fixed and working)
2. Labor costs
3. Material costs
4. Maintenance costs
5. Property taxes and insurance
6. Quality (and scrap) costs
7. Overhead costs
8. Disposal costs
9. Revenues
10. Salvage or market values

7.2.3 Estimating Techniques (Models)

The third basic component of the integrated approach (Figure 7-1) involves estimating techniques (models). These techniques, together with the detailed cost and revenue data, are used to develop individual cash-flow estimates and the net cash-flow for each alternative.

> The purpose of estimating is to develop cash-flow projections—*not to produce exact data* about the future, which is virtually impossible. Neither a preliminary estimate nor a final estimate is expected to be exact; rather, it should adequately suit the need at a reasonable cost and is often presented as a range of numbers.

Cost and revenue estimates can be classified according to detail, accuracy, and their intended use as follows:

1. *Order-of-magnitude estimates:* used in the planning and initial evaluation stage of a project.
2. *Semidetailed, or budget, estimates:* used in the preliminary or conceptual design stage of a project.
3. *Definitive (detailed) estimates:* used in the detailed engineering/construction stage of a project.

Order of magnitude estimates are used in selecting the feasible alternatives for the study. They typically provide accuracy in the range of ±30 to 50% and are developed through semiformal means such as conferences, questionnaires, and generalized equations applied at Level 1 or 2 of the WBS.

Budget (semidetailed) estimates are compiled to support the preliminary design effort and decision making during this project period. Their accuracy usually lies in the range of ±15%. These estimates differ in the fineness of cost and revenue breakdowns and the amount of effort spent on the estimate. Estimating equations applied at Levels 2 and 3 of the WBS are normally used.

Detailed estimates are used as the basis for bids and to make detailed design decisions. Their accuracy is about ±5%. They are made from specifications, drawings, site surveys, vendor quotations, and in-house historical records and are usually done at Level 3 and successive levels in the WBS.

Thus, it is apparent that a cost or revenue estimate can vary from a "back of the envelope" calculation by an expert to a very detailed and accurate prediction of the future prepared by a project team. The level of detail and accuracy of estimates should depend on

1. Time and effort available as justified by the importance of the study.
2. Difficulty of estimating the items in question.
3. Methods or techniques employed.
4. Qualifications of the estimator(s).
5. Sensitivity of study results to particular factor estimates.

As estimates become more detailed, accuracy typically improves, but the cost of estimating increases dramatically. This general relationship is shown in Figure 7-4 and illustrates the idea that cost and revenue estimates should be prepared with full recognition of how accurate a particular study requires them to be.

Figure 7-4

Accuracy of Cost and
Revenue Estimates
versus the Cost of
Making Them

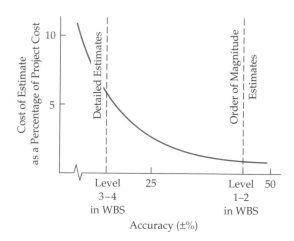

Regardless of how estimates are made, individuals who use them should recognize that the estimates will be in error to some extent, even if sophisticated estimation techniques are used. However, estimation errors can be minimized by using reliable data and the appropriate estimating method.

7.2.3.1 Sources of Estimating Data The number of information sources useful in cost and revenue estimating is too great to list completely. The following four major sources of information are listed roughly in order of importance:

1. Accounting records.
2. Other sources within the firm.
3. Sources outside the firm.
4. Research and development.

1. *Accounting records.* A prime source of information for economic analyses; however, they are often not suitable for direct, unadjusted use.

   A brief discussion of the accounting process and information is given in Appendix A. In its most basic sense, accounting consists of a series of procedures for keeping a detailed record of monetary transactions between established categories of assets, each of which has an accepted interpretation useful for its own purposes. The data generated by the accounting function are often inherently misleading for engineering economic analyses, not only because they are based on past results, but also because of the following limitations:

   (a) The accounting system is rigidly categorized. Categories of various types of assets, liabilities, net worth, income, and expenses for a given firm may be perfectly appropriate for operating decisions and financial summaries, but rarely are they fully appropriate to the needs of economic analyses and decision making involving engineering design and project alternatives.

   (b) Standard accounting conventions cause misstatements of some types of financial information to be built into the system. These misstatements tend to be based on the philosophy that management should avoid overstating the value of its assets or understating the value of its liabilities and should therefore assess them very conservatively.

(c) Accounting data often have illusory precision and implied authoritativeness. Although it is customary to present data to the nearest dollar or the nearest cent, the records are not nearly that accurate in general.

In summary, accounting records are a good source of historical data, but have some limitations when used in making prospective estimates for engineering economic analyses. Moreover, accounting records rarely contain direct statements of incremental costs or opportunity costs, both of which are essential in most engineering economic analyses.

2. *Other sources within the firm.* The typical firm has a number of people and records that may be excellent sources of estimating information. Examples of functions within firms that keep records useful to economic analyses are engineering, sales, production, quality, purchasing, and personnel.

3. *Sources outside the firm.* There are numerous sources outside the firm that can provide helpful information. The main problem is in determining those that are most beneficial for particular needs. The following is a listing of some commonly used outside sources:

(a) *Published information.* Technical directories, buyer indexes, U.S. government publications, reference books, and trade journals offer a wealth of information. For instance, *Standard and Poor's Industry Surveys* gives monthly information regarding key industries. *The Statistical Abstract of the United States* is a remarkably comprehensive source of cost indexes and data. The Bureau of Labor Statistics publishes many periodicals that are good sources of labor costs, such as the *Monthly Labor Review, Employment and Earnings, Current Wage Developments, Handbook of Labor Statistics*, and the *Chartbook on Wages, Prices and Productivity.* An annual construction cost handbook, *Building Construction Cost Data*, is published by the R. S. Means Company in Kingston, Massachusetts. It includes standard crew sizes, unit prices, and prevailing wage rates for various regions of the country.

(b) *Personal contacts* are excellent potential sources. Vendors, salespeople, professional acquaintances, customers, banks, government agencies, chambers of commerce, and even competitors are often willing to furnish needed information on the basis of a serious and tactful request.

4. *Research and Development (R&D).* If the information is not published and cannot be obtained by consulting someone, the only alternative may be to undertake R&D to generate it. Classic examples are developing a pilot plant and undertaking a test market program. These activities are usually expensive and may not always be successful; thus, this final step is taken only in connection with very important decisions and when the sources mentioned earlier are known to be inadequate.

The evaluation of the market and the business environment for large new capital projects is, along with the related estimating of project sales, product prices, and so on, a major area of analysis. R. F. de la Mare provides a good summary discussion of economic forecasting and market analysis related to large investment projects, and of the incorporation of revenue estimates into cash flows.*

7.2.3.2 How Estimates Are Accomplished

Estimates can be prepared in a number of ways—for example, by

1. A *conference* of various people who are thought to have good information or bases for estimating the quantity in question. A special version of this is the *Delphi method*, which involves cycles of questioning and feedback in which the opinions of individual participants are kept anonymous.
2. *Comparison* with similar situations or designs about which there is more information and from which estimates for the alternatives under consideration can be extrapolated. This is sometimes called *estimating by analogy*. The comparison method may be used to approximate the cost of a design or product that is new. This is done by taking the cost of a more complex design for a similar item as an upper bound and the cost of a less complex item of similar design as a lower bound. The resulting approximation may not be very accurate, but the comparison method does have the virtue of setting bounds that might be useful for decision making.
3. *Using quantitative techniques,* which do not always have standardized names. Some selected techniques, with the names used being generally suggestive of the approaches, are discussed in the next section.

7.3 Selected Estimating Techniques (Models)

The estimating models discussed in this section are applicable for order-of-magnitude estimates and for many semidetailed or budget estimates. They are useful in the initial selection of feasible alternatives for further analysis and in the conceptual or preliminary design phase of a project. Sometimes, these models can be used in the detailed design phase of a project to reduce the number of engineering estimates based on bills of material, standard costs, and other detailed information.

7.3.1 Indexes

Costs and prices† vary with time for a number of reasons, including 1) technological advances; 2) availability of labor and materials; and 3) inflation. An *index* is a dimensionless number that indicates how a cost or a price has changed with time (typically escalated) with respect to a base year. Indexes provide a convenient means for developing present and future cost and price estimates from historical data. An estimate of the cost or selling price of an item in year n can be obtained

---

* R. F. de la Mare, *Manufacturing Systems Economics: The Life-Cycle Cost and Benefits of Industrial Assets* (London: Holt, Rinehart and Winston, 1982), pp. 123–149.

† The terms *cost* and *price* are often used together. The cost of a product or service is the total of the resources, direct and indirect, required to produce it. The price is the value of the good or service in the marketplace. In general, price is equal to cost plus a profit.

by multiplying the cost or price of the item at an earlier point in time (year k) by the ratio of the index value in year n ($\overline{I}_n$) to the index value in year k ($\overline{I}_k$);* that is,

$$C_n = C_k \left(\frac{\overline{I}_n}{\overline{I}_k} \right), \tag{7-1}$$

where k = reference year (e.g., 1996) for which cost or price of item is known;

n = year for which cost or price is to be estimated $(n > k)$;

C_n = estimated cost or price of item in year n;

C_k = cost or price of item in reference year k.

Equation (7-1) is sometimes referred to as the *ratio technique* of updating costs and prices. Use of this technique allows the cost or potential selling price of an item to be taken from historical data with a specified base year and updated with an index. This concept can be applied at the lower levels of a WBS to estimate the cost of equipment, materials, and labor, as well as at the top level of a WBS to estimate the total project cost of a new facility, bridge, and so on.

EXAMPLE 7-2

A certain index for the cost of purchasing and installing utility boilers is keyed to 1974, where its baseline value was arbitrarily set at 100. Company XYZ installed a 50,000-lb/hr boiler in 1996 for $525,000 when the index had a value of 468. This same company must install another boiler of the same size in 1999. The index in 1999 is 542. What is the approximate cost of the new boiler?

SOLUTION
In this example, n is 1999 and k is 1996. From Equation (7-1), an approximate cost of the boiler in 1999 is

$$C_{1999} = \$525,000(542/468) = \$608,013.$$

Indexes can be created for a single item or for multiple items. For a single item, the index value is simply the ratio of the cost of the item in the current year to the cost of the same item in the reference year, multiplied by the reference year factor (typically, 100). A composite index is created by averaging the ratios of selected item costs in a particular year to the cost of the same items in a reference year. The developer of an index can assign different weights to the items in the index

---

* In this section only, k is used to denote the reference year.

according to their contribution to total cost. For example, a general weighted index is given by

$$\bar{I}_n = \frac{W_1(C_{n1}/C_{k1}) + W_2(C_{n2}/C_{k2}) + \cdots + W_M(C_{nM}/C_{kM})}{W_1 + W_2 + \cdots + W_M} \times \bar{I}_k, \qquad (7\text{-}2)$$

where M = total number of items in the index ($1 \le m \le M$);

C_{nm} = unit cost (or price) of the mth item in year n;

C_{km} = unit cost (or price) of the mth item in year k;

W_m = weight assigned to the mth item;

$\bar{I}_k$ = composite index value in year k.

The weights $W_1, W_2, \ldots, W_M$ can sum to any positive number, but typically sum to 1.00 or 100. Almost any combination of labor, material, products, services, and so on can be used for a composite cost or price index.

EXAMPLE 7-3

Based on the following data, develop a weighted index for the price of a gallon of gasoline in 1999, when 1986 is the reference year having an index value of 99.2. The weight placed on *regular unleaded* gasoline is three times that of either premium or unleaded plus, because roughly three times as much regular unleaded is sold compared with premium or unleaded plus.

	Price (Cents/Gal) in year		
	1986	1992	1999
Premium	114	138	120
Unleaded plus	103	127	109
Regular unleaded	93	117	105

SOLUTION
In this example, k is 1986 and n is 1999. From Equation (7-2), the value of $\bar{I}_{1999}$ is

$$\frac{(1)(120/114) + (1)(109/103) + (3)(105/93)}{1+1+3} \times 99.2 = 109.$$

Now if the index in 2004, for example, is estimated to be 189, it is a simple matter to determine the corresponding prices of gasoline from $\bar{I}_{1999} = 109$:

$$\text{Premium:} \quad 120 \text{ cents/gal} \left(\frac{189}{109}\right) = 208 \text{ cents/gal},$$

$$\text{Unleaded plus:} \quad 109 \text{ cents/gal} \left(\frac{189}{109}\right) = 189 \text{ cents/gal},$$

$$\text{Regular unleaded:} \quad 105 \text{ cents/gal} \left(\frac{189}{109}\right) = 182 \text{ cents/gal}.$$

Many indexes are periodically published, including the *Engineering News Record* Construction Index, which incorporates labor and material costs, and the Marshall and Stevens cost index. The *Statistical Abstract of the United States* publishes government indexes on yearly materials, labor, and construction costs. The Bureau of Labor Statistics publishes the *Producer Prices and Price Indexes* and the *Consumer Price Index Detailed Report.* Indexes of cost and price changes are frequently used in engineering economy studies.

7.3.2 Unit Technique

The *unit technique* involves using a "per unit factor" that can be estimated effectively. Examples are as follows:

Capital cost of plant per kiloWatt of capacity

Revenue per mile

Fuel cost per kilowatt-hour generated

Annual savings per 500 operating hours

Capital cost per installed telephone

Revenue per customer served

Temperature loss per 1,000 feet of steam pipe

Operating cost per mile

Revenue per case

Maintenance cost per hour

Construction cost per square foot

Revenue per thousand pounds

Such factors, when multiplied by the appropriate unit, give a total estimate of cost, savings, or revenue.

As a simple example, suppose that we need a preliminary estimate of the cost of a particular house. Using a unit factor of, say, $55 per square foot and knowing that the house is approximately 2,000 square feet, we estimate its cost to be $55 × 2,000 = $110,000.

While the unit technique is very useful for preliminary estimating purposes, such average values can be misleading. In general, more detailed methods will result in greater estimation accuracy.

7.3.3 Factor Technique

The *factor technique* is an extension of the unit method, within a basic segmenting strategy, in which one sums the product of several quantities or components and adds these to any components estimated directly. That is,

$$C = \sum_d C_d + \sum_m f_m U_m, \tag{7-3}$$

where C = cost being estimated;

C_d = cost of the selected component d that is estimated directly;

f_m = cost per unit of component m;

U_m = number of units of component m.

As a simple example, suppose that we need a slightly refined estimate of the cost of a house consisting of 2,000 square feet, two porches, and a garage. Using a unit factor of $50 per square foot, and $5,000 per porch and $8,000 per garage for the two directly estimated components, we can calculate the total estimate as

$$(\$5,000 \times 2) + \$8,000 + (\$50 \times 2,000) = \$118,000.$$

The factor technique is particularly useful when the complexity of the estimating situation does not require a WBS, but several different parts are involved. Example 7-4 and the product cost-estimating example to be presented in Section 7.5.1 further illustrate this technique.

EXAMPLE 7-4

The detailed design of the commercial building described in Example 7-1 affects the utilization of the gross square feet (and, thus, the net rentable space) available on each floor. Also, the size and location of the parking lot and the prime road frontage available along the property may offer some additional revenue sources. As project manager, analyze the potential revenue impacts of the following considerations.

The first floor of the building has 15,000 gross square feet of retail space, and the second floor has the same amount planned for office use. Based on discussions with the sales staff, you develop the following additional information:

1. The retail space should be designed for two different uses—60% for a restaurant operation (utilization = 79%) and 40% for a retail clothing store (utilization = 83%).
2. There is a high probability that all the office space on the second floor will be leased to one client (utilization = 89%).
3. An estimated 20 parking spaces can be rented on a long-term basis to two existing businesses that adjoin the property. Also, one spot along the road frontage can be leased to a sign company for erection of a billboard without impairing the primary use of the property.

SOLUTION
Based on this information, you estimate annual project revenue ($\hat{R}$) as

$$\hat{R} = W(r_1)(12) + Y(r_2)(12) + \sum_{j=1}^{3} S_j(u_j)(d_j),$$

where　W = number of parking spaces;

Y = number of billboards;

r_1 = rate per month per parking space = $22;

r_2 = rate per month per billboard = $65;

j = index of type of building space use;

S_j = space (gross square feet) being used for purpose j;

u_j = space j utilization factor (% net rentable);

d_j = rate per (rentable) square foot per year of building space used for purpose j.

Then

$$\hat{R} = [20(\$22)(12) + 1(\$65)(12)] + [9{,}000(0.79)(\$23)$$

$$+ 6{,}000(0.83)(\$18) + 15{,}000(0.89)(\$14)]$$

$$\hat{R} = \$6{,}060 + \$440{,}070 = \$446{,}130.$$

A breakdown of the annual estimated project revenue in Example 7-4 shows that

1.4% is from miscellaneous revenue sources,

98.6% is from leased building space.

From a detailed design perspective, changes in annual project revenue due to changes in building space utilization factors can be easily calculated. For example, an average 1% improvement in the ratio of rentable space to gross square feet would change the annual revenue as follows:

$$\Delta\hat{R} = \sum_{j=1}^{3} S_j(u_j + 0.01)(d_j) - (\$446{,}130 - \$6{,}060)$$

$$= \$445{,}320 - \$440{,}070$$

$$= \$5{,}250 \text{ per year.}$$

7.4 Parametric Cost Estimating

Parametric cost estimating is the use of historical cost data and statistical techniques to predict future costs. Statistical techniques are used to develop cost estimating relationships (CERs) that tie the cost or price of an item (e.g., a product, good, service, or activity) to one or more independent variables (i.e., cost drivers). Recall from Chapter 2 that cost drivers are design variables that account for a large portion of total cost behavior. Table 7-1 lists a variety of items and associated cost

TABLE 7-1 Examples of Cost Drivers Used in Parametric Cost Estimates

Product	Cost Driver (Independent Variable)
Construction	Floor space, roof surface area, wall surface area
Trucks	Empty weight, gross weight, horsepower
Passenger car	Curb weight, wheel base, passenger space, horsepower
Turbine engine	Maximum thrust, cruise thrust, specific fuel consumption
Reciprocating engine	Piston displacement, compression ratio, horsepower
Sheetmetal	Net weight, number of holes drilled, number of rivets placed
Aircraft	Empty weight, speed, wing area
Diesel locomotive	Horsepower, weight, cruising speed
Pressure vessels	Volume
Spacecraft	Weight
Electrical power plants	KiloWatts
Motors	Horsepower
Computers	Megabytes
Software	Number of lines of code
Documentation	Pages
Jet engines	Pounds of thrust

drivers. The unit technique described in the previous section is a simple example of parametric cost estimating.

Parametric models are used in the early design stages to get an idea of how much the product (or project) will cost based on a few physical attributes (such as weight, volume, and power). The output of the parametric models (an estimated cost) is used to gauge the impact of design decisions on the total cost. Awareness of the effect our engineering design decisions have on total cost is essential to developing a product that is both economically and technically sound.

Various statistical and other mathematical techniques are used to develop the cost estimating relationships. For example, simple linear regression and multiple linear regression models, which are standard statistical methods for estimating the value of a dependent variable (the unknown quantity) as a function of one or more independent variables, are often used to develop estimating relationships. This section describes two commonly used estimating relationships, the power-sizing technique and the learning curve, followed by an overview of the procedure used to develop CERs.

7.4.1 Power-Sizing Technique

The *power-sizing technique*, which is sometimes referred to as an *exponential model*, is frequently used for developing capital investment estimates for industrial plants and equipment. This CER recognizes that cost varies as some power of the change in capacity or size. That is,

$$\frac{C_A}{C_B} = \left(\frac{S_A}{S_B}\right)^X,$$

$$C_A = C_B \left(\frac{S_A}{S_B}\right)^X, \tag{7-4}$$

where
$$\left.\begin{array}{l} C_A = \text{cost for plant A} \\ C_B = \text{cost for plant B} \end{array}\right\} \quad \begin{array}{l}\text{(both in \$ as of the point in} \\ \text{time for which the estimate is} \\ \text{desired)}\end{array} \quad ;$$

$$\left.\begin{array}{l} S_A = \text{size of plant A} \\ S_B = \text{size of plant B} \end{array}\right\} \quad \text{(both in same physical units)} \quad ;$$

$$X = \textit{cost-capacity factor} \text{ to reflect economies of scale.*}$$

The value of the cost-capacity factor will depend on the type of plant or equipment being estimated. For example, $X = 0.68$ for nuclear generating plants and 0.79 for fossil-fuel generating plants. Note that $X < 1$ indicates decreasing economies of scale (each additional unit of capacity costs less than the previous unit), $X > 1$[†] indicates increasing economies of scale (each additional unit of capacity costs more than the previous unit), and $X = 1$ indicates a linear cost relationship with size.

EXAMPLE 7-5

Suppose that it is desired to make a preliminary estimate of the cost of building a 600-MW fossil-fuel plant. It is known that a 200-MW plant cost $100 million 20 years ago when the approximate cost index was 400, and that cost index is now 1,200. The cost capacity factor for a fossil-fuel power plant is 0.79.

SOLUTION
Before using the power-sizing model to estimate the cost of the 600-MW plant (C_A), we must first use the cost index information to update the known cost of the 200-MW plant 20 years ago to a current cost. Using Equation (7-1), we find that the cost now of a 200-MW plant is

$$C_B = \$100 \text{ million} \left(\frac{1,200}{400}\right) = \$300 \text{ million.}$$

---

[*] May be calculated or estimated from experience using statistical techniques. See W. R. Park, *Cost Engineering Analysis* (New York: John Wiley & Sons, 1973), p. 137, for typical factors.

[†] Precious gems are an example of increasing economies of scale. For example, a one-carat diamond typically costs more than four quarter-carat diamonds.

Now, using Equation (7-4), we obtain the following estimate for the 600-MW plant:

$$C_A = \$300 \text{ million} \left(\frac{600\text{-MW}}{200\text{-MW}}\right)^{0.79}$$

$$C_A = \$300 \text{ million} \times 2.38 = \$714 \text{ million.}$$

Note that Equation (7-4) can be used to estimate the cost of a larger plant (as in Example 7-5) or the cost of a smaller plant. For example, suppose we need to estimate the cost of building a 100-MW plant. Using Equation (7-4) and the data for the 200-MW plant in Example 7-5, we find that the cost now of a 100-MW plant is

$$C_A = \$300 \text{ million} \left(\frac{100\text{-MW}}{200\text{-MW}}\right)^{0.79}$$

$$C_A = \$300 \text{ million} \times 0.58 = \$174 \text{ million.}$$

7.4.2 Learning and Improvement

A *learning curve* is a mathematical model that explains the phenomenon of increased worker efficiency and improved organizational performance with repetitive production of a good or service. The learning curve is sometimes called an *experience curve*, or a *manufacturing progress function;* fundamentally, it is an estimating relationship. The learning (improvement) curve effect was first observed in the aircraft and aerospace industries with respect to labor hours per unit.[*] However, it applies in many different situations. For example, the learning curve effect can be used in estimating the professional hours expended by an engineering staff to accomplish successive detailed designs within a family of products, as well as in estimating the labor hours required to assemble automobiles.

The basic concept of learning curves is that some input resources (e.g., energy costs, labor hours, material costs, engineering hours) decrease, on a per-output-unit basis, as the number of units produced increases. Most learning curves are based on the assumption that a constant percentage reduction occurs in, say, labor hours, as the number of units produced is *doubled*. For example, if 100 labor hours are required to produce the first output unit and a 90% learning curve is assumed, then $100(0.9) = 90$ labor hours would be required to produce the second unit. Similarly, $100(0.9)^2 = 81$ labor hours would be needed to produce the fourth unit, $100(0.9)^3 = 72.9$ hours to produce the eighth unit, and so on. Therefore, a 90% learning curve results in a 10% reduction in labor hours each time the production quantity is doubled.

---

[*] T. P. Wright, "Factors Affecting the Cost of Airplanes," *Journal of Aeronautical Sciences,* vol. 3, no.4 (February 1936).

The assumption of a constant percentage reduction in the amount of an input resource used (per output unit) each time the number of output units is doubled can be used to develop a mathematical model for the learning (improvement) function. Let

$u =$ the output unit number,

$Z_u =$ the number of input resource units needed to produce output unit number u,

$K =$ the number of input resource units needed to produce the first output unit,

$s =$ the learning curve slope parameter expressed as a decimal (for a 90% learning curve, $s = 0.9$).

Then

$$Z_u = K(s^a), \quad \text{where } a = 0, 1, 2, 3, \ldots.$$

Thus,

$$\log Z_u - \log K = a(\log s).$$

Because $u = 2^a$,

$$\log u = a(\log 2),$$

or,

$$a = \frac{\log Z_u - \log K}{\log s} = \frac{\log u}{\log 2}$$

and

$$\log Z_u - \log K = n(\log u), \quad \text{where } n = \frac{\log s}{\log 2}.$$

Now, by taking the antilog of both sides, we have

$$\frac{Z_u}{K} = u^n,$$

or

$$Z_u = K(u^n). \tag{7-5}$$

EXAMPLE 7-6

The Mechanical Engineering department has a student team that is designing a formula car for national competition. The time required for the team to assemble the first car is 100 hours. Their improvement (or learning rate) is 0.8, which means that as output is doubled, their time to assemble a car is reduced by 20%. Use this information to determine (a) the time it will take the team to assemble the 10th car, (b) the *total time* required to assemble the first 10 cars, and (c) the estimated *cumulative average* assembly time for the first 10 cars.

SOLUTION

(a) From Equation (7-5) and assuming a proportional decrease in assembly time for output units between doubled quantities, we have,

$$Z_{10} = 100(10)^{\log 0.8/\log 2}$$

$$= 100(10)^{-0.322}$$

$$= \frac{100}{2.099} = 47.6 \text{ hr.}$$

(b) The total time to produce x units, T_x, is given by

$$T_x = \sum_{u=1}^{x} Z_u = \sum_{u=1}^{x} K(u^n) = K \sum_{u=1}^{x} u^n. \qquad (7\text{-}6)$$

Using Equation (7-6), we see that

$$T_{10} = 100 \sum_{u=1}^{10} u^{-0.322} = 100[1^{-0.322} + 2^{-0.322} + \cdots + 10^{-0.322}] = 631 \text{ hr.}$$

(c) The cumulative average time for x units, C_x, is given by

$$C_x = T_x/x. \qquad (7\text{-}7)$$

Using Equation (7-7), we get

$$C_{10} = T_{10}/10 = 631/10 = 63.1 \text{ hr.}$$

EXAMPLE 7-7

The Betterbilt Construction Company designs and builds residential family homes. The purchasing manager for the company has developed a strategy whereby all the construction materials for a home are purchased from the same large supplier, but competitive bidding among a few firms is used to select the supplier for each home.

The company is ready to construct, in sequence, 16 new homes of 2,400 square feet each. The same basic design, with minor changes, will be used for each home. The successful bid for the construction materials in the first home is $64,800, or $27 per square foot. The purchasing manager believes, based on past experience, that several actions can be taken to reduce material costs by 8% each time the number of homes constructed doubles. Based on this information, (a) what is the estimated cumulative average material cost per square foot for the first five homes, and (b) what is the estimated material cost per square foot for the last (sixteenth) home?

SOLUTION

(a) Based on the constant reduction rate of 8% each time the number of homes constructed doubles, a 92% learning curve applies to the situation. The cumulative average material cost for the first five homes is developed in the following table (assuming a proportional decrease in material costs for homes between doubled quantities):

(A)	(B)	(C)	(D) = (C) ÷ (A)
Home	Material Cost[a] per Ft²	Cumulative Sum	Cumulative Average Cost per Ft²
1	$27.00	$ 27.00	$27.00
2	24.84	51.84	25.92
3	23.66	75.50	25.17
4	22.85	98.35	24.59
5	22.25	120.60	24.12

[a] From Equation (7-5); for example $Z_3 = \$27(3)^{\log 0.92/\log 2} = \$23.66$.

(b) From Equation (7-5),

$$Z_{16} = \$27(16)^{\log 0.92/\log 2}$$

$$= \$27(16)^{-0.1203}$$

$$= \frac{\$27}{1.3959} = \$19.34 \text{ per square foot.}$$

7.4.3 Developing a Cost Estimating Relationship (CER)

A cost estimating relationship (CER) is a mathematical model that describes the cost of an engineering project as a function of one or more design variables. CERs are useful tools because they allow the estimator to develop a cost estimate quickly and easily. Furthermore, estimates can be made early in the design process before detailed information is available. As a result, engineers can use CERs to make design decisions that are cost effective in addition to meeting technical requirements.

There are four basic steps in developing a CER:

1. Problem definition
2. Data collection and normalization
3. CER equation development
4. Model validation and documentation

7.4.3.1 Problem Definition The first step in any engineering analysis is to define the problem to be addressed. A well-defined problem is much easier to solve. For the purposes of cost estimating, developing a WBS is an excellent way of describing the elements of the problem. A review of the completed WBS can also help identify potential cost drivers for the development of CERs.

7.4.3.2   Data Collection and Normalization The collection and normalization of data is the most critical step in the development of a CER. We're all familiar with the adage "garbage in—garbage out." Without reliable data, the cost estimates obtained using the CER would be meaningless. The WBS is also helpful in the data collection phase. The WBS helps to organize the data and ensure that no elements are overlooked.

Data can be obtained from both internal and external sources. Costs of similar projects in the past are one source of data. Published cost information is another source of data. Regardless of the source, it is important that technical noncost data describing the physical and performance characteristics of the system be available. For example, if product weight is a potential cost driver, it is essential that we know the weights associated with the cost data.

Once collected, data must be normalized to account for differences due to inflation, geographical location, labor rates, and so on. For example, cost indexes or the techniques to be described in Chapter 8 can be used to normalize costs that occurred at different times. Consistent definition of the data is another important part of the normalization process.

7.4.3.3   CER Equation Development The next step in the development of a CER is to formulate an equation that accurately captures the relationship between the selected cost driver(s) and project cost. Table 7-2 lists four general equation types commonly used in CER development. In these equations, a, b, c, and d are constants, while x_1, x_2, and x_3 represent design variables.

A simple, yet very effective, way to determine an appropriate equation form for the CER is to plot the data. If a plot of the data on regular graph paper appears to follow a straight line, then a linear relationship is suggested. If a curve is suggested, then try plotting the data on semilog or log-log paper. If a straight line results using semilog paper, then the relationship is logarithmic or exponential. If a straight line results using log-log paper, the relationship is a power curve.

Once we have determined the basic equation form for the CER, the next step is to determine the values of the coefficients in the CER equation. The most common technique used to solve for the coefficient values is the method of least squares. Basically, this method seeks to determine a straight line through the data that minimizes the total deviation of the actual data from the predicted values. (The line itself represents the CER.) This method is relatively easy to apply manually and is also available in many commercial software packages. (Most spreadsheet packages are

TABLE 7-2   Typical Equation Forms

Type of Relationship	Generalized Equation
Linear	$\text{Cost} = a + bx_1 + cx_2 + dx_3 + \dots$
Power	$\text{Cost} = a + bx_1^c x_2^d \dots$
Logarithmic	$\text{Cost} = a + b\log(x_1) + c\log(x_2) + \dots$
Exponential	$\text{Cost} = a + b\exp^{cx_1}\exp^{dx_2} \dots$

capable of performing a least-squares fit of data.) The primary requirement for the use of the least-squares method is a linear relationship between the independent variable (the cost driver) and the dependent variable (project cost).*

All of the equation forms presented in Table 7-2 can easily be transformed into a linear form. The following equations can be used to calculate the values of the coefficients a and b in the simple linear equation $y = a + bx$:

$$b = \frac{n \sum_{i=1}^{n} x_i y_i - \left(\sum_{i=1}^{n} x_i \right) \left(\sum_{i=1}^{n} y_i \right)}{n \sum_{i=1}^{n} x_i^2 - \left(\sum_{i=1}^{n} x_i \right)^2}, \tag{7-8}$$

$$a = \frac{\sum_{i=1}^{n} y_i - b \sum_{i=1}^{n} x_i}{n}. \tag{7-9}$$

Note that the variable n in the foregoing equations is equal to the number of data sets used to estimate the values of a and b.

EXAMPLE 7-8

In the early stages of design, it is believed that the cost of a spacecraft is related to its weight. Cost and weight data for six spacecraft have been collected and normalized and are shown in the next table. A plot of the data suggests a linear relationship. Determine the values of the coefficients for the CER.

Spacecraft	Weight (lb)	Cost ($millions)
i	x_i	y_i
1	400	278
2	530	414
3	750	557
4	900	689
5	1,130	740
6	1,200	851

SOLUTION
In this problem, $n = 6$. The following table facilitates the intermediate calculations needed to compute the values of a and b using Equations (7-8) and (7-9).

---

*In addition, the observations should be independent. The difference between predicted and actual values is assumed to be normally distributed with an expected value of zero. Furthermore, the variance of the dependent variable is assumed equal for each value of the independent variable.

i	x_i	y_i	x_i^2	$x_i y_i$
1	400	278	160,000	111,200
2	530	414	280,900	219,420
3	750	557	562,500	417,750
4	900	689	810,000	620,100
5	1,130	740	1,276,900	836,200
6	1,200	851	1,440,000	1,021,200
Totals	4,910	3,529	4,530,300	3,225,870

$$b = \frac{(6)(3,225,870) - (4,910)(3,529)}{(6)(4,530,300) - (4,910)^2} = \frac{2,027,830}{3,073,700} = 0.6597,$$

$$a = \frac{3,529 - (0.6597)(4,910)}{6} = 48.31.$$

The resulting CER relating spacecraft cost (in millions of dollars) to spacecraft weight is

$$\text{Cost} = 48.31 + 0.6597x,$$

where x represents the weight of the spacecraft in pounds, and $400 \leq x \leq 1,200$.

7.4.3.4 Model Validation and Documentation Once the CER equation has been developed, we need to determine how well the CER can predict cost (i.e. model validation) and document the development and appropriate use of the CER. Validation can be accomplished using statistical "goodness of fit" measures such as standard error and the correlation coefficient. Analysts must use goodness of fit measures to infer how well the CER predicts cost as a function of the selected cost driver(s). Documenting the development of the CER is important for future use of the CER. It is important to include in the documentation the data used to develop the CER and the procedures used for normalizing the data.

The standard error (SE) measures the average amount by which the actual cost values and the predicted cost values vary. The SE is calculated by

$$\text{SE} = \sqrt{\frac{\sum_{i=1}^{n}(y_i - \text{Cost}_i)^2}{n}}, \tag{7-10}$$

where Cost_i is the cost predicted using the CER with the independent variable values for data set i and y_i is the actual cost. A small value of standard error is preferred.

The correlation coefficient (R) measures the closeness of the actual data points to the regression line ($y = a + bx$). It is simply the ratio of explained deviation to total deviation.

$$R = \frac{\sum_{i=1}^{n}(x_i - \bar{x})(y_i - \bar{y})}{\sqrt{\left[\sum_{i=1}^{n}(x_i - \bar{x})^2\right]\left[\sum_{i=1}^{n}(y_i - \bar{y})^2\right]}}, \qquad (7\text{-}11)$$

where $\bar{x} = \frac{1}{n}\sum_{i=1}^{n} x_i$ and $\bar{y} = \frac{1}{n}\sum_{i=1}^{n} y_i$. The sign $(+/-)$ of R will be the same as the sign of the slope (b) of the regression line. Values of R close to one (or minus one) are desirable in that they indicate a strong linear relationship between the dependent and independent variables.

In cases where it is not clear which is the "best" cost driver to select or which equation form is best, you can use the goodness of fit measures to make a selection. In general, all other things being equal, the CER with better goodness of fit measures should be selected.

EXAMPLE 7-9

Compute the SE and the correlation coefficient for the CER developed in Example 7-8.

SOLUTION

The CER developed in Example 7-8 related the cost of a spacecraft to its weight using the equation

$$\text{Cost} = 48.31 + 0.6597x.$$

Using this equation, we can predict the cost of the six spacecraft given their weights:

i	x_i	y_i	Cost$_i$	$(y_i - \text{Cost}_i)^2$	$(x_i - \bar{x})(y_i - \bar{y})$	$(x_i - \bar{x})^2$	$(y_i - \bar{y})^2$
1	400	278	312.19	1,168.96	129,753.42	174,999.99	96,205.43
2	530	414	397.95	257.60	50,218.44	83,134.19	30,335.19
3	750	557	543.09	193.49	2,129.85	4,668.99	971.57
4	900	689	642.04	2,205.24	8,234.79	6,669.99	10,166.69
5	1,130	740	793.77	2,891.21	47,320.86	97,138.19	23,052.35
6	1,200	851	839.95	122.10	100,314.33	145,671.99	69,079.61
Totals	4,910	3,529	3,528.99	6,838.60	337,971.69	512,283.34	229,810.84

Note that $\bar{x} = \frac{1}{6}(4,910) = 818.33$ and $\bar{y} = \frac{1}{6}(3,529) = 588.17$. Using Equations (7-10) and (7-11), we can compute the SE and correlation coefficient for the CER:

$$SE = \sqrt{\frac{6,638.60}{6}} = 33.76,$$

$$R = \frac{337,971.69}{\sqrt{(512,283.34)(229,810.84)}} = 0.985.$$

The value of the correlation coefficient is close to one, indicating a strong positive linear relationship between the cost of the spacecraft and the spacecraft's weight.

In summary, CERs are useful for a number of reasons. First, given the required input data, they are quick and easy to use. Second, a CER usually requires very little detailed information making it possible to use the CER early in the design process. Finally, a CER is an excellent predictor of cost if correctly developed using good historical data.

7.5 Cost Estimation in the Design Process

Today's companies are faced with the problem of providing quality goods and services at competitive prices. The price of their product is based on the overall cost of making the item plus a built-in profit. To ensure that products can be sold at competitive prices, cost must be a major factor in the design of the product. As was discussed in the preface of this book, a functionally well-designed product can be worthless if it is not economically feasible. (Recall the thousand-dollar composite tricycle.) For a product to have value to a customer, the benefits must outweigh the costs.

In this section, we will discuss both a "bottom-up" approach and a "top-down" approach to determining product costs and selling price. Used together with the concepts of target costing, design-to-cost, and value engineering, these techniques can assist engineers in the design of cost effective systems and competitively priced products.

Companion Web Site (http://www.prenhall.com/sullivan_engineering/):   Cost estimation for a heat exchanger involves the computation of a base cost, as well as installation, operation, and maintenance costs—a *life cycle cost*. Visit the Web site to view comparisons of cost estimates for alternative types of heat exchangers. A spreadsheet calculator is included so that you may develop cost estimates for your own designs.

7.5.1 The Elements of Product Cost and "Bottom-Up" Estimating

As discussed in Chapter 2, product costs are classified as direct or indirect. Direct costs are easily assignable to a specific product, while indirect costs are not easily allocated to a certain product. For instance, direct labor would be the wages of a machine operator, indirect labor would be supervision.

Manufacturing costs have a distinct relationship to production volume in that they may be fixed, variable, or step-variable. Generally, administrative costs are fixed, regardless of volume; material costs vary directly with volume; and equipment cost is a step function of production level.

The primary costs within the manufacturing expense category include engineering and design, development costs, tooling, manufacturing labor, materials, supervision, quality control, reliability and testing, packaging, plant overhead, general and administrative, distribution and marketing, financing, taxes, and insurance. Where do we start?

Engineering and design costs consist of design, analysis, and drafting, together with miscellaneous charges such as reproductions. The engineering cost may be allocated to a product on the basis of how many engineering labor hours are involved. Other major types of costs that must be estimated are as follows:

- Tooling costs, which consist of repair and maintenance plus the cost of any new equipment.
- Manufacturing labor costs, which are determined from standard data, historical records, or the accounting department. Learning curves are often used for estimating direct labor.
- Materials costs, which can be obtained from historical records, vendor quotations, and the bill of materials. Scrap allowances must be included.
- Supervision, which is a fixed cost based on the salaries of supervisory personnel.
- Factory overhead, which includes utilities, maintenance, and repairs. As discussed in Chapter 2 and Appendix A, there are various methods used to allocate overhead, such as in proportion to direct labor hours or machine hours.
- General and administrative costs, which are sometimes included with the factory overhead.

A "bottom-up" procedure for estimating total product cost is commonly used by companies to help them make decisions about what to produce and how to price their products. The term "bottom-up" is used because the procedure requires estimates of cost elements at the lower levels of the cost structure which are then added together to obtain the total cost of the product. The following simple example shows the general bottom-up procedure for making a per unit product cost estimate and illustrates the use of a typical spreadsheet form of the cost structure for preparing the estimate.

The spreadsheet in Figure 7-5 shows the determination of the cost of a throttle assembly. The EXCEL cell formulas are provided in Appendix 7-A. Column A shows typical cost elements that contribute to total product cost. The list of cost elements can easily be modified to meet a company's needs. This spreadsheet allows for per unit estimates (column B), factor estimates (column C), and direct estimates (column D). The shaded rows are selected subtotals.

Typically, direct labor costs are estimated via the unit technique. The manufacturing process plan is used to estimate the total number of direct labor hours required per output unit. This quantity is then multiplied by the composite labor rate to obtain the total direct labor cost. In this example, 36.48 direct labor hours are required to produce 50 throttle assemblies and the composite labor rate is $10.54 per hour, which yields a total direct labor cost of $384.50.

Indirect costs, such as quality control and planning labor, are often allocated to individual products using factor estimates. Estimates are obtained by expressing the cost as a percentage of another cost. In this example, planning labor and quality control are expressed as 12% and 11% of direct labor cost (row A), respectively. This gives a total labor cost of $472.93. Factory overhead and general and administrative expenses are estimated as percentages of the total labor cost (row D).

Entries for cost elements for which direct estimates are available are placed in column D. The total production materials cost for the 50 throttle assemblies is $167.17. A direct estimate of $28.00 applies to the outside manufacture of required components. The subtotal of cost elements at this point is $1,235.62.

Column A	Column B		Column C		Column D	Column E
	Unit Estimate		Factor Estimate		Direct	Row
	Unit	Cost/Unit	Factor	of Row	Estimate	Total
A: Factory Labor	36.48	$ 10.54				$ 384.50
B: Planning Labor			12%	A		46.14
C: Quality Control			11%	A		42.29
D: TOTAL LABOR						472.93
E: Factory Overhead			105%	D		496.58
F: General & Admin. Expense			15%	D		70.94
G: Production Material					$ 167.17	167.17
H: Outside Manufacture					28.00	28.00
I: SUBTOTAL						1,235.62
J: Packing Costs			5%	I		61.78
K: TOTAL DIRECT CHARGE						1,297.41
L: Other Direct Charge			1%	K		12.97
M: Facility Rental						--
N: TOTAL MANUFACTURING COST						1,310.38
O: Quantity (lot size)						50
P: MANUFACTURING COST / UNIT						26.21
Q: Profit/Fee			10%	P		2.62
R: UNIT SELLING PRICE						$ 28.83

Figure 7-5 Manufacturing Cost-Estimating Spreadsheet

Packing costs are estimated as 5% of all previous costs (row I), giving a total direct charge of $1,297.41. The cost of other miscellaneous direct charges are figured in as 1% of the current subtotal (row K). This results in a total manufacturing cost of $1,310.38 for the entire lot of 50 throttle assemblies. The manufacturing cost per assembly is $1,310.38/50 = $26.21.

As mentioned earlier in this section, the price of a product is based on the overall cost of making the item plus a built-in profit. The bottom of the spreadsheet in Figure 7-5 shows the computation of unit selling price based on this strategy. In this example, the desired profit (often called the profit margin) is 10% of the unit manufacturing cost, which corresponds to a profit of $2.62 per throttle assembly. The total selling price of a throttle assembly is then $26.21 + $2.62 = $28.83.

As mentioned earlier, learning curves are often used when estimating direct labor costs. The following example illustrates how learning curves were used to obtain the factory labor hours for the throttle assemblies.

EXAMPLE 7-10

Let a batch of 50 throttle assemblies define one output unit. The 36.48 factory labor hours used to estimate the cost of the throttle assembly is based on the sixteenth output unit. Assuming a 90% learning curve, what was the number of factory labor hours required for the first batch of 50 throttle assemblies? What is your estimate of the labor hours needed for the 64th and 100th output units (i.e., batches)?

SOLUTION

Let $K =$ the number of labor hours required for the first batch of throttle assemblies. From Equation (7-5), we have

$$Z_{16} = K(16)^{\log 0.9/\log 2};$$

$$36.48 = K(16)^{-0.152};$$

$$K = 55.6 \text{ hr.}$$

Thus, the estimate of 36.48 hr was obtained based on the fact that it took 55.6 hours to assemble the first batch of throttle assemblies. Using $K = 55.6$, we can easily estimate the time required for the 64th and 100th output units:

$$Z_{64} = 55.6(64)^{-0.152} = 29.54 \text{ hr;}$$

$$Z_{100} = 55.6(100)^{-0.152} = 27.61 \text{ hr.}$$

7.5.2 Target Costing and Design to Cost: A "Top-Down" Approach

Traditionally, American firms determine an initial estimate of a new product's selling price using the bottom-up approach described in the previous section. That is, the estimated selling price is obtained by accumulating relevant fixed and variable costs and then adding a profit margin, which is a percentage of total manufacturing costs. This process is often termed *design to price*. The estimated selling price is then used by the marketing department to determine whether the new product can be sold.

In contrast, Japanese firms apply the concept of *target costing*, which is a top-down approach. The focus of target costing is "what *should* the product cost" instead of "what *does* the product cost." The objective of target costing is to design costs out of products before those products enter the manufacturing process. In this top-down approach, cost is viewed as an input to the design process, not as an outcome of it.

As shown in Figure 7-6, target costing is initiated by conducting market surveys to determine the selling price of the best competitor's product. A target cost is obtained by deducting the desired profit from the best competitor's selling price:

$$\text{Target Cost} = \text{Competitor's Price} - \text{Desired Profit}. \tag{7-12}$$

As discussed in the previous section, desired profit is often expressed as a percentage of the total manufacturing cost called the profit margin. For a specific profit margin (e.g., 10%), the target cost can be computed using the following equation:

$$\text{Target Cost} = \frac{\text{Competitor's Price}}{(1 + \text{profit margin})}. \tag{7-13}$$

This target cost is obtained prior to the design of the product and is used as a goal for engineering design, procurement, and production.

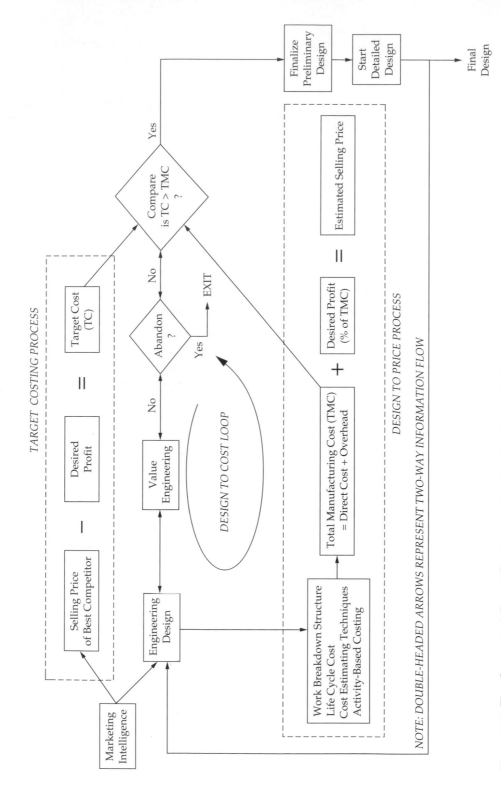

Figure 7-6 The Concept of Target Costing and Its Relationship to Design

EXAMPLE 7-11

Recall the throttle assemblies discussed in the previous example. Suppose that a market survey has shown that the best competitor's selling price is $27.50 per assembly. If a profit margin of 10% (based on total manufacturing cost) is desired, determine a target cost for the throttle assembly.

SOLUTION

Since the desired profit has been expressed as a percentage of total manufacturing costs, we can use Equation (7-13) to determine the target cost:

$$\text{Target Cost} = \frac{\$27.50}{(1 + 0.10)} = \$25.00.$$

Note that the total manufacturing cost calculated in Figure 7-5 was $26.21 per assembly. Since this cost exceeds the target cost, a redesign of the product itself or of the manufacturing process is needed to achieve a competitive selling price.

Companion Web Site (http://www.prenhall.com/sullivan_engineering/): Demanufacturing of computers involves disassembly of older computers, refurbishing, and either donating or reselling the units. However, there are some residuals, or remaining components, that cannot be reused, are harmful to the environment, and contribute to the cost of demanufacturing. Visit the Web site to view two applications of *target costing* that examine issues involved with the recycling of computers.

As mentioned before, target costing takes place before the design process begins. Engineers use the target cost as a cost performance requirement. Now the final product must satisfy both technical and cost performance requirements. The practice of considering cost performance as important as technical performance during the design process is termed *design to cost*. The design to cost procedure begins with the target cost as the cost goal for the entire product. This target cost is then broken down into cost goals for major subsystems, components, and subassemblies. These cost goals would cover targets for direct material costs and direct labor. Cost goals are not typically established for indirect cost categories such as overhead and administrative costs. It is important to note that reasonable cost goals must be established. If cost goals are too easy to achieve, designers have little motivation for seeking a better alternative. If the cost goals are too hard to achieve, people become discouraged.

Once cost goals have been established, the preliminary engineering design process is initiated. Conventional tools such as work breakdown structure and cost estimating are utilized to prepare a bottom-up total manufacturing cost projection discussed in the previous section. The total manufacturing cost represents an *initial* appraisal of what it would cost the firm to design and manufacture the product being considered. The total manufacturing cost is then compared with the top-down target cost. If the total manufacturing cost is more than the target cost, then the design must be fed back into the *value engineering process* (to be discussed in

Section 7.5.3) to challenge the functionality of the design and attempt to reduce the cost of the design. This iterative process is a key feature of the design to cost procedure. If the total manufacturing cost can be made less than the target cost, the design process continues into detailed design, culminating in the final design to be produced. If the total manufacturing cost cannot be reduced to the target cost, the firm should seriously consider abandoning the product.

The spreadsheet in Figure 7-7 illustrates the use of the manufacturing cost estimating spreadsheet to compute both a target cost and the cost reductions necessary to achieve the target cost. As computed in Example 7-11, the target cost for the throttle assembly is $25.00. Since the initial total manufacturing cost (determined to be $26.21 in Figure 7-5) is greater than the target cost, we must work backward from the total manufacturing cost, changing values of a chosen (single) cost element to the level required to reduce the cost to the desired target. This method of establishing new cost targets for individual elements can be accomplished by trial and error (manually manipulating the spreadsheet values) or by making use of the "solver" feature of the software package (if one is available). Figure 7-7 shows one possible result of this process. If the process of assembling the throttles could be made more efficient such that total direct labor requirements are reduced to 34.48 hr (instead of

Column A		Column B		Column C		Column D	Column E
		Unit Estimate		Factor Estimate		Direct	Row
MANUFACTURING COST ELEMENTS		Units	Cost/Unit	Factor	of Row	Estimate	Total
A:	Factory labor	34.48	$ 10.54				$ 363.42
B:	Planning labor			12%	A		43.61
C:	Quality control			11%	A		39.98
D:	TOTAL LABOR						447.01
E:	Factory overhead			105%	D		469.36
F:	General & admin. expense			15%	D		67.05
G:	Production material					$ 167.17	167.17
H:	Outside manufacture					28.00	28.00
I:	SUBTOTAL						1,178.58
J:	Packing costs			5%	I		58.93
K:	TOTAL DIRECT CHARGE						1,237.51
L:	Other direct charge			1%	K		12.38
M:	Facility rental						-
N:	TOTAL MANUFACTURING COST						1,249.89
O:	Quantity (lot size)						50
P:	MANUFACTURING COST / UNIT						25.00
	Competitor's selling price	$ 27.50					
	Desired return on sales	10%					
	Target cost	$ 25.00					

Figure 7-7 Manufacturing Cost Estimating and Target Costing

36.48), the target cost could be realized. Now the challenge is to find a way, either through product or process redesign, to reduce the direct labor requirements.

EXAMPLE 7-12

Given the current estimated total manufacturing cost of $26.21 as shown in Figure 7-5, determine a cost goal for production material that would allow us to achieve a target cost of $25.00.

SOLUTION

Using the spreadsheet of Figure 7-5 as a starting point, one approach to determining a cost goal for production material would be to iteratively change the value entered in Row G and Column D until the desired total manufacturing cost of $25.00 is obtained. The following table shows a series of costs for production material and the resulting total manufacturing cost per assembly:

Production Material Cost per 50 Assemblies	Total Manufacturing Cost per Assembly
$167.17	$26.21
150.00	25.84
140.00	25.63
130.00	25.42
120.00	25.21
110.00	25.00

As seen in the table, a production material cost of $110 per batch of 50 assemblies would result in a total manufacturing cost of $25.00, the target cost. Now it is left to design engineers to determine if a different, less costly material could be used or if process improvements could be made to reduce material scrap. Another possibility would be to negotiate a new purchase price with the supplier of the material or seek a new supplier.

We could also have made use of the "solver" feature included in most spreadsheet packages. Figure 7-8 shows the result of this approach.

An additional example of target costing is provided in Appendix 7-B.

7.5.3 Value Engineering

This section introduces the topic of value engineering (VE). The objective of VE is very similar to that of design to cost. The VE objective is to provide the required product functions at a minimum cost. VE necessitates a detailed examination of a product's functions, and the cost of each, in addition to a thorough review of product specifications. VE is performed by a team of specialists from a variety

Column A		Column B		Column C		Column D	Column E
		Unit Estimate		Factor Estimate		Direct	Row
MANUFACTURING COST ELEMENTS		Units	Cost/Unit	Factor	of Row	Estimate	Total
A:	Factory labor	36.48	$ 10.54				$ 384.50
B:	Planning labor			12%	A		46.14
C:	Quality control			11%	A		42.29
D:	TOTAL LABOR						472.93
E:	Factory overhead			105%	D		496.58
F:	General & admin. expense			15%	D		70.94
G:	Production material					$ 110.23	110.23
H:	Outside manufacture					28.00	28.00
I:	SUBTOTAL						1,178.69
J:	Packing costs			5%	I		58.93
K:	TOTAL DIRECT CHARGE						1,237.62
L:	Other direct charge			1%	K		12.38
M:	Facility rental						-
N:	TOTAL MANUFACTURING COST						1,250.00
O:	Quantity (lot size)						50
P:	MANUFACTURING COST / UNIT						25.00
	Competitor's selling price	$ 27.50					
	Desired return on sales	10%					
	Target cost	$ 25.00					

Figure 7-8 Production Material Cost Goal for Example 7-12

of disciplines (design, manufacturing, marketing, etc.), and the team focuses on determining the most cost-effective way to provide high value at an acceptable cost to the customer. Value engineering is most appropriately applied early in the life cycle where there is more potential for cost savings. VE is applied repeatedly during the design phase as new information becomes available. Note that in Figure 7-6, the VE function appears within the design-to-cost loop and is a critical part of obtaining a total manufacturing cost that is less than the target cost.

The key to successful VE is to ask critical questions and seek creative answers. Table 7-3 lists some sample questions that should be included in a VE study. It is important to question everything and not take anything for granted. Cost-reduction opportunities are sometimes so simple that they get overlooked. Creative solutions can be obtained using classical brainstorming or the Nominal Group Technique (discussed in Chapter 1). Alternatives that seem promising should be analyzed to determine if cost reduction is possible without compromising functionality.

The examples that follow illustrate how VE has been used to obtain enhanced functionality and improved value. The redesign of the remote control console (for a TV/VCR) is a classic example of VE. In Figure 7-9, the product on the left looks more like a calculator than a remote control console. VE was used to identify many

TABLE 7-3 Value Engineering Checklist

Are all the functions provided required by the customer?
Can less expensive material be used?
Can the number of different materials used be reduced?
Can the design be simplified to reduce the number of parts?
Can a part designed for another product be used?
Are all the machined surfaces and finishes necessary?
Can redundant quality inspections be eliminated?
Would product redesign eliminate a quality problem?
Is the current level of packaging necessary?

unnecessary functions that customers did not want and would not pay for. By eliminating these unwanted functions, we were able to cut significant production costs. The redesigned remote control on the right-hand side projects a highly simplified, user-friendly appearance. This design, therefore, enhances use value and esteem value even before considering the effect of production cost savings on the product's selling price.

In Figure 7-10, a connector used in an electrical regulator was subject to value engineering. The present assembly (on the left) has nine different parts, of which five are specially made and four can be bought off the shelf. Unit material costs amount to $2.34, and the labor cost is $2.93. The basic function of the assembly is "conduct current," and the secondary functions are "create seal" and "establish connection." After applying the VE methodology, we found that the "create seal" function was unnecessary. By eliminating this redundant function, we reduced the number of parts required for the assembly to three (as shown on the right-hand

Figure 7-9
Remote Control
Console

Present   Recommended

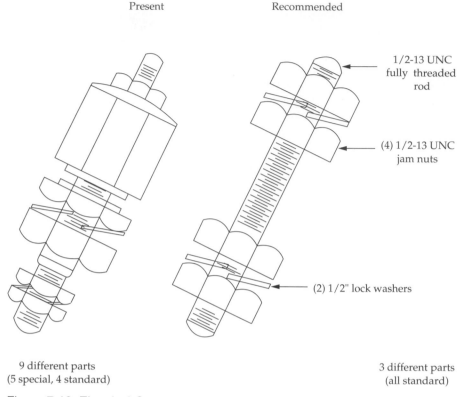

1/2-13 UNC
fully threaded
rod

(4) 1/2-13 UNC
jam nuts

(2) 1/2" lock washers

9 different parts   3 different parts
(5 special, 4 standard)   (all standard)

Figure 7-10 **Electrical Connecter**

side of Figure 7-10). Unit material cost was reduced to $1.99, a 15% reduction; the overall cost reduction was 38%. Other indirect benefits included lower inventory costs (less parts count) and shorter manufacturing and assembly times.

7.6 Estimating Cash Flows for a Typical Small Project

We will consider a small project typical of those often encountered in practice. To what extent does the integrated approach (Figure 7-1) apply when the project is not large and complex? The answer is that it applies regardless of the size and complexity of the project. However, several adjustments can be made to reduce the level of detail to fit the specific situation.

1. *WBS.* The number of levels and scope of the WBS can normally be significantly reduced for a small project. Sometimes the WBS can be combined with the cost and revenue structure into a worksheet for developing the estimates. (See Example 7-13.) The important point is that this initial component of the integrated approach needs to be explicitly evaluated for the specific project. A WBS in the proper form and scope will facilitate the economic analysis of any project.
2. *Cost and revenue structure.* The number of cost and revenue categories and elements required can be reduced for most small projects. This second component,

however, still needs to be considered in detail. For example, the number of operating and maintenance cost elements that may need to be included, even in a small project, can be quite extensive.

3. *Estimating techniques (models).* Estimating future costs and revenues is usually less complex with small projects. The techniques discussed in Sections 7.2.3 through 7.4, however, will still need to be used.

These three basic components of the integrated approach apply regardless of the size of the project. Their application to small projects, however, is reduced in scope along with the information database required. In any engineering economy study, it is necessary to (1) define the cash flow perspective, (2) determine the estimating baseline, and (3) establish the length of the analysis (study) period. These parts of the approach do not vary with project size.

EXAMPLE 7-13

Your company is involved in the manufacture of transmission components and axles for heavy-duty trucks and is a major supplier to three truck-manufacturing plants. Just-in-time inventory concepts are used at each of the three manufacturing plants. Therefore, price competitiveness, reliable delivery to meet plant production schedules, and quality of the delivered product are essential to maintain the company's position as a supplier to the three plants. Meeting such customer expectations is critical to increasing the company's market share. Consequently, a project is being considered to replace some existing equipment with new automated equipment for the production of axles.

One of the feasible alternatives involves new equipment manufactured by Company *A*. Describe the development of its BTCF using the integrated approach in Figure 7-1. Discuss potential sources and compilation of necessary data as appropriate (all details do not need to be given). Some basic data related to the project are the following:

1. The equipment acquisition cost is $2,650,000 (including computer software and primary installation costs) if purchased from Company *A*. Other miscellaneous installation costs of $83,000 would be expensed in the first year of operation (i.e., not included in the cost basis of the equipment).
2. The analysis (study) period established by the company for this type of investment is six years.

SOLUTION
The new automated equipment, if purchased from Company *A*, is a complete system; that is, the hardware and software do not need to be broken down further to define the system explicitly for cost and revenue estimating. Therefore, the WBS level that can be used for estimating is the total project (i.e., Level 1 of the WBS). As a result, the WBS and the cost and revenue structure can be combined in a single worksheet. Thus, in this situation a separate, detailed WBS is not required.

The cash-flow perspective that should be used in this project is that of the company (owners). Since this is a project that will upgrade an existing operation, the best estimating baseline is the current operation, and the differential approach (Section 7.2) should be used. Thus, cost data from the present operation plus those obtained from the manufacturer (Company A) are the primary sources of data for estimating purposes. The estimating techniques to be used are determined by the database that is available.

A representative worksheet is shown in Figure 7-11 for summarizing the costs and revenues needed to develop the net cash flow over six years when equipment is purchased from Company A. The estimate of capital investment is based primarily on data from the manufacturer (cost of the equipment and computer software). Internal estimates developed by the project engineering group are used for the other cost elements (installation costs, working capital, etc.).

A. Nonrecurring Costs and Revenues	Costs	Revenues
1. Capital Investment		
a. Hardware (including computer equipment)	$2,195,000	
b. Computer software	185,700	
c. Primary installation	269,300	
d. Other installation costs	83,000	
e. Working capital	28,400	
f. Project engineering and management	172,400	
	Total: $2,933,900	
2. Revenue		
a. Sale of present equipment (year 0)		$185,000
b. Sale of new equipment (year 6)		310,000

B. Recurring Annual Costs and Revenues	Costs	Revenues or Reduced Costs
1. Operational and Maintenance (O&M) Costs		
a. Direct Costs		
Labor		$201,000
Material		58,000
Other direct costs		44,600
b. Indirect Costs		
Labor/overtime		14,300
Materials and supplies		
Cost of quality (during production)		32,000
Tooling/fixtures		11,500
Maintenance	$18,600	
Utilities	4,200	
Property taxes and insurance	28,900	
Other indirect costs		5,900
2. Revenue		
Increased sales		525,000
	Total: $51,700	$892,300

Figure 7-11 Project Cost and Revenue Estimating Worksheet for Example 7-13

The increased revenue estimate, based on additional market share (sales volume) as a result of the project, would be developed by the sales staff. The estimated MV for the equipment being replaced and the new equipment at the end of six years could be developed using data obtained from firms involved in the resale of this type of equipment. The operational and maintenance costs would be estimated from present operating experience and expected new equipment performance data supplied by Company A.

Based on the cost and revenue estimates shown on the worksheet, the estimated six-year BTCF for the alternative involving purchase of the new equipment from Company A is shown in the following table:

End of Year	BTCF (Company A)
0	$-\$2,748,900$ $(= -\$2,933,900 + \$185,000)$
1	840,600
2	840,600
3	840,600
4	840,600
5	840,600
6	1,179,000 $(= \$892,300 - \$51,700 + \$310,000 + \$28,400)$

The cash-flow amounts in year zero and year six include revenue from the disposal of assets of $185,000 and $310,000, respectively, as indicated on the worksheet. Also, the amount at the end of year six includes recovery of working capital.

▼ 7.7 Summary

Developing the cash flow for each alternative in a study is a pivotal step in the engineering economic analysis procedure. An integrated approach for developing cash flows includes three major components: (1) a WBS definition of the project, (2) a cost and revenue structure that identifies all the cost and revenue elements involved in the study, and (3) estimating techniques (models). Other considerations such as the length of the analysis period, the perspective and estimating baseline for the cash flows, and a cost and revenue database are illustrated in Figure 7-1 and discussed in the chapter.

The WBS is a powerful technique for defining all the work elements and their interrelationships for a project. It is a basic tool in project management and is a vital aid in an engineering economy study. Understanding this technique and its applications are important in engineering practice.

The development of a cost and revenue structure will help to ensure that a cost element or a source of revenue is not overlooked in the analysis. The life-cycle concept and the WBS are used in developing this structure for a project.

Estimating techniques (models) are used to develop the cash flows for the alternatives as they are defined by the WBS. Thus, the estimating techniques form a bridge between the WBS and detailed cost and revenue data and the estimated cash flows for the alternatives.

7.8 References

Engineering News-Record. Published monthly by McGraw-Hill Book Co., New York.

JELEN, F. C., and J. H. BLACK. *Cost and Optimization Engineering,* 2d ed. (New York: McGraw-Hill Book Co., 1983).

MATTHEWS, L. M. *Estimating Manufacturing Costs: A Practical Guide for Managers and Estimators* (New York: McGraw-Hill Book Co., 1983).

MICHAELS, J. V., and W. P. WOOD. *Design to Cost* (New York: John Wiley & Sons, 1989).

OSTWALD, P. F. *Engineering Cost Estimating,* 3d ed. (Englewood Cliffs, NJ: Prentice Hall, 1992).

PARK, W. R., and D. E. JACKSON. *Cost Engineering Analysis: A Guide to Economic Evaluation of Engineering Projects,* 2nd ed. (New York: John Wiley & Sons, 1984).

STEWART, R. D. *Cost Estimating* (New York: John Wiley & Sons, 1982).

STEWART, R. D., R. M. WYSKIDA, and J. D. JOHANNES, eds. *Cost Estimators' Reference Manual,* 2d ed. (New York: John Wiley & Sons, 1995).

7.9 Problems

The number in parentheses ( ) that follows each problem refers to the section from which the problem is taken.

7-1. Visually examine a lawnmower for home use that is (a) nonriding, (b) approximately 21 inches in cutting width, and (c) powered with a 3.5- to 5.0-hp, air-cooled engine. Develop a WBS for this product through Level 3. (7.2)

7-2. You are planning to build a new home with approximately 2,000 to 2,500 gross square feet of living space on one floor. In addition, you are planning an attached two-car garage (with storage space) of approximately 450 gross square feet. Develop a cost and revenue structure for designing and constructing, operating (occupying) for 10 yr, and then selling the home at the end of the tenth year. (7.2)

7-3. Suppose that your brother-in-law has decided to start a company that produces synthetic lawns for senior homeowners. He anticipates starting production in 18 months. In estimating future cash flows of the company, which of the following would be relatively easy versus relatively difficult to obtain? Also, suggest how each might be estimated with reasonable accuracy. (7.2)

a. Cost of land for a 10,000-square-foot building.

b. Cost of the building (cinder block construction).

c. Initial working capital.

d. Total capital investment cost.

e. First year's labor and material costs.

f. First year's sales revenues.

7-4. Manufacturing equipment that was purchased in 2000 for $200,000 must be replaced at the end of 2005. What is the estimated cost of the replacement based on the following equipment cost index? (7.3)

Year	Index	Year	Index
2000	223	2003	257
2001	238	2004	279
2002	247	2005	293

7-5. Prepare a composite (weighted) index for housing construction costs in 2004 using the following data: (7.3)

Type of Housing	Percent	Reference Year ($I = 100$)	2004
Single Units	70	41 $\}$ $/ft^2$	62 $\}$ $/ft^2$
Duplex Units	5	38	57
Multiple Units	25	33	53

7-6. The major manufacturing cost elements for a model of electronic process control equipment are shown in the accompanying table. The related cost indexes for a reference year and for the current year are also shown in the accompanying table. (7.3)

Cost Element	Percent of Manufacturing Cost	Index (Reference Year)	Index (Current Year)
Factory Labor	13%	131	176
Direct material	20	150	210
Purchased components	32	172	231
Indirect Costs	21	160	190
Engineering	8	135	180
Other costs	6	140	172

a. Based on this information, develop a composite (weighted) cost index for the reference year and the current year.

b. If the manufacturing cost for one unit of the equipment was $314,300 in the reference year, what would be your semidetailed estimate of the manufacturing cost in the current year?

7-7. A microbrewery was built in 2000 at a total cost of $650,000. Additional information is given in the accompanying table (all 1995 indices = 100). (7.3)

Cost Element	Average Percentage of Total Brewery Cost	Index (2000)	Index (2004)
Labor	30	160	200
Materials	20	145	175
Equipment	50	135	162

a. Calculate a weighted index for microbrewery construction in 2004.

b. Prepare a budget estimate for a microbrewery in 2004.

7-8. The purchase price of a natural gas–fired commercial boiler (capacity X) was $181,000 eight years ago. Another boiler of the same basic design, except with capacity $1.42X$, is currently being considered for purchase. If it is purchased, some optional features presently costing $28,000 would be added for your application. If the cost index was 162 for this type of equipment when the capacity X boiler was purchased, and is 221 now, and the applicable cost capacity factor is 0.8, what is your estimate of the purchase price for the new boiler? (7.3, 7.4)

7-9. Use the factor technique to estimate the cost of installing a local-area network in a factory environment having the following characteristics. One large building on a single level will require a total of 3,000 ft of coaxial (broadband) cable to network its six departments. Six network interface units (NIUs) will be required, and a total of 50 taps will have to be made to connect all the anticipated workstations and programmable devices. Two modems are needed in addition to one network manager/analyzer that costs $30,000. The information necessary to make the estimate may be obtained from the worksheet shown on the following page. (See Table P7-9.) How accurate do you think such an estimate would be? (7.3)

7-10. If an ammonia plant that produces 500,000 pounds per year cost $2,500,000 to construct eight years ago, what would a 1,500,000-pound-per-year plant cost now? Suppose that the construction cost index has increased an average rate of 12% per year for the past eight years and that the cost-capacity factor (X) to reflect economy of scale is 0.65. (7.4)

7-11. Based on driving 15,000 miles per year, the average annual cost of owning and operating a four-cylinder automobile in 2002 is estimated to be $0.42 per mile. The cost breakdown is shown in the following table: (7.3)

Cost Element	Cost per mile
Depreciation	$0.210
Gasoline and oil	0.059
Finance charges*	0.065
Insurance costs (including collision)	0.060
Taxes, license and registration fees	0.015
Tires	0.011

*Based on 20% down and 48 month financing at an A.P.R of 10%.

a. If a person who owns this "average" automobile plans to drive 15,000 miles during 2002, how much would it cost to own and operate the automobile?

TABLE P7-9	Worksheet for Problem 7-9			
1.	Interbuilding connections	$100–$150 per foot	×⬚	=⬚
2.	Intrabuilding connections	$20–$50 per foot	×⬚	=⬚
3.	Cable installation	$20 per foot	×⬚	=⬚
4.	Equipment			
	a. Broadband CATV amplifier	$500–$1,500	×⬚	=⬚
	Taps	$17–$20 each	×⬚	=⬚
	Splitters	$5–$15	×⬚	=⬚
	NIUs	$500–$1,000 per port	×⬚	=⬚
	Modems	$1,000 each	×⬚	=⬚
	b. Basebands			
	NIUs	$600 per port	×⬚	=⬚
	Repeaters	$1,200–$1,500 each	×⬚	=⬚
	Taps/transceivers	$200–$300 each	×⬚	=⬚
	c. Network manager	$10,000–$30,000		=⬚
	Network analyzer	$30,000		=⬚

b. If the person actually drives 30,000 miles in 2002, give some reasons why his or her actual cost may not be twice the answer obtained in Part (a).

7-12. Six years ago, an 80-kW diesel electric set cost $160,000. The cost index for this class of equipment six years ago was 187 and is now 194. The cost-capacity factor is 0.6. (7.4)

a. The plant engineering staff is considering a 120-kW unit of the same general design to power a small isolated plant. Assume we want to add a precompressor, which (when isolated and estimated separately) currently costs $18,000. Determine the total cost of the 120-kW unit.

b. Estimate the cost of a 40-kW unit of the same general design. Include the cost of the $18,000 precompressor.

7-13. The plant manager of MOMAX, Inc. decides that a new hydraulic lift is needed in the production section. If the cost of a 150,000-lb lift 10 years ago was $200.000, and a lift with a capacity of 125,000 lb is needed now, what will be the cost of the new lift? The cost index now is 343.8 and the index 10 yr ago was 171.6. The cost capacity factor for his type of equipment is 0.80.

7-14. A 250-ft^2 shell and tube heat exchanger was purchased for $13,500 in 1994 when the index value was 830. Estimate the cost of a 150-ft^2 shell and tube heat exchanger in 2006 when the index

value is 964 and the appropriate cost-capacity factor is 0.6. (7.4)

7-15. The structural engineering design section within the engineering department of a regional electrical utility corporation has developed several standard designs for a group of similar transmission line towers. The detailed design for each tower is based on one of the standard designs. A transmission line project involving 50 towers has been approved. The estimated number of engineering hours needed to accomplish the first detailed tower design is 126. Assuming a 95% learning curve, (a) what is your estimate of the number of engineering hours needed to design the eighth tower and to design the last tower in the project, and (b) what is your estimate of the cumulative average hours required for the first five designs? (7.4)

7-16. The time to produce the 32nd unit of a particular product is 8.74 hr. If the learning curve, based on previous experience with similar products, is 85%, (a) what was the number of hours required for the first unit and (b) what is the estimated number of hours needed for the 1,000th output unit? (7.4)

7-17. The overhead costs for a company are presently $X per month. The management team of the company, in cooperation with the employees, is ready to implement a comprehensive improvement program to reduce these costs. If you (a) consider an observation of actual overhead costs for one month analogous to an output unit, (b) estimate the overhead costs for the first month of program implementation to be $1.15X$ due to extra front-end effort, and (c) consider a 90% improvement curve applicable to the situation, what is your estimate of the percentage reduction in present overhead costs per month after 30 months of program implementation? (7.4)

7-18. Refer to Problem 7-2. You have decided to build a one-floor home with 2,450 gross square feet of living space. Also, the attached two-car garage (with storage space) will have 450 gross square feet of area.

   a. Develop a WBS (through Level 3) defining the work elements involved in the design and construction of the home. (7.2)

   b. Develop a semidetailed estimate of your capital investment cost associated with the project until the time of your initial occupancy of the home. (*Note:* Your instructor will provide you with additional information to assist with this part of the problem.) (7.3)

7-19. The basic power-sizing model [Equation (7-4)] can be modified to better represent a specific estimating situation. Consider the situation of an automated warehousing system for a new distribution center handling case goods (e.g., an area distribution center for a supermarket company). Equation (7-4) can be modified to improve its capability to estimate the capital investment required for this project (system) by (a) separating the equipment and installation part of the capital investment (which can be estimated based on the exponential model) from the other project and support cost part (engineering, purchasing, project management, etc.) of the capital investment, and (b) adjusting both parts of the initial cost with price index changes from the previous comparable system installation (in the reference year). That is, the modified form of Equation (7-4) would be

$$C_A = C_{B1}(S_A/S_B)^X(\overline{I}_{B1}) + C_{B2}(\overline{I}_{B2}),$$

where

C_A = estimated cost of new automated warehousing system;

C_{B1} = equipment and installation cost of previous comparable system;

C_{B2} = other project and support costs of previous comparable system;

S_A = capacity of new automated warehousing system;

S_B = capacity of previous comparable system;

X = cost-capacity factor to reflect economies of scale;

$\overline{I}_{B1}$ = composite cost index ratio (current/reference) for equipment and installation costs;

$\overline{I}_{B2}$ = composite cost index ratio (current/reference) for other project and support costs.

Equipment and Installation Costs			
		Index	Index
Cost Element	Weight	(Ref. Yr)	(This Yr)
Mechanical equipment	0.41	122	201
Automation equipment	0.22	131	212
Installation hardware	0.09	118	200
Installation labor	0.28	135	184

Other Project and Support Costs			
		Index	Index
Cost Element	Weight	(Ref. Yr)	(This Yr)
Engineering	0.38	136	206
Project management	0.31	128	194
Purchasing	0.11	105	162
Other support	0.20	113	179

   a. Develop the cost index ratios for $\overline{I}_{B1}$ and $\overline{I}_{B2}$ based on the above data. (7.3)

   b. Develop the estimated capital investment cost for the new automated warehousing system when the equipment and installation cost of the previous comparable system was $1,226,000; the capacity of the new system is 11,000 cases of goods per eight-hour shift; the capacity of the previous comparable system is 5,800 cases per eight-hour shift; the cost capacity factor is 0.7; and other project and support costs for the previous comparable system were $234,000. (7.3, 7.4)

7-20. The cost of building a supermarket is related to the total area of the building. Data for the last 10 supermarkets built for Regork, Inc., are shown in the accompanying table.

Building	Area (ft^2)	Cost
1	14,500	$800,000
2	15,000	825,000
3	17,000	875,000
4	18,500	972,000
5	20,400	1,074,000
6	21,000	1,250,000
7	25,000	1,307,000
8	26,750	1,534,000
9	28,000	1,475,500
10	30,000	1,525,000

a. Develop a CER for the construction of supermarkets. Use the CER to estimate the cost of Regork's next store which has a planned area of 23,000 ft^2. (7.4)

b. Compute the standard error and correlation coefficient for the CER developed in part (a). (7.4)

7-21. In the packaging department of a large automotive parts distributor, a fairly reliable estimate of packaging and processing costs can be determined by knowing the weight of an order. Thus, the weight is a cost driver that accounts for a sizable fraction of the packaging and processing costs at this company. Data for the past 10 orders are given as follows: (7.4)

Packaging and Processing Costs ($) y	Weight (Pounds) x
97	230
109	280
88	210
86	190
123	320
114	300
112	280
102	260
107	270
86	190

a. Estimate the a and b coefficients, and determine the linear regression equation to fit these data.

b. We is the correlation coefficient (r)?

c. If an order weighs 250 lb, how much should it cost to package and process it?

7-22. Using the costing worksheet provided in this chapter (Figure 7-5), estimate the unit cost and selling price of manufacturing metal wire cutters in lots of 100 when these data have been obtained: (7.5)

Factory (direct) labor :	4.2 hr at $11.15/hr
Factory overhead :	150% of factory labor
Outside manufacture :	$74.87
Production material :	$26.20
Packing costs :	7% of factory labor
Desired profit :	12% of total manufacturing cost

7-23. You have been asked to *estimate the per unit selling price* of a new line of widgets. Pertinent data are as follows:

Direct labor rate:	$15.00 per hour
Production material:	$375 per 100 widgets
Factory overhead:	125% of direct labor
Packing costs:	75% of direct labor
Desired profit:	20% of total manufacturing cost

Past experience has shown that an 80% learning curve applies to the labor required for producing widgets. The time to complete the first widget has been estimated to be 1.76 hr. Use the estimated time to complete the 50th widget as your standard time for the purpose of estimating the unit selling price. (7.4, 7.5)

7-24. An electronics manufacturing company is planning to introduce a new product in the market. The best competitor sells a similar product at $420/unit. Other pertinent data are as follows:

Direct labor cost:	$15.00/hour
Factory overhead:	120% of direct labor
Production materials:	$300/unit
Packaging costs:	20% of direct labor

It has been found that an 85% learning curve applies to the labor required. The time to complete the first unit has been estimated to be 5.26 hr. The company decides to use the time required to complete the 20th unit as a standard for cost estimation purposes. The profit margin is based on the total manufacturing costs.

a. Based on the above information, determine the maximum profit margin that the company can have so as to remain competitive. (7.4, 7.5)

b. If the company desires a profit margin of 15%, can the target cost be achieved? If not, suggest two ways in which the target cost can be achieved.

7-25. Given the following information, how many units must be sold to achieve a profit of $25,000? [Note that the units sold must account for total production costs (direct and overhead) plus desired profit.] (7.4, 7.5)

Direct labor hours:	0.2 hr/unit
Direct labor costs:	$21.00/hr
Direct materials cost:	$4.00/unit
Overhead costs:	120% of direct labor
Packaging and shipping:	$1.20/unit
Selling price:	$20.00/unit

7-26. A personal computer (PC) company is trying to bring a new model of a PC to the market. According to the marketing department, the best selling price for a similar model from a world-class competitor is $2,500 per computer. The company wants to sell at the same price as its best competitor. The cost breakdown of the new model is shown below:

Assembling time for the first unit:	1.00 hr
Handling time:	10% of assembling time
Direct labor rate:	$15/hr
Planning labor:	10% of direct labor
Quality control:	50% of direct labor
Factory overhead:	200% of total labor
General & Admin. expense:	300% of total labor
Direct material cost:	$200/computer
Outside manufacture:	$2,000/computer
Packing cost:	10% of total labor
Facility rental:	10% of total labor
Profit:	20% of total manufacturing cost
Number of units:	20,000

Since the company mainly produces subassemblies purchased from other manufacturers and repackages the product, the direct material cost

is estimated at only $200 per computer. Direct labor time consists of handling time and assembling time. The company estimates the learning curve for assembling the new model is 95%. Compute the total manufactured cost for 20,000 of these PCs and determine the unit selling price. How can the company reduce its costs to meet target cost based on Equation (7-13)? (7.3, 7.4)

7-27. Use Figure 7-1 and the information and results of Problem 7-18 to develop an estimated before-tax net cash flow for 10 years of ownership of the new home. Assume the sale (disposal) of the home at the end of 10 yr. Obtain (locally) representative operating, repair, resale, and other data related to home ownership as needed to support development of the 10-yr cash-flow estimates. Indicate the cost-estimating techniques used in estimating cash flows. State any assumptions you make. (7.6)

7-28. A small plant has been constructed and the costs are known. A new plant is to be estimated using the exponential (power sizing) costing model. Major equipment, costs, and factors are as show in Table P8-28 on the following page. (Note mW = 10^6 Watts.)

   If ancillary equipment will cost an additional $200,000, find the cost for the proposed plant. (7.4)

7-29. If 846.2 labor hours are required for the third production unit, and 873.0 labor hours are required for the fifth production unit, determine the learning curve parameter s. (7.4)

7-30. Your company is planning to produce and sell high-density double-sided computer disks with 2 MB storage capacity. The disks are produced by installing a magnetic film into a plastic cartridge. A total of three operations need to be performed:

   (1) Cut out disks from magnetic film. The magnetic film is bought in rolls that cost $90 each. From each roll, 2,000 circular disks can be cut out. One person is needed to operate and supervise the cut-out machine. Installing a new roll takes 8 min, and cutting out 2,000 circular disks takes 25 min.

   (2) Apply disk control center-pieces. The disk control center-pieces cost $0.12 per unit. One person is required to apply the center-pieces to the magnetic disks. Applying the first center-piece takes 3 sec, and for the remaining center-pieces an 80% learning curve is applicable.

TABLE P7-28 Table for Problem 7-28

Equipment	Reference Size	Unit Reference Cost	Cost-Capacity Factors	New Design Size
Two boilers	6 mW	$300.000	0.80	10 mW
Two generators	6 mW	400,000	0.60	9 mW
Tank	80,000 gal	106,000	0.66	91,500 gal

(3) Insert into plastic cartridges. The plastic cartridges cost $0.15 per unit. One person is needed to supervise the disk-insertion operation. This operation is done automatically by a machine which can insert 1,500 disks per hour. The film, center-pieces, and cartridges are purchased from an outside manufacturer. A total of 10,000 disks are to be produced. Other relevant cost data are as follows:

The direct labor rate is $15.00 per hour.

Planning labor is 15% of factory labor.

Quality control is 30% of factory labor.

Factory overhead is 80% of total labor.

General and administrative expenses are 50% of total labor.

Packing costs are 100% of total labor.

The profit margin is 15% of total manufacturing cost.

a. Based on this information, estimate the unit selling price for a single disk. (7.5)
b. Compute the target cost when the best competitor's selling price is $0.50 per disk and a 15% profit margin is desired. (7.5)
c. Investigate and report on any cost reduction alternatives that can be implemented to reach the target cost. (7.5)

7-31. *Brain Teaser* You have been asked to prepare a quick estimate of the construction cost for a coal-fired electricity generating plant and facilities. A work breakdown structure (levels one through three) is shown in Table P7-31. You have the following information available.

A coal-fired generating plant twice the size of the one you are estimating was built in 1977. The 1977 boiler (1.2) and boiler support system (1.3) cost $110 million. The cost

index for boilers was 110 in 1977; it is 492 in 2000. The cost capacity factor for similar boilers and support systems is 0.9.

The 600-acre site is on property you already own, but improvements (1.1.1) and roads (1.1.2), will cost $2,000 per acre and railroads (1.1.3) will cost $3,000,000. Project integration (1.9) is projected to cost 3% of all other construction costs.

The security systems (1.5.4) are expected to cost $1,500 per acre, based on recent (2000) construction of similar plants. All other support facilities and equipment (1.5) elements are to be built by Viscount Engineering. Viscount Engineering has built the support facilities and equipment elements for two similar generating plants. Their experience is expected to reduce labor requirements substantially; a 90% learning curve can be assumed. Viscount built the support facilities and equipment on their first job in 95,000 hours. For this project, Viscount's labor will be billed to you at $60/hour. Viscount estimates materials for the construction of the support facilities and equipment elements (except 1.5.4) will cost you $15,000,000.

The coal storage facility (1.4) for the coal-fired generating plant built in 1977 cost $5 million. Although your plant is smaller, you require the same size coal storage facility as the 1977 plant. You assume you can apply the cost index for similar boilers to the coal storage facility.

What is your estimated 2000 cost for building the coal-fired generating facility? Summarize your calculations in a cost estimating worksheet, and state the assumptions you make.

TABLE P7-31	**Work Breakdown Structure for Problem P7-31**

PROJECT: Coal-Fired Electricity Generating Plant and Facilities

Line No.	Title	WBS Element Code
001	Coal-Fired Power Plant	1.
002	Site	1.1
003	Land Improvements	1.1.1
004	Roads, Parking, and Paved Areas	1.1.2
005	Railroads	1.1.3
006	Boiler	1.2
007	Furnace	1.2.1
008	Pressure Vessel	1.2.2
009	Heat Exchange System	1.2.3
010	Generators	1.2.4
011	Boiler Support System	1.3
012	Coal Transport System	1.3.1
013	Coal Pulverizing System	1.3.2
014	Instrumentation & Control	1.3.3
015	Ash Disposal System	1.3.4
016	Transformers & Distribution	1.3.5
017	Coal Storage Facility	1.4
018	Stockpile Reclaim System	1.4.1
019	Rail Car Dump	1.4.2
020	Coal Handling Equipment	1.4.3
021	Support Facilities & Equipment	1.5
022	Hazardous Waste Systems	1.5.1
023	Support Equipment	1.5.2
024	Utilities & Communications System	1.5.3
025	Security Systems	1.5.4
026	Project Integration	1.9
027	Project Management	1.9.1
028	Environmental Management	1.9.2
029	Project Safety	1.9.3
030	Quality Assurance	1.9.4
031	Test, Start-Up, & Transition Management	1.9.5

Appendix 7-A   EXCEL Spreadsheet for Figure 7-5

Table 7-A-1 contains the cell formulas for Column E (Row Total) of Figure 7.5. Note that in the actual spreadsheet, Column E is spreadsheet column H and Row A is spreadsheet row 6.

TABLE 7-A-1 Cell Formulas for Column E of Figure 7-5. Note that in the actual spreadsheet, Column E is spreadsheet column H and Row A is spreadsheet row 6.

Column E
Row
Total

A: = IF($C6 > 0, $C6 * $D6, $G6)

B: = IF($C7 > 0, $C7 * $D7, IF($F7 = "A", $E7 * $H6, $G7))

C: = IF($C8 > 0, $C8 * $D8, IF($F8 = "A", $E8 * $H6, IF($F8 = "B", $E8 * $H7, $G8)))

D: = SUM(I6 : J8)

E: = IF($C10 > 0, $C10 * $D10, IF($F10 = "A", $E10 * $H$6, IF($F10 = "B", $E10 * $H$7, IF($F10 = "C", $E10 * $H$8, IF($F10 = "D", $E10 * $H$9, $G10)))))

F: = IF($C11 > 0, $C11 * $D11, IF($F11 = "A", $E11 * $H$6, IF($F11 = "B", $E11 * $H$7, IF($F11 = "C", $E11 * $H$8, IF($F11 = "D", $E11 * $H$9, $G11)))))

G: = IF($C12 > 0, $C12 * $D12, IF($F12 = "A", $E12 * $H$6, IF($F12 = "B", $E12 * $H$7, IF($F12 = "C", $E12 * $H$8, IF($F12 = "D", $E12 * $H$9,
       IF($F12 = "E", $E12 * $H$11, IF($F12 = "F", $E12 * $H$11, $G12)))))))

I: = SUM($H9 : $H13)

J: = IF($C15 > 0, $C15 * $D15, IF($F15 = "A", $E15 * $H$6, IF($F15 = "B", $E15 * $H$7, IF($F15 = "C", $E15 * $H$8, IF($F15 = "D", $E15 * $H9,
       IF($F15 = "G", $E15 * $H$12, IF($F15 = "T", $E15 * $H$14, $G15)))))))

K: = SUM($H14 : $H15)

L: = IF($C17 > 0, $C17 * $D17, IF($F17 = "A", $E·7 * $H$6, IF($F17 = "B", $E17 * $H$7, IF($F17 = "C", $E17 * $H$8, IF($F17 = "D", $E17 * $H$9,
       IF($F17 = "G", $E17 * $H$12, IF($F17 = "T", $E17 * $H$14, IF($F17 = "K", $E17 * $H$16, $G17)))))))

M: = IF($C18 > 0, $C18 * $D18, IF($F18 = "A", $E18 * $H$6, IF($F18 = "B", $E18 * $H$7, IF($F18 = "C", $E18 * $H$8, IF($F18 = "D", $E18 * $H$9,
       IF($F18 = "G", $E18 * $H$12, IF($F18 = "T", $E18 * $H$14, IF($F18 = "K", $E18 * $H$16, $G18)))))))

N: = SUM($H16 : $H18)

O:

Q: = IF($C22 > 0, $C22 * $D22, IF($F22 = "A", $E22 * $H$6, IF($F22 = "C", $E22 * $H$8, IF($F22 = "D", $E22 * $H$9, IF($F22 = "G", $E22 * $H$12,
       IF($F22 = "T", $E22 * $H$14, IF($F22 = "K", $E22 * $H$16, IF($F22 = "P", $E22 * $H$21, $G22)))))))

R: = SUM($H21 : $H22)

The nested IF statements are used to determine what type of estimate has been provided for each row. The cell formula first looks to see whether a unit estimate has been provided. If no value has been entered in the "units" column (column C in the formula), the formula seeks to determine whether a factor estimate has been provided (signaled by the presence of a row letter in column F). If neither a unit estimate or a factor estimate has been provided, the row total is set equal to the value in column G, which corresponds to the column for direct estimates.

Appendix 7-B An Additional Example of Target Costing

The purpose of this appendix is to provide an additional illustration of the iterative use of "top-down" and "bottom-up" estimating together with the concepts of target costing, design to cost, and value engineering.

The problem is to estimate the cost and selling price for a metallic handle. The purpose of the product is to serve as a common handle for a series of handheld tools (hammer, chisel, etc.). A market survey has shown that the best competitor's selling price for a similar product is $10.00. Your company sets a profit margin of 10% (based on total manufacturing cost) for this type of product. Thus, the target cost for the handle is

$$\text{Target Cost} = \frac{\$10.00}{1.10} = \$9.09$$

A preliminary design calls for the handle to be machined from an aluminum rod. A total of 13 machining operations are specified. A total of 1,000 handles are to be produced. The following data are used to obtain an initial estimate of the total manufacturing cost:

Handling time (first unit):	15 min
Machining time:	12 min/unit
Tool changing time:	3.4 min/unit
Direct material cost:	$1.40/unit
Tool material cost:	$5.00/tool
Average tool life:	300 min/tool
Direct labor rate:	$8.00/hr
Planning labor:	9% of factory labor
Quality control:	15% of factory labor
Factory overhead:	90% of total labor
General and administrative:	25% of total labor
Packing costs:	$0.80/unit

Estimating Direct Labor Time

In the manufacture of the handle, direct labor time consists of handling time, machining time, and tool-changing time. Handling time is the time spent by the machinist loading and unloading the part from the machines as well as the time required for machine adjustments. Machining time is the actual time spent machining the part. Tool-changing time is the time spent by the machinist changing tools.

It is felt that a 90% learning curve applies to the handling time component. As the machinist gets into the rhythm of the work, less time should be spent handling the part. For estimating purposes, the cumulative average handling time over the 1,000 handles to be produced will be used. To obtain this estimate, a short computer program is written to obtain the total handling time for the 1,000 units:

$$
\begin{aligned}
&K = 15 \\
&n = \log(0.9)/\log(2) \\
&T = 0 \\
&\text{FOR } I = 1 \text{ to } 1000 \\
&\qquad T = T + K * I\hat{\ }n \\
&\text{NEXT } I \\
&C = T/1000
\end{aligned}
$$

The results obtained are $T = 6{,}180$ min and $C = 6.18$ min. The estimate of total factory labor hours is given by

Factory Labor = Handling time + Machining time + Tool Changing time

$$
= (6.18 \text{ min} + 12 \text{ min} + 3.4 \text{ min}) \left(\frac{\text{hour}}{60 \text{ min}} \right) = 0.36 \text{ hr per unit}
$$

Estimating Production Material Costs

Production material costs in this example consist of the direct material used for the handle and the tools used to machine the handle. The tool cost per unit is based on the expected tool life and the machining time for an individual unit. The formula used to estimate tool cost is:

$$
C_t = C_{tm}(t_m/T)
$$

where C_t = tool cost ($/unit)

C_{tm} = tool material cost ($/tool)

t_m = machining time (min/unit)

T = average tool life (min/tool)

Our estimate of tool cost per handle is therefore

$$
C_t = \left(\frac{\$5.00}{\text{tool}} \right) \left(\frac{12 \text{ min/unit}}{300 \text{ min/tool}} \right) = \frac{\$0.20}{\text{unit}}.
$$

Total production material cost is therefore $1.40/unit + $0.20/unit = $1.60/unit.

Column A		Column B		Column C		Column D	Column E
		Unit Estimate		Factor Estimate		Direct	Row
MANUFACTURING COST ELEMENTS		Units	Cost/Unit	Factor	Row	Estimate	Estimate
A:	Factory Labor	0.36	$8.00				$2.88
B:	Planning Labor			9%	A		0.26
C:	Quality Control			15%	A		0.43
D:	TOTAL LABOR						3.57
E:	Factory Overhead			90%	D		3.21
F:	General & Admin. Expense			25%	D		0.89
G:	Production Material					$1.60	1.60
H:	Outside Manufacture						0.00
I:	SUBTOTAL						9.28
J:	Packing Costs					$0.80	0.80
K:	TOTAL DIRECT CHARGE						10.08
L:	Other Direct Charge						0.00
M:	Facility Rental						0.00
N:	TOTAL MANUFACTURING COST						10.08
O:	Quantity (lot size)						1
P:	MANUFACTURING COST/UNIT						10.08
Q:	Profit/Fee			10%	P		1.01
R:	UNIT SELLING PRICE						$11.09

Figure 7-B-1 Initial Estimate of Manufacturing Cost and Selling Price

Figure 7-B-1 shows the completed manufacturing cost estimating spreadsheet. Our current estimate of total manufacturing cost is $10.08, which exceeds our target cost of $9.09. We will now identify some areas for cost reduction and apply value engineering to obtain the necessary reductions.

Setting Cost Goals for Potential Cost Reduction Areas

Making use of the solver feature of our spreadsheet package, we can quickly obtain cost goals for specific cost categories. Figure 7-B-2 shows that reducing factory labor to 0.3135 hr per unit would enable us to meet our target cost. As shown in Figure 7-B-3, reducing packaging cost alone would not allow us to achieve the target cost. This doesn't mean that we shouldn't seek to reduce packaging at all, just that cost reductions in other areas will also be required.

Application of Value Engineering to Obtain Cost Reduction

Three potential areas for cost reduction have been identified: factory labor, production material, and packaging. An in-depth study of these areas resulted in these suggested changes:

1. Part redesign reduced the number of machining operations (and therefore machining time) required. The new estimate of machining time is 10.8 minutes.

Column A	Column B		Column C		Column D	Column E
	Unit Estimate		Factor Estimate		Direct	Row
MANUFACTURING COST ELEMENTS	Units	Cost/Unit	Factor	Row	Estimate	Estimate
A: Factory Labor	0.3135	$ $8.00				$ 2.51
B: Planning Labor			9%	A		0.23
C: Quality Control			15%	A		0.38
D: TOTAL LABOR						3.11
E: Factory Overhead			90%	D		2.80
F: General & Admin. Expense			25%	D		0.78
G: Production Material					$ 1.60	1.60
H: Outside Manufacture						0.00
I: SUBTOTAL						8.29
J: Packing Costs					$ 0.80	0.80
K: TOTAL DIRECT CHARGE						9.09
L: Other Direct Charge						0.00
M: Facility Rental						0.00
N: TOTAL MANUFACTURING COST						9.09
O: Quantity (lot size)						1
P: MANUFACTURING COST/UNIT						9.09
Competitor's Selling Price	$ 10.00					
Desired Return on Sales	10%					
Target Cost	$ 9.09					

Figure 7-B-2 Cost Goal for Factory Labor

Column A	Column B		Column C		Column D	Column E
	Unit Estimate		Factor Estimate		Direct	Row
MANUFACTURING COST ELEMENTS	Units	Cost/Unit	Factor	Row	Estimate	Estimate
A: Factory Labor	0.36	$ $8.00				$ 2.88
B: Planning Labor			9%	A		0.26
C: Quality Control			15%	A		0.43
D: TOTAL LABOR						3.57
E: Factory Overhead			90%	D		3.21
F: General & Admin. Expense			25%	D		0.89
G: Production Material					$ 1.60	1.60
H: Outside Manufacture						0.00
I: SUBTOTAL						9.28
J: Packing Costs					($0.19)	(0.19)
K: TOTAL DIRECT CHARGE						9.09
L: Other Direct Charge						0.00
M: Facility Rental						0.00
N: TOTAL MANUFACTURING COST						9.09
O: Quantity (lot size)						1
P: MANUFACTURING COST/UNIT						9.09
Competitor's Selling Price	$ 10.00					
Desired Return on Sales	10%					
Target Cost	$ 9.09					

Figure 7-B-3 Cost Goal for Packaging Cost

Handling time and tool-changing time are unaffected by this change. The new estimate of total factory labor hours is:

$$\text{Factory Labor} = (6.18 \text{ min} + 10.8 \text{ min} + 3.4 \text{ min}) \left(\frac{\text{hour}}{60 \text{ min}} \right) = 0.34 \text{ hr per unit}$$

A reduction in machining time will also result in a reduction in tool cost. The new tool cost would be:

$$C_t = \left(\frac{\$5.00}{\text{tool}} \right) \left(\frac{10.8 \text{ min/unit}}{300 \text{ min/tool}} \right) = \frac{\$0.18}{\text{unit}}$$

2. Negotiations with the supplier of aluminum bar stock has led to a reduction in raw material cost. This is due to an agreement to return all scrap material to the supplier. Scrap material is significant in this case—approximately 60% of the original material is removed during the machining operations. The new direct material cost is $1.10 per unit. Combined with the new tool cost, the new estimate of production material is $1.28 per unit.
3. An analysis of the packing material requirements revealed that a lesser grade of cardboard would provide the needed protection during shipment. The new estimate of packing costs is $0.55 per unit.

The impact of these changes on total manufacturing cost of the handle is shown in Figure 7-B-4. None of these changes would have the desired impact on cost individually; however, collectively, the reductions are enough to meet the target cost.

Column A		Column B		Column C		Column D	Column E
		Unit Estimate		Factor Estimate		Direct	Row
MANUFACTURING COST ELEMENTS		Units	Cost/Unit	Factor	Row	Estimate	Estimate
A:	Factory Labor	0.34	$ 8.00				$ 2.72
B:	Planning Labor			9%	A		0.24
C:	Quality Control			15%	A		0.41
D:	TOTAL LABOR						3.37
E:	Factory Overhead			90%	D		3.04
F:	General & Admin. Expense			25%	D		0.84
G:	Production Material					$ 1.28	1.28
H:	Outside Manufacture						0.00
I:	SUBTOTAL						8.53
J:	Packing Costs					$ 0.55	0.55
K:	TOTAL DIRECT CHARGE						9.08
L:	Other Direct Charge						0.00
M:	Facility Rental						0.00
N:	TOTAL MANUFACTURING COST						9.08
O:	Quantity (lot size)						1
P:	MANUFACTURING COST/UNIT						9.08
Q:	Profit/Fee			10%	P		0.91
R:	UNIT SELLING PRICE						$ 9.99

Figure 7-B-4 Final Estimate of Manufacturing Cost and Selling Price

CHAPTER

8

*P*rice Changes and Exchange Rates

*W*hen the monetary unit does not have a constant value in exchange for goods and services in the marketplace, and when future price changes are expected to be significant, an undesirable choice among competing alternatives can be made if price change effects are not included in an engineering economy study (before taxes and after taxes). The objectives of this chapter are to (1) introduce a methodology for dealing with price changes caused by inflation and deflation, (2) develop and illustrate proper techniques to account for these effects, and (3) discuss the relationship of these concepts to foreign exchange rates and the analysis of engineering projects in currencies other than the U.S. dollar.

The following topics are discussed in this chapter:

Price changes

The Consumer Price Index and the Producer Price Index

Terminology and basic concepts

The relationship between actual (current) dollars and real (constant) dollars

Use of combined (market) interest rates versus real interest rates

Differential price inflation or deflation

Modeling price changes with geometric sequences of cash flows

Application strategy for use of actual- and real-dollar analysis

A comprehensive example

Foreign exchange rates

8.1 Price Changes

In earlier chapters, we assumed that prices for goods and services in the market-place remain relatively unchanged over extended periods. Unfortunately, this is not generally a realistic assumption.

> *General price inflation,* which is defined here as an increase in the average price paid for goods and services bringing about a reduction in the purchasing power of the monetary unit, is a business reality that can affect the economic comparison of alternatives. The history of price changes shows that price inflation is much more common than general price *deflation,* which involves a decrease in the average price for goods and services with an increase in the purchasing power of the monetary unit. The concepts and methodology discussed in this chapter, however, apply to any price changes.

One measure of price changes in our economy (and an estimate of general price inflation or deflation for the average consumer) is the Consumer Price Index (CPI). The CPI is a composite price index that measures average change in the prices paid for food, shelter, medical care, transportation, apparel, and other selected goods and services used by individuals and families.

Another measure of price changes in the economy (and also an estimate of general price inflation or deflation) is the Producer Price Index (PPI). In actuality, a number of different indexes are calculated covering most areas of the U.S. economy. These indexes are composite measures of average changes in the selling prices of items used in the production of goods and services. These different indexes are calculated by stage of production [crude materials (e.g., iron ore), intermediate materials (rolled sheet steel), and finished goods (automobiles)], by Standard Industrial Classification (SIC), and by the census product code extension of the SIC areas. Thus, PPI information is available to meet the needs of most engineering economy studies.

The CPI and PPI indexes are calculated monthly from survey information by the Bureau of Labor Statistics in the U.S. Department of Labor. These indexes are based on current and historical information and may be used, as appropriate, to represent future economic conditions or for short-term forecasting purposes only. Also, longer-term forecasts of price changes may be purchased from private firms engaged in the business of providing economic forecasting services.

The annual CPI and PPI (finished goods) end-of-year values for 1988–2001 are shown in Table 8-1. Also, the annual inflation or deflation rates are shown for each of the two indexes. Since we have used end-of-year index values, the annual change rates reflect what occurred over the 12-month calendar year. The annual change rates (%) are calculated as follows:

$$(\text{CPI or PPI Annual Change Rate, \%})_k = \frac{(\text{Index})_k - (\text{Index})_{k-1}}{(\text{Index})_{k-1}}(100\%).$$

For example, the PPI (finished goods) annual change rate (%) for 1996 is

$$\frac{(\text{PPI})_{1996} - (\text{PPI})_{1995}}{(\text{PPI}_{1995})}(100\%) = \frac{133.0 - 129.3}{129.3}(100\%) = 2.86\%.$$

TABLE 8-1 CPI and PPI Values and Annual Change Rates, 1988–2001

Year	End-of-Year CPI Value	Annual Change Rate (%)	End-of-Year PPI (Finished Goods) Value	Annual Change Rate (%)
1988	120.5	4.42	110.0	3.87
1989	126.1	4.65	115.4	4.91
1990	133.8	6.10	122.0	5.72
1991	137.9	3.06	121.9	−0.01
1992	141.9	2.90	124.2	1.89
1993	145.8	2.75	124.5	0.02
1994	149.7	2.67	126.5	1.60
1995	153.5	2.54	129.3	2.21
1996	158.6	3.32	133.0	2.86
1997	161.3	1.70	131.4	−1.95
1998	163.9	1.6	131.1	−0.2
1999	168.3	2.7	134.9	2.9
2000	174.0	3.4	139.7	3.6
2001	178.2 (est.)	2.4	143.4 (est.)	2.6

Source: *CPI and PPI Detailed Reports,* U.S. Department of Labor, Bureau of Labor Statistics (U.S. Government Printing Office, Washington, D.C.).

For the years 1991, 1997, and 1998, deflation occurred as measured by the PPI for finished goods [i.e., an estimate of the general inflation or deflation rate (f)].

8.2 Terminology and Basic Concepts

To facilitate the development and discussion of the methodology for including price changes of goods and services in engineering economy studies, we need to define and discuss some terminology and basic concepts. The dollar is used as the monetary unit in this book except when discussing foreign exchange rates.

1. *Actual dollars (A$):* The number of dollars associated with a cash flow (or a noncash flow amount such as depreciation) as of the time it occurs. For example, people typically anticipate their salaries two years in advance in terms of actual dollars. Sometimes A$ are referred to as *nominal* dollars, *current* dollars, *then-current* dollars, and *inflated* dollars, and their relative purchasing power is affected by general price inflation or deflation.
2. *Real dollars (R$):* Dollars expressed in terms of the same purchasing power relative to a particular time. For instance, the future unit prices of goods or services that are changing rapidly are often estimated in real dollars (relative to some base year) to provide a consistent means of comparison. Sometimes R$ are termed *constant* dollars.

3. *General price inflation (or deflation) rate* (f): A measure of the average change in the purchasing power of a dollar during a specified period of time. The general price inflation or deflation rate is defined by a selected, broadly based index of market price changes. In engineering economic analysis, the rate is projected for a future time interval and usually is expressed as an effective annual rate. Many large organizations have their own selected index that reflects the particular business environment in which they operate.

4. *Combined (market) interest rate* (i_c): The money paid for the use of capital, normally expressed as an annual rate (%) that includes a market adjustment for the anticipated general price inflation rate in the economy. Thus, it is a *market interest rate* and represents the time value change in future actual dollar cash flows that takes into account both the potential real earning power of money and the estimated general price inflation or deflation in the economy. It is sometimes called the *nominal* interest rate.

5. *Real interest rate* (i_r): The money paid for the use of capital, normally expressed as an annual rate (%) that does *not* include a market adjustment for the anticipated general price inflation rate in the economy. It represents the time value change in future real-dollar cash flows based only on the potential real earning power of money. It is sometimes called the *inflation-free* interest rate.

6. *Base time period* (b): The reference or base time period used to define the constant purchasing power of real dollars. Often, in practice, the base time period is designated as the time of the engineering economic analysis, or reference time 0 (i.e., $b = 0$). However, b can be any designated point in time.

With an understanding of these definitions, we can delineate and illustrate some useful relationships that are important in engineering economy studies.

8.2.1 The Relationship between Actual Dollars and Real Dollars

The relationship between actual dollars (A$) and real dollars (R$) is defined in terms of the general price inflation (or deflation) rate; that is, it is a function of f.

Actual dollars as of any period (e.g., a year), k, can be converted into real dollars of *constant* market purchasing power as of any base period, b, by the relationship

$$(R\$)_k = (A\$)_k \left(\frac{1}{1+f} \right)^{k-b} = (A\$)_k (P/F, f\%, k-b), \qquad (8\text{-}1)$$

for a given b value. This relationship between actual dollars and real dollars applies to the unit prices, or costs of fixed amounts of individual goods or services, used to develop (estimate) the individual cash flows related to an engineering project. The designation for a specific type of cash flow, j, would be included as

$$(R\$)_{k,j} = (A\$)_{k,j} \left(\frac{1}{1+f} \right)^{k-b} = (A\$)_{k,j} (P/F, f\%, k-b), \qquad (8\text{-}2)$$

for a given b value, where the terms $R\$_{k,j}$ and $A\$_{k,j}$ are the unit price, or cost for a fixed amount, of goods or services j in period k in real dollars and actual dollars, respectively.

EXAMPLE 8-1

Suppose that your salary is $35,000 in year one, will increase at 6% per year through year four, and is expressed in actual dollars as follows:

End of Year, k	Salary (A$)
1	$35,000
2	37,100
3	39,326
4	41,685

If the general price inflation rate (f) is expected to average 8% per year, what is the real-dollar equivalent of these actual dollar salary amounts? Assume that the base time period is year one $(b = 1)$.

SOLUTION

By using Equation (8-2), we see that the real-dollar salary equivalents are readily calculated relative to the base time period, $b = 1$:

Year	Salary (R$, $b = 1$)
1	$35,000(P/F, 8\%, 0) = \$35,000$
2	$37,100(P/F, 8\%, 1) = 34,351$
3	$39,326(P/F, 8\%, 2) = 33,714$
4	$41,685(P/F, 8\%, 3) = 33,090$

In year one (the designated base time period for the analysis), the annual salary in actual dollars remained unchanged when converted to real dollars. *This illustrates an important point: In the base time period (b), the purchasing power of an actual dollar and a real dollar is the same; that is, $R\$_{b,j} = A\$_{b,j}$.* This example also illustrates the results when the actual annual rate of increase in salary (6% in this example) is less than the general price inflation rate (f). As you can see, the actual-dollar salary cash flow shows some increase, but a decrease in the real-dollar salary cash flow occurs (and thus a decrease in total market purchasing power). This is the situation when people say their salary increases have not kept pace with market inflation.

EXAMPLE 8-2

An engineering project team is analyzing the potential expansion of an existing production facility. Different design alternatives are being considered. The estimated after-tax cash flow (ATCF) in actual dollars for one alternative is shown in

TABLE 8-2	ATCFs for Example 8-2		
(1) End-of-Year, k	(2) ATCF (A\$)	(3 $(P/F, f\%, k - b)$ $= [1/(1.052)^{k-0}]$	(4) ATCF (R\$), $b = 0$
0	−\$172,400	1.0	−\$172,400
1	−21,000	0.9506	−19,963
2	51,600	0.9036	46,626
3	53,000	0.8589	45,522
4	58,200	0.8165	47,520
5	58,200	0.7761	45,169
6	58,200	0.7377	42,934
7	58,200	0.7013	40,816
8	58,200	0.6666	38,796

column 2 of Table 8-2. If the general price inflation rate (f) is estimated to be 5.2% per year during the eight-year analysis period, what is the real dollar ATCF that is equivalent to the actual dollar ATCF? The base time period is year zero ($b = 0$).

SOLUTION
The application of Equation (8-1) is shown in column 3 of Table 8-2. The ATCF in real dollars shown in column 4 has purchasing power in each year equivalent to the original ATCF in actual dollars (column 2).

8.2.2 The Correct Interest Rate to Use in Engineering Economy Studies

In general, the interest rate that is appropriate for equivalence calculations in engineering economy studies depends on whether actual-dollar or real-dollar cash flow estimates are used:

Method	If Cash Flows Are in Terms of	Then the Interest Rate to Use Is
A	Actual dollars (A\$)	Combined (market) interest rate, i_c
B	Real dollars (R\$)	Real interest rate, i_r

This table should make intuitive sense as follows: If one is estimating cash flows in terms of actual (inflated) dollars, the combined interest rate (market interest rate with inflation/deflation component) is used. Similarly, if one is estimating cash flows in terms of real dollars, the real (inflation-free) interest rate is used. Thus, one can make economic analyses in either the actual- or real-dollar domain with equal validity, provided that the appropriate interest rate is used for equivalence calculations.

It is important to be consistent in using the correct interest rate for the type of analysis (actual or real dollars) being done. The two mistakes commonly made are as follows:

Interest Rate	Type of Analysis	
(MARR)	A$	R$
i_c	$\checkmark$(Correct)	Mistake 1 Bias is against capital investment
i_r	Mistake 2 Bias is toward capital investment	$\checkmark$(Correct)

In Mistake 1, the combined interest rate (i_c), which includes an adjustment for the general price inflation rate (f), is used in equivalent worth calculations for cash flows estimated in real dollars. Because real dollars have constant purchasing power expressed in terms of the base time period (b) and do not include the effect of general price inflation, we have an inconsistency. There is a tendency to develop future cash flow estimates in terms of dollars with purchasing power at the time of the study (i.e., real dollars with $b = 0$), and then to use the combined interest rate in the analysis [a firm's MARR is normally a combined (market) interest rate]. The result of Mistake 1 is a bias against capital investment. The cash-flow estimates in real dollars for a project are numerically lower than actual-dollar estimates with equivalent purchasing power (assuming that $f > 0$). Additionally, the i_c value (which is greater than the i_r value that should be used) further reduces (understates) the equivalent worth of the results of a proposed capital investment.

In Mistake 2, the cash-flow estimates are in actual dollars, which include the effect of general price inflation (f), but the real interest rate (i_r) is used for equivalent worth calculations. Since the real interest rate does not include an adjustment for general price inflation, we again have an inconsistency. The effects of this mistake, in contrast to those in Mistake 1, result in a bias toward capital investment by overstating the equivalent worth of future cash flows.

8.2.3 The Relationship among i_c, i_r, and f

Equation (8-1) showed that the relationship between an actual-dollar amount and a real-dollar amount of equal purchasing power in period k is a function of the general inflation rate (f). It is desirable to do engineering economy studies in terms of either actual dollars or real dollars. Thus, the relationship between the two dollar domains is important, as well as the relationship among i_c, i_r, and f, so that the equivalent worth of a cash flow is equal in the base time period when either an actual or a real-dollar analysis is used. The relationship among these three factors is (derivation not shown)

$$1 + i_c = (1 + f)(1 + i_r); \tag{8-3}$$

$$i_c = i_r + f + i_r(f); \tag{8-4}$$

$$i_r = \frac{i_c - f}{1 + f}. \tag{8-5}$$

Thus, the combined (market) interest rate [Equation (8-4)] is the sum of the real interest rate (i_r) and the general price inflation rate (f), plus the product of those two terms. Also, as shown in Equation (8-5), the real interest rate (i_r) can be calculated from the combined interest rate and the general price inflation rate. Similarly, based on Equation (8-5), the IRR of a real-dollar cash flow is related to the IRR of an actual-dollar cash flow (with the same purchasing power each period) as follows: $\text{IRR}_r = (\text{IRR}_c - f)/(1 + f)$.

EXAMPLE 8-3

If a company borrowed $100,000 today to be repaid at the end of three years at a combined (market) interest rate of 11%, what is the actual dollar amount owed at the end of three years, the real IRR to the lender, and the real-dollar amount equivalent in purchasing power to the actual-dollar amount at the end of the third year? Assume that the base or reference period is now $(b = 0)$ and that the general price inflation rate (f) is 5% per year.

SOLUTION

In three years, the company will owe the original $100,000 plus interest that has accumulated, in actual dollars:

$$(A\$)_3 = (A\$)_0(F/P, i_c\%, 3) = \$100,000(F/P, 11\%, 3) = \$136,763.$$

With this payment, the actual internal rate of return, IRR_c, to the lender is 11%. Thus, the real rate of return to the lender can be calculated based on Equation (8-5):

$$\text{IRR}_r = \frac{0.11 - 0.05}{1.05} = 0.05714, \text{ or } 5.714\%.$$

The real interest rate in this example is the same as IRR_r. Using this value for i_r, the real-dollar amount equivalent in purchasing power to the actual-dollar amount owed is

$$(R\$)_3 = (R\$)_0(F/P, i_r\%, 3) = \$100,000(F/P, 5.714\%, 3) = \$118,140.$$

This amount can be verified by the following calculation based on Equation (8-1):

$$(R\$)_3 = (A\$)_3(P/F, f\%, 3) = \$136,763(P/F, 5\%, 3) = \$118,140.$$

EXAMPLE 8-4

In Example 8-1, your salary was projected to increase at the rate of 6% per year, and the general price inflation rate was expected to be 8% per year. Your resulting estimated salary for the four years in actual and real dollars was as follows:

End of Year, k	Salary (A$)	Salary (R$), $b = 1$
1	$35,000	$35,000
2	37,100	34,351
3	39,326	33,714
4	41,685	33,090

What is the equivalent worth (EW) of the four-year actual- and real-dollar salary cash flows at the end of year one (base year) if your personal MARR is 10% per year (i_c)?

SOLUTION

(a) Actual-dollar salary cash flow:

$$EW(10\%)_1 = \$35,000 + \$37,100(P/F, 10\%, 1) + \$39,326(P/F, 10\%, 2) + \$41,685(P/F, 10\%, 3)$$

$$= \$132,545.$$

(b) Real-dollar salary cash flow:

$$i_r = \frac{i_c - f}{1 + f} = \frac{0.10 - 0.08}{1.08} = 0.01852, \text{ or } 1.852\%;$$

$$EW(1.852\%)_1 = \$35,000 + \$34,351\left(\frac{1}{1.01852}\right)^1 + \$33,714\left(\frac{1}{1.01852}\right)^2 + \$33,090\left(\frac{1}{1.01852}\right)^3$$

$$= \$132,545.$$

Thus, we obtain the same equivalent worth at the end of year one (the base time period) for both the actual-dollar and real-dollar four-year salary cash flows when the appropriate interest rate is used for the equivalence calculations.

8.2.4 Fixed and Responsive Annuities

Whenever future cash flows are predetermined by contract, as in the case of a bond or a fixed annuity, these amounts do not respond to general price inflation or deflation. In cases where the future amounts are not predetermined, however, they may respond to general price changes. The degree of response varies from case to case. To illustrate the nature of this situation, let us consider two annuities. The first annuity is fixed (unresponsive to general price inflation) and yields $2,000 per year in actual dollars for 10 years. The second annuity is of the same duration and yields enough future actual dollars to be equivalent to $2,000 per year in real dollars (purchasing power). Assuming a general price inflation rate of 6% per year, pertinent values for the two annuities over a 10-yr period are as shown in Table 8-3.

TABLE 8-3 Illustration of Fixed and Responsive Annuities with General Price Inflation Rate of 6% per Year

End of Year k	Fixed Annuity		Responsive Annuity	
	In Actual Dollars	In Equivalent Real Dollars[a]	In Actual Dollars	In Equivalent Real Dollars[a]
1	$2,000	$1,887	$2,120	$2,000
2	2,000	1,780	2,247	2,000
3	2,000	1,679	2,382	2,000
4	2,000	1,584	2,525	2,000
5	2,000	1,495	2,676	2,000
6	2,000	1,410	2,837	2,000
7	2,000	1,330	3,007	2,000
8	2,000	1,255	3,188	2,000
9	2,000	1,184	3,379	2,000
10	2,000	1,117	3,582	2,000

[a] See Equation (8-1)

Thus, when the amounts are constant in actual dollars (unresponsive to general price inflation), their equivalent amounts in real dollars decline over the 10-yr interval to $1,117 in the final year. When the future cash-flow amounts are fixed in real dollars (responsive to general price inflation), their equivalent amounts in actual dollars rise to $3,582 by year 10.

Included in engineering economy studies are certain quantities unresponsive to general price inflation, such as depreciation, or lease fees and interest charges based on an existing contract or loan agreement. For instance, depreciation amounts, once determined, do not increase (with present accounting practices) to keep pace with general price inflation; lease fees and interest charges typically are contractually fixed for a given period of time. Thus, it is important when doing an actual-dollar analysis to recognize the quantities that are unresponsive to general price inflation, and when doing a real-dollar analysis to convert these A$ quantities to R$ quantities using Equation (8-2).

If this is not done, not all cash flows will be in the same dollar domain (A$ or R$), and the analysis results will be distorted. Specifically, the equivalent worths of the cash flows for an A$ and a R$ analysis will not be the same in the base year, b, and the A$ IRR and the R$ IRR for the project will not have the proper relationship based on Equation (8-5); that is, $IRR_r = (IRR_c - f)/(1 + f)$.

8.2.5 The Impact of Price Changes on After-Tax Analysis

Engineering economy studies that include the effects of price changes caused by inflation or deflation may also include such items as interest charges, depreciation amounts, lease payments, and other contract amounts that are actual dollar cash flows based on past commitments. They are generally *unresponsive* to further price changes. At the same time, many other types of cash flows (e.g., labor, materials) are *responsive* to market price changes. In Example 8-5, an *after-tax analysis* is presented that shows the correct handling of these different situations.

EXAMPLE 8-5

The cost of some new and more efficient electrical circuit switching equipment is $180,000. It is estimated (in base year dollars, $b = 0$) that the equipment will reduce current net operating expenses by $36,000 per year (for 10 yr) and will have a $30,000 market value at the end of the 10th year. For simplicity, these cash flows are estimated to increase at the general price inflation rate ($f = 8\%$ per year). Due to new computer control features on the equipment, it will be necessary to contract for some maintenance support during the first three years. The maintenance contract will cost $2,800 per year. This equipment will be depreciated under the MACRS (GDS) method, and it is in the five-year property class. The effective income tax rate (t) is 38%; the selected analysis period is 10 yr; and the MARR (after taxes) is $i_c = 15\%$ per year.

(a) Based on an actual-dollar after-tax analysis, is this capital investment justified?

(b) Develop the ATCF in real dollars.

SOLUTION

(a) The actual-dollar after-tax economic analysis of the new equipment is shown in Table 8-4 (columns 1–7). The capital investment, savings in operating expenses, and market value (in the 10th year) are estimated in actual dollars (column 1) using the general price inflation rate and Equation (8-1). The maintenance contract amounts for the first three years (column 2) are already in actual dollars. (They are unresponsive to further price changes.) The algebraic sum of columns 1 and 2 equals the before-tax cash flow (BTCF) in actual dollars (column 3).

In columns 4, 5, and 6, the depreciation and income tax calculations are shown. The depreciation deductions in column 4 are based on the MACRS (GDS) method and, of course, are in actual dollars. The entries in columns 5 and 6 are calculated as discussed in Chapter 6. The effective income tax rate (t) is 38% as given. The entries in column 6 are equal to the entries in column 5 multiplied by $-t$. The algebraic sum of columns 3 and 6 equals the ATCF in actual dollars (column 7). The present worth of the actual-dollar ATCF, using $i_c = 15\%$ per year, is

$$PW(15\%) = -\$180,000 + \$36,050(P/F, 15\%, 1) + \cdots + \$40,156(P/F, 15\%, 10)$$

$$= \$33,790.$$

Therefore, the project is economically justified.

TABLE 8-4 Example 8-5 when the General Price Inflation Rate Is 8% Per Year

End of Year (k)	(1) A$ Cash Flows	(2) Contract (A$)	(3) BTCF (A$)	(4) Depreciation (A$)	(5) Taxable Income	(6) Income Taxes (t = 0.38)	(7) ATCF (A$)	(8) R$ Adjustment $[1/(1+f)]^{k-b}$	(9) ATCF (R$)
0	−$180,000		−$180,000				−$180,000	1.0000	−$180,000
1	38,880[a]	−$2,800	36,080	$36,000	$80	−$30	36,050	0.9259	33,379
2	41,990	−2,800	39,190	57,600	−18,410	+6,996	46,186	0.8573	39,595
3	45,349	−2,800	42,549	34,560	7,989	−3,036	39,513	0.7938	31,366
4	48,978		48,978	20,736	28,242	−10,732	38,246	0.7350	28,111
5	52,895		52,895	20,736	32,159	−12,220	40,675	0.6806	27,683
6	57,128		57,128	10,368	46,760	−17,769	39,359	0.6302	24,804
7	61,697		61,697		61,697	−23,445	38,252	0.5835	22,320
8	66,632		66,632		66,632	−25,320	41,312	0.5403	22,320
9	71,964		71,964		71,964	−27,346	44,618	0.5003	22,320
10	77,720		77,720		77,720	−29,534	48,186	0.4632	22,320
10	64,767[b]		64,767		64,767	−24,611	40,156	0.4632	18,600

[a] $(A\$)_k = \$36,000(1.08)^{k-0}$, $k = 1, \cdots, 10$.

[b] $MV_{10,A\$} = \$30,000(1.08)^{10} = \$64,767$.

(b) Next, Equation (8-1) is used to calculate the ATCF in real dollars from the entries in column 7. The real-dollar ATCF (column 9) shows the estimated economic consequences of the new equipment in dollars that have the constant purchasing power of the base year. The actual-dollar ATCF (column 7) is in dollars that have the purchasing power of the year in which the cost or saving occurs. The comparative information provided by the ATCF in both actual dollars and real dollars is helpful in interpreting the results of an economic analysis. Also, as illustrated in this example, the conversion between actual dollars and real dollars can easily be done. The PW of the real-dollar ATCF (column 9) using $i_r = (i_c - f)/(1 + f) = (0.15 - 0.08)/1.08 = 0.06481$, or 6.48%, is

$$PW(6.48\%) = -\$180,000 + \$33,379(P/F, 6.48\%, 1) + \cdots + \$18,600(P/F, 6.48\%, 10)$$

$$= \$33,790.$$

Thus, the PW (equivalent worth in the base year with $b = 0$) of the real-dollar ATCF is the same as the present worth calculated previously for the actual-dollar ATCF.

8.2.5.1 Variation in the Average Rate of Price Change In Example 8-5, the general price inflation rate (f) was projected to be 8% each year during the 10-yr analysis period. In the case where the estimated annual rates vary during the analysis period, these varying rates would be applied successively to the costs and revenues for the years involved. For example, assume that the annual rates of inflation for Example 8-5 were estimated to vary as shown in column 1 of Table 8-5. Then the annual operating cost savings and market value originally estimated in base-year dollars in the example, would be projected in actual dollars (column 3) by successive application of the annual rates, as shown in column 2 (where the symbol $\prod$ means the product).

8.2.5.2 Calculating an Effective General Price Inflation Rate In Table 8-5, the projected annual general price inflation rates varied during the 10-yr analysis period. Suppose that these rates are the best estimates of future price changes in your company. However, for a study of a small capital investment project, you consider the successive application of the varying annual rates in the analysis to be a refinement not justified by the situation. In this case, the analysis can be simplified by using an annual effective rate ($\overline{f}$) in the same way that $f = 8\%$ per year was used in the original solution to Example 8-5. Assume that the analysis period is 10 yr for the small project. The calculation of $\overline{f}$ (based on the entries in column 1 of Table 8-5) would be

$$\overline{f} = \left[\prod_{k=1}^{N} (1 + f_k) \right]^{1/N} - 1 = \left[\prod_{k=1}^{10} (1 + f_k) \right]^{1/10} - 1 \tag{8-6}$$

$$= [(1.04)^1 (1.055)^2 (1.07)^3 (1.08)^4]^{0.1} - 1 = (1.9292)^{0.1} - 1$$

$$= 0.067917, \text{ or } 6.7917\%.$$

TABLE 8-5 Variation in the General Price Inflation Rate

End of Year, k	(1) General Price Inflation Rate (f_k)	(2) A\$ Price Change Factors $\left[\prod_{l=1}^{k}(1+f_l)\right]$; for $b=0$	(3) Estimated A\$ Cash Flows
0	—		$-\$180{,}000$
1	4.0	1.04	$37{,}440^a$
2	5.5	$(1.04)(1.055) = 1.0972$	$39{,}499$
3	5.5	$(1.04)(1.055)^2 = 1.1576$	$41{,}674$
4	7.0	$(1.04)(1.055)^2(1.07) = 1.2386$	$44{,}590$
5	7.0	$(1.04)(1.055)^2(1.07)^2 = 1.3253$	$47{,}711$
6	7.0	$(1.04)(1.055)^2(1.07)^3 = 1.4180$	$51{,}048$
7	8.0	$(1.04)(1.055)^2(1.07)^3(1.08) = 1.5315$	$55{,}134$
8	8.0	$(1.04)(1.055)^2(1.07)^3(1.08)^2 = 1.6540$	$59{,}544$
9	8.0	$(1.04)(1.055)^2(1.07)^3(1.08)^3 = 1.7863$	$64{,}307$
10⎱ 10⎰	8.0	$(1.04)(1.055)^2(1.07)^3(1.08)^4 = 1.9292$	$\begin{cases} 69{,}451 \\ 57{,}876^b \end{cases}$

$^a(A\$)_k = \$36{,}000$ [Col. (2)].
$^bMV_{10} = \$30{,}000(1.9292) = \$57{,}876$.

If this approach were applied to the original calculations in Table 8-5, the entries in column 3 would be slightly different for years 1 through 9. However, the operating cost savings in the 10th year would be $\$36{,}000(1.067917)^{10} = \$69{,}451$, the same value calculated by successive application of the varying annual rates originally used in the table.

8.3 Differential Price Inflation or Deflation

The general price inflation (or deflation) rate (f) may not be the best estimate of future price changes for one or more cost and revenue cash flows in an engineering economy study. The variation between the general price inflation rate and the best estimate of future price changes for specific goods and services is called *differential price inflation* (or *deflation*), and it is caused by factors such as technological improvements, and changes in productivity, regulatory requirements, and so on. Also, a restriction in supply, an increase in demand, or a combination of both may change the market value of a particular good or service relative to others. Price changes caused by some combination of general price inflation and differential price inflation (or deflation) can be represented by a *total price escalation* (or *de-escalation*) *rate*. These rates are further defined as follows:

1. Differential price inflation (or deflation) rate (e_j'): the increment (%) of price change (in the unit price, or cost for a fixed amount), above or below the general price inflation rate, during a period (normally a year) for good or service j.

2. Total price escalation (or de-escalation) rate (e_j): The total rate (%) of price change (in the unit price, or cost for a fixed amount) during a period (normally a year), for good or service j. The total price escalation rate for a good or service includes the effects of both the general price inflation rate (f) and the differential price inflation rate (e_j') on price changes.

8.3.1 The Relationship among e_j, e_j', and f

The differential price inflation rate (e_j') is a price change in good or service j in *real dollars* caused by various factors in the marketplace. Similarly, the total price escalation rate (e_j) is a price change in *actual dollars*. The relationship among these two factors (e_j, e_j') and f is (derivation not shown)

$$1 + e_j = (1 + e_j')(1 + f); \tag{8-7}$$

$$e_j = e_j' + f + e_j'(f); \tag{8-8}$$

$$e_j' = \frac{e_j - f}{1 + f}. \tag{8-9}$$

Thus, as shown in Equation (8-8), the total price escalation rate (e_j) for good or service j in actual dollars is the sum of the general price inflation rate and the differential price inflation rate plus their product. Also, as shown in Equation (8-9), the differential price inflation rate (e_j') in real dollars can be calculated from the total price escalation rate and the general price inflation rate.

In practice, the general price inflation rate (f) and the total price escalation rate (e_j) for each good and service involved are usually estimated for the study period. For each of these rates, different values may be used for subsets of periods within the analysis period if justified by the available data. This was illustrated in Table 8-5 for the general price inflation rate. The differential price inflation rates (e_j'), when needed, normally are not estimated directly but are calculated using Equation (8-9).

EXAMPLE 8-6

The prospective maintenance expenses for a commercial heating, ventilating, and air-conditioning (HVAC) system are estimated to be $12,200 per year in base-year dollars (assume that $b = 0$). The total price escalation rate is estimated to be 7.6% for the next three years ($e_{1,2,3} = 7.6\%$), and for years four and five it is estimated to be 9.3% ($e_{4,5} = 9.3\%$). The general price inflation rate (f) for this five-year period is estimated to be 4.7% per year. Develop the maintenance expense estimates for years one through five in actual dollars and in real dollars, using e_j and e_j' values, respectively.

TABLE 8-6 Example 8-6 Calculations				
(1) End of Year, k	(2) A\$ Adjustment (e_j)	(3) Maintenance Expenses, A\$	(4) R\$ Adjustment (e'_j)	(5) Maintenance Expenses, R\$
1	$12,200(1.076)^1$	\$13,127	$12,200(1.0277)^1$	\$12,538
2	$12,200(1.076)^2$	14,125	$12,200(1.0277)^2$	12,885
3	$12,200(1.076)^3$	15,198	$12,200(1.0277)^3$	13,242
4	$12,200(1.076)^3(1.093)^1$	16,612	$12,200(1.0277)^3(1.0439)^1$	13,823
5	$12,200(1.076)^3(1.093)^2$	18,157	$12,200(1.0277)^3(1.0439)^2$	14,430

SOLUTION

The development of the annual maintenance expenses in actual dollars is shown in column 2 of Table 8-6. In this example, the general price inflation rate is not the best estimate of changes in future maintenance expenses. The five-year period is divided into two subperiods corresponding to the two different price escalation rates ($e_{1,2,3} = 7.6\%$; $e_{4,5} = 9.3\%$). These rates are then used with the estimated expenses in the base year; $(A\$)_0 = (R\$)_0 = \$12,200$.

The development of the maintenance expenses in real dollars is shown in column 4. This development is the same as for actual dollars, except that the e'_j values [Equation (8-9)] are used instead of the e_j values. The e'_j values in this example are as follows:

$$e'_{1,2,3} = \frac{0.076 - 0.047}{1.047} = 0.0277, \text{ or } 2.77\%;$$

$$e'_{4,5} = \frac{0.093 - 0.047}{1.047} = 0.0439, \text{ or } 4.39\%.$$

This illustrates that differential inflation, or deflation, also results in market price changes in real dollars, as well as in actual dollars.

8.3.2 Modeling Price Changes with Geometric Cash Flow Sequences

In Chapter 3, we discussed equivalence calculations involving projected cash-flow patterns that are increasing at a rate of $\overline{f}\%$ per period. When total price escalation is included in an engineering economic analysis, projected prices of goods and services can be modeled as increasing at a constant rate per period. Thus, the resulting end-of-period cash-flow pattern is often a geometric sequence.

In Section 8.2.2, the correct interest rate to use in an engineering economic analysis was shown to depend on the dollar terms of the cost and revenue cash flows; specifically, the combined interest rate (i_c) is used in an actual dollar analysis, and the real interest rate (i_r) is used in a real dollar analysis. An additional question is, "what $\overline{f}$ value is used for each method of analysis when cost escalation is included and a geometric sequence cash-flow model is appropriate?" In the following table, we see that $\overline{f}$ is equal to e_j in an A\$ analysis and equal to e'_j in a R\$ analysis:

Method	Cash flows	Interest Rate (i)	Geometric Gradient ($\bar{f}$)
A	Actual dollars (A\$)	i_c	e_j
B	Real dollars (R\$)	i_r	e'_j

From this it follows that the "convenience rate" (Chapter 3) needed to evaluate a geometric cash-flow sequence involving price escalation would be as follows:

$$A\$ \ Analysis \qquad\qquad R\$ \ Analysis$$

$$i_{CR} = \frac{i_c - e_j}{1 + e_j} \qquad\qquad i_{CR} = \frac{i_r - e'_j}{1 + e'_j}. \qquad\qquad (8\text{-}10)$$

EXAMPLE 8-7

A water-service public utility district is considering some new replacement pumping equipment to reduce operating expenses and improve service reliability. The base time period is the present, or year zero ($b = 0$). The estimated annual savings in year zero dollars is \$78,000. The utility district uses an eight-year study period for this type of equipment replacement study; the general price inflation rate is projected to be 4.6% per year; the total price escalation rate (e_j) for operating expenses is projected to be 6.2% per year; the utility district is using a MARR (which includes the effect of general price inflation) of 9.5% per year; the old equipment has no net market value; and no income taxes are involved. Based on these estimates, calculate the maximum amount that could be paid for the equipment now (a) using an actual-dollar analysis and (b) using a real-dollar analysis.

SOLUTION

(a) The actual-dollar analysis is

$$i_c = \text{MARR (as given)} = 9.5\%; \ \ \overline{f} = e_j = 6.2\%; \ \ N = 8;$$

$$i_{CR} = \frac{i_c - e_j}{1 + e_j} = \frac{0.095 - 0.062}{1.062} = 0.03107, \ \text{or } 3.11\%.$$

From the data given, the annual savings of $\$78,000(1.062)^k$ for $1 \leq k \leq 8$ constitute a geometric cash-flow sequence. By using Equation (3-27), we can write

$$\text{PW}(3.11\%) \text{ of savings} = \$78,000(P/A, 3.11\%, 8)$$

$$= \$78,000 \left[\frac{(1.0311)^8 - 1}{0.0311(1.0311)^8} \right] = \$545,000,$$

which is the maximum amount that should be paid for the equipment.

(b) The real-dollar analysis is

$$i_r = \text{MARR(Real)} = \frac{i_c - f}{1 + f} = \frac{0.095 - 0.046}{1.046} = 0.04685;$$

$$\overline{f} = e'_j = \frac{e_j - f}{1 + f} = \frac{0.062 - 0.046}{1.046} = 0.01530;$$

$$i_{CR} = \frac{i_r - e'_j}{1 + e'_j} = \frac{0.04685 - 0.0153}{1.0153} = 0.03107, \text{ or } 3.11\%.$$

At this point in the real-dollar analysis, we see that the convenience rate is the same value calculated for the actual-dollar analysis, and the PW of the savings would thus be the same. This should make intuitive sense, because we know that the equivalent worth of a cash flow is the same in the base period using either actual-dollar or real-dollar analysis. Thus, the PW of the annual savings for the new pump would equal $545,000 when using method A or B, since the base period is the present ($b = 0$).

EXAMPLE 8-8

An air-pollution control project is being evaluated by a chemical processing company. The estimated initial capital investment required for the project is $1,240,000 ($1,100,000 in depreciable assets and $140,000 in additional working capital). The annual expenses have been divided into two categories, labor and other operating and maintenance (O & M) expenses. In the first year, the estimated annual labor expenses are $42,000 (estimated to increase $2,000 per year thereafter), and the estimated other annual O & M expenses are $68,000 (estimated to decrease 3.2% per year thereafter; that is, *de-escalation* in these costs is expected to occur). Assume that the analysis period is 10 years; the company's after-tax market MARR is 12% per year; the base period is the present ($b = 0$); the company's effective income tax rate (t) is 40%; the general inflation rate (f) is 2.6% per year; and, for simplicity, straight-line depreciation will be used over the 10-yr analysis period with an estimated salvage value of zero at the end of 10 yr ($SV_{10} = 0$). Based on this information and an after-tax analysis, (a) what is the equivalent future worth of project costs at the end of 10 years in actual dollars and in real dollars, and (b) if a by-product were produced by the air-pollution control process that had commercial value, what would the equivalent annual worth of revenues need to be in actual dollars and in real dollars to break even relative to the costs of the project?

SOLUTION

(a) Step 1: Determine the PW of the project's ATCF using an actual-dollar analysis. We thus have

$$PW(12\%)_{\text{ATCF}} = -\$1,240,000 + \$140,000(P/F, \ 12\%, \ 10)$$

$$- (1-0.4)[\$42,000(P/A, \ 12\%, \ 10) + \$2,000(P/G, \ 12\%, \ 10)]$$

$$- (1-0.4)\left[\frac{\$68,000}{1+(-0.032)}(P/A, \ 15.7\%, \ 10)\right]$$

$$+ 0.4\left[\frac{\$1,100,000-0}{10}(P/A, \ 12\%, \ 10)\right]$$

$$= -\$1,319,012,$$

where the \$140,000 initial investment in working capital is assumed to be recovered at the end of the analysis period, the convenience rate [Equation (8-10)] for the after-tax calculation related to the other annual O & M expenses is $i_{CR} = [0.12-(-0.032)]/[1+(-0.032)] = 0.157$ (15.7%), and the annual straight-line depreciation amounts are a reduction in after-tax project costs.

Step 2: Convert the PW of the ATCF to a FW equivalent at the end of the analysis period in both dollar domains. (Note that in this example the PW is the equivalent worth of the ATCF in the base year.) In actual dollars,

$$FW(12\%)_{\text{ATCF}} = -\$1,319,012(F/P, \ 12\%, \ 10) = -\$4,096,051.$$

In the real-dollar domain, however, we need the after-tax real interest rate for the FW calculation: $i_r = (0.12 - 0.026)/1.026 = 0.091618$, or 9.1618%. Then

$$FW(9.1618\%)_{\text{ATCF}} = -\$1,139,012(F/P, \ 9.1618\%, \ 10) = -\$3,169,244.$$

From these FW calculations, we see that (with $f = 2.6\%$ per year) it takes 4,096,051 equivalent actual dollars at the end of 10 years to have the same purchasing power of 3,169,244 equivalent real dollars that have constant purchasing power as measured by a present (base-year) dollar.

(b) Based on the PW calculation in Part (a), the required annual equivalent revenue to break even in actual dollars, is

$$AW(12\%)_{\text{ATCF}} = \$1,319,012(A/P, \ 12\%, \ 10) = \$233,465,$$

and, in real dollars, it is

$$AW(9.1618\%)_{\text{ATCF}} = \$1,319,012(A/P, \ 9.1618\%, \ 10) = \$206,953.$$

In Example 8-9, we look at a bond (Chapter 4), which is a fixed-income asset and see how its current value is affected by a period of projected deflation.

EXAMPLE 8-9

Suppose that deflation occurs in the U.S. economy and that the CPI (as a measure of f) is expected to decrease an average of 2% per year for the next five years. A bond with a face (par) value of $10,000 and a life of five years (i.e., it will be redeemed in five years) pays an interest (bond) rate of 5% per year. The interest is paid to the owner of the bond once each year. If an investor expects a *real* rate of return of 4% per year, what is the maximum amount that should be paid now for this bond?

SOLUTION

The cash flows over the life of the bond are 0.05($10,000) = $500 per year in interest (actual dollars) for years 1–5, plus the redemption of $10,000 (the face value of the bond), also in actual dollars, at the end of year five. To determine the current value of this bond (i.e., the maximum amount an investor should pay for it), these cash flows must be discounted to the present using the combined (market) interest rate. From Equation (8-4), we can compute i_c (where $f = -2\%$ per year) as follows:

$$i_c = i_r + f + i_r(f) = 0.04 - 0.02 - 0.04(0.02)$$

$$= 0.0192, \text{ or } 1.92\% \text{ per year.}$$

Therefore, the current market value of the bond is

$$PW = \$500(P/A, \ 1.92\%, \ 5) + \$10,000(P/F, \ 1.92\%, \ 5)$$

$$= \$500(4.7244) + \$10,000(0.9093)$$

$$= \$11,455.$$

As additional information, if we had made the mistake of discounting the future cash flows over the life of the bond at the bond rate of 5% per year, the current value would have been $10,000, the face value of the bond. Also, in general, if the rate used to discount the future cash flows over the life of a bond is *less* than the bond rate (the situation in this example), then the current (market) value will be greater than the bond's face value. Therefore, during periods of deflation, owners of bonds (or other types of fixed-income assets) need to monitor their market values closely because a favorable "sell situation" may occur.

8.4 Application Strategy

In practice, should actual-dollar or real-dollar analysis be used, and when should price changes be included in an engineering economy study? Judgment based on the prospective price change estimates, and sensitivity analysis, are used in practice. However, either the actual dollar or the real dollar analysis method may be used. Both methods, properly applied, result in the same equivalent worth for a cash flow in the base time period, require the same amount of information, and have no practical difference in application effort.

There is some difference, however, in the information available for interpreting the economic results. The results from an actual-dollar analysis are in market purchasing power that varies with time, while the results from a real-dollar analysis are in constant market purchasing power defined by the base period (*b*). *Thus, real-dollar analysis provides information in terms of a constant purchasing power unit of measure, while actual-dollar analysis provides information on the money quantities that will occur during the study period.*

An analysis or *application strategy* that works well in engineering practice is to use actual dollar analysis for both before-tax and after-tax studies, and then, at the end of an analysis, to use Equation (8-1) or Equation (8-2) to provide selected cash flows (particularly the before-tax or after-tax net cash flows) in real-dollar terms. With little effort, this strategy provides additional information that is useful. In some organizations, a particular method of analysis may be specified; but even then, selected cash flows can be easily converted to the other dollar domain to assist with interpretation of results.

8.5 A Comprehensive Example

In many engineering economy studies of projects conducted in industry, price changes have to be considered in addition to the applicable income tax provisions. To illustrate this situation, a fairly comprehensive engineering project analysis is presented.

EXAMPLE 8-10

A company is considering an investment opportunity that requires the investment of $20,000 in production control equipment to increase the output of an assembly line. As a result, the revenue obtained from the modified assembly line is expected to increase. The following information applies to the investment opportunity:

Analysis period	10 yr
Base time period	Present ($b = 0$)
Estimated useful life of the equipment	10 yr
MACRS (GDS) property class	5 yr
Effective income tax rate (t)	39%
Real (after-tax) MARR (i_r)	6%
General price inflation rate (f)	8% per year
Combined (after-tax) MARR (i_c)	$14.48\% = [0.06 + 0.08 + (0.06)(0.08)]100\%$
Increased revenue (assume that revenue escalates at the general price inflation rate of 8% per year)	$15,000 per year in year zero dollars
Market value in 10 yr	10% of the capital investment ($e_{MV} = 8\%$)

	Annual Expenses	
Category	Estimate (Year 0 Dollars)	Price Escalation Rate per Year (e_j)
Material	$1,200	10%
Labor	2,500	5.5%
Energy	2,500	15%
Other expenses	500	8%

Leased equipment is also required, which can be obtained for the first five years at a rate of $800 per year. The contract will be renegotiated at the beginning of the sixth year at an escalated value based on the general price inflation rate.

Perform an after-tax analysis of this project using the PW method and including the effects of total price escalation: (a) Conduct an actual-dollar analysis (and calculate the PW of the ATCF); (b) convert the actual-dollar ATCF to a real-dollar ATCF; and (c) calculate the PW of the real-dollar ATCF and show that it is identical to the PW of the actual-dollar ATCF.

SOLUTION

(a) Actual-dollar analysis: The preliminary calculations required are as follows:

1. **Revenue:** The estimated revenue of $15,000 per year in year zero dollars must be increased each year by the general price inflation rate.

$$(\text{Revenue})_k = \$15,000(1.08)^k.$$

2. **Material, labor, energy, and other annual expenses:** These annual expenses, estimated in year-zero dollars, are increased each year by the appropriate total price escalation rate (e_j):

$$(\text{Material})_k = \$1,200(1.1)^k;$$

$$(\text{Labor})_k = \$2,500(1.055)^k;$$

$$(\text{Energy})_k = \$2,500(1.15)^k;$$

$$(\text{Other expenses})_k = \$500(1.08)^k.$$

3. **Leased property:** The lease will be adjusted at the end of year five to account for five years of general price inflation at 8% per year:

$$\text{Lease expense (years 6–10)} = \$800(1.08)^5 = \$1,175.$$

4. **Depreciation:** The MACRS (GDS) depreciation amounts are as follows:

End of Year k	Cost Basis	MACRS (GDS) Recovery Rates	MACRS (GDS) Depreciation (A$)
1	$20,000	0.2000	$4,000
2	20,000	0.3200	6,400
3	20,000	0.1920	3,840
4	20,000	0.1152	2,304
5	20,000	0.1152	2,304
6	20,000	0.0576	1,152

5. *Market value:* The 10% MV, based on the capital investment, is a year-zero amount and must be increased to account for the estimated 8% per year total price escalation rate ($e_{MV} = f$).

$$MV_{10} = 0.1(\$20,000)(1.08)^{10} = \$4,318$$

6. *Gain on disposal:* The market value in actual dollars of $4,318 represents a gain on disposal (Chapter 6) and is taxed like ordinary income at the 39% rate.

The actual-dollar after-tax analysis is shown in Table 8-7. The PW of the actual-dollar ATCF, using $i_c = 14.48\%$, is $16,780.

(b) Real-dollar ATCF: The conversion of the actual-dollar ATCF to a real-dollar ATCF is shown in the last two columns in Table 8-7. Equation (8-1), with $b = 0$, was used to make the conversion.

TABLE 8-7 Actual-Dollar Cash Flow Analysis (with Conversion to Real-Dollar ATCF) for Example 8-10

End of Year, k	Capital Investment	Revenue	Material	Labor	Energy	Other Expenses	Leased Equipment
0	−$20,000						
1		$16,200	−$1,320	−$2,638	−$2,875	−$540	−$800
2		17,496	−1,452	−2,783	−3,306	−583	−800
3		18,896	−1,597	−2,936	−3,802	−630	−800
4		20,407	−1,757	−3,097	−4,373	−680	−800
5		22,040	−1,933	−3,267	−5,028	−735	−800
6		23,803	−2,126	−3,447	−5,783	−793	−1,175
7		25,707	−2,338	−3,637	−6,650	−857	−1,175
8		27,764	−2,572	−3,837	−7,648	−925	−1,175
9		29,985	−2,830	−4,048	−8,795	−1,000	−1,175
10		32,384	−3,112	−4,270	−10,114	−1,079	−1,175
10		4,318[a]					

[a] Estimated MV.

TABLE 8-7 (*continued*) Actual-Dollar Cash Flow Analysis (with Conversion to Real-Dollar ATCF) for Example 8-10

	A$ After-Tax Analysis					R$ ATCF	
End of Year, k	Before-Tax Cash Flow	Depreciation	Taxable Income	Income Taxes	A$ After-Tax Cash Flow	R$ Adjustment Factor $(1/1.08)^{k-0}$	R$ After-Tax Cash Flow
0	−$20,000				−$20,000	1.0	−$20,000
1	8,028	$4,000	$4,028	$1,571	6,457	0.92593	5,979
2	8,572	6,400	2,172	847	7,725	0.85734	6,623
3	9,131	3,840	5,291	2,063	7,068	0.79383	5,611
4	9,700	2,304	7,396	2,884	6,816	0.73503	5,010
5	10,277	2,304	7,973	3,109	7,168	0.68058	4,878
6	10,479	1,152	9,327	3,638	6,841	0.63017	4,311
7	11,050		11,050	4,310	6,740	0.58349	3,933
8	11,607		11,607	4,527	7,080	0.54027	3,825
9	12,137		12,137	4,733	7,404	0.50025	3,704
10	12,634		12,634	4,927	7,707	0.46319	3,570
10	4,318		4,318[b]	1,684	2,634	0.46319	1,220
			PW(i_c = 14.48%) = $16,780			PW(i_r = 6%) = $16,780	

[b] Recovery of depreciation (gain on disposal)—taxed as ordinary income.

This solution method implements the strategy recommended in Section 8.4 of using an actual-dollar analysis and then converting selected cash flows into real dollars. Reviewing the actual-dollar ATCF in this case indicates an equivalent annual positive cash flow during the analysis period of approximately $7,184 from the $20,000 investment in new equipment. However, the real-dollar ATCF shows that in terms of dollars with constant purchasing power ($b = 0$), the net positive cash flow from the investment (except for year 2) decreases from $5,979 in year 1 to $3,570 in year 10.

(c) The PW of the real-dollar ATCF, using $i_r = 6\%$, is $16,780. This is the same value as the PW of the actual-dollar ATCF calculated in (a) using $i_c = 14.48\%$.

8.6 Foreign Exchange Rates and Purchasing Power Concepts

When domestic corporations make foreign investments, the resultant cash flows that occur over time are in a different currency from U.S. dollars. Typically, foreign investments are characterized by two (or more) translations of currencies: (1) when the initial investment is made and (2) when cash flows are returned to U.S.-based corporations. Exchange rates between currencies fluctuate, sometimes dramatically, over time, so a typical question that can be anticipated is "What return (profit) did we make on our investment in the synthetic fiber plant in Thatland?"

For the engineer who is designing another plant in Thatland, the question might be "What is the PW (or IRR) that our firm will obtain by constructing and operating this new plant in Thatland?"

> Observe that changes in the exchange rate between two currencies over time are analogous to changes in the general inflation rate because the relative purchasing power between the two currencies is changing similar to the relative purchasing power between actual dollar amounts and real dollar amounts.

Assume the following:

i_{us} = rate of return in terms of a combined (market) interest rate relative to U.S. dollars.

i_{fc} = rate of return in terms of a combined (market) interest rate relative to the currency of a foreign country.

f_e = annual devaluation rate (rate of annual change in the exchange rate) between the currency of a foreign country and the U.S. dollar. In the following relationships, a positive f_e is used when the foreign currency is being devalued relative to the dollar, and a negative f_e is used when the dollar is being devalued relative to the foreign currency.

Then (derivation not shown)

$$1 + i_{us} = \frac{1 + i_{fc}}{1 + f_e},$$

or

$$i_{fc} = i_{us} + f_e + f_e(i_{us}), \tag{8-11}$$

and

$$i_{us} = \frac{i_{fc} - f_e}{1 + f_e}. \tag{8-12}$$

EXAMPLE 8-11

The CMOS Electronics Company is considering a capital investment of 50,000,000 pesos in an assembly plant located in a foreign country. Currency is expressed in pesos, and the exchange rate is now 100 pesos per U.S. dollar.

The country has followed a policy of devaluing its currency against the dollar by 10% per year to build up its export business to the United States. This means that each year the number of pesos exchanged for a dollar increases by 10% ($f_e = 10\%$), so in two years $(1.10)^2(100) = 121$ pesos would be traded for one dollar. Labor is quite inexpensive in this country, so management of CMOS Electronics feels that the proposed plant will produce the following rather attractive ATCF, stated in pesos:

End of Year	0	1	2	3	4	5
ATCF	-50	+20	+20	+20	+30	+30
(millions of pesos)						

If CMOS Electronics requires a 15% IRR per year, after taxes, in U.S. dollars (i_{us}) on its foreign investments, should this assembly plant be approved? Assume that there are no unusual risks of nationalization of foreign investments in this country.

SOLUTION

To earn a 15% annual rate of return in U.S. dollars, the foreign plant must earn, based on Equation (8-11), $0.15 + 0.10 + 0.15(0.10) = 0.265$, which is 26.5% on its investment in pesos (i_{fc}). As shown next, the PW (at 26.5%) of the ATCF, in pesos, is 9,165,236, and its IRR is 34.6%. Therefore, investment in the plant appears to be economically justified. We can also convert pesos into dollars when evaluating the prospective investment:

End of Year	ATCF (Pesos)	Exchange Rate	ATCF (Dollars)
0	-50,000,000	100 pesos per $1	-500,000
1	20,000,000	110 pesos per $1	181,818
2	20,000,000	121 pesos per $1	165,289
3	20,000,000	133.1 pesos per $1	150,263
4	30,000,000	146.4 pesos per $1	204,918
5	30,000,000	161.1 pesos per $1	186,220
IRR:	34.6%	IRR:	22.4%
PW(26.5%):	9,165,236 pesos	PW(15%):	$91,632

The PW (at 15%) of the ATCF, in dollars, is $91,632, and its IRR is 22.4%. Therefore, the plant again appears to be a good investment in economic terms. Notice that the two IRRs can be reconciled by using Equation (8-12):

$$i_{us}(\text{IRR in \$}) = \frac{i_{fc}(\text{IRR in pesos}) - 0.10}{1.10}$$

$$= \frac{0.346 - 0.10}{1.10}$$

$$= 0.224, \text{ or } 22.4\%.$$

Remember that devaluation of a foreign currency relative to the U.S. dollar produces less expensive exports to the United States. Thus, the devaluation means that the U.S. dollar is stronger relative to the foreign currency. That is, fewer dollars are needed to purchase a fixed amount (barrels, tons, items) of goods and services from the foreign source or, stated differently, more units of the foreign currency are required to purchase U.S. goods. This phenomenon was observed in Example 8-11.

Conversely, when exchange rates for foreign currencies gain strength against the U.S. dollar (f_e has a negative value and the U.S. dollar is weaker relative to the foreign currency), the prices for imported goods and services into the United States rise. In such a situation, U.S. products are less expensive in foreign markets. For example, in 1986 the U.S. dollar was exchanged for approximately 250 Japanese yen, but in 1999 a weaker U.S. dollar was worth about 110 Japanese yen. As a consequence, U.S. prices on Japanese goods and services almost doubled in theory (but in actuality increased in the United States by lesser amounts). One explanation for this anomaly is that Japanese companies were willing to reduce their profit margins to retain their share of the U.S. market.

> In summary, if the average devaluation of currency A is f_e% per year relative to currency B, then each year it will require f_e% more of currency A to exchange for the same amount of currency B.

EXAMPLE 8-12

The monetary (currency) unit of another country, A, has a present exchange rate of 10.7 units per U.S. dollar. If (a) the average devaluation of currency A in the international market is estimated to be 4.2% per year (for the next five years) relative to the U.S. dollar, what would be the exchange rate three years from now; and if, instead, (b) the average devaluation of the U.S. dollar (for the next five years) is estimated to be 3% per year relative to currency A, what would be the exchange rate three years from now?

SOLUTION

(a) $$10.7(1.042)^3 = 12.106 \text{ units of } A \text{ per U.S. dollar.}$$

(b) $$10.7 \text{ units of } A = 1(1.03)^3 \text{ U.S. dollars, and;}$$

$$1 \text{ U.S. dollar} = \frac{10.7}{1.09273} = 9.792 \text{ units of } A.$$

EXAMPLE 8-13

A U.S. firm is analyzing a potential investment project in another country. The present exchange rate is 425 of their currency units (A) per U.S. dollar. The best estimate is that currency A will be devalued in the international market an average of 2% per year relative to the U.S. dollar over the next several years. The estimated before-tax net cash flow (in currency A) of the project is as follows:

End of Year	Net Cash Flow (Currency A)
0	−850,000,000
1	270,000,000
2	270,000,000
3	270,000,000
4	270,000,000
5	270,000,000
6	270,000,000
6 (MV)[a]	120,000,000

[a] Estimated market value at the end of year six.

If (a) the MARR value of the U.S. firm (before taxes and based on the U.S. dollar) is 20% per year, is the project economically justified; and if, instead, (b) the U.S. dollar is estimated to be devalued in the international market an average of 2% per year over the next several years, what would be the rate of return based on the U.S. dollar? Is the project economically justified?

SOLUTION

(a)
$$PW(i'\%) = 0 = -850,000,000 + 270,000,00(P/A, \ i'\%, \ 6)$$
$$+ 120,000,000(P/F, \ i', \ 6).$$

By trial and error (Chapter 4), we determine that $i'\% = i_{fc} = IRR_{fc} = 24.01\%$. Now, using Equation (8-12), we have

$$i_{us} = IRR_{us} = \frac{0.2401 - 0.02}{1.02} = 0.2158, \text{ or } 21.58.\%$$

Since this rate of return in terms of the U.S. dollar is greater than the firm's MARR (20% per year), the project is economically justified (but very close to the minimum rate of return required on the investment).

A SECOND SOLUTION TO PART (A)
Using Equation (8-11), we can determine the firm's MARR value in terms of currency A as follows:

$$i_{fc} = MARR_{fc} = 0.20 + 0.02 + 0.02(0.20) = 0.244, \text{ or } 22.4\%.$$

Using this MARR$_{fc}$ value, we can calculate the PW (22.4%) of the project's net cash flow (which is estimated in units of currency A); that is, PW(22.4%) = 32,597,000 units of currency A. Since this PW is greater than 0, we confirm that the project is economically justified.

   A third solution approach (not shown) would use projected exchange rates for each year to convert the estimated annual net cash flow amounts in currency A to U.S. dollar amounts. Then, the PW of the U.S. dollar net cash flow at MARR$_{us}$ = 20% would be calculated to determine whether the project is economically acceptable. This solution method was demonstrated in Example 8-11.

(b) Using Equation (8-12) and the IRR_{fc} value (24.01% calculated in the solution to Part (a), we have

$$i_{us} = IRR_{us} = \frac{0.2401 - (-0.02)}{1 - 0.02} = 0.2654, \text{ or } 26.54\%.$$

Since the rate of return in terms of the U.S. dollar is greater than the required $MARR_{us} = 20\%$, the project is economically justified.

As additional information, notice in the first solution to Part (a) when currency A is estimated to be devalued relative to the U.S. dollar, $IRR_{us} = 21.58\%$, while in Part (b), when the U.S. dollar is estimated to be devalued relative to currency A, $IRR_{us} = 26.54\%$. What is the relationship between the different currency devaluations in the two parts of the problem and these results?

Answer: Since the annual earnings to the U.S. firm from the investment are in units of currency A originally, and currency A is being devalued relative to the U.S. dollar in Part (a), these earnings would be exchanged for decreasing annual amounts of U.S. dollars causing an unfavorable impact on the project's IRR_{us}. In Part (b), however, the U.S. dollar is being devalued relative to currency A and these annual earnings would be exchanged for increasing annual amounts of U.S. dollars causing a favorable impact on the project's IRR_{us}.

8.7 Spreadsheet Applications

The following example illustrates the use of spreadsheets to convert actual dollars to real dollars and vice versa.

EXAMPLE 8-14

Sara B. Goode wishes to retire in the year 2022 with personal savings of $500,000 (1997 spending power). Assume that the expected inflation rate in the economy will average 3.75% per year during this period. Sara plans to invest in a 7.5% per year savings account, and her salary is expected to increase by 8.0% per year between 1997 and 2022. Assume that Sara's 1997 salary was $60,000 and that the first deposit took place at the end of 1997. What percent of her yearly salary must Sara put aside for retirement purposes to make her retirement plan a reality?

This example demonstrates the flexibility of a spreadsheet, even in instances where all of the calculations are based on a piece of information (% of salary to be saved) that we do not yet know. If we deal in actual dollars, the cash-flow relationships are straightforward. The formula in cell F7 in Figure 8-1 converts the desired ending balance into actual dollars. The salary is paid at the end of the year, at which point some percentage is placed in a bank account. The interest calculation is based on the cumulative deposits and interest in the account at the beginning of the year, but not on the deposits made at the end of the year. The salary is increased and the cycle repeats.

	A	B	C	D	E	F
1	Base Rate	Figure Entry		Starting Salary in 1997		$ 60,000
2	10.000%	1		Annual Salary Increase		8.00%
3	1.000%	2		Savings Interest Rate		7.50%
4	0.100%	4		Average Inflation Rate		3.75%
5	0.010%	5		Desired 2022 Amount (R$)		$ 500,000
6						
7	Savings Rate	12.45%		Desired 2022 Amount (A$)		$ 1,255,084
8				Year 2022 Bank Balance		$ 1,255,913
9						
10					Bank	
11			Salary	Savings	Balance	
12		Year	(A$)	(A$)	(A$)	
13		1997	$ 60,000	$ 7,470	$ 7,470	
14		1998	64,800	8,068	16,098	
15		1999	69,984	8,713	26,018	
16		2000	75,583	9,410	37,380	
17		2001	81,629	10,163	50,346	
18		2002	88,160	10,976	65,098	
19		2003	95,212	11,854	81,834	
20		2004	102,829	12,802	100,774	
21		2005	111,056	13,826	122,158	
22		2006	119,940	14,933	146,253	
23		2007	129,535	16,127	173,349	
24		2008	139,898	17,417	203,767	
25		2009	151,090	18,811	237,861	
26		2010	163,177	20,316	276,016	
27		2011	176,232	21,941	318,658	
28		2012	190,330	23,696	366,253	
29		2013	205,557	25,592	419,314	
30		2014	222,001	27,639	478,402	
31		2015	239,761	29,850	544,132	
32		2016	258,942	32,238	617,180	
33		2017	279,657	34,817	698,286	
34		2018	302,030	37,603	788,261	
35		2019	326,192	40,611	887,991	
36		2020	352,288	43,860	998,450	
37		2021	380,471	47,369	1,120,703	
38		2022	$ 410,909	$ 51,158	$ 1,255,913	

Figure 8-1 **Spreadsheet for Example 8-14**

The spreadsheet model is shown in Figure 8-1. We can enter the formulas for the geometric gradient representing the salary increase (column C), the percentage of the salary (cell B7) that goes into savings (column D), and the bank balance at the end of the year (column E) without knowing the percent of salary being saved.

Most spreadsheets have a *solver* feature that will automatically determine the desired savings strategy. This example illustrates an approach that is not as elegant, but is nonetheless fast and will work for software that does not have this solver feature.

The approach is to revise the base savings rate systematically and compare the ending bank balance (copied to cell F8 for ease of viewing on the screen) with the desired year-2022 balance. To save keystrokes, the base rate is broken down by powers of 10 into separate cells, in the range B2:B5. The base rate is recombined with the formula in cell B7. Starting with the highest power of 10 (cell B2), we bracket the savings rate that will set cells F7 and F8 equal (or nearly equal). The highlighted cell formulas in Figure 8-1 are as follows.

Cell	Contents
B7	$= A2 * B2 + A3 * B3 + A4 * B4 + A5 * B5$
F7	$= F5 * (1 + F4)\,\hat{}\,25$
F8	$= E38$
C13	$= F1$
C16	$= C15 * (1 + \$F\$2)$
D16	$= C16 * \$B\$7$
E16	$= E15 * (1 + \$F\$3) + D16$

Once the problem has been formulated in a spreadsheet, we can determine the impact of different interest rates, inflation rates, and so on, on the retirement plan with minimal changes and effort.

▼ 8.8 Summary

Price changes caused by inflation or deflation are an economic and business reality that can affect the comparison of alternatives. In fact, since the 1920s, the historical U.S. inflation rate has averaged approximately 4% per year. Much of this chapter has dealt with incorporating price changes into before-tax and after-tax engineering economy studies.

In this regard, it must be ascertained whether cash flows have been estimated in actual dollars or real dollars. The appropriate interest rate to use when discounting or compounding actual dollar amounts is a combined, or marketplace, rate, while the corresponding rate to apply in real dollar analysis is the firm's real interest rate.

Engineering economy studies often involve quantities that do not respond to inflation, such as depreciation amounts, interest charges, and lease fees and other amounts established by contract. Identifying these quantities and handling them properly in an analysis are necessary to avoid erroneous economic results. The use of the chapter's basic concepts in dealing with foreign exchange rates has also been demonstrated.

8.9 References

Freidenfelds, J., and M. Kennedy. "Price Inflation and Long-Term Present Worth Studies," *The Engineering Economist,* vol. 24, no. 3, Spring 1979, pp. 143–160.

Industrial Engineering, vol. 12, no. 3, March 1980. The entire issue is devoted to "The Industrial Engineer and Inflation." Of particular interest are the following articles:

(a) Estes, C. B., W. C. Turner, and K. E. Case. "Inflation—Its Role in Engineering-Economic Analysis," pp. 18–22.

(b) Sullivan, W. G., and J. A. Bontadelli. "How an IE Can Account for Inflation in Decision-Making," pp. 24–33.

(c) Ward, T. L. "Leasing During Inflation: A Two-Edged Sword," pp. 34–37.

Jones, B. W., *Inflation in Engineering Economic Analysis* (New York: John Wiley & Sons, 1982).

Lee, P. M., and W. G. Sullivan. "Considering Exchange Rate Movements in Economic Evaluation of Foreign Direct Investments," *The Engineering Economist,* vol. 40, no. 2, Winter 1995, pp. 171–199.

Watson, F. A., and F. A. Holland. "Profitability Assessment of Projects Under Inflation," *Engineering and Process Economics,* vol. 2, no. 3, 1976, pp. 207–221.

8.10 Problems

The number in parentheses ( ) that follows each problem refers to the section from which the problem is taken.

8-1. Your rich aunt is going to give you an end-of-year gift of $1,000 for each of the next 10 years.

a. If general price inflation is expected to average 6% per year during the next 10 years, what is the equivalent value of these gifts at the present time? The real interest rate is 4% per year.

b. Suppose that your aunt specified that the annual gifts of $1,000 are to be increased by 6% each year to keep pace with inflation. With a real interest rate of 4% per year, what is the current PW of the gifts? (8.2)

8-2. Because of general price inflation in our economy, the purchasing power of the dollar shrinks with the passage of time. If the average general price inflation rate is expected to be 4% per year into the foreseeable future, how many years will it take for the dollar's purchasing power to be one-half of what it is now? (That is, at what future point in time will it take $2 to buy what can be purchased today for $1?) (8.2)

8-3. Which of these situations would you prefer? (8.2)

a. You invest $2,500 in a certificate of deposit that earns an effective interest rate of 8% per year. You plan to leave the money alone for five years, and the general price inflation rate is expected to average 5% per year. Income taxes are ignored.

b. You spend $2,500 on a piece of antique furniture. You believe that in five years the furniture can be sold for $4,000. Assume that the average general price inflation rate is 5% per year. Again, income taxes are ignored.

8-4. Annual expenses for two alternatives have been estimated on different bases as follows:

End of Year	Alternative A Annual Expenses Estimated in Dollars	Alternative B Annual Expenses Estimated in Real Dollars with b = 0
1	$120,000	$100,000
2	132,000	110,000
3	148,000	120,000
4	160,000	130,000

If the average general price inflation rate is expected to be 6% per year and the real rate of interest is 9% per year, show which alternative has the least negative equivalent worth in the base period. (8.2)

8-5. A firm desires to determine the most economic equipment-overhauling schedule alternative to provide for service for the next nine years of operation. The firm's real minimum attractive rate of return is 8% per year, and the inflation rate is estimated at 7% per year. The following are alternatives with all the costs expressed in real (constant worth) dollars. (8.2)

   a. Completely overhaul for $10,000 now.
   b. A major overhaul for $7,000 now that can be expected to provide six years of service and then a minor overhaul costing $5,000 and the end of six years.
   c. A minor overhaul costing $5,000 now as well as at the end of three years and six years from now.

8-6. A recent engineering graduate has received the annual salaries shown in the following table over the past four years. During this time, the CPI has performed as indicated. Determine the engineer's annual salaries in *year-0 dollars* ($b = 0$) using the CPI as the indicator of general price inflation. (8.2)

End of Year	Salary (A$)	CPI
1	$34,000	7.1%
2	36,200	5.4%
3	38,800	8.9%
4	41,500	11.2%

8-7. A large corporation's electricity bill amounts to $400 million. During the next 10 yr, electricity usage is expected to increase by 75%, and the estimated electricity bill 10 yr hence has been projected to be $920 million. Assuming electricity usage and rates increase at uniform annual rates over the next 10 yr, what is the annual rate of inflation of electricity prices expected by the corporation? (8.2)

8-8. A high school graduate has decided to invest 5% of her first-year's salary in a mutual fund. This amounts to $1,000 in the first year. She has been told that her savings should keep up with expected salary increases, so she plans to invest an *extra* 8% each year over a 10-yr period. Thus, at the end of year one she invests $1,000; in year

two, $1,080; in year three, $1,166.40; and so on through year 10. If the average rate of inflation is expected to be 5% over the next 10 yr and if she expects a 2% *real* return on this investment, what is the future worth of the mutual fund at the end of the 10th year? (8.2)

8-9. A commercial building design cost $89/ft^2 to construct eight years ago (for an 80,000-ft^2 building). This construction cost has escalated 5.4% per year since then. Presently, your company is considering construction of a 125,000-ft^2 building of the same design. The cost capacity factor is $X = 0.92$. In addition, it is estimated that working capital will be 5% of construction costs, and that project management, engineering services, and overhead will be 4.2%, 8%, and 31%, respectively, of construction costs. Also, it is estimated that annual expenses in the first year of operation will be $5/ft^2, and these are estimated to increase 5.66% per year thereafter. The future general inflation rate is estimated to be 7.69% per year, and the market-based MARR$_c$ = 12% per year. (Chapter 7 and 8.2)

   a. What is the estimated capital investment for the 125,000-ft^2 building?
   b. Based on a before-tax analysis, what is the PW for the first 10 yr of ownership of the building?
   c. What is the AW for the first 10 yr of ownership in real dollars (R$)?

8-10. The operating budget estimate for an engineering staff for fiscal year 2004 is $1,780,000. The actual budget expenditures of the staff for the previous two fiscal years, as well as estimates for the next two years, are as shown on the following page. These are actual-dollar amounts. Management, however, also wants annual budget amounts for these years, using a constant-dollar perspective. The 2004 fiscal year is to be used for this purpose ($b = 2004$). The estimated annual general price inflation rate is 5.6%. What are the annual constant- (real-) dollar budget amounts? (8.2)

Fiscal Year	Budget Amount (A$)
2002	$1,615,000
2003	1,728,000
2004	1,780,000
2005	1,858,300
2006	1,912,200

8-11. An individual wishes to have a preplanned amount in a savings account for retirement in 20 years. This amount is to be equivalent to $30,000 in today's purchasing power. If the expected average inflation rate is 7% per year and the savings account earns 5% interest, what lump sum should be deposited now in the savings account? (8.2)

8-12. The AZROC Corporation needs to acquire a computer system for one of its regional engineering offices. The purchase price of the system has been quoted at $50,000, and the system will reduce annual expenses by $18,000 per year in real dollars. Historically, these annual expenses have escalated at an average rate of 8% per year, and this is expected to continue into the future. Maintenance services will be contracted for, and their cost per year (in actual dollars) is constant at $3,000. Also, assume $f = 8\%$ per year.

What is the minimum (integer-valued) life of the system such that the new computer can be economically justified? Assume that the computer's market value is zero at all times. The firm's MARR is 25% per year (which includes an adjustment for anticipated inflation in the economy). Show all calculations. (8.2)

8-13. An investor lends $10,000 today, to be repaid in a lump sum at the end of 10 years with interest at 10% ($= i_c$) compounded annually. What is the real rate of return, assuming that the general price inflation rate is 8% annually? (8.2)

8-14. An investor established an individual savings account in 1991 that involves a series of 20 deposits as shown in the accompanying figure.

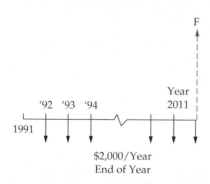

$2,000/Year
End of Year

The account is expected to compound at an average interest rate of 12% per year through the year 2011. General price inflation is expected to average 6% per year during this time. (8.2)

a. What is the FW of the savings account at the end of year 2011?

b. What is the FW of the savings account in 1991 (base time period) spending power?

8-15. Consider a project requiring an investment of $20,000 that is expected to return, in actual dollars, $6,000 at the end of the first year, $8,000 at the end of the second year, and $12,000 at the end of the third year. The general price inflation rate is 5% per year, and the real interest rate is 10% per year. Compare the PW of this project using before-tax actual dollar and real dollar analysis (assume $b = 0$). (8.2)

8-16. Listed in the following table are the estimated annual price changes, in percentages, for two products over the next seven years. You want to simplify the modeling of these price change effects in some cost analyses being done. What single annual rate for each product would you use in your simplified model for this seven-year period? (8.2)

Year	Price Change (%)	
	Product A	Product B
1	4.6	8.3
2	4.8	7.5
3	6.1	9.0
4	6.9	8.0
5	5.8	7.0
6	7.2	9.0
7	6.6	9.5

8-17. Your company *must* obtain some new production equipment for the next six years and is considering leasing. You have been directed to perform an actual-dollar after-tax study of the leasing approach. The pertinent information for the study is as follows:

Lease costs: First year, $80,000; second year, $60,000; third through sixth years, $50,000 per year. Assume that a six-year contract has been offered by the lessor that fixes these costs over the six-year period.

Other costs (not covered under contract): $4,000 in year-0 dollars, and estimated to increase 10% each year.

Effective income tax rate: 40%. (8.5)

a. Develop the actual-dollar ATCF for the leasing alternative.

b. If the real MARR (i_r) after taxes is 5% per year and the annual inflation rate (f) is 9.524% per year, what is the actual-dollar after-tax equivalent annual cost for the leasing alternative?

8-18. The capital investment for a new highway paving machine is $838,000. The estimated annual expense, in year zero dollars, is $92,600. This expense is estimated to increase at the rate of 6.3% per year. Assume that $f = 4.5\%$, $N = 7$ years, MV at the end of year seven is 15% of the capital investment, and the MARR (in real terms) is 10.05% per year. What uniform annual revenue (before taxes), in actual dollars, would the machine need to generate to break even? (8.3)

8-19. A gas-fired heating unit is expected to meet an annual demand for thermal energy of 500 million Btu, and the unit is 80% efficient. Assume that each 1,000 cubic feet of natural gas, if burned at 100% efficiency, can deliver one million Btu. Suppose further that natural gas is now selling for $2.50 per 1,000 cubic feet. What is the PW of fuel cost for this heating unit over a 12-yr period if natural gas prices are expected to escalate at an average rate of 10% per year? The firm's MARR ($= i_c$) is 18% per year. (8.3)

8-20. Your company uses the same large electric drive motor at several locations in its electric power generating plants. A new motor, which is much more efficient, is available. The market price of the new model is $71,000. Assume the following:

- Analysis period is 10 years.
- General inflation rate is 3.2%.
- The total price escalation rate on annual savings in operating expenses is 5.7% per year. Assume any savings in year one would escalate at this rate thereafter.
- Before-tax MARR = 12% per year (does not include inflation component).
- Base time period is year zero ($b = 0$).

If you ignore any market value or income taxes, what would the *annual savings in year one* need to be per motor to break even, if purchased at the $71,000 market price? (Use an actual dollar analysis). (8.3)

8-21. A small heat pump, including the duct system, now costs $2,500 to purchase and install. It

has a useful life of 15 yr and incurs annual maintenance expenses of $100 per year in real (year-zero) dollars over its useful life. A compressor replacement is required at the end of the 8th year at a cost of $500 in real dollars. The annual cost of electricity for the heat pump is $680 based on present prices. Electricity prices are projected to escalate at an annual rate of 10%. All other costs are expected to escalate at 6%, which is the projected general price inflation rate. The firm's MARR, which includes an allowance for general price inflation, is 15% per year. No market value is expected from the heat pump at the end of 15 yr. (8.3)

a. What is the AW, expressed in actual dollars, of owning and operating the heat pump?

b. What is the AW in real dollars of owning and operating the heat pump?

8-22. A company has two different machines it can purchase to perform a specified task. Both machines will perform the same job. Machine *A* costs $150,000 initially, while machine *B* (the deluxe model) costs $200,000. It has been estimated that costs will be $1,000 for machine *A* and $500 for machine *B* in the first year. Management expects these costs to increase with inflation, which is expected to average 10% per year. The company uses a 10-yr study period, and its effective income tax rate is 50%. Both machines qualify as 5-yr MACRS (GDS) property. Which machine should the company choose? (8.5)

8-23. An electric utility company in the Northeast is trying to decide whether to switch from oil to coal at one of its generating stations. After much investigation, the problem has been reduced to these economic trade-offs:

	Oil	Coal
Cost of retrofitting boilers to burn coal	—	?
Annual fuel expense (year 0 dollars)	25×10^6	17×10^6
Escalation rate (e_j)	10%/yr	6%/yr
Life of plant	25 yr	25 yr

Determine the cost of retrofitting the boilers (to burn coal) that could be justified at this generating station. The utility's real MARR is 3% per year, and the general price inflation rate in the economy will average 6% per year over the next 25 yr. (8.2, 8.3)

a. Solve using an actual-dollar analysis.

b. Solve using a real-dollar analysis.

8-24. Suppose that Problem 5-2 is modified such that the estimated annual revenues less expenses for the three designs are as follows:

Design	Annual Revenues Less Expenses
1	$5,500 in year one and increasing $300 per year thereafter.
2	$3,300 in years one and two and increasing at the rate of 10% per year thereafter.
3	$4,800 in years one through four and increasing at the rate of 7% per year thereafter.

Rework Problem 5-2 using the PW method to determine the preferred design. (5.4, 8.3)

8-25. Because of tighter safety regulations, an improved air filtration system must be installed at a plant that produces a highly corrosive chemical compound. The capital investment in the system is $260,000 in present-day dollars. The system has a useful life of 10 yr and is in the MACRS (GDS) five-year property class. It is expected that the MV of the system at the end of its 10-yr life will be $50,000 in present-day dollars. Annual expenses, estimated in present-day dollars, are expected to be $6,000 per year, not including an annual property tax of 4% of the investment cost (*does not inflate*). Assume that the *plant* has a remaining life of 20 yr, and that replacement costs, annual expenses, and MV escalate at 6% per year.

If the effective income tax rate is 40%, set up a table to determine the ATCF for the system over a 20-yr period. The after-tax market rate of return desired on investment capital is 12% per year. What is the PW of costs of this system after income taxes have been taken into account? Develop the real-dollar ATCF. (Assume that the annual general price inflation rate is 4.5% over the 20-yr period.) (Chap. 6 and Secs. 8.4, 8.5)

8-26. A certain engine lathe can be purchased for $150,000 and depreciated over three years to a zero salvage value with the SL method. This machine will produce metal parts that will generate revenues of $80,000 (time 0 dollars) per year. It is a policy of the company that the annual revenues will be escalated each year to keep pace with the general inflation rate, which is expected to average 5%/yr ($\bar{f} = 0.05$). Labor, materials, and utilities totaling $20,000 (time 0 dollars) per year are all expected to increase at 9% per year. The firm's effective income tax rate is 50%, and its after-tax MARR (i_c) is 26% per year.

Perform an actual-dollar ($A\$$) analysis, and determine the annual ATCFs of the preceding investment opportunity. Use a life of three years and work to the nearest dollar. What interest rate would be used for discounting purposes? (8.5)

8-27. Your company manufactures circuit boards and other electronic parts for various commercial products. Design changes in part of the product line, which are expected to increase sales, will require changes in the manufacturing operation. The cost basis of new equipment required is $220,000 (MACRS five-year property class). Increased annual revenues, in year zero dollars, are estimated to be $360,000. Increased annual expenses, in year zero dollars, are estimated to be $239,000. The estimated market value of equipment in actual dollars at the end of the six-year analysis period is $40,000. General price inflation is estimated at 4.9% per year; the total escalation rate on annual revenues is 2.5%, and for annual expenses it is 5.6%; the after-tax MARR (in market terms) is 10% per year; and $t = 39\%$. (Chap. 6 and Secs. 8.4, 8.5)

a. Based on an after-tax, actual-dollar analysis, what is the maximum amount that your company can afford to spend on the total project (i.e., changing the manufacturing operations)? Use the PW method of analysis.

b. Develop (show) your ATCF in real dollars.

8-28. You have been assigned the task of analyzing for your company whether to purchase or lease some transportation equipment. The analysis period is six years, and the base year is year zero ($b = 0$). Other pertinent information is given in Table P8-28. Also,

a. The contract terms for the lease specify a cost of $300,000 in the first year and $200,000 annually in years two through six (the contract, i.e., these rates, do not cover the annual expense items).

TABLE P8-28 Table for Problem P8-28

Cash Flow Item	Estimate in Year-0 Dollars		Best Estimate of Price Change (% per year, e_j)
	Purchase	Lease	
Capital investment	$600,000	—	—
MV at end of six years	90,000	—	2%
Annual operating, insurance and other expenses	26,000	$26,000	6
Annual maintenance expenses	32,000	32,000	9

b. The after-tax MARR (*not* including inflation) is 13.208% per year.
c. The general inflation rate (f) is 6%.
d. The effective income tax rate (t) is 34%.
e. Assume the equipment is in the MACRS (GDS) five-year property class.

Which alternative is preferred (use an after-tax, actual dollar analysis and the FW criterion)? (Chap. 6 and Secs. 8.4, 8.5)

8-29. An international corporation located in Country *A* is considering a project in the United States. The currency in Country *A*, say *X*, has been strengthening relative to the U.S. dollar; specifically, the average devaluation of the U.S. dollar has been 2.6% per year (which is projected to continue). Assume the present exchange rate is 6.4 units of *X* per U.S. dollar. (a) What is the estimated exchange rate two years from now, and (b) if, instead, currency *X* was devaluing at the same rate (2.6% per year) relative to the U.S. dollar, what would be the exchange rate three years from now?

8-30. A company requires a 26% internal rate of return (before taxes) in U.S. dollars on project investments in foreign countries. (8.6)

a. If the currency of Country *A* is projected to average an 8% annual devaluation relative to the dollar, what rate of return (in terms of the currency there) would be required for a project?
b. If the dollar is projected to devaluate 6% annually relative to the currency of Country *B*, what rate of return (in terms of the currency there) would be required for a project?

8-31. A U.S. company is considering a high-technology project in a foreign country. The es-

timated economic results for the project (after taxes) in the foreign currency (T-marks), is shown in the following table for the seven-year analysis period being used. The company requires an 18% rate of return in U.S. dollars (after taxes) on any investments in this foreign country. (8.6)

End of year	Cash Flow (T-marks after Taxes)
0	−3,600,000
1	450,000
2	1,500,000
3	1,500,000
4	1,500,000
5	1,500,000
6	1,500,000
7	1,500,000

a. Should the project be approved, based on a PW analysis in U.S. dollars, if the devaluation of the T-mark, relative to the U.S. dollar, is estimated to average 12% per year and the present exchange rate is 20 T-marks per dollar?
b. What is the IRR of the project in T-marks?
c. Based on your answer in (b), what is the IRR in U.S. dollars?

8-32. An automobile manufacturing company in Country *X* is considering the construction and operation of a large plant on the eastern seaboard of the United States. Their MARR = 20% per year on a before-tax basis. (*This is a market rate relative to their currency in Country X.*) The study period used by the company for this type of investment is 10 yr. Additional information is provided as follows:

- The currency in Country X is the Z-Kron.
- It is estimated that the U.S. dollar will become weaker relative to the Z-Kron during the next 10 yr. Specifically, the dollar is estimated to be devalued at an average rate of 2.2% per year.
- The present exchange rate is 92 Z-Krons per U.S. dollar.
- The estimated before-tax net cash flow (in U.S. dollars) is as follows:

EOY	Net Cash Flow (U.S. Dollars)
0	$-\$168,000,000$
1	$-32,000,000$
2	$69,000,000$
⋮	⋮
10	$69,000,000$

Based on a before-tax analysis, will this project meet the company's economic decision criterion? (8.6)

8-33. XYZ rapid prototyping (RP) software costs $20,000, lasts one year, and will be expensed (i.e., written off in one year). The cost of the upgrades will increase 10% per year starting at the beginning of year two. How much can be spent now for a RP software upgrade agreement that lasts three years and must be depreciated with the SL method to zero over three years? The MARR is 20% per year (i_c), and the effective income tax rate (t) is 34%. (8.5)

8-34. *Brain Teaser* This case study is a justification of a computer system for a theoretical firm, the ABC Manufacturing Company. The following are known data:

- The combined initial hardware and software cost is $80,000.
- Contingency costs have been set at $15,000 (these are not necessarily incurred).
- A service contract on the hardware costs $500 per month.
- The effective income tax rate (t) is 38%.
- Company management has established a 15% $(= i_c)$ per year hurdle rate (MARR).

In addition, the following assumptions and projections have been made:

- In order to support the system on an ongoing basis, a programmer/analyst will be required. The starting salary (first year) is $28,000, and fringe benefits amount to 30% of the base salary. Salaries are expected to increase by 6% each year thereafter.
- The system is expected to yield a staff savings of three persons (to be reduced through normal attrition) at an average salary of $16,200 per year per person (base salary plus fringes) in year-0 (base year) dollars. It is anticipated that one person will retire during the second year, another in the third year, and a third in the fourth year.
- A 3% reduction in purchased material costs is expected; first year purchases are $1,000,000 in year-0 dollars and are expected to grow at a compounded rate of 10% per year.
- The project life is expected to be six years, and the computer capital investment will be fully depreciated over that time period [MACRS (GDS) five-year property class].

Based on this information, perform an actual-dollar ATCF analysis. Is this investment acceptable based on economic factors alone? (8.2, 8.5)

*R*eplacement Analysis

*R*eplacement decisions are critically important to an operating organization. The objectives of this chapter are (1) to discuss the considerations involved in replacement studies and (2) to address the key question of whether an asset should be kept one or more years or immediately replaced with the best available challenger.

The following topics are discussed in this chapter:

Reasons for the replacement of assets

Factors that must be considered in replacement studies

Typical replacement problems

Determining the economic life of a challenger

Determining the economic life of the defender

Comparisons when the useful life of the defender and the challenger differ

Retirement without replacement (abandonment)

After-tax replacement studies

A comprehensive example (including augmentation)

9.1 Introduction

A decision situation often encountered in business firms and government organizations, as well as by individuals, is whether an existing asset should be retired from use, continued in service, or replaced with a new asset. As the pressures of worldwide competition continue to increase, requiring higher quality goods and services, shorter response times, and other changes, this type of decision is occurring more frequently. Thus, the *replacement problem*, as it is commonly called, requires careful engineering economy studies to provide the information needed to make sound decisions that improve the operating efficiency and the competitive position of an enterprise.

> Engineering economy studies of replacement situations are performed using the same basic methods as other economic studies involving two or more alternatives. The specific decision situation, however, occurs in different forms. Sometimes it may be whether to retire an asset without replacement (*abandonment*) or whether to retain the asset for backup rather than primary use. Also, the decision may consider if changed production requirements can be met by *augmenting* the capacity or capability of the existing asset(s). Often, however, the decision is whether to replace an existing (old) asset, descriptively called the *defender*, with a new asset. The one or more alternative replacement (new) assets are then called *challengers*.

9.2 Reasons for Replacement Analysis

The need to evaluate the replacement, retirement, or augmentation of assets results from changes in the economics of their use in an operating environment. Various reasons can underlie these changes, and unfortunately they are sometimes accompanied by unpleasant financial facts. The following are the four major reasons that summarize most of the factors involved:

1. *Physical Impairment (Deterioration).* These are changes that occur in the physical condition of the asset. Normally, continued use (aging) results in the less efficient operation of an asset. Routine maintenance and breakdown repair costs increase, energy use may increase, more operator time may be required, and so forth. Or, some unexpected incident such as an accident occurs that affects the physical condition and the economics of ownership and use of the asset.
2. *Altered Requirements.* Capital assets are used to produce goods and services that satisfy human wants. When the demand for a good or service either increases or decreases or the design of a good or service changes, the related asset(s) may have the economics of its use affected.
3. *Technology.* The impact of changes in technology varies among different types of assets. For example, the relative efficiency of heavy highway construction equipment is impacted less rapidly by technological changes than automated manufacturing equipment. In general, the costs per unit of production, as well as quality and other factors, are favorably impacted by changes in technology, which result in more frequent replacement of existing assets with new and better challengers.
4. *Financing.* Financial factors involve economic opportunity changes external to the physical operation or use of assets and may involve income tax considerations.* For example, the rental (lease) of assets may become more attractive than ownership.

---

*In this chapter, we often refer to Chapter 6 for details on depreciation methods and after-tax analysis.

Reason 2 (altered requirements) and Reason 3 (technology) are sometimes referred to as different categories of *obsolescence*. Even financial changes (Reason 4) could be considered a form of obsolescence. In any replacement problem, however, factors from more than one of these four major areas may be involved. Regardless of the specific considerations, and even though there is a tendency to regard it with some apprehension, the replacement of assets often represents an economic opportunity for the firm.

For the purposes of our discussion of replacement studies, the following is a distinction between various types of lives for typical assets.

Economic life is the period of time (years) that results in the minimum equivalent uniform annual cost (EUAC) of owning and operating an asset.* If we assume good asset management, economic life should coincide with the period of time extending from the date of acquisition to the date of abandonment, demotion in use, or replacement from the primary intended service.

Ownership life is the period between the date of acquisition and the date of disposal by a specific owner. A given asset may have different categories of use by the owner during this period. For example, a car may serve as the primary family car for several years and then serve only for local commuting for several more years.

Physical life is the period between original acquisition and final disposal of an asset over its succession of owners. For example, the car just described may have several owners over its existence.

Useful life is the time period (years) that an asset is kept in productive service (either primary or backup). It is an estimate of how long an asset is expected to be used in a trade or business to produce income.

9.3 Factors That Must Be Considered in Replacement Studies

There are several factors that must be considered in replacement studies. Once a proper perspective has been established regarding these factors, little difficulty should be experienced in making replacement studies. Six factors and related concepts are discussed in this section:

1. Recognition and acceptance of past errors
2. Sunk costs
3. Existing asset value and the *outsider viewpoint*
4. Economic life of the proposed replacement asset (challenger)
5. Remaining (economic) life of the old asset (defender)
6. Income tax considerations

---

*The AW of a primarily cost cash flow pattern is sometimes called the equivalent uniform annual cost (EUAC). Because this term is commonly used in the definition of the economic life of an asset, we will often use EUAC in this chapter.

9.3.1 Past Estimation Errors

The economic focus in a replacement study is the future. Any *estimation errors* made in a previous study related to the defender are not relevant (unless there are income tax implications). For example, when an asset's book value (BV) is greater than its current market value (MV), the difference frequently has been designated as an estimation error. Such "errors" also arise when capacity is inadequate, maintenance expenses are higher than anticipated, and so forth.

This implication is unfortunate because in most cases these differences are not the result of errors, but of the inability to foresee future conditions better at the time of the original estimates. Acceptance of unfavorable economic realities may be made easier by posing a hypothetical question: "What will be the costs of my competitor, who has no past errors to consider?" In other words, we must decide whether we wish to live in the *past*, with its errors and discrepancies, or to be in a sound competitive position in the *future*. A common reaction is "I can't afford to take the loss in value of the existing asset that will result if the replacement is made." The fact is that the loss already has occurred, whether or not it could be afforded, and it exists whether or not the replacement is made.

9.3.2 The Sunk-Cost Trap

Only present and future cash flows should be considered in replacement studies. Any unamortized values (i.e., unallocated value of an asset's capital investment) of an existing asset under consideration for replacement are strictly the result of *past* decisions—that is, the initial decision to invest in that asset and decisions as to the method and number of years to be used for depreciation purposes. For purposes of this chapter, we define a *sunk cost* to be the difference between an asset's BV and its MV at a particular point in time. Sunk costs have no relevance to the replacement decisions that must be made (*except to the extent that they affect income taxes*). When income tax considerations are involved, we must include the sunk cost in the engineering economy study. Clearly, serious errors can be made in practice when sunk costs are incorrectly handled in replacement studies.

9.3.3 Investment Value of Existing Assets and the Outsider Viewpoint

Recognition of the nonrelevance of BVs and sunk costs leads to the proper viewpoint to use in placing value on existing assets for replacement study purposes. In this chapter, we use the so-called *"outsider viewpoint"* for approximating the investment amount of an existing asset (defender). In particular, the outsider viewpoint* is the perspective that would be taken by an impartial third party to establish the fair MV of a used (secondhand) asset. This viewpoint forces the analyst to focus on present and future cash flows in a replacement study, thus avoiding the temptation to dwell on past (sunk) costs.

---

*The outsider viewpoint is also known as the opportunity cost approach to determining the value of the defender.

The *present realizable* MV is the correct capital investment amount to be assigned to an existing asset in replacement studies.* A good way to reason that this is true is to use the *opportunity cost* or *opportunity foregone principle*. That is, if it should be decided to keep the existing asset, we are giving up the opportunity to obtain its net realizable MV at that time. Therefore, this represents the *opportunity cost* of keeping the defender.

There is one addendum to this rationale: If any new investment expenditure (such as for overhaul) is needed to upgrade the existing asset so that it will be competitive in level of service with the challenger, the extra amount should be added to the present realizable MV to determine the total investment in the existing asset for replacement study purposes.

> When using the outsider viewpoint, the total investment in the defender is the opportunity cost of not selling the existing asset for its current MV, *plus* the cost of upgrading it to be competitive with the best available challenger (all feasible challengers are to be considered).

Clearly, the MV of the defender must not also be claimed as a reduction in the challenger's capital investment because doing so would provide an unfair advantage to the challenger due to double counting the defender's selling price.

EXAMPLE 9-1

The purchase price of a certain new automobile (challenger) being considered for use in your business is $21,000. Your firm's present automobile (defender) can be sold on the open market for $10,000. The defender was purchased with cash three years ago, and its current book value is $12,000. To make the defender comparable in continued service to the challenger, your firm would need to make some repairs at an estimated cost of $1,500.

Based on this information, what are (a) the total capital investment in the defender using the outsider viewpoint and (b) the unamortized value of the defender?

SOLUTION

(a) The total capital investment in the defender (if kept) is its current market value (an opportunity cost) plus the cost of upgrading the car to make it comparable in service to the challenger. Hence, the total capital investment in the defender is $10,000 + $1,500 = $11,500 (from an outsider's viewpoint). This represents a good starting point for estimating the cost of keeping the defender.

(b) The unamortized value of the defender is the book loss (if any) associated with disposing of it. Given that the defender is sold for $10,000, the unamortized

---

*In after-tax replacement studies, the before-tax MV is modified by income tax effects related to potential gains (losses) foregone if the defender is kept in service.

value (loss) is $12,000 − $10,000 = $2,000. This is the difference between the current market value and the current book value of the defender. As discussed in Section 9.3.2, this amount represents a sunk cost and has no relevance to the replacement decision, except to the extent that it may impact income taxes (to be discussed in Section 9.9).

9.3.4 Economic Life of the Challenger

The economic life of an asset minimizes the EUAC of owning and operating the asset, and it is often shorter than the useful or physical life. It is essential to know a challenger's economic life in view of the principle that new and existing assets should be compared over their economic (optimum) lives. Economic data regarding challengers are periodically updated (often annually) and replacement studies are then repeated to ensure an on-going evaluation of improvement opportunities.

9.3.5 Economic Life of the Defender

As we shall see later in this chapter, the economic life of the defender is often one year. Consequently, care must be taken when comparing the defender asset with a challenger asset, because *different lives* are involved in the analysis. We shall see that the defender should be kept longer than its apparent economic life as long as its *marginal cost* is less than the minimum EUAC of the challenger over its economic life. What *assumptions* are involved when two assets having different apparent economic lives are compared, knowing that the defender is a nonrepeating asset? These concepts will be discussed in Section 9.7.

9.3.6 The Importance of Income Tax Consequences

The replacement of assets often results in gains or losses from the sale of *depreciable property*, as discussed in Chapter 6. Consequently, to perform an accurate economic analysis in such cases, the studies must be made on an *after-tax basis*. It is evident that the existence of a taxable gain or loss, in connection with replacement, can have a considerable effect on the results of an engineering study. A prospective gain from the disposal of assets can be reduced by as much as 40% or 50%, depending on the effective income tax rate used in a particular study. Hence, the decision to dispose of or retain an existing asset can be influenced by income tax considerations.

9.4 Typical Replacement Problems

The following typical replacement situations are used to illustrate several of the factors that must be considered in replacement studies. These analyses use the outsider viewpoint to determine the investment in the defenders.

EXAMPLE 9-2

A firm owns a pressure vessel that it is contemplating replacing. The old pressure vessel has annual operating and maintenance expenses of $60,000 per year and it can be kept for five more years, at which time it will have zero market value. It is believed that $30,000 could be obtained for the old pressure vessel if it were sold now.

A new pressure vessel can be purchased for $120,000. The new pressure vessel will have a market value of $50,000 in five years and will have annual operating and maintenance expenses of $30,000 per year. Using a before-tax MARR of 20% per year, determine whether or not the old pressure vessel should be replaced. A study period of five years is appropriate.

SOLUTION

The first step in the analysis is to determine the investment value of the defender (old pressure vessel). Using the outsider viewpoint, the investment value of the defender is $30,000, its present MV. We can now compute the PW (or FW or AW) of each alternative and decide whether the old pressure vessel should be kept in service or replaced immediately.

$Defender:$ $\quad PW(20\%) = -\$30,000 - \$60,000(P/A, 20\%, 5)$

$$= -\$209,436.$$

$Challenger:$ $\quad PW(20\%) = -\$120,000 - \$30,000(P/A, 20\%, 5) + \$50,000(P/F, 20\%, 5)$

$$= -\$189,623.$$

The PW of the challenger is greater (less negative) than the PW of the defender. Thus, the old pressure vessel should be replaced immediately. (The EUAC of the defender is $70,035, and that of the challenger is $63,410.)

EXAMPLE 9-3

The manager of a carpet manufacturing plant became concerned about the operation of a critical pump in one of the processes. After discussing this situation with the supervisor of plant engineering, they decided that a replacement study should be done, and that a nine-year study period would be appropriate for this situation. The company that owns the plant is using a before-tax MARR of 10% per year for its capital investment projects.

The existing pump, Pump A, including driving motor with integrated controls, cost $17,000 five years ago. An estimated market value of $750 could be obtained for the pump if it were sold now. Some reliability problems have been experienced with Pump A, including annual replacement of the impeller and bearings at a cost of $1,750. Annual operating and maintenance (O&M) expenses have been

TABLE 9-1 Summary of Information for Example 9-3

MARR (before taxes) = 10% per year

Existing Pump A *(defender)*

Capital investment when purchased five years ago		$17,000
Annual expenses:		
Replacement of impeller and bearings	$1,750	
Operating and maintenance	3,250	
Taxes and insurance: $17,000 \times 0.02$	340	
Total annual expenses		$5,340
Present market value		$750
Estimated market value at the end of nine additional years		$200

Replacement Pump B *(challenger)*

Capital investment		$16,000
Annual expenses:		
Operating and maintenance	$3,000	
Taxes and insurance: $16,000 \times 0.02$	320	
Total annual expenses		$3,320
Estimated market value at the end of nine years: $16,000 \times 0.20$		$3,200

averaging $3,250. Annual insurance and property tax expenses are 2% of the initial capital investment. It appears that the pump will provide adequate service for another nine years if the present maintenance and repair practice is continued. It is estimated that if this pump is continued in service, its final MV after nine more years will be about $200.

An alternative to keeping the existing pump in service is to sell it immediately and to purchase a replacement pump, Pump *B*, for $16,000. An estimated market value at the end of the nine-year study period would be 20% of the initial capital investment. O&M expenses for the new pump are estimated to be $3,000 per year. Annual taxes and insurance would total 2% of the initial capital investment. The data for Example 9-3 are summarized in Table 9-1.

Based on these data, should the defender (Pump *A*) be kept [and the challenger (Pump *B*) not purchased] or should the challenger be purchased now (and the defender sold)? Use a before-tax analysis and the outsider viewpoint in the evaluation.

SOLUTION

In an analysis of the defender and challenger, care must be taken to correctly identify the investment amount in the existing pump. Based on the outsider viewpoint, this would be the current MV of $750; that is, the *opportunity cost* of keeping the defender. Note that the investment amount of Pump *A* ignores the original purchase price of $17,000. Using the principles discussed thus far, a before-tax analysis of EUAC of Pump *A* and Pump *B* can now be made.

The solution of Example 9-3 using EUAC (before taxes) as the decision criterion follows:

Study Period = 9 Years	Keep Old Pump *A*	Replacement Pump *B*
EUAC(10%):		
Annual expenses	$5,340	$3,320
Capital recovery cost (Equation 4-7):		
$[(\$750 - \$200)(A/P, 10\%, 9) + \$200(0.10)]$	115	
$[(\$16,000 - \$3,200)(A/P, 10\%, 9) + \$3,200(0.10)]$		2,542
Total EUAC(10%)	$5,455	$5,862

Because Pump *A* has the smaller EUAC ($5,455 < $5,862), the replacement pump is apparently not justified and the defender should be kept at least one more year. We could also make the analysis using other methods (e.g., PW), and the indicated choice would be the same.

9.5 Determining the Economic Life of a New Asset (Challenger)

Sometimes in practice the useful lives of the defender and the challenger(s) are not known and cannot be reasonably estimated. The time an asset is kept in productive service might be extended indefinitely with adequate maintenance and other actions, or it might be suddenly jeopardized by an external factor such as technological change. Under this situation, it is important to know the economic life, minimum EUAC, and total year-by-year (marginal) costs for both the best challenger and the defender *so they can be compared based on an evaluation of their economic lives and the costs most favorable to each.*

The economic life of an asset was defined in Section 9.2 as the time that results in the minimum EUAC of owning and operating the asset. Also, the economic life is sometimes called the minimum-cost life or optimum replacement interval. For a new asset, its EUAC can be computed if the capital investment, annual expenses, and year-by-year market values are known or can be estimated. The apparent difficulties of estimating these values in practice may discourage performing the economic life and equivalent cost calculations. Similar difficulties, however, are encountered in most engineering economy studies when estimating the *future* economic consequences of alternative courses of action. Therefore, the estimating problems in replacement analysis are not unique and can be overcome in most application studies.

The estimated initial capital investment, as well as the annual expense and market value estimates, may be used to determine the PW through year k of total costs, PW_k. That is, on a *before-tax* basis,

$$PW_k(i\%) = I - MV_k(P/F, i\%, k) + \sum_{j=1}^{k} E_j(P/F, i\%, j), \qquad (9\text{-}1)$$

which is the sum of the initial capital investment, I, (PW of the initial investment amounts if any occur after time zero) adjusted by the PW of the MV at the end of year k, and of the PW of annual expenses (E_j) through year k. The *total marginal cost* for each year k, TC_k, is calculated using Equation (9-1) by finding the increase in the PW of total cost from year $k-1$ to year k, and then determining the equivalent worth of this increase at the end of year k. That is, $TC_k = (PW_k - PW_{k-1})(F/P, i\%, k)$. The algebraic simplification of this relationship results in the formula

$$TC_k(i\%) = MV_{k-1} - MV_k + iMV_{k-1} + E_k, \qquad (9\text{-}2)$$

which is the sum of the loss in MV during the year of service, the opportunity cost of capital invested in the asset at the beginning of year k, and the annual expenses incurred in year $k(E_k)$. These total marginal (or year-by-year) costs, based on Equation (9-2), are then used to find the EUAC of each year prior to and including year k. The minimum $EUAC_k$ value during the useful life of the asset identifies its economic life, N_C^*. This procedure is illustrated in Example 9-4.

EXAMPLE 9-4

A new forklift truck will require an investment of $20,000 and is expected to have year-end market values and annual expenses as shown in columns 2 and 5, respectively, of Table 9-2. If the before-tax MARR is 10% per year, how long should the asset be retained in service?

TABLE 9-2 Determination of the Economic Life N^* of a New Asset (Example 9-4)

(1) End of Year, k	(2) MV, End of Year k	(3) Loss in Market Value (MV) during Year k	(4) Cost of Capital = 10% of Beginning of Year MV	(5) Annual Expenses (E_k)	(6) [=(3)+(4)+(5)] Total (Marginal) Cost for Year, (TC_k)	(7) EUAC[a] through Year k
0	$20,000	—	—	—	—	—
1	15,000	$5,000	$2,000	$2,000	$9,000	$9,000
2	11,250	3,750	1,500	3,000	8,250	8,643
3	8,500	2,750	1,125	4,620	8,495	8,598 ($N_C^* = 3$)
4	6,500	2,000	850	8,000	10,850	9,084
5	4,750	1,750	650	12,000	14,400	9,954

[a] $EUAC_k = \left[\sum_{j=1}^{k} TC_j(P/F, 10\%, j)\right](A/P, 10\%, k)$

SOLUTION

The solution to this problem is obtained by completing Columns 3, 4, 6 [Equation (9-2)], and 7 of Table 9-2. In the solution, the customary year-end occurrence of all cash flows is assumed. The loss in market value during year k is simply the difference between the beginning-of-year market value, MV_{k-1}, and the end-of-year market value, MV_k. The opportunity cost of capital in year k is 10% of the capital unrecovered (invested in the asset) at the beginning of each year. The values in column 7 are the equivalent uniform annual costs that would be incurred each year (1 to k) if the asset were retained in service through year k and then replaced (or retired) at the end of the year. The minimum EUAC occurs at the end of year N_C^*.

It is apparent from the values shown in column 7 that the new forklift truck will have a minimum EUAC if it is kept in service for only three years (i.e., $N_C^* = 3$).

The computational approach in the preceding example, as shown in Table 9-2, was to determine the total marginal cost for each year and then to convert these into an EUAC through year k. The before-tax EUAC for any life can also be calculated using the more familiar capital recovery formulas presented in Chapter 4. For example, for a life of two years, the EUAC can be calculated with the help of Equation (4-5) as follows:

$$EUAC_2(10\%) = \$20,000(A/P, 10\%, 2) - \$11,250(A/F, 10\%, 2)$$

$$+ [\$2,000(P/F, 10\%, 1) + \$3,000(P/F, 10\%, 2)](A/P, 10\%, 2)$$

$$= \$8,643.$$

This agrees with the corresponding row in column 7 of Table 9-2.

9.6 Determining the Economic Life of a Defender

In replacement analyses, we must also determine the economic life (N_D^*) that is most favorable to the defender. This gives us the choice of keeping the defender as long as its EUAC at N_D^* is less than the minimum EUAC of the challenger. When a major outlay for defender alteration or overhaul is needed, the life that will yield the minimum EUAC is likely to be the period that will elapse before the next major alteration or overhaul is needed. Alternatively, *when there is no defender MV now or later (and no outlay for alteration or overhaul) and when the defender's operating expenses are expected to increase annually, the remaining life that will yield the minimum EUAC will be one year.*

When MVs are greater than zero and expected to decline from year to year, it is necessary to calculate the apparent remaining economic life, which is done in the same manner as in Example 9-4 for a new asset. Using the outsider viewpoint, the investment value of the defender is considered to be its present realizable MV.

Regardless of how the remaining economic life for the defender is determined, a decision to keep the defender does not mean that it should be kept only for this period of time. Indeed, the defender should be kept longer than the apparent economic life as long as its *marginal* cost (total cost for an additional year of service) is less than the minimum EUAC for the best challenger.

This important principle of replacement analysis is illustrated in Example 9-5.

EXAMPLE 9-5

It is desired to determine how much longer a forklift truck should remain in service before it is replaced by the new truck (challenger) for which data were given in Example 9-4 and Table 9-2. The defender in this case is two years old, originally cost $13,000, and has a present realizable MV of $5,000. If kept, its market values and annual expenses are expected to be as follows:

End of Year, k	MV, End of Year k	Annual Expenses, E_k
1	$4,000	$5,500
2	3,000	6,600
3	2,000	7,800
4	1,000	8,800

Determine the most economical period to keep the defender before replacing it (if at all) with the present challenger of Example 9-4. The before-tax cost of capital (MARR) is 10% per year.

SOLUTION

Table 9-3 shows the calculation of total cost for each year (marginal cost) and the EUAC at the end of each year for the defender based on the format used in Table 9-2. Note that the minimum EUAC of $7,000 corresponds to keeping the defender for one more year. However, the marginal cost of keeping the truck for the second year is $8,000, which is still less than the minimum EUAC for the challenger (i.e., $8,598, from Example 9-4). The *marginal cost* for keeping the defender the third year and beyond is greater than the $8,598 minimum EUAC for the challenger. Based on the available data shown, it would be most economical to keep the defender for two more years and then to replace it with the challenger. This situation is portrayed graphically in Figure 9-1.

Example 9-5 assumes that a comparison is being made with the best challenger alternative available. In this situation, if the defender is retained beyond the point where its marginal costs exceed the minimum EUAC for the challenger, the difference in costs continues to grow and replacement becomes more urgent. This is illustrated to the right of the intersection in Figure 9-1.

TABLE 9-3 Determination of the Economic Life N^* of an Old Asset (Example 9-5)

(1) End of Year, k	(2) MV, End of Year k	(3) Loss in Market Value (MV) during Year k	(4) Cost of Capital = 10% of Beginning of Year MV	(5) Annual Expenses (E_k)	(6) [=(3)+(4)+(5)] Total (Marginal) Cost for Year, (TC_k)	(7) EUAC[a] through Year k	
0	$5,000	—	—	—	—	—	
1	4,000	$1,000	$500	$5,500	$7,000	$7,000	$(N_D^* = 1)$
2	3,000	1,000	400	6,600	8,000	7,476	
3	2,000	1,000	300	7,800	9,100	7,966	
4	1,000	1,000	200	8,800	10,000	8,405	

[a] $\text{EUAC}_k = \left[\sum_{j=1}^{k} \text{TC}_j (P/F, 10\%, j) \right] (A/P, 10\%, k)$

Figure 9-2 illustrates the effect of improved new challengers in the future. If an improved challenger X becomes available before replacement with the new asset of Figure 9-1, then a new replacement study should take place to consider the improved challenger. If there is a possibility of a further-improved challenger Y as of, say, four years later, it may be better still to postpone replacement until that challenger becomes available. Although retention of the old asset beyond its break-even point with the best available challenger has a cost that may well grow with time, this cost of waiting can, in some instances, be worthwhile if it permits the purchase of an improved asset having economies that offset the cost

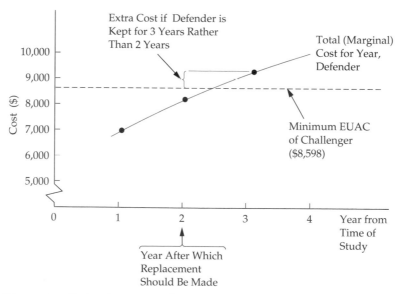

Figure 9-1 Defender versus Challenger Forklift Trucks (Based on Examples 9-4 and 9-5)

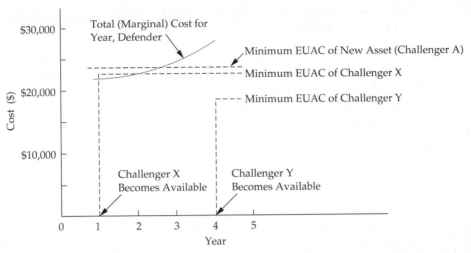

Figure 9-2 Old versus New Asset Costs with Improved Challengers Becoming Available in the Future

of waiting. Of course, a decision to postpone a replacement may also "buy time and information." Because technological change tends to be sudden and dramatic rather than uniform and gradual, new challengers with significantly improved features can arise sporadically and can change replacement plans substantially.

When replacement is not signaled by the engineering economy study, more information may become available before the next analysis of the defender. Hence, the next study should include any additional information. *Postponement* generally should mean a postponement of the decision on when to replace, not the decision to postpone replacement until a specified future date.

9.7 Comparisons in Which the Defender's Useful Life Differs from That of the Challenger

In Section 9.4, we discussed a typical replacement situation in which the useful lives of the defender and the challenger were known and were the same, as well as equal to the study period. When this situation occurs, any of the analysis methods, properly applied, may be used.

In the previous two sections (9.5 and 9.6), we discussed the economic lives of a new asset and of a defender and how these results (along with the related cost information) are used in replacement analysis when the useful lives of the assets may or may not be known.

A third situation occurs when the useful lives of the best challenger and the defender are known, or can be estimated, but are not the same. The comparison of the challenger and the defender under these circumstances is the topic of this section.

In Chapter 5, two assumptions used for the economic comparisons of alternatives, including those having different useful lives, were described: (1) *repeatability* and (2) *cotermination*. Under either assumption, the analysis period used is the same for all alternatives in the study. The repeatability assumption, however, involves two main stipulations:

1. The period of needed service for which the alternatives are being compared is either indefinitely long or a length of time equal to a common multiple of the useful lives of the alternatives.
2. What is estimated to happen in the first useful life span will happen in all succeeding life spans, if any, for each alternative.

For replacement analyses, the first of these conditions may be acceptable, but normally the second is not reasonable for the defender. The defender is typically an older and used piece of equipment. An identical replacement, even if it could be found, probably would have an installed cost in excess of the current MV of the defender.

Failure to meet the second stipulation can be circumvented if the period of needed service is assumed to be indefinitely long and *if we recognize that the analysis is really to determine if* **now** *is the time to replace the defender.* When the defender is replaced, either now or at some future date, it will be by the challenger—the best available replacement.

Example 9-5, involving a before-tax analysis of the defender versus a challenger forklift truck, made implicit use of the *repeatability* assumption. That is, it was assumed that the particular challenger analyzed in Table 9-2 would have a minimum EUAC of $8,598 regardless of when it replaces the defender. Figure 9-3 shows time diagrams of the cost consequences of keeping the defender for two more years versus adopting the challenger now, with the challenger costs to be repeated into the indefinite future. Recall that the economic life of the challenger is three years. *It can be seen in Figure 9-3 that the only difference between the alternatives is in years 1 and 2.*

The repeatability assumption, applied to replacement problems involving assets with different useful or economic lives, often simplifies the economic comparison of the alternatives. For example, the comparison of the PW values of the alternatives in Figure 9-3 *over an infinite analysis period* (refer to the capitalized worth calculation in Chapter 5) will confirm our previous answer to Example 9-5 that Alternative *A* (keep defender for two more years) is preferred to Alternative *B* (replace with challenger now). Using a MARR = 10% per year, we have

$$\mathrm{PW}_A(10\%) = -\$7,000(P/F, 10\%, 1) - \$8,000(P/F, 10\%, 2) - \frac{\$8,598}{0.10}(P/F, 10\%, 2)$$

$$= -\$84,029;$$

$$\mathrm{PW}_B(10\%) = -\frac{\$8,598}{0.10} = -\$85,980.$$

The difference ($\mathrm{PW}_B - \mathrm{PW}_A$) is $-\$1,951$, which confirms that the additional cost of the challenger over the next two years is not justified and it is best to keep the defender for two more years before replacing it with the challenger.

Figure 9-3
Effect of the
Repeatability
Assumption Applied
to Alternatives in
Example 9-5

A: Keep Defender for Two More Years

Defender Challenger 1 Challenger 2
          (Years 3–5)   (Years 6–8)

$7,000
$8,000
A = $8,598

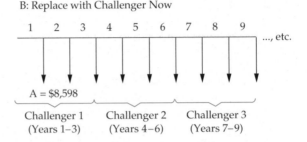

B: Replace with Challenger Now

A = $8,598

Challenger 1   Challenger 2   Challenger 3
(Years 1–3)   (Years 4–6)   (Years 7–9)

> Whenever the *repeatability* assumption is not applicable, the *coterminated* assumption may be used; it involves using a finite study period for all alternatives. As described in Chapter 5, use of the *coterminated* assumption requires detailing what and when cash flows are expected to occur for each alternative and then determining which is most economical, using any of the correct economic analysis methods. *When the effects of price changes and taxes are to be considered in replacement studies, it is recommended that the coterminated assumption be used.*

EXAMPLE 9-6

Suppose that we are faced with the same replacement problem as in Example 9-5, except that the period of needed service is (a) three years or (b) four years. That is, a finite analysis period under the coterminated assumption is being used. In each case, which alternative should be selected?

SOLUTION

(a) For a planning horizon of three years, we might intuitively think that either the defender should be kept for three years or it should be replaced immediately by the challenger to serve for the next three years. From Table 9-3, the EUAC for the defender for three years is $7,966, and from Table 9-2 the EUAC for the challenger for three years is $8,598. Thus, following this reasoning, the defender would be kept for three years. However, this is not quite right. Focusing on the "Total (Marginal) Cost for Each Year" columns, we can see that the defender has the lowest cost in the first two years, but in the third year

TABLE 9-4 Determination of When to Replace the Defender for a Planning Horizon of Four Years [Example 9-6, Part (b)]

Keep Defender for	Keep Challenger for	Total (Marginal) Costs for Each Year				EUAC at 10% for 4 Years	
		1	2	3	4		
0 years	4 years	$9,000[a]	$8,250[a]	$8,495[a]	$10,850[a]	$9,084	
1	3	7,000	9,000	8,250	8,495	8,140	
2	2	7,000	8,000	9,000	8,250	8,005	←Least cost
3	1	7,000	8,000	9,100	9,000	8,190	alternative
4	0	7,000[b]	8,000[b]	9,100[b]	10,000[b]	8,405	

[a] Column 6 of Table 9-2.
[b] Column 6 of Table 9-3.

its cost is $9,100; the EUAC of one year of service for the challenger is only $9,000. Hence, it would be more economical to replace the defender after the second year. This conclusion can be confirmed by enumerating all replacement possibilities and their respective costs and then computing the EUAC for each, as will be done for the four-year planning horizon in part (b).

(b) For a planning horizon of four years, the alternatives and their respective costs for each year and the EUAC of each are given in Table 9-4. Thus, the most economical alternative is to keep the defender for two years and then replace it with the challenger, to be kept for the next two years. The decision to keep the defender for two years happens to be the same as when the repeatability assumption was used, which, of course, would not be true in general.

When a replacement analysis involves a defender that cannot be continued in service because of changes in technology, service requirements, and so on, a choice among two or more new challengers must be made. Under this situation the repeatability assumption may be a convenient economic modeling approach for comparing the alternatives and making a present decision. Note that when the defender is not a feasible alternative, the replacement problem is no different than any other analysis involving mutually exclusive alternatives.

9.8 Retirement without Replacement (Abandonment)

Consider a project for which the period of required service is finite and that has *positive* net cash flows following an initial capital investment. Market values, or abandonment values, are estimated for the end of each remaining year in the project's life. In view of an opportunity cost (MARR) of $i\%$ per year, should the project be undertaken? Given that we have decided to implement the project, what is the best year to abandon the project? In other words, what is the "economic life" of the project?

For this type of problem, the following assumptions are applicable:

1. Once a capital investment has been made, the firm desires to postpone the decision to abandon a project as long as its present equivalent value (PW) is not decreasing.
2. The existing project will be terminated at the best abandonment time and will not be replaced by the firm.

Solving the abandonment problem is similar to determining the economic life of an asset. In abandonment problems, however, annual benefits (cash inflows) are present but in economic life analysis, costs (cash outflows) are dominant. In both cases, the objective is to increase the overall wealth of the firm by finding the life that maximizes profits or, equivalently, minimizes costs.

EXAMPLE 9-7

A $50,000 baling machine for recycled paper is being considered by the XYZ Company. Annual revenues less expenses and end-of-year abandonment (market) values for the machine have been estimated for the project. The firm's MARR is 12% per year. When is the best time to abandon the project if the firm has already decided to acquire the baling machine and use it for no longer than seven years?

	End of Year						
	1	2	3	4	5	6	7
Annual revenues less expenses	$10,000	$15,000	$18,000	$13,000	$9,000	$6,000	$5,000
Abandonment value of machine[a]	40,000	32,000	25,000	21,000	18,000	17,000	15,000

[a] Estimated market value

SOLUTION

The PWs that result from deciding now to keep the machine exactly one, two, three, four, five, six, and seven years are as follows:

Keep for one year:

$$PW(12\%) = -\$50,000 + (\$10,000 + \$40,000)(P/F, 12\%, 1)$$

$$= -\$5,355$$

Keep for two years:

$$PW(12\%) = -\$50,000 + \$10,000(P/F, 12\%, 1) + (\$15,000 + \$32,000)(P/F, 12\%, 2)$$

$$= -\$3,603$$

In the same manner, the PW for years three through seven can be computed. The results are as follows:

Keep for three years:	PW(12%) = $1,494
Keep for four years:	PW(12%) = $5,306
Keep for five years:	PW(12%) = $7,281
Keep for six years:	PW(12%) = $8,719
Keep for seven years:	PW(12%) = $9,153

As you can see, PW is maximized ($9,153) by retaining the machine for the total seven years. Thus, the best abandonment time would be in seven years.

In some cases, management may decide that although an existing asset is to be retired from its current use, it will not be replaced or removed from all service. Although the existing asset may not be able to compete economically at the moment, it may be desirable and even economical to retain the asset as a standby unit or for some different use. The cost to retain the defender under such conditions may be quite low, because of its relatively low realizable MV and perhaps low annual expenses. Often income tax considerations (to be discussed in the next section) also bear on the true cost of retaining the defender.

9.9 After-Tax Replacement Studies

As discussed in Chapter 6, income taxes associated with a proposed project may represent a major cash outflow for a firm. Therefore, income taxes should be considered along with all other relevant cash flows when assessing the economic profitability of a project. This fact also holds true for replacement decisions. The replacement of an asset often results in gains or losses from the sale of the existing asset (defender). The income tax consequence resulting from the gain (loss) associated with the sale (or retention) of the defender has the potential to impact the decision to keep the defender or to sell it and purchase the challenger. The remainder of this section is devoted to demonstrating the procedure for performing replacement analyses on an after-tax basis. Note that after-tax replacement analyses require knowledge of the depreciation schedule already in use for the defender, as well as the appropriate depreciation schedule to be used for the challenger.

9.9.1 After-Tax Economic Life

In earlier sections, the economic life of a new asset (Example 9-4) and the economic life of an existing asset (Example 9-5) were determined on a before-tax basis. However, an *after-tax* analysis can also be used to determine the economic life of an asset by extending Equation (9-1) to account for income tax effects:

$$PW_k(i\%) = I + \sum_{j=1}^{k} [(1-t)E_j - td_j](P/F, i\%, j) - [(1-t)MV_k + t(BV_k)](P/F, i\%, k).$$

(9-3)

This computation finds the PW of the after-tax cash flows (expressed as costs) through year k, PW_k, by (1) adding the initial capital investment, I, (PW of investment amounts if any occur after time zero) and the sum of the after-tax PW of annual expenses through year k, including adjustments for annual depreciation amounts (d_j), and then (2) adjusting this total after-tax PW of costs by the after-tax consequences of gain or loss on disposal of the asset at the end of year k. Similar to the previous before-tax analysis using Equation (9-1), Equation (9-3) is used to determine the after-tax total marginal cost for each year k, TC_k. That is, $TC_k = (PW_k - PW_{k-1})(F/P, i\%, k)$. The algebraic simplification of this relationship results in Equation (9-4):

$$TC_k(i\%) = (1-t)(MV_{k-1} - MV_k + iMV_{k-1} + E_k) + i(t)(BV_{k-1}).$$ (9-4)

Equation (9-4) is $(1-t)$ times Equation (9-2) plus interest on the tax adjustment from the book value of the asset at the beginning of year k. A tabular format incorporating Equation (9-4) is used in the solution of the next example to find the economic life of a new asset on an after-tax basis (N_{AT}^*). This same procedure can also be used to find the after-tax economic life of an existing asset.

EXAMPLE 9-8

Find the economic life on an after-tax basis for the new forklift truck (challenger) described in Example 9-4. Assume that the new forklift is depreciated as a MACRS (GDS)*three-year property class asset, the effective income tax rate is 40%, and the after-tax MARR is 6% per year.

SOLUTION

The calculations using Equation (9-4) are shown in Table 9-5. The expected year-by-year market values and annual expenses are repeated from Example 9-4 in columns 2 and 5, respectively. In column 6, the *sum* of the loss in MV during year k, cost of capital based on the MV at the beginning-of-year (BOY) k, and annual expenses in year k are multiplied by $(1-t)$ to determine an *approximate* after-tax total marginal cost in year k.

---

*In Chapter 6, the GDS (general depreciation system) and ADS (alternative depreciation system) are discussed.

TABLE 9-5 Determination of the After-Tax Economic Life for the Asset Described in Example 9-4

(1) End of Year, k	(2) MV, End of Year k	(3) Loss in MV during Year k	(4) Cost of Capital = 6% of BOY MV in Col. 2	(5) Annual Expenses	(6) Approximate After-Tax Total (Marginal) Cost for Year k $(1-t) \cdot (\text{Col. } 3+4+5)$
0	$20,000	0	0	0	0
1	15,000	$5,000	$1,200	$2,000	$4,920
2	11,250	3,750	900	3,000	4,590
3	8,500	2,750	675	4,620	4,827
4	6,500	2,000	510	8,000	6,306
5	4,750	1,750	390	12,000	8,484

(1) End of Year, k	(7) MACRS BV at End of Year k	(8) Interest on Tax Adjustment = $6\% \cdot t \cdot BOY\ BV$ in Col. 7	(9) Adjusted After-Tax Total (Marginal) Cost (TC_k) (Col. 6 + Col. 8)	(10) EUAC[a] (After Tax) through Year k	
0	$20,000	0	0	0	
1	13,334	$480	$5,400	$5,400	
2	4,444	320	4,910	5,162	
3	1,482	107	4,934	5,090	$N^*_{AT} = 3$
4	0	36	6,342	5,377	
5	0	0	8,484	5,928	

[a] $\text{EUAC}_k = [\sum_{j=1}^{k}(\text{Col.9})_j \cdot (P/F, 6\%, j)](A/P, 6\%, k)$

The BV amounts at the end of each year, based on the new forklift truck being a MACRS (GDS) three-year property class asset, are shown in column 7. These amounts are then used in column 8 to determine an annual tax adjustment [last term in Equation (9-4)], based on the BOY book values (BV_{k-1}). This annual tax adjustment is algebraically added to the entry in column 6 to obtain an *adjusted* after-tax total marginal cost in year k, TC_k. The total marginal cost amounts are used in column 10 to calculate, successively, the equivalent uniform annual cost, $EUAC_k$, of retirement of the asset at the end of year k. In this case, the after-tax economic life (N^*_{AT}) is three years, the same result obtain on a before-tax basis in Example 9-4.

It is not uncommon for the before-tax and after-tax economic lives of an asset to be the same (as occurred in Examples 9-4 and 9-8).

9.9.2 After-Tax Investment Value of the Defender

The outsider viewpoint has been used in this chapter to establish a before-tax investment value of an existing asset. Using this viewpoint, the present realizable MV of the defender is the appropriate before-tax investment value. This value (although not an actual cash flow) represents the opportunity cost of keeping the defender. In determining the after-tax investment value, we must also include the opportunity cost of gains (losses) not realized if the defender is kept.

Consider, for example, a printing machine that was purchased three years ago for $30,000. It has a present market value of $5,000 and a current book value of $8,640. If the printing machine were sold now, the company would experience a loss on disposal of $5,000 - $8,640 = -$3,640. Assuming a 40% effective income tax rate, this loss would translate into a $(-0.40)(-\$3,640) = \$1,456$ tax savings. Thus, if it is decided to keep the printing machine, not only would the company be giving up the opportunity to obtain the $5,000 market value, it would also be giving up the opportunity to obtain the $1,456 tax credit that would result from selling the printing machine at a price less than its current BV. Thus, the total after-tax investment value of the existing printing machine is $5,000 + $1,456 = $6,456.

The computation of the after-tax investment value of an existing asset is quite straightforward. Using the general format for computing after-tax cash flows (ATCF) presented previously in Figure 6-5, we would have the following entries if the defender were sold now (year 0). Note that MV_0 and BV_0 represent the MV and BV, respectively, of the defender at the time of the analysis.

End of Year, k	BTCF	Depreciation	Taxable Income	Cash Flow for Income Taxes	ATCF (if defender is sold)
0	MV_0	None	$MV_0 - BV_0$	$-t(MV_0 - BV_0)$	$MV_0 - t(MV_0 - BV_0)$

Now, if it was decided to keep the asset, the preceding entries become the opportunity costs associated with keeping the defender. The appropriate year 0 entries for analyzing the after-tax consequences of keeping the defender are shown in Figure 9-4. Note that the entries in Figure 9-4 are simply the same values shown above only reversed in sign to account for the change in perspective (keep versus sell).

	(A)	(B)	(C)	$(D) = -t(C)$	$(E) = (A) + (D)$
End of Year, k	BTCF	Depreciation	Taxable Income	Cash Flow for Income Taxes	ATCF (if defender is kept)
0	$-MV_0$	None	$-(MV_0 - BV_0)$	$-t[-(MV_0 - BV_0)]$ $= t(MV_0 - BV_0)$	$-MV_0 + t(MV_0 - BV_0)$

Figure 9-4 **General Procedure for Computing the After-Tax Investment Value of a Defender**

EXAMPLE 9-9

An existing asset being considered for replacement has a current market value of $12,000 and a current book value of $18,000. Determine the after-tax investment value of the existing asset (if kept) using the outsider viewpoint and an effective income tax rate of 34%.

SOLUTION

Given that $MV_0 = \$12,000$, $BV_0 = \$18,000$ and $t = 0.34$, we can easily compute the ATCF associated with keeping the existing asset by using the format of Figure 9-4:

End of Year, k	BTCF	Depreciation	Taxable Income	Cash Flow for Income Taxes	ATCF
0	−$12,000	None	−($12,000 − $18,000) = $6,000	(−0.34)($6,000) = −$2,040	−$12,000 − $2,040 = −$14,040

The appropriate after-tax investment value for the existing asset is $14,040. Note that this is higher than the before-tax investment value of $12,000, due to the tax credit given up by *not* selling the existing machine at a loss.

EXAMPLE 9-10

An engineering consulting firm is considering the replacement of its CAD workstation. The workstation was purchased four years ago for $20,000. Depreciation deductions have followed the MACRS (GDS) five-year property class schedule. The workstation can be sold now for $4,000. Assuming the effective income tax rate is 40%, compute the after-tax investment value of the CAD workstation if it is kept.

SOLUTION

To compute the ATCF associated with keeping the defender, we must first compute the current BV, BV_0. The workstation has been depreciated for four years under the MACRS (GDS) system with a five-year property class. Thus,

$$BV_0 = \$20,000(1 - 0.2 - 0.32 - 0.192 - 0.1152) = \$3,456.^*$$

Using the format presented in Figure 9-4, we find that the ATCF associated with keeping the defender can be computed as follows:

---

*Current tax law dictates that gains and losses be taxed as ordinary income. As a result, it is not necessary to explicitly account for the MACRS half-year convention when computing the "if sold" BV (the increase in taxable income due to a higher BV is offset by the half-year of depreciation that could be claimed if the defender is kept). This allows us to simplify the procedure for computing the after-tax investment value of the defender.

End of Year, k	BTCF	Depreciation	Taxable Income	Cash Flow for Income Taxes	ATCF
0	−$4,000	None	−($4,000 − $3,456) = −$544	(−0.4)(−$544) = $218	−$4,000 + $218 = −$3,782

The after-tax investment value of keeping the existing CAD workstation is $3,782. Note that in the case where MV_0 is higher than BV_0, the after-tax investment value is lower than the before-tax investment value. This is because the gain on disposal (and resulting tax liability) does not occur at this time if the defender is retained.

9.9.3 Illustrative After-Tax Replacement Analyses

The following examples represent typical after-tax replacement analyses. They illustrate the appropriate method for including the effect of income taxes, as well as several of the factors that must be considered in general replacement studies.

EXAMPLE 9-11 **(Restatement of Example 9-3 with Tax Information)**

The manager of a carpet manufacturing plant became concerned about the operation of a critical pump in one of the processes. After discussing this situation with the supervisor of plant engineering, they decided that a replacement study should be done and that a nine-year study period would be appropriate for this situation. The company that owns the plant is using an after-tax MARR of 6% per year for its capital investment projects. The effective income tax rate is 40%.

The existing pump, Pump A, including driving motor with integrated controls, cost $17,000 five years ago. The accounting records show the depreciation schedule to be following that of an asset with a MACRS (ADS) recovery period of nine years. Some reliability problems have been experienced with Pump A, including annual replacement of the impeller and bearings at a cost of $1,750. Annual expenses have been averaging $3,250. Annual insurance and property tax expenses are 2% of the initial capital investment. It appears that the pump will provide adequate service for another nine years if the present maintenance and repair practice is continued. An estimated market value of $750 could be obtained for the pump if it is sold now. It is estimated that if this pump is continued in service, its final MV after nine more years will be about $200.

An alternative to keeping the existing pump in service is to sell it immediately and to purchase a replacement pump, Pump B, for $16,000. A nine-year class life (MACRS five-year property class) would be applicable to the new pump under the GDS. An estimated market value at the end of the nine-year study period would be 20% of the initial capital investment. O&M expenses for the new pump are estimated to be $3,000 per year. Annual taxes and insurance would total 2% of the initial capital investment. The data for Example 9-11 are summarized in Table 9-6.

TABLE 9-6 Summary of Information for Example 9-11

MARR (after taxes) = 6% per year
Effective income tax rate = 40%

Existing Pump A (defender)	
MACRS (ADS) recovery period	9 years
Capital investment when purchased five years ago	$17,000
Total annual expenses	$5,340
Present market value	$750
Estimated market value at the end of nine additional years	$200
Replacement Pump B (challenger)	
MACRS (GDS) property class	5 years
Capital investment	$16,000
Total annual expenses	$3,320
Estimated market value at the end of nine years	$3,200

Based on these data, should the defender (Pump A) be kept [and the challenger (Pump B) not purchased], or should the challenger be purchased now (and the defender sold)? Use an after-tax analysis and the outsider viewpoint in the evaluation.

SOLUTION

The after-tax computations for keeping the defender (Pump A) and not purchasing the challenger (Pump B) are shown in Table 9-7. *Year 0* of the *analysis period* is at the *end of the current (fifth) year of service* of the defender. The year-0 entries of Table 9-7 are computed using the general format presented in Figure 9-4 and are further explained in the following:

TABLE 9-7 ATCF Computations for the Defender (Existing Pump A) in Example 9-11

End of Year, k	(A) BTCF[a]	(B) MACRS (ADS) Depreciation	$(C) = (A) - (B)$ Taxable Income	$(D) = -0.4(C)$ Income Taxes at 40%	$(E) = (A) + (D)$ ATCF
0	−$750	None	$7,750	−$3,100	−$3,850
1–4	−5,340	$1,889	−7,229	2,892	−2,448
5	−5,340	944	−6,284	2,514	−2,826
6–9	−5,340	0	−5,340	2,136	−3,204
9	200		200[b]	−80	120

[a] Before-Tax Cash Flow (BCTF).

[b] Gain on disposal (taxable at the 40% rate).

1. BTCF ($-\$750$): The same amount used in the before-tax analysis of Example 9-3. This amount is based on the outsider viewpoint and is the opportunity cost of keeping the defender instead of replacing it (and selling it for the estimated present MV of $750).

2. Taxable income ($7,750): This amount is the result of an increase in taxable income of $7,750 due to the tax consequences of keeping the defender instead of selling it. Specifically, *if we sold the defender now*, the loss on disposal would be as follows:

$$\text{Gain or loss on disposal (if sold now)} = MV_0 - BV_0;$$

$$BV_0 = \$17,000[1 - 0.0556 - 4(0.1111)] = \$8,500;$$

$$\text{Loss on disposal (if sold now)} = \$750 - \$8,500 = -\$7,750.$$

But *since we are keeping the defender (Pump A) in this alternative,* we have the reverse effect on taxable income, an increase of $7,750 due to an opportunity foregone.

3. Cash flow for income taxes ($-\$3,100$): The increase in taxable income due to the tax consequences of keeping the defender results in an increased tax liability (or tax credit foregone) of $-0.4(\$7,750) = -\$3,100$.

4. ATCF ($-\$3,850$): The total after-tax investment value of the defender is the result of two factors: the present MV ($750) and the tax credit ($3,100) foregone by keeping the existing Pump A. Therefore, the ATCF representing the investment in the defender (based on the outsider viewpoint) is $-\$750 - \$3,100 = -\$3,850$.

The remainder of the ATCF computations over the nine-year analysis period for the alternative of keeping the defender are shown in Table 9-7. The after-tax computations for the alternative of purchasing the challenger (Pump B) are shown in Table 9-8.

The next step in an after-tax replacement study involves equivalence calculations using an after-tax MARR. The following is the after-tax EUAC analysis for Example 9-11:

$$\text{EUAC}(6\%) \text{ of Pump } A \text{ (defender)} = \$3,850(A/P, 6\%, 9)$$

$$+ \$2,448(P/A, 6\%, 4)(A/P, 6\%, 9)$$

$$+ [\$2,826(F/P, 6\%, 4)$$

$$+ \$3,204(F/A, 6\%, 4) - \$120](A/F, 6\%, 9)$$

$$= \$3,332;$$

TABLE 9-8 ATCF Computations for the Challenger (Replacement Pump B) in Example 9-11

End of Year, k	(A) BTCF	(B) MACRS (GDS) Depreciation	$(C) = (A) - (B)$ Taxable Income	$(D) = -0.4(C)$ Income Taxes at 40%	$(E) = (A) + (D)$ ATCF
0	−$16,000	None			−$16,000
1	−3,320	$3,200	−$6,520	$2,608	−712
2	−3,320	5,120	−8,440	3,376	56
3	−3,320	3,072	−6,392	2,557	−763
4	−3,320	1,843	−5,163	2,065	−1,255
5	−3,320	1,843	−5,163	2,065	−1,255
6	−3,320	922	−4,242	1,697	−1,623
7-9	−3,320	0	−3,320	1,328	−1,992
9	3,200		3,200[a]	−1,280	1,920

[a] Gain on disposal (taxable at the 40% rate).

$$\text{EUAC}(6\%) \text{ of Pump } B \text{ (challenger)} = \$16,000(A/P, 6\%, 9)$$
$$+ [\$712(P/F, 6\%, 1) - \$56(P/F, 6\%, 2)$$
$$+ \$763(P/F, 6\%, 3)$$
$$+ \cdots + \$1,992(P/F, 6\%, 9)](A/P, 6\%, 9)$$
$$- \$1,920(A/F, 6\%, 9)$$
$$= \$3,375.$$

Because the EUACs of both pumps are very close, other considerations, such as the improved reliability of the new pump, could detract from the slight economic preference for Pump A. The after-tax annual costs of both alternatives are considerably less than their before-tax annual costs.

The after-tax analysis does not *reverse* the results of the before-tax analysis for this problem (see Example 9-3). However, due to income tax considerations, identical before-tax and after-tax recommendations should not necessarily be expected.

The next example involves the determination of the economic life of a defender on an after-tax basis and the use of after-tax marginal costs to determine the most economical time to replace the defender.

EXAMPLE 9-12

The Hokie Metal Stamping Company is considering the replacement of a spray system. The new improved system will cost $60,000 installed and will have an estimated economic life of 12 yr. The market value of the new system at the end of 12 yr is expected to be $6,000. Further, it is estimated that annual operating and maintenance expenses will average $32,000 per year for the new system and that straight-line depreciation (with a $6,000 terminal market value) will be used.

The present system has a remaining useful life of three years. It has a current book value of $12,000 and a present realizable market value of $8,000. The estimated operating expenses, market values, and book values of the present system for the next three years are as follows:

Year	Market Value at End of Year	Book Value at End of Year	Operating Expenses during Year
1	$6,000	$9,000	$40,000
2	5,000	6,000	50,000
3	4,000	3,000	60,000

A spray system will be needed for as long as the company remains in business (which it hopes is a long, long time). Perform an after-tax analysis to determine the most economical period to keep the defender *before* replacing it with the new system. The after-tax MARR is 15% per year and the effective income tax rate is 50%.

SOLUTION

This analysis begins with the determination of the after-tax economic life of the present system (the economic life of the challenger was given to be 12 yr). Table 9-9 shows the calculations of the year-by-year after-tax marginal costs [Equation (9-4)] of the defender and the corresponding EUAC. It can be seen in column 10 that the economic life of the defender is one year.

Table 9-10 contains the ATCF calculations for the challenger. The ATCFs are used to compute the after-tax EUAC of the challenger as follows:

$$EUAC = \$60,000(A/P, 15\%, 12) + \$13,750 - \$6,000(A/F, 15\%, 12) = \$24,613.$$

Comparing just the EUACs of the defender with the EUAC of the challenger, you may be tempted to conclude that the old system should be kept at least one more year, perhaps even two more years. However, in this situation you should examine marginal costs. The valid economic criterion when operating expenses are increasing over time is to keep the old system as long as the marginal cost of an additional year of service is less than the equivalent uniform annual cost of the new system. The marginal cost of keeping the old system for the first year is $22,500. This $22,500 is less than the $24,613 EUAC of the new system, thus justifying keeping the old system for the first year. The marginal cost of keeping the old system for the second year is $26,625. The $26,625 is greater than the $24,613 average annual cost of the new system, thus indicating that the old system should not be kept the second year, but rather that it be replaced at the end of the first year.

TABLE 9-9 Determination of the After-Tax Economic Life for the Defender Described in Example 9-12

(1) End of Year, k	(2) MV, End of Year k	(3) Loss in MV during Year k	(4) Cost of Capital = 15% of BOY MV in Col. 2	(5) Annual Expenses	(6) Approximate After-Tax Total (Marginal) Cost for Year k $(1 - t) \cdot (\text{Col. } 3 + 4 + 5)$
0	$8,000	0	0	0	0
1	6,000	$2,000	$1,200	$40,000	$21,600
2	5,000	1,000	900	50,000	25,950
3	4,000	1,000	750	60,000	30,875

TABLE 9-9 (cont'd.)

(7) End of Year, k	BV at End of Year k	(8) Interest on Tax Adjustment = $15\% \cdot t \cdot$ BOY BV In Col. 7	(9) Adjusted After-Tax Total (Marginal) Cost (TC_k) (Col. 6 + Col. 8)	(10) EUAC[a] (After-Tax) through Year k	
0	$12,000	0	0	0	
1	9,000	$900	$22,500	$22,500	$N_{AT}^* = 1$
2	6,000	675	26,625	24,418	
3	3,000	450	31,325	26,408	

[a] $\text{EUAC}_k = [\sum_{j=1}^{k} (\text{Col. } 9)_j \cdot (P/F, 15\%, j)](A/P, 15\%, k)$

TABLE 9-10 ATCF Computations for the Challenger in Example 9-12

End of Year, k	(A) BTCF	(B) Straight-line Depreciation	(C) = (A) − (B) Taxable Income	(D) = −0.4(C) Income Taxes at 40%	(E) = (A) + (D) ATCF
0	−$60,000	None			−$60,000
1–12	−32,000	$4,500[a]	−$36,500	$18,250	−13,750
12	6,000		0[b]	0	6,000

[a] Straight-line depreciation amount = ($60,000 − $6,000)/12 = $4,500.

[b] BV_{12} = $60,000 − (12)($4,500) = $6,000; $MV_{12} - BV_{12}$ = 0.

9.10 A Comprehensive Example

Sometimes in engineering practice, a replacement analysis involves an existing asset that cannot meet future service requirements without *augmentation* of its capabilities. When this is the case, the defender with increased capability should be competitive with the best available challenger. The analysis of this situation is included in the following comprehensive example.

EXAMPLE 9-13

The emergency electrical supply system of a hospital owned by a medical service corporation is presently supported by an 80-kW diesel-powered electrical generator that was put into service five years ago [capital investment = $210,000; MACRS (GDS) seven-year property class]. An engineering firm is designing modifications to the electrical and mechanical systems of the hospital as part of an expansion project. The redesigned emergency electrical supply system will require 120-kW of generating capacity to serve the increased demand. Two preliminary designs for the system are being considered. The first involves the augmentation of the existing 80-kW generator with a new 40-kW diesel-powered unit (GDS seven-year property class). This alternative represents the augmented defender. The second design includes replacement of the existing generator with the best available alternative, a new turbine-powered unit with 120-kW of generating capacity (the challenger). Both alternatives will provide the same level of service to the operation of the emergency electrical supply system.

The challenger, if selected, will be leased by the hospital for a ten-year period. At that time, the lease contract would be renegotiated either for the original piece of equipment or for a replacement generator with the same capacity. The following additional estimates have been generated for use in the replacement analysis.

| | Alternative | | |
| | Defender | | |
	80-kW	40-kW	Challenger
Capital investment	$90,000[a]	$140,000	$10,000[b]
Annual lease amount	0	0	$39,200
Operating hours / year	260	260	260
Annual expenses (year zero $):			
Operating and maintenance (O&M) expense per hour	$80	$35	$85
Other expenses	$3,200	$1,000	$2,400
Useful life	10 years	15 years	15 years

[a] Opportunity cost based on present market value of the defender (outsider viewpoint).

[b] Deposit required by the terms of the contract to lease the challenger. It is refundable at the end of the study period.

The annual lease amount for the challenger will not change over the ten-year contract period. The operating and maintenance (O&M) expense per hour of operation and the other annual expense amounts for both alternatives are estimated in year zero dollars, and are expected to escalate at the rate of 4% per year. (Assume that base year, b, is year 0; see Chapter 8 for dealing with price changes).

The present estimated market value of the 80-kW generator is $90,000 and its estimated market value at the end of an additional 10 years, in year-zero dollars, is $30,000. The estimated market value of the new 40-kW generator, 10 years from now in year-zero dollars, is $38,000. Both future market values are estimated to escalate at the rate of 2% per year.

The corporation's after-tax, market-based MARR (i_c) is 12% per year, and its effective income tax rate is 40%. A 10-year planning horizon (study period) is considered appropriate for this decision situation (note that with income tax considerations and price changes in the analysis, a study period based on the coterminated assumption is being used).

Based on an after-tax, actual dollar analysis, which alternative (augmentation of the defender or lease of the challenger) should be selected as part of the design of the modified emergency electrical power system?

SOLUTION

The after-tax analysis of the first alternative (defender), keeping the existing 80-kW generator and augmenting its capacity with a new 40-kW generator is shown in Table 9-11. The initial $230,000 before-tax capital investment amount is the sum of (1) the present $90,000 market value of the existing 80-kW generator, which is

TABLE 9-11 Defender Augmented with a New 40-kW Generator (Example 9-13)

End of Year, k	BTCF	Depreciation 80-kW	Depreciation 40-kW	Taxable Income	Cash Flow for Income Taxes	ATCF
0	−$230,000	None		−$43,149[c]	$17,260	−$212,740
1	−35,464	$18,732	$20,006	−74,202	29,681	−5,783
2	−36,883[a]	18,753	34,286	−89,922	35,969	−914
3	−38,358	9,366	24,486	−72,210	28,884	−9,474
4	−39,892		17,486	−57,378	22,951	−16,941
5	−41,488		12,502	−53,990	21,596	−19,892
6	−43,147		12,488	−55,635	22,254	−20,893
7	−44,873		12,502	−57,375	22,950	−21,923
8	−46,668		6,244	−52,912	21,165	−25,503
9	−48,535			−48,535	19,414	−29,121
10	−50,476			−50,476	20,190	−30,286
10	82,892[b]			82,892	−33,157	49,735

[a] $-[260(\$80 + \$35) + (\$3,200 + \$1,000)](1.04)^2 = -\$36,883$

[b] $MV_{10} = (\$30,000 + \$38,000)(1.02)^{10} = \$82,892$

[c] If the defender was sold now, gain on disposal = $90,000 − \$46,851 = \$43,149$; where $BV_0 = \$46,851$.

an opportunity cost based on an outsider viewpoint and (2) the $140,000 capital investment for the new 40-kW generator. The −$43,149 of taxable income at year zero is due to the gain on disposal, *which is not incurred* when the 80-kW generator is kept instead of sold.

The after-tax PW of keeping the defender and augmenting its capacity is

$$PW_D(12\%) = -\$212,740 - \$5,783(P/F, 12\%, 1) - \cdots$$

$$+ (\$49,735 - \$30,286)(P/F, 12\%, 10)$$

$$= -\$282,468.$$

Under the contract terms for leasing the challenger, there is an initial $10,000 deposit, which is fully refundable at the end of the 10-year period. There are no tax consequences associated with the deposit transaction. The annual before-tax cash flow (BTCF) for the challenger is the sum of (1) the annual lease amount, which stays constant over the 10-yr contract period, and (2) the annual O&M and other expenses, which escalate at the rate of 4% per year. For example, the BTCF for the challenger in year one is −$39,200 − [$85(260) + $2,400](1.04) = −$64,680. These annual BTCF amounts for years one through ten are fully deductible from taxable income by the corporation, and they are also the taxable income amounts for the alternative (the corporation cannot claim any depreciation on the challenger because it does not own the equipment). Hence, the after-tax PW of selecting the challenger, assuming that it is leased under these contract terms, is

$$PW_C(12\%) = -\$10,000 + \$10,000(P/F, 12\%, 10)$$

$$- (1 - 0.4)(\$39,200)(P/A, 12\%, 10)$$

$$- (1 - 0.4)[\$85(260) + \$2,400](P/A, i_{CR} = 7.69\%, 10)$$

$$= -\$239,705,$$

where $i_{CR} = (0.12 - 0.04)/(1.04) = 0.0769$, and $(P/A, 7.69\%, 10) = 6.8049$.

Based on an after-tax analysis, the challenger is economically preferable for use in the emergency electrical supply system because its PW has the least negative value.

9.11 Spreadsheet Applications

A vital ingredient of many replacement studies is an asset's economic life. The next example provides a general spreadsheet model that can be used to determine the economic life of an asset given the initial capital investment, year-by-year market values, and the annual operating expenses. This spreadsheet can also be used to determine the best time to abandon a project.

	A	B	C	D	E	F	G	H
1	MARR	15%						
2								
3					Net	Total (Marginal)		
4			Loss in MV		Cash Flow	Cash Flow	Equivalent	
5	End of	MV at End	During	Cost of	for Year	for Year	Annual Worth	
6	Year	of Year	Year k	Capital	(R-E)	(R-E-CR)	Through Year	
7	0	$ 15,000						
8	1	$ 12,000	$ 3,000	$ 2,250	$ (1,000)	$ (6,250)	($6,250)	
9	2	$ 10,000	$ 2,000	$ 1,800	$ (1,000)	$ (4,900)	($5,622)	Economic Life
10	3	$ 7,000	$ 3,000	$ 1,500	$ (1,300)	$ (5,800)	($5,673)	
11	4	$ 3,000	$ 4,000	$ 1,050	$ (2,000)	$ (7,050)	($5,949)	
12	5	$ 500	$ 2,500	$ 450	$ (2,500)	$ (5,450)	($5,875)	

Figure 9-5 Spreadsheet for Determining Economic Life in Example 9-14.

EXAMPLE 9-14

The estimated year-by-year market values and operating expenses for a replacement piece of equipment are shown in Figure 9-5 (column B and column E, respectively). The market values are used to compute year-by-year loss in value (column C) and the cost of capital (column D). The resulting capital recovery amount is combined with the expenses for the year (shown as a net cash flow in column E) to determine the total marginal cost for the year (column F). Column G calculates the equivalent annual worth of the cash flows in column F successively through each year. Column H contains an IF( ) function that places the label "Economic Life" next to the maximum equivalent annual worth (which corresponds to the minimum equivalent uniform annual cost) found in Column G. The formulas for the highlighted cells in Figure 9-5 are given in the following table:

Cell	Contents
C10	$= B9 - B10$
D10	$= B9 * \$B\$1$
E10	User input of net cash flow for year
F10	$= E10 - (C10 + D10)$
G10	$= -PMT(\$B\$1, A10, NPV(\$B\$1, F\$8 : F10))$
H10	$= IF(G10 = MAX(G\$8: G\$12), \text{"Economic Life, "})$

▼ 9.12 Summary

In summary, there are several important points to keep in mind when conducting a replacement or retirement study:

The MV of the defender must *not* also be deducted from the purchase price of the challenger *when using the outsider viewpoint* to analyze a replacement problem. This error double-counts the defender's MV and biases the comparison toward the challenger.

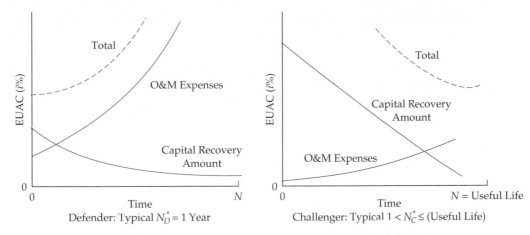

Figure 9-6 Typical Pattern of the EUAC for a Defender and a Challenger.

A sunk cost (i.e., $MV - BV < 0$) associated with keeping the defender must *not* be added to the purchase price of the best available challenger. This error results in an incorrect penalty that biases the analysis in favor of retaining the defender.

In Section 9.6, we observed that the economic life of the defender is often one year, which is generally true if annual expenses are high relative to the defender's investment cost when using the outsider viewpoint. Hence, the marginal cost of the defender should be compared with the EUAC at the *economic life* of the challenger to answer the fundamental question "Should the defender be kept for one or more years or disposed of now?" The typical pattern of the EUAC for a defender and a challenger is illustrated in Figure 9-6.

The income tax effects of replacement decisions should not be ignored. The foregone income tax credits associated with keeping the defender may swing the economic preference away from the defender, thus making the challenger the better choice.

The *best available* challenger(s) must be determined. Failure to do so represents unacceptable engineering practice.

Any excess capacity, reliability, flexibility, safety, and so on of the challenger may have value to the owner and should be claimed as a dollar benefit if a dollar estimate can be placed on it. Otherwise, this value would be treated as a *nonmonetary* benefit.

9.13 References

BARISH, N. N., and S. KAPLAN. *Economic Analysis for Engineering and Managerial Decision Making* (New York: McGraw-Hill Book Co., 1978).

BEAN, J. C., J. R. LOHMANN, and R. L. SMITH. "A Dynamic Infinite Horizon Replacement Economy Decision Model," *The Engineering Economist*, vol. 30, no. 2, 1985, pp. 99–120.

BERNHARD, R. H. "Improving the Economic Logic Underlying Replacement Age Decisions for Municipal Garbage Trucks: Case Study," *The Engineering Economist*, vol. 35, no. 2, Winter 1990, pp. 129–147.

Hartman, J. C. "A General Procedure for Incorporating Asset Utilization Decisions into Replacement Analysis," *The Engineering Economist*, vol. 44, no. 3, 1999, pp. 217–238.

Lake, D. H., and A. P. Muhlemann. "An Equipment Replacement Problem," *Journal of the Operational Research Society*, vol. 30, no. 5, 1979, pp. 405–411.

Leung, L. C., and J. M. A. Tanchoco. "Multiple Machine Replacement within an Integrated Systems Framework," *The Engineering Economist*, vol. 32, no. 2, 1987, pp. 89–114.

Matsuo, H. "A Modified Approach to the Replacement of an Existing Asset," *The Engineering Economist*, vol. 33, no. 2, Winter 1988, pp. 109–120.

Morris, W. T. *Engineering Economic Analysis* (Reston, VA: Publishing Co., 1976).

Naik, M. D., and K. P. Nair. "Multistage Replacement Strategies," *Journal of the Operations Research Society of America*, vol. 13, no. 2, March–April 1965, pp. 279–290.

Oakford, R. V., J. R. Lohmann, and A. Salazar. "A Dynamic Replacement Economy Decision Model," *IIE Transactions*, vol. 16, no. 1, 1984, pp. 65–72.

Park, C. S., and G. P. Sharp-Bette. *Advanced Engineering Economics* (New York: John Wiley & Sons, 1990).

9.14 Problems

The number in parentheses ( ) that follows each problem refers to the section from which the problem is taken.

9-1. An industrial lift truck has been in service for several years and management is contemplating replacing it. A planning horizon of five years is to be used in the replacement study. The old lift truck (defender) has a current market value of $1,500. If the defender is retained, it is anticipated to have annual operating and maintenance costs of $7,300. It will have a zero market value at the end of five additional years of service. The new lift truck (challenger) will cost $10,000 and will have operating and maintenance costs of $5,100. It will have a market value of $2,500 at the end of the planning horizon. Determine the preferred alternative using a present worth comparison and a minimum attractive rate of return (before taxes) of 20% per year. (9.4)

9-2. Suppose that you have an old car, which is a real gas-guzzler. It is 10 years old and could be sold to a local dealer for $400 cash. Assume that its MV two years from now is zero. The annual maintenance expenses will average $800 into the foreseeable future, and the car averages only 10 miles per gallon. Gasoline costs $1.50 per gal-

lon, and you average 15,000 miles per year. You now have an opportunity to replace the old car with a better one that costs $8,000. If you buy it, you will pay cash. Because of a two-year warranty, the maintenance expenses are expected to be negligible. This car averages 30 miles per gallon. Use the IRR method to determine which alternative you should select. Utilize a two-year analysis period and assume that the new car can be sold for $5,000 at the end of year two. Let your MARR be 15% per year. State any other assumptions you make. (9.4)

9-3. The Ajax Corporation has an overhead crane that has an estimated remaining life of 10 yr. The crane can be sold now for $8,000. If the crane is kept in service, it must be overhauled immediately at a cost of $4,000. Operating and maintenance costs will be $3,000 per year after the crane is overhauled. The overhauled crane will have zero market value at the end of the 10-year study period. A new crane will cost $18,000, will last for 10 years, and will have a $4,000 market value at that time. Operating and maintenance costs are $1,000 per year for the new crane. The company uses a before-tax interest rate of 10% per year in evaluating investment alternatives. Should the company replace the old crane? (9.4)

9-4.

a. Find the economic life of an asset having the following projected cash flows:

Capital investment =$5,000

MV =$0(at all times)

Annual expenses =$3,000 (EOY 1),

$4,000 (EOY 2),

$5,000 (EOY 3),

and $6,000 (EOY 4).

The MARR is 0% per year. (9.5)

b. Find the economic life of another asset having these cash flow estimates:

Capital investment =$10,000

MV =$10,000 (at all times)

Annual expenses =$3,000 (EOY 1),

$4,000 (EOY 2),

$5,000 (EOY 3),

and $6,000 (EOY 4).

The MARR is 12% per year. (9.5)

c. Repeat part (b), except that MV = $0 at all times. (9.5)

9-5. Robert Roe has just purchased a four-year-old used car, paying $3,000 for it. A friend has suggested that he determine in advance how long he should keep the car so as to ensure the greatest overall economy. Robert has decided that, because of style changes, he would not want to keep the car longer than four years, and he has estimated the annual expenses and market values for years 1 through 4 as follows:

	Year 1	Year 2	Year 3	Year 4
Annual expenses	$950	$1,050	$1,100	$1,550
Market value at end of year	2,250	1,800	1,450	1,160

If Robert's capital is worth 12% per year, at the end of which year should he dispose of the car? (9.5)

9-6. A present asset (defender) has a current market value of $87,000 ($MV_0$). Based on the used equipment market, the estimated market values at the end of the next three years are $MV_1 = \$76,000$, $MV_2 = \$60,000$, $MV_3 = \$40,000$. The annual expenses are $18,000 in present (year-0) dollars, and these expenses are estimated to increase at 4.1% per year. The before-tax MARR is 10% per year. The best challenger available has an economic life of six years, and its EUAC over this period is $44,210. Based on this information and a before-tax analysis, when should you plan to replace the defender with the challenger? (9.6, 9.7)

9-7. The replacement of a planing machine is being considered by the Reardorn Furniture Company. (There is an indefinite future need for this type of machine.) The best challenger will cost $30,000 installed, and will have an estimated economic life of 12 yr and a $2,000 MV at that time. It is estimated that annual expenses will average $16,000 per year. The defender has a present BV of $6,000 and a present MV of $4,000. Data for the defender for the next three years are as follows:

Year	MV at End of Year	BV at End of Year	Expenses During The Year
1	$3,000	$4,500	$20,000
2	2,500	3,000	25,000
3	2,000	1,500	30,000

a. Using a before-tax interest rate of 15% per year, make a comparison to determine whether it is economical to make the replacement now.

b. If the annual expenses for the present machine had been estimated to be $15,000, $18,000, and $23,000 in years one, two, and three, respectively, what replacement strategy should be recommended? (9.6, 9.7)

9-8. A construction firm currently owns a heavy-duty tractor that has a present market value (MV) of $80,000. Estimates of the tractor's operating and maintenance (O&M) expenses and MV at the end of each of the remaining six years of useful life are in Table P9-8a:

TABLE P9-8a Tractor Operating and Maintenance Expenses for Problem P9-8

	End of year k					
	1	2	3	4	5	6
O&M expenses	$20,000	$25,000	$38,000	$45,000	$47,000	$50,000
Market value	70,000	60,000	50,000	40,000	30,000	20,000

TABLE P9-8b New Purchase Price and O&M and MV for Problem P9-8

	End of year k					
	1	2	3	4	5	6
O&M expenses	$10,000	$12,000	$16,000	$17,000	$20,000	$25,000
Market value	180,000	150,000	120,000	100,000	90,000	75,000

The firm is considering a new heavy-duty tractor to replace the one presently owned. The new tractor's purchase price is $220,000 and its estimated O & M and MV for each of the next six years of the study period are in Table P9-8b.

If the MARR = 0% per year, should the new tractor be purchased? If so, when? (9.5, 9.6)

9-9. An existing robot is used in a commercial material laboratory to handle ceramic samples in the high-temperature environment that is part of several test procedures. Due to changing customer needs, the robot will not meet future service requirements unless it is upgraded at a cost of $2,000. Because of this situation, a new advanced technology robot has been selected for potential replacement of the existing robot. The accompanying estimates have been developed from information provided by some current users of the new robot and data obtained from the manufacturer. The firm's before-tax MARR is 25% per year. Based on this information, should the existing robot be replaced? Assume that a robot will be needed for an indefinite period. (9.4, 9.7)

Defender	
Current market value	$38,200
Upgrade cost (year 0)	2,000
Annual expenses	$1,400 in year one, and increasing at the rate of 8%/yr thereafter.
Useful life (years)	6
Market value at end of useful life	−$1,500

Challenger	
Purchase price	$51,000
Installation cost	5,500
Annual expenses	$1,000 in year one, and increasing by $150/yr thereafter.
Useful life (years)	10
Market value at end of useful life	$7,000

9-10. A diesel engine (defender) was installed 10 years ago at a cost of $50,000. It has a present realizable MV of $14,000. If kept, it can be expected to last for five more years, have annual expenses of $14,000, and have a market value of $8,000 at the end of the five years. This engine can be replaced with an improved version costing $65,000 and having an expected life of 20 yr. The challenger will have estimated annual expenses of $9,000 and a final market value of $13,000. It is thought that an engine will be needed indefinitely and that the results of the economy study would not be affected by the consideration of income taxes. Using a before-tax MARR of 15% per year, perform an analysis to determine whether to keep or replace the old engine. (9.4, 9.7)

9-11. A steel pedestrian overpass must either be reinforced or replaced. Reinforcement would cost $22,000 and would make the overpass adequate for an additional five years of service. If the overpass is torn down now, the scrap value of the steel would exceed the removal cost by $14,000. If it is reinforced, it is estimated that its net salvage (market) value would be $16,000 at the time it is retired from service. A new pre-stressed concrete overpass would cost $140,000 and would meet the foreseeable requirements of the next 40 yr. Such a design would have no net scrap or market value. It is estimated that the annual expenses of the reinforced overpass would exceed those of the concrete overpass by $3,200. Assume that money costs the state 10% per year and that the state pays no taxes. What would you recommend? (9.4, 9.7)

9-12. A small high-speed commercial centrifuge has the following net cash flows and abandon-ment values over its useful life (see Table P9-12).

The firm's MARR is 10% per year. Deter-mine the optimal time for the centrifuge to be abandoned if it is acquired for $7,500 and not to be used for more than five years. (9.8)

9-13. Consider a piece of equipment that initially cost $8,000 and has these estimated annual ex-penses and market value:

End of Year, k	Annual Expenses	MV at End of Year
1	$3,000	$4,700
2	3,000	3,200
3	3,500	2,200
4	4,000	1,450
5	4,500	950
6	5,250	600
7	6,250	300
8	7,750	0

If the after-tax MARR is 7% per year, determine the after-tax economic life of this equipment. MACRS (GDS) depreciation is being used (five-year property class). The effective income tax rate is 40%. (9.9)

9-14. A current asset (defender) is being evaluated for potential replacement. It was purchased four years ago at a cost of $62,000. It has been depre-ciated as a MACRS (GDS) five-year property-class asset. The present market value of the de-fender is $12,000. Its remaining useful life is es-timated to be four years, but it will require ad-ditional repair work now (a one-time $4,000 ex-pense) to provide continuing service equivalent to the challenger. The current effective income tax rate is 39% and the after-tax MARR = 15% per year. Based on an outsider viewpoint, what is the after-tax initial investment in the defender if it is kept (*not* replaced now)? (9.9)

9-15. The present worth of the after-tax cash flows through year k, PW_k, for a defender (three-year remaining useful life) and a challenger (five-year useful life) are given in the following table:

	PW of ATCF through Year k, PW_k	
Year	Defender	Challenger
1	$-$14,020	$-$18,630
2	$-$28,100	$-$34,575
3	$-$43,075	$-$48,130
4		$-$65,320
5		$-$77,910

Assume the after-tax MARR is 12% per year. Based on this information,

a. What is the economic life and the related minimum equivalent uniform annual cost, EUAC, when $k = N_{AT}^*$, for both the defender and the challenger? (9.5, 9.6)

b. When should the challenger (based on the present analysis) replace the defender? Why? (9.6, 9.7)

c. What assumption(s) have you made in an-swering Part (b)?

TABLE P9-12 Cash Flows and Abandonment Values for Problem P9-12

	End of Year				
	1	2	3	4	5
Annual revenues less expenses	$2,000	$2,000	$2,000	$2,000	$2,000
Abandonment value of machine[a]	$6,200	$5,200	$4,000	$2,200	0

[a] Estimated market value.

9-16. Four years ago, the Attaboy Lawn Mower Company purchased a piece of equipment for its assembly line. Because of increasing maintenance costs for this equipment, a new piece of machinery is being considered. Information about the defender (present equipment) and the challenger are shown in the following table:

Defender	Challenger
Original cost = $9,000	Purchase cost = $13,000
Maintenance = $300 in first year of use four years ago, increasing by 10% per year thereafter	Maintenance = $100 in year one, increasing by 10% per year thereafter
MACRS (ADS) with a nine-year recovery period	MACRS (GDS) five-year property class
MV = 0 five years from now	MV = $3,000 at the end of year five

Assume that a $3,200 MV is available now for the defender. Perform an after-tax analysis using an after-tax MARR of 10% per year and a five-year analysis period to determine which alternative to select. The effective income tax rate is 40%. (9.9)

9-17. It is being decided whether or not to replace an existing piece of equipment with a newer, more productive one that costs $80,000 and has an estimated MV of $20,000 at the end of its useful life of six years. Installation charges for the new equipment will amount to $3,000; this is not added to the capital investment, but will be an expensed item during the first year of operation. MACRS (GDS) depreciation (five-year property class) will be used. The new equipment will reduce direct costs (labor, maintenance, rework, etc.) by $10,000 in the first year, and this amount is expected to increase by $500 each year thereafter during its six-year life. It is also known that the BV of the fully depreciated old machine is $10,000 but that its present fair MV is $14,000. The MV of the old machine will be zero in six years. The effective income tax rate is 40%. (9.9)

a. Determine the prospective after-tax *incremental* cash flow associated with the new equipment if it is believed that the existing machine could perform satisfactorily for six more years.

b. Assume the after-tax MARR is 12% per year. Based on the ERR method, should you replace the defender with the challenger? Assume ε = MARR.

9-18. Ten years ago, a corporation built a facility at a cost of $400,000 in an area that has since developed into a major retail location. At the time the facility was constructed, it was estimated to have a depreciable life of 20 yr with no market value and straight-line depreciation has been used. The corporation now finds it would be more convenient to be in a less congested area and can sell the old facility for $250,000. A new facility in the desired location would cost $500,000 and have a MACRS (GDS) property class of 10 yr. There would be annual savings of $4,000 per year in expenses. Taxes and insurance on the old facility have been 5% of the initial capital investment per year, while for the new facility they are estimated to be only 3% of the initial capital investment per year. The study period is 10 yr and the estimated MV of the new facility at the end of 10 yr is $200,000. The corporation has a 40% income tax rate and capital is worth 12% per year after taxes. What would you recommend on the basis of an after-tax IRR analysis? (9.9)

9-19. Use the PW method to select the better of the following alternatives:

Annual Expenses	Defender: Alternative *A*	Challenger: Alternative *B*
Labor	$300,000	$250,000
Material	250,000	100,000
Insurance and property taxes	4% of initial capital investment	None
Maintenance	$8,000	None
Leasing cost	None	$100,000

Assume that the defender was installed five years ago and that its MACRS (GDS) property class is seven years. The after-tax MARR is 10% per year, and the effective income tax rate is 40%. (9.9, 9.10)

Definition of alternatives:
A: Retain an already owned machine (defender) in service for eight more years.
B: Sell the defender and lease a new one (challenger) for eight years.

Alternative *A* (additional information):
    Cost of defender five years ago = $500,000
    BV now = $111,550
    Estimated market value eight years from now = $50,000
    Present MV = $150,000

9-20. Suppose that it is desired to make an after-tax analysis for the situation posed in Problem 9-10. The defender is being depreciated by the straight-line method over 15 yr; an estimated market value of $8,000 is being used for depreciation purposes. Assume that if the replacement is made, the challenger will be depreciated as a MACRS (GDS) five-year property class asset. Also, assume that the effective income tax rate is 40%. Use the AW method to determine if the replacement is justified by earning an after-tax MARR of 10% per year or more. (9.7, 9.9)

9-21. You have a machine that was purchased four years ago and depreciated under MACRS (ADS) using a five-year recovery period. The original cost was $150,000 and the machine can last for 10 years or more in its current application. A new machine is now available at a cost of only $100,000. It can be depreciated with the MACRS (GDS) method (five-year property class). The annual expenses of the challenger are only $5,000, while those of the defender are $20,000. The challenger has a useful life greater than 10 years. You find that $40,000 is the best price you can get if you sell the present machine now. Your best projection for the future is that you will need the service provided by one of the two machines for the next five years. The MV of the defender is estimated at $2,000 in five years, but that of the challenger is estimated at $5,000 in five years. If the after-tax MARR is 10% per year, should you sell the defender and purchase the challenger? You do not need both. Assume that the company is in the 40% income tax bracket. (9.9)

9-22. Five years ago, an airline installed a baggage conveyor system in a terminal knowing that within a few years it would have to be moved. The original cost of the installation was $120,000 and through accelerated depreciation methods the company has been able to depreciate the entire cost. It now finds that it will cost $40,000 to move and upgrade the conveyor. This capitalized cost would be recovered over the next

six years [the MACRS (ADS) with half-year convention and a five-year recovery period would be used], which the airline believes is a good estimate of the remaining useful life of the system if moved. As an alternative, the airline finds that it can purchase a somewhat more efficient conveyor system for an installed cost of $120,000. The new system would result in an estimated reduction in annual expenses of $6,000 in year-0 dollars. Annual expenses are expected to escalate by 6% per year. The new system is in the five-year MACRS (GDS) property class, and its estimated market value six years from now, in year-0 dollars, is 50% of its installed cost. This MV is estimated to escalate 3% per year. A small airline company, which will occupy the present space, has offered to buy the old conveyor for $90,000.

Annual property taxes and insurance on the present equipment have been $1,500, but it is estimated that they would increase to $1,800 if the equipment is moved and upgraded. For the new system it is estimated that these costs would be about $2,750 per year. All other expenses would be about equal for the two alternatives. The company is in the 40% income tax bracket. It wishes to obtain at least 10% per year, after taxes, on any invested capital. What would you recommend? (9.9, 9.10)

9-23. A manufacturing company has some existing semiautomatic production equipment that it is considering replacing. This equipment has a present MV of $57,000 and a BV of $27,000. It has five more years of depreciation available under MACRS (ADS) of $6,000 per year for four years and $3,000 in year five. (The original recovery period was nine years.) The estimated MV of the equipment five years from now (in year-zero dollars) is $18,500. The MV escalation rate on this type of equipment has been averaging 3.2% per year. The total annual expenses are averaging $27,000 per year.

New automated replacement equipment would then be leased. Estimated annual expenses for the new equipment are $12,200 per year. The *annual* leasing costs would be $24,300. The MARR (after taxes) is 9% per year, $t = 40\%$, and the analysis period is five years. (*Remember*: The owner claims depreciation and the leasing cost is an operating expense.)

Based on an after-tax, actual-dollar analysis, should the new equipment be leased? Base your answer on the IRR of the incremental cash flow. (9.9, 9.10)

9-24. A company is considering replacing a turret lathe (defender) with a single-spindle screw machine (challenger). The turret lathe was purchased four years ago at a cost of $80,000 and depreciation has been based on MACRS (GDS) five-year property class calculations. It can be sold now for $15,000, but if it is retained, it would operate satisfactorily for four more years and have zero market value. The screw machine is estimated to have a useful life of 10 yr. MACRS (GDS) depreciation would be used (five-year property class). It would require only 50% attendance of an operator who earns $12.00 per hour. The machines would have equal capacities and would be operated eight hours per day, 250 days per year. Maintenance on the turret lathe has been $3,000 per year; for the screw machine it is estimated to be $1,500 per year. Taxes and insurance on each machine would be 2% annually of the initial capital investment. If capital is worth 10% per year to the company after taxes, and the company has a 40% income tax rate, what is the maximum price it can afford to pay for the screw machine? Assume a four-year analysis period and an imputed market value (Chapter 5) for the challenger at the end of four years. (9.9)

9-25. *Brain Teaser* There are two customers requiring three-phase electrical service, one existing at location A and a new customer at location B. The load at location A is known to be 110 KVA, and at location B it is contracted to be 280 KVA. Both loads are expected to remain constant indefinitely into the future. Already in service at A are three 100-KVA transformers that were installed some years ago when the load was much greater. Thus, the alternatives are as follows:

> Alternative A: Install three 100-KVA transformers (new) at B now and replace those at A with three 37.5-KVA transformers only when the existing ones must be retired.

> Alternative B: Remove the three 100-KVA transformers now at A and relocate them at B. Then install three 37.5-KVA transformers (new) at A.

Data for both alternatives are provided in Table P9-25. The existing transformers have 10 yr of life remaining. Suppose that the before-tax MARR = 8% per year. Recommend which action to follow after calculating an appropriate criterion for comparing these alternatives. List all assumptions necessary and ignore income taxes. (9.7)

TABLE P9-25 Table for Problem P9-25		
	Existing and New Transformers	
	Three 37.5-KVA	Three 100-KVA
Capital Investment:		
Equipment	$900	$2,100
Installation	$340	$475
Property tax	2% of capital investment	2% of capital investment
Removal cost	$100	$110
Market value	$100	$110
Useful life (years)	30	30

10

Dealing with Uncertainty

*T*he objective of this chapter is to present and discuss nonprobabilistic methods that are helpful in analyzing the economic consequences of engineering projects where uncertainty exists.

The following topics are discussed in this chapter:

The nature of risk, uncertainty, and sensitivity

Sources of uncertainty

Sensitivity analysis:

Break-even analysis

Sensitivity graphs

Combinations of factors

Optimistic-most likely-pessimistic estimating

Risk-adjusted MARR

Reduction in useful life

▼ 10.1 Introduction

In previous chapters, we stated specific assumptions concerning applicable revenues, costs, and other quantities important to an engineering economic analysis. It was assumed that a high degree of confidence could be placed in all estimated values. That degree of confidence is sometimes called *assumed certainty*. Decisions made solely on the basis of this kind of analysis are sometimes called *decisions under certainty*. The term is rather misleading in that there rarely is a case in which the best of estimates of quantities can be assumed as certain.

In virtually all situations there is doubt as to the ultimate economic results that will be obtained from an engineering project. We now examine techniques applicable to Step 5 of the seven-step procedure for conducting engineering economy studies (Chapter 1). The motivation for dealing with risk and uncertainty is to establish the bounds of error in our estimates such that another alternative being considered may turn out to be a better choice than the one we recommended under assumed certainty.

10.2 What Are Risk, Uncertainty, and Sensitivity?

Both *risk* and *uncertainty* in decision-making activities are caused by lack of precise knowledge regarding future business conditions, technological developments, synergies among funded projects, and so on. *Decisions under risk* are decisions in which the analyst models the decision problem in terms of assumed possible future outcomes, or scenarios, whose probabilities of occurrence can be estimated. A *decision under uncertainty,* by contrast, is a decision problem characterized by several unknown futures for which probabilities of occurrence cannot be estimated.

In reality, the difference between risk and uncertainty is somewhat arbitrary. One contemporary school of thought posits that representative and likely future outcomes and their probabilities can always be subjectively developed.*Hence, it is not unreasonable to suggest that decision making under risk is the more plausible and tractable framework for dealing with lack of perfect knowledge concerning the future. Although we may make a technical distinction between risk and uncertainty, both can cause study results to vary from predictions, and there seldom is anything significant to be gained by attempting to treat them separately. Therefore, in the remainder of this book, the terms *risk* and *uncertainty* are used interchangeably.

PRINCIPLE 6—MAKE UNCERTAINTY EXPLICIT (CHAPTER 1)

In dealing with uncertainty, it is often helpful to determine to what degree changes in an estimate would affect a capital investment decision, that is, how *sensitive* a given investment is to changes in particular factors that are not known with certainty. If a factor such as project life or annual revenue can be varied over a wide range without causing much effect on the investment decision, the decision under consideration is said *not* to be sensitive to that particular factor. Conversely, if a small change in the relative magnitude of a factor will reverse an investment decision, the decision is highly sensitive to that factor.

In this chapter, we discuss nonprobabilistic techniques for considering uncertainty in engineering economic analysis. The use of probability models is introduced in Chapter 13.

---

*R. Schlaifer, *Analysis of Decisions Under Uncertainty* (New York: McGraw-Hill, 1969).

10.3 Sources of Uncertainty

It is useful to consider some of the factors that affect the uncertainty involved in the analysis of the future economic consequences of an engineering project. It would be almost impossible to list and discuss all of the potential factors. There are four major sources of uncertainty, however, which are nearly always present in engineering economy studies.

The first source that is always present is the *possible inaccuracy of the cash-flow estimates used in the study.* If representative information is available regarding the items of revenue and expense, the resulting accuracy should be good. If, on the other hand, little factual information is available on which to base the estimates, their accuracy may be high or low.

The accuracy of the cash-inflow estimates is often difficult to determine. If they are based on past experience or have been determined by adequate market surveys, a fair degree of reliance may be placed on them. On the other hand, if they are based on limited information with a considerable element of hope thrown in, they probably contain a sizable element of uncertainty.

A saving in existing operating expenses, however, should involve less uncertainty. It is usually easier to determine what the saving will be because there is considerable experience and past history on which to base the estimates. Similarly, there should be no large error in most estimates of capital required. Uncertainty in capital investment requirements is often reflected as a *contingency* above the actual cost of plant and equipment.

The second major source affecting uncertainty is the *type of business involved in relation to the future health of the economy.* Some types of business operations are less stable than others. For example, most mining enterprises are more risky than one engaged in manufactured homes. However, we cannot arbitrarily say that an investment in the latter operations always involves less uncertainty than investment in mining. Whenever capital is to be invested in an engineering project, the nature of the business as well as expectations of future economic conditions (e.g., interest rates) should be considered in deciding what risk is present.

A third source affecting uncertainty is the *type of physical plant and equipment involved.* Some types of structures and equipment have rather definite economic lives and market values. Little is known of the physical or economic lives of others, and they have almost no resale value. A good engine lathe generally can be used for many purposes in nearly any fabrication shop. Quite different would be a special type of lathe that was built to do only one unusual job. Its value would be dependent almost entirely upon the demand for the special task that it can perform. Thus, the type of physical property involved affects the accuracy of the estimated cash-flow patterns. Where money is to be invested in specialized plant and equipment, this factor should be considered carefully.

The fourth important source of uncertainty that must always be considered is the *length of the study period* used in the analysis. The conditions that have been assumed in regard to cash inflows and outflows must exist throughout the study period in order to obtain a satisfactory return on the capital investment. A long study period naturally decreases the probability of all the factors turning out as estimated. Therefore, a long study period, all else being equal, generally increases the uncertainty of a capital investment.

◥ 10.4 Sensitivity Analysis

In the economic analysis of most engineering projects, it is helpful to determine how sensitive the situation is to the several factors of concern so that proper consideration may be given to them in the decision process. *Sensitivity*, in general, means the relative magnitude of change in the measure of merit (such as PW or IRR) caused by one or more changes in estimated study factor values. Sometimes sensitivity is more specifically defined to mean the relative magnitude of the change in one or more factors that will reverse a decision among project alternatives or a decision about the economic acceptability of a project.

> In engineering economy studies, sensitivity analysis is a general nonprobabilistic methodology, readily available, to provide information about the potential impact of uncertainty in selected factor estimates. Its routine use is fundamental to developing economic information useful in the decision process.

As we discussed in the previous section (10.3), there are several potential sources contributing to uncertainty in the cash-flow estimates of an engineering project. The specific factors of concern will vary with each project, but one or more of them will normally need to be further analyzed before the best decision can be made. Simply stated, engineering economy studies focus on the future, and uncertainty about the estimated prospective economic results cannot be avoided.

Several techniques are usually included in a discussion of sensitivity analysis in engineering economy. We will discuss the topic in terms of the following three techniques:

1. *Break-even analysis.* This technique is commonly used when the selection among project alternatives or the economic acceptability of an engineering project is heavily dependent upon a single factor, such as capacity utilization, which is uncertain.
2. *Sensitivity graph (spiderplot).* This approach is used when two or more project factors are of concern and an understanding of the sensitivity of the economic measure of merit to changes in the value of *each* factor is needed.
3. *Combination of factors.* When the combined effects of uncertainty in two or more project factors need to be examined, this analysis approach may be used. Normally, a spiderplot is developed first to identify the most sensitive factors and to aid in determining what combination(s) of factors to analyze.

10.4.1 Break-Even Analysis

When the selection between two engineering project alternatives is heavily dependent on a single factor, we can solve for the value of that factor at which the conclusion is a standoff. That value is known as the *break-even point*, that is, the value at which we are indifferent between the two alternatives. (The use of break-even points with respect to production and sales volumes was discussed in Chapter 2.) Then, if the best estimate of the actual outcome of the common factor is higher or lower than the break-even point, and assumed certain, the best alternative becomes apparent.

In mathematical terms, we have

$$EW_A = f_1(y) \text{ and } EW_B = f_2(y),$$

where EW_A = an equivalent worth calculation for the net cash flow of Alternative A;

EW_B = the same equivalent worth calculation for the net cash flow of Alternative B;

y = a common factor of interest affecting the equivalent worth values of Alternative A and Alternative B.

Therefore, the break-even point between Alternative A and Alternative B is the value of factor y for which the two equivalent worth values are equal. That is, $EW_A = EW_B$, or $f_1(y) = f_2(y)$, which may be solved for y.

Similarly, when the economic acceptability of an engineering project depends upon the value of a single factor, say, z, mathematically we can set an equivalent worth of the project's net cash flow for the analysis period equal to zero [$EW_p = f(z) = 0$] and solve for the break-even value of z. That is, the value of z at which we would be indifferent (economically) between implementing and rejecting the project. Then, if the best estimate of the value of z is higher or lower than the break-even point value, and assumed certain, the economic acceptability of the project is known.

The following are examples of common factors for which break-even analyses might provide useful insights into the decision problem:

1. *Annual revenue and expenses.* Solve for the annual revenue required to equal (break even with) annual expenses. Break-even annual expenses of an alternative can also be determined in a pairwise comparison when revenues are identical for both alternatives being considered.
2. *Rate of return.* Solve for the rate of return on the increment of invested capital at which two given alternatives are equally desirable.
3. *Market (or salvage) value.* Solve for the future resale value that would result in indifference as to preference for an alternative.
4. *Equipment life.* Solve for the useful life required for an engineering project to be economically justified.
5. *Capacity utilization.* Solve for the hours of utilization per year, for example, at which an alternative is justified or at which two alternatives are equally desirable.

The usual break-even problem involving two alternatives can be most easily approached mathematically by equating an equivalent worth of the two alternatives expressed as a function of the factor of interest. Using the same approach for the economic acceptability of an engineering project, we can mathematically equate an equivalent worth of the project to zero as a function of the factor of concern. In break-even studies, project lives may or may not be equal, so care should be taken to determine whether the coterminated or repeatability assumption best fits the situation.

The following examples illustrate both mathematical and graphical solutions to typical break-even problems.

EXAMPLE 10-1

Suppose that there are two alternative electric motors that provide 100-hp output. An Alpha motor can be purchased for $12,500 and has an efficiency of 74%, an estimated useful life of 10 yr, and estimated maintenance expenses of $500 per year. A Beta motor will cost $16,000 and has an efficiency of 92%, a useful life of 10 yr, and annual maintenance expenses of $250. Annual taxes and insurance expenses on either motor will be $1\frac{1}{2}$% of the investment. If the MARR is 15% per year, how many hours per year would the motors have to be operated at full load for the annual costs to be equal? Assume that market values at the end of 10 yr for both are negligible and that electricity costs $0.05 per kiloWatt-hour.

SOLUTION BY MATHEMATICS

Note: 1 hp = 0.746 kW and input = output/efficiency. If X = number of hours of operation per year, components of the equivalent annual worth, AW_α, for the Alpha motor would be as follows:

Capital recovery amount:

$$\$12,500(A/P, 15\%, 10) = \$12,500(0.1993) = \$2,490 \text{ per year};$$

Operating expense for power:

$$(100)(0.746)(\$0.05)X/0.74 = \$5.04X \text{ per year};$$

Maintenance expense:

$$\$500 \text{ per year};$$

Taxes and insurance:

$$\$12,500(0.015) = \$187 \text{ per year}.$$

Similarly, components of the equivalent annual worth, AW_β, for the Beta motor would be as follows:

Capital recovery amount:

$$\$16,000(A/P, 15\%, 10) = \$16,000(0.1993) = \$3,190 \text{ per year};$$

Operating expense for power:

$$(100)(0.746)(\$0.05)X/0.92 = \$4.05X \text{ per year};$$

Maintenance expense:
$$\$250 \text{ per year};$$

Taxes and insurance:

$$\$16,000(0.015) = \$240 \text{ per year};$$

Because we are dealing with costs only in this example (revenues are assumed to be equal), we use the equivalent uniform annual cost (EUAC) metric to solve for the break-even point.

At the break-even point, $\text{EUAC}_\alpha = \text{EUAC}_\beta$. Thus,

$$\$2,490 + \$5.04X + \$500 + \$187 = \$3,190 + \$4.05X + \$250 + \$240$$

$$\$5.04X + \$3,177 = \$4.05X + \$3,680$$

$$\hat{X} \simeq 508 \text{ hr/yr}.$$

PLOT OF THE MATHEMATICAL SOLUTION

Figure 10-1 shows a plot of the total EUAC of each motor as a function of the number of hours of operation per year. The constant annual costs (EUAC intercepts) are $3,177 and $3,680 for Alpha and Beta, respectively, and the expenses that vary directly with hours of operation per year (slopes of lines) are $5.04 and $4.05 for

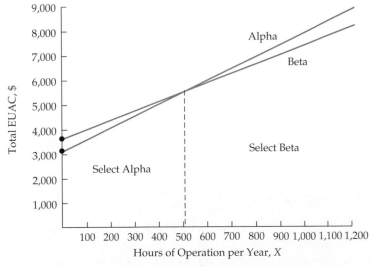

Figure 10-1 Graphical Plot of the Break-Even Point for Example 10-1

Alpha and Beta, respectively. (Refer to the previous mathematical solution.) Of course, the break-even point is the value of the independent variable, X, at which the linear EUAC functions for the two alternatives intersect (at approximately 508 hr per year). Therefore, if the best estimate of hours of operation per year is > 508, the Beta motor is preferred.

Hours of operation per year was the measure of business activity in Example 10-1 and was used as the variable for which a break-even value was desired. Also, in Example 10-1, break-even analysis was applied where only two alternatives were involved. However, break-even analysis can be extended to multiple alternatives, which is demonstrated in Example 10-2.

EXAMPLE 10-2

The Universal Postal Service is considering the possibility of putting wind deflectors on the tops of 500 of their long-haul tractors. Three types of deflectors, with the following characteristics, are being considered (MARR = 10% per year):

	Windshear	Blowby	Air-vantage
Capital investment	$1,000	$400	$1,200
Drag reduction	20%	10%	25%
Maintenance/year	$10	$5	$5
Useful life	10 yr	10 yr	5 yr

If 5% in drag reduction means 2% in fuel savings per mile, how many miles do the tractors have to be driven per year before the Windshear deflector is favored over the other deflectors? Over what range of miles driven per year is Air-vantage the best choice? (*Note:* Fuel cost is expected to be $1.00 per gallon and average fuel consumption is five miles per gallon without the deflectors.) State any assumptions you make.

SOLUTION

The annual operating expenses of long-haul tractors equipped with the various deflectors are calculated as a function of mileage driven per year, X:

$$\text{Windshear: } [(X \text{ miles/yr})(0.92)(0.2 \text{ gal/mi})(\$1.00/\text{gal})] = \$0.184X/\text{yr};$$

$$\text{Blowby: } [(X \text{ miles/yr})(0.96)(0.2 \text{ gal/mi})(\$1.00/\text{gal})] = \$0.192X/\text{yr};$$

$$\text{Air-vantage: } [(X \text{ miles/yr})(0.90)(0.2 \text{ gal/mi})(\$1.00/\text{gal})] = \$0.180X/\text{yr}.$$

Plotting EUAC of the deflectors yields the break-even values of X shown in Figure 10-2. In summary, when $X \leq 12{,}831$, Blowby would be selected. If $X \geq 37{,}203$, the Air-vantage deflector would be chosen; otherwise, Windshear is the preferred alternative. Mathematically, the exact values can be calculated for each pair of

EUAC equations (Windshear versus Blowby, Windshear versus Air-vantage, and Blowby versus Air-vantage). For example, the break-even value between the Windshear deflector and the Blowby deflector is

$$\$1{,}000(A/P, 10\%, 10) + \$10 + \$0.184X = \$400(A/P, 10\%, 10) + \$5 + \$0.192X$$

$$\$172.75 + \$0.184X = \$70.1 + \$0.192X$$

$$X = \frac{102.65}{0.008} = 12{,}831 \text{ miles/year.}$$

The repeatability assumption, which is appropriate in this situation, allows the EUACs to be compared over different periods of time.

It is often helpful to know at what future date a deferred investment will be needed so that an alternative permitting deferred investment will break even with one that provides immediately for all future requirements. Where only the costs of acquiring the assets by the two alternatives need to be considered or where the annual expenses throughout the entire life are not affected by the date of acquisition of the deferred asset, the break-even point may be determined very easily and may be helpful in arriving at a decision between alternatives. Example 10-3 illustrates this type of break-even study.

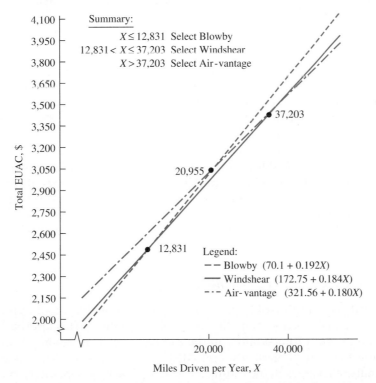

Figure 10-2 Graphical Plot of the Break-Even Analysis of Example 10-2

EXAMPLE 10-3

In planning a small two-story office building, the architect has submitted two designs. The first provides foundation and structural details so that two additional stories can be added to the required initial two stories at a later date and without modifications to the original structure. This building would cost $1,400,000. The second design, without such provisions, would cost only $1,250,000. If the first plan is adopted, an additional two stories could be added at a later date at a cost of $850,000. If the second plan is adopted, however, considerable strengthening and reconstruction would be required, which would add $300,000 to the cost of a two-story addition. Assuming that the building is expected to be needed for 75 yr, by what time would the additional two stories have to be built to make the adoption of the first design justified? (The MARR is 10% per year.)

SOLUTION

The break-even deferment period, $\hat{T}$, is determined as follows:

	Provide Now	No Provision
PW cost:		
First unit	$1,400,000	$1,250,000
Second unit	$850,000(P/F, 10%, \hat{T})	$1,150,000(P/F, 10%, \hat{T})
Equating total PW of costs:		
$1,400,000 + $850,000(P/F, 10%, \hat{T}) = $1,250,000 + $1,150,000(P/F, 10%, \hat{T})$		

If the *difference* between the two alternatives is examined, it can be seen that $150,000 now is being traded off against $300,000 at a later date. The question is, what "later date" constitutes the break-even point?

Solving, we have

$$(P/F, 10\%, \hat{T}) = 0.5.$$

From the 10% interest table in Appendix C, $\hat{T}$ = seven years (approximately). Thus, if the additional space will be required in less than seven years, it would be more economical to make immediate provision in the foundation and structural details. If the addition would not likely be needed until after seven years, greater economy would be achieved by making no such provision in the first structure.

10.4.2 Sensitivity Graph (Spiderplot)

The sensitivity graph (spiderplot) technique is an analysis tool applicable when break-even analysis does not "fit" the project situation. This approach makes explicit the impact of uncertainty in the estimates of each factor of concern on the economic measure of merit. Example 10-4 demonstrates this technique by plotting the results of changes in the estimates of several factors, separately, on the present worth of an engineering project.

EXAMPLE 10-4

The best (most likely) cash-flow estimates for a new piece of equipment being considered for immediate installation are as follows:

Capital investment, I	$11,500
Revenues/yr $\Big\}$ A	5,000
Expenses/yr	2,000
Market value, MV	1,000
Useful life, N	6 years

Because of the new technology built into this machine, it is desired to investigate its PW over a range of ±40% changes in the estimates for (a) capital investment, (b) annual net cash flow, (c) market value, and (d) useful life. Based on these best estimates, plot a diagram that summarizes the sensitivity of present worth to percent deviation changes in each separate factor estimate when the MARR=10% per year.

SOLUTION

The PW of this project (installation of the new equipment) based on the best estimates of the factors given previously is

$$PW(10\%) = -\$11,500 + (\$5,000 - \$2,000)(P/A, 10\%, 6) + \$1,000(P/F, 10\%, 6)$$

$$= \$2,130.$$

This value of the PW in Figure 10-3 occurs at the common intersection point of the percent deviation graphs for the four separate project factors (I, A, N, and MV).

(a) When the capital investment (I) varies by ±$p\%$ the PW is

$$PW(10\%) = -(1 \pm p\%/100)(\$11,500) + \$3,000(P/A, 10\%, 6) + \$1,000(P/F, 10\%, 6).$$

If we let $p\%$ vary in increments of 10% to ±40%, the resultant calculations of PW(10%) can be plotted as shown in Figure 10-3.

(b) The equation for PW can be modified to reflect ±$a\%$ changes in net annual cash flow, A:

$$PW(10\%) = -\$11,500 + (1 \pm a\%/100)(\$3,000)(P/A, 10\%, 6) + \$1,000(P/F, 10\%, 6).$$

Results are plotted in Figure 10-3 for 10% increments in A within the prescribed ±40% interval.

(c) When market value (MV) varies by ±$s\%$, the PW is

$$PW(10\%) = -\$11,500 + \$3,000(P/A, 10\%, 6)$$

$$+ (1 \pm s\%/100)(\$1,000)(P/F, 10\%, 6).$$

Results are shown in Figure 10-3 for changes in MV over the ±40% internal.

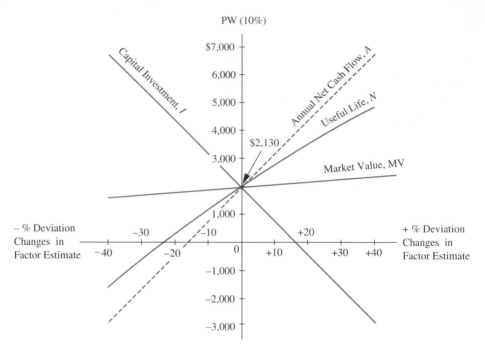

Figure 10-3 Sensitivity Graph (Spiderplot) of Four Factors in Example 10-4

(d) Plus and minus $n\%$ changes in useful life (N), as they affect PW(10%), can be represented by this equation:

$$PW(10\%) = -\$11,500 + \$3,000[P/A, 10\%, 6(1 \pm n\%/100)]$$

$$+ \$1,000[P/F, 10\%, 6(1 \pm n\%/100)].$$

When $n\%$ varies by 10% increments within the desired ±40% interval, resultant changes in PW(10%) can be calculated and plotted as shown in Figure 10-3.

In summary, the spiderplot in Figure 10-3 shows the sensitivity of the present worth to percent deviation changes in each factor's best estimate. The other factors are assumed to remain at their best estimate values. *The relative degree of sensitivity of the present worth to each factor is indicated by the slope of the curves* (the "steeper" the slope of a curve, the more sensitive the present worth is to the factor). Also, the intersection of each curve with the abscissa shows the percent change in each factor's best estimate at which the present worth is zero.

Based on the spiderplot, we see that the present worth is insensitive to MV, but quite sensitive to changes in I, A, and N. For example, it is clear that the capital investment can increase approximately $2,130 (to $13,630) without causing the project's present worth to become negative. This is an 18.5% increase, which can be approximated from Figure 10-3.

As additional information, consider using the sensitivity graph technique to compare two or more mutually exclusive project alternatives. If only two alternatives are being compared, a spiderplot based on the incremental cash flow between the alternatives can be used to aid in the selection of the preferred alternative. Extending this approach to three alternatives, two sequential paired comparisons can be used to help select the preferred alternative. Another approach is to plot (overlay) in the same figure a sensitivity graph for each alternative. Obviously, if this later approach is used for the comparison of more than, say, three alternatives (with two or three factors each), interpreting the results may become a problem.

10.4.3 Combinations of Factors

We are often concerned about the *combined effects* of uncertainty in two or more project factors on the economic measure of merit. When this situation occurs, the following approach should be used in developing additional information to assist decision making:

1. Develop a sensitivity graph for the project as discussed in Section 10.4.2. Also, for the most sensitive factors, try to develop improved estimates and reduce the range of uncertainty before proceeding further with the analysis.
2. Select the most sensitive project factors based on the information in the sensitivity graph. Analyze the combined effects of these factors on the project's economic measure of merit by (a) using an additional graphical technique to make the combined impact of the two most sensitive factors more explicit and (b) determining the impact of selected combinations of three or more factors (these combinations are sometimes called *scenarios*).

The first technique is demonstrated in Example 10-5, and the second technique is illustrated in Example 10-6.

EXAMPLE 10-5

Refer to the engineering project in Example 10-4. This situation, with additional assumptions, will be used to demonstrate a graphical technique which makes the combined impact of the two most sensitive factors on the present worth more explicit.

In Example 10-4, a common range of uncertainty ($\pm 40\%$ of each factor's best estimate) was used for the four project factors of concern: capital investment, I; annual net cash flow, A; useful life, N; and market value, MV. Assume the following new estimate ranges for this example: capital investment, -10% to $+15\%$; annual net cash flow, -40% to $+25\%$; and useful life, -10% to $+20\%$. Market value is removed as a factor of concern and the best estimate value of $1,000 will be used. Also, instead of replotting the sensitivity graph in Figure 10-3, we will use the parts of the curves that are within the new estimated ranges of uncertainty. This is feasible since the best estimate values remain the same. The present worth of the project still remains the most sensitive to I and A, and slightly less sensitive to N.

Therefore, in this example, we will focus on the combined impact of these factors (I, A) on the present worth value, PW(10%).

SOLUTION

We will plot the present worth of the project, PW(10%), as a function of both factors (I, A) assuming useful life and market value remain at their best estimate values of six years and $1,000, respectively. The following information is needed:

Project Factor (Variable)	Deviation Range[a]	Best Estimate	Range Estimate[b] Minimum	Maximum
Capital investment, I	−10% to +15%	$11,500	$10,350	$13,225
Annual net cash flow, A	−40% to +25%	3,000	1,800	3,750

[a] New estimated range of percent deviations from the best estimate value.
[b] Based on the minimum and maximum percent deviation values and the best estimate value.

Using this information, the two-dimensional graph shown in Figure 10-4 is plotted. Annual net cash flow (A) is represented as a variable on the abscissa. The other variable, capital investment (I), is represented by a set of curves, and the ordinate reflects the present worth of the project. The two curves plotted in the graph are based on the minimum and maximum values of capital investment and delineate the boundaries of the set of curves representing this variable. The two curves are the plot of

$$PW(10\%) = -\$10,350 + A(P/A, 10\%, 6) + \$1,000(P/F, 10\%, 6)$$

and

$$PW(10\%) = -\$13,225 + A(P/A, 10\%, 6) + \$1,000(P/F, 10\%, 6).$$

The shaded area shown in Figure 10-4 reflects the present worth values resulting from all combinations of I and A values and defines the region of uncertainty. Also shown is the maximum value ($6,547) and the minimum value (−$4,820) of the present worth. Since the shaded region is not all above the abscissa, PW(10%) > 0, nor all below the abscissa, PW(10%) < 0, the decision is sensitive to the combined impact of these two factors. It is not correct, however, to interpret the proportions of the shaded region above and below the abscissa as the probability of the present worth being greater than or less than zero unless it can be assumed that all combinations of the I and A values are equally likely, which is very improbable.

The combined impact of changes in the best estimate values for three or more factors on the economic measure of merit for an engineering project can be analyzed using selected combinations of the changes. This approach and the *Optimistic-Most Likely-Pessimistic (O-ML-P)* technique of estimating factor values are illustrated in Example 10-6.

An optimistic estimate for a factor is one that is in the favorable direction (for example, the minimum capital investment cost in Example 10-5). The most likely

Figure 10-4

Combined Impact of
Two Factors (I, A)
on PW Value in
Example 10-5

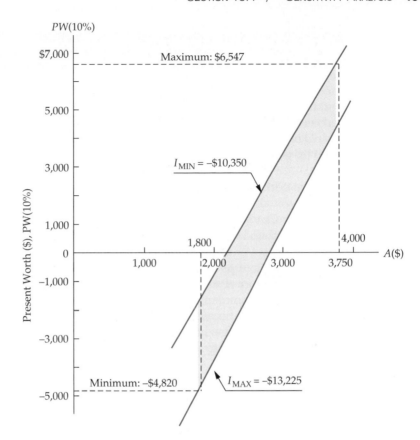

value for a factor is defined for our purposes as the best estimate value. This
definition was used in Example 10-4. The pessimistic estimate for a factor is one
that is in the unfavorable direction (say, the maximum capital investment cost in
Example 10-5). In applications of this technique, the optimistic condition for a
factor is often specified as a value that has 19 chances out of 20 of being better than
the actual outcome. Similarly, the pessimistic condition has 19 chances out of 20 of
being worse than the actual outcome. In operational terms, the optimistic condition
for a factor is the value when things occur as well as can be reasonably expected,
and the pessimistic estimate is the value when things occur as detrimentally as can
be reasonably expected.

EXAMPLE 10-6

Consider a proposed ultrasound inspection device for which the optimistic, pessimistic, and most likely estimates are given in Table 10-1. The MARR is 8% per
year. Also shown at the end of Table 10-1 are the AWs for all three estimation
conditions. Based on this information, analyze the combined effects of uncertainty
in the factors on the AW value.

TABLE 10-1 Optimistic, Most Likely, and Pessimistic Estimates and AWs for Proposed Ultrasound Device (Example 10-6)

	Estimation Condition		
	Optimistic (O)	Most Likely (M)	Pessimistic (P)
Capital investment, I	$150,000	$150,000	$150,000
Useful life, N	18 yr	10 yr	8 yr
Market value, MV	0	0	0
Annual revenues, R	$110,000	$70,000	$50,000
Annual expenses, E	20,000	43,000	57,000
AW (8%):	+$73,995	+$4,650	−$33,100

SOLUTION

Step 1: Before proceeding further with the solution, we need to evaluate the two extreme values of the AW. As shown at the end of Table 10-1, the AW for the optimistic estimates is very favorable ($73,995), while the AW for the pessimistic estimates is quite unfavorable (−$33,100). If both extreme AW values were positive, we would make a "go" decision with respect to the device without further analysis because no combination of factor values based on the estimates will result in AW < 0. By similar reasoning, if both AW values were negative, a "no-go" decision would be made regarding the device. In this example, however, the decision is sensitive to other combinations of outcomes and we proceed to Steps 2 and 3.

Step 2: A sensitivity graph (spiderplot) for this situation is needed to show explicitly the sensitivity of the AW to the three factors of concern: useful life, N; annual revenues, R; and annual expenses, E. The spiderplot is shown in Figure 10-5. The curves for N, R, and E plot percent deviation changes from the most likely (best) estimate, over the range of values defined by the optimistic and pessimistic estimates for each factor, versus AW. As additional information, a curve is also shown for the MARR versus the AW. Based on the spiderplot, the AW for the proposed ultrasound device appears very sensitive to annual revenues and quite sensitive to annual expenses and reductions in useful life. However, even if we were to significantly change the MARR (8%), it would have little impact on the AW.

Step 3: The various combinations of the optimistic, most likely, and pessimistic factor values (outcomes) for annual revenues, useful life, and annual expenses need to be analyzed for their combined impacts on the AW. The results for these 27 (3 × 3 × 3) combinations are shown in Table 10-2.

MAKING THE AW RESULTS EASIER TO INTERPRET

Since the AW values in Table 10-2 result from estimates subject to varying degrees of uncertainty, little information of value would be lost if the numbers were rounded to the nearest thousand dollars. Further, suppose that management is

Figure 10-5
Sensitivity Graph for
Proposed Ultrasound
Device (Example 10-6)

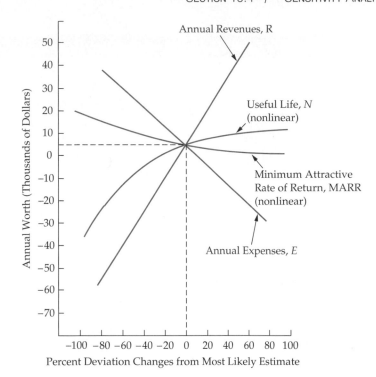

most interested in the number of combinations of outcomes in which the AW is, say, (1) more than $50,000 and (2) less than $0. Table 10-3 shows how Table 10-2 might be changed to make it easier to interpret and use in communicating the AW results to management.

From Table 10-3, it is apparent that four combinations result in AW > $50,000, while nine produce AW < $0. Each combination of conditions is not necessarily equally likely. Therefore, again, statements such as "There are 9 chances out of 27 that we will lose money on this project" are not appropriate.

TABLE 10-2 AWs ($) for All Combinations of Estimated Outcomes[a] for Annual Revenues, Annual Expenses, and Useful Life: Proposed Ultrasound Device (Example 10-6)

Annual Revenues, R	Annual Expenses, E								
	O			M			P		
	Useful Life, N			Useful Life, N			Useful Life, N		
	O	M	P	O	M	P	O	M	P
O	73,995	67,650	63,900	50,995	44,650	40,900	36,995	30,650	26,900
M	34,000	27,650	23,900	10,995	4,650	900	−3,005	−9,350	−13,100
P	14,000	7,650	3,900	−9,005	−15,350	−19,100	−23,005	−29,350	−33,100

[a] Estimates: O, optimistic; M, most likely; P, pessimistic.

TABLE 10-3 Results in Table 10-2 Made Easier to Interpret (AWs in $000s)[a,b]

Annual Revenues, R	Annual Expenses, E								
	O			M			P		
	Useful Life, N			Useful Life, N			Useful Life, N		
	O	M	P	O	M	P	O	M	P
O	74	68	64	51	45	41	37	31	27
M	34	28	24	11	5	1	−3	−9	−13
P	14	8	4	−9	−15	−19	−23	−29	−33

[a] Estimates: O, optimistic; M, most likely; P, pessimistic.

[b] Boxed entries, AW > $50,000 (4 out of 27 combinations); underscored entries, AW < $0 (9 out of 27 combinations).

Another form of sensitivity results that is often quite valuable is to determine the relative (or absolute) change in one or more factors that will just reverse the decision. Even though the change can be estimated on a spiderplot, it is better to calculate it for each factor of concern. Applied to Example 10-6, this means determining the relative change in each factor that will decrease the AW by $4,650 so that it reaches $0. Table 10-4 shows this application by using a table and bars of varying lengths to emphasize that the AW of the device is (1) most sensitive to changes in the estimated annual revenues and (2) least sensitive to changes in the MARR value.

It is clear that even with a few factors, and using the O-ML-P estimating technique, the number of possible combinations of conditions in a sensitivity analysis can become quite large, and the task of investigating all of them might be quite time-consuming. One goal of progressive sensitivity analysis is to eliminate from detailed consideration those factors for which the measure of merit is quite insensitive and highlighting the conditions for other factors to be studied further in accordance with the degree of sensitivity of each. Thus, the number of combinations of conditions included in the analysis can perhaps be kept to a manageable size.

TABLE 10-4 Sensitivity of Decision Reversal to Changes in Selected Estimates

	Most Likely Estimate	Required Outcome[a]	Amount of Change	Change Amount as Percentage of Most Likely Estimate	
Capital investment	$150,000	$181,000	$31,200	+20.8%	_____
Useful life	10 yr	7.3 yr	−2.7 yr	−27.0%	_____
Annual revenues	70,000	65,350	−4,650	−6.6%	___
Annual expenses	43,000	47,650	4,650	+10.8%	____
MARR	8%	12.5%	+4.5%	+56%	_____

[a] To reverse decision (decrease AW to $0). Notice that reversal of AW is most sensitive to change in annual revenues.

Companion Web Site (http://www.prenhall.com/sullivan_engineering/):   Many mechanical insulation companies are using computerized systems for cost estimation; however, an investment decision is required before purchasing such a system. Visit the Web site to view an economic comparison of manual versus computerized cost-estimating methods that demonstrates *sensitivity analysis* using various money–time relationships.

10.5 Analyzing a Proposed Business Venture

Another illustration of the use of sensitivity analysis is provided in Example 10-7 in which a new business venture is analyzed. This example includes several factors whose outcomes are believed to be crucial to the success of the venture. Tabular displays are used to summarize results of the various analyses.

EXAMPLE 10-7

A small group of investors is considering starting a small premixed-concrete plant in a rapidly developing suburban area about 15 miles from a large city. The group believes that there will be a good market for premixed concrete in this area for at least the next 10 yr, and that if they establish such a local plant, it will be unlikely that another local plant would be established. Existing plants in the adjacent large city would, of course, continue to serve this new area. The investors believe that the plant could operate at about 75% of capacity 250 days per year, because it is located in an area where the weather is mild throughout the year.

The plant will cost $100,000 and will have a maximum capacity of 72 cubic yards of concrete per day. Its market value at the end of 10 yr is estimated to be $20,000, which is the value of the land. To deliver the concrete, four secondhand trucks would be acquired, costing $8,000 each, having an estimated life of five years and a market value of $500 each at the end of that time. In addition to the four truck drivers, who would be paid $50.00 per day each, four people would be required to operate the plant and office, at a total cost of $175.00 per day. Annual operating and maintenance expenses for the plant and office are estimated at $7,000 and for each truck at $2,250, both in view of 75% capacity utilization. Raw material costs are estimated to be $27.00 per cubic yard of concrete. Payroll taxes, vacations, and other fringe benefits would amount to 25% of the annual payroll. Annual taxes and insurance on each truck would be $500, and taxes and insurance on the plant would be $1,000 per year. The investors would not contribute any labor to the business, but a manager would be employed at an annual salary of $20,000.

Delivered, premixed concrete currently is selling for an average of $45 per cubic yard. A useful plant life of 10 years is expected, and capital invested elsewhere by these investors is earning about 15% per year before income taxes. It is desired to find the AW for the expected conditions described and to perform sensitivity analyses for certain factors.

SOLUTION BY THE AW METHOD
Annual revenue:

$$72 \times 250 \times \$45 \times 0.75 = \$607,500$$

Annual expenses:

1. Capital recovery amount
   Plant: $\$100,000(A/P, 15\%, 10)$
      $-\$20,000(A/F, 15\%, 10)$ =   \$18,940
   Trucks: $4[\$8,000(A/P, 15\%, 5)$
      $-\$500(A/F, 15\%, 5)]$ =   9,250
                                \$28,190

2. Labor:
   Plant and office: $\$175 \times 250$ =   43,750
   Truck drivers: $4 \times \$50 \times 250$ =   50,000
   Manager   =   20,000
                                113,750

3. Payroll taxes, fringe benefits, etc.:
   $\$113,750 \times 0.25$ 28,438

4. Taxes and insurance:
   Plant   =   1,000
   Trucks: $\$500 \times 4$ =   2,000
                                3,000

5. Operations and maintenance at 75% capacity:
   Plant and office   =   7,000
   Trucks: $\$2,250 \times 4$ =   9,000
                                16,000

6. Materials: $72 \times 0.75 \times 250 \times \$27.00$ 364,500
                Total expenses   \$553,878

The net AW for these most likely (best) estimates is $\$607,500 - \$553,878 = \$53,622.$ Apparently, the project is an attractive investment opportunity.

In Example 10-7, there are three factors that are of great importance and that must be estimated: *capacity utilization,* the *selling price of the product,* and the *useful life of the plant.* A fourth factor—raw material costs—is important, but any significant change in this factor would probably have an equal impact on competitors and probably would be reflected in a corresponding change in the selling price of mixed concrete. The other cost elements should be determinable with considerable accuracy. Therefore, we need to investigate the effect of variations in the plant utilization, selling price, and useful life. Sensitivity analysis is needed in this situation.

10.5.1 Sensitivity to Capacity Utilization

As a first step, we will determine how expenses would vary, if at all, as capacity utilization is varied in Example 10-7. In this case, it is probable that the annual

TABLE 10-5 AW at $i = 15\%$ per year for the Premixed-Concrete Plant in Example 10-7 for Various Capacity Utilizations (Average Selling Price Is $45 per Cubic Yard)			
	50% Capacity	65% Capacity	90% Capacity
Annual revenue	$405,000	$526,500	$729,000
Annual expenses:			
Capital recovery	28,190	28,190	28,190
Labor	113,750	113,750	113,750
Payroll taxes and similar items	28,438	28,438	28,438
Taxes and insurance	3,000	3,000	3,000
Operations and maintenance[a]	13,715	15,086	17,372
Materials	243,000	315,900	437,400
Total expenses	$430,093	$504,364	$628,150
AW (15%)	−$25,093	+$22,136	+$100,850

[a] Let $x =$ annual operations and maintenance expenses and assume that 50% of the cost varies directly with capacity utilization. At 75% capacity utilization, $x/2 + (x/2)(0.75) = \$16,000$, so that $x = \$18,286$ at 100% capacity utilization. Therefore, at 50% utilization, the operations and maintenance expense would be $\$9,143 + 0.5(\$9,143) = \$13,715$.

expense items listed under groups 1, 2, 3, and 4 in the previous tabulation would be virtually unaffected if capacity utilization should vary over a quite wide range—from 50% to 90%, for example. To meet peak demands, the same amount of plant, trucks, and personnel probably would be required. Operations and maintenance expenses (group 5) would be affected somewhat. For this factor we must try to determine what the variation would be or make a reasonable assumption as to the probable variation. For this case, it will be *assumed* that one half of these expenses would be fixed and the other half would vary with capacity utilization by a straight-line relationship. Certain other factors, such as the cost of materials in this case, will vary in direct proportion to capacity utilization.

Using these assumptions, Table 10-5 shows how the revenue, expenses, and net AW would change with different capacity utilizations. It will be noted that the AW is moderately sensitive to capacity utilization. The plant could be operated at a little less than 65% of capacity, instead of the assumed 75%, and still produce an AW greater than zero. Also, quite clearly, if they should be able to operate above the assumed 75% of capacity, the AW would be very good. This type of analysis gives the analyst a good idea of how much leeway the company can have in capacity utilization and still have an acceptable venture.

10.5.2 Sensitivity to Selling Price

Examination of the sensitivity of the project to the selling price of the concrete reveals the situation shown in Table 10-6. The values in this table assume that the plant would operate at 75% of capacity; the expenses would thus remain constant,

TABLE 10-6	Effect of Various Selling Prices on the AW for the Premixed-Concrete Plant in Example 10-7 Operating at 75% of Capacity			
	Selling Price			
	$45.00	$43.65(3%)[a]	$42.75(5%)[a]	$40.50(10%)[a]
Annual revenue	$607,500	$589,275	$577,125	$546,750
Annual expenses	553,878	553,878	553,878	553,878
AW(15%)	$53,622	$35,397	$23,247	−$7,128

[a] Percentage values shown in parentheses are reductions in price below $45.

with only the selling price varying. Here it will be noted that the project is quite sensitive to price. A decrease in price of 10% would reduce the IRR to less than 15% (i.e., AW < 0). Since a decrease of 10% is not very large, the investors would want to make a thorough study of the price structure of concrete in the area of the proposed plant, particularly with respect to the possible effect of the increased competition that the new plant would create. If such a study reveals price instability in the market for concrete, the plant could be a risky investment.

10.5.3 Sensitivity to Useful Life

The effect of the third factor, assumed useful life of the plant, can be investigated readily. If a life of five years were assumed for the plant, instead of the assumed value of ten years, the only factor in the study that would be changed would be the cost of capital recovery. If the market value is assumed to remain constant, the capital recovery amount over a five-year period is

$$\$100,000(A/P, 15\%, 5) - \$20,000(A/F, 15\%, 5) = \$26,866 \text{ per year,}$$

which is $7,926 more expensive than the initial value of $18,940. In this case, the AW would be reduced to $45,696—a decline of 14.8%. Hence, a 50% reduction in useful life causes only a 14.8% reduction in AW. Clearly, the venture is rather insensitive to the assumed useful life of the plant.

With the added information supplied by the sensitivity analyses that have just been described, those who make the investment decision concerning the proposed concrete plant would be in a much better position than if they had only the initial study results, based on an assumed utilization of 75% of capacity, available to them.

Also, additional information useful to the investors (about the combined effects of various outcomes of the three factors) can be obtained by using the graphical technique illustrated in Example 10-5, and by analyzing selected combinations of the factor outcomes explained in Example 10-6. Accomplishing this additional analysis is the task in Problem 10-23 in Section 10.11.

10.6 Risk-Adjusted Minimum Attractive Rates of Return

Uncertainty causes factors inherent to engineering economy studies, such as cash flows and project life, to become random variables in the analysis. (Simply stated, a random variable is a function that assigns a unique numerical value to each possible outcome of a probabilistic quantity.) A widely used industrial practice for including some consideration of uncertainty is to increase the MARR when a project is thought to be relatively uncertain. Hence, a procedure has emerged that employs *risk-adjusted* interest rates. It should be noted, however, that many pitfalls of performing studies of financial profitability with risk-adjusted MARRs have been identified.*Also, the procedure does not make the uncertainty in the project estimates explicit.

In general, the preferred practice to account for uncertainty in estimates (cash flows, project life, etc.) is to deal directly (explicitly) with their suspected variations in terms of probability assessments (Chapter 13) rather than to manipulate the MARR as a means of reflecting the virtually certain versus highly uncertain status of a project. Intuitively, the risk-adjusted interest-rate procedure can be defended because more certainty regarding the overall profitability of a project exists in the early years compared to, say, the last two years of its life. Increasing the MARR places emphasis on early cash flows rather than on longer-term benefits, and this would appear to help compensate for time-related project uncertainties. But the question of uncertainty in cash-flow amounts is not directly addressed. The following example illustrates how this method of dealing with uncertainty can lead to an illogical recommendation.

EXAMPLE 10-8

The Atlas Corporation is considering two alternatives, both affected by uncertainty to different degrees, for increasing the recovery of a precious metal from its smelting process. The following data concern capital investment requirements and estimated annual savings of both alternatives:

End of Year, k	Alternative	
	P	Q
0	−$160,000	−$160,000
1	120,000	20,827
2	60,000	60,000
3	0	120,000
4	60,000	60,000

The firm's MARR for its risk-free investments is 10% per year. Because of the technical considerations involved, Alternative P is thought to be *more uncertain*

---

* A. A. Robichek and S. C. Myers, "Conceptual Problems in the Use of Risk-Adjusted Discount Rates," *Journal of Finance*, vol. 21, December 1966, pp. 727–730.

than Alternative Q. Therefore, according to the Atlas Corporation's engineering economy handbook, the risk-adjusted MARR applied to P will be 20% per year and the risk-adjusted MARR for Q has been set at 17% per year. Which alternative should be recommended?

SOLUTION

At the risk-free MARR of 10%, both alternatives have the same PW of $39,659. All else being equal, Alternative Q would be chosen because it is less uncertain than Alternative P. Now a PW analysis is performed for the Atlas Corporation, using its prescribed risk-adjusted MARRs for the two options:

$$PW_P(20\%) = -\$160,000$$

$$+ \$120,000(P/F, 20\%, 1) + \$60,000(P/F, 20\%, 2)$$

$$+ \$60,000(P/F, 20\%, 4) = \$10,602;$$

$$PW_Q(17\%) = -\$160,000 + \$20,827(P/F, 17\%, 1)$$

$$+ \$60,000(P/F, 17\%, 2)$$

$$+ \$120,000(P/F, 17\%, 3)$$

$$+ \$60,000(P/F, 17\%, 4) = \$8,575.$$

Without considering economic uncertainty (i.e., MARR=10% per year), and based on technical considerations, the selection was seen to be Alternative Q. But when Alternative P is "penalized" due to the technical considerations by applying a higher risk-adjusted MARR to compute its PW, the comparison of alternatives favors Alternative P. One would expect to see Alternative Q recommended with this procedure. This contradictory result can be seen clearly in Figure 10-6, which demonstrates the general situation in which contradictory results might be expected.

Even though the intent of the risk-adjusted MARR is to make more uncertain projects appear less economically attractive, the opposite was shown to be true in Example 10-8. Furthermore, a related shortcoming of the risk-adjusted MARR procedure is that cost-only projects are made to appear more desirable (to have a less negative PW, for example) as the interest rate is adjusted upward to account for uncertainty. At extremely high interest rates, the alternative having the lowest investment requirement would be favored, regardless of subsequent cost cash flows. Because of difficulties such as those illustrated, this procedure is not generally recommended as an acceptable means of dealing with uncertainty.

▼ 10.7 Reduction of Useful Life

Some of the methods for dealing with uncertainty that have been discussed to this point have attempted to compensate for potential losses that could be incurred

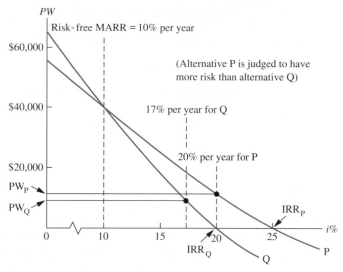

Figure 10-6 Graphical Portrayal of Risk-Adjusted Interest Rates (Example 10-8)

if conservative decision-making practices are not followed. Thus, dealing with uncertainty in an engineering economy study tends to lead to the adoption of conservative estimates of factors so as to reduce downside risks of making a wrong decision.

> The method considered in this section makes use of a truncated project life that is often considerably less than the estimated useful life. By dropping from consideration those revenues (savings) and expenses that may occur after the reduced study period, heavy emphasis is placed on rapid recovery of investment capital in the early years of a project's life. Consequently, this method is closely related to the discounted payback technique discussed in Chapter 4, and it suffers from most of the same deficiencies that beset the payback method.

EXAMPLE 10-9

Suppose that the Atlas Corporation referred to in Example 10-8 decided not to utilize risk-adjusted interest rates as a means of recognizing uncertainty in their engineering economy studies. Instead, they have decided to truncate the study period at 75% of the most likely estimate of useful life. Hence, all cash flows past the third year would be ignored in the analysis of alternatives. By using this method, should Alternative P or Q be selected when MARR=10% per year?

SOLUTION
Based on the PW criterion, it is apparent that neither alternative would be the choice with this procedure for recognizing uncertainty:

$$PW_P(10\%) = -\$160,000 + \$120,000(P/F, 10\%, 1)$$

$$+ \$60{,}000(P/F, 10\%, 2) = -\$1{,}324;$$

$$PW_Q(10\%) = -\$160{,}000 + \$20{,}827(P/F, 10\%, 1)$$

$$+ \$60{,}000(P/F, 10\%, 2)$$

$$+ \$120{,}000(P/F, 10\%, 3) = -\$1{,}324.$$

EXAMPLE 10-10

A proposed new product line requires $2,000,000 in capital investment over a two-year construction period. Projected revenues and expenses over this product's anticipated eight-year commercial life, along with its capital requirements are as follows:

Type of Cash Flow	End of Year (Millions of $)									
	−1	0	1	2	3	4	5	6	7	8
Capital investment	0.9	1.1	0	0	0	0	0	0	0	0
Revenues	0	0	1.8	2.0	2.1	1.9	1.8	1.8	1.7	1.5
Expenses	0	0	0.8	0.9	0.9	0.9	0.8	0.8	0.8	0.7

The company's maximum simple payback period is four years (after taxes), and its after-tax MARR is 15% per year. This investment will be depreciated using the MACRS (GDS) method and a five-year property class (Chapter 6). An effective income-tax rate of 40% applies to taxable income produced by this new product.

The company's management is quite concerned about the financial attractiveness of this venture if unforeseen circumstances (e.g., loss of market or technological breakthroughs) occur. They are cautious of investing a large sum of capital in this product because competition is quite keen and companies that wait to enter the market may be able to purchase more cost-efficient technology. You have been given the assignment of assessing the downside profitability of the product when the primary concern is its staying power (life) in the marketplace. That is, they must determine the minimum life of the product that will produce an acceptable after-tax IRR. Draw a graph of your results and list all appropriate assumptions.

SOLUTION

An analysis of after-tax cash flows is shown in Table 10-7 for the most likely product life of eight years.

TABLE 10-7 After-Tax Analysis of Example 10-10

End of Year, k	(A) BTCF	(B) Depreciation Deduction	(C) = (A) − (B) Taxable Income	(D) = −0.4(C) Cash Flow for Income Taxes	(E) = (A) + (D) ATCF
−1	−900,000	—	—	—	−900,000
0	−1,100,000	—	—	—	−1,100,000
1	1,000,000	$400,000	$600,000	240,000	760,000
2	1,100,000	640,000	460,000	184,000	916,000
3	1,200,000	384,000	816,000	326,400	873,600
4	1,000,000	230,400	769,600	307,840	692,160
5	1,000,000	230,400	769,600	307,840	692,160
6	1,000,000	115,200	884,800	353,920	646,080
7	900,000	0	900,000	360,000	540,000
8	800,000	0	800,000	320,000	480,000

It has been assumed that the residual (market) value of the investment is zero. Moreover, the MACRS depreciation deductions are assumed to be unaffected by the useful life of this product, and they begin in the first year of commercial operation (year one). A plot of after-tax IRR versus actual life of the product line is shown in Figure 10-7. To make at least 15% per year after taxes on this venture, the product's life must be four years or more. It can be quickly determined from Table 10-7 that the after-tax *simple* payback period is three years. Consequently, this new product would appear to be a judicious investment as long as its actual life turns out to be four years or greater.

Figure 10-7
IRR for Different
Product Lives in
Example 10-10

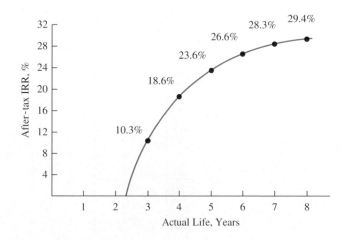

10.8 Spreadsheet Applications

Spreadsheet applications provide an excellent capability to answer "what-if" questions. In the following example, a spreadsheet is used to determine the sensitivity of a project's present worth to several factors.

EXAMPLE 10-11

In this example, we explore the impact on the present worth of an engineering project relative to changes in the capital investment, annual savings, market value, study period, and MARR.

Figure 10-8 shows the resulting table of present-worth values as each factor (variable) of the present-worth calculation is varied over a range of ±50% from the most likely estimate. Each column has a unique formula that refers to the factors located in the range C2: C6 to determine present worth. The particular factor of interest, for example the study period in column E, is multiplied by the factor (1 + % change) to create the table. You can verify your formulas by noting that all

Figure 10-8
Spreadsheet for
Performing a Sensitivity
Analysis (PW shown in
the table)

	A	B	C	D	E	F
1	Most Likely Estimates					
2	Capital Investment (I):		($50,000)			
3	Annual Savings (A):		$12,000			
4	Market Value (MV):		$5,000			
5	Study Period (N):		8			
6	MARR (i):		10%			
7						
8	% Change	I	A	MV	N	i
9						
10	-50%	$ 41,352	$ (15,658)	$ 15,185	$ (8,547)	$ 30,943
11	-40%	$ 36,352	$ (9,256)	$ 15,419	$ (2,780)	$ 27,655
12	-30%	$ 31,352	$ (2,854)	$ 15,652	$ 2,563	$ 24,566
13	-20%	$ 26,352	$ 3,548	$ 15,885	$ 7,514	$ 21,661
14	-10%	$ 21,352	$ 9,950	$ 16,118	$ 12,101	$ 18,927
15	0%	$ 16,352	$ 16,352	$ 16,352	$ 16,352	$ 16,352
16	10%	$ 11,352	$ 22,754	$ 16,585	$ 20,290	$ 13,923
17	20%	$ 6,352	$ 29,155	$ 16,818	$ 23,940	$ 11,631
18	30%	$ 1,352	$ 35,557	$ 17,051	$ 27,321	$ 9,466
19	40%	$ (3,648)	$ 41,959	$ 17,285	$ 30,454	$ 7,419
20	50%	$ (8,648)	$ 48,361	$ 17,518	$ 33,357	$ 5,482

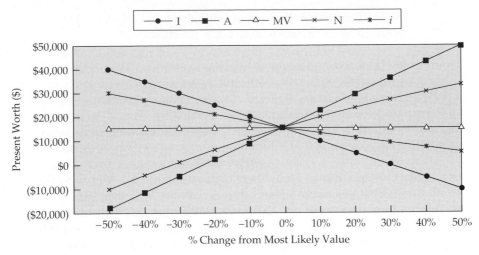

Figure 10-9 Sensitivity Analysis of Five Factors in Example 10-11

columns are equal at the most likely value (% Change = 0). The formulas in the highlighted cells in Figure 10-8 are as follows:

Cell	Contents
B10	= $C$2 ∗ (1+A10)+PV($C$6, $C$5, −$C$3)+$C$4/(1+$C$6)^$C$5
C10	= $C$2+PV($C$6, $C$5, −$C$3 ∗ (1+A10))+$C$4/(1+$C$6)^$C$5
D10	= $C$2+PV($C$6, $C$5, −$C$3)+$C$4 ∗ (1+A10)/(1+$C$6)^$C$5
E10	= $C$2+PV($C$6, $C$5 ∗ (1+A10), −$C$3)+$C$4/(1+$C$6)^($C$5 ∗ (1+A10))
F10	= $C$2+PV($C$6 ∗ (1+A10), $C$5, −$C$3)+$C$4/(1+$C$6 ∗ (1+A10))^$C$5

For ease of interpretation, it is helpful to graph the results of the sensitivity analysis, which is easily accomplished using the charting feature that exists in most spreadsheet packages. The graphed results (spiderplot) of this analysis are shown in Figure 10-9. The graph uses the "% Change" column as the independent (X) axis and columns B through F.

This graph indicates that present worth is most sensitive to annual savings. The next most influential factor is the capital investment, and the least sensitive is the market value (which is as expected because it is a small dollar amount and is more heavily discounted because it occurs at the end of the study period).

10.9 Summary

Engineering economy involves decision making among competing uses of scarce capital resources. The consequences of resultant decisions usually extend far into the future. In this chapter, we have used nonprobabilistic techniques to deal with the realization that the consequences (cash flows, useful lives, etc.) of engineering projects can never be known with absolute certainty. This situation is generally referred to as *decision making under uncertainty.*

Several of the most commonly applied and useful nonprobabilistic procedures for dealing with uncertainty in engineering economy studies have been presented in this chapter: (1) sensitivity analysis–break-even analysis, sensitivity graphs, and combinations of factors; (2) optimistic–pessimistic estimates; (3) risk-adjusted MARRs; and (4) reduction in useful life. Break-even analysis determines the value of a key common factor, such as utilization of capacity, at which the economic desirability of two alternatives is equal or a project is economically justified. This break-even point is then compared to an independent estimate of the factor's most likely (best estimate) value to assist with the selection between alternatives or to make a decision about a project. The sensitivity graph technique makes explicit the impact of uncertainty in the estimates of each project factor of concern on the economic measure of merit, and it is a valuable analysis tool. The techniques discussed in Section 10.4.3 for evaluating the combined impact of changes in two or more factors are important when the additional information provided is needed to assist decision making. The remaining procedures for dealing with uncertainty are aimed at selecting the best course of action when one or more consequences of the alternatives being evaluated lack estimation precision.

Regrettably, there is no quick and easy answer to the question "How should uncertainty best be considered in an engineering economic analysis?" Generally, simple procedures (e.g., sensitivity analysis) allow reasonable discrimination among alternatives to be made or the acceptability of a project to be determined on the basis of the uncertainties present, and they are relatively inexpensive to apply. Additional discrimination among alternatives or determining the acceptability of a project is possible with more complex procedures that utilize probabilistic concepts (Chapter 13), but their difficulty of application and expense may be prohibitive.

10.10 References

CANADA, J. R., W. G. SULLIVAN, and J. A. WHITE. *Capital Investment Decision Analysis for Engineering and Management,* 2nd ed. (Englewood Cliffs, NJ: Prentice Hall, Inc., 1996).

CHURCHMAN, C. W., R. L. ACKOFF, and E. L. ARNOFF. *Introduction to Operations Research* (New York: John Wiley & Sons, 1957).

FLEISCHER, G.A. *Introduction to Engineering Economy* (Boston; PWS Publishing Company, 1994).

GRANT, E. L., W. G. IRESON, and R. S. LEAVENWORTH. *Principles of Engineering Economy* (New York: John Wiley & Sons, 1990).

MORRIS, W. T. *The Analysis of Management Decisions* (Homewood, IL: Richard D. Irwin Co., 1964).

▼10.11 Problems

The number in parentheses ( ) that follows each problem refers to the section from which the problem is taken.

10-1. Why should the effects of uncertainty be considered in engineering economy studies? What are some likely sources of uncertainty in these studies? (10.3)

10-2. Construct your own *nonlinear* break-even analysis problem, develop a solution for it, and bring a one-page summary of your problem and solution to class for discussion. (10.4)

10-3. Refer to Example 10-3. This question is identical to Example 10-3 where the *extra* cost of strengthening the structure to accommodate two more floors is $300,000. This extra cost is uncertain. All other costs are assumed to be certain. For Design 2, there is no provision now for extra floors to be added later.

What is the sensitivity of the choice of Design 1 and Design 2 to ±30% changes in this uncertain cost estimate? Express this sensitivity in terms of $\hat{T}$. Draw a graph to illustrate your answer. The MARR is 10% per year.

10-4. A certain potential investment project is critical to a firm. The following are "best" or "most likely" estimates

Investment:	$100,000
Life:	10 yr
Salvage value:	$20,000
Net annual cash flow:	$30,000
MARR:	10%

It is desired to show the sensitivity of a measure of merit (net annual worth) to variation, over a range of ±50% of the expected values, in the following elements: (a) life, (b) net annual cash flow, and (c) interest rate. Graph the results. To which element is the decision most sensitive? (10.4)

10-5. Consider these two alternatives:

	Alternative 1	Alternative 2
Capital investment	$4,500	$6,000
Annual revenues	$1,600	$1,850
Annual expenses	$400	$500
Estimated market value	$800	$1,200
Useful life	8 yr	10 yr

a. Suppose that the market value of Alternative 1 is known with certainty. By how much would the estimate of market value for Alternative 2 have to vary so that the *initial* decision based on these data would be reversed? The annual MARR is 15% per year. (10.4.1)

b. Determine the life of Alternative 1 for which the AWs are equal. (10.4.1)

10-6. Two 100-horsepower motors are being considered for use in the accompanying table.

	ABC Brand	XYZ Brand
Purchase price	$1,900	$6,200
Useful life in years	10	10
Market value	none	none
Annual maintenance expense	$170	$310
Efficiency	80%	90%

a. If power cost is $0.10 per kWh, and the interest rate is 12% per year, how many hours of operation per year are required to justify the purchase of XYZ brand motor? (1 hp = 0.746 kW) (10.5)

b. Given your answer in Part (a), which motor would you select if the motor is expected to operate 2,000 hr per year? Explain why. (10.4.1)

10-7. The following alternatives are available to fill a given need that is expected to exist indefinitely:

	Plan A	Plan B	Plan C
Initial investment:	$2,000	$6,000	$12,000
Useful life:	6 yr	3 yr	4 yr
Annual expenses:	$3,500	$1,000	$400

Each plan is expected to have $0 market value at the end of each life cycle.

a. Analyze the sensitivity of the preferred plan due to ±30% errors in estimating the annual expenses. Use a MARR of 10%. (10.4)

b. Analyze the sensitivity of the preferred plan due to ±50% errors in estimating the MARR (i.e., the MARR will vary form 5% to 15%).

10-8. Two electric motors are being considered to power an industrial hoist. Each is capable of providing 90 hp. Pertinent data for each motor are as follows:

	Motor	
	D-R	Westhouse
Capital investment	$2,500	$3,200
Electrical efficiency	0.74	0.89
Maintenance per year	$40	$60
Useful life	10 yr	10 yr

If the expected usage of the hoist is 500 hr per year, what would the cost of electrical energy have to be (in cents per kilowatt-hour) before the D-R motor is favored over the Westhouse motor? The MARR is 12% per year. [*Note:* 1 hp = 0.746 kW.] (10.4.1)

10-9. Your company operates a fleet of light trucks that are used to provide contract delivery services. As the engineering and technical manager, you are analyzing the purchase of 55 new trucks as an addition to the fleet. These trucks would be used for a new contract the sales staff is trying to obtain. If purchased, the trucks would cost $21,200 each; estimated use is 20,000 miles per year per truck; estimated operation and maintenance and other related expenses (year zero dollars) are $0.45 per mile, which is forecasted to escalate (increase) at the rate of 5% per year; and the trucks are MACRS (GDS) three-year property class assets. The analysis period is four years; $t = 38\%$; MARR = 15% per year (after taxes; includes an inflation component); and the estimated MV at the end of four years (in year zero dollars) is 35% of the purchase price of the vehicles. This estimate is expected to escalate at the rate of 2% per year.

Based on an after-tax, actual dollar analysis, what is the annual revenue required by your company from the contract to justify these expenditures before any profit is considered? This calculated amount for annual revenue is the break-even point between purchasing the trucks and which other alternative? (10.4.1)

10-10. A nationwide motel chain is considering locating a new motel in Bigtown, USA. The cost of building a 150-room motel (excluding furnishings) is $5 million. The firm uses a 15-yr planning horizon to evaluate investments of this type. The furnishings for this motel must be replaced every five years at an estimated cost of $1,875,000 (at $k = 0, 5$, and 10). The old furnishings have no market value. Annual operating and maintenance expenses for the facility are estimated to be $125,000. The market value of the motel after 15 yr is estimated to be 20% of the original building cost.

Rooms at the motel are projected to be rented at an average rate of $45 per night. On the average, the motel will rent 60% of its rooms each night. Assume the motel will be open 365 days per year. The MARR is 10% per year. (10.4)

a. Using an annual-worth measure of merit, is the project economically attractive?

b. Investigate sensitivity to decision reversal for the following three factors: (1) capital investment, (2) MARR, and (3) occupancy rate (average percent of rooms rented per night). To which of these factors is the decision most sensitive?

c. Graphically investigate the sensitivity of the annual worth to changes in these three factors. Investigate changes over the interval $\pm 40\%$. On your graph, use percent change as the x-axis and annual worth as the y-axis.

10-11. An improved facility costing $50,000 has been proposed. Construction time will be two years with capital expenditures of $20,000 the first year and $30,000 the second year. Cash flows are as follows:

Year	Savings
−1	−$20,000
0	−$30,000
1	10,000
2	14,000
3	18,000
4	22,000
5	26,000

The facility will not be required after five years and will have a market value of $5,000. Analyze the sensitivity of annual worth due to errors in estimating both the savings in the first year and magnitude of the gradient amount. Use a table to show results of ±50% changes in both variables. The MARR is 10% per year. (10.4)

10-12. Consider the following cash-flow diagram:

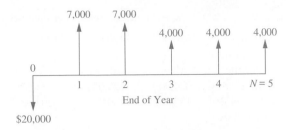

Insulation, in.	Heat loss, Btu per hr
3	4,400
4	3,400
5	2,800
6	2,400
7	2,000
8	1,800

Plot changes in PW to $\pm20\%$ and $\pm40\%$ changes in the project's life, N. Let $i = 10\%$ per year, and assume that MV = 0. State any other assumptions you make. (10.4.2)

10-13. A new steam-flow monitoring device must be purchased immediately by a local municipality. These most likely (best) estimates have been developed by a group of engineers:

Capital investment	$140,000
Annual savings	$25,000
Useful life	12 yr
Market value (end of year 12)	$40,000
MARR	10%/yr

Because considerable uncertainty surrounds these estimates, it is desired to evaluate the sensitivity of PW to $\pm50\%$ changes in the most likely estimates of (a) annual savings, (b) useful life, and (c) interest rate (MARR). Graph the results and determine to which factor the decision is most sensitive. (10.4.2)

10-14. It is desired to determine the most economic thickness of insulation for a large cold-storage room. Insulation is expected to cost $150 per 1,000 sq ft of wall area per inch of thickness installed and to require annual property taxes and insurance of 5% of the capital investment. It is expected to have $0 market value after a 20-yr life. The following are estimates of the heat loss per 1,000 sq ft of wall area for several thicknesses:

The cost of heat removal (loss) is estimated at $0.02 per 1,000 Btu. The MARR is 20% per year. Assuming continuous operation throughout the year, analyze the sensitivity of the optimal thickness to errors in estimating the cost of heat removal. Use the AW technique. (A computer spreadsheet should be considered here.) (10.4, 10.9)

10-15. An industrial machine costing $10,000 will produce net cash savings of $4,000 per year. The machine has a five-year useful life, but must be returned to the factory for major repairs after three years of operation. These repairs cost $5,000. The company's MARR is 10% per year. What internal rate of return will be earned on the purchase of this machine? Analyze the sensitivity of the internal rate of return to $\pm$2,000 changes in the repair cost. (10.4)

10-16. It is desired to determine the optimal height for a proposed building that is expected to last 40 years and then be demolished at zero net market value. Pertinent data are shown in Table P10-16. In addition to the building capital investment, the land requires an investment of $50,000 and is expected to retain that value throughout the useful life period. Analyze the sensitivity of the decision due to changes in estimates of the MARR between 10%, 15%, and 20%. Use the PW method and ignore income taxes. (10.4)

TABLE P10-16 Data for Problem P10-16				
	Number of floors			
	2	3	4	5
Capital investment	$200,000	$250,000	$320,000	$400,000
Annual revenue	40,000	60,000	85,000	100,000
Annual expenses	15,000	25,000	25,000	45,000

10-17. An office building is considering converting from a coal-burning furnace to one that burns either fuel oil or natural gas. The cost of converting to fuel oil is estimated to be $80,000 initially; annual operating expenses are estimated to be $4,000 less than that experienced using the coal furnace. Approximately 140,000 Btus are produced per gallon of fuel oil; fuel oil is anticipated to cost $1.10 per gallon.

The cost of converting to natural gas is estimated to be $60,000 initially; additionally, annual operating and maintenance expenses are estimated to be $6,000 less than that for the coal-burning furnace. Approximately 1,000 Btus are produced per cubic foot of natural gas; it is estimated natural gas will cost $0.02 per cu ft.

A planning horizon of 20 years is to be used. Zero market values and a MARR of 10% per year are appropriate. Perform a sensitivity analysis for the annual Btu requirement for the heating system. (*Hint:* First calculate the break-even number of Btus (in thousands). Then determine AWs if Btu requirement varies over ±30% of the break-even amount.) (10.4)

10-18. Suppose that for an engineering project the optimistic, most likely, and pessimistic estimates are as shown in the accompanying table. (10.4.3)

	Optimistic	Most Likely	Pessimistic
Capital investment	$80,000	$95,000	$120,000
Useful life	12 yr	10 yr	6 yr
Market value	$30,000	$20,000	$0
Net annual cash flow	$35,000	$30,000	$20,000
MARR	12%/yr	12%/yr	12%/yr

a. What is the AW for each of the three estimation conditions?
b. It is thought that the most critical elements are useful life and net annual cash flow. Develop a table showing the AW for all combinations of the estimates for these two factors, assuming that all other factors remain at their most likely values.

10-19. Suppose that, for a certain potential investment project, the optimistic, most likely, and pessimistic estimates are as shown in the accompanying table. (10.4.3)

	Optimistic	Most Likely	Pessimistic
Capital investment	$90,000	$100,000	$120,000
Useful life	12 yr	10 yr	6 yr
Market value	$30,000	$20,000	$0
Net annual cash flow	$35,000	$30,000	$20,000
MARR (per year)	10%	10%	10%

a. What is the annual worth for each of the three estimation conditions?
b. It is thought that the most critical factors are useful life and net annual cash flow. Develop a table showing the net annual worth for all combinations of the estimates for these two factors assuming all other factors to be at their "most likely" values.

10-20. A bridge is to be constructed now as part of a new road. Engineers have determined that traffic density on the new road will justify a two-lane road and a bridge at the present time. Because of uncertainty regarding future use of the road, the time at which an extra two lanes will be required is currently being studied.

The two-lane bridge will cost $200,000 and the four-lane bridge, if built initially, will cost $350,000. The future cost of widening a two-lane bridge to four lanes will be an extra $200,000 plus $25,000 for every year that widening is delayed. The MARR used by the highway department is 12% per year. The following estimates have been made of the times at which the four-lane bridge will be required:

Pessimistic estimate	4 yr
Most likely estimate	5 yr
Optimistic estimate	7 yr

In view of these estimates, what would you recommend? What difficulty, if any, do you have in interpreting your results? List some advantages and disadvantages of this method of preparing estimates. (10.4.3)

10-21. Individual industries will use energy as efficiently as it is economical to do so, and there are several incentives to improve the efficiency of energy consumption.

To illustrate, consider the selection of a new electric motor-driven water pump. The pump is to operate 800 hours per year. Pump A costs $2,000, has an overall efficiency of 82.06%, and delivers 11 hp. The other available alternative, Pump B, costs $1,000, has an overall efficiency of 45.13%, and delivers 12.1 hp. Both pumps have a useful life of five years and will be sold at that time. (Remember, 1 hp = 0.746 kW.) Assume that there is no use for the extra pumping capacity of Pump B.

    Pump A will use SL depreciation over five years, with an estimated SV of zero. Pump B will use the MACRS depreciation method with a property class of three years. After five years, Pump A has an MV $400 and Pump B has an MV of $200.

    Using the IRR method based on ACTFs and a *before-tax* MARR of 16.667%, is the *increased investment* in Pump A economically justifiable? The effective income-tax rate is 40%. The cost of electricity is $0.05 per kWh, and the pumps are subject to a study period of five years. Work this problem on an after-tax basis. (10.7)

10-22. Motor XYZ in Problem 10-6 is manufactured in a foreign country and is believed to be less reliable than motor ABC. To cope with this uncertainty, a risk-adjusted MARR of 20% is utilized to calculate its AW. When hours of operation per year total 1,000, which motor would be selected? What difficulty is encountered with this method? (10.6)

10-23. Refer to Example 10-7, Section 10.5. For this small premixed-concrete plant, do the following: (10.4.2, 10.4.3)

a. Develop an appropriate sensitivity graph (spiderplot). Include any additional values of the factors you consider necessary. Also, include raw material costs as an additional factor in the sensitivity graph under the assumption in this problem that all competitors may not respond to changes in these costs the same way.

b. For the two project factors that individually impact the AW the most (the two most sensitive factors), use the graphical technique applied in Example 10-5 to show more explicitly their combined impact on the AW.

c. For the three most sensitive factors, further analyze the combined effect of their outcomes on the AW. (You determine how to best formulate the combinations of factor outcomes—Example 10-6 illustrated one approach with O-ML-P estimating; however, you can develop three or four scenarios with selected changes in specific factors.)

10-24. *Brain Teaser* (10-4, 10-7, 10-8)
Consider these two alternatives for solid-waste removal:

Alternative A: Build a solid-waste processing facility. Financial variables are as follows:

Capital investment	$108 million in 2004 (commercial operation starts in 2004)
Expected life of facility	20 yr
Annual operating expenses	$3.46 million (expressed in 2004 dollars)
Estimated market value	40% of initial capital cost at all times

Alternative B: Contract with vendors for solid-waste disposal after intermediate recovery. Financial variables are as follows:

Capital investment	$17 million in 2004 (This is for *intermediate* recovery from the solid-waste stream.)
Expected contract period	20 yr
Annual operating expenses	$2.10 million (in 2004 dollars)
Repairs costs to intermediate recovery system every five years	$3.0 million (in 2004 dollars)
Annual fee to vendors	$10.3 million (in 2004 dollars)
Estimated market value at all times	$0

Related Data:

MACRS (GDS) property class	15 yr (Chapter 6)
Study period	20 yr
Effective income-tax rate	40%
Company MARR (after-tax)	10% per year
Inflation rate	0% (ignore inflation)

a. How much more expensive (in terms of capital investment only) could Alternative B be in order to break even with Alternative A?

b. How sensitive is the after-tax PW of Alternative B to cotermination of both alternatives at the end of *year 10*?

c. Is the initial decision to adopt Alternative B in (a) reversed if our company's annual operating expenses for Alternative B only ($2.10 million per year) unexpectedly double? Explain why (or why not).

d. Use a computer spreadsheet available to you to solve this problem.

*A*dditional Topics
in Engineering Economy

Money is the seed of money, and the first guinea is sometimes more difficult to acquire than the second million
—Jean Jacques Rousseau, "A Discourse on Political Economy," in *The Social Contract*, 1762

Evaluating Projects with the Benefit–Cost-Ratio Method

*T*he objectives of this chapter are (1) to describe many of the unique characteristics of public projects and (2) to learn how to use the benefit–cost (B–C) ratio as a criterion for project selection. Consideration of both independent projects and mutually exclusive projects is presented.

The following topics are discussed in this chapter:

> The perspective and terminology associated with public projects
> Self-liquidating and multipurpose projects
> Difficulties in evaluating public-sector projects
> The interest rate that should be used for public projects
> The benefit–cost-ratio method
> Evaluating independent projects by B–C ratios
> Comparison of mutually exclusive alternatives
> Criticisms and shortcomings of the B–C-ratio method

▼ 11.1 Introduction

Public projects are those authorized, financed, and operated by federal, state, or local governmental agencies. Such public works are numerous, and although they may be of any size, they are frequently much larger than private ventures. Since they require the expenditure of capital, such projects are subject to the principles of engineering economy with respect to their design, acquisition, and operation. However, because they are public projects, a number of important special factors exist that are not ordinarily found in privately financed and operated businesses. The differences between public and private projects are listed in Table 11-1.

TABLE 11-1 Some Basic Differences between Privately Owned and Publicly Owned Projects

	Private	Public
Purpose	Provide goods or services at a profit; Maximize profit or minimize cost	Protect health; Protect lives and property; Provide services (at no profit); Provide jobs
Sources of capital	Private investors and lenders	Taxation; Private lenders
Method of financing	Individual ownership; Partnerships; Corporations	Direct payment of taxes; Loans without interest; Loans at low interest; Self-liquidating bonds; Indirect subsidies; Guarantee of private loans
Multiple purposes	Moderate	Common (e.g., reservoir project for flood control, electrical power generation, irrigation, recreation, education)
Project life	Usually relatively short (5 to 20 yr)	Usually relatively long (20 to 60 yr)
Relationship of suppliers of capital to project	Direct	Indirect, or none
Nature of "benefits"	Monetary or relatively easy to equate to monetary terms	Often nonmonetary, difficult to quantify, difficult to equate to monetary terms
Beneficiaries of project	Primarily, entity undertaking project	General public
Conflict of purposes	Moderate	Quite common (dam for flood control vs. environmental preservation)
Conflict of interests	Moderate	Very common (between agencies)
Effect of politics	Little to moderate	Frequent factors; Short-term tenure for decision makers; Pressure groups; Financial and residential restrictions; etc.
Measurement of efficiency	Rate of return on capital	Very difficult; No direct comparison with private projects

As a consequence of these differences, it is often difficult to make engineering economy studies and investment decisions for public-works projects in exactly the same manner as for privately owned projects. Different decision criteria are often used, which creates problems for the public (which pays the bill), for those who must make the decisions, and for those who must manage public-works projects.

The benefit–cost-ratio method, which is normally used for the evaluation of public projects, has its roots in federal legislation. Specifically, the Flood Control Act of 1936 requires that for a federally financed project to be justified, its benefits must be in excess of its costs. In general terms, benefit–cost analysis is a systematic method of assessing the desirability of government projects or policies when it is important to take a long-term view of future effects and a broad view of possible side effects. In meeting the requirements of this mandate, the B–C method evolved into the calculation of a ratio of project benefits to project costs. Rather than allowing the analyst to apply criteria more commonly used for evaluating private projects (IRR, PW, etc.), most governmental agencies require the use of the B–C method.

11.2 Perspective and Terminology for Analyzing Public Projects

Before applying the benefit–cost-ratio method to evaluate a public project, the appropriate perspective must be established. In conducting an engineering economic analysis of any project, whether it is a public or a private undertaking, the proper perspective is to maximize the net benefits to the owners of the enterprise considering the project. This process requires that the question of who owns the project be addressed. Consider, for example, a project involving the expansion of a section of I-80 from four to six lanes. Because the project is paid for primarily with federal funds channeled through the Department of Transportation, we might be inclined to say that the federal government is the "owner." These funds, however, originated from tax dollars—thus, the true owners of the project are the taxpayers.

As mentioned previously, the benefit–cost method requires that a ratio of benefits to costs be calculated. Project *benefits* are defined as the favorable consequences of the project to the public, but project *costs* represent the monetary disbursement(s) required of the government. It is entirely possible, however, for a project to have unfavorable consequences to the public. Considering again the widening of I-80, some of the *owners* of the project—farmers along the interstate—would lose a portion of their arable land, along with a portion of their annual revenues. Because this negative financial consequence is borne by (a segment of) the public, it cannot be classified as either a benefit or a cost. The term *disbenefits* is generally used to represent the negative consequences of a project to the public.

EXAMPLE 11-1

A new convention center and sports complex has been proposed to the Gotham City Council. This public-sector project, if approved, will be financed through the issue of municipal bonds. The facility will be located in the City Park near downtown Gotham City, in a wooded area, which includes a bike path, a nature trail, and a pond. Because the city already owns the park, no purchase of land is necessary. List separately the project's *benefits, costs,* and any *disbenefits.*

SOLUTION

BENEFITS
Improvement of the image of the downtown area of Gotham City
Potential to attract conferences and conventions to Gotham City
Potential to attract professional sports franchises to Gotham City
Revenues from rental of the facility
Increased revenues for downtown merchants of Gotham City
Use of facility for civic events

COSTS:	Architectural design of the facility
	Construction of the facility
	Design and construction of parking garage adjacent to the facility
	Facility operating and maintenance costs
	Facility insurance costs
DISBENEFITS:	Loss of use of a portion of the City Park to Gotham City residents, including the bike path, the nature trail, and the pond
	Loss of wildlife habitat in urban area

▼ 11.3 Self-Liquidating Projects

The term *self-liquidating project* is applied to a governmental project that is expected to earn direct revenue sufficient to repay its cost in a specified period of time. Most of these projects provide utility services—for example, the fresh water, electric power, irrigation water, and sewage disposal provided by a hydroelectric dam. Other examples of self-liquidating projects include toll bridges and highways.

As a rule, self-liquidating projects are expected to earn direct revenues that offset their costs, but they are not expected to earn profits or pay income taxes. Although they also do not pay property taxes, in some cases in-lieu payments are made to state, county, or municipal governments in place of the property or franchise taxes that would have been paid had the project been under private ownership. For example, the U.S. government agreed to pay the states of Arizona and Nevada $300,000 each annually for 50 yr in-lieu of taxes that would have accrued if Hoover Dam had been privately constructed and operated. These in-lieu payments are usually considerably less than the actual property and franchise taxes would have been. Furthermore, once such payments are agreed upon, usually at the origination of the project, they are virtually never changed thereafter. These unchanging payments are not the case with property taxes, which are based upon the appraised value of the property.

▼ 11.4 Multiple-Purpose Projects

An important characteristic of public-sector projects is that many such projects have multiple purposes or objectives. One example of this would be the construction of a dam to create a reservoir on a river. (See Figure 11-1.) This project would have multiple purposes: (1) assist in flood control, (2) provide water for irrigation, (3) generate electric power, (4) provide recreational facilities, and (5) provide drinking water. Developing such a project to meet more than one objective ensures that greater overall economy can be achieved. Because the construction of a dam involves very large sums of capital and the use of a valuable natural resource—a river—it is likely that the project could not be justified unless it served multiple purposes. This type of situation is generally desirable, but, at the same time, it creates economic and managerial problems due to the overlapping utilization of facilities and the possibility of a conflict of interest between the several purposes and the agencies involved.

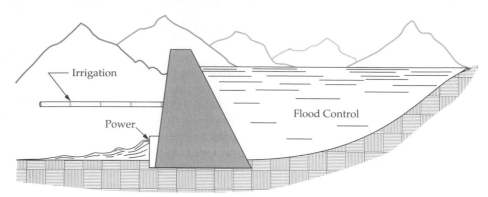

Figure 11-1 Schematic Representation of a Multiple-Purpose Project Involving Flood Control, Irrigation, and Power

The basic problems that often arise in evaluating public projects can be illustrated by returning to the dam shown in Figure 11-1. The project under consideration is to be built in the semi-arid central portion of California, primarily to provide control against spring flooding resulting from the melting snow in the Sierra Nevadas. If a portion of the water impounded behind the dam could be diverted onto the adjoining land below the dam, the irrigation water would greatly increase the productivity, and thus the value, of that land, which would result in an increase in the nation's resources. So the objectives of the project should be expanded to include both flood control and irrigation.

The existence of a dam with a high water level on one side and a much lower level on the other side also suggests that some of the nation's resources will be wasted unless a portion of the water is diverted to run through turbines, generating electric power. This electricity can be sold to customers in the areas surrounding the reservoir, giving the project the third purpose of generating electric power.

In this semiarid region, the creation of a large reservoir behind the dam would provide valuable facilities for hunting, fishing, boating, swimming, and camping. Thus, the project has the fourth purpose of providing recreation facilities. A fifth purpose would be the provision of a steady, reliable supply of drinking water.

Each of the above-mentioned objectives of the project has desirable economic and social value, so what started out a single-purpose project now has five purposes. The failure to meet all five objectives would mean that valuable national resources are being wasted. On the other hand, there are certain *disbenefits* to the public that must also be considered. Most apparent of these is the loss of farm land above the dam in the area covered by the reservoir. Other disbenefits might include (1) the loss of a white-water recreational area enjoyed by canoeing, kayaking, and rafting enthusiasts, (2) the loss of annual deposits of fertile soil in the river basin below the dam due to spring flooding, and (3) the negative ecological impact of obstructing the flow of the river.

If the project is built to serve five purposes, the fact that one dam will serve all of them leads to at least three basic problems. The *first* of these is the allocation of the cost of the dam to each of its intended purposes. Suppose, for example, that the estimated costs of the project are $35,000,000. This figure includes costs incurred for the purchase and preparation of land to be covered by water above the dam site; the actual construction of the dam, the irrigation system, the power generation plant, and the purification and pumping stations for drinking water; and the design/development of recreational facilities. The allocation of some of these costs to specific purposes is obvious (e.g., the cost of constructing the irrigation system), but what portion of the cost of purchasing and preparing the land should be assigned to flood control? What amounts should be assigned to irrigation, power generation, drinking water, and recreation?

The *second* basic problem is the conflict of interest among the several purposes of the project. Consider the decision as to the water level to be maintained behind the dam. In meeting the first purpose—flood control—the reservoir should be maintained at a near-empty level to provide the greatest storage capacity during the months of the spring thaw. This lower level would be in direct conflict with the purpose of power generation, which could be maximized by maintaining as high a level as possible behind the dam at all times. Further, maximizing the recreational benefits would suggest that a constant water level be maintained throughout the year. Thus, conflicts of interest arise among the multiple purposes, and compromise decisions must be made. These decisions ultimately affect the magnitude of benefits resulting from the project.

A *third* problem with multiple-purpose public projects is political sensitivity. Because each of the various purposes, or even the project itself, is likely to be desired or opposed by some segment of the public and by various interest groups that may be affected, inevitably such projects frequently become political issues.*This conflict often has an effect on cost allocations and thus on the overall economy of these projects.

> The net result of these three factors is that the cost allocations made in multiple-purpose public sector projects tend to be arbitrary. As a consequence, production and selling costs of the services provided also are arbitrary. Because of this fact, they cannot be used as valid yardsticks with which similar private-sector projects can be compared to determine the relative efficiencies of public and private ownership.

11.5 Difficulties in Evaluating Public-Sector Projects

With all of the difficulties that have been cited in evaluating public-sector projects, we may wonder whether engineering economy studies of such projects should be attempted. In most cases, economy studies cannot be made in as complete, comprehensive, and satisfactory a manner as in the case of studies of privately financed projects. In the private sector, the *costs* are borne by the firm undertaking the project, and the *benefits* are the favorable outcomes of the project accrued by the

---

*The construction of the Tellico Dam on the Little Tennessee River was considerably delayed due to two important *disbenefits*: (1) concerns over the impact of the project on the environment of a small fish, the snail darter, and (2) the flooding of burial grounds considered sacred by the Cherokee Nation.

firm. Any costs and benefits that occur outside of the firm are generally ignored in evaluations unless it is anticipated that those external factors will indirectly impact the firm. But the opposite is true in the case of public-sector projects. In the wording of the Flood Control Act of 1936, "if the benefits to whomsoever they may accrue are in excess of the estimated costs," all of the potential benefits of a public project are relevant and should be considered. Simply enumerating all of the benefits for a large-scale public project is a formidable task! Further, the monetary value of these benefits to all of the affected segments of the public must somehow be estimated. Regardless, decisions about the investment of capital in public projects must be made by elected or appointed officials, by managers, or by the general public in the form of referendums. Because of the magnitude of capital and the long-term consequences associated with many of these projects, following a systematic approach for evaluating their worthiness is vital.

There are a number of difficulties inherent in public projects that must be considered in conducting engineering economy studies and making economic decisions regarding those projects. Some of these are as follows:

1. There is no profit standard to be used as a measure of financial effectiveness. Most public projects are intended to be nonprofit.
2. The monetary impact of many of the benefits of public projects is difficult to quantify.
3. There may be little or no connection between the project and the public, which is the owner of the project.
4. There is often strong political influence whenever public funds are used. When decisions regarding public projects are made by elected officials who will soon be seeking reelection, *the immediate benefits and costs are stressed, often with little or no consideration for the more important long-term consequences.*
5. The usual profit motive as a stimulus to promote effective operation is absent, which is not intended to imply that all public projects are ineffective or that managers and employees are not attempting to do their jobs efficiently. But the direct profit stimuli present in privately owned firms are considered to have a favorable impact on project effectiveness in the private sector.
6. Public projects are usually much more subject to legal restrictions than are private projects. For example, the area of operations for a municipally owned power company may be restricted such that power can be sold only within the city limits, regardless of whether a market for any excess capacity exists outside the city.
7. The ability of governmental bodies to obtain capital is much more restricted than with private enterprises.
8. The appropriate interest rate for discounting the benefits and costs of public projects is often controversially and politically sensitive. Clearly, lower interest rates favor long-term projects having major social or monetary benefits in the future, whereas higher interest rates promote a short-term outlook whereby decisions are based mostly on initial investments and immediate benefits.

A discussion of several viewpoints and considerations that are often used to establish an appropriate interest rate for public projects is included in the next section.

▼ 11.6 What Interest Rate Should Be Used for Public Projects?

When public-sector projects are evaluated, interest rates play the same role of accounting for the time value of money as in the evaluation of projects in the private sector. The rationale for the use of interest rates, however, is somewhat different. The choice of an interest rate in the private sector is intended to lead directly to a selection of projects to maximize profit or minimize cost. In the public sector, on the other hand, projects are not usually intended as profit-making ventures. Instead, the goal is the *maximization of social benefits,* assuming that these have been appropriately measured. The choice of an interest rate in the public sector is intended to determine how available funds should best be allocated among competing projects to achieve social goals. The relative differences in magnitude of interest rates between governmental agencies, regulated monopolies, and private enterprises are illustrated in Figure 11-2.

Three main considerations bear on what interest rate to use in engineering economy studies of public-sector projects:

1. The interest rate on borrowed capital
2. The opportunity cost of capital to the governmental agency
3. The opportunity cost of capital to the taxpayers

As a general rule, it is appropriate to use the interest rate on borrowed capital as the interest rate for cases in which money is borrowed specifically for the project(s) under consideration. For example, if municipal bonds are issued specifically for the financing of a new school, the effective interest rate on those bonds should be the interest rate.

For public-sector projects, the opportunity cost of capital to a governmental agency encompasses the annual rate of *benefit* to either the constituency served by that agency or the composite of taxpayers who will eventually pay for the project. If projects are selected such that the estimated *return* (in terms of benefits) on all accepted projects is higher than that on any of the rejected projects, then the interest

Figure 11-2
Relative Differences
in Interest Rates
for Governmental
Agencies, Regulated
Monopolies, and
Private Enterprise

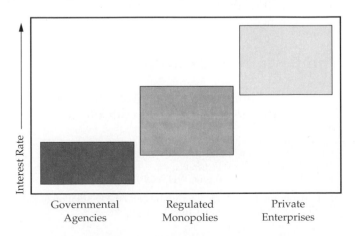

rate used in economic analyses is that associated with the best opportunity forgone. If this process is done for all projects and investment capital available within a governmental agency, the result is *an opportunity cost of capital for that governmental agency.* A strong argument against this philosophy, however, is that the different funding levels of the various agencies and the different nature of projects under the direction of each agency would result in different interest rates for each of the agencies, even though they all share a common primary source of funds—taxation of the public.

The third consideration—the opportunity cost of capital to the taxpayers—is based on the philosophy that all government spending takes potential investment capital away from the taxpayers. The taxpayers' opportunity cost is generally greater than either the cost of borrowed capital or the opportunity cost to governmental agencies, and there is a compelling argument for applying the largest of these three rates as the interest rate for evaluating public projects; it is not economically sound to take money away from a taxpayer to invest in a government project yielding benefits at a rate less than what could have been earned by that taxpayer.

This argument was supported by a federal government directive issued in 1992—and still in force—by the Office of Management and Budget (OMB).* According to this directive, a 7% interest rate should be used in economic evaluations for a wide range of federal projects, with certain exceptions (e.g., a lower rate can be applied in evaluating water resource projects). This 7%, it can be argued, is at least a rough approximation of the real-dollar return that taxpayers could earn from the use of that money for private investment. This corresponds to an approximate nominal (market) return of 10% per year.

One additional theory on establishing interest rates for federal projects advocates that the "social discount rate" used in such analyses should be the market-determined *risk-free* rate for private investments.[†] According to this theory, a nominal interest rate on the order of 3–4% per year should be used.

The preceding discussion focuses on the *considerations* that should play a role in establishing an interest rate for public projects. As in the case of the private sector, there is no simple formula for determining the appropriate interest rate for public projects. With the exception of projects falling under the 1992 OMB directive, setting the interest rate is ultimately a policy decision at the discretion of the governmental agency conducting the analysis.

▼ 11.7 The Benefit–Cost-Ratio Method

As the name implies, the benefit–cost-ratio method involves the calculation of a ratio of benefits to costs. Whether evaluating a project in the private sector or in

---

[*] Office of Management and Budget, "Guidelines and Discount Rates for Benefit–Costs Analysis of Federal Programs," *OMB Circular No. A-94 (revised),* February 21, 1997. The OMB home page is http://www.whitehouse.gov/WH/EOP/omb.

[†] K. J. Arrow and R. C. Lind, "Uncertainty and the Evaluation of Public Investment Decisions," *American Economic Review,* vol. 60, June 1970, pp. 364–378.

the public sector, the time value of money must be considered to account for the timing of cash flows (or benefits) occurring after the inception of the project. Thus, the B–C ratio is actually a ratio of *discounted benefits* to *discounted costs*.

Any method for formally evaluating projects in the public sector must consider the worthiness of allocating resources to achieve social goals. For over 60 years, the B–C-ratio method has been the accepted procedure for making go/no-go decisions on independent projects and for comparing mutually exclusive projects in the public sector, even though the other methods discussed in Chapter 4 (PW, AW, IRR, etc.) will lead to identical recommendations, *assuming all these procedures are properly applied.*

Accordingly, the purpose of this section is to describe and illustrate the mechanics of the B–C ratio method for evaluating projects. Two different B–C ratios will be presented because they are used in practice by various government agencies and municipalities. Both ratios lead to the *identical choice* of which project is best when comparing mutually exclusive alternatives.

The B–C ratio is defined as the ratio of the equivalent worth of benefits to the equivalent worth of costs. The equivalent-worth measure applied can be present worth, annual worth, or future worth, but customarily, either PW or AW is used. An interest rate for public projects, as discussed in the previous section, is used in the equivalent worth calculations. The benefit–cost ratio is also known as the *Savings-Investment Ratio (SIR)* by some governmental agencies.

Several different formulations of the B–C ratio have been developed. Two of the more commonly used formulations are presented in this section, illustrating the use of both present worth and annual worth.

Conventional B–C ratio with PW:

$$\text{B–C} = \frac{\text{PW(benefits of the proposed project)}}{\text{PW(total costs of the proposed project)}} = \frac{\text{PW}(B)}{I + \text{PW(O\&M)}} \qquad (11\text{-}1)$$

where   PW($\cdot$) = present worth of ($\cdot$);

B = benefits of the proposed project;

I = initial investment in the proposed project;

O&M = operating and maintenance costs of the proposed project.

Modified B–C ratio with PW:

$$\text{B–C} = \frac{\text{PW}(B) - \text{PW(O\&M)}}{I} \qquad (11\text{-}2)$$

The numerator of the modified benefit–cost ratio expresses the equivalent worth of the benefits minus the equivalent worth of the O&M costs, and the denominator includes only the initial investment costs. A project is acceptable when the B–C ratio, as defined in either Equation (11-1) or (11-2), is greater than or equal to 1.0.

Equations (11-1) and (11-2) can be rewritten in terms of equivalent annual worth as follows:

Conventional B–C ratio with AW:

$$B\text{–}C = \frac{AW(\text{benefits of the proposed project})}{AW(\text{total costs of the proposed project})} = \frac{AW(B)}{CR + AW(O\&M)} \qquad (11\text{-}3)$$

where $AW(\cdot)$ = annual worth of $(\cdot)$;

$\quad B$ = benefits of the proposed project;

$\quad CR$ = capital-recovery amount (i.e., the equivalent annual cost of the initial investment, I, including an allowance for market, or salvage value, if any)

$\quad O\&M$ = operating and maintenance costs of the proposed project

Modified B–C ratio with AW:

$$B\text{–}C = \frac{AW(B) - AW(O\&M)}{CR} \qquad (11\text{-}4)$$

Note that when using the annual-worth approach, the annualized equivalent of any *market value* associated with the investment is effectively subtracted from the denominator in the calculation of the capital-recovery amount (CR) in Equations (11-3) and (11-4). Similarly, when using the present-worth approach to calculate a benefit–cost ratio, it is customary to reduce the investment in the denominator by the discounted equivalent of any market value. Equations (11-1) and (11-2) are rewritten as follows to incorporate the market value of an investment:

Conventional B–C ratio with PW, Market Value included:

$$B\text{–}C = \frac{PW(\text{benefits of the proposed project})}{PW(\text{total costs of the proposed project})} = \frac{PW(B)}{I - PW(MV) + PW(O\&M)}$$
$$(11\text{-}5)$$

where

$\quad PW(\cdot)$ = present worth of $(\cdot)$;

$\quad B$ = benefits of the proposed project;

$\quad I$ = initial investment in the proposed project;

$\quad MV$ = market value of investment;

$\quad O\&M$ = operating and maintenance costs of the proposed project.

Modified B–C ratio with PW, Market Value included:

$$B\text{–}C = \frac{PW(B) - PW(O\&M)}{I - PW(MV)} \qquad (11\text{-}6)$$

The resulting B–C ratios for all the previous formulations will give identical results in determining the acceptability of a project (i.e., either B–C $\geq$ 1.0 or B–C < 1.0). The conventional B–C ratio will give identical numerical results for both PW and AW formulations; similarly, the modified B–C ratio gives identical numerical results whether PW or AW is used. Although the magnitude of the B–C ratio will differ between conventional and modified B–C, go/no-go decisions are not affected by the choice of approach, as shown in Example 11-2.

In examples presented in the remainder of Chapter 11, we assume a nominal (market) interest rate is used to discount "actual dollar" cash flows. The curious reader is referred to Chapter 8 for definitions of these terms.

EXAMPLE 11-2

The city of Bugtussle is considering extending the runways of its municipal airport so that commercial jets can use the facility. The land necessary for the runway extension is currently farmland, which can be purchased for $350,000. Construction costs for the runway extension are projected to be $600,000, and the additional annual maintenance costs for the extension are estimated to be $22,500. If the runways are extended, a small terminal will be constructed at a cost of $250,000. The annual operating and maintenance costs for the terminal are estimated at $75,000. Finally, the projected increase in flights will require the addition of two air traffic controllers, at an annual cost of $100,000. Annual *benefits* of the runway extension have been estimated as follows:

$325,000	Rental receipts from airlines leasing space at the facility
$65,000	Airport tax charged to passengers
$50,000	Convenience benefit for residents of Bugtussle
$50,000	Additional tourism dollars for Bugtussle

Apply the B–C-ratio method with a study period of 20 yr and a nominal interest rate of 10% per year to determine whether the runways at Bugtussle Municipal Airport should be extended.

SOLUTION

Conventional B–C: *Eqn. 11-1*	B–C = PW(B)/[I + PW(O&M)] B–C = $490,000 (P/A, 10%, 20)/[$1,200,000 + $197,500 (P/A, 10%, 20)] **B–C = 1.448 > 1; extend runways.**
Modified B–C: *Eqn. 11-2*	B–C = [PW(B) − PW(O&M)]/I B–C = [$490,000 (P/A, 10%, 20) − $197,500 (P/A, 10%, 20)]/$1,200,000 **B–C = 2.075 > 1; extend runways.**
Conventional B–C: *Eqn. 11-3*	B–C = AW(B)/[CR + AW(O&M)] B–C = $490,000/[$1,200,000 (A/P, 10%, 20) + $197,500] **B–C = 1.448 > 1; extend runways.**
Modified B–C: *Eqn. 11-4*	B–C = [AW(B) − AW(O&M)]/CR B–C = [$490,000 − $197,500]/[$1,200,000 (A/P, 10%, 20)] **B–C = 2.075 > 1; extend runways.**

As can be seen in the preceding example, the difference between conventional and modified B–C ratios is essentially due to subtracting the equivalent-worth measure of operating and maintenance costs from both the numerator and the denominator of the B–C ratio. In order for the B–C ratio to be greater than 1.0, the numerator must be greater than the denominator. Similarly, the numerator must be less than the denominator for the B–C ratio to be less than 1.0. Subtracting a constant (the equivalent worth of O&M costs) from both numerator and denominator does not alter the *relative* magnitudes of the numerator and denominator. Thus, project acceptability is not affected by the choice of conventional versus modified B–C ratio. This information is stated mathematically as follows for the case of B–C > 1.0:

Let N = **the numerator of the conventional B–C ratio;**

D = **the denominator of the conventional B–C ratio;**

$O\&M$ = **the equivalent worth of operating and maintenance costs.**

If **B–C** $= \dfrac{N}{D}$ *> 1.0, then N > D.*

If N > D, and $[N - O\&M] > [D - O\&M]$, *then* $\dfrac{N - O\&M}{D - O\&M}$ *> 1.0.*

Note that $\dfrac{N - O\&M}{D - O\&M}$ *is the modified* **B–C** *ratio, thus if conventional* **B–C** *> 1.0, then modified* **B–C** *> 1.0.*

Two additional issues of concern are the treatment of *disbenefits* in benefit–cost analyses and the decision as to whether certain cash flow items should be treated as *additional benefits* or as *reduced costs.* The first concern arises whenever disbenefits are formally defined in a B–C evaluation of a public-sector project. An example of the second concern would be a public-sector project proposing to replace an existing asset having high annual operating and maintenance costs with a new asset having lower O&M costs. As will be seen in Sections 11.7.1 and 11.7.2, the final recommendation on a project is not altered by either the approach to incorporating disbenefits or the classification of an item as a reduced cost or an additional benefit.

11.7.1 Disbenefits in the B–C Ratio

In a previous section, disbenefits were defined as negative consequences to the public resulting from the implementation of a public-sector project. The traditional approach for incorporating disbenefits into a benefit–cost analysis is to reduce benefits by the amount of disbenefits (i.e., to subtract disbenefits from benefits in the numerator of the B–C ratio). Alternatively, the disbenefits could be treated as costs (i.e., add disbenefits to costs in the denominator). Equations (11-7) and (11-8) illustrate the two approaches for incorporating disbenefits in the conventional B–C ratio, with benefits, costs, and disbenefits in terms of equivalent AW. (Similar equations could also be developed for the modified B–C ratio or for PW as the measure of equivalent worth.) Again, the *magnitude* of the B–C ratio will be different depending upon which approach is used to incorporate disbenefits, but project acceptability—that is, whether the B–C ratio is $>$, $<$, or $= 1.0$—will not be affected, as shown in Example 11-3.

Conventional B–C ratio with AW, *benefits* reduced by amount of *disbenefits*:

$$\text{B–C} = \frac{\text{AW(benefits)} - \text{AW(disbenefits)}}{\text{AW(costs)}} = \frac{\text{AW}(B) - \text{AW}(D)}{\text{CR} + \text{AW(O\&M)}}. \quad (11\text{-}7)$$

Here, $\text{AW}(\cdot)$ = annual worth of $(\cdot)$;

$\quad\quad\quad B$ = benefits of the proposed project;

$\quad\quad\quad D$ = disbenefits of the proposed project;

$\quad\quad\text{CR}$ = capital recovery amount (i.e., the equivalent annual cost of the initial investment, I, including an allowance for market value, if any);

$\quad\text{O\&M}$ = operating and maintenance costs of the proposed project.

Conventional B–C ratio with AW, *Costs* increased by amount of *disbenefits*:

$$\text{B–C} = \frac{\text{AW(benefits)}}{\text{AW(costs)} + \text{AW(disbenefits)}} = \frac{\text{AW}(B)}{\text{CR} + \text{AW(O\&M)} + \text{AW}(D)} \quad (11\text{-}8)$$

EXAMPLE 11-3

Refer back to Example 11-2. In addition to the benefits and costs, suppose that there are disbenefits associated with the runway extension project. Specifically, the increased noise level from commercial jet traffic will be a serious nuisance to homeowners living along the approach path to the Bugtussle Municipal Airport. The annual disbenefit to citizens of Bugtussle caused by this "noise pollution" is estimated to be $100,000. Given this additional information, reapply the conventional B–C ratio, with equivalent annual worth, to determine whether this disbenefit affects your recommendation on the desirability of this project.

SOLUTION

Disbenefits Reduce Benefits Eqn. 11-7	B–C = [AW(B) − AW(D)]/[CR + AW(O&M)] B–C = [$490,000 − $100,000]/[$1,200,000 $(A/P, 10\%, 20)$ + $197,500] **B–C= 1.152 >1; extend runways.**
Disbenefits Treated as Additional Costs Eqn. 11-8	B–C = AW(B)/[CR + AW(O&M) + AW(D)] B–C = $490,000/[$1,200,000 $(A/P, 10\%, 20)$ + $197,500 + $100,000] **B–C= 1.118 >1; extend runways.**

As in the case of conventional and modified B–C ratios, the treatment of dis-benefits may affect the magnitude of the B–C ratio, but it has no effect on project desirability in go/no-go decisions. It is left to the reader to develop a mathematical rationale for this, similar to that included in the discussion of conventional versus modified B–C ratios.

11.7.2 Added Benefits versus Reduced Costs in B–C Analyses

The analyst often needs to classify certain cash flows as either added benefits or reduced costs in calculating a B–C ratio. The questions arise, "How critical is the proper assignment of a particular cash flow as an added benefit or a reduced cost?" and "Is the outcome of the analysis affected by classifying a reduced cost as a benefit?" *An arbitrary decision as to the classification of a benefit or a cost has no impact on project acceptability.* The mathematical rationale for this information is presented below and in Example 11-4.

Let B = the equivalent annual worth of project benefits;

C = the equivalent annual worth of project costs;

X = the equivalent annual worth of a cash flow (either an added benefit or a reduced cost) not included in either B or C.

If X is classified as an added benefit, then $B–C = \dfrac{B + X}{C}$. Alternatively,

if X is classified as a reduced cost, then $B–C = \dfrac{B}{C - X}$.

Assuming that the project is acceptable, that is, B–C ≥ 1.0,

$$\frac{B + X}{C} \geq 1.0, \text{ which indicates that } B + X \geq C, \text{ and}$$

$$\frac{B}{C - X} \geq 1.0, \text{ which indicates that } B \geq C - X,$$

which can be restated as $B + X \geq C$.

EXAMPLE 11-4

A project is being considered by the Tennessee Department of Transportation to replace an aging bridge across the Cumberland River on a state highway. The existing two-lane bridge is expensive to maintain and creates a traffic bottleneck because the state highway is four lanes wide on either side of the bridge. The new bridge can be constructed at a cost of $300,000, and estimated annual maintenance costs are $10,000. The existing bridge has annual maintenance costs of $18,500. The annual benefit of the new four-lane bridge to motorists, due to the removal of the traffic bottleneck, has been estimated to be $25,000. Conduct a benefit–cost analysis, using a nominal interest rate of 8% and a study period of 25 yr, to determine whether the new bridge should be constructed.

SOLUTION

Treating the reduction in annual maintenance costs as a *reduced cost*:

$$B–C = \$25,000/[\$300,000(A/P, 8\%, 25) - (\$18,500 - \$10,000)]$$

$$B–C = 1.275 > 1; \text{construct new bridge.}$$

Treating the reduction in annual maintenance costs as an *increased benefit*:

$$B–C = [\$25,000 + (\$18,500 - \$10,000)]/[\$300,000(A/P, 8\%, 25)]$$

$$B–C = 1.192 > 1; \text{construct new bridge.}$$

Therefore, the decision to classify a cash flow item as an additional benefit or as a reduced cost will affect the magnitude of the calculated B–C ratio, but it will have no effect on project acceptability.

▼ 11.8 Evaluating Independent Projects by B–C Ratios

Independent projects are categorized as groupings of projects for which the choice to select any particular project in the group is *independent* of choices regarding any and all other projects within the group. Thus, it is permissible to select none of the projects, any combination of projects, or all of the projects from an independent group. (Note that this does not hold true under conditions of *capital rationing*. Methods of evaluating otherwise independent projects under capital rationing are discussed in a later section of this chapter.) Because any or all projects from an independent set can be selected, formal comparisons of independent projects are unnecessary. The issue of whether one project is *better* than another is unimportant if those projects are independent; the only criterion for selecting each of those projects is whether their respective B–C ratios are equal to or greater than 1.0.

A typical example of an economy study of a federal project—using the conventional B–C-ratio method—is a study of a flood control and power project on the White River in Missouri and Arkansas. Considerable flooding and consequent damage had occurred along certain portions of this river, as shown in Table 11-2. In addition, the uncontrolled water flow increased flood conditions on the lower Mississippi River. In this case, there were independent options of building a reservoir *and/or* a channel improvement to alleviate the problem. The cost and benefit summaries for the Table Rock reservoir and the Bull Shoals channel improvement are shown in Table 11-3. The fact that the Bull Shoals channel improvement project

TABLE 11-2 Annual Loss as a Result of Floods on Three Stretches of the White River

Item	Annual Value of Loss	Annual Loss per Acre of Improved Land in Floodplain	Annual Loss per Acre for Total Area in Floodplain
Crops	$1,951,714	$6.04	$1.55
Farm (other than crops)	215,561	0.67	0.17
Railroads and highways	119,800	0.37	0.09
Levees[a]	87,234	0.27	0.07
Other losses	168,326	0.52	0.13
TOTALS:	$2,542,635	$7.87	$2.01

[a] Expenditures by the United States for levee repairs and high-water maintenance.

has the higher benefit cost ratio is irrelevant; both options are acceptable because their B–C ratios are greater than one.

Several interesting facts may be noted concerning this study. *First,* there was no attempt to allocate the cost of the projects between flood control and power production. *Second,* very large portions of the flood-control benefits were shown to be in connection with the Mississippi River and are not indicated in Table 11-2; these were not detailed in the main body of the report, but were shown in an appendix. Only a moderate decrease in the value of these benefits would have changed the B–C ratio considerably. *Third,* without the combination of flood-control and power-generation objectives, neither project would have been economical for either purpose. These facts point to the advantages of multiple purposes for making flood-control projects economically feasible and to the necessity for careful enumeration and evaluation of the prospective benefits of a public-sector project.

▼ 11.9 Comparison of Mutually Exclusive Projects by B–C Ratios

Recall that a group of *mutually exclusive projects* was defined as *a group of projects from which, at most, one project may be selected.* When using an equivalent-worth method to select from among a set of mutually exclusive alternatives (MEAs), the "best" alternative can be selected by maximizing the PW (or AW, or FW). Because the benefit–cost method provides a *ratio* of benefits to costs rather than a direct measure of each project's *profit potential,* selecting the project that maximizes the B–C ratio does not guarantee that the best project is selected. In addition to the fact that *maximizing the B–C ratio for mutually exclusive alternatives is **incorrect**,* any attempt to do so would be further confounded by the potential for inconsistent ranking of projects by the conventional B–C ratio versus the modified B–C ratio (i.e., the conventional B–C ratio might favor a different project than would the modified B–C

TABLE 11-3 Estimated Costs, Annual Charges, and Annual Benefits for the Table Rock Reservoir and Bull Shoals Channel Improvement Projects

Item	Table Rock Reservoir	Bull Shoals Channel Improvement
Cost of dam and appurtenances, and reservoir:		
Dam, including reservoir-clearing, camp, access railroads and highways, and foundation exploration and treatment	$20,447,000	$25,240,000
Powerhouse and equipment	6,700,000	6,650,000
Power-transmission facilities to existing load-distribution centers	3,400,000	4,387,000
Land	1,200,000	1,470,000
Highway relocations	2,700,000	140,000
Cemetery relocations	40,000	18,000
Damage to villages	6,000	94,500
Damage to miscellaneous structures	7,000	500
Total Construction Cost (estimated appropriation of public funds necessary for the execution of the project)	$34,500,000	$38,000,000
Federal investment:		
Total construction cost	$34,500,000	$38,000,000
Interest during construction	$1,811,300	1,995,000
Total	$36,311,300	$39,995,000
Present value of federal properties	1,200	300
Total Federal Investment	$36,312,500	$39,995,300
Total Annual Costs	$1,642,200	$1,815,100
Annual benefits:		
Prevented direct flood losses in White River basin:		
Present conditions	60,100	266,900
Future developments	19,000	84,200
Prevented indirect flood losses owing to floods in White River basin	19,800	87,800
Enhancement in property values in White River valley	7,700	34,000
Prevented flood losses on Mississippi River	220,000	980,000
Annual Flood Benefits	326,000	1,452,900
Power value	1,415,600	1,403,400
Total Annual Benefits	$1,742,200	$2,856,300
Conventional B–C Ratio = Total Annual Benefits ÷ Annual Costs	1.06	1.57

ratio). This phenomenon is illustrated in Example 11-5. (The approach to handling *disbenefits* or classifications of cash-flow items as *added benefits* versus *reduced costs* could also change the preference for one MEA over another.) As with the rate of return procedures in Chapter 5, an evaluation of mutually exclusive alternatives by the B–C ratio requires that an *incremental* benefit–cost analysis be conducted.

EXAMPLE 11-5

The required investments, annual operating and maintenance costs, and annual benefits for two mutually exclusive alternative projects are shown subsequently. Both conventional and modified B–C ratios are included for each project. Note that **Project A** has the greater *conventional* B–C, but **Project B** has the greater *modified* B–C. Given this information, which project should be selected?

	Project A	Project B	
Capital investment	$110,000	$135,000	**Nominal interest rate = 10% per year**
Annual O&M cost	12,500	45,000	**Study period = 10 yr**
Annual benefit	37,500	80,000	
Conventional B–C	**1.475**	1.315	
Modified B–C	1.935	**2.207**	

SOLUTION

The B–C analysis has been conducted improperly. Although each of the B–C ratios shown is *numerically correct*, a comparison of mutually exclusive alternatives requires that an incremental analysis be conducted.

When comparing mutually exclusive alternatives with the B–C-ratio method, *they are first ranked in order of increasing total equivalent worth of costs.* This rank ordering will be identical whether the ranking is based on PW, AW, or FW of costs. The "do-nothing" alternative is selected as a baseline alternative. The B–C ratio is then calculated for the alternative having the lowest equivalent cost. If the B–C ratio for this alternative is equal to or greater than 1.0, then that alternative becomes the new baseline; otherwise, "do-nothing" remains as the baseline. The next least equivalent cost alternative is then selected, and the difference (Δ) in the respective benefits and costs of this alternative and the baseline is used to calculate an incremental B–C ratio ($\Delta B / \Delta C$). If that ratio is equal to or greater than 1.0, then the higher equivalent cost alternative becomes the new baseline; otherwise the last baseline alternative is maintained. Incremental B–C ratios are determined for each successively higher equivalent cost alternative until the last alternative has been compared. The flowchart of this procedure is included as Figure 11-3, and the procedure is illustrated in Example 11-6.

Figure 11-3
The Incremental
Benefit–Cost-Ratio
Procedure

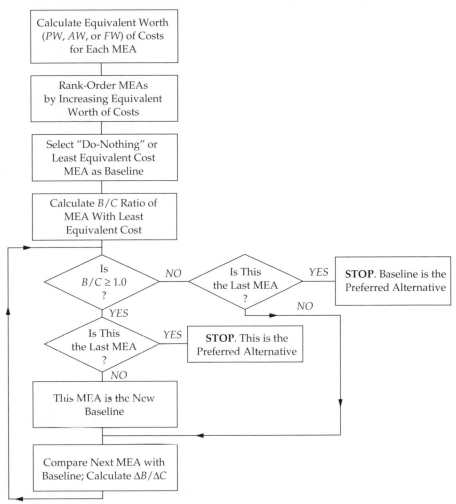

EXAMPLE 11-6

Three mutually exclusive alternative public-works projects are currently under consideration. Their respective costs and benefits are included in the table that follows. Each of the projects has a useful life of 50 yr, and the nominal interest rate is 10% per year. Which, if any, of these projects should be selected?

	A	B	C
Capital investment	$8,500,000	$10,000,000	$12,000,000
Annual oper. & maint. costs	750,000	725,000	700,000
Market value	1,250,000	1,750,000	2,000,000
Annual benefit	2,150,000	2,265,000	2,500,000

SOLUTION

$$\text{PW(Costs, } A) = \$8,500,000 + \$750,000(P/A, 10\%, 50)$$

$$- \$1,250,000(P/F, 10\%, 50) = \$15,925,463$$

$$\text{PW(Costs, } B) = \$10,000,000 + \$725,000(P/A, 10\%, 50)$$

$$- \$1,750,000(P/F, 10\%, 50) = \$17,173,333$$

$$\text{PW(Costs, } C) = \$12,000,000 + \$700,000(P/A, 10\%, 50)$$

$$- \$2,000,000(P/F, 10\%, 50) = \$18,923,333$$

$$\text{PW(Benefit, } A) = \$2,150,000(P/A, 10\%, 50) = \$21,316,851$$

$$\text{PW(Benefit, } B) = \$2,265,000(P/A, 10\%, 50) = \$22,457,055$$

$$\text{PW(Benefit, } C) = \$2,750,000(P/A, 10\%, 50) = \$24,787,036$$

$$\text{B–C}(A) = \$21,316,851/\$15,925,463$$

$$= \mathbf{1.3385 > 1.0} \therefore \textbf{Project A is acceptable}$$

$$\Delta B/\Delta C(B - A) = (\$22,457,055 - \$21,316,851)/(\$17,173,333 - \$15,925,463)$$

$$= \mathbf{0.9137 < 1.0} \therefore \textbf{increment required for Project B is not acceptable}$$

$$\Delta B/\Delta C(C - A) = (\$24,787,036 - \$21,316,851)/(\$18,923,333 - \$15,925,463)$$

$$= \mathbf{1.1576 > 1.0} \therefore \textbf{increment required for Project C is acceptable}$$

Decision: *Recommend Project C.*

It is not uncommon for some of the projects in a set of mutually exclusive public-works projects to have different lives. Recall from Chapter 5 that the AW criterion can be used to select from among alternatives with different lives as long as the assumption of *repeatability* is valid. Similarly, if a mutually exclusive set of public-works projects includes projects with varying useful lives, it may be possible to conduct an incremental B–C analysis using the AW of benefits and costs of the various projects. This analysis is illustrated in Example 11-7.

EXAMPLE 11-7

Two mutually exclusive alternative public-works projects are under consideration. Their respective costs and benefits are included in the table that follows. Project I has an anticipated life of 35 yr, and the useful life of Project II has been estimated to be 25 yr. If the nominal interest rate is 9%, which, if either, of these projects should be selected? The effect of inflation is negligible.

	Project I	Project II
Capital investment	$750,000	$625,000
Annual oper. & maint. costs	120,000	110,000
Annual benefit	245,000	230,000
Useful life of project (years)	35	25

SOLUTION

AW(Costs, I) = $750,000$(A/P, 9\%, 35)$ + $120,000 = $190,977$

AW(Costs, II) = $625,000$(A/P, 9\%, 25)$ + $110,000 = $173,629$

B–C(II) = $230,000/$173,629 = **1.3247** > **1.0** ∴ **Project II is acceptable**

ΔB/ΔC(I–II) = ($245,000 − $230,000)/($190,977 − $173,629)

$= 0.8647 < 1.0$ ∴ **increment required for Project I is not acceptable**

Decision: *Project II should be selected.*

In a previous section, the calculation of B–C ratios for independent projects was presented, and it was stated that the issue of whether one project was superior to another is irrelevant. How, then, can we select from among a set of independent public-works projects under conditions of capital rationing? Recall that in Chapter 5 mutually exclusive combinations (MECs) of independent projects were developed when budgetary constraints prevented all of the economically viable projects from being selected. Such an analysis can also be conducted with the B–C ratio, but as with mutually exclusive alternatives, the procedure must be applied incrementally, as illustrated in Example 11-8.

EXAMPLE 11-8

A governmental agency is considering four independent projects, each having 30-year projected useful lives. The current budget for this agency allows not more than $35,000,000 to be spent, in terms of initial investments, and the nominal interest rate is 10% per year. Using the B–C-ratio method, which of the projects shown below should be selected?

Project	Initial Investment	Annual Costs	Annual Benefits
A	$12,000,000	$1,250,000	$3,250,000
B	20,000,000	4,500,000	8,000,000
C	10,000,000	750,000	1,250,000
D	14,000,000	1,850,000	4,050,000

SOLUTION

Project C is first removed from further consideration because its B–C ratio is < 1.0. From the remaining three projects, a total of $2^3 = 8$ mutually exclusive combinations can be developed. One of these, the combination of all three remaining projects, is infeasible due to the budget constraint. Comparing the mutually exclusive combinations incrementally, beginning with the combination having the least PW of costs, it is shown that the combination of Projects A and B should be selected.

Project	PW (Costs)	PW (Benefits)	B–C Ratio	Acceptable?
A	$23,783,643	$30,637,472	1.2882	YES
B	62,421,115	75,415,316	1.2082	YES
C	17,070,186	11,783,643	0.6903	NO
D	31,439,792	38,179,004	1.2144	YES

MEC	Projects	Total Investment	PW of Costs	PW of Benefits	Feasible?
1	do nothing	0	0	0	YES
2	A	$12,000,000	$23,783,643	$30,637,472	YES
3	B	20,000,000	62,421,115	75,415,316	YES
4	D	14,000,000	31,439,792	38,179,004	YES
5	AB	32,000,000	86,204,758	106,052,788	YES
6	AD	26,000,000	55,223,435	68,816,476	YES
7	BD	34,000,000	93,860,907	113,594,319	YES
8	ABD	46,000,000	117,644,550	144,231,791	NO

Incremental Comparison of MEC	Δ PW(Costs)	Δ PW (Benefits)	Δ B–C Ratio	Increment Justified?
1 → 2	$23,783,643	$30,637,472	1.2882	MEC 2 accepted
2 ⇒ 4	7,656,149	7,541,532	0.9850	MEC 4 not accepted
2 ⇒ 6	31,439,792	38,179,004	1.2144	MEC 6 accepted
6 ⇒ 3	7,197,680	6,598,840	0.9168	MEC 3 not accepted
6 ⇒ 5	30,981,323	37,236,312	1.2019	MEC 5 accepted
5 ⇒ 7	7,656,149	7,541,532	0.9850	MEC 7 not accepted

By applying the B–C ratio incrementally to mutually exclusive combinations of independent projects, the optimal set of projects is shown to be MEC 5. Note that although this approach is acceptable and, when properly applied, will lead to the selection of the "best" set of projects, the same result can be accomplished in a more straightforward manner by using PW (or AW or FW) and selecting the feasible MEC that maximizes the equivalent-worth criterion without the need of an incremental analysis.

▼11.10 Criticisms and Shortcomings of the Benefit–Cost-Ratio Method*

Although the benefit–cost-ratio method is firmly entrenched as *the* procedure used by most governmental agencies for the evaluation of public-sector projects, it has been widely criticized over the years. Among the criticisms are that (1) it is often used as a tool for after-the-fact justifications rather than for project evaluation, (2) serious distributional inequities (i.e., one group reaps the benefits while another incurs the costs) may not be accounted for in B–C studies, and (3) qualitative information is often ignored in B–C studies.

Regarding the first criticism, the benefit–cost ratio is considered by many to be a method of using the numbers to support the views and interests of the group paying for the analysis. A congressional subcommittee once supported this critical view with the following conclusion:

> ... the most significant factor in evaluating a benefit-cost study is the name of the sponsor. Benefit–cost studies generally are formulated after basic positions on an issue are taken by the respective parties. The results of competing studies predictably reflect the respective positions of the parties on the issue (p. 55 of Campen).

One illustration of such a flawed B–C study is the Bureau of Reclamation's 1967 analysis supporting the proposed Nebraska Mid-State Project. The purpose of this project was to divert water from the Platte River to irrigate crop land, and the computed B–C ratio for this project was 1.24, indicating that the project should be undertaken. This favorable ratio was based, in part, on the following fallacies (p. 53):

1. An unrealistically low interest rate of only 3.125% per year was used.
2. The analysis used a project life of 100 yr, rather than the generally more accepted and supportable assumption of a 50-yr life.
3. The analysis counted wildlife and fish benefits among the project's benefits, although in reality, the project would have been devastating to wildlife. During a 30-yr history of recorded water flow (1931–1960), the proposed water diversion would have left a substantial stretch of the Platte River dry for over half of that time, destroying fish and waterfowl habitats on 150 miles of the river. This destruction of wildlife habitats would have adversely affected the populations of endangered species including the bald eagle, the whooping crane, and the sandhill crane.
4. The benefits of increased farm output were based on government support prices that would have required a substantial federal subsidy.

Campen states that "the common core of these criticisms is not so much the fact that benefit–cost analysis is used to justify particular positions but that it is presented as a scientific unbiased method of analysis" (pp. 52–53). In order for an analysis to be fair and reliable, it must be based on an accurate and realistic assessment of

---

*J. T. Campen, *Benefit, Cost, and Beyond: The Political Economy of Benefit-Cost Analysis* (Cambridge, MA: Ballinger, 1986). All quotes in Section 11.10 are taken from this source.

all the relevant benefits and costs. Thus, the analysis must be performed either by an unbiased group or by a group that includes representatives from each of the interest groups involved. When the Nebraska Mid-State Project was reevaluated by an impartial third party, for example, a more realistic B–C ratio of only 0.23 was determined. Unfortunately, analyses of public-sector projects are sometimes performed by parties who have already formed strong opinions on the worth of these projects.

Another shortcoming of the B–C-ratio method is that benefits and costs effectively cancel each other out without regard for who reaps the benefits or who incurs the costs, which does not cause significant difficulties in the private sector where the benefits accrue to and the costs are borne by the owners of the firm. Recall that in the public sector, however, "the benefits to whomsoever they accrue" must be considered, which may result in serious distributional inequities in benefit–cost studies. Two reasons for being critical of this potential lack of distributional equity are that (1) public policy generally "operates to reduce economic inequality by improving the well-being of disadvantaged groups" and (2) there is little concern for the equality or inequality of people who are in the same general economic circumstances (p. 56 of Campen).

Public policy is generally viewed as one method of reducing the economic inequalities of the poor, residents of underdeveloped regions, and racial minorities. Of course, there are many cases where it is not possible to make distributional judgments, but there are others where it is clear. Conceivably, a project could have adverse consequences for Group A, a disadvantaged group, but the benefits to another group, Group B, might far outweigh the disbenefits to Group A. If the project has a B–C ratio > 1, it will likely be accepted without regard for the consequences to Group A, particularly if Group B includes wealthy, politically influential members.

A lack of consideration of the distributional consequences of a project might also produce inequities for individuals who are in the same approximate economic circumstances. Consider, for example, the inequity shown in the following example:

> Consider a proposal to raise property taxes by 50 percent on all properties with odd-numbered street addresses and simultaneously to lower property taxes by 50 percent on all properties with even-numbered addresses. A conventional benefit-cost analysis of this proposal would conclude that its net benefits were approximately zero, and an analysis of its impact on the overall level of inequality in the size distribution of income would also show no significant effects. Nevertheless, such a proposal would be generally, and rightly, condemned as consisting of an arbitrary and unfair redistribution of income (p. 56 of Campen).

Another, more realistic, example of adverse distributional consequences would involve a project to construct a new chemical plant in *Town A*. The chemical plant would employ hundreds of workers in an economically depressed area, but, according to one group of concerned citizens, it would also release hazardous byproducts that could pollute the groundwater and a nearby stream that provides most of the drinking water for *Town B*. Thus, the benefits of this project would be the addition of jobs and a boost to the local economy for *Town A*, but *Town B* would have increased water-treatment costs and residents may be subject to long-term health hazards and increased medical bills. Unfortunately, a B–C-ratio analysis would show only the net monetary effect of the project, without regard to the serious distributional inequities.

In the previously cited example of the Bureau of Reclamation's analysis of the Nebraska Mid-State Project, the problem of using unreliable monetary values for nonmonetary considerations (irreducibles) was highlighted in the discussion of wildlife and fish "benefits." But the outcome of the B–C analysis might also be highly suspect if no attempt had been made to quantify these aspects of the project. When only easily quantifiable information is included in the analysis, the importance of the other factors is totally neglected, which results in projects with mostly monetary benefits being favored, and projects with less tangible benefits being rejected without due consideration. Unfortunately, busy decision makers want a single number upon which to base their acceptance or rejection of a project. No matter how heavily the nonmonetary aspects of a project are emphasized during the discussion, most managers will go directly to the bottom line of the report to get a single number and use that to make their decision. A 1980 congressional committee concluded that "whenever some quantification is done—no matter how speculative or limited—the number tends to get into the public domain and the qualifications tend to get forgotten.... The number is the thing" (p. 68 of Campen).

Although these criticisms have often been leveled at the benefit–cost method itself, problems in the use (and abuse) of the B–C procedure are largely due to the inherent difficulties in evaluating public projects (see Section 11.5) and the manner in which the procedure is applied. Note that these same criticisms might be in order for a poorly conducted analysis that relied on an equivalent worth or rate-of-return method.

11.11 Spreadsheet Applications

To demonstrate the use of spreadsheets in a benefit–cost analysis, we consider the three mutually exclusive public-works projects introduced in Example 11-6. Their respective benefits and costs are inputs to the spreadsheet model shown in Figure 11-4. The B–C ratio is computed for each alternative, using both conventional and modified forms of the ratio. Because all of the projects have a B–C > 1, we must perform an incremental analysis to determine the best public-works project.

The incremental analysis is shown in the lower portion of Figure 11-4. Project A is our base alternative because it has the smallest annual equivalent of costs and a B–C ratio ≥ 1. The first comparison then is between A and B. The incremental benefits and costs are found by subtracting the estimates for Project A from the estimates for Project B. The ratio of the incremental benefits to the incremental costs, $\Delta B/\Delta C$, is less than one, indicating that the increment is not justified.

Project C is then compared to Project A in a similar fashion. The resulting $\Delta B/\Delta C > 1$ tells us that the incremental benefits of Project C outweigh the incremental costs. Because Project C is the last alternative, it is the recommended alternative. Note that (1) the same conclusion is reached regardless of whether we

	A	B	C	D
1	MARR	10%		
2	Study Period	50		
3				
4		Project A	Project B	Project C
5	Initial Costs	$ 8,500,000	$ 10,000,000	$ 12,000,000
6	Annual O&M Costs	$ 750,000	$ 725,000	$ 700,000
7	Market Value	$ 1,250,000	$ 1,750,000	$ 2,000,000
8	Annual Benefit	$ 2,150,000	$ 2,265,000	$ 2,500,000
9				
10	CR Amount	$ 856,229	$ 1,007,088	$ 1,208,592
11				
12	Conventional B/C Ratio	1.3385	1.3077	1.3099
13	Modified B/C Ratio	1.6351	1.5292	1.4893
14				
15	Incremental Analysis			
16		Δ(B–A)	Δ(C–A)	
17	Δ Initial Costs	$ 1,500,000	$ 3,500,000	
18	Δ Annual O&M Costs*	$ 25,000	$ 50,000	*Savings
19	Δ Market Value	$ 500,000	$ 750,000	
20	Δ Annual Benefit	$ 115,000	$ 350,000	
21				
22	Δ CR Amount	$ 150,859	$ 352,363	
23				
24	ΔB/ΔC (Conventional)	0.9137	1.1576	
25	ΔB/ΔC (Modified)	0.9280	1.1352	
26	Increment Justified?	No	Yes	

Figure 11-4 **Spreadsheet for Comparing MEAs Using an Incremental B–C Ratio**

use the conventional or modified B–C ratio, and (2) Project A, which had the highest B–C ratio, is not the recommended project. The formulas for the highlighted cells are given in the following table:

Cell	Contents
B10	$= -\text{PMT}(\$B\$1, \$B\$2, B5 - B7/(1 + \$B\$1)^{\wedge}\$B\$2)$
B12	$= B8/(B10 + B6)$
B13	$= (B8 - B6)/B10$
B17	$= C5 - B5$
C17	$= D5 - B5$
B22	$= -\text{PMT}(\$B\$1, \$B\$2, B17 - B19/(1 + \$B\$1)^{\wedge}\$B\$2)$
B24	$= B20/(B22 - B18)$
B25	$= (B20 + B18)/B22$
B26	$= \text{IF} (B24 >= 1, \text{"Yes"}, \text{"No"})$

11.12 Summary

From the discussion and examples of public projects presented in this chapter, it is apparent that because of the methods of financing, the absence of tax and profit requirements, and political and social factors, the criteria used in evaluating privately financed projects frequently cannot be applied to public works. Nor should public projects be used as yardsticks with which to compare private projects. Nevertheless, whenever possible, public works should be justified on an economic basis to ensure that the public obtains the maximum return from the tax money that is spent. Whether an engineer is working on such projects, is called upon to serve as a consultant, or assists in the conduct of the benefit–cost analysis, he or she is bound by professional ethics to do his or her utmost to see that the projects and the associated analyses are carried out in the best possible manner and within the limitations of the legislation enacted for their authorization.

The B–C ratio has remained a popular method for evaluating the financial performance of public projects. Both the conventional and modified B–C-ratio methods have been explained and illustrated for the case of *independent* and *mutually exclusive* projects. A final note of caution: The best project among a *mutually exclusive* set of projects is not necessarily the one that maximizes the B–C ratio. In this chapter we have seen that an incremental analysis approach to evaluating benefits and costs is necessary to ensure the correct choice.

11.13 References

CAMPEN, J. T. *Benefit, Cost, and Beyond* (Cambridge, MA: Ballinger, 1986).

DASGUPTA, AGIT K., and D. W. PEARCE. *Cost-Benefit Analysis: Theory & Practice* (New York: Harper & Row, 1972).

MISHAN, E. J. *Cost-Benefit Analysis* (New York: Praeger, 1976).

OFFICE OF MANAGEMENT AND BUDGET, "Guidelines and Discount Rates for Benefit-Cost Analysis of Federal Programs," OMB Circular A-94 (revised), February 21, 1997.

PREST, A. R., and R. TURVEY, "Cost-Benefit Analysis: A Survey," *The Economic Journal,* vol. 75, no. 300, December 1965, pp. 683–735.

SASSONE, PETER G., and WILLIAM A. SCHAFFER. *Cost-Benefit Analysis: A Handbook* (New York: Academic Press, 1978).

SCHWAB, B., and P. LUSZTIG. "A Comparative Analysis of the Net Present Value and the Benefit-Cost Ratio as Measure of Economic Desirability of Investment," *Journal of Finance,* vol. 24, 1969, pp. 507–516.

11.14 Problems

The number in parentheses () that follows each problem refers to the section from which the problem is taken.

11-1. A government agency is considering buying a piece of land for $500,000 and constructing an office building on it. Three different building designs are being analyzed (see Table P11-1).

TABLE P11-1 Different Building Design for Problem P11-1

	Design A 2 stories	Design B 5 stories	Design C 10 stories
Cost of building (excluding cost of land)	$800,000	$1,200,000	$3,000,000
Resale value* of land and building at end of 20-yr analysis period	500,000	900,000	2,000,000
Annual rental income after deducting all operating expenses	120,000	300,000	450,000

*Resale value is considered as a reduction in cost, rather than a benefit.

Using the modified benefit–cost-ratio method and a MARR of 10% per year, determine which alternative, if any, should be selected. (11.9)

11-2. Five mutually exclusive alternatives are being considered for providing a sewage-treatment facility. The annual equivalent costs and estimated benefits of the alternatives are as follows:

	Annual Equivalent (in thousands)	
Alternative	Cost	Benefits
A	$1,050	$1,110
B	900	810
C	1,230	1,390
D	1,350	1,500
E	990	1,140

Which plan, if any, should be adopted, if the Sewage Authority wishes to invest if, and only if, the B–C ratio on any cost is at least 1.0.? (11.9)

11-3. Perform a conventional B–C analysis for the six mutually exclusive alternatives in Table P11-3. Also, compute the conventional B–C values for each individual alternative and compare the values obtained with the modified B–C values. The MARR is 10% per year. (11.9)

11-4. Five mutually exclusive machines are being considered for a particular job. Each is expected to have a salvage value of 50% of the investment amount at the end of the four-year study period. Using a B–C analysis, which machine should be selected on the basis of the data in Table P11-4? (11.9)

11-5. A nonprofit government corporation is considering two alternatives for generating power:

Alternative A. Build a coal-powered generating facility at a cost of $20,000,000. Annual power sales are expected to be $1,000,000 per year. Annual O&M costs are $200,000 per year. A benefit of this alternative is that it is expected to attract new industry, worth $500,000 per year, to the region.

Alternative B. Build a hydroelectric generating facility. The capital investment, power sales, and operating costs are $30,000,000, $800,000, and $100,000 per year, respectively. Annual benefits of this alternative are as follows:

TABLE P11-3 Six Mutually Exclusive Alternatives for Problem P11-3

	Alternative Project					
	A	B	C	D	E	F
Investment	$1,000	$1,500	$2,500	$4,000	$5,000	$7,000
Annual savings in cash disbursements	150	375	500	925	1,125	1,425
Salvage Value	1,000	1,500	2,500	4,000	5,000	7,000

TABLE P11-4 Data for Problem P11-4

	Alternatives				
	A	B	C	D	E
Investment	$2,100	$3,400	$1,000	$2,700	$1,400
Net cash flow per year	280	445	110	340	180
MARR = 12% per year					

Flood-control savings	$600,000
Irrigation	$200,000
Recreation	$100,000
Ability to attract new industry	$400,000

The useful life of both alternatives is 50 yr. Using an interest rate of 5%, determine which alternative (if either) should be selected according to the conventional B–C-ratio method. (11.7, 11.9)

11-6. Five independent projects are available for funding by a certain public agency. The following tabulation shows the equivalent annual benefits and costs for each: (11.8)

Project	Annual Benefits	Annual Costs
A	$1,800,000	$2,000,000
B	5,600,000	4,200,000
C	8,400,000	6,800,000
D	2,600,000	2,800,000
E	6,600,000	5,400,000

a. Assume that the projects are of the type for which the benefits can be determined with considerable certainty and that the agency is willing to invest money as long as the B–C ratio is at least one. Which alternatives should be selected for funding?

b. What is the rank ordering of projects from best to worst?

c. If the projects involved intangible benefits that required considerable judgment in assigning their values, would your recommendation be affected?

11-7. In the development of a publicly owned, commercial waterfront area, three possible inde-

pendent plans are being considered. Their costs and estimated benefits are as follows (11.8)

	PW ($000s)	
Plan	Costs	Benefits
A	$123,000	$139,000
B	135,000	150,000
C	99,000	114,000

a. Which plan(s) should be adopted, if any, if the controlling board wishes to invest any amount required, provided that the B–C ratio on the required investment is at least 1.0?

b. Suppose that 10% of the costs of each plan are reclassified as "disbenefits." What percentage change in the B–C ratio of each plan results from the reclassification?

c. Comment on why the rank orderings in (a) are unaffected by the change in (b).

11-8. Consider the two types of equipment in the following table and determine which choice is better if a firm desires to invest as long as the B–C ratio is greater than or equal to one. The firm's MARR is 10% per year. Assume repeatability and show all work. (11.9)

	Equipment Type	
	RS-422	RS-511
Capital investment	$500	$1,750
Useful life (years)	6	12
Market (salvage) value	$125	$375
Annual benefits	$238	$388
Annual O&M costs	$108	$113

TABLE P11-9 Mutually Exclusive Alternatives for Problem P11-9

Alternative	Equivalent Annual Cost of Project	Expected Annual Flood Damage	Annual Benefits
I. No flood control	0	$100,000	0
II. Construct levees	$30,000	80,000	$112,000
III. Build small dam	$100,000	5,000	110,000

11-9. Consider the mutually exclusive alternatives in Table P11-9. Which alternative would be chosen according to these decision criteria?

a. Maximum benefit

b. Minimum cost

c. Maximum benefits minus costs

d. Largest investment having an incremental B–C ratio larger than one

e. Largest B–C ratio

Which project should be chosen? (11.9)

11-10. *A river that passes through private lands is formed from four branches of water that flow from a national forest. Some flooding occurs each year, and a major flood generally occurs every few years. If small earthen dams are placed on each of the four branches, the chances of major flooding would be practically eliminated. Construction of one or more dams would reduce the amount of flooding by varying degrees.

Other potential benefits from the dams are the reduction of damages to fire and logging roads in the forest, the value of the dammed water for protection against fires, and the increased recreational use. The table that follows contains the estimated benefits and costs associated with building one or more dams.

The equation used to calculate B–C ratios is as follows:

$$\text{B–C ratio} = \frac{\text{annual flood and fire savings} + \text{recreational benefits}}{\text{equivalent annual construction costs} + \text{maintenance}}$$

Benefits and costs are to be compared using the AW method with an interest rate of 8%; useful lives of the dams are 100 years. See Table P11-10. (11.7, 11.9)

a. Which of the four options would you recommend? Show why.

b. If fire benefits are reclassified as reduced costs, would the choice in part (a) be affected? Show your work.

11-11. A state-sponsored Forest Management Bureau is evaluating alternative routes for a new road into a formerly inaccessible region. Three mutually exclusive plans for routing the road provide different benefits, as indicated in Table P11-11. The roads are assumed to have an

TABLE P11-10 Costs and Benefits for Problem P11-10

Option	Dam Sites	Costs		Benefits		
		Construction	Annual Maintenance	Annual Flood	Annual Fire	Annual Recreation
A	1	3,120,000	52,000	520,000	52,000	78,000
B	1 & 2	3,900,000	91,000	630,000	104,000	78,000
C	1, 2 & 3	7,020,000	130,000	728,000	156,000	156,000
D	1, 2, 3 & 4	9,100,000	156,000	780,000	182,000	182,000

*Fashioned after a problem from James L. Riggs, *Engineering Economics* (New York: McGraw–Hill, 1977), pp. 432–434.

TABLE P11-11 Mutually Exclusive Plans for Problem P11-11

Route	Construction Costs	Annual Maintenance Cost	Annual Savings in Fire Damage	Annual Recreational Benefit	Annual Timber Access Benefit
A	185,000	2,000	5,000	3,000	500
B	220,000	3,000	7,000	6,500	1,500
C	290,000	4,000	12,000	6,000	2,800

economic life of 50 yr, and the nominal interest rate is 8% per year. Which route should be selected according to the B–C-ratio method? (11.7, 11.9)

11-12. An area on the Colorado River is subject to periodic flood damage that occurs, on the average, every two years and results in a $2,000,000 loss. It has been proposed that the river channel should be straightened and deepened, at a cost of $2,500,000, to reduce the probable damage to not over $1,600,000 for each occurrence during a period of 20 yr before it would have to be deepened again. This procedure would also involve annual expenditures of $80,000 for minimal maintenance. One legislator in the area has proposed that a better solution would be to construct a flood-control dam at a cost of $8,500,000, which would last indefinitely, with annual maintenance costs of not over $50,000. He estimates that this project would reduce the probable annual flood damage to not over $450,000. In addition, this solution would provide a substantial amount of irrigation water that would produce annual revenue of $175,000 and recreational facilities, which he estimates would be worth at least $45,000 per year to the adjacent populace. A second legislator believes that the dam should be built and that the river channel also should be straightened and deepened, noting that the total cost of $11,000,000 would reduce the probable annual flood loss to not over $350,000 while

providing the same irrigation and recreational benefits. If the state's capital is worth 10%, determine the B–C ratios and the incremental B–C ratio. Recommend which alternative should be adopted. (11.9)

11-13. Ten years ago the port of Secoma built a new pier containing a large amount of steel work, at a cost of $300,000, estimating that it would have a life of 50 yr. The annual maintenance cost, much of it for painting and repair caused by environmental damage, has turned out to be unexpectedly high, averaging $27,000. The port manager has proposed to the port commission that this pier be replaced immediately with a reinforced concrete pier at a construction cost of $600,000. He assures them that this pier will have a life of at least 50 yr, with annual maintenance costs of not over $2,000. He presents the information in Table P11-13 as justification for the replacement, having determined that the net market value of the existing pier is $40,000.

He has stated that because the port earns a net profit of over $3,000,000 per year, the project could be financed out of annual earnings. Thus, there would be no interest cost, and an annual savings of $19,000 would be obtained by making the replacement. (11.9)

a. Comment on the port manager's analysis.

b. Make your own analysis and recommendation regarding the proposal.

TABLE P11-13 Pier Replacement Cost for Problem P11-13

Annual Cost of Present Pier		Annual Cost of Proposed Pier	
Depreciation ($300,000/50)	$6,000	Depreciation ($600,000/50)	$12,000
Maintenance cost	27,000	Maintenance cost	2,000
Total	$33,000	Total	$14,000

11-14. A toll bridge across the Mississippi River is being considered as a replacement for the current I-40 bridge linking Tennessee to Arkansas. Because this bridge, if approved, will become a part of the U.S. Interstate Highway system, the B–C-ratio method must be applied in the evaluation. Investment costs of the structure are estimated to be $17,500,000, and $325,000 per year in operating and maintenance costs are anticipated. In addition, the bridge must be resurfaced every fifth year of its 30-yr projected life, at a cost of $1,250,000 per occurrence (no resurfacing cost in year 30). Revenues generated from the toll are anticipated to be $2,500,000 in its first year of operation, with a projected annual rate of increase of 2.25% per year due to the anticipated annual increase in traffic across the bridge. Assuming zero market (salvage) value for the bridge at the end of 30 yr and an MARR of 10% per year, should the toll bridge be constructed? (11.7)

11-15. Refer back to **Problem 11-14.** Suppose that the toll bridge can be redesigned such that it will have a (virtually) infinite life. The MARR remains at 10% per year. Revised costs and revenues (benefits) are given as follows: (11.7, 11.9)

Capital investment: $22,500,000
Annual operating and maintenance costs: $250,000
Resurface cost every 7th year: $1,000,000
Structural repair cost, every 20th year: $1,750,000
Revenues (treated as constant—no rate of increase): $3,000,000

a. What is the capitalized worth of the bridge?
b. Determine the B–C ratio of the bridge over an infinite time horizon.
c. Should the initial design (*Problem 11-14*) or the new design be selected?

11-16. In the aftermath of Hurricane Thelma, the U.S. Army Corps of Engineers is considering two alternative approaches to protect a fresh-water wetland from the encroaching seawater during high tides. The first alternative, the construction of a 5-mile long, 20-foot high levee, would have an investment cost of $25,000,000 with annual upkeep costs estimated at $725,000. A new roadway along the top of the levee would provide two major "benefits": (1) improved recreational access for fishermen and (2) reduction of

the driving distance between the towns at opposite ends of the proposed levee by 11 miles. The annual benefit for the levee has been estimated at $1,500,000. The second alternative, a channel dredging operation, would have an investment cost of $15,000,000. The annual cost of maintaining the channel is estimated at $375,000. There are no documented "benefits" for the channel-dredging project. Using a MARR of 8% and assuming a 25-yr life for either alternative, apply the incremental benefit–cost-ratio ($\Delta B/\Delta C$) method to determine which alternative should be chosen. (NOTE: The null alternative, "Do Nothing," is not a viable alternative.) (11.9)

11-17. A tunnel through a mountain is being considered as a replacement for an existing stretch of highway in southeastern Kentucky. The existing road is a steep, narrow, winding two-lane highway that has been the site of numerous fatal accidents, with an average of 2.05 fatalities and 3.35 "serious injuries" per year. It has been projected that the tunnel will significantly reduce the frequency of accidents, with estimates of not more than 0.15 fatalities and 0.35 "serious injuries" per year. Initial capital investment requirements, including land acquisition, tunnel excavation, lighting, roadbed preparation, etc., have been estimated to be $45,000,000. Annualized upkeep costs for the tunnel will be significantly less than for the existing highway, resulting in an annual savings of $85,000. For purposes of this analysis, a "value per life saved" of $1,000,000 will be applied, along with an estimate of $750,000 for medical costs, disability, etc., per "serious injury." (11.7)

a. Apply the benefit–cost-ratio method, with an anticipated life of the tunnel project of 50 yr and an interest rate of 8% per year, to determine whether tunnel should be constructed.
b. Assuming that the cost per "serious injury" is unchanged, determine the "value per life saved" at which the tunnel project is marginally justified (i.e., B–C = 1.0).

11-18. You have been assigned the task of comparing the economic results of three alternative designs for a state government public-works project. The estimated values for various economic factors related to the three designs are given in Table P11-18. The MARR being used is 9% and the analysis period is 15 yr. (11.9)

TABLE P11-18 Estimated Values for Problem P11-18

	Alternative Design		
Factor	1	2	3
Capital investment	$1,240,000	$1,763,000	$1,475,000
Market value (end of year 15)	90,000	150,000	120,000
Annual O & M costs	215,000	204,000	201,000
Annual benefits to user group A	315,000	367,000	355,000
Annual benefits to other user groups	147,800	155,000	130,500

a. Use the conventional benefit–cost-ratio method, with AW as the equivalent worth measure, to select the preferred design for the project.

b. Use the modified B–C-ratio method, with PW as the equivalent-worth measure, to select the preferred design for the project.

11-19. The Fox River is bordered on the east by Illinois Route 25 and on the west by Illinois Route 31. Along one stretch of the river, there is a distance of 16 miles between adjacent crossings. An additional crossing in this area has been proposed, and three alternative bridge designs are under consideration. Two of the designs have 25-yr useful lives, and the third has a useful life of 35 yr. Each bridge must be resurfaced periodically, and the roadbed of each bridge will be replaced at the end of its useful life, at a cost significantly less than initial construction costs. The annual benefits of each design differ on the basis of disruption to normal traffic flow along Routes

25 and 31. Given the information in Table P11-19, use the B–C-ratio method to determine which bridge design should be selected. Assume that the selected design will be used indefinitely, and use a nominal interest rate of 10% per year. (11.9)

11-20. Jackson County is planning to make road improvements along a section of one of the county highways. Two alternatives have been identified. Alternative A requires an initial outlay of $100,000 at the end of year 0 and yearly (end of year) maintenance costs of $15,200 each year thereafter. It will produce benefits to the public valued at $34,400 per year (end of year 1 and each year thereafter). Alternative B requires an initial outlay of $210,000 at the end of year 0 and yearly (end of year) maintenance costs of $10,600 each year thereafter. It will produce benefits to the public valued at $36,500 per year (end of year 1 and each year thereafter). It would also be possible to do nothing, which

TABLE P11-19 Bridge Design Information for Problem P11-19

	Bridge Design		
	A	B	C
Capital investment	$17,000,000	$14,000,000	$12,500,000
Annual maintenance cost*	12,000	17,500	20,000
Resurface (every fifth year)*	—	40,000	40,000
Resurface (every seventh year)*	40,000	—	—
Bridge replacement cost	3,000,000	3,500,000	3,750,000
Annual benefit	2,150,000	1,900,000	1,750,000
Useful life of bridge (yr)**	35	25	25

* Cost not incurred in last year of bridge's useful life.

** Applies to roadbed only; Structural portion of bridge has indefinite useful life

would cost nothing and would produce no benefits to the public. The county uses an annual 12% MARR for decision making, and it is willing to make this decision on the basis of cost. Analyze this problem *using the benefit–cost-ratio method* and explain what you would recommend that the county do. You must show the ratios that you use to make your recommendation. (11.9)

12

Engineering Economy Studies in Investor-Owned Utilities

*O*ur objective in this chapter is to present an economic evaluation technique, called the revenue requirement method, *that is widely used by utility companies to select from among mutually exclusive projects. Because utilities are expected to minimize the revenues required from customers paying for services, the project having the smallest revenue requirement while providing an acceptable level of service will normally be recommended.*

The following topics are discussed in this chapter:

General characteristics of investor-owned utilities

Development of the revenue requirement method

Assumptions of the revenue requirement method

Utility rate regulation

Illustration of the revenue requirement method

Immediate versus deferred investment

Revenue requirement analysis under conditions of inflation

12.1 Introduction

Investor-owned utilities provide services such as gas, electric power, water, telephone communications, environmental protection, and some types of transportation. Because public utilities are usually monopolistic, their financing and management have customarily been a government responsibility. However, the last three decades have been characterized by a strong privatization movement that has included power generation and transportation. For example, in the United Kingdom, the British Electric Board has been completely sold to private investors.

In the United States, the government has urged and helped private investors to enter the public utilities sector via the Private Producers Act.

Because city, state, and federal governments, acting on behalf of the public, have historically granted a monopolistic position to utility companies, they retain the right to regulate and control the utility's prices. Therefore, a regulatory body, generally in the form of a *public utilities commission*, is established by the public to exercise the desired regulation and control functions. Although public utilities commissions were originally established to prevent utilities from discriminating among customers with respect to services provided and prices charged, their functions have been expanded. They encompass the setting of rates, so that excessive profits are eliminated, and the establishment and maintenance of standards of service. For instance, during the 1970s, energy issues became critical and the major piece of legislation addressing the issue of efficient energy use was the 1979 Public Utilities Regulatory Policies Act (PURPA). This act required utilities to purchase power from available industrial sources and to pay as much for it per kiloWatt-hour as they would have if they themselves had generated it.

A recent trend in the United States is to deregulate electric power companies so that, for example, an electric utility company in New York can sell its services (electricity) and own plants in Oklahoma and California. An important issue facing the Federal Energy Regulatory Commission is how to maintain the high reliability of services among such deregulated electric power companies.[*]

In the remainder of Chapter 12, we discuss the revenue requirement method of capital investment evaluation. This method has traditionally been employed by utility companies to minimize the life cycle costs of providing their services. Even if deregulation of utility companies accelerates in the years ahead, it is likely that the traditionally accepted revenue requirement method will remain in widespread use for gauging the economic profitability of proposed capital investments. In fact, results provided by the revenue requirement method are generally equivalent to those obtained from a present worth analysis of a project's after-tax cash flows.[†] (See Chapter 6).

12.2 General Characteristics of Investor-Owned Utilities

Because of the nature of the services they render, their monopolistic position, and the regulation to which they are subject, investor-owned utilities have a number of unique economic characteristics that must be taken into account in making engineering economy studies. Some of these characteristics are discussed in the following paragraphs:

---

[*] "De-regulation Puts Electricity Reliability in Question," *USA Today*, July 10, 1998, p. B-1.

[†] T. L. Ward and W. G. Sullivan, "Equivalence of the Present Worth and Revenue Requirement Method of Capital Investment Analysis," *AIIE Transactions*, vol. 13, no. 1, pp. 29–40.

1. The capital investment per worker and the ratio of fixed costs to variable costs are very high. This means that careful attention must be given to investment problems and to ensuring an adequate flow of capital for expansion.

2. Utilities *must* render whatever service is demanded by customers within established rate schedules. Subject to regulatory safeguards, a utility must expand to meet the growth of the community it serves.

3. A utility is *required* to keep abreast of technical developments in its field that would permit reduction in the cost of its services and improve the reliability of the service. This must be done, even though not demanded immediately by customers, to maintain public goodwill and to protect the utility's monopolistic position.

4. The rates charged for a utility's services are based on total costs, including a fair return after income taxes, on the rate-base value of its property. The so-called rate base is roughly equal to the book value of a utility's plant and equipment in service.

5. A basic concept in setting rates for utility services is that the companies must be able to earn enough profits to pay sufficient dividends to attract the capital necessary for rendering the service. If an adequate rate of profit is not obtained, the necessary capital will not be forthcoming from investors, and, as a result, the public will not be able to have the desired utility service.

6. The earnings of a utility are limited by the rate base. As a result, profit on sales is of very little significance. If sales income increases as a result of decreased operating costs—for example, by more effective energy conservation—it may not produce any long-term profit increase. Profits for the current year might be realized, but if the increase were to result in a return that was judged by the regulatory commission to be greater than necessary, a rate reduction would be ordered. Thus, the benefits of the improved operation in terms of financial gain to the company would be eliminated.

7. Utilities have much greater stability of income than other companies. The upper limit of earnings, after income taxes, is not usually permitted to exceed about 12% to 16% on equity capital. It should be noted that even though a maximum limit is put on earnings, there is no guarantee of any profit, and there is no assurance against loss. However, if the utility can show that it is operating efficiently, it can usually obtain permission to increase its rates when needed to produce a fair profit, and may thus be able to attract needed capital.

8. Because of the stable nature of their business and earnings, utilities commonly finance their capital expenditures with a higher percentage of *borrowed* capital than do nonutility companies. Most nonutilities seldom use more than about 30% debt capital, but many utilities use 50% to 60% borrowed capital as a percentage of their total capitalization.

9. The assets of utilities, on the average, involve longer write-off periods than those of nonutilities. This is caused by the physical nature of the assets and by the fact that the monopolistic situation results in less functional depreciation.

10. Utilities are much less limited in terms of the availability of capital than are nonutility companies, due to their greater stability of revenues and earnings and

to the fact that regulatory agencies recognize that utilities must be permitted to earn a return that will ensure an adequate flow of capital.

12.3 General Concepts of Utility Economy Studies

There are several concepts that are usually inherent to engineering economy studies in regulated utility companies. These are as follows:

1. Economy studies of regulated utilities usually reflect the interests of the customer, whereas those of nonutility companies generally reflect the viewpoint of the owner.
2. Investor-owned utility economy studies usually involve alternative ways of, or alternative programs for, *doing* something. Because a utility is obligated to provide the service demanded by its customers, studies are seldom made of the economy of doing versus not doing. Instead, it is more often a matter of how to do something most economically.
3. Administrative and general supervision expenses frequently are not included. Because these expenses will be about the same for each alternative, they usually may be omitted.
4. The costs of money, depreciation, income taxes, and property taxes are usually expressed in terms of the capital invested.

12.4 Methods of Engineering Economy for Investor-Owned Utility Projects

The economic evaluation method most widely used by privately owned, regulated utilities is the *revenue requirement method*. This method provides a basis for comparing mutually exclusive alternatives. It can be applied to a wide spectrum of regulated businesses that have the characteristics discussed in the previous sections.

> In essence, the revenue requirement method calculates the revenues that a given project must provide just to meet all the costs associated with it, including a fair return to investors.

The relationship between a project's revenue requirements and its costs is shown in Figure 12-1. Because regulatory commissions act on behalf of the consumers of a utility's services, investment project selection should be made in such a way that revenue requirements are minimized.

The following sections develop and illustrate the revenue requirement method. Examples will be used to explain various facets of this method as it applies to privately owned, regulated utilities.

12.5 Development of the Revenue Requirement Method[*]

As shown in Figure 12-1, the minimum revenue requirement consists of carrying charges resulting from capital investments that must be recovered, plus all associated expenses that occur periodically (i.e., fuel, O&M expenses, property taxes, and insurance). Carrying charges are also called *total fixed charges*. They include the following:

- Interest on bonds used to partially finance the project
- Equity return requirements for the stockholders
- Income taxes to be paid to state and local governments
- Depreciation charges on the investment

The concept of a fixed-charge rate is widely used in the utility industry. The fixed-charge rate is defined as the annual owning cost of an investment (carrying charges) expressed as a percentage of the investment.

The equation used to find the annual carrying charges in year k, CC_k, is

$$CC_k = D_{B_k} + [(1 - \lambda)e_a + \lambda i_b] \cdot UI_k + T_k,$$ (12-1)

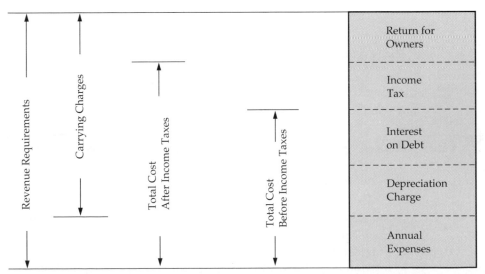

Figure 12-1 Relationship of Revenue Requirements and Costs for an Investor-Owned Utility

---

[*]The notation in Chapter 12 differs from that used in the rest of the book because various costs of debt and equity capital and resultant costs of total capital are required here. Many of these same concepts are utilized in other chapters, but are not developed and employed as rigorously as in Chapter 12.

where D_{B_k} = book depreciation taken in year k, $1 \leq k \leq N$;

λ = fraction of borrowed money in a utility's total capitalization;

e_a = return to equity capital (as a decimal);

i_b = cost of borrowed capital (as a decimal);

UI_k = unrecovered investment at the beginning of year k;

$$\text{UI}_k = \begin{cases} \text{I (initial investment)}, k = 1 \\ \text{UI}_{k-1} - D_{B_{k-1}}, \ 2 \leq k \leq N \end{cases} ;$$

T_k = income taxes paid in year k.

Since depreciation claimed for income-tax purposes and interest paid on debt are tax deductible, the income tax in any given year is determined by the following equation:

$$T_k = t\left(\text{CC}_k - \lambda \cdot i_b \cdot \text{UI}_k - D_{T_k}\right), \tag{12-2}$$

where D_{T_k} is the tax depreciation for income-tax purposes in year k and t is the effective income-tax rate.

Note that carrying charges (CC_k) are a function of income taxes (T_k) in Equation (12-1) and that income taxes (T_k) are a function of carrying charges (CC_k) in Equation (12-2). This can be seen clearly in Figure 12-1. The revenue requirement can be determined if the income taxes are known and, conversely, the income taxes can be computed if the revenue requirement is known. There are two equations and two unknowns (i.e., CC_k and T_k). Solving for T_k yields

$$T_k = [t/(1-t)]\left[(1-\lambda)e_a \cdot \text{UI}_k + D_{B_k} - D_{T_k}\right]. \tag{12-3}$$

The revenue requirement in year k, RR_k, is

$$\text{RR}_k = \text{CC}_k + C_k, \tag{12-4}$$

where C_k represents all recurring annual expenses in year k.

12.6 Assumptions of the Revenue Requirement Method

The following assumptions are common when using the revenue requirement method:

1. The total investment in an asset during any year is equal to its beginning-of-year book value.
2. The amount of debt capital invested in an asset during any year is a constant fraction of its book value during that year, and this fraction remains constant throughout the asset's life.
3. Equity and debt capital involve constant rates of return throughout the life of the project.
4. Book depreciation charges are used to retire capital stock and bonds each year in proportion to the debt–equity mix of financing employed.
5. The effective income-tax rate is constant over the life of the project.

12.7 Utility Rate Regulation

Utility rates are established during a regulatory rate proceeding. When changes in a utility's cost or income occur because of a change in the company's physical plant, a regulatory rate proceeding takes place to consider whether a new rate is warranted. First, an acceptable return on investors' equity is determined based on factors such as what is required to instill financial confidence in the utility, what other utilities are allowed when operating in the same business risk environment, and what is fair and reasonable. Revenues are then calculated to yield the required return on equity.

H. G. Stoll of the General Electric Company distinguishes these two criteria for electric rate regulation: the return on equity and the return on the rate base.* The return on common equity criterion calculates the ratio of net income available for common stock (from the utility's income statement) to the average or end-of-year common equity (from the utility's balance sheet). The revenue requirement is then increased or decreased so that the return-on-equity target is achieved. Alternatively, return on the rate base is a more traditional rate regulation criterion. The rate base is defined as follows:

Rate base = total plant in service

− accumulated depreciation reserve

+ materials and supplies (optional)

+ fossil fuel inventory (optional)

+ working capital allowance (optional)

− deferred income taxes (optional)

− deferred investment tax credit (optional)

+ construction work in progress (optional).

Because of the role and importance of a utility's cost of capital and capitalization structure to the minimum revenue requirement, we discuss selected aspects of an investor-owned utility's financing operation at this point. First, the interest paid on borrowed capital (debt) is tax deductible. Therefore, the after-tax cost of debt is

$$i_a' = i_b' - t i_b'$$

$$= (1 - t)\left[(1 + i_b)\left(1 + \bar{f}\right) - 1\right], \tag{12-5}$$

where i_b' = inflation-adjusted cost of borrowed capital;

$= \left[(1 + i_b)\left(1 + \bar{f}\right) - 1\right]$;

t = effective income-tax rate;

$\bar{f}$ = average annual inflation rate.

---

*Stoll, H. G., *Least-Cost Electric Utility Planning* (New York: John Wiley & Sons, 1987).

Second, the firm's cost of capital depends on the proportion and cost of both debt and equity capital. The *after-tax cost of capital,* including an adjustment for inflation, is

$$K'_a = \lambda i'_a + (1 - \lambda)e'_a$$

$$= \lambda(1 - t)i'_b + (1 - \lambda)e'_a, \tag{12-6}$$

where λ = fraction of borrowed money in the utility's total capitalization;

$(1 - \lambda)$ = fraction of equity capital in total capitalization;

e'_a = inflation-adjusted equity rate = $\left[(1 + e_a)(1 + \bar{f}) - 1\right]$.

The real (inflation-free) after-tax cost of capital is

$$K_a = \frac{1 + K'_a}{1 + \bar{f}} - 1$$

$$= \frac{\lambda(1 - t)i_b + (1 - \lambda)e_a - \lambda t \bar{f}}{1 + \bar{f}}, \tag{12-7}$$

where e_a is the real equity rate.

▼ 12.8 Flow-Through and Normalized Accounting

The revenue requirement method of project comparison presented in Section 12.9 uses the *flow-through* accounting method. Flow-through accounting requires income-tax savings (credits) resulting from (1) accelerated depreciation, (2) investment credits (when applicable), and (3) interest paid on funds used during construction to be passed on to a utility's customers in the year that they occur. For instance, straight-line depreciation for rate-setting purposes and accelerated depreciation for determining federal income taxes owed typically combine to *reduce* the revenue requirements for a project when using flow-through accounting methods. However, this accounting method is also frequently employed to compare the relative economics of competing projects, and it produces revenue requirements that are equivalent to those of the after-tax discounted cash-flow methods illustrated in Chapter 6.

On the other hand, normalized accounting requires the income-tax savings listed previously to be amortized (depreciated) over the life of a project. Normalized accounting is utilized by most investor-owned utilities as a way of protecting the company against unforeseen changes in future income-tax rates and state/federal laws that govern their operation. In addition, normalized accounting is almost exclusively used for setting rates on services provided to customers. This accounting method produces revenue requirements that are frequently higher than those resulting from the flow-through method. Because of the additional details associated with normalized accounting, we have elected in this chapter to discuss only the flow-through method for determining revenue requirements.

12.9 Illustration of the Revenue Requirement Method: A Tabular Procedure

Using a tabular form to calculate the annual revenue requirements for a utility project offers an easy-to-manipulate and understandable computational format. The analyst can use tabular columns, as required by a given problem, to account for the various components of revenue requirement shown in Figure 12-1.

EXAMPLE 12-1

This example, evaluating a single investment project, utilizes the following project data and the column-by-column operations shown in Table 12-1 to determine RR_k:

Project life, N = book life = 4 yr ;

Initial capital investment, I = $7,500;

Market value, MV = $1,500;

Annual O&M expenses, C = $500;

Real (inflation-free) cost of borrowed money, i_b = 5% per year;

Real (inflation-free) return on equity, e_a = 16.07% per year;

Debt ratio, λ = 0.3;

Effective income-tax rate, t = 50%;

Book depreciation method = straight line;

Tax depreciation method = straight line;

Average annual inflation rate, $\bar{f}$ = 0%.

For each operating year, k, $1 \leq k \leq 4$, the revenue requirement in year k, RR_k, is calculated by using Equation (12-4). One column is reserved for each term of the carrying charges [see Equation (12-1)], and an additional column is used for the recurring annual expenses associated with the project.

TABLE 12-1 Annual Revenue Requirements for Example 12-1

Year, k	(1) Unrecovered Invest., UI_k	(2) Book Depr., D_{B_k}	(3) Tax Depr., D_{T_k}	(4) Debt Return, $\lambda i_b UI_k$	(5) Equity Return, $(1-\lambda)e_a UI_k$	(6) Income Tax, T_k	(7) Annual Expenses, C_k	(8) RR_k = Cols. $2+4+5$ $+6+7$
1	$7,500	$1,500	$1,500	$113	$844	$844	$500	$3,780
2	6,000	1,500	1,500	90	675	675	500	3,440
3	4,500	1,500	1,500	68	506	506	500	3,080
4	3,000	1,500	1,500	45	337	337	500	2,720

For instance, RR_2 is calculated as follows:

Column 1: $UI_2 = UI_1 - D_{B_1}$
                  $= \$7{,}500 - \$1{,}500 = \$6{,}000$

Columns 2 $D = (I - MV)/N$
and 3: $= (\$7{,}500 - \$1{,}500)/4 = \$1{,}500$

Column 4: $\lambda i_b \cdot UI_2 = 0.3(0.05)(\$6{,}000) = \$90$

Column 5: $(1 - \lambda)e_a \cdot UI_2 = 0.7(0.1607)(\$6{,}000) = \$674.94$

Column 6: $T_2 = [t/(1 - t)][(1 - \lambda)e_a \cdot UI_2 + D_{B_2} - D_{T_2}]$
                $= [0.5/(1 - 0.5)] \cdot [0.7 \cdot 0.1607 \cdot \$6{,}000 + \$1{,}500 - \$1{,}500]$
                $= \$674.94$

Column 7: $C_2 = \$500$

Column 8: $RR_2 = \$1{,}500 + \$500 + \$90 + \$674.94 + \$674.94$
                $= \$3{,}439.88$

Calculations for the remaining years are performed similarly. Table 12-1 provides a summary of the results in Example 12-1. Note that no unrecovered investment is left after the end of year 4.

It is customary to express the yearly revenue requirements (column 8) as a single measure of worth for the project under study.

Cumulative present worth, equivalent annual worth (also called *levelized revenue requirement*, $\overline{RR}$), and capitalized worth are the three measures most often used by utilities to report the economic merit of a given project. To calculate these quantities, a discounting factor is needed to account for the time value of money. The utility's real after-tax cost of capital, K_a, is the interest rate employed for such calculations when $\overline{f} = 0$.

In Example 12-1, K_a is determined using Equation (12-7) with an inflation rate of $\overline{f} = 0$:

$$K_a = 0.3 \cdot (1 - 0.5) \cdot 0.05 + (1 - 0.3) \cdot 0.1607 - 0.3 \cdot 0.5 \cdot 0/(1 + 0)$$

$$= 0.12 \quad (12\%).$$

Hence, the present worth of RR as a function of K_a is

$$PWRR(K_a) = \sum_{k=1}^{N} RR_k \cdot (P/F, K_a\%, k)$$

$$= [\$3{,}799.86(P/F, 12\%, 1) + \$3{,}439.88(P/F, 12\%, 2)$$

$$+ \$3{,}079.92(P/F, 12\%, 3) + \$2{,}719.94(P/F, 12\%, 4)]$$

$$= \$10{,}055.59.$$

The levelized revenue requirement is then

$$\overline{RR}(K_a) = PWRR(K_a) \cdot (A/P, K_a\%, N)$$

$$= \$10,055.59 \cdot (A/P, 12\%, 4)$$

$$= \$3,310.70.$$

Finally, the capitalized revenue requirement is

$$CRR(K_a) = \overline{RR}(K_a) \div K_a$$

$$= \$3,310.70 \div 0.12$$

$$= \$27,589.17.$$

In selecting among alternative investment projects, these three quantitative measures are equivalent. The alternative that minimizes the selected revenue requirement measure represents the most economical choice. Because the utility is obligated to provide services to the public, a request for a rate increase can be presented to the regulatory commission if the investors feel that revenues generated from a project are unsatisfactory.

EXAMPLE 12-2

A public utility must extend electric power service to a small shopping center. A decision must be made as to whether a pole line or an underground system should be used. The pole-line system would cost only $158,000 to install, but because of numerous changes that are anticipated in the development and use of the shopping center, it is estimated that annual maintenance expenses would be $29,000. An underground system would cost $315,000 to install, but the annual maintenance expenses would not exceed $5,500. Annual property taxes are 1.5% of the capital investment. The company operates with 33% borrowed capital, on which it pays an interest rate of 8% per year. Capital should earn about 11% per year after taxes. For this problem, the after-tax return of 11% is interpreted as the value of K_a. A 20-yr study period is to be used, and the effects of inflation on cash flows are to be ignored. The straight-line method is to be used for both book and tax depreciation purposes. Finally, the effective income-tax rate is 39.94%.

Before column 5 (equity return) in the tabular procedure can be computed, the value of the return on equity, e_a, has to be determined. By using Equation (12-7), one can write

$$e_a = \{K_a - \lambda \cdot [(1-t) \cdot i_b - t \cdot \bar{f} \cdot (1 - \bar{f})]\}/(1 - \lambda)$$

$$= \{0.11 - 0.33 \cdot [(1 - 0.3994) \cdot 0.08 - 0.3994 \cdot 0 \cdot (1 - 0)]\}/(1 - 0.33)$$

$$= 0.1405.$$

TABLE 12-2 Pole-Line System Calculations for Example 12-2

	(1)	(2)	(3)	(4)	(5)	(6)	(7)	(8)
								RR_k = Cols.
	Unrecovered	Book	Tax	Debt	Equity	Income	Annual	2 + 4 + 5
Year, k	Investment	Deprec.	Deprec.	Return	Return	Tax	Expenses	+ 6 + 7
1	$158,000	$7.900	$7.900	$4,171	$14,875	$9,892	$31,370	$68,208
2	150,100	7,900	7,900	3,963	14,131	9,397	31,370	66,761
3	142,200	7,900	7,900	3,754	13,387	8,902	31,370	65,313
4	134,300	7,900	7,900	3,546	12,644	8,408	31,370	63,867
5	126,400	7,900	7,900	3,337	11,900	7,914	31,370	62,420
6	118,500	7,900	7,900	3,128	11,156	7,419	31,370	60,974
7	100,600	7,900	7,900	2,920	10,412	6,924	31,370	59,526
8	102,700	7,900	7,900	2,711	9,669	6,430	31,370	58,080
9	94,800	7,900	7,900	2,503	8,925	5,935	31,370	56,632
10	86,900	7,900	7,900	2,294	8,181	5,440	31,370	55,185
11	79,000	7,900	7,900	2,086	7,437	4,946	31,370	53,739
12	71,000	7,900	7,900	1,877	6,694	4,452	31,370	52,292
13	63,200	7,900	7,900	1,668	5,950	3,957	31,370	50,845
14	55,300	7,900	7,900	1,460	5,206	3,462	31,370	49,399
15	47,400	7,900	7,900	1,251	4,462	2,967	31,370	47,951
16	39,500	7,900	7,900	1,043	3,719	2,473	31,370	46,504
17	31,600	7,900	7,900	834	2,975	1,978	31,370	45,057
18	23,700	7,900	7,900	626	2,231	1,484	31,370	43,611
19	15,800	7,900	7,900	417	1,487	989	31,370	42,163
20	7,900	7,900	7,900	209	744	495	31,370	40,717
								$\overline{RR}$ = $59,497

Tables 12-2 and 12-3 display the yearly RR results for the pole-line and underground systems, respectively. The values of $\overline{RR}$ are given for each alternative at the bottom of the corresponding RR table. Accordingly, the pole-line system shows a lower $\overline{RR}$ and is, therefore, the system to be chosen based on financial considerations only.

12.10 Immediate versus Deferred Investment

Because utilities must always be prepared to meet the demands for service placed on them, many engineering economy studies in utility companies involve immediate versus deferred investment to meet future demands. The following is an example.

TABLE 12-3 Underground System Calculations for Example 12-2

	(1)	(2)	(3)	(4)	(5)	(6)	(7)	(8)
								RR_k = Cols.
	Unrecovered	Book	Tax	Debt	Equity	Income	Annual	2 + 4 + 5
Year, k	Investment	Deprec.	Deprec.	Return	Return	Tax	Expenses	+ 6 + 7
1	$315,000	$15,750	$15,750	$8,316	$29,655	$19,721	$10,225	$83,667
2	299,250	15,750	15,750	7,900	28,173	18,735	10,225	80,783
3	283,500	15,750	15,750	7,484	26,690	17,749	10,225	77,899
4	267,750	15,750	15,750	7,069	25,207	16,763	10,225	75,014
5	252,000	15,750	15,750	6,653	23,724	15,777	10,225	72,129
6	236,250	15,750	15,750	6,237	22,242	14,791	10,225	69,245
7	220,500	15,750	15,750	5,821	20,759	13,805	10,225	66,360
8	204,750	15,750	15,750	5,405	19,276	12,819	10,225	63,476
9	189,000	15,750	15,750	4,990	17,793	11,832	10,225	60,590
10	173,250	15,750	15,750	4,574	16,310	10,846	10,225	57,705
11	157,500	15,750	15,750	4,158	14,828	9,861	10,225	54,822
12	141,750	15,750	15,750	3,742	13,345	8,874	10,225	51,936
13	126,000	15,750	15,750	3,326	11,862	7,888	10,225	49,052
14	110,250	15,750	15,750	2,911	10,379	6,902	10,225	46,167
15	94,500	15,750	15,750	2,495	8,897	5,917	10,225	43,283
16	78,750	15,750	15,750	2,079	7,414	4,930	10,225	40,398
17	63,000	15,750	15,750	1,663	5,931	3,944	10,225	37,513
18	47,250	15,750	15,750	1,247	4,448	2,958	10,225	34,629
19	31,500	15,750	15,750	832	2,966	1,972	10,225	31,744
20	15,750	15,750	15,750	416	1,483	986	10,225	28,859
								$\overline{RR}$ = $66,305

EXAMPLE 12-3

A water company must decide whether to install a new pumping plant now and abandon a gravity-feed system, which has been fully depreciated, or wait five years to install the new plant because of deteriorated piping in the gravity-feed system. Annual O&M expenses and taxes for the gravity system are $45,000. The pumping plant will cost $375,000 to install, and it is estimated that it would have a market value of 5% of its capital investment at the time of removal from service 20 yr hence, when a new and larger system will be installed. Annual O&M expenses and property taxes for the proposed plant would be $30,000. The gravity-feed system has no market value now or later.

If the pumping plant is installed now, it would have a useful life of 20 yr. If installed 5 yr hence, its useful life would be only 15 yr, but its market value would still be 5% of its capital investment. Using the revenue requirement method, determine which alternative is better. Straight-line depreciation is assumed for both book and tax purposes. The company operates with 50% borrowed capital, on which it pays an interest rate of 7% per year. The equity rate is expected to be about 14% per year, and the company pays an effective income-tax rate of 50%.

SOLUTION
First, we determine K_a from Equation (12-7) to be $0.5[(1 - 0.5)(0.07)] + 0.5(0.14) = 0.0875$. Next, from Table 12-4, the levelized revenue requirement of the new pumping plant using the tax-adjusted cost of capital is found to be

$$\overline{RR}(8.75\%) = \$92,135.$$

From Table 12-5, the levelized revenue requirement of the deferred installation is

$$\overline{RR}(8.75\%) = \$74,876.$$

TABLE 12-4 Install New Pumping Plant Now for Example 12-3

Year, k	(1) Unrecovered Investment	(2) Book Deprec.	(3) Tax Deprec.	(4) Debt Return	(5) Equity Return	(6) Income Tax	(7) Annual Expenses	(8) RR_k = Cols. 2 + 4 + 5 + 6 + 7
1	$375,000.00	$17,812.50	$17,812.50	$13,125.00	$26,250.00	$26,250	$30,000	$113,438
2	357,187.50	17,812.50	17,812.50	12,501.56	25,003.13	25,003	30,000	110,320
3	339,375.00	17,812.50	17,812.50	11,878.13	23,756.25	23,756	30,000	107,203
4	321,562.50	17,812.50	17,812.50	11,254.69	22,509.38	22,509	30,000	104,086
5	303,750.00	17,812.50	17,812.50	10,631.25	21,262.50	21,263	30,000	100,970
6	285,937.50	17,812.50	17,812.50	10,007.81	20,015.62	20,016	30,000	97,852
7	268,125.00	17,812.50	17,812.50	9,384.38	18,768.76	18,769	30,000	94,735
8	250,312.50	17,812.50	17,812.50	8,760.94	17,521.88	17,522	30,000	91,618
9	232,500.00	17,812.50	17,812.50	8,137.50	16,275.00	16,275	30,000	88,501
10	214,687.50	17,812.50	17,812.50	7,514.06	15,028.12	15,028	30,000	85,383
11	196,875.00	17,812.50	17,812.50	6,890.63	13,781.26	13,781	30,000	82,266
12	179,062.50	17,812.50	17,812.50	6,267.19	12,534.38	12,534	30,000	79,148
13	161,250.00	17,812.50	17,812.50	5,643.75	11,287.50	11,288	30,000	76,033
14	143,437.50	17,812.50	17,812.50	5,020.31	10,040.62	10,041	30,000	72,915
15	125,625.00	17,812.50	17,812.50	4,396.88	8,793.76	8,794	30,000	69,798
16	107,812.50	17,812.50	17,812.50	3,773.44	7,546.88	7,547	30,000	66,680
17	90,000.00	17,812.50	17,812.50	3,150.00	6,300.00	6,300	30,000	63,563
18	72,187.50	17,812.50	17,812.50	2,526.56	5,053.12	5,053	30,000	60,446
19	54,375.00	17,812.50	17,812.50	1,903.31	3,806.26	3,806	30,000	57,328
20	36,562.50	17,812.50	17,812.50	1,279.69	2,559.38	2,559	30,000	54,210
							$\overline{RR}$ =	$92,135

TABLE 12-5 Defer Installation of New Pump for Five Years for Example 12-3

Year, k	(1) Unrecovered Investment	(2) Book Deprec.	(3) Tax Deprec.	(4) Debt Return	(5) Equity Return	(6) Income Tax	(7) Annual Expenses	(8) RR_k = Cols. 2 + 4 + 5 + 6 + 7
1	$ 0	$ 0	$ 0	$ 0	$ 0	$ 0	$45,000	$ 45,000
2	0	0	0	0	0	0	45,000	45,000
3	0	0	0	0	0	0	45,000	45,000
4	0	0	0	0	0	0	45,000	45,000
5	0	0	0	0	0	0	45,000	45,000
6	375,000	23,750	23,750	13,125.00	26,250	26,250	30,000	119,375
7	351,250	23,750	23,750	12,293.75	24,588	24,588	30,000	115,220
8	327,500	23,750	23,750	11,462.50	22,925	22,925	30,000	111,063
9	303,750	23,750	23,750	10,631.25	21,263	21,263	30,000	106,907
10	280,000	23,750	23,750	9,800.00	19,600	19,600	30,000	102,750
11	256,250	23,750	23,750	8,968.75	17,938	17,938	30,000	98,595
12	232,500	23,750	23,750	8,137.50	16,275	16,275	30,000	94,438
13	208,750	23,750	23,750	7,306.25	14,673	14,673	30,000	90,402
14	185,000	23,750	23,750	6,475.00	12,950	12,950	30,000	86,125
15	161,050	23,750	23,750	5,643.75	11,288	11,288	30,000	81,970
16	137,500	23,750	23,750	4,812.50	9,625	9,625	30,000	77,813
17	113,750	23,750	23,750	3,981.25	7,963	7,963	30,000	73,657
18	90,000	23,750	23,750	3,150.00	6,300	6,300	30,000	69,500
19	66,250	23,750	23,750	2,318.75	4,638	4,638	30,000	65,345
20	42,500	23,750	23,750	1,487.50	2,975	2,975	30,000	61,188

$$\overline{RR} = \$74,876$$

Finally, a comparison of the levelized revenue requirements for both alternatives shows that it is more economical to defer the new pumping plant for five years.

▼ 12.11 Revenue Requirement Analysis under Conditions of Inflation*

As discussed in Chapter 8, considerable confusion frequently arises when we consider inflation in engineering economy studies because of depreciation and other fixed actual dollar annuities that are not sensitive to inflation. This same difficulty also arises in connection with the revenue requirement method. Example 12-4 illustrates the correct treatment of inflation in revenue requirement studies.

---

*Remaining examples include MACRS depreciation under the "Tax Depreciation" column to illustrate the situation where book depreciation and tax depreciation are different.

EXAMPLE 12-4

We now re-evaluate Example 12-1 when annual expenses inflate at 10% per year and when the cost of borrowed money and return to equity also increase due to this rate of inflation. It is assumed further that the market value is *not* responsive to inflation. Furthermore, amounts estimated for annual expenses are expressed in year zero purchasing power.

The revenue requirement table is obtained by using the equations employed in Example 12-1, *except* that all year 0 (real) quantities are replaced by their equivalent inflation-adjusted values and MACRS (GDS property class is three years) depreciation is included in column 3. In particular, the inflation-adjusted cost of borrowed capital is calculated as

$$i'_b = (1 + i_b) \cdot (1 + \bar{f}) - 1$$

$$= (1 + 0.05) \cdot (1 + 0.1) - 1$$

$$= 0.155,$$

and the inflation-adjusted equity rate is

$$e'_a = (1 + e_a) \cdot (1 + \bar{f}) - 1$$

$$= (1 + 0.1607) \cdot (1 + 0.1) - 1$$

$$= 0.27677.$$

Similarly, annual expenses in year k are

$$C_k = \$500 \cdot (1 + \bar{f})^k, \quad 1 \le k \le 4.$$

The results of the revenue requirement analysis are summarized in Table 12-6, and representative calculations are the following:

Column 4:   Debt return in year k $= \lambda i'_b \cdot \text{UI}_k$
            Debt return in year 1 $= (0.3)(0.155)(\$7,500)$
                                  $= \$348.75;$

Column 5:   Equity return in year k $= (1 - \lambda)e'_a \cdot \text{UI}_k$
            Equity return in year 1 $= (1 - 0.3)(0.27677)(\$7,500)$
                                  $= \$1,453.04;$

Column 6:   Income tax $T_k = \left[t/(1 - t) \right]\left[(1 - \lambda)e'_a \cdot \text{UI}_k + D_{B_k} - D_{T_k} \right]$
            Income tax in year 1 $= [0.5/(1 - 0.5)][(1 - 0.3)(0.27677)(\$7,500)$
                                  $+ \$1,500 - \$2,500] = \$453.$

The inflation-adjusted after-tax cost of capital calculated from Equation (12-6) is

$$K'_a = \lambda(1 - t)i'_b + (1 - \lambda)e'_a$$

$$= 0.3(1 - 0.5)(0.155) + (1 - 0.3)(0.27677)$$

$$= 0.216989$$

$$\cong 21.7\%.$$

TABLE 12-6 Solution of Example 12-1 with $\overline{f}$ = 10% (Example 12-4)

Year, k	(1) Unrecovered Investment	(2) Book Deprec.	(3) Tax Deprec.	(4) Debt Return	(5) Equity Return	(6) Income Tax	(7) Annual Expenses	(8) RR_k = Cols. 2 + 4 + 5 + 6 + 7
1	$7,500	$1,500	$2,500	$349	$1,453	$ 453	$550	$4,305
2	6,000	1,500	3,334	279	1,162	−671	605	2,875
3	4,500	1,500	1,111	209	872	1,261	666	4,508
4	3,000	1,500	556	140	581	1,525	733	4,479

The levelized revenue requirement of the project under conditions of inflation is then

$$\overline{RR}\left(K'_a\right) = [\$4,305.08 \cdot (P/F, 21.7\%, 1) + \$2,875.11 \cdot (P/F, 21.7\%, 2)$$

$$= +\$4,507.66 \cdot (P/F, 21.7\%, 3) + \$4,478.69 \cdot (P/F, 21.7\%, 4)]$$

$$= \$3,996.43.$$

12.12 Summary

Because of the special characteristics of utilities, privately owned utilities are usually granted a monopolistic franchise by the public. In return, these utilities are expected to satisfy their customers' expressed demands under the control of a regulatory body acting on behalf of the public.

The revenue requirement method was introduced as an appropriate economic evaluation technique for public-utility projects. A fundamental principle underlying utility-rate regulation is the fact that revenues should be generated so as just to cover the utility-service expenses and provide a fair return on equity to investors.

> Recommended choices with the revenue requirement method are the same as those made when using conventional PW and AW methods at a discounting rate equal to the utility's weighted after-tax cost of capital.

The revenue requirement method is equivalent to an after-tax PW or AW analysis of competing alternatives. Only the perspective is different. That is, PW and AW methods evaluate the project from the shareholder's viewpoint, while the revenue requirement method uses the utility customer's viewpoint because rates are regulated by representatives of the public.

▼12.13 References

Commonwealth Edison Company. *Engineering Economics* (Chicago: Commonwealth Edison Company, 1975).

Jeynes, P. H. *Profitability and Economic Choice* (Ames: Iowa State University Press, 1968).

Mayer, R. R. "Finding Your Minimum Revenue Requirements," *Industrial Engineering*, vol. 9, no. 4, April 1977, pp. 16–22.

Stoll, H. G. *Least-Cost Electric Utility Planning* (New York: John Wiley & Sons, 1987).

Ward, T. L., and W. G. Sullivan. "Equivalence of the Present Worth and Revenue Requirements Method of Capital Investment Analysis," *AIIE Transactions*, vol. 13, no. 1, pp. 29–40.

▼12.14 Problems

The number in parentheses () that follows each problem refers to the section from which the problem is taken.

12-1.

 a. Describe the types of regulation to which investor-owned utilities, but not private industries, are subject. Why is regulation necessary? (12.1)

 b. How would economy studies differ in a government-owned utility as opposed to an investor-owned utility? (12.2)

12-2.

 a. What advantages to the public result from utility companies? (12.1)

 b. What disadvantages might there be to utilities? (12.1)

 c. How is regulation of utilities that operate solely within an individual state achieved as opposed to those that provide services to many states (e.g., telephone companies and gas pipelines)? (12.2)

12-3. Briefly summarize the basic characteristics that distinguish investor-owned utilities from nonregulated industries such as steel, automobile, and chemical manufacturing. (12.3)

12-4. Why are most utilities heavily financed with borrowed capital? What characteristics of this industry make it possible to attract large amounts of borrowed capital, and what advantages (disadvantages) are associated with the use of borrowed money? (12.2)

12-5. Explain why it may be in the best interest of the consuming public for a regulatory agency to permit a utility to charge sufficiently high rates to allow it to earn an adequate return on its capital. (12.7)

12-6.

 a. In a certain state, a member of the Public Utilities Commission said, "I will oppose all rate increases. I am interested only in the rates the customers have to pay today." Comment on the results that could follow if all members of the Commission rigidly followed this philosophy. (12.7)

 b. Comment on this statement: "No company that provides an exclusive and required service, such as electric power, should be permitted to make a profit." (12.7)

12-7. Is there justification for a privately owned, regulated utility being permitted to include in its rates the cost of advertising (to encourage the public to increase utilization of its service)? (12.7)

> *Note:* Solve the remaining problems by using the after-tax cost of capital, K_a (or K'_a).

12-8. A telephone company can provide certain facilities having a 10-yr life and zero market value by either of two alternatives. Alternative A requires a capital investment of $70,000 and $3,000 per year for maintenance. Alternative B will have a capital investment of $48,000 and will require $6,000 annually for maintenance. Property taxes and insurance would be 4% of the capital investment per year for either alternative. The after-tax cost of capital is 10%, with 30% being borrowed at a 6% interest rate. The effective income-tax rate is 50%. Which alternative will provide the lower annual equivalent revenue requirement? The MACRS (GDS) recovery period is five years, and book depreciation is computed with the straight-line method over 10 years. (12.5, 12.8)

12-9. A gas company must decide whether to build a new meter-repair and testing facility now or wait three years before doing so. It estimates that until the new facility is built, its annual expenses for these functions will be $90,000 greater than when the new facility is completed. The new facility would cost $900,000 and would not be needed after 20 yr (which is the analysis period). The ultimate market value would be $200,000 at that time. The company uses 40% borrowed capital, paying 8% per year interest (before taxes) for it, and the regulatory body permits it to earn 13.8% per year on its equity capital. Assuming that the company has a 46% income-tax rate, determine the equivalent annual revenue requirement for both options and recommend which is better. Assume that book (and tax) depreciation over 20 yr is computed with the straight-line method. (12.5, 12.8)

12-10. A utility company can construct a modern power plant that can generate power at 24 mil ($0.024) per kilowatt-hour at a 70% load factor. The 24 mil covers all expenses, including profit on capital and also income taxes. A large industrially owned power plant will soon make wholesale power available. To take advantage of the wholesale power, it will cost $180 per kilowatt of capacity to build the necessary transmission line, which will experience a 70% load factor. Annual maintenance expenses for this line will be $0.90 per kilowatt of capacity, and it is to be fully depreciated for book purposes over a 30-yr period. A 15-yr MACRS recovery period (GDS) will be used for calculating depreciation for income-tax purposes. The cost of money for the company is 12% per year, with 40% borrowed capital at an interest rate of 7% per year. The effective income-tax rate is 50%. The study period is 30 years. At what price must the company be able to purchase the power in order for it to be as economical as if it were generated with the new modern plant? (12.5, 12.8)

12-11. Determine the annual revenue requirements for the following proposed 280-KVA transformer bank. (12.5)

Installed cost	= $240,000
Property taxes and insurance/yr	= 2% of installed cost
Market value	= 0
Tax life = book life	= 4 yr
Depreciation method (for book purposes)	= Straight line
Depreciation method (for tax purposes)	= MACRS (GDS, 3-yr recovery period)
Effective income-tax rate	= 0.40
Cost of equity capital	= 20% per year, $(1 - \lambda)$ = 0.60
Cost of borrowed funds	= 12% per year, $\lambda = 0.40$

Fill in Table P12-11 in completing this problem.

TABLE P12-11 Table for Problem 12-11

EOY, k	Unrecovered Investment	Depreciation: Book	Tax	Recurring Annual Expenses	Return to Debt	Return to Equity	Income Tax	Revenue Requirement
1								
2								
3								
4								

12-12. A telephone company must provide a direct current battery unit to a new service region in 2002. The expected useful life of the equipment is seven years. Alternative A requires a capital investment of $75,000 and has O&M expenses of $8,000 per year. The market value for tax purposes is zero, and this is also the expected realizable market value. A tax life of five years will be claimed, and MACRS (GDS, five-year property) depreciation will be used for tax purposes. However, straight-line depreciation over seven years will be taken for rate-setting purposes (i.e., book depreciation).

The after-tax cost of capital (K_a') is 12% per year, with 40% being borrowed at 8% per year. The effective income-tax rate is 40%, and the general inflation rate is 6% per year. Only O&M expenses are affected by inflation, and the costs of capital given previously include an allowance for anticipated inflationary pressures in the economy.

For Alternative A, answer the following questions. Be sure to state any assumptions you feel are appropriate and necessary. (12.10)

a. What is the actual dollar ATCF in year five of this alternative's useful life?

b. What is the *income tax* entry in an RR table for year five?

12-13. In 2002, the installed cost of a new transformer at the OPEC Utility Company is $50,000. Annual maintenance expenses, which are expected to escalate by 5% each year, are $1,500 in today's dollars. A five-year MACRS (GDS) recovery period is to be used for tax depreciation purposes, and the expected life of the transformer is eight years. The terminal MV is negligible. Straight-line depreciation is used for determining BV for rate-setting purposes. Borrowed capital represents 40% of the company's capitalization, and it costs 10% per year before taxes. The return to equity is approximately 15% per year. (12.10)

a. If the firm's effective income-tax rate is 40%, calculate the RR in year three.

b. If the firm's effective income-tax rate is 50%, by how much does the RR in year three increase?

12-14. An electric utility company has an opportunity to build a small hydroelectric generating plant, of 20,000-kW capacity, on a mountain stream where the flow is seasonal. As a consequence, the annual output of energy would be only 40,000,000 kWh. The capital investment would be $2,000,000, and it is estimated that the annual operation and maintenance expenses would be $32,000 during its estimated 30-yr economic life. It is believed that the property would have a market value of $200,000 at the end of the 30-yr period. An alternative is to build a geothermal generating plant, which would have the same annual capacity, at a cost of $1,600,000. Because the company would have to pay the owners of the property for the geothermal steam, the estimated annual expense for the steam and operation and maintenance is $120,000. A 30-yr contract can be obtained on the steam supply, and it is believed that this period is realistic for the economic life of the plant but that the market value at that time would be little more than zero. Property taxes and insurance on either plant would be 2% of the capital investment per year. The company employs 40% borrowed capital, for which it pays 8.5% interest per year. It earns 13% per year after taxes on total capital, and it has a 50% effective income-tax rate. Which development should be undertaken? State all assumptions that you make. (12.8)

12-15. Use the RR method to compare alternatives A and B in Problem 12-8 when the average inflation rate on maintenance is 6% per year. Assume that property taxes do not respond to inflation, and adjust the cost of capital to account for inflation. (12.10)

12-16. A natural-gas pipeline company is considering two plans to provide service required by present demand and the forecasted growth of demand for the coming 18 yr. Alternative A requires an immediate investment of $700,000 in property that has an estimated life of 18 yr, with 10% of the capital investment as the terminal market value. Annual expenses will be $25,000. Annual property taxes will be 2% of the capital investment. Alternative B requires an immediate investment of $400,000 in property that has an estimated life of 18 years, with 20% of the capital investment as the terminal market value. Annual O&M expenses during the first eight years will be $42,000. After eight years, an additional investment of $450,000 will be required in property having an estimated life of 10 yr with 50% of the additional investment as the terminal market value. After this additional property is installed, annual O&M expenses (for years 9 to 18) of the combined properties will be $72,000.

Annual property taxes will be 2% of the initial capital investment of the property in service at any time. The regulatory commission is allowing a 10% per year fair return on depreciated BV to cover the cost of money (K_a) to the utility. Assume that this rate of return will continue throughout the 18 yr. The utility company's effective tax rate is 50%. Straight-line depreciation is to be used for book purposes in setting rates, and MACRS (GDS) depreciation is to be used for income-tax purposes. The MACRS recovery period is seven years for all depreciable assets. Half of the utility's financing is by debt with interest at 8% per year. Determine which plan minimizes equivalent annual revenue requirements after property and income taxes have been taken into account. (12.9)

12-17. In making its forecast of requirements in a certain area for the next 30 yr, a telephone company has determined that a 600-pair cable is required immediately and a total of 1,000 pairs will be required by the end of 15 yr. An underground conduit of sufficient size to handle the cable needs is being installed now at a cost of $10,000. If a 1,000-pair cable is installed now, it will cost $30,000. As an alternative, it can install a 600-pair cable immediately at a cost of $20,000 and install an additional 400-pair cable at the end of 15 yr at an estimated cost of $16,000. Because of technical obsolescence, it is company policy to consider the economic life of either installation to be 30 yr from the present time. Annual property taxes on either alternative would be 2% of the installed cost, and the market value of all cable and conduit at the end of the 30-yr period is estimated to be 10% of the installed cost. The company uses 40% borrowed capital, for which it pays 8% per year. It earns 12% per year after taxes on total capital and has a 50% effective income-tax rate. Which alternative would you recommend? Assume that depreciation for tax and book purposes is computed with the straight-line method over 15 yr for both alternatives. (12.9)

13

Probabilistic Risk Analysis

*T*he objectives of this chapter are to (1) introduce the use of statistical and probability concepts in decision situations involving risk and uncertainty, (2) illustrate how they can be applied in engineering economic analysis, and (3) discuss the considerations and limitations relative to their application.

The following topics are discussed in this chapter:

The distribution of random variables

The basic characteristics of probability distributions

Evaluation of projects with discrete random variables

Probability trees

Evaluation of projects with continuous random variables

A description of Monte Carlo simulation

Accomplishing Monte Carlo simulation with a computer

Decision tree analysis

13.1 Introduction

In this chapter, selected statistical and probability concepts are used to analyze the economic consequences of some decision situations involving risk and uncertainty and requiring engineering knowledge and input. The probability that a cost, revenue, useful life, or other factor value will occur, or that a particular equivalent-worth or rate-of-return value for a cash flow will occur, is usually considered to be the long-run relative frequency with which the event (value) occurs or the subjectively estimated likelihood that it will occur. Factors such as these, having probabilistic outcomes, are called *random variables.*

As discussed in Chapter 1, a decision situation such as a design task, a new venture, an improvement project, or any similar effort requiring engineering knowledge, has two or more alternatives associated with it. The cash-flow amounts for each alternative often result from the sum, difference, product, or quotient of random variables such as initial capital investments, operating expenses, revenues, changes in working capital, and other economic factors. Under these circumstances, the measures of profitability (e.g., equivalent-worth and rate-of-return values) of the cash flows will also be random variables.

> The information about these random variables that is particularly helpful in decision making are their expected values and variances, especially for the economic measures of merit of the alternatives. These derived quantities for the random variables are used to make the uncertainty associated with each alternative more explicit, including any probability of loss. Thus, when uncertainty is considered, the variability in the economic measures of merit and the probability of loss associated with the alternatives are normally used in the decision-making process.

13.2   The Distribution of Random Variables

Capital letters such as X, Y, and Z are usually used to represent random variables and lowercase letters (x, y, z) to denote the particular values that these variables take on in the sample space (i.e., in the set of all possible outcomes for each variable). When a random variable X is considered to follow some *discrete* probability distribution, its *probability mass function* is usually indicated by $p(x)$ and its *cumulative distribution function* by $P(x)$. When a random variable is considered to follow a *continuous* probability distribution, its *probability density function* and its cumulative distribution function are usually indicated by $f(x)$ and $F(x)$, respectively.

13.2.1   Discrete Random Variables

A random variable X is said to be *discrete* if it can take on at most a countable (finite) number of values $(x_1, x_2, \ldots, x_L)$. The probability that a discrete random variable X takes on the value x_i is given by

$$\Pr\{X = x_i\} = p(x_i) \text{ for } i = 1, 2, \ldots, L \text{ (i is a \textit{sequential index} of}$$

the discrete values, x_i, that the variable takes on),

where $p(x_i) \geq 0$ and $\sum_i p(x_i) = 1$.

The probability of events about a discrete random variable can be computed from its probability mass function $p(x)$. For example, the probability of the event that the value of X is contained in the closed interval $[a, b]$ is given by (where the colon is read "such that")

$$\Pr\{a \leq X \leq b\} = \sum_{i:a \leq X_i \leq b} p(x_i). \tag{13-1}$$

The probability that the value of X is less than or equal to $x = h$, the cumulative distribution function $P(x)$ for a discrete case, is given by

$$\Pr\{X \leq h\} = P(h) = \sum_{i:X_i \leq h} p(x_i). \tag{13-2}$$

In most practical applications, discrete random variables represent *countable* data such as the useful life of an asset in years, number of maintenance jobs per week, or number of employees as positive integers.

13.2.2 Continuous Random Variables

A random variable X is said to be continuous if there exists a nonnegative function $f(x)$ such that for any set of real numbers $[c, d]$, where $c < d$, the probability of the event that the value of X is contained in the set is given by

$$\Pr\{c \leq X \leq d\} = \int_c^d f(x)dx \tag{13-3}$$

and

$$\int_{-\infty}^{\infty} f(x)dx = 1.$$

Thus, the probability of events about the continuous random variable X can be computed from its probability density function, and the probability that X assumes exactly any one of its values is 0. Also, the probability that the value of X is less than or equal to a value $x = k$, the cumulative distribution function $F(x)$ for a continuous case, is given by

$$\Pr\{X \leq k\} = F(k) = \int_{-\infty}^{k} f(x)dx. \tag{13-4}$$

Also, for a continuous case,

$$\Pr\{c \leq X \leq d\} = \int_c^d f(x)dx = F(d) - F(c). \tag{13-5}$$

In most practical applications, continuous random variables represent *measured* data such as time, cost, and revenue on a continuous scale. Depending upon the situation, the analyst decides to model random variables in engineering economic analysis as either discrete or continuous.

13.2.3 Mathematical Expectation and Selected Statistical Moments

The expected value of a single random variable X, $E(X)$, is a weighted average of the distributed values x that it takes on and is a measure of the central location of the distribution (central tendency of the random variable). The $E(X)$ is the first moment of the random variable about the origin and is called the *mean* (central moment) of the distribution. The expected value is

$$E(X) = \begin{cases} \sum_i x_i p(x_i) & \text{for } x \text{ discrete and } i = 1, 2, \ldots, L \\ \int_{-\infty}^{\infty} x f(x) dx & \text{for } x \text{ continuous.} \end{cases} \qquad (13\text{-}6)$$

While $E(X)$ provides a measure of central tendency, it does not measure how the distributed values x cluster around the mean. The *variance*, $V(X)$, which is nonnegative, of a single random variable X is a measure of the dispersion of the values it takes on around the mean. It is the expected value of the square of the difference between the values x and the mean, which is the second moment of the random variable about its mean:

$$E\{[X - E(X)]^2\} = V(X) = \begin{cases} \sum_i [x_i - E(X)]^2 p(x_i) & \text{for } x \text{ discrete} \\ \int_{-\infty}^{\infty} [x - E(X)]^2 f(x) dx & \text{for } x \text{ continuous.} \end{cases} \qquad (13\text{-}7)$$

From the binomial expansion of $[X - E(X)]^2$, it can be easily shown that $V(X) = E(X^2) - [E(X)]^2$. That is, $V(X)$ equals the second moment of the random variable around the origin, which is the expected value of X^2, minus the square of its mean. This is the form often used for calculating the variance of a random variable X:

$$V(X) = \begin{cases} \sum_i x_i^2 p(x_i) - [E(X)]^2 & \text{for } x \text{ discrete} \\ \int_{-\infty}^{\infty} x_i^2 f(x) dx - [E(X)]^2 & \text{for } x \text{ continuous.} \end{cases} \qquad (13\text{-}8)$$

The *standard deviation* of a random variable, $\text{SD}(X)$, is the positive square root of the variance; that is, $\text{SD}(X) = [V(X)]^{1/2}$.

13.2.4 Multiplication of a Random Variable by a Constant

A common operation performed on a random variable is to multiply it by a constant, for example, the estimated maintenance labor expense for a time period, $Y = cX$, when the number of labor hours per period (X) is a random variable, and the cost per labor hour (c) is a constant. Another example is the PW calculation for a project when the before-tax or after-tax net cash-flow amounts, F_k, are random variables, and then each F_k is multiplied by a constant $(P/F, i\%, k)$ to obtain the PW value.

When a random variable, X, is multiplied by a constant, c, the expected value, $E(cX)$, and the variance, $V(cX)$, are

$$E(cX) = cE(X) = \begin{cases} \sum_i cx_i p(x_i) & \text{for } x \text{ discrete} \\ \int_{-\infty}^{\infty} cxf(x)dx & \text{for } x \text{ continuous} \end{cases} \tag{13-9}$$

and

$$\begin{aligned} V(cX) &= E\{[cX - E(cX)]^2\} \\ &= E\{c^2X^2 - 2c^2X \cdot E(X) + c^2[E(X)]^2\} \\ &= c^2E\{[X - E(X)]^2\} \\ &= c^2V(X). \end{aligned} \tag{13-10}$$

13.2.5 Multiplication of Two Independent Random Variables

A cash-flow random variable, say Z, may result from the product of two other random variables, $Z = XY$. Sometimes, X and Y can be treated as statistically independent random variables. For example, consider the estimated annual expenses, $Z = XY$, for a repair part repetitively procured during the year on a competitive basis, when the unit price (X) and the number of units used per year (Y) are modeled as independent random variables.

When a random variable, Z, is a product of two independent random variables, X and Y, the expected value, $E(Z)$, and the variance, $V(Z)$, are

$$Z = XY$$

$$E(Z) = E(X)E(Y); \tag{13-11}$$

$$\begin{aligned} V(Z) &= E[XY - E(XY)]^2 \\ &= E\{X^2Y^2 - 2XYE(XY) + [E(XY)]^2\} \\ &= E(X^2)E(Y^2) - [E(X)E(Y)]^2. \end{aligned}$$

But the variance of any random variable, $V(\text{RV})$, is

$$V(\text{RV}) = E[(\text{RV})^2] - [E(\text{RV})]^2,$$

$$E[(\text{RV})^2] = V(\text{RV}) + [E(\text{RV})]^2.$$

Then

$$V(Z) = \{V(X) + [E(X)]^2\}\{V(Y) + [E(Y)]^2\} - [E(X)]^2[E(Y)]^2$$

or

$$V(Z) = V(X)[E(Y)]^2 + V(Y)[E(X)]^2 + V(X)V(Y). \tag{13-12}$$

13.3 Evaluation of Projects with Discrete Random Variables

Expected value and variance concepts apply theoretically to long-run conditions in which it is assumed that the event is going to occur repeatedly. However, application of these concepts is often useful even when investments are not going to be made repeatedly over the long run. In this section, several examples are used to illustrate these concepts with selected economic factors modeled as discrete random variables.

EXAMPLE 13-1

We now apply the expected value and variance concepts to the small premixed-concrete plant project discussed in Example 10-7. Suppose that the estimated probabilities of attaining various capacity utilizations are as follows:

Capacity (%)	Probability
50	0.10
65	0.30
75	0.50
90	0.10

It is desired to determine the expected value and variance of *annual revenue*. Subsequently, the expected value and variance of AW for the project can be computed. By evaluating both $E(AW)$ and $V(AW)$ for the concrete plant, indications of the venture's average profitability and its uncertainty are obtained. The calculations are shown in Tables 13-1 and 13-2.

SOLUTION

Expected value of annual revenue: $\sum(A \times B) = \$575,100$.

Variance of annual revenue: $\sum(A \times C) - (575,100)^2 = 6,360 \times 10^6(\$)^2$.

TABLE 13-1 Solution for Annual Revenue (Example 13-1)

i	Capacity (%)	(A) Probability, $p(x_i)$	(B) Revenue[a] x_i	(A) × (B) Expected Revenue	(C) = (B)² x_i^2	(A) × (C)
1	50	0.10	$405,000	$40,500	1.64×10^{11}	0.164×10^{11}
2	65	0.30	526,500	157,950	2.77×10^{11}	0.831×10^{11}
3	75	0.50	607,500	303,750	3.69×10^{11}	1.845×10^{11}
4	90	0.10	729,000	72,900	5.31×10^{11}	0.531×10^{11}
				$575,100		$3.371 \times 10^{11}(\$)^2$

[a] From Table 10-5 with revenue for Capacity = 75% added.

	Capacity (%)	(A) $p(x_i)$	(B) AW,[a] x_i	$(A) \times (B)$ Expected AW	$(C) = (B)^2$ $(AW)^2$	$(A) \times (C)$
i						
1	50	0.10	−$25,093	−$2,509	0.63×10^9	0.063×10^9
2	65	0.30	22,136	6,641	0.49×10^9	0.147×10^9
3	75	0.50	53,622	26,811	2.88×10^9	1.440×10^9
4	90	0.10	100,850	10,085	10.17×10^9	1.017×10^9
				$41,028		$2.667 \times 10^9 (\$)^2$

TABLE 13-2 Solution for AW (Example 13-1)

[a] From Table 10-5 with AW for Capacity = 75% added.

$$\text{Expected value of AW: } \sum (A \times B) = \$41,028$$
$$\text{Variance of AW: } \sum (A \times C) - (41,028)^2 = 9,837 \times 10^5 (\$)^2$$
$$\text{Standard deviation of AW: } \$31,364$$

The standard deviation of AW, SD(AW), is less than the expected AW, $E(\text{AW})$, and only the 50% capacity utilization situation results in a negative AW. Consequently, with this additional information, the investors in this undertaking may well judge the venture to be an acceptable one.

There are projects, such as the flood-control situation in the next example, in which future losses due to natural or human-made risks can be decreased by increasing the amount of capital that is invested. Drainage channels or dams, built to control floodwaters, may be constructed in different sizes, costing different amounts. If they are correctly designed and used, the larger the size, the smaller will be the resulting damage loss when a flood occurs. As we might expect, the most economical size would provide satisfactory protection against most floods, although it could be anticipated that some overloading and damage might occur at infrequent periods.

EXAMPLE 13-2

A drainage channel in a community where flash floods are experienced has a capacity sufficient to carry 700 cubic feet per second. Engineering studies produce the following data regarding the probability that a given water flow in any one year will be exceeded and the cost of enlarging the channel:

Water Flow (ft^3/sec)	Probability of a Greater Flow Occurring in Any One Year	Capital Investment to Enlarge Channel to Carry This Flow
700	0.20	—
1,000	0.10	$20,000
1,300	0.05	30,000
1,600	0.02	44,000
1,900	0.01	60,000

TABLE 13-3 Expected Equivalent Annual Cost (Example 13-2)

Water Flow (ft³/sec)	Capital Recovery Amount	Expected Annual Property Damage[a]	Total Expected Equivalent Uniform Annual Cost
700	None	$20,000(0.20) = $4,000	$4,000
1,000	$20,000(0.0839) = $1,678	20,000(0.10) = 2,000	3,678
1,300	30,000(0.0839) = 2,517	20,000(0.05) = 1,000	3,517
1,600	44,000(0.0839) = 3,692	20,000(0.02) = 400	4,092
1,900	60,000(0.0839) = 5,034	20,000(0.01) = 200	5,234

[a] These amounts are obtained by multiplying $20,000 by the probability of greater water flow occurring.

Records indicate that the average property damage amounts to $20,000 when serious overflow occurs. It is believed that this would be the average damage whenever the storm flow is *greater* than the capacity of the channel. Reconstruction of the channel would be financed by 40-year bonds bearing 8% interest per year. It is thus computed that the capital recovery amount for debt repayment (principal of the bond plus interest) would be 8.39% of the capital investment, because $(A/P, 8\%, 40) = 0.0839$. It is desired to determine the most economical channel size (water-flow capacity).

SOLUTION
The total expected equivalent uniform annual cost for the structure and property damage for all alternative channel sizes would be as shown in Table 13-3. These calculations show that the minimum expected annual cost would be achieved by enlarging the channel so that it would carry 1,300 cubic feet per second, with the expectation that a greater flood might occur in 1 year out of 20 on the average and cause property damage of $20,000.

Note that when loss of life or limb might result, as in Example 13-2, there usually is considerable pressure to disregard pure economy and build such projects in recognition of the nonmonetary values associated with human safety.

The following example illustrates the same principles as in Example 13-2, except that it applies to safety alternatives involving electrical circuits.

EXAMPLE 13-3

Three alternatives are being evaluated for the protection of electrical circuits, with the following required investments and probabilities of failure:

Alternative	Capital Investment	Probability of Loss in Any Year
A	$90,000	0.40
B	100,000	0.10
C	160,000	0.01

TABLE 13-4 Expected Equivalent Annual Worth (Example 13-3)

Alternative	Capital Recovery Amount = Capital Investment × (A/P, 12%, 8)	Annual Maintenance Expense = Capital Investment × (0.10)	Expected Annual Cost of Failure	Total Expected Equivalent Annual Cost
A	$90,000(0.2013) = $18,117	$9,000	$94,000(0.40) = $37,600	$64,717
B	100,000(0.2013) = 20,130	10,000	94,000(0.10) = 9,400	39,530
C	160,000(0.2013) = 32,208	16,000	94,000(0.01) = 940	49,148

If a loss does occur, it will cost $80,000 with a probability of 0.65, and $120,000 with a probability of 0.35. The probabilities of loss in any year are independent of the probabilities associated with the resultant cost of a loss if one does occur. Each alternative has a useful life of eight years and no estimated market value at that time. The MARR is 12% per year, and annual maintenance expenses are expected to be 10% of the capital investment. It is desired to determine which alternative is best based on expected total annual costs (Table 13-4).

SOLUTION

The expected value of a loss, if it occurs, can be calculated as follows:

$$\$80,000(0.65) + \$120,000(0.35) = \$94,000.$$

Thus, Alternative B is the best based on total expected equivalent uniform annual cost, which is a long-run average cost. However, one might rationally choose Alternative C to reduce significantly the chance of an $80,000 or $120,000 loss occurring in any year in return for a 24.3% increase in the total expected equivalent uniform annual cost.

In Examples 13-1 through 13-3, a revenue or cost factor was modeled as a discrete random variable, with project life assumed certain. A second type of situation involves the cash-flow estimates being certain, but the project life being modeled as a random variable. This is illustrated in Example 13-4, where project life is modeled as a discrete random variable.

EXAMPLE 13-4

The heating, ventilating, and air-conditioning (HVAC) system in a commercial building has become unreliable and inefficient. Rental income is being hurt, and the annual expenses of the system continue to increase. Your engineering firm has been hired by the owners to (1) perform a technical analysis of the system, (2) develop a preliminary design for rebuilding the system, and (3) accomplish an engineering economic analysis to assist the owners in making a decision. The estimated capital-investment cost and annual savings in O&M expenses, based on the preliminary design, are shown in the following table. The estimated annual increase in rental income with a modern HVAC system has been developed by the owner's marketing staff and is also provided in the following table. These estimates are considered reliable because of the extensive information available. The useful life of the rebuilt system, however, is quite uncertain. The estimated probabilities of various useful lives are provided. Assume that the MARR = 12% per year and the estimated market value of the rebuilt system at the end of its useful life is zero. Based on this information, what is the E(PW), V(PW), and SD(PW) of the project's cash flows? Also, what is the probability of the PW $\geq$ 0? What decision would you make regarding the project, and how would you justify your decision using the available information?

Economic Factor	Estimate	Useful Life, Year (N)	p (N)	
Capital investment	$521,000	12	0.1	
Annual savings	48,600	13	0.2	
Increased annual revenue	31,000	14	0.3	
		15	0.2	$\sum = 1.00$
		16	0.1	
		17	0.05	
		18	0.05	

SOLUTION

The PW of the project's cash flows, as a function of project life (N), is

$$\text{PW}(12\%)_N = -\$521,000 + \$79,600(P/A, 12\%, N).$$

The calculation of the value of E(PW) = $9,984, and the value of $E[(\text{PW})^2] = 577.527 \times 10^6 (\$)^2$, are shown in Table 13-5. Then, by using Equation (13-8), the variance of the PW is

TABLE 13-5 Calculation of E(PW) and $E[(\text{PW})^2]$ (Example 13-4)

(1) Useful Life (N)	(2) PW(N)	(3) $p(N)$	(4) = (2) × (3) $E[\text{PW}(N)]$	(5) = $(2)^2$ $[\text{PW}(N)]^2$	(6) = (3) × (5) $p(N)[\text{PW}(N)]^2$
12	−$27,926	0.1	−$2,793	779.86 × 10^6	77.986 × 10^6
13	−9,689	0.2	−1,938	93.88 × 10^6	18.776 × 10^6
14	6,605	0.3	1,982	43.63 × 10^6	13.089 × 10^6
15	21,148	0.2	4,230	447.24 × 10^6	89.448 × 10^6
16	34,130	0.1	3,413	1,164.86 × 10^6	116.486 × 10^6
17	45,720	0.05	2,286	2,090.32 × 10^6	104.516 × 10^6
18	56,076	0.05	2,804	3,144.52 × 10^6	157.226 × 10^6
		E(PW) =	$9,984	$E[(\text{PW})^2]$ =	577.527 × 10^6 ($)^2

$$V(\text{PW}) = E[(\text{PW})^2] - [E(\text{PW})]^2$$

$$= 577.527 \times 10^6 - (\$9,984)^2$$

$$= 477.847 \times 10^6 (\$)^2.$$

The SD(PW) is equal to the positive square root of the variance, $V(\text{PW})$:

$$\text{SD(PW)} = [V(\text{PW})]^{1/2} = (477.847 \times 10^6)^{1/2}$$

$$= \$21,859.$$

Based on the PW of the project as a function of N (column 2), and the probability of each PW(N) value occurring (column 3), the probability of the PW being ≥ 0 is

$$\Pr\{\text{PW} \geq 0\} = 1 - (0.1 + 0.2) = 0.7.$$

The results of the engineering economic analysis indicate that the project is a questionable business action. The $E(\text{PW})$ of the project is positive (\$9,984) but small relative to the large capital investment. Also, even though the probability of the PW being greater than zero is somewhat favorable (0.7), the SD(PW) value is large [over two times the $E(\text{PW})$ value].

13.3.1 Probability Trees

The discrete distribution of cash flows sometimes occurs in each time period. A *probability tree diagram* is useful in describing the prospective cash flows, and the probability of each value occurring, for this situation. Example 13-5 is a problem of this type.

EXAMPLE 13-5

The uncertain cash flows for a small improvement project are described by the probability tree diagram in Figure 13-1. (Note that the probabilities emanating from each node sum to unity.) The analysis period is two years, and the MARR = 12% per year. Based on this information, (a) what are the $E(\text{PW})$, $V(\text{PW})$, and SD(PW) of the project, (b) what is the probability that PW ≤ 0, and (c) which analysis result(s) favor approval of the project and which ones appear unfavorable?

SOLUTION

(a) The calculation of the values for $E(\text{PW})$ and $E[(\text{PW})^2]$ is shown in Table 13-6. In column 2, PW_j, is the PW of branch j in the tree diagram. The probability of each branch occurring, $p(j)$, is shown in column 3. For example, proceeding from the right node for each cash flow in Figure 13-1 to the left node, we have $p(1) = (0.3)(0.2) = 0.06$ and $p(9) = (0.5)(0.3) = 0.15$. Hence,

$$E(\text{PW}) = \sum_j (PW_j) p(j) = \$39.56.$$

Figure 13-1
Probability Tree
Diagram for
Example 13-5

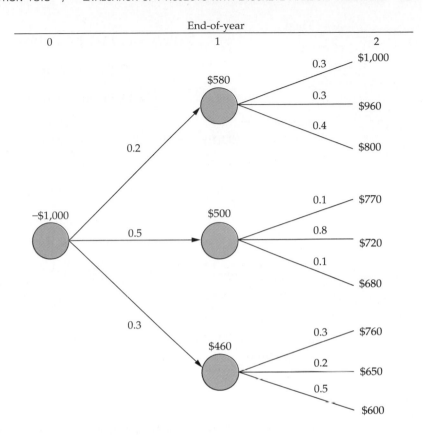

TABLE 13-6 Calculation of $E(\text{PW})$ and $E[(\text{PW})^2]$ (Example 13-5)

	(1)			(2)	(3)	(4) = (2) × (3)	(5) = (2)²	(6) = (3) × (5)
	Net Cash Flow							
	EOY							
j	0	1	2	PW_j	$p(j)$	$E(\text{PW}_j)$	$(\text{PW}_j)^2$	$E[(\text{PW}_j)^2]$
1	−$1,000	$580	$1,000	$315	0.06	$18.90	99,225$²	5,953$²
2	−1,000	580	960	283	0.06	16.99	80,089	4,805
3	−1,000	580	800	156	0.08	12.45	24,336	1,947
4	−1,000	500	770	60	0.05	3.04	3,600	180
5	−1,000	500	720	20	0.40	8.17	400	160
6	−1,000	500	680	−11	0.05	−0.57	121	6
7	−1,000	460	760	17	0.09	1.49	289	26
8	−1,000	460	650	−71	0.06	−4.27	5,044	302
9	−1,000	460	600	−111	0.15	−16.64	12,321	1,848
						$E(\text{PW}) = \$39.56$		$E[(\text{PW})^2] = 15,227\$^2$

Then

$$V(\text{PW}) = E[(\text{PW})^2] - [E(\text{PW})]^2$$

$$= 15{,}227 - (\$39.56)^2$$

$$= 13{,}662(\$)^2,$$

and

$$\text{SD}(\text{PW}) = [V(\text{PW})]^{1/2} = (13{,}662)^{1/2} = \$116.88.$$

(b) Based on the entries in column 2, PW_j, and column 3, $p(j)$, we have

$$\Pr\{\text{PW} \leq 0\} = p(6) + p(8) + p(9)$$

$$= 0.05 + 0.06 + 0.15$$

$$= 0.26.$$

(c) The analysis results that favor approval of the project are $E(\text{PW}) = \$39.56$, which is greater than zero only by a small amount, and $\Pr\{\text{PW} > 0\} = 1 - 0.26 = 0.74$. However, the $\text{SD}(\text{PW}) = \$116.92$ is approximately three times the $E(\text{PW})$. This indicates a relatively high variability in the measure of economic merit, (PW) of the project, and is usually an unfavorable indicator of project acceptability.

13.3.2 An Application Perspective

One of the major problems in computing expected values is the determination of the probabilities. In many situations, there is no precedent for the particular venture being considered. Therefore, probabilities seldom can be based on historical data and rigorous statistical procedures. In most cases, the analyst, or person making the decision, must make a judgment based on all available information in estimating the probabilities. This fact makes some people hesitate to use the expected-value concept, because they cannot see the value of applying such a technique to improve the evaluation of uncertainty when so much apparent subjectivity is present.

Although this argument has merit, the fact is that engineering economy studies deal with future events and there must be an extensive amount of estimating. Furthermore, even if the probabilities could be based accurately on past history, there rarely is any assurance that the future will repeat the past. Thus, structured methods for assessing subjective probabilities are often used in practice.* Also, even if we must estimate the probabilities, the very process of doing so requires us to make explicit the uncertainty that is inherent in all estimates going into the analysis. Such structured thinking is likely to produce better results than little or no thinking about such matters.

---

*For further information, see W. G. Sullivan and W. W. Claycombe, *Fundamentals of Forecasting* (Reston, VA: Reston Publishing Co., 1977), Chapter 6.

▼ 13.4 Evaluation of Projects with Continuous Random Variables

In Section 13.1, we discussed using the variance of a random variable, in addition to its expected value, in decision making. Thus, the uncertainty associated with an alternative can be represented more realistically. This was demonstrated in Examples 13-1, 13-4, and 13-5, when a revenue factor, a cost factor, and the project life, respectively, were modeled as *discrete* random variables. In each of these examples, the expected value and the variance of the project's equivalent worth were determined and used in the evaluation. Also, in the latter two examples, the probability of the PW being greater than or less than zero was computed.

In this section, we continue to compute mathematically the expected values and the variances of probabilistic factors, but we model the selected probabilistic factors as *continuous* random variables. In each example, simplifying assumptions are made about the distribution of the random variable and the statistical relationship among the values it takes on. When the situation is more complicated, such as a problem that involves probabilistic cash flows and probabilistic project lives, a second general procedure that utilizes Monte Carlo simulation is normally used. This is the subject of Section 13.5.

> Two frequently used assumptions about uncertain cash-flow amounts are that they are distributed according to a normal distribution* and are statistically independent. Underlying these assumptions is a general characteristic of many cash flows in that they result from a number of different and independent factors.

The advantage of using statistical independence as a simplifying assumption, when appropriate, is that no correlation between the cash-flow amounts (e.g., the net annual cash-flow amounts for an alternative) is being assumed. Consequently, if we have a linear combination of two or more independent cash-flow amounts, say $PW = c_0 F_0 + \cdots + c_N F_N$, where the c_k values are coefficients and the F_k values are periodic net cash flows, the expression for the $V(PW)$, based on Equation (13-10), reduces to

$$V(PW) = \sum_{k=0}^{N} c_k^2 V(F_k). \qquad (13\text{-}13)$$

And, based on Equation (13-9), we have

$$E(PW) = \sum_{k=0}^{N} c_k E(F_k). \qquad (13\text{-}14)$$

EXAMPLE 13-6

For the following annual cash-flow estimates, find the $E(PW)$, $V(PW)$, and SD(PW) of the project. Assume that the annual net cash-flow amounts are normally dis-

---

*This frequently encountered continuous probability function is discussed in any good statistics book, such as R. E. Walpole and R. H. Myers, *Probability and Statistics for Engineers and Scientists* (New York: Macmillan Publishing Co., 1989), pp. 139–154.

tributed with the expected values and standard deviations as given and statistically independent, and that the MARR = 15% per year.

End of Year, k	Expected Value of Net Cash Flow, F_k	SD of Net Cash Flow, F_k
0	−$7,000	0
1	3,500	$600
2	3,000	500
3	2,800	400

A graphical portrayal of these normally distributed cash flows is shown in Figure 13-2.

SOLUTION

The expected PW, based on Equation (13-14), is calculated as follows where $E(F_k)$ is the expected net cash flow in year k ($0 \leq k \leq N$) and c_k is the single-payment PW factor ($P/F, 15\%, k$):

$$E(\text{PW}) = \sum_{k=0}^{3}(P/F, 15\%, k)E(F_k)$$

$$= -\$7,000 + \$3,500(P/F, 15\%, 1) + \$3,000(P/F, 15\%, 2)$$

$$+ \$2,800(P/F, 15\%, 3)$$

$$= \$153.$$

To determine $V(\text{PW})$, we use the relationship in Equation (13-13). Thus,

$$V(\text{PW}) = \sum_{k=0}^{3}(P/F, 15\%, k)^2 V(F_k)$$

$$= 0^2 1^2 + 600^2 (P/F, 15\%, 1)^2 + 500^2 (P/F, 15\%, 2)^2$$

$$+ 400^2 (P/F, 15\%, 3)^2$$

$$= 484,324\$^2$$

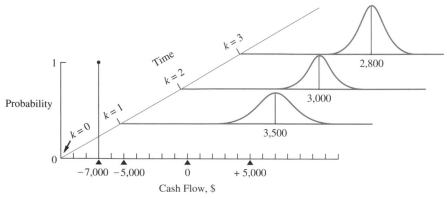

Figure 13-2 Probabilistic Cash Flows over Time (Example 13-6)

and

$$SD(PW) = [V(PW)]^{1/2} = \$696.$$

When we can assume that a random variable, say the PW of a project's cash flow, is normally distributed with a mean, $E(PW)$, and a variance, $V(PW)$, we can compute the probability of events about the random variable occurring. This assumption can be made, for example, when we have some knowledge of the shape of the distribution of the random variable, and when it is appropriate to do so. Also, this assumption may be supportable when a random variable, such as a project's PW value, is a linear combination of other independent random variables (say, the cash-flow amounts, F_k), regardless of whether the form of the probability distribution(s) of these variables is known.*

EXAMPLE 13-7

Refer to Example 13-6. For this problem, what is the probability that the IRR of the cash-flow estimates is less than the MARR, Pr{IRR < MARR}? Assume that the PW of the project is a normally distributed random variable, with its mean and variance equal to the values calculated in Example 13-6.

SOLUTION

For a decreasing PW(i) function having a unique IRR, the probability that the IRR is less than the MARR is the same as the probability that PW is less than zero. Consequently, by using the standardized normal distribution in Appendix E, we can determine the probability that PW is less than zero:[†]

$$Z = \frac{PW - E(PW)}{SD(PW)} = \frac{0 - 153}{696} = -0.22;$$

$$Pr\{PW \le 0\} = Pr\{Z \le -0.22\}.$$

From Appendix E, we find that Pr$\{Z \le -0.22\} = 0.4129.$

---

*The theoretical basis of this assumption is the Central Limit Theorem of statistics. For a summary discussion of the supportability of this assumption under different conditions, see C. S. Park and G. P. Sharpe-Bette, *Advanced Engineering Economics* (New York: John Wiley & Sons, 1990), pp. 420–421.

[†] A random variable, X, is normally distributed with mean μ and standard deviation σ in accordance with the following equation:

$$f(X) = \frac{1}{\sigma\sqrt{2\pi}}\exp\left\{-\left[\frac{(X-\mu)^2}{2\sigma^2}\right]\right\}.$$

The standardized normal distribution, $f(Z)$, of the variable $Z = (X - \mu)/\sigma$ has a mean of 0 and a standard deviation of 1.

EXAMPLE 13-8

The estimated cash-flow data for a project are shown in the following table for the five-year study period being used. Each annual net cash-flow amount, F_k, is a linear combination of two statistically independent random variables, X_k and Y_k, where X_k is a revenue factor and Y_k is an expense factor. The X_k cash-flow amounts are statistically independent of each other, and the same is true of the Y_k amounts. Both X_k and Y_k are continuous random variables, but the form of their probability distributions is not known. The MARR = 20% per year. Based on this information, (a) what are the $E(\text{PW})$, $V(\text{PW})$, and SD(PW) values of the project's cash flows, and (b) what is the probability that the PW is less than zero, that is, $\Pr\{PW \le 0\}$, and the project is economically attractive?

End of Year, k	Net Cash Flow $F_k = a_k X_k - b_k Y_k$	Expected Value X_k	Expected Value Y_k	Standard Deviation (SD) X_k	Standard Deviation (SD) Y_k
0	$F_0 = X_0 + Y_0$	$0	-$100,000	$0	$10,000
1	$F_1 = X_1 + Y_1$	60,000	-20,000	4,500	2,000
2	$F_2 = X_2 + 2Y_2$	65,000	-15,000	8,000	1,200
3	$F_3 = 2X_3 + 3Y_3$	40,000	-9,000	3,000	1,000
4	$F_4 = X_4 + 2Y_4$	70,000	-20,000	4,000	2,000
5	$F_5 = 2X_5 + 2Y_5$	55,000	-18,000	4,000	2,300

SOLUTION

(a) The calculation of the $E(F_k)$ and $V(F_k)$ values of the project's annual net cash flows are shown in Table 13-7. The $E(\text{PW})$ is calculated using Equation (13-14) as follows:

TABLE 13-7 Calculation of $E(F_k)$ and $V(F_k)$ (Example 13-8)

End of Year, k	F_k	$E(F_k) = a_k E(X_k) + b_k E(Y_k)$		$V(F_k) = a_k^2 V(X_k) + b_k^2 V(Y_k)$
0	F_0	$0 - $100,000 =	-$100,000	$0 + (1)^2(10,000)^2 = 100.0 \times 10^6 \$^2$
1	F_1	60,000 - 20,000 =	40,000	$(4,500)^2 + (1)^2 (2,000)^2 = 24.25 \times 10^6$
2	F_2	65,000 - 2(15,000) =	35,000	$(8,000)^2 + (2)^2 (1,200)^2 = 69.76 \times 10^6$
3	F_3	2(40,000) - 3(9,000) =	53,000	$(2)^2(3,000)^2 + (3)^2 (1,000)^2 = 45.0 \times 10^6$
4	F_4	70,000 - 2(20,000) =	30,000	$(4,000)^2 + (2)^2 (2,000)^2 = 32.0 \times 10^6$
5	F_5	2(55,000) - 2(18,000) =	74,000	$(2)^2(4,000)^2 + (2)^2 (2,300)^2 = 85.16 \times 10^6$

$$E(\text{PW}) = \sum_{k=0}^{5}(P/F, 20\%, k)E(F_k)$$

$$= -\$100{,}000 + \$40{,}000(P/F, 20\%, 1) + \cdots$$

$$+ \$74{,}000(P/F, 20\%, 5)$$

$$= \$32{,}517.$$

Then the $V(\text{PW})$ is calculated using Equation (13-13) as follows:

$$V(\text{PW}) = \sum_{k=0}^{5}(P/F, 20\%, k)^2 V(F_k)$$

$$= 100.0 \times 10^6 + (24.25 \times 10^6)(P/F, 20\%, 1)^2 + \cdots$$

$$+ (85.16 \times 10^6)(P/F, 20\%, 5)^2$$

$$= 186.75 \times 10^6 (\$)^2.$$

Finally,

$$\text{SD(PW)} = [V(\text{PW})]^{1/2}$$

$$= [186.75 \times 10^6]^{1/2}$$

$$= \$13{,}666.$$

(b) The PW of the project's net cash flow is a linear combination of the annual net cash-flow amounts, F_k, that are independent random variables. Each of these random variables, in turn, is a linear combination of the independent random variables X_k and Y_k. We can also observe in Table 13-7 that the $V(\text{PW})$ calculation does not include any dominant $V(F_k)$ value. Therefore, we have a reasonable basis on which to assume that the PW of the project's net cash-flow is approximately normally distributed, with $E(\text{PW}) = \$32{,}517$ and $\text{SD(PW)} = \$13{,}666$.

Based on this assumption, we have

$$Z = \frac{\text{PW} - E(\text{PW})}{\text{SD(PW)}} = \frac{0 - \$32{,}517}{\$13{,}666} = -2.3794;$$

$$\Pr\{\text{PW} \le 0\} = \Pr\{Z \le -2.3794\}.$$

From Appendix E, we find that $\Pr\{Z \le -2.3794\} = 0.0087$. Therefore, the probability of loss on this project is negligible. Based on this result, the $E(\text{PW}) > 0$, and the $\text{SD(PW)} = 0.42[E(\text{PW})]$; therefore, the project is economically attractive and has low risk of failing to add value to the firm.

▼ 13.5 Evaluation of Uncertainty Using Monte Carlo Simulation*

The modern development of computers and related software has resulted in the increased use of Monte Carlo simulation as an important tool for analysis of project uncertainties. For complicated problems, Monte Carlo simulation generates random outcomes for probabilistic factors so as to imitate the randomness inherent in the original problem. In this manner, a solution to a rather complex problem can be inferred from the behavior of these random outcomes.

> To perform a simulation analysis, the first step is to construct an analytical model that represents the actual decision situation. This may be as simple as developing an equation for the PW of a proposed industrial robot in an assembly line or as complex as examining the economic effects of proposed environmental regulations on typical petroleum refinery operations. The second step is to develop a probability distribution from subjective or historical data for each uncertain factor in the model. Sample outcomes are randomly generated by using the probability distribution for each uncertain quantity and are then used to determine a *trial* outcome for the model. Repeating this sampling process a large number of times leads to a frequency distribution of trial outcomes for a desired measure of merit, such as PW or AW. The resulting frequency distribution can then be used to make probabilistic statements about the original problem.

To illustrate the Monte Carlo simulation procedure, suppose that the probability distribution for the useful life of a piece of machinery has been estimated as shown in Table 13-8. The useful life can be simulated by assigning random numbers to each value such that they are proportional to the respective probabilities. (A random number is selected in a manner such that each number has an equal probability of occurrence.) Because two-digit probabilities are given in Table 13-8, random numbers can be assigned to each outcome, as shown in Table 13-9. Next, a single outcome is simulated by choosing a number at random from a table of random numbers.[†] For example, if any random number between and including 00

TABLE 13-8 Probability Distribution for Useful Life

Number of Years, N		p(N)	
3 ⎫		0.20 ⎫	
5 ⎬ possible values		0.40 ⎬ $\sum p(N) = 1.00$	
7 ⎪		0.25 ⎪	
10 ⎭		0.15 ⎭	

---

*Adapted from W. G. Sullivan and R. Gordon Orr, "Monte Carlo Simulation Analyzes Alternatives in Uncertain Economy," *Industrial Engineering* vol. 14, no. 11, November 1982. Reprinted with permission from *Industrial Engineering* magazine. Copyright Institute of Industrial Engineers, Inc., 25 Technology Park/Atlanta, Norcross, Georgia, 30092-2988.

[†] The last two digits of randomly chosen telephone numbers in a telephone directory are usually quite close to being random numbers.

TABLE 13-9 Assignment of Random Numbers

Number of Years, N	Random Numbers
3	00–19
5	20–59
7	60–84
10	85–99

TABLE 13-10 Random Normal Deviates (*RND*s)

−1.565	0.690	−1.724	0.705	0.090
0.062	−0.072	0.778	−1.431	0.240
0.183	−1.012	−0.844	−0.227	−0.448
−0.506	2.105	0.983	0.008	0.295
1.613	−0.225	0.111	−0.642	−0.292

and 19 is selected, the useful life is three years. As a further example, the random number 74 corresponds to a life of seven years.

If the probability distribution that describes a random variable is *normal*, a slightly different approach is followed. Here the simulated outcome is based on the mean and standard deviation of the probability distribution and on a random normal deviate, which is a random number of standard deviations above or below the mean of a standardized normal distribution. An abbreviated listing of typical random normal deviates is shown in Table 13-10. For normally distributed random variables, the simulated outcome is based on Equation (13-15):

$$\text{Outcome value} = \text{mean} + [\text{random normal deviate} \times \text{standard deviation}]. \tag{13-15}$$

For example, suppose that an *annual* net cash flow is assumed to be normally distributed, with a mean of $50,000 and a standard deviation of $10,000, as shown in Figure 13-3.

Figure 13-3
**Normally Distributed
Annual Cash Flow**

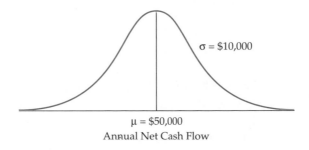

σ = $10,000

μ = $50,000
Annual Net Cash Flow

		TABLE 13-11 Example of the Use of *RND*s

Year	*RND*	Annual Net Cash Flow [$50,000 + *RND* ($10,000)]
1	0.090	$50,900
2	0.240	52,400
3	−0.448	45,520
4	0.295	52,950
5	−0.292	47,080

Simulated cash flows for a period of five years are listed in Table 13-11. Notice that the average annual net cash flow is $248,850/5, which equals $49,770. This approximates the known mean of $50,000 with an error of 0.46%.

If the probability distribution that describes a random event is *uniform* and continuous, with a minimum value of *A* and a maximum value of *B*, another procedure should be followed to determine the simulated outcome. Here the outcome can be computed with the formula

$$\text{Simulation outcome} = A + \frac{\text{RN}}{\text{RN}_m}[B - A], \tag{13-16}$$

where RN_m is the maximum possible random number (9 if one digit is used, 99 if two are used, etc.) and RN is the random number actually selected. This equation should be used when the minimum outcome, *A*, and the maximum outcome, *B*, are known.

For example, suppose that the market value in year *N* is assumed to be uniformly and continuously distributed between $8,000 and $12,000. A value of this random variable would be generated as follows with a random number of 74:

$$\text{Simulation outcome} = \$8,000 + \frac{74}{99}(\$12,000 - \$8,000) = \$10,990.$$

Proper use of these procedures, coupled with an accurate model, will result in an approximation of the actual outcome. But how many simulation trials are necessary for an *accurate* approximation of, for example, the average outcome? In general, the greater the number of trials, the more accurate the approximation of the mean and standard deviation will be. One method of determining whether a sufficient number of trials has been conducted is to keep a running average of results. At first, this average will vary considerably from trial to trial. The amount of change between successive averages should decrease as the number of simulation trials increases. Eventually, this running (cumulative) average should level off at an accurate approximation.

EXAMPLE 13-9

Monte Carlo simulation can also simplify the analysis of a more complex problem. The following estimates relate to an engineering project being considered by a large manufacturer of air-conditioning equipment. Subjective probability functions have been estimated for the four independent uncertain factors as follows:

CAPITAL INVESTMENT   Normally distributed with a mean of $50,000 and a standard deviation of $1,000.

USEFUL LIFE   Uniformly and continuously distributed with a minimum life of 10 years and a maximum life of 14 years.

ANNUAL REVENUE

$35,000 with a probability of 0.4

$40,000 with a probability of 0.5

$45,000 with a probability of 0.1

ANNUAL EXPENSE   Normally distributed, with a mean of $30,000 and a standard deviation of $2,000.

The management of this company wishes to determine whether the capital investment in the project will be a profitable one. The interest rate is 10% per year. In order to answer this question, the PW of the venture will be simulated.

SOLUTION
To illustrate the Monte Carlo simulation procedure, five trial outcomes are computed manually in Table 13-12. The estimate of the average present worth based on this very small sample is $19,010/5 = $3,802. For more accurate results, hundreds or even thousands of repetitions would be required.

The applications of Monte Carlo simulation for investigating uncertainty are many and varied. However, remember that the results can be no more accurate than the model and the probability estimates used. In all cases, the procedure and rules are the same: careful study of the problem and development of the model; accurate assessment of the probabilities involved; true randomization of outcomes as required by the Monte Carlo simulation procedure; and calculation and analysis of the results. Furthermore, a sufficiently large number of Monte Carlo trials should always be used to reduce the estimation error to an acceptable level.

13.6   Performing Monte Carlo Simulation with a Computer*

It is apparent from the preceding section that a Monte Carlo simulation of a complex project requiring several thousand trials can be accomplished only with the help of a computer. Indeed, there are numerous simulation programs that can be obtained from software companies and universities. To illustrate the computational features and output of a typical simulation program, Example 13-9 has been evaluated

---

*MS-Excel™ has a simulation feature that can generate random data for seven different probability distributions.

TABLE 13-12	Monte Carlo Simulation of PW Involving Four Independent Factors (Example 13-9)				
Trial Number	Random Normal Deviate (RND_1)	Capital Investment, I [$50,000 + RND_1($1,000)$]	Three-Digit RNs	Project Life, N $[10 + \frac{RN}{999}(14 - 10)]$	Project Life, N (Nearest Integer)
1	-1.003	$48,997	807	13.23	13
2	-0.358	49,642	657	12.63	13
3	$+1.294$	51,294	488	11.95	12
4	-0.019	49,981	282	11.13	11
5	$+0.147$	50,147	504	12.02	12

	One-Digit RN	Annual Revenue, R $35,000$ for 0–3 $40,000$ for 4–8 $45,000$ for 9	RND_2	Annual Expense, E [$30,000 + RND_2 ($2,000)$]	$PW = -I$ $+(R - E)(P/A, 10\%, N)$
1	2	$35,000	-0.036	$29,928	$-$12,969
2	0	35,000	$+0.605$	31,210	$-22,720$
3	4	40,000	$+1.470$	32,940	$-3,189$
4	9	45,000	$+1.864$	33,728	$+23,232$
5	8	40,000	-1.223	27,554	$+34,656$
				Total	$+$19,010

with a computer program. (A spreadsheet Monte Carlo simulation example is shown in Section 13.8.) The computer queries and user responses (in boxes) are shown in Figure 13-4. Simulation results for 3,160 trials are shown in Figure 13-5. (This number of trials was needed for the cumulative average PW to stabilize to a variation of ±0.5%.)

The average PW is $7,759.60, which is larger than the $3,801 obtained from Table 13-12. This underscores the importance of having a sufficient number of simulation trials to ensure reasonable accuracy in Monte Carlo analyses.

The histogram in Figure 13-5 indicates that the *median* PW of this investment is $6,700 and that the dispersion of PW trial outcomes is considerable. The standard deviation of simulated trial outcomes is one way to measure this dispersion. Based on Figure 13-5, 59.5% of all simulation outcomes have a PW of $0 or greater. Consequently, this project may be too risky for the company to undertake because the down-side risk of failing to realize at least a 10% per year return on the capital investment is about four chances out of 10. Perhaps another project investment should be considered.

A typical application of simulation involves the analysis of several mutually exclusive alternatives. In such studies, how can one compare alternatives that have different expected values and standard deviations of, for instance, PW? One approach is to select the alternative that *minimizes* the probability of attaining a PW

THE FOLLOWING PROGRAM USES MONTE CARLO SIMULATION
TECHNIQUES AS APPLIED TO RISK ANALYSIS PROBLEMS OF
ENGINEERING ECONOMY.

WILL YOU BE USING A REMOTE PRINTER FOR OUTPUT ? (Y OR
N) Y

INPUT A RANDOM NUMBER BETWEEN 1 AND 1000. 199

MAXIMUM NUMBER OF ITERATIONS YOU WISH TO RUN ? 1000

WHAT INTEREST RATE (PERCENT) IS TO BE USED ? 10

THE DATA FOR EACH RANDOM VARIABLE INVOLVED MAY BE
FORMULATED AS FOLLOWS:

1.   SINGLE VALUE OR ANNUITY
2.   SINGLE VALUE WITH UNIFORM GRADIENT
3.   SINGLE VALUE WITH GEOMETRIC GRADIENT
4.   DISCRETE DISTRIBUTION
5.   UNIFORM DISTRIBUTION
6.   NORMAL DISTRIBUTION
7.   A SERIES OF YEARLY CASH FLOWS
8.   SALVAGE VALUE DEPENDENT ON PROJECT LIFE
9.   TRIANGULAR DISTRIBUTION

INFORMATION FOR INITIAL CASH FLOW:

     DISTRIBUTION IDENTIFICATION NUMBER =   6

     MEAN VALUE =   -50000

     STANDARD DEVIATION =   1000

INFORMATION FOR YEARLY CASH FLOW:

     THIS CASH FLOW MAY CONSIST OF A NUMBER OF
     DIFFERENT ELEMENTS WHICH MAY FOLLOW DIFFERENT
     DISTRIBUTIONS.
     PLEASE INPUT THE DATA ONE ELEMENT AT AS TIME AND
     YOU WILL BE PROMPTED FOR ADDITIONAL INFORMATION.
     DISTRIBUTION IDENTIFICATION NUMBER =   4

NUMBER OF VALUES =   3   *(continued)*

Figure 13-4 Example Monte Carlo Simulation—Computer Queries and User
Responses

```
    INPUT VALUES IN ASCENDING ORDER:

    VALUE 1 =    35000
         WITH PROBABILITY    0.4

    VALUE 2 =    4000
       WITH PROBABILITY    0.5

    VALUE 3 =    45000
       WITH PROBABILITY    0.1

    IS THERE ADDITIONAL ANNUAL CASH FLOW DATA? (Y OR N)    Y

    DISTRIBUTION IDENTIFICATION NUMBER =    6

    MEAN VALUE =    -30000

    STANDARD DEVIATION =    2000

    IS THERE ADDITIONAL ANNUAL CASH FLOW DATA? (Y OR N)    N
INFORMATION FOR SALVAGE VALUE:

    DISTRIBUTION IDENTIFICATION NUMBER =    1

    CASH VALUE =    0
    INFORMATION FOR PROJECT LIFE:
       DISTRIBUTION IDENTIFICATION NUMBER =    5

    MINIMUM VALUE =    10

    MAXIMUM VALUE =    14

    EXPECTED VALUE OF PRESENT WORTH =          7759.60
    VARIANCE OF PRESENT WORTH =         680623960.00
    STANDARD DEVIATION OF PRESENT WORTH =     26088.77
    PROBABILITY THAT PRESENT WORTH IS GREATER THAN
    ZERO =                                    0.595

    EXPECTED VALUE OF ANNUAL WORTH =          1114.15
    VARIANCE OF ANNUAL WORTH =          14611587.00
    STANDARD DEVIATION OF ANNUAL WORTH =      3822.51
    PROBABILITY THAT ANNUAL WORTH IS GREATER THAN
    ZERO =                                    0.595
```

Figure 13-4 *(continued)* Example Monte Carlo Simulation—Computer Queries and User Responses

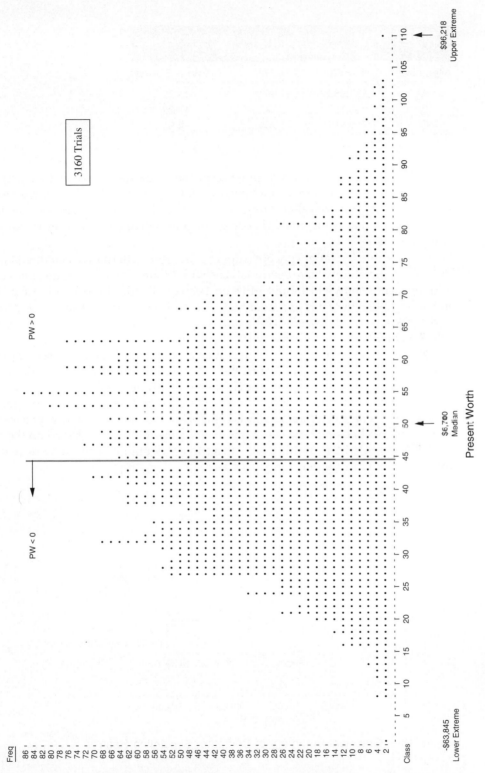

Figure 13-5 Histogram of PW for Example 13-9

TABLE 13-13	Simulation Results for Three Mutually Exclusive Alternatives		
Alternative	E(PW)	SD(PW)	E(PW) ÷ SD(PW)
A	$37,382	$1,999	18.70
B	49,117	2,842	17.28
C	21,816	4,784	4.56

that is less than zero. Another popular response to this question utilizes a graph of expected value (a measure of the reward) plotted against standard deviation (an indicator of risk) for each alternative. An attempt is then made to assess subjectively the trade-offs that result from choosing one alternative over another in pairwise comparisons.

To illustrate the latter concept, suppose that three alternatives having varying degrees of uncertainty have been analyzed with Monte Carlo computer simulation, and the results shown in Table 13-13 have been obtained. These results are plotted in Figure 13-6, where it is apparent that Alternative C is inferior to Alternatives A and B because of its lower E(PW) and larger standard deviation. Therefore, C offers a smaller PW that has a greater amount of risk associated with it! Unfortunately, the choice of B versus A is not as clear because the increased expected PW of B has to be balanced against the increased risk of B. This trade-off *may or may not* favor B, depending on management's attitude toward accepting the additional uncertainty associated with a larger expected reward. The comparison also presumes that Alternative A is acceptable to the decision maker. One simple procedure for choosing between A and B is to rank alternatives based on the ratio of E(PW) to SD(PW). In this case, Alternative A would be chosen because it has the more favorable (larger) ratio.

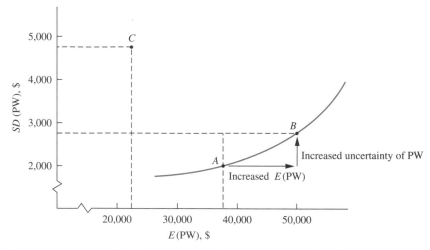

Figure 13-6 Graphical Summary of Computer Simulation Results

▼ 13.7 Decision Trees*

> Decision trees, also called *decision flow networks* and *decision diagrams,* are powerful means of depicting and facilitating the analysis of important problems, especially those that involve sequential decisions and variable outcomes over time. Decision trees are used in practice because they make it possible to break down a large, complicated problem into a series of smaller simple problems, and they enable objective analysis and decision making that includes explicit consideration of the risk and effect of the future.

The name *decision tree* is appropriate, because it shows branches for each possible alternative for a given decision and branches for each possible outcome (event) that can result from each alternative. Such networks reduce abstract thinking to a logical visual pattern of cause and effect. When costs and benefits are associated with each branch and probabilities are estimated for each possible outcome, analysis of the decision flow network can clarify choices and risks.

13.7.1 Deterministic Example

The most basic form of a decision tree occurs when each alternative can be assumed to result in a single outcome—that is, when certainty is assumed. The replacement problem in Figure 13-7 illustrates this. The problem as shown reflects that the decision about whether to replace the defender (old machine) with the new machine (challenger) is not just a one-time decision, but rather one that recurs periodically. That is, if the decision is made to keep the old machine at decision point one, then later, at decision point two, a choice again has to be made. Similarly, if the old machine is chosen at decision point two, then a choice again has to be made at decision point three. For each alternative, the cash inflow and duration of the project are shown above the arrow and the capital investment is shown below the arrow.

Figure 13-7
Deterministic
Replacement Example

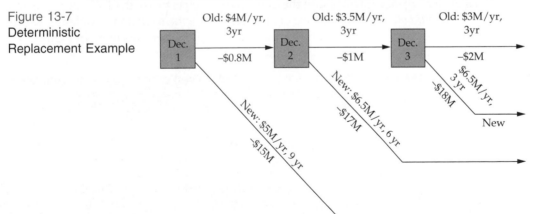

---

*Adapted (except Section 13.7.3) from John R. Canada and William G. Sullivan, *Economic and Multiattribute Evaluation of Advanced Manufacturing Systems,* 1989, pp. 341–343, 347. Reprinted by permission of Prentice Hall, Upper Saddle River, NJ.

For this problem, the question initially seems to be which alternative to choose at decision point one. But an intelligent choice at decision point one should take into account the later alternatives and decisions that stem from it. Hence, the correct procedure in analyzing this type of problem is to start at the most distant decision point, determine the best alternative and quantitative result of that alternative, and then "roll back" to each preceding decision point, repeating the procedure until finally the choice at the initial or present decision point is determined. By this procedure, one can make a present decision that directly takes into account the alternatives and expected decisions of the future.

For simplicity in this example, timing of the monetary outcomes will first be neglected, which means that a dollar has the same value regardless of the year in which it occurs. Table 13-14 shows the necessary computations and decisions using a nine-year study period. Note that the monetary outcome of the best alternative at decision point three ($7.0M for the *old*) becomes part of the outcome for the old alternative at decision point two. Similarly, the best alternative at decision point two ($22.0M for the *new*) becomes part of the outcome for the defender alternative at decision point one.

The computations in Table 13-14 show that the answer is to keep the old alternative now and plan to replace it with the new one at the end of three years (at decision point two). But this does not mean that the old machine should necessarily be kept for a full three years and a new machine bought without question at the end of that period. Conditions may change at any time, necessitating a fresh analysis—probably a decision tree analysis—based on estimates that are reasonable in light of conditions at that time.

13.7.1.1 Deterministic Example Considering Timing For decision tree analyses, which involve working from the most distant to the nearest decision point, the easiest way to take into account the timing of money is to use the PW approach and thus *discount all monetary outcomes to the decision points in question*. To demonstrate, Table 13-15 shows computations for the same replacement problem of Figure 13-7 using an interest rate of 25% per year.

TABLE 13-14 Monetary Outcomes and Decisions at Each Point—Deterministic Replacement Example of Figure 13-7[a]

Decision Point	Alternative	Monetary Outcome		Choice
3	Old	$3M(3) − $2M	= $7.0M	Old
	New	$6.5M(3) − $18M	= $1.5M	
2	Old	$7M + $3.5M(3) − $1M	= $16.5M	
	New	$6.5M(6) − $17M	= $22.0M	New
1	Old	$22.0M + $4M(3) − $0.8M	= $33.2M	Old
	New	$5M(9) − $15M	= $30.0M	

[a] Interest = 0% per year, that is, ignoring timing of cash flows.

TABLE 13-15 Decision at Each Point with Interest = 25% Per Year for Deterministic Replacement Example of Figure 13-7

Decision Point	Alternative	PW of Monetary Outcome		Choice
3	Old	$3M(P/A, 3) - $2M$ $3M(1.95) - $2M$	= $\underline{\$3.85M}$	Old
	New	$6.5M(P/A, 3) - $18M$ $6.5M(1.95) - $18M$	= $-\$5.33M$	
2	Old	$3.85M(P/F, 3) + $3.5M(P/A, 3) - $1M$ $3.85M(0.512) + $3.5M(1.95) - $1M$	= $\underline{\$7.79M}$	Old
	New	$6.5M(P/A, 6) - $17M$ $6.5M(2.95) - $17M$	= $\$2.18M$	
1	Old	$7.79M(P/F, 3) + $4M(P/A, 3) - $0.8M$ $7.79M(0.512) + $4M(1.95) - $0.8M$	= $\underline{\$10.99M}$	Old
	New	$5.0M(P/A, 9) - $15M$ $5.0M(3.46) - $15M$	= $\$2.30M$	

Note from Table 13-15 that when taking into account the effect of timing by calculating PWs at each decision point, the indicated choice is not only to keep the old alternative at decision point one, but also to keep the old alternative at decision points two and three as well. This result is not surprising since the high interest rate tends to favor the alternatives with lower capital investments, and it also tends to place less weight on long-term returns (benefits).

13.7.2 General Principles of Diagramming

The proper diagramming of a decision problem is, in itself, generally very useful to the understanding of the problem, and it is essential to correct subsequent analysis.

The placement of decision points (nodes) and chance outcome nodes from the initial decision point to the base of any later decision point should give an accurate representation of the information that will and will not be available when the choice represented by the decision point in question actually has to be made. The decision tree diagram should show the following (normally a square symbol is used to depict a decision node, and a circle symbol is used to depict a chance outcome node):

1. all initial or immediate alternatives among which the decision maker wishes to choose;
2. all uncertain outcomes and future alternatives that the decision maker wishes to consider because they may directly affect the consequences of initial alternatives;
3. all uncertain outcomes that the decision maker wishes to consider because they may provide information that can affect his or her future choices among alternatives and hence indirectly affect the consequences of initial alternatives.

Note that the alternatives at any decision point and the outcomes at any chance outcome node must be

1. mutually exclusive, (i.e., no more than one can be chosen);
2. collectively exhaustive, (i.e., one event must be chosen or something must occur if the decision point or outcome node is reached).

13.7.3 Decision Trees with Random Outcomes

The deterministic replacement problem discussed in Section 13.7.1 introduced the concept of sequential decisions using assumed certainty for alternative outcomes. However, an engineering problem requiring sequential decisions often includes random outcomes, and decision trees are very useful in structuring this type of situation. The decision tree diagram helps to make the problem explicit and assists in its analysis. This is illustrated in Examples 13-10 through 13-12.

EXAMPLE 13-10

The Ajax Corporation manufactures compressors for commercial air-conditioning systems. A new compressor design is being evaluated as a potential replacement for the most frequently used unit. The new design involves major changes that have the expected advantage of better operating efficiency. From the perspective of a typical user, the new compressor (as an assembled component in an air-conditioning system) would have an increased investment of $8,600 relative to the present unit and an annual expense saving dependent upon the extent to which the design goal is met in actual operations.

Estimates by the multidisciplinary design team of the new compressor achieving four levels (percentages) of the efficiency design goal and the probability and annual expense saving at each level are as follows:

Level (Percentage) of Design Goal Met (%)	Probability $p(L)$	Annual Expense Saving
90	0.25	$3,470
70	0.40	2,920
50	0.25	2,310
30	0.10	1,560

Based on a before-tax analysis (MARR = 18% per year, analysis period = 6 years, and market value = 0) and $E(\text{PW})$ as the decision criterion, is the new compressor design economically preferable to the current unit?

SOLUTION

The single-stage decision tree diagram for the design alternatives is shown in Figure 13-8. The PWs associated with each of the efficiency design goal levels being

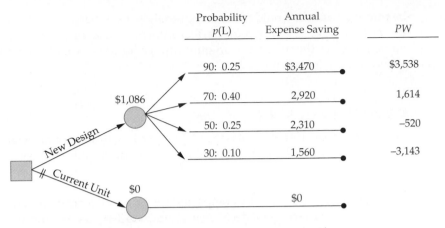

Probability p(L)	Annual Expense Saving	PW
90: 0.25	$3,470	$3,538
70: 0.40	2,920	1,614
50: 0.25	2,310	−520
30: 0.10	1,560	−3,143
	$0	$0

Figure 13-8 Single-State Decision Tree (Example 13-10)

met are as follows:

$$PW(18\%)_{90} = -\$8,600 + \$3,470(P/A, 18\%, 6) = \$3,538$$

$$PW(18\%)_{70} = -\$8,600 + \$2,920(P/A, 18\%, 6) = \$1,614$$

$$PW(18\%)_{50} = -\$8,600 + \$2,310(P/A, 18\%, 6) = -\$520$$

$$PW(18\%)_{30} = -\$8,600 + \$1,560(P/A, 18\%, 6) = -\$3,143$$

Based on these values, the $E(PW)$ of each installed unit of the new compressor is as follows:

$$E(PW) = 0.25(\$3,538) + 0.40(\$1,614) + 0.25(-\$520) + 0.10(-\$3,143)$$

$$= \$1,086$$

The $E(PW)$ of the current unit is zero since the cash-flow estimates for the new design are incremental amounts relative to the present design. Therefore, the analysis indicates that the new design is economically preferable to the present design. (The parallel lines across the current unit path on the diagram indicate that it was not selected.)

13.7.3.1 The Expected Value of Perfect Information (EVPI) The probability estimates of achieving the efficiency design goal levels, $p(L)$, developed by the design team in Example 13-10 reflect the uncertainty about the future operational performance of the new compressor. These probabilities are based on present information and are prior to obtaining any additional test data.

If more test data were obtained to reduce the uncertainty, additional costs would be incurred. Therefore, these additional costs must be balanced against the value of reducing the uncertainty. Obviously, if perfect information were available about the future operational efficiency of the new compressor, all uncertainty would be removed and we could make an optimal decision between the current unit and the new design. Even though perfect information is unobtainable, its expected value indicates the maximum amount (upper limit) we should consider spending for additional information.

EXAMPLE 13-11

Refer to Example 13-10. What is the EVPI regarding future operational performance of the new compressor to the typical user of an air-conditioning system?

SOLUTION

We can calculate the EVPI by comparing the optimal decision based on perfect information with the original decision in Example 13-10 to select the new compressor. This comparison is shown in Table 13-16. Based on this comparison, the EVPI to the typical user of a unit is

$$\text{EVPI} = \$1,530 - \$1,086 = \$444.$$

13.7.3.2 The Use of Additional Information to Reduce Uncertainty The solution to Example 13-11 shows that there is some potential value to be obtained from additional test information about the operational performance of the new compressor. From the viewpoint of the typical user of a new unit versus the current unit, the maximum estimated value of additional information is $444.

The members of the management team of Ajax Corporation are strongly customer focused and want to exceed customers' expectations about the performance of their products. Thus, they asked the design team to estimate the value of data that could be obtained from an additional comprehensive test of prototypes of the new compressor. The information from the test will not be perfect, because the test cannot determine exactly the long-term operational performance of the new design

TABLE 13-16 Expected Value of Perfect Information (Example 13-11)

Level (%)of Design Goal Met	Probability $p(L)$	Decision with Perfect Information		Prior Decision (New Design)
		Decision	Outcome	
90	0.25	New	$3,538	$3,538
70	0.40	New	1,614	1,614
50	0.25	Current	0	−520
30	0.10	Current	0	−3,143
		Expected value:	$1,530	$1,086

under diverse customer applications. However, the imperfect test information may reduce uncertainty and merit the additional cost of obtaining it.

We can evaluate the value of additional information before obtaining it only if we can estimate the reliability of the experiment being used. Therefore, the design team addressed the reliability of additional test information in predicting the future operational performance of the new compressor. The estimates developed by the design team, the calculated *revised probabilities* of achieving the new efficiency design goal levels, and the value of additional test information from a user's perspective are discussed in Example 13-12.

EXAMPLE 13-12

Refer to Examples 13-10 and 13-11. The members of the design team are confident that data from the additional comprehensive test of prototype compressors will show whether future operational performance will be favorable (60% or more of the design goal met) or not favorable (60% of the design goal not met). Based on obtaining these results from the test, and by using current engineering data within Ajax Corporation, the design team developed the following *conditional probability* estimates:

Comprehensive Test Outcome	Conditional Probabilities of Test Outcome Given the Level (%) of the Design Goal Met			
	90	70	50	30
Favorable (F)	0.95	0.85	0.30	0.05
Not favorable (NF)	0.05	0.15	0.70	0.95
Sum:	1.00	1.00	1.00	1.00

For example, given that the future operational performance of the new compressor is 90% of the design goal, the conditional probability of the comprehensive test results being favorable is estimated to be 0.95, and the conditional probability of the test results being not favorable is estimated to be 0.05. That is to say, $p(F \mid 90) = 0.95$, and $p(NF \mid 90) = 0.05$, where the | means "given."

Based on these conditional probabilities (reliability estimates of test results) and the choice of either doing the test or not doing it, (a) calculate the revised probabilities of the four efficiency design goal levels being met and (b) estimate the value of performing the comprehensive test to the typical user of a new compressor unit.

SOLUTION

(a) The two-stage decision tree diagram, which includes an initial decision on whether to do the test, is shown in Figure 13-9. In order to calculate the revised probabilities, we need to determine the joint probabilities of each level of the efficiency design goal being met and each test outcome occurring, as well as the marginal probability of each test outcome. These probabilities are shown

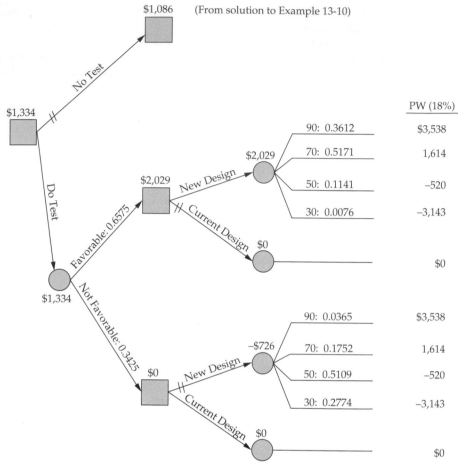

Figure 13-9 **Two-Stage Decision Tree (Example 13-12)**

in Table 13-17. As an example, the joint probabilities of the efficiency design goal being met at the 90% level and each test outcome occurring are calculated as follows:

$$p(90, \text{F}) = p(\text{F} \mid 90) \cdot p(90) = (0.95)(0.25) = 0.2375;$$

$$p(90, \text{NF}) = p(\text{NF} \mid 90) \cdot p(90) = (0.05)(0.25) = 0.0125.$$

The remaining six joint probabilities are determined in a similar manner. The sum of the joint probabilities over the four design goal levels gives the marginal probability of each test outcome occurring. That is to say, $p(\text{F}) = 0.6575$ and $p(\text{NF}) = 0.3425$. Similarly, the sums of the joint probabilities over the test outcomes are the marginal probabilities of achieving the new efficiency design goal levels [and are the same as the prior probabilities, $p(\text{L})$, in Example 13-10]. The revised probabilities of each level of the design goal being met, based on Table 13-17 [e.g., $p(90) = p(90, \text{F})/p(\text{F}) = 0.2375/0.6575 = 0.3612$], are also shown in Figure 13-9.

TABLE 13-17 Joint and Marginal Probabilities (Example 13-12)

Level (%) of Design Goal Met	Joint Probabilities		Marginal Probabilities, $p(L)$
	Favorable (F)	Not Favorable (NF)	
90	0.2375	0.0125	0.25
70	0.3400	0.0600	0.40
50	0.0750	0.1750	0.25
30	0.0050	0.0950	0.10
Marginal probabilities of test outcome:	0.6575	0.3425	1.00 (Sum)

(b) The estimated value (to the typical user of a new compressor unit) of completing the additional comprehensive test can be determined by using the data shown in Figure 13-9. By starting on the right side of Figure 13-9 and working toward the left side, we can calculate the E(PW) of the new design for both a favorable test outcome ($2,029) and an unfavorable test outcome (−$726). Based on these results, the choices at the two decision nodes are the new design alternative for a favorable test outcome and the current design alternative for a not favorable test outcome, respectively.

Based on these alternative design selections at the two decision nodes, the E(PW) at the chance node for the "do test" option is $1,334. The expected value of the comprehensive test, before the additional cost is considered, is $1,334 − $1,086 = $248, where $1,086 is the E(PW) of the new design without additional test information (Example 13-10).

The management team at Ajax Corporation used this information to help make a final decision about the additional comprehensive test of the new compressor design. Since the estimated total cost of the test was less than the expected value to the typical user of a unit ($248) times the estimated number of units to be sold in one year, and because of their strong customer focus, management decided to proceed with the additional test.

13.8 Spreadsheet Applications

Earlier in this chapter, we learned how Monte Carlo simulation can simplify the analysis of relatively complex problems. To reduce the estimation error, it is recommended that a large number of trials (as many as several thousand) be used. This is a formidable task if you were to perform this analysis using hand calculations. In this section, a spreadsheet model for Monte Carlo simulation is presented.

At the heart of Monte Carlo simulation is the generation of random numbers. Most spreadsheet packages include a RAND() function that returns a random number between zero and one. Other advanced statistical functions, such as NORMSINV(), will return the inverse of a cumulative distribution function (the standard normal distribution in this case). This function can be used to generate random normal deviates. The spreadsheet model shown in Figure 13-10 makes use of these functions in performing a Monte Carlo simulation for the project evaluated in Example 13-9.

The probabilistic functions for the four independent uncertain factors are included in the spreadsheet model. The required capital investment and annual expenses are normally distributed with the mean and standard deviations shown. The project life is expected to be uniformly distributed between 10 and 14 years. A discrete probability distribution has been compiled for annual revenues. The associated cumulative probability distribution (shown in column I, rows 4–6) is computed by the spreadsheet model.

Random normal deviates are generated in column B and column H to compute each trial's capital investment and annual expenses values. A uniform random number is generated in column D for the purposes of obtaining the project life. The trial value of project life makes use of the function ROUND() to return integer values. A second uniform random number is generated to obtain the annual revenues. The random number is compared to the cumulative distribution function for annual revenues and the appropriate value is placed in column G. The present worth of each trial is computed in column J.

Figure 13-10 shows only 10 trials. The average present worth over these trials was computed to be $6,164. More trials can be created simply by copying blocks of cells. The average present worth found using this spreadsheet model over 1000 trials was $7,949 (which is very close to the expected value of the present worth). The formulas for the highlighted cells are given in the following table:

Cell	Contents
I5	= I4 + H5
B11	= NORMSINV(RAND())
C11	= ($D$3 + $E$3 * B11)
D11	= RAND()
E11	= ROUND($D$7 + D11($E$7 − $D$7))
F11	= RAND()
G11	= IF(F11 <= I$4,G$4,IF(F11 <= I$5,G$5,G$6))
H11	= NORMSINV(RAND())
I11	= ($D$4 + $E$4 * H11)
J11	= −C11 − PV($B$1, E11, G11 − I11)
J22	= AVERAGE(J11 : J20)

▼ 13.9 Summary

Engineering economy involves decision making among competing uses of scarce capital resources. The consequences of resultant decisions usually extend far into the future. In this chapter, we have presented various statistical and probability concepts that address the fact that the consequences (cash flows, project lives, etc.) of engineering alternatives can never be known with certainty, including Monte Carlo computer simulation techniques and decision tree analysis. Cash inflow and cash outflow factors, as well as project life, were modeled as discrete and continuous random variables. The resulting impact of uncertainty on the economic measures

	A	B	C	D	E	F	G	H	I	J
1	MARR	10%								
2				Mean	Std. Dev.		Annual		Cumulative	
3			Capital Investment	$50,000	$1,000		Revenues	Prob.	Probability	
4			Annual Expenses	$30,000	$2,000		$35,000	0.4	0.4	
5							$40,000	0.5	0.9	
6				Minimum	Maximum		$45,000	0.1	1	
7			Project Life	10	14					
8										
9			Capital	Uniform	Project	Uniform	Annual		Annual	
10	Trial	RND1	Investment	RN [0,1]	Life	RN [0,1]	Revenues	RND2	Expenses	PW
11	1	+0.346	50346	0.912	14	0.315	35,000	+0.268	30536	$ (17,461)
12	2	−0.453	49547	0.213	11	0.413	40,000	−1.397	27206	$ 33,551
13	3	−1.092	48908	0.992	14	0.146	35,000	−0.382	29236	$ (6,446)
14	4	+0.064	50064	0.688	13	0.898	40,000	−0.807	28386	$ 32,434
15	5	−0.983	49017	0.638	13	0.212	35,000	−0.158	29684	$ (11,256)
16	6	−1.274	48726	0.477	12	0.895	40,000	−0.189	29622	$ 21,986
17	7	+1.083	51083	0.157	11	0.766	40,000	+0.003	30006	$ 13,829
18	8	−0.535	49465	0.771	12	0.239	35,000	+0.513	31026	$ (21,236)
19	9	−0.167	49833	0.488	12	0.470	40,000	+1.168	32336	$ 2,387
20	10	+0.499	50499	0.073	10	0.982	45,000	−1.061	27878	$ 54,708
21										
22									Average	$ 6,164

Figure 13-10 Spreadsheet Model for Monte Carlo Simulation

of merit for an alternative was analyzed. Included in the discussion were several considerations and limitations relative to the use of these methods in application.

Regrettably, there is no quick and easy answer to the question "How should uncertainty best be considered in an engineering economic evaluation?" Generally, simple procedures (e.g., breakeven analysis and sensitivity analysis, discussed in Chapter 10) allow some discrimination among alternatives to be made on the basis of the uncertainties present, and they are relatively inexpensive to apply. Additional discrimination among alternatives is possible with more complex procedures that utilize probabilistic concepts. These procedures, however, are more difficult to apply and require additional time and expense.

13.10 References

Bonini, C. P. "Risk Evaluation of Investment Projects," *OMEGA*, vol. 3, no. 6, 1975, pp. 735–750.

Hertz, D. B., and H. Thomas. *Risk Analysis and Its Applications* (New York: John Wiley & Sons, 1983).

Hillier, F. S. *The Evaluation of Risky Interrelated Investments* (Amsterdam: North-Holland, 1969).

Hull, J. C. *The Evaluation of Risk in Business Investment* (New York: Pergamon Press, 1980).

Magee, J. F. "Decision Trees for Decision Making," *Harvard Business Review*, vol. 42, no. 4, July-August 1964, pp. 126–138.

Park, C. S., and G. Sharpe-Bette. *Advanced Engineering Economics.* (New York: John Wiley & Sons, 1990).

Rose, L. M. *Engineering Investment Decisions: Planning Under Uncertainty.* (Amsterdam: Elsevier, 1976).

Walpole, R. E., and R. H. Meyers. *Probability and Statistics for Engineers*, 4th ed. (New York: Macmillan Publishing Company, 1989).

13.11 Problems

The number in parentheses () that follows each problem refers to the section from which the problem is taken.

13-1. Assume that the estimated net annual benefits for a project during each year of its life have the following probabilities:

Net Annual Benefits (NAB)	p(NAB)
$2,000	0.40
3,000	0.50
4,000	0.10

The life is three years for certain and the initial capital investment is $7,000, with negligible salvage value. The MARR is 15% per year. Determine $E(\text{PW})$ and the probability that PW is greater than zero [i.e., $\Pr(\text{PW} \geq 0)$]. (13.3)

13-2. A bridge is to be constructed now as part of a new road. An analysis has shown that traffic density on the new road will justify a two-lane bridge at the present time. Because of uncertainty regarding future use of the road, the time at which an extra two lanes will be required is currently being studied. The estimated probabilities of having to widen the bridge to four lanes at various times in the future are as follows:

Widen Bridge In	Probability
3 years	0.1
4 years	0.2
5 years	0.3
6 years	0.4

The present estimated cost of the two-lane bridge is $2,000,000. If constructed now, the four-lane bridge will cost $3,500,000. The future cost of widening a two-lane bridge will be an extra $2,000,000 plus $250,000 for every year that widening is delayed. If money can earn 12% per year, what would you recommend? (13.3)

13-3. In Problem 13-2, perform an analysis to determine how sensitive the choice of a four-lane bridge built now versus a four-lane bridge that is constructed in two stages is to the interest rate. Will an interest rate of 15% per year reverse the initial decision? At what interest rate would constructing the two-lane bridge now be preferred? (13.3)

13-4. In a building project, the amount of concrete to be poured during the next week is uncertain. The foreman has estimated the following probabilities:

Amount (cubic yards)	Probability
1,000	0.1
1,200	0.3
1,300	0.3
1,500	0.2
2,000	0.1

Determine the expected value (amount) of concrete to be poured next week. Also compute the variance and standard deviation of the amount of concrete to be poured. (13.3)

13-5. Consider the two random variables P and Q given in the following table:

Price, P	$p(P)$	Quantity Sold, Q	$p(Q)$
$6	$\frac{1}{3}$	10	$\frac{1}{3}$
5	$\frac{1}{3}$	15	$\frac{1}{3}$
4	$\frac{1}{3}$	20	$\frac{1}{3}$

Assume that P and Q are independent. What are the mean, variance, and standard deviation of the probability distribution for revenue? (13.3)

13-6. A small dam is being planned for a river tributary that is subject to frequent flooding. From past experience, the probabilities that water flow will exceed the design capacity of the dam during a year, plus relevant cost information, are as follows:

	Probability of Greater Flow during a Year	Capital Investment
A	0.100	$180,000
B	0.050	195,000
C	0.025	208,000
D	0.015	214,000
E	0.006	224,000

Estimated annual damages that occur if water flows exceed design capacity are $150,000, $160,000, $175,000, $190,000, and $210,000 for design A, B, C, D, and E, respectively. The life of the dam is expected to be 50 years, with negligible market value. For an interest rate of 8% per year, determine which design should be implemented. What nonmonetary considerations might be important to the selection? (13.3)

13-7. A diesel generator is needed to provide auxiliary power in the event that the primary source of power is interrupted. Various generator designs are available, and more expensive generators tend to have higher reliabilities should they be called on to produce power. Estimates of reliabilities, capital investment costs, O&M expenses, market value and damages resulting from a complete power failure (i.e., the standby generator fails to operate) are given in Table P13-7 for three alternatives. If the life of each generator is 10 years and the MARR = 10% per year, which generator should be chosen if you assume one main power failure per year? Does your choice change if you assume two main power failures per year (O&M expenses remain the same)? (13.3)

13-8. The owner of a ski resort is considering installing a new ski lift, which will cost $900,000. Expenses for operating and maintaining the lift are estimated to be $1,500 per day when operating. The U.S. Weather Service estimates that there is a 60% probability of 80 days of skiing weather per year, a 30% probability of 100 days per year, and a 10% probability of 120 days

TABLE P13-7 Three Generator Designs for Problem P13-7

Alternative	Capital Investment	O&M Expenses/Year	Reliability	Cost of Power Failure	Market Value
R	$200,000	$5,000	0.96	$400,000	$40,000
S	170,000	7,000	0.95	400,000	25,000
T	214,000	4,000	0.98	400,000	38,000

per year. The operators of the resort estimate that during the first 80 days of adequate snow in a season, an average of 500 people will use the lift each day, at a fee of $10 each. If 20 additional days are available, the lift will be used by only 400 people per day during the extra period, and if 20 more days of skiing are available, only 300 people per day will use the lift during those days. The owners wish to recover any invested capital within five years and want at least a 25% per year rate of return before taxes. Based on a before-tax analysis, should the lift be installed? (13.3)

13-9. Refer to Problem 13-8. Assume the following changes: The study period is eight years; the ski lift will be depreciated using the MACRS Alternative Depreciation System (ADS); the ADS recovery period is seven years; the MARR = 15% per year (after-tax); and the effective income-tax rate (t) is 40%. Based on this information, what is the $E(PW)$ and SD(PW) of the ATCF? Interpret the analysis results and make a recommendation on installing the ski lift. (13.3)

13-10. An energy conservation project is being evaluated. Four levels of performance are considered feasible. The estimated probabilities of each performance level and the estimated before-tax cost savings in the first year are shown in the following table:

Performance Level (L)	$p(L)$	Cost Savings (1st yr; before taxes)
1	0.15	$22,500
2	0.25	35,000
3	0.35	44,200
4	0.25	59,800

Assume the following:

- Initial capital investment: $100,000 [80% is depreciable property and the rest (20%) are

costs that can be immediately expensed for tax purposes].
- The ADS under MACRS is being used. The ADS recovery period is four years.
- The before-tax cost savings are estimated to increase 6% per year after the first year.
- $MARR_{AT} = 12\%$ per year; the analysis period is five years; $MV_5 = 0$.
- The effective income-tax rate is 40%.

Based on $E(PW)$ and after-tax analysis, should the project be implemented? (13.3)

13-11. The purchase of a new piece of electronic measuring equipment for use in a continuous metal-forming process is being considered. If this equipment were purchased, the capital cost would be $418,000, and the estimated savings are $148,000 per year. The useful life of the equipment in this application is uncertain. The estimated probabilities of different useful lives occurring are shown in the following table. Assume that the MARR = 15% per year before taxes, and the market value at the end of its useful life is equal to zero. Based on a before-tax analysis, (a) what are the $E(PW)$, $V(PW)$, and SD(PW) associated with the purchase of the equipment, and (b) what is the probability that the PW is less than zero? Make a recommendation and give your supporting logic based on the analysis results. (Chapter 8 and 13.3)

Useful Life, Years (N)	$p(N)$
3	0.1
4	0.1
5	0.2
6	0.3
7	0.2
8	0.1

13-12. The tree diagram in Figure P13-12 describes the uncertain cash flows for an engineering

Figure P13-12
Probability Tree Diagram
for Problem P13-12

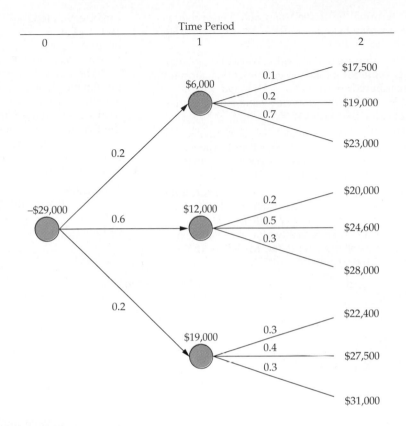

Time Period

project. The analysis period is two years, and the MARR = 15% per year. Based on this information, (a) what are the $E(PW)$, $V(PW)$, and SD(PW) of the project, and (b) what is the probability that PW ≥ 0? (13.3)

13-13. A potential project has an initial capital investment of $100,000. Net annual revenues minus expenses are estimated to be $40,000 (A$) in the first year and to increase at the rate of 6.48% per year. However, the useful life of the primary equipment is uncertain, as shown in the following table:

Useful Life, Years (N)	$p(N)$
1	0.03
2	0.10
3	0.30
4	0.30
5	0.17
6	0.10

(Assume that i_c = MARR = 15% per year and f = 4% per year.) Based on this information, (a) what are the $E(PW)$ and SD(PW) for this project, (b) what is the Pr{PW ≥ 0}, and (c) what is the $E(AW)$ in R$? Do you consider the project economically acceptable, questionable, or not acceptable, and why? (Chapter 8, 13.3)

13-14. A proposed venture has an initial capital investment of $80,000, annual revenues minus expenses of $30,000, and an uncertain useful life, N, as follows:

N	Probability of N
1	0.05
2	0.15
3	0.20
4	0.30
5	0.20
6	0.05
7	0.05

Determine the $E(PW)$ and $SD(PW)$ of this investment when the MARR is 20% per year. Also, what is the $Pr\{PW \leq 0\}$? (13.3)

13-15. Suppose that a random variable (e.g., market value for a piece of equipment) is normally distributed, with mean = $175 and variance = 25^2. What is the probability that the actual market value is *at least* $171? (13.4)

13-16. The AW of project R-2 is normally distributed, with a mean of $1,500 and a variance of 810,000($)^2$. Determine the probability that this project's AW is less than $1,700. (13.4)

13-17. For the following cash-flow estimates, determine the $E(PW)$ and $V(PW)$. Also, find the probability that the PW will exceed $0. The cash flows are statistically independent and normally distributed, and the MARR is 12% per year. (13.4)

End of Year, k	Expected Value of Cash Flow	Standard Deviation of Cash Flow
0	−$14,000	0
1	6,000	$800
2	4,000	400
3	4,000	400
4	8,000	1,000

13-18. The use of three estimates (defined here as H = high, L = low, and M = most likely) for random variables is a practical technique for modeling uncertainty in some engineering economy studies. Assume that the mean and variance of the random variable, X_k, in this situation

can be estimated by $E(X_k) = (1/6)(H + 4M + L)$ and $V(X_k) = [(H - L)/6]^2$. The estimated net cash-flow data for one alternative associated with a project are shown in Table P13-18.

The random variables, X_k, are assumed to be statistically independent, and the applicable MARR = 15% per year. Based on this information, (a) what are the mean and variance of the PW, (b) what is the probability that PW ≥ 0 (state any assumptions that you make), and (c) is this the same as the probability that the IRR is acceptable? Explain. (13.4)

13-19. Consider Problem 13-8 when, in addition to uncertainty regarding number of skiing days per year, the useful life of the venture is *also* uncertain as follows:

Useful Life, Years (N)	$p(N)$
4	0.2
5	0.6
6	0.2

Finally, the market value (MV) of the ski lift is a function of the venture's life:

$$MV = \$10,000(7 - N)$$

a. Set up a table and use Monte Carlo simulation to determine five trial outcomes of the venture's before-tax AW. Recall that the MARR is 25% per year.

b. Based on your simulation outcomes, should the lift be installed? State any assumptions you make. (13.5, 13.6)

TABLE P13-18 Estimates for Problem P13-18

End of Year, k	Net Cash Flow	Three-Point Estimates for X_k		
		L	M	H
0	$F_0 = X_0$	−$38,000	−$41,000	−$45,000
1	$F_1 = 2X_1$	−1,900	−2,200	−2,550
2	$F_2 = X_2$	9,800	10,600	11,400
3	$F_3 = 4X_3$	5,600	6,100	6,400
4	$F_4 = 5X_4$	4,600	4,800	5,100
5	$F_5 = X_5$	16,500	17,300	18,300

TABLE P13-20 Equipment Estimates for Problem P13-20

Factor	Expected Value	Type of Probability Distribution
Capital investment	$150,000	Known with certainty
Market value	$2,000 $(13 - N)$	Normal, $\sigma = \$500$
Annual savings	$70,000	Normal, $\sigma = \$4,000$
Annual expenses	$43,000	Normal, $\sigma = \$2,000$
Useful life, N	13 years	Uniform in [8, 18]
MARR	8%/year	Known with certainty

13-20. Consider the estimates for a new piece of manufacturing equipment shown in Table P13-20. (13.5, 13.6)

a. Set up a table and simulate five trials of the equipment's PW.

b. Compute the mean of the five trials and recommend whether the equipment should be purchased.

13-21. Simulation results are available for two mutually exclusive alternatives. A large number of trials have been run with a computer, with the results shown in Figure P13-21.

Discuss the issues that may arise when attempting to decide between these two alternatives. (13.5, 13.6)

13-22. A large mudslide caused by heavy rains will cost Sabino County $1,000,000 per occurrence in lost property-tax revenues. In any given year there is one chance in 100 that a major mudslide will occur.

A civil engineer has proposed constructing a culvert on a mountain where mudslides are likely. The culvert will reduce the likelihood of a mudslide to almost zero. The capital investment would be $50,000, and annual maintenance expenses would be $2,000 in the first year, increasing by 5% per year thereafter.

If the life of the culvert is expected to be 20 years and the cost of capital to Sabino County is 7% per year, should the culvert be built? (13.3)

13-23. A company is considering an engineering improvement project with uncertain outcomes. The best present estimates, including prior probabilities of success, are as follows:

Success Category	Probability of Success	Net Annual Benefits
A	0.35	$200,000
B	0.35	100,000
C	0.30	20,000

The estimated net annual benefits are relative to current operations. Assume the project's initial investment is $280,000; taxes are not being

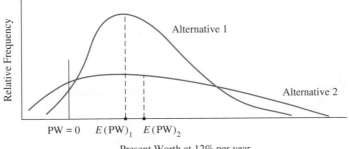

Figure P13-21 Simulation Results for Problem P13-21

considered; the MARR (before-taxes) is 15% per year; and a six-year analysis period applies to this type of project.

Due to the uncertain outcomes, the responsible manager has directed that a potential test experiment be evaluated prior to further consideration of the project. The estimated reliability of the test experiment is as follows:

Test Outcome	Conditional Probabilities of Test Outcome Given the Success Category		
	A	B	C
Good (G)	0.90	0.25	0.05
Quite Poor (P)	0.10	0.75	0.95
Sum:	1.0	1.0	1.0

Based on a decision tree analysis and the E(PW) as the economic measure of interest, what is the estimated value of the additional information that would be provided by the test experiment in this case? (13.7)

13-24. An improved design of a computerized piece of continuous quality measuring equipment used to control the thickness of rolled sheet products is being developed. It is estimated to sell for $125,000 more than the current design. However, based on present test data, the typical user has the following probabilities of achieving different performance results and cost savings (relative to the current unit) in the first year of operation (assume these annual cost savings would escalate 5% per year thereafter; a five-year analysis period is appropriate for this situation; the before-tax market MARR is 18% per year; and the net MV at the end of five years is 0):

Performance Results	Probability	Cost Savings in First Year
Optimistic	0.30	$60,000
Most likely	0.55	40,000
Pessimistic	0.15	18,000

Based on the E(PW), is the new design preferable to the current unit (show a single-stage decision tree diagram for this situation)? What is the EVPI? What does the EVPI tell you? (13.7)

13-25. If the interest rate is 8% per year, what decision would you make based on the decision tree diagram in Figure P13-25(13.7)

13-26. The vice-president of operations at a plant that manufactures components for hydraulic systems is considering an improvement to present production capability. The decision situation has been reduced to three alternatives. The first alternative would result in significant changes in present operations, including increased automation. The second alternative would involve fewer changes in present operations and not include any new automation. The third alternative is to make no changes (do nothing).

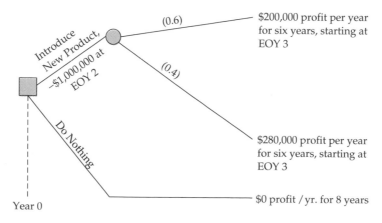

Figure P13-25 Decision Tree Diagram for Problem P13-25

The incremental capital investment and incremental annual revenue for the first two alternatives, relative to present operations, are shown in the accompanying table. The annual revenue estimates are based on future sales of the components. The sales department estimates the probability of good, average, and poor future sales as 0.30, 0.60, and 0.10, respectively.

Alternative	Capital Investment	Future Sales	Annual Revenue
1	$300,000	Good	$142,000
		Average	119,000
		Poor	50,000
2	85,000	Good	66,000
		Average	46,000
		Poor	17,000

Draw a single-stage decision tree for this situation. Then, based on a before-tax analysis (MARR = 20% per year, analysis period equal to five years, and zero market value for all alternatives) and the $E(PW)$ as a decision criterion, determine which alternative is preferred. What would be the expected value of perfect information (EVPI) about future sales in this case? (13.7)

13-27. Refer to Problem 13-26. At the end of the analysis of the single-stage decision tree diagram

by the vice-president of operations, the plant management team realized that additional information about future sales of the hydraulic components should reduce the uncertainty involved. Therefore, they asked the sales department to survey their customers and improve the information about future sales conditions. The management team's estimates of the *conditional probabilities* of survey outcomes for each potential sales condition are shown in the accompanying table.

	Conditional Probabilities of Survey Outcome Given the Future Sales Condition		
Survey Outcome	Good (G)	Average (A)	Poor (P)
Optimistic (O)	0.85	0.60	0.10
Not Favorable (NF)	0.15	0.40	0.90
Sum:	1.00	1.00	1.00

Based on this information, develop a two-stage decision tree for this situation. Calculate the revised probabilities of the three future sales conditions occurring. What is the estimated value to the plant of conducting the sales survey (before considering any additional cost involved)? (13.7)

Capital Financing and Allocation

*F*or ease of presentation and discussion, we have divided this chapter into two main areas: (1) the long-term sourcing of capital for a firm (capital financing) and (2) the expenditure of capital through development, selection, and implementation of specific projects (capital allocation). Our aim is to give the student an understanding of these basic components of the capital budgeting process so that the important role of the engineer in this complex and strategic function will be made clear.

The following topics are discussed in this chapter:

The capital financing and allocation functions

Differences between sources of capital

Cost of debt capital

Cost of equity capital

Weighted average cost of capital

Leasing as a source of capital

Capital allocation

An overview of a typical corporate capital budgeting process

14.1 Introduction

An entrepreneurial firm must obtain capital funds from investors and lenders (*capital financing*) and then invest these funds in equipment, tools, and other resources (*capital allocation*) to produce goods and services for sale. The revenues from the engineering and other capital projects involved must earn an adequate return on the funds invested in terms of profit (additional wealth) if the firm is to achieve economic growth and be competitive in the future. Thus, the decision by a com-

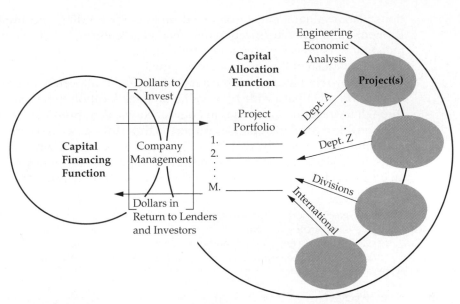

Figure 14-1 Overview of Capital Financing and Allocation Activities in a Typical Organization

pany to implement an engineering project involves the expenditure of current capital funds to obtain future economic benefits, or to meet safety, regulatory, or other operating requirements. This implementation decision is normally made, in a well-managed company, as part of a capital budgeting process (Section 14.8). The capital financing and allocation functions are primary components of this process.

The capital financing and allocation functions are closely linked, as illustrated in Figure 14-1, and they are simultaneously managed as part of the capital budgeting process. The amounts of new funds needed from investors and lenders, as well as the amounts of funds available from internal sources (depreciation and retained earnings*), to support new capital projects are determined in the capital financing function. Also, the *sources* of any new externally acquired funds—issuing additional stock, selling bonds, obtaining loans, and so on—are decided upon. These amounts, in total, as well as the ratio of debt to equity capital, must be commensurate with the financial status of the firm and balanced with the current and future capital investment requirements.

The selection of the engineering projects for implementation occurs in the capital allocation function. The total capital investment in new projects is constrained by the amount decided upon for this purpose during the capital financing considerations. The capital allocation activities begin in the various organizations in the company—departments (say, engineering), operating divisions, research and development, and so on. During each capital budgeting cycle, these organiza-

---

*Retained earnings are the portion of a firm's after-tax accumulated earnings not paid out as dividends to the owners (stockholders) and reinvested into the company.

tions plan, evaluate, and recommend projects for funding and implementation. Economic and other justification information is required with each project recommendation. Engineering economy studies are accomplished as part of this process to develop much of the information required.

As shown in Figure 14-1, the available capital is allocated among the projects selected on a companywide basis in the capital allocation function. Management, through its integrated activities in both functions, is responsible for ensuring that a reasonable return (in dollars) is earned on these investments so providers of debt and equity capital will be motivated to furnish more capital when the need arises. Thus, it should be apparent why the informed practice of engineering economy is an essential element in the foundation of an organization's competitive culture.

> In summary, the capital financing and allocation functions are closely linked decision processes regarding *how much* and *where* financial resources will be obtained and expended on future engineering and other capital projects to achieve economic growth and to improve the competitiveness of a firm. The scope of these activities encompasses
>
> 1. how the financial resources are acquired from equity, debt, and other sources;
> 2. how minimum requirements of economic acceptability are established;
> 3. how the capital projects are identified and evaluated;
> 4. how final project selections are made for implementation;
> 5. how post-audit reviews are conducted.

14.2 Differences between Sources of Capital

As discussed in the previous section, capital plays an essential role in engineering and business projects. Although engineers seldom engage in obtaining the capital needed for projects, the methods by which it is obtained (equity or borrowed capital, funds from internal sources, or indirectly through the leasing of assets) will impact the minimum required rate of return, some income-tax considerations, and possibly other related factors.

Most engineering economy studies are concerned with the *total* capital used, without regard for source; this approach, in effect, evaluates the *project* rather than the interests of any particular group of capital suppliers. The examples and problems in this book normally evaluate the project because in most analyses the choice between alternatives can be made independently of the sources of funds to be used. Hence, up to this point, the firm's overall pool of investment funds has been regarded as the source of capital needed for projects. The various sources of capital that may be available to a firm, and their differences, are summarized as follows:

1. *Debt capital* involves both short-term and long-term borrowing of funds. Interest must be paid to the suppliers of the capital, and the debt must be repaid at a specified time. The suppliers of debt capital do not share in the profits resulting from the use of the capital; the interest they receive, of course, comes out of the firm's revenues. In many instances, the borrower pledges some type of

security to ensure that the money will be repaid. Sometimes the terms of the loan may place limitations on the uses to which the funds may be put, and in some cases restrictions may be put on further borrowing. Interest paid for the use of borrowed funds is a tax-deductible expense for the firm.

2. *Equity capital* is supplied and used by its owners in the expectation that a *profit* will be earned. There is no assurance that a profit will, in fact, be gained or that the investment capital will be recovered. Similarly, there are no limitations placed on the use of the funds except those imposed by the owners themselves. There is no *explicit* cost for the use of such capital in the ordinary sense of a tax-deductible expense. Equity capital cannot be obtained, however, unless the expected rate of return is high enough, at an acceptable risk, to be attractive to potential investors.

3. *Retained earnings* are an important source of internal capital. Simply stated, retained earnings are profits that are reinvested in the business instead of being paid as dividends to the owners. This method of financing capital projects is used by most companies, but a deterrent is the fact that owners usually expect and demand that they receive some profits in the form of dividends from their investment. Therefore, it is *usually* necessary for a large portion (maybe 50% or more) of the profits to be paid to the owners in the form of dividends. Retention of the remaining profits reduces the immediate amount of dividends per share of stock, increases the book value of the stock, and results in greater future dividends or market resale value for the stock. Many investors prefer to have some of the profits retained and reinvested to help increase the value of their stock.

4. *Depreciation reserves* are set aside out of revenue as an allowance for the replacement of equipment and other depreciable assets and are an additional source of internal capital. In effect, the depreciation funds provide a revolving investment fund that may be used to the best possible advantage. The funds are thus an important source of capital for financing new projects within an existing firm. Obviously, depreciation funds must be managed so that required capital is available for replacing essential equipment when the time for replacement arrives.

5. *Leasing* of an asset is a way of acquiring the use of the asset without the capital expenditure of purchasing it. A lease is a form of contract that establishes the conditions under which the lessor (owner of an asset) conveys to the lessee the use of the asset, including the cost involved. Therefore, leasing is a method of achieving the benefits of capital investment without actually acquiring additional debt or equity capital. In addition, leasing costs are deductible from operating income for tax purposes.

In this chapter, we assume that the *capital structure* for a firm (the overall pool of investment funds referred to previously) is a relatively constant mix of various components of debt and equity capital. It is beyond the scope of our discussion, however, to address the issues related to the mix of debt and equity capital that will optimize the future value of the firm to its owners. Therefore, we focus on the after-tax cost (to the firm) of the primary components of both types of capital, and then on their combined effects in terms of the total weighted cost of capital for a given mix of these primary components.

▼ 14.3 Cost of Debt Capital

The debt part of the capital structure leverages the equity part by increasing the total funds available for capital projects as well as the potential wealth of a firm. The proportion of debt capital, however, must be maintained below a level that would adversely affect the market value of the firm's common stock (Section 14.4). This level will vary by the type of company, say, approximately 30% for a medium to large consumer goods company to more than 50% for a public utility company. The primary components of debt capital are short-term loans and long-term bonds (introduced in Chapter 4). Each of these components is further discussed in the sections that follow.

14.3.1 Loans (Short-Term Debt)

Short-term loans are usually for periods less than five years and more frequently less than two years. The sources of funds are banks, insurance companies, retirement systems, and other lending agencies. A financial instrument such as a line of credit or a short-term note is used that involves a promise to repay the amount of funds borrowed, with interest, on some prearranged schedule. The lending agency may require something of a tangible value (such as a mortgage on physical assets or a current asset such as accounts receivable) as security for the loan, or at least it will make certain the financial position of the borrowing company is such that there is minimal risk involved. For simplicity, we will assume that all interest payments on loans, as well as income taxes paid by a firm, are on an annual basis. Based on this assumption, the after-tax cost of capital for a short-term loan from a lending agency is

$$c_L = i_L(1 - t), \tag{14-1}$$

where
c_L = after-tax cost of capital for a loan;

i_L = rate of interest per year paid on the loan;

t = effective (marginal) income-tax rate (Chapter 6).

EXAMPLE 14-1

The chief financial officer for the Interstate Products Company has acquired a three-year loan for $3,600,000 from a regional bank. The funds from this loan are the short-term debt part of the company's capital structure. The financial arrangements require the payment of interest at the end of each year based on the outstanding principal at the beginning of the year, and annual payments on the principal amount. The interest rate on the loan is 8.3% per year and the company's marginal income-tax rate is 42%. (a) Based on this information, what is the after-tax cost of capital to the company of this short-term loan? (b) If the company repaid $500,000 of the principal at the end of year 1, what is the after-tax cash flow for interest on the loan at the end of year 2?

SOLUTION

(a) After-tax cost of capital on the loan is

$$c_L = 0.083(1 - 0.42) = 0.0481, \text{ or } 4.81\% \text{ per year.}$$

(b) Principal at beginning of year 2 = $3,600,000 - $500,000 = $3,100,000;

$$\text{ATCF}_{\text{Int}}(\text{year } 2) = \$3,100,000(0.0481) = \$149,110.$$

14.3.2 Bonds (Long-Term Debt)

A *bond* is essentially a long-term note given to the lender by the borrower, stipulating the terms of repayment and other conditions. In return for the money loaned, the company promises to repay the loan (bond) and interest upon it at a specified rate. Because the bond represents company indebtedness, the bondholder has no voice in the affairs of the business, at least for as long as the interest is paid, and of course is not entitled to any share of the profits.

Bonds are issued in units such as $1,000, $10,000, and so on, which is known as the *face value*, or *par value*, of the bond. This amount is to be repaid to the lender at the end of a specified period of time. When the face value has been repaid, the bond is said to have been *retired* or *redeemed*. The interest rate quoted on the bond is called the *bond rate*, and the periodic interest payment due is computed as the face value times the bond interest rate per period. A description of what happens during the normal life cycle of a bond is presented in Figure 14-2.

The annual after-tax cost of capital for a bond can be estimated as[*]

$$C_B = \frac{[Zr + (Z - P + S_e)/N + A_e](1 - t)}{(Z + P - S_e)/2}, \tag{14-2}$$

where Z = face (par) value of bond;

r = bond rate (nominal interest) per year;

N = number of years until the bond is retired (redeemed);

S_e = initial selling expenses associated with the bond;

P = actual selling price of the bond [if $P < Z$, the bond sold at a discount (to the par value), and if $P > Z$, the bond sold at a premium];

A_e = annual administrative expenses associated with the bond;

t = effective (marginal) income-tax rate.

[*]Based on approximative formula 5.8 in C. S. Park, and G. P. Sharp-Bette, *Advanced Engineering Economics* (New York: John Wiley & Sons, 1990), p. 178.

Figure 14-2
Life Cycle of Financing
a Bond

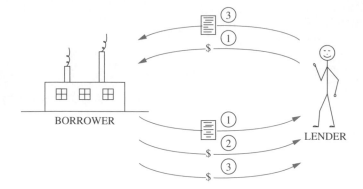

Description of Step	Supplementary Comments
① Borrower sells a bond to the lender. Lender gets bond certificate.	Bonds are issued in even denominations (face values) such as $1,000, $10,000, etc., but amount paid by lender is determined by market supply and demand. The transaction is usually done through a broker.
② Periodic bond interest payments are made to lender.	The amount of each bond interest payment is computed as face value times bond interest rate.
③ Borrower redeems bond by paying principal and getting bond certificate back.	Usually done at end of the stated bond life, the amount paid back is usually the face value.

The numerator of Equation (14-2) is the *annual after-tax cost of the bond* based on the annual interest expenses plus the annualized amount (over the life of the bond) of any discount or premium and the initial selling expenses plus the annual administrative expenses. The denominator is the *average investment in the bond* over its life. As additional information, notice that Equation (14-2), as the terms are defined, is solved on a per bond basis. However, each term in the equation can be "scaled up" in value for a total bond issue and the equation solved on that basis.

EXAMPLE 14-2

The Interstate Products Company recently sold (with the assistance of an investment bank) a $10,000,000 issue of eight-year bonds in which each bond had a par value of $5,000. These funds are the long-term debt part of the company's capital structure. The bond interest rate is 6.6% per year and the interest is paid on an annual basis to the bondholders (owners) of record. The initial selling cost charged by the investment bank was 0.94% of the face value of the bond. Due to the market prime interest rate at the time of sale (the prime rate is the interest rate charged on loans to firms with the best credit ratings), each $5,000 bond (in par value) actually sold for $4,870. That is, the prime interest rate at that time was greater than the bond interest rate (6.6%) and the bonds sold at a discount. Also, as part of the bond issue transaction, a contract was made with a separate (service) bank to maintain the required records on the bonds, make annual interest payments to the bondholders, and accomplish other administrative tasks. The annual cost for this service is 2% of the annual interest expense on the bond. The company's marginal income-tax rate is 42%.

Based on this information, what is the annual after-tax cost of capital to the company of this long-term debt part of their capital structure?

SOLUTION
Equation (14-2) is applied on a per bond basis to estimate the after-tax cost of capital to the company of the bond issue. That cost is

$$C_B = \frac{\{\$5,000(0.066) + [\$5,000 - \$4,870 + 0.0094(\$5,000)]/8 + 0.02(0.066)(\$5,000)\}(1 - 0.42)}{[\$5,000 + \$4,870 - 0.0094(\$5,000)]/2}$$

$$= \frac{\$359(0.58)}{\$4,912} = 0.0424, \text{ or } 4.24\% \text{ per year,}$$

where $\quad Z = \$5,000;$

$\quad\quad\quad\quad r = 6.6\%;$

$\quad\quad\quad\quad N = 8 \text{ years};$

$\quad\quad\quad\quad P = \$4,870;$

$\quad\quad\quad\quad t = 42\%.$

14.3.3 Bond Retirement

The interest paid on bonds is a cost of doing business. In addition to this periodic cost, the company must look forward to the day when the bonds are redeemed and the principal (par value) must be repaid to the bondholders.

When it is desired to repay long-term loans and thus reduce company indebtedness, a systematic program frequently is adopted for repayment of a bond issue when it becomes due. Some such provision, planned in advance, gives assurance to the bondholders and makes the bonds more attractive to the investing public; it may also allow the bonds to be issued at a lower rate of interest.

In many cases, the company periodically sets aside definite sums that, with the interest they earn, will accumulate to the amount needed to retire the bonds at the time they are due. Because it is convenient to have these periodic deposits equal in amount, the retirement procedure becomes a *sinking fund*. This is one of the most common uses of a sinking fund. By its use, the bondholders know that adequate provision is being made to safeguard their investment, and the company knows in advance what the annual cost for bond retirement will be.

If a bond issue of $100,000 in 10-year bonds, in $1,000 units, paying 10% nominal interest in semiannual payments must be retired by the use of a sinking fund that earns 8% compounded semiannually, the semiannual cost for retirement will be as follows:

$$A = F(A/F, i\%, N);$$

$$F = \$100,000;$$

$$i = 8\%/2 = 4\% \text{ per interest period;}$$

$$N = 2 \times 10 = 20 \text{ interest periods.}$$

Thus,

$$A = \$100,000(0.0336) = \$3,360.$$

In addition, the semiannual interest on the bonds must be paid, which would be calculated as follows:

$$\text{Interest} = \$100,000 \times \frac{0.10}{2} = \$5,000;$$

$$\text{Total semiannual cost} = \$3,360 + \$5,000 = \$8,360;$$

$$\text{Annual cost} = \$8,360 \times 2 = \$16,720.$$

The total cost for interest and retirement of the entire bond issue over 20 periods (10 years) will be

$$\$8,360 \times 20 = \$167,200.$$

14.4 Cost of Equity Capital

We have referred to the form of organization so far in our discussion of the capital financing function as the "firm" or the "company." However, private-sector companies of any relative size are usually organized as a *corporation*. A corporation is a fictitious being, recognized by law, that can pursue almost any type of business transaction in which a real person can engage. It operates under a charter that is granted by a state and is endowed by this charter with certain rights and privileges, such as perpetual life without regard to any change in the persons of its owners. In payment for these privileges and the enjoyment of being a legal entity, the corporation is subject to certain restrictions. It is limited in its field of action by the provisions of its charter. In order to enter new fields of enterprise, it must apply for a revision of its charter or obtain a new one. Special taxes may be assessed against it.

The equity capital of a corporation is acquired through the sale of stock. The purchasers of the stock are part owners, usually called *stockholders*, of the corporation and its assets. In this manner, ownership may be spread throughout the world, and as a result, enormous sums of capital can be accumulated. With few exceptions, the stockholders of a corporation, although they are the owners and are entitled to share in the profits, are not liable for the debts of the corporation. *They are thus never compelled to suffer any loss beyond the value of their stock.* Because the life of a corporation is continuous or indefinite, long-term investments can be made and the future faced with some degree of certainty, which also makes debt capital (particularly long-term) easier to obtain, and generally at a lower interest cost.

There are many types of stock, but two are of primary importance. These are *common stock*, which represents ordinary ownership without special guarantees of return on investment, and *preferred stock*, which has certain privileges and restrictions not available for common stock. For instance, dividends on common stock are *not* paid until the fixed percentage return on preferred stock has been paid.

14.4.1 Common Stock

The issuance of common stock is a primary source of equity capital utilized to finance the capital projects of a corporation. Other sources of equity capital include preferred stock, retained earnings, and depreciation reserves.

Establishing value for a share of common stock is not as straightforward as placing value on a bond and its after-tax cost of capital. Valuation and the after-tax cost of common stock is actually a controversial subject because of numerous assumptions regarding the future of dividend growth rates, future stock prices, perceived riskiness of the investment, projected after-tax earnings, and so on.[*] The value of common stock must be a measure of the earnings that will be received through ownership of the stock, and it is dependent upon several factors, which can probably all be summed up under two headings—dividends and market price.

We present a very simple approach for the valuation of common stock and the estimation of per share rate of return expected by the investor. The approach is termed the *dividend valuation model*. Other approaches, such as the earnings model and the investment-opportunities model, are discussed in any good finance textbook.

The owner of a share of common stock in a corporation is entitled to receive cash dividends declared by the company, as well as the price of the stock at the time it is sold. If we let the after-tax value of cash dividends (dividends are paid out of after-tax earnings) received during year k equal Div_k, the current value of a share of common stock in the dividend valuation model can be approximated by the PW of future cash receipts during an N-year ownership period. That is,

$$P_0 \simeq \frac{\text{Div}_1}{(1+e_a)} + \frac{\text{Div}_2}{(1+e_a)^2} + \cdots + \frac{\text{Div}_N}{(1+e_a)^N} + \frac{P_N}{(1+e_a)^N}, \qquad (14\text{-}3)$$

where e_a = rate of return per year (expressed as a decimal) required by common stockholders (*after-tax cost of equity to the corporation*);

       P_0 = current value of a share of common stock;

       P_N = selling price of a share of common stock at the end of N years.

The value of e_a must be sufficient to compensate the shareholder for his or her time value of money and for the risk that is believed to be associated with the investment. How we estimate the value of P_N is a further complication in determining P_0.

---

[*]See, for example, Franco Modigliani and Merton H. Miller, "The Cost of Capital, Corporation Finance, and the Theory of Investment," *American Economic Review,* vol. 48, no. 3, June 1958, pp. 261–297; and D. Durand, "The Cost of Capital in an Imperfect Market: A Reply to Modigliani and Miller," *American Economic Review,* vol. 49, no. 4, September 1959, pp. 639–655.

The dividend valuation model incorporates the two conservative assumptions that dividends are *constant* over the long term and that $P_0 = P_N$. In this case, the current price of a share of common stock equals the PW of an assumed infinite series of dividend receipts that remain constant in amount:

$$P_0 = \text{Div}\,(P/A,\ e_a,\ \infty) = \frac{\text{Div}}{e_a}. \qquad (14\text{-}4)$$

Thus, if the current selling price of a share of common stock is known and the annual dividend for the past year is also known, the after-tax cost of equity (common stock) is conservatively estimated to be

$$e_a = \frac{\text{Div}}{P_0}. \qquad (14\text{-}5)$$

When the future price of the security is expected to grow at a rate of g (expressed as a decimal) each year, the cost of equity can be approximated by adding a growth factor to the dividend valuation model [Equation (14-5)]:

$$e_a = \frac{\text{Div}}{P_0} + g. \qquad (14\text{-}6)$$

Suppose that a share of common stock is priced at $100 and a dividend of $8 is currently paid annually. The expected annual growth in price is 4% per year. If an investor is willing to purchase this security based on the assumption that dividends remain constant and the price grows at 4% annually, the expected return is about $8/$100 + 0.04 = 0.12, or 12% per year. A second, less risky security being considered may sell for $100 and pay a dividend of $10 annually, with $g = 0$. In this case, $e_a = 10\%$ per year. If the investor is indifferent between the two securities, an additional expected return of 2% per year is required to compensate for the extra risk associated with the first investment.

The determination of the after-tax cost of all types of equity is difficult in practice. For the purposes of this book, the opportunity cost principle and Equations (14-5) and (14-6) provide a basic, though simplified, approach for approximating this quantity.

EXAMPLE 14-3

The Interstate Products Company (IPC) is expected to generate perpetual after-tax net earnings of $2,700,000 per year with its existing assets. This firm produces a stable product and has been in business for 15 years. Furthermore, it has 1,000,000 shares of common stock outstanding and has a long-standing policy of declaring an annual dividend that is 50% of its after-tax earnings. The remaining 50% of earnings is retained for cash reserves, equipment replacement, and so on.

(a) If investors require a 4% per year return on their investment from dividends only, how much would they be willing to pay for a share of IPC common stock if dividends remain constant?

(b) An investor who owns 1,000 shares of IPC stock believes that its share price will grow at a rate of 6% per year in the future. What is the rate of return on IPC stock expected by this investor (that is, what is the after-tax cost of capital for common stock based on the growth valuation model)?

SOLUTION

(a) From Equation (14-4), the estimated current selling price of a share of IPC common stock should be [\$2,700,000(0.5)/1,000,000 shares]/0.04 = \$33.75.

(b) The return to equity based on Equation (14-6) would be approximately (\$1.35/\$33.75) + 0.06 = 0.10, or 10% per year.

14.4.2 Preferred Stock

Preferred stock also represents ownership, but the owner has certain additional privileges and restrictions not assigned to the holder of common stock. Preferred stockholders are guaranteed a dividend on their stock, usually a percentage of its par value, before the holders of the common stock may receive any return. In case of dissolution of the corporation, the assets must be used to satisfy the claims of the preferred stockholders before those of the holders of common stock. Preferred stockholders usually, but not always, have voting rights. Occasionally they are granted certain privileges, such as the election of special representatives on the board of directors, if their preferred dividends are not paid for a specified period.

Because the dividend rate is fixed, preferred stock is a more conservative investment than common stock and has many of the features of long-term bonds. For this reason, the market value of such stock is less likely to fluctuate. Therefore, the after-tax cost of capital for preferred stock (e_p) can be approximated by dividing the guaranteed dividend (Div_p, which is paid out of after-tax earnings) by the original par value of the stock (P_p):

$$e_p = \frac{\text{Div}_p}{P_p}. \tag{14-7}$$

EXAMPLE 14-4

The Interstate Products Corporation previously issued 80,000 shares of preferred stock at a par value of \$25 per share. The guaranteed annual dividend is \$2 per share. What is the after-tax cost of the preferred stock part of IPC's capital structure?

SOLUTION
Based on Equation (14-7), we have $e_p = \$2/\$25 = 0.08$, or 8% per year.

14.4.3 Retained Earnings

The after-tax cost of retained earnings is normally assumed to be the same as for common stock (the rate of return expected by common stockholders). It may appear that retained earnings are a free resource to the corporation, but this is not the case.

These earnings, which are equity funds, do not belong to the corporation, but rather to the stockholders. They have been retained and reinvested in the firm for the purpose of enhancing future growth and revenues and increasing stockholders wealth. Thus, there is the same opportunity cost for these funds as would occur if the shareholders received them originally and then invested them in additional common shares of the corporation.

14.5 Weighted Average Cost of Capital

The after-tax weighted average cost of capital (WACC) for a firm can be determined once the amount and explicit cost is established for each debt and equity component in the capital structure. We illustrate the calculation for the Interstate Products Corporation (IPC) in the next section.

14.5.1 The Interstate Products Corporation Case

The amount and after-tax cost of the individual short-term debt, bond, common stock, and preferred stock components of IPC's capital structure were established in Examples 14-1 through 14-4, respectively. Retained earnings are also part of its overall pool of investment funds. As discussed in Section 14.4.3, the cost of these internal equity funds should be the same as the cost of common stock.

We will assume that on its most recent financial statement date, IPC had retained earnings of $4,300,000. This amount ($4,300,000) and the information from Examples 14-1 through 14-4 can be combined into a weighted average cost of capital for the firm. The weighting of each of the capital components should be proportional to its part in the total pool of funds. The calculations for the Interstate Products Corporation case are shown in Table 14-1.

As additional information, notice that *depreciation funds (reserves)*, another source of internal funds for investment, are not explicitly included in the weighted average cost of capital calculation. This does not mean, however, that these funds are a free resource to the corporation. This would be false logic. Instead, we assume that these funds replace the need for additional debt and equity capital in the same proportions as the present capital structure and have an opportunity cost equal to the weighted average cost of capital (8.2% per year in the IPC case).

14.5.2 Relationship with the Minimum Attractive Rate of Return

What is the relationship between the WACC value and the MARR? Assume, for example, that the rate of return of an engineering project is estimated to be less than the WACC. Then, if the project was implemented, the subsequent economic results would decrease the value of the firm since there would not be any surplus earned above the cost of capital invested in the project. That is, the project is estimated to result in a negative impact on the wealth of the firm. Obviously, we do not want this situation to occur. Therefore, the WACC should be the minimum value used for the MARR.

Extending this logic leads to another important consideration. Assume that the current MARR value being used by a firm is greater than the WACC (for example,

TABLE 14-1	Calculation of After-Tax Weighted Average Cost of Capital (IPC Case)			
Source of Financing	Amount	Proportion	After-Tax Cost (decimal)	Weighted Cost
Short-term debt	$3,600,000[b]	0.0809	0.0481[b]	0.0039
Bonds	10,000,000[c]	0.2247	0.0424[c]	0.0095
Common stock[a]	24,600,000	0.5528	0.1000[d]	0.0553
Preferred stock	2,000,000[e]	0.0449	0.0800[e]	0.0036
Retained earnings	4,300,000	0.0967	0.1000	0.0097
	$44,500,000	1.0000		WACC = 0.0820
				or 8.2% per year

[a] The 1,000,000 shares of common stock were originally sold for an average price of $24.60 per share.

[b] Refer to Example 14-1.

[c] Refer to Example 14-2.

[d] Refer to Example 14-3.

[e] Refer to Example 14-4.

it was established using the opportunity cost approach discussed in Chapter 4). Then the *best economic measure* of the equivalent current value that would be added to the firm by the project *remains the PW value calculated at $i = WACC$*. Therefore, regardless of the current MARR value, this information is important and should be available for use in decision making.

14.5.3 Marginal Weighted Average Cost of Capital

A logical argument is sometimes made that the current (historical) WACC is not the best value to use for *new* projects. The viewpoint presented is that new debt and equity capital to fund these projects or to replace existing funds later will normally be at a higher cost, and the weighted average cost of these additional (marginal) funds should be used.

We take the viewpoint in this book that the most representative cost of capital is situation dependent. That is, if a firm must acquire additional capital resources to fund new projects, then the best value to use is the after-tax marginal weighted average cost based on the mix of new funding sources. However, if the existing pool of investment funds, including depreciation reserves, is adequate to meet the firm's future capital financing requirements, then the current after-tax WACC is the best value to use.

14.6 Leasing as a Source of Capital

As mentioned in Section 14.2, leasing is a business arrangement that makes assets available for use without incurring the initial capital investment costs of purchase.

The decision to lease or purchase an asset represents a situation in which the source of capital may affect which alternative is eventually chosen. Leasing is a source of capital generally regarded as a long-term liability similar to a mortgage, whereas the purchase of an asset typically uses funds from the firm's capital structure (much of which is equity). Before considering examples of lease–buy problems, we provide some background information about leases.

For corporations, the rent paid on leased assets used in their trade or business is generally deductible as a business expense. To make lease payments deductible as rent, the contract must be for a true leasing arrangement instead of a conditional sales agreement. In a true lease, the corporation using the property (lessee) does not acquire ownership in or title to the asset, whereas a conditional sales contract will transfer to the lessee an equity interest in or title to the asset being leased. Hence, the test of whether lease payments qualify as business expenses lies in distinguishing between a true lease and a conditional sale.* For our purposes, it is assumed that a true lease exists and that an asset may be acquired through *leasing* or *purchasing*.

A number of studies have shown that there is no real income-tax advantage in leasing. This is particularly true since accelerated methods (e.g., MACRS) have been permitted for depreciation. Assuming a given purchase price, the firm offering a lease contract (lessor) can charge no more for depreciation than can the owner of assets. If assets are leased, the annual lease payments are deducted in computing income taxes; if the assets are purchased, annual depreciation is deducted. Most companies have now come to realize that leasing may not offer major tax advantages.

There may or may not be savings in maintenance expenses through leasing. Any savings will depend on the actual circumstances, which should be carefully evaluated in each case. There is no doubt that leasing usually does simplify maintenance problems, which may be an important factor. Also, many indirect costs, which frequently are difficult to determine, are usually associated with ownership.

These same studies have concluded that a true advantage of leasing lies in allowing a firm to obtain modern equipment that is subject to rapid technological change. Further, leasing for this purpose typically provides an effective hedge against obsolescence and inflation.

The following example illustrates correct methods of handling a lease versus purchase study on an after-tax basis; the analysis uses the tabular format presented in Chapter 6 (Figure 6-5).

EXAMPLE 14-5

An industrial forklift truck can be purchased for $30,000 or leased for a fixed amount of $9,200 per year payable at the *beginning* of each year. The lease contract provides that maintenance expenses are borne by the lessor. Regardless of whether the

---

*For further information, see *Tax Guide for Small Business,* U.S. Internal Revenue Service Publication 334, published annually.

TABLE 14-2 ATCFs for Example 14-5

Year	(A) BTCF	(B) Depreciation[a]	(C) = (A) − (B) Taxable Income	(D) = −0.4(C) Cash Flow for Income Taxes	(E) = (A) + (D) ATCF
Purchase Truck (A $ Study)[b]					
0	−$30,000				−$30,000
1	−1,050	$6,000	−$7,050	$2,820	1,770
2	−1,102	9,600	−10,702	4,281	3,179
3	−1,158	5,760	−6,918	2,767	1,609
4	−1,216	3,456	−4,672	1,869	653
5	−1,276	3,456	−4,732	1,893	617
6	−1,340	1,728	−3,058	1,227	−113
Lease Truck (A $ Study)[c]					
0	−$9,200		−$9,200	$3,680	−$5,520
1–5	−9,200		−9,200	3,680	−5,520
6	0	0	0	0	0

[a] MACRS rates are given in Table 6-3.
[b] The AW at MARR = 15% is −$6,439.
[c] The AW at MARR = 15% is −$6,348.

truck is purchased or leased, the study period is six years. If purchased, annual maintenance expenses are expected to be $1,000 in year 0 purchasing power, and they will inflate at 5% per year over the study period. The MV of the truck is expected to be negligible after six years of normal use. Depreciation is determined with the MACRS (GDS) method using a five-year recovery period (deductions occur over six years). The effective income-tax rate is 40%, and the after-tax MARR, which includes an allowance for general price inflation, is 15% per year.

Use the AW method, and determine whether the forklift truck should be purchased or leased. This firm is profitable in its overall business activity.

SOLUTION
The effects of general price inflation and income taxes on the ATCFs of both alternatives are shown in Table 14-2. The lease alternative is less costly than the purchase alternative (AW = −$6,348 > −$6,439) and would probably be selected. Moreover, if capital is not readily available, the firm would prefer to lease the forklift truck. Furthermore, if the estimates of maintenance expenses and general price inflation are believed to be relatively uncertain, the firm would tend to favor leasing as a hedge against the future.

Rather than make use of the tabular procedures illustrated in Example 14-5, models may be developed that yield the same equivalent worths (e.g., PWs) for the lease and purchase alternatives. These are summarized as follows.

14.6.1 Cost of the Lease Alternative

The after-tax cost of a lease during year k is given by

$$l_k = L_k(1 - t),$$

where l_k = after-tax lease expense during year k;

L_k = before-tax lease expense during year k;

t = effective income-tax rate.

If i, the after-tax MARR the firm expects from the use of its money, is known and fixed, the PW of the after-tax *cost* of the lease during its life of N years is given by

$$\text{PW}_{\text{Lease}}(i\%) = \sum_{k=1}^{N} \frac{L_k(1 - t)}{(1 + i)^k}. \tag{14-8}$$

It should be noted that annual maintenance expenses are not included in Equation (14-8) because they are assumed to be borne by the equipment supplier and are included in the annual lease cost L_k. Furthermore, the standard end-of-year cash-flow convention is assumed.

14.6.2 Cost of the Purchase Alternative

The after-tax cost of equipment when it is purchased is a function of the expected annual expenses during the life of the equipment, as well as the purchase price, book value, and expected market value. The PW of the after-tax *cost* of purchased equipment is given by

$$\text{PW}_{\text{Buy}}(i\%) = I - \frac{\text{MV}_N(1 - t) + t\text{BV}_N}{(1 + i)^N} + \sum_{k=1}^{N} \frac{\text{O\&M}_k(1 - t) - d_k(t)}{(1 + i)^k}, \tag{14-9}$$

where I = capital investment;

MV_N = expected market value at end of year N;

BV_N = book value at end of year N;

i = interest rate per year;

N = life of equipment in years;

$\text{O\&M}_k$ = operating and maintenance expense during year k;

t = effective income-tax rate;

d_k = depreciation during year k.

It should be noted that the market value, book value, and depreciation amounts of Equation (14-9) are negative because they reduce costs. Again, the end-of-year cash-flow convention is assumed.

14.7 Capital Allocation

We have been discussing capital financing topics in Sections 14.2 through 14.6 that deal with (1) how a company obtains capital (and from what sources) and (2) how much capital the company has available, and at what cost, to maintain a successful business enterprise in the years ahead.

An outstanding phenomenon of present-day industrialized civilizations is the extent to which engineers and managers, by using capital (money and property), are able to create wealth through activities that transform various types of resources into goods and services. Historically, the largest industrialized nations in the world consume a significant portion of their gross national product each year by investing in wealth-creating assets such as equipment and machinery (so-called intermediate goods of production).

The remainder of this section examines the capital-expenditure decision-making process, also referred to as *capital allocation.* This process involves the planning, evaluation, and management of capital projects. In fact, much of this book has dealt with concepts and techniques required to make correct capital-expenditure decisions involving engineering projects. Now our task is to place them in the broader context of upper management's responsibility for proper planning, measurement, and control of the firm's overall portfolio of capital investments.

14.7.1 Allocating Capital among Independent Projects

Companies are constantly presented with independent opportunities in which they can invest capital across the organization. These opportunities usually represent a collection of the best projects for improving operations in all areas of the company (e.g., manufacturing, research and development, etc.). In most cases, the amount of available capital is limited, and additional capital can be obtained only at increasing incremental cost. Thus, companies have a problem of budgeting, or allocating, available capital to numerous possible uses.

One popular approach to capital allocation among projects uses the PW criterion and was discussed in Chapter 5. If project risks are about equal, the procedure is to compute the PW for each investment opportunity and then to determine the combination of projects that maximizes PW, subject to various constraints on the availability of capital. The following example provides a general review of this procedure.

EXAMPLE 14-6

Consider these four independent projects and determine the best allocation of capital among them if no more than $300,000 is available to invest:

Independent Project	Initial Capital Outlay	PW
A	$100,000	$25,000
B	125,000	30,000
C	150,000	35,000
D	75,000	40,000

TABLE 14-3 Project Combinations for Example 14-6

Combination	Total PW ($\times 10^3$)	Total Capital Outlay ($\times 10^3$)
AB	$55	$225
AC	60	250
AD	65	175
BC	65	275
BD	70	200
CD	75	225
ABC	90	375
ACD	100	325
BCD	105	350
ABD	95	300* **Best**
ABCD	130	450

SOLUTION

All possible combinations of these independent projects taken two, three, and four at a time are shown in Table 14-3 together with the total PW and initial capital outlay of each. After eliminating those combinations that violate the $300,000 funds constraint, the proper selection of projects would be ABD, and the maximum PW is $95,000. The process of enumerating combinations of projects having nearly identical risks is best accomplished with a computer when large numbers of projects are being evaluated.

Methods for determining which possible projects should be allocated available funds seem to require the exercise of judgment in most realistic situations. Example 14-7 illustrates such a problem and possible methods of solution.

EXAMPLE 14-7

Assume that a firm has five investment opportunities (projects) available, which require the indicated amounts of capital and which have economic lives and prospective after-tax IRRs as shown in Table 14-4. Further, assume that the five ventures are independent of each other, investment in one does not prevent investment in any other, and none is dependent upon the undertaking of another.

Now suppose that the company has unlimited funds available, or at least sufficient funds to finance all these projects, and that capital funds cost the company 6% per year after taxes. For these conditions, the company probably would decide to undertake all projects that offered a return of at least 6% per year, and thus projects A, B, C, and D would be financed. However, such a conclusion would assume that the risks associated with each project are reasonable in light of the prospective IRR or are no greater than those encountered in the normal projects of the company.

Unfortunately, in most cases the amount of capital is limited, either by absolute amount or by increasing cost. If the total of capital funds available is $60,000, the

	TABLE 14-4 Prospective Projects for a Firm[a]		
Project	Capital Investment	Life (years)	Rate of Return (% per year)
A	$40,000	5	7
B	15,000	5	10
C	20,000	10	8
D	25,000	15	6
E	10,000	4	5

[a] Here we assume that the indicated rates of return for these projects can be repeated indefinitely by subsequent "replacements."

decision becomes more difficult. Here it would be helpful to list the projects in order of *decreasing* profitability in Table 14-5 (omitting the undesirable project E). Here it is clear that a complication exists. We naturally would wish to undertake those ventures that have the greatest profit potential. However, if projects B and C are undertaken, there will not be sufficient capital for financing project A, which offers the next greatest rate of return. Projects B, C, and D could be undertaken and would provide an approximate annual return of $4,600 (= $15,000 × 10% + $20,000 × 8% + $25,000 × 6%). If project A were undertaken, together with either B or C, the total annual return would not exceed $4,600.* A further complicating factor is the fact that project D involves a longer life than the others. It is thus apparent that we might *not* always decide to adopt the alternative that offers the greatest profit potential.

The problem of allocating limited capital becomes even more complex when the risks associated with the various available projects are not the same. Assume that the risks associated with project B are determined to be higher than the average risk associated with projects undertaken by the firm and that those associated with project C are lower than average. The company thus might rank the projects according to their overall desirability, as in Table 14-6. Under these conditions, the company might decide to finance projects C and A, thus avoiding one project with a higher than average risk and another having the lowest prospective return and longest life of the group.

	TABLE 14-5 Prospective Projects of Table 14-4 Ordered by IRR		
Project	Capital Investment	Life (years)	Rate of Return (%)
B	$15,000	5	10
C	20,000	10	8
A	40,000	5	7
D	25,000	15	6

* This return amount is given assuming that the leftover capital could earn no more than 6% per year.

TABLE 14-6 Prospective Projects of Table 14-5 Ordered According to Overall Desirability

Project	Capital Investment	Life (years)	Rate of Return (%)	Risk Rating
C	$20,000	10	8	Lower
A	40,000	5	7	Average
B	15,000	5	10	Higher
D	25,000	15	6	Average

14.7.2 Linear Programming Formulations of Capital Allocation Problems

For large numbers of independent or interrelated investments, the "brute force" enumeration and evaluation of all combinations of projects, as illustrated in Example 14-7, is impractical. This section describes a mathematical procedure for efficiently determining the optimal *portfolio* of projects in industrial capital allocation problems (Figure 14-1). Only the formulations of these problems will be presented in this section; their solution is beyond the scope of the book.

Suppose that the goal of a firm is to maximize its net PW by adopting a capital budget that includes a large number of mutually exclusive combinations of projects. When the number of possible combinations becomes fairly large, manual methods for determining the optimal investment plan tend to become complicated and time consuming, and it is worthwhile to consider linear programming as a solution procedure. The remainder of this section describes how simple capital allocation problems can be formulated as linear programming problems. Linear programming is a mathematical procedure for maximizing (or minimizing) a linear objective function, subject to one or more linear constraint equations. Hopefully, the reader will obtain some feeling for how more involved problems might also be modeled.

Linear programming is a useful technique for solving certain types of multiperiod *capital allocation problems* when a firm is not able to implement all projects that may increase its PW. For example, constraints often exist on how much investment capital can be committed during each fiscal year, and interdependencies among projects may affect the extent to which projects can be successfully carried out during the planning period.

The *objective function* of the capital allocation problem can be written as

$$\text{Maximize net PW} = \sum_{j=1}^{m} B_j^* X_j,$$

where $B_j^* =$ net PW of investment opportunity (project) j during the planning period being considered;

X_j = fraction of project j that is implemented during the planning period (*Note:* In most problems of interest, X_j will be either 0 or 1; the X_j values are the decision variables);

m = number of mutually exclusive combinations of projects under consideration.

In computing the net PW of each mutually exclusive combination of projects, a MARR must be specified.

The following notation is used in writing the constraints for a linear programming model:

c_{kj} = cash outlay (e.g., initial capital investment or annual operating budget) required for project j in time period k;

C_k = maximum cash outlay that is permissible in time period k.

Typically, two types of constraints are present in capital budgeting problems:

1. *Limitations on cash outlays for period k of the planning horizon:*

$$\sum_{j=1}^{m} c_{kj} X_j \leq C_k.$$

2. *Interrelationships among the projects.* The following are examples:

   (a) If projects p, q, and r are mutually exclusive, then

   $$X_p + X_q + X_r \leq 1.$$

   (b) If project r can be undertaken only if project s has already been selected, then

   $$X_r \leq X_s \quad \text{or} \quad X_r - X_s \leq 0.$$

   (c) If projects u and v are mutually exclusive and project r is dependent (contingent) on the acceptance of u or v, then

   $$X_u + X_v \leq 1$$

   and $$X_r \leq X_u + X_v.$$

To illustrate the formulation of linear programming models for capital allocation problems, Example 14-8 and Example 14-9 are presented.

EXAMPLE 14-8

Five engineering projects are being considered for the upcoming capital budget period. The interrelationships among the projects and the estimated net cash flows of the projects are summarized in the following table:

	Cash Flow ($000s) for End of Year, k					PW at
Project	0	1	2	3	4	MARR = 10% per year
B1	−50	20	20	20	20	13.4
B2	−30	12	12	12	12	8.0
C1	−14	4	4	4	4	−1.3
C2	−15	5	5	5	5	0.9
D	−10	6	6	6	6	9.0

Projects B1 and B2 are mutually exclusive. Projects C1 and C2 are mutually exclusive *and* dependent on the acceptance of B2. Finally, project D is dependent on the acceptance of C1.

Using the PW method and MARR = 10% per year, determine which combination (portfolio) of projects is best if the availability of capital is limited to $48,000.

SOLUTION

The objective function and constraints for this problem are written as follows: Maximize

$$\text{Net PW} = 13.4X_{B1} + 8.0X_{B2} - 1.3X_{C1} + 0.9X_{C2} + 9.0X_D,$$

subject to

$$50X_{B1} + 30X_{B2} + 14X_{C1} + 15X_{C2} + 10X_D \leq 48;$$

(constraint on investment funds)

$$X_{B1} + X_{B2} \leq 1;$$

($B1$ and $B2$ are mutually exclusive)

$$X_{C1} + X_{C2} \leq X_{B2};$$

($C1$ or $C2$ is contingent on $B2$)

$$X_D \leq X_{C1};$$

(D is contingent on $C1$)

$$X_j = 0 \text{ or } 1 \text{ for } j = \text{B1, B2, C1, C2, D.}$$

(no fractional projects are allowed)

A problem such as this could be solved readily by using the simplex method of linear programming if the last constraint ($X_j = 0$ or 1) were not present. With that constraint included, the problem is classified as a linear *integer* programming problem. (Many computer programs are available for solving large linear integer programming problems.)

TABLE 14-7 Project Interrelationships and PW Values (Example 14-9)

Project		Net Cash Flow ($000s), End of Year[a]				Net PW ($000s) at 12% per year[b]
		0	1	2	3	
A1		−225	150 (60)	150 (70)	150 (70)	+135.3
A2	mutually exclusive	−290	200 (180)	180 (80)	160 (80)	+146.0
A3		−370	210 (290)	200 (170)	200 (170)	+119.3
B1		−600	100 (100)	400 (200)	500 (300)	+164.1
B2	independent	−1,200	500 (250)	600 (400)	600 (400)	+151.9
C1		−160	70 (80)	70 (50)	70 (50)	+8.1
C2	mutually exclusive and dependent on acceptance of A1 or A2	−200	90 (65)	80 (65)	60 (65)	−13.1
C3		−225	90 (100)	95 (60)	100 (70)	+2.3

[a] Estimates in parentheses are annual operating expenses (which have already been subtracted in the determination of net cash flows).
[b] For example, net PW for A1 = −$225,000 + $150,000($P/A$, 12%, 3) = +$135,300.

EXAMPLE 14-9

Consider a three-period capital allocation problem having the net cash-flow estimates and PW values shown in Table 14-7. The MARR is 12% per year and the ceiling on investment funds available is $1,200,000. In addition, there is a constraint on operating funds for support of the combination of projects selected—$400,000 in year 1. From these constraints on funds outlays and the interrelationships among projects indicated in Table 14-7, we shall formulate this situation in terms of a linear integer programming problem.

SOLUTION

First, the net PW of each investment opportunity at 12% per year is calculated (Table 14-7). The objective function then becomes

$$\text{Maximize net PW} = 135.3X_{A1} + 146.0X_{A2} + 119.3X_{A3} + 164.1X_{B1}$$

$$+ 151.9X_{B2} + 8.7X_{C1} - 13.1X_{C2} + 2.3X_{C3}.$$

The budget constraints are the following:

Investment funds constraint:

$$225X_{A1} + 290X_{A2} + 370X_{A3} + 600X_{B1} + 1,200X_{B2}$$

$$+ 160X_{C1} + 200X_{C2} + 225X_{C3} \leq 1,200$$

First year's operating cost constraint:

$$60X_{A1} + 180X_{A2} + 290X_{A3} + 100X_{B1} + 250X_{B2}$$

$$+ 80X_{C1} + 65X_{C2} + 100X_{C3} \leq 400$$

Interrelationships among the investment opportunities give rise to these constraints on the problem:

$$
\begin{array}{lll}
X_{A1} + X_{A2} + X_{A3} & \leq 1 & \text{A1, A2, A3 are mutually exclusive} \\
X_{B1} & \leq 1 \\
X_{B2} & \leq 1 & \text{B1, B2 are independent} \\
X_{C1} + X_{C2} + X_{C3} & \leq X_{A1} + X_{A2} & \text{accounts for dependence of} \\
& & \text{C1, C2, C3 (which are mutually exclusive)} \\
& & \text{on A1 } or \text{ A2}
\end{array}
$$

Finally, if all decision variables are required to be either 0 (not in the optimal solution) or 1 (included in the optimal solution), the last constraint on the problem would be written

$$X_j = 0, 1 \quad \text{for } j = \text{A1, A2, A3, B1, B2, C1, C2, C3.}$$

As can be seen, a fairly simple problem such as this would require a large amount of time to solve by listing and evaluating all mutually exclusive combinations, as suggested in Chapter 5. Consequently, it is recommended that a suitable computer program be used to obtain solutions for all but the most simple capital allocation problems.

14.8 An Overview of a Typical Corporate Capital Budgeting Process

There is always the possibility that a student of engineering economy has been immersed in such a maze of details by this point that he or she has lost sight of the "enterprise context" in which various types of calculations are performed to evaluate proposed capital expenditures. Therefore, our objective in the remainder of this chapter is to focus on how the results of engineering economic analyses are used in the corporate capital budgeting process. The student should pay close attention to how measures of financial merit, such as present worth and internal rate of return, are used in the corporate capital appropriation process.

A typical corporate capital budgeting process consists of several interrelated steps:

1. preliminary planning and cost of capital;
2. annual capital budget and proposed project portfolio;

3. capital expenditure policies and evaluation procedures;
4. project implementation and postaudit review;
5. communication.

14.8.1 Preliminary Planning and Cost of Capital

A considerable amount of planning must be accomplished before capital financing and allocation decisions can be made. The main purpose of capital expenditure planning is to make sure that long-term goals of the organization can be attained. These long-term goals and strategic plans directly tie profit plans to capital budgets. Although budget periods normally range from 3 to 10 years, most large- and medium-sized firms use a five-year period and small firms use a three- to five-year period.

In long-range planning, a company decides how big it wants to be, how fast it wants to grow, how much capital it needs, and how it will acquire the capital funds needed. As discussed earlier, the acquisition of these funds from either internal or external sources determines the cost of capital. Also, as illustrated earlier, the most common approach to determining the cost of capital is the after-tax weighted average of the debt and equity components in the capital structure.

Some firms use the weighted average cost of capital as the MARR for capital expenditure planning, but others use this as a starting point in developing the MARR for each division. The latter is more prominent in medium-size firms, although most firms tend to use one companywide rate. A firm's cost of capital is periodically updated as the mix and amounts of debt and equity capital change.

14.8.2 Annual Capital Projects Budget and Proposed Project Portfolio

A normal procedure for developing an annual capital projects budget within a firm is for division and functional-level managers to develop the list of proposed projects. As these project proposals go up the ranks of the organizational hierarchy, some are eliminated and others are added. To assist management in the capital budgeting process, the proposed projects must be classified in some manner. No matter what size the firm is, the two most common methods of classifying proposed projects are by operating division (project type and purpose) and by project size in dollars.

Once the proposed projects are classified, it is necessary to rank them within the portfolio according to various selection criteria. The profitability of invested capital and adherence to the long-term strategy and goals of the business are often ranked as the two top criteria. Three methods frequently used by companies to measure economic merit in the planning stages of a project are the payback period, IRR, and PW methods. Projects with long payback periods, low IRRs, or undesirable PWs are dropped from further consideration unless there are extenuating circumstances to retain them in the project portfolio (e.g., projects that must be funded to ensure compliance with legal requirements).

Every year, a company will have some projects that can be called *noneconomic*. A noneconomic project is one that requires capital investment, but provides little or no monetary return. Most companies separate economic and noneconomic projects when requesting funding, and some firms further divide noneconomic projects into various categories such as sustaining, regulatory and environmental, safety and health, and administrative.

For various reasons, not all profitable projects are accepted. A project may be turned down at two stages of the capital budgeting process, the first being at the planning and selection stage and the second at the implementation stage. Although productivity of capital is important, the two major reasons for rejecting a proposed project at either stage are incompatibility with company goals and objectives and unavailability of capital.

As might be expected, especially in large companies, top management and the board of directors will usually approve the overall capital budget; division and functional managers will decide on the allocation of capital to most of the individual projects.

14.8.3 Capital Expenditure Policies and Evaluation Procedures

Capital expenditure policies and procedures can be subdivided into two broad parts: (1) management approval levels for projects of different sizes and (2) management control over specific capital expenditures.

Three typical plans for delegating management responsibility for project approvals are as follows:

1. Whenever proposed projects are clearly good in terms of economic desirability according to operating division analysis, the division is given approval power as long as control can be maintained over the total amount invested by each division and as long as the division analyses are considered reliable.
2. Whenever projects represent the execution of policies already established by headquarters, such as routine replacements, the division is given the power to commit funds within the limits of appropriate controls.
3. Whenever a project requires a total commitment of more than a certain amount, this request is sent to higher levels within the organization. The request is often coupled with a budget limitation regarding the maximum total investment that a division may undertake in a budget period.

To illustrate the idea of a larger investment requiring higher administrative approval, the limitations for a particular firm might be as follows:

If the Total Capital Investment Is. . .		
More Than	But Less Than or Equal to	Then Approval Is Required through
$5,000	$100,000	Plant Manager
100,000	1,000,000	Division Vice-President
1,000,000	2,500,000	President
2,500,000	—	Board of Directors

The purpose of these policies is to streamline the capital expenditure planning and control process by delegating authority to various management levels to approve projects that can be handled effectively at these levels. This streamlining permits top management to concentrate on the most significant capital demands.

Establishing capital expenditure policies is a major responsibility of top management, but responsibility for developing sound economic selection criteria tends to vary by organization. However, regardless of which group develops these criteria, they are applied when a project is proposed and again when it is ready for implementation.

14.8.4 Project Implementation and Postaudit Review

The implementation time for a project may be either short or very long, and responsibility for project implementation customarily rests with division management and the project sponsor. Approximately two to six months before implementation, an appropriations request (AR) must be submitted and approved. During the time that a project is being implemented, a periodic progress report is usually submitted to the proper levels of management. This report is used to ensure that the project is on schedule and that management is aware of any problems that may have arisen. Often a project will have a cost overrun due to the difficulty in estimating future cash flows. Most companies allow some overrun (say, 10%) without requiring that a new AR be submitted.

In most firms, division management is responsible for conducting a postaudit review after a project has attained operational status. (See Step 7 of the engineering economic analysis procedure discussed in Section 1.4.) This review is usually a constructive learning experience that includes a review of project operations and financial performance. The primary objectives of the postaudit appraisal are (1) to determine whether project objectives have been achieved, (2) to discover the degree of conformance to the plan and ascertain where variance occurred, (3) to encourage more careful estimates in the original proposal, and (4) to learn from the results, to identify problems, and to promote better estimates in the future. The postaudit appraisal varies from three months to two years after start-up, but is normally done after one year of operation.

14.8.5 Communication

If project proposals are to be transmitted from one organizational unit to another for review and approval, there must be effective means for communication that may range from standard forms to personal appearances. In communicating proposed projects to higher levels in the management structure, it is desirable to use a format that is as standardized as possible to help ensure uniformity and completeness of information and evaluation. In general, the technical and marketing aspects of each proposed project should be completely described in the manner most appropriate to each individual case. However, financial summaries of all proposals should be standardized so that they may be consistently and fairly evaluated.

▼ 14.9 Summary

This chapter has provided an overview of the capital financing and capital allocation functions as well as the total capital budgeting process. Our discussion of capital financing has dealt with where companies get money to continue to grow and prosper and how much it costs to obtain this capital. Also included was a discussion of the weighted average cost of capital. In this regard, differences between debt capital and owner's (equity) capital were made clear. Leasing as a source of capital was also described, and a lease-versus-purchase example was analyzed.

Our treatment of capital allocation among independent investment opportunities has been built on two important observations. First, the primary concern in capital expenditure activity is to ensure the survival of the company by implementing ideas to maximize future shareholder wealth, which is equivalent to maximization of shareholder PW. Second, engineering economic analysis plays a vital role in deciding which projects are recommended for funding approval and included in a company's overall capital investment portfolio.

▼ 14.10 References

BAUMOL, W. J., and R. E. QUANDT. "Investment and Discount Rates Under Capital Rationing—A Programming Approach," *Economic Journal,* vol. 75, no. 298, June 1965, pp. 317–329.

BERNARD, R. H. "Mathematical Programming Models for Capital Budgeting—A Survey, Generalization, and Critique," *Journal of Financial and Quantitative Analysis,* vol. 4, no. 2, 1969, pp. 111–158.

BUSSEY, L. E., and T. G. ESCHENBACH. *The Economic Analysis of Industrial Projects,* 2nd ed. (Englewood Cliffs, NJ: Prentice Hall, 1992).

GURNANI, C., "Capital Budgeting: Theory and Practice," *The Engineering Economist,* vol. 3, no. 1 (Fall 1984), pp. 19–46.

LEVY, H., and M. SARNAT. *Capital Investment and Financial Decisions,* 2nd ed. (Englewood Cliffs, NJ: Prentice-Hall, 1983).

PARK, C. S., and G. P. SHARPE-BETTE. *Advanced Engineering Economics,* (New York: John Wiley & Sons, Inc., 1990).

WEINGARTNER, H. M. *Mathematical Programming and the Analysis of Capital Budgeting Problems* (Englewood Cliffs, NJ: Prentice-Hall, 1963).

▼ 14.11 Problems

The number in parentheses () that follows each problem refers to the section from which the problem is taken.

14-1. Describe how an organization's capital financing and allocation activities affect the practice of engineering economy. (14.1)

14-2. Why do most engineering economic analyses normally assume that a company's pool of capital is being used to fund a capital project instead of a specific source of capital (e.g., equity versus borrowed funds)? (14.2)

14-3. List five possible sources of funds to a corporation for funding capital projects and continuing operations. (14.2)

14-4. Briefly describe the five basic steps associated with a company's capital budgeting process. (14.8)

14-5.

   a. What is equity capital, and how is it different from debt capital? (14.2)

   b. Why do bondholders, on the average, receive a lower rate of return than do holders of common stock in the same corporation? (14.2, 14.3)

14-6.

   a. List at least four characteristics of a corporation. (14.4)

   b. What may be the advantages to a company of leasing assets? (14.2, 14.6)

14-7.

   a. What is the cost of capital of retained earnings? Why? (14.4)

   b. How should we view the cost of depreciation funds (reserves)? Why? (14.5)

14-8. A corporation sold an issue of 20-year bonds, having a total face value of $5,000,000, for $4,750,000. The bonds bear interest at 10%, payable semiannually. The company wishes to establish a sinking fund for retiring the bond issue and will make semiannual deposits that will earn 8%, compounded semiannually. Compute the semiannual cost for interest and redemption of these bonds. (14.3)

14-9. The Yog Manufacturing Company's common stock is presently selling for $32 per share, and annual dividends have been constant at $2.40 per share. If an investor believes that the price of a share of common stock will grow at 5% per year into the foreseeable future, what is the approximate cost of common stock equity to Yog? What assumptions did you make? (14.4)

14-10. A small corporation having a capitalization of $200,000, represented by 2,000 shares of common stock, has been in operation for five years. During this time, it has paid no dividends in order to be able to finance its growth through retained earnings. It now needs $100,000 in additional capital to finance expansion. It is considering three methods of obtaining the capital: (1) attempting to issue $100,000 in new common stock; (2) borrowing from a bank at 8% interest; and (3) selling five-year bonds bearing interest at 7%, with the restriction that no further indebtedness can be incurred during the life of the bond issue. Discuss briefly the advantages and disadvantages of each method of financing. (14.2, 14.3, 14.4)

14-11. Refer to Problem 14-8. Assume the initial selling expenses of the bond issue is 1.17% of the par (face) value; the annual administrative expense of servicing the bond issue is 3.1% of the annual interest costs; and the corporation's marginal (effective) income-tax rate is 39.6%. Based on this additional information, what is the after-tax cost of capital to the corporation of the bond issue? (14.3)

14-12. Refer to Example 14-5. If annual maintenance expenses can range from $800 to $1,300 per year and inflation can vary from 3% to 8% per year (as shown in the table that follows), determine whether the forklift truck should be purchased or leased for each combination of extreme values. (14.6)

Annual Maintenance	Annual Inflation Rate (%)	Recommendation
$800	3	?
1,300	8	?

14-13. An existing piece of equipment has been performing poorly and needs replacing. More modern equipment can be *purchased* or it can be *leased*. If purchased, the equipment will cost $20,000 and have a depreciable life of five years with no market value. For simplicity, assume straight-line depreciation is used by the firm. Because of improved operating characteristics of the equipment, raw materials savings of $5,000 per year are expected to result relative to continued use of the present equipment. However, annual labor expenses for the new equipment will most likely increase by $2,000 and annual maintenance will go up by $1,000. To lease the new equipment requires a refundable deposit of $2,000, and the end-of-year leasing fee is $6,000. Annual materials savings and extra labor expenses will be the same when purchasing or leasing the equipment, but the lessee company will provide maintenance for its equipment as part of the leasing fee. The after-tax MARR is 15% per year and the effective income-tax rate is 50%. If purchased, it is believed that the equipment can be sold at the end of five years for $1,500 even though $0 was used in calculating straight-line depreciation. Determine whether the company should buy or lease the new equipment, assuming that it has been decided to replace the present equipment. (14.6)

14-14. Determine the more economical means of acquiring a business machine if you may either (a) purchase the machine for $5,000 with a probable resale value of $1,000 at the end of five years or (b) lease the machine at an annual rate of $900 per year for five years with an initial deposit of $500, refundable upon returning the machine in good condition. If you own the machine, assume (for simplicity) you will depreciate it at an annual rate of $800. All lease charges are deductible for income-tax purposes. As the owner or lessor, you will pay all expenses associated with the operation of the machine.

a. Compare these alternatives by using the AW method. The after-tax MARR is 10% per year and the effective income-tax rate is 50%. Do *not* use a tabular method of solution.

b. How high could the annual leasing fees be such that leasing remains the more desirable alternative? (14.6)

14-15. A firm is considering the development of several new products. The products under consideration are listed in the next table. Products in each group are mutually exclusive. At most, one product from each group will be selected. The firm has an MARR of 10% per year and a budget limitation on development costs of $2,100,000. The life of all products is assumed to be 10 years, with no salvage value. Formulate this capital allocation problem as a linear integer programming model. (14.7)

Group	Product	Development Cost	Annual Net Cash Income
A	A1	$500,000	$90,000
	A2	650,000	110,000
	A3	700,000	115,000
B	B1	600,000	105,000
	B2	675,000	112,000
C	C1	800,000	150,000
	C2	1,000,000	175,000

14-16. Four proposals are under consideration by your company. Proposals A and C are mutually exclusive; proposals B and D are mutually exclusive and cannot be implemented unless proposal A *or* C has been selected. No more

than $140,000 can be spent at time zero. The before-tax MARR is 15% per year. The estimated cash flows are shown in the accompanying table. Form all mutually exclusive combinations in view of the specified contingencies, and formulate this problem as an integer linear programming model. (14.7)

End	Proposal			
of Year	A	B	C	D
0	−$100,000	−$20,000	−$120,000	−$30,000
1	40,000	6,000	25,000	6,000
2	40,000	10,000	50,000	10,000
3	60,000	10,000	85,000	19,000

14-17. Three alternatives are being considered for an engineering project. Their cash-flow estimates are shown in the accompanying table. A and B are mutually exclusive, and C is an optional add-on feature to alternative A. Investment funds are limited to $5,000,000. Another constraint on this project is the engineering personnel needed to design and implement the solution. No more than 10,000 person-hours of engineering time can be committed to the project. Set up a linear integer programming formulation of this resource allocation problem. (14.7)

	Alternative		
	A	B	C
Initial investment (10^6)	4.0	4.5	1.0
Personnel requirement (hours)	7,000	9,000	3,000
After-tax annual savings, years one through four (10^6)	1.3	2.2	0.9
PW at 10% per year (10^6)	0.12	2.47	1.85

14-18. Four proposals are under consideration by your company. Proposals A and C are mutually exclusive; proposals B and D are mutually exclusive and cannot be implemented unless proposal A *or* C has been selected. No more than $140,000

TABLE P14-18 Four Proposals for Problem P14-18

	Estimated Cash Flow for Proposals			
End of Year	A	B	C	D
0	−50,000	−20,000	−120,000	−30,000
1	0	10,000	55,000	15,000
2	0	10,000	55,000	15,000
3	83,000	10,000	55,000	15,000
PW(15%)	4,574	2,832	5,577	4,248
IRR	18.4%	23.4%	17.8%	23.4%

can be spent at time zero. The before-tax MARR is 15% per year. Formulate this situation in terms of a linear integer programming problem. Relevant data are provided in Table P14-18. (14.7)

14-19. Refer to the IPC weighted average cost of capital case (Section 14.5.1) and Examples 14-1 through 14-4. Assume IPCs retained earnings in its capital structure remain $4,300,000, and the changes to Examples 14-1 through 14-4 are as shown in Table P14-19. Based on this information, what is the updated after-tax WACC for the Interstate Products Corporation?

TABLE P14-19 Example 14-1 Changes for Problem P14-19

Example	Change(s)
14-1	The three-year loan was for $4,800,000 at an interest rate of 9.1% per year.
14-2	A $15,000,000 issue of 12-year bonds; par value of $10,000 per bond; $r = 5.92\%$ per year; and each bond sold for $10,430.
14-3	After-tax earnings of $1,650,000 per year were realized, and the 1,000,000 shares were originally sold at an average price of $18.40. The future share price is expected to grow 8% per year.
14-4	100,000 shares of preferred stock were sold at a par value of $29 per share.

Dealing with Multiattributed Decisions

*T*he objective of this chapter is to discuss how several relatively straightforward models can be used to evaluate alternatives in a manner that encompasses the monetary and nonmonetary attributes that are integral to almost all real-life decisions.

The following topics are discussed in this chapter:

Examples of multiattributed decisions

Choice of attributes

Selection of a measurement scale

Dimensionality of the problem

Noncompensatory models

Compensatory models

▼ 15.1 Introduction

All chapters up to Chapter 15 have dealt principally with the assessment of equivalent monetary worth of competing alternatives and proposals. As you know, *few* decisions are based strictly on dollars and cents. In this chapter, our attention is directed at how diverse nonmonetary considerations (*attributes*) that arise from multiple objectives can be explicitly included in the evaluation of engineering and business ventures. By *nonmonetary*, we mean that a formal market mechanism does not exist in which value can be established for various aspects of a venture's performance such as aesthetic appeal, employee morale, and environmental enhancement.

Value is difficult to define because it is used in a variety of ways. In fact, in 350 B.C., Aristotle conceived seven classes of value that are still recognized today:

(1) economic, (2) moral, (3) aesthetic, (4) social, (5) political, (6) religious, and (7) judicial. Of these classes, only *economic value* can be measured in terms of (hopefully) objective monetary units such as dollars, yen, or pesos. However, economic value is also established through an item's *use value* (properties that provide a unit of use, work, or service) and *esteem value* (properties that make something desirable). In highly simplified terms, we can say that use values cause a product to perform (e.g., a car serves as a reliable means of transportation) and esteem values cause it to sell (e.g., a convertible automobile has a look of sportiness). Use value and esteem value defy precise quantification in monetary terms, so we often resort to multiattribute techniques for evaluating the total value of complex designs and complicated systems or machinery.

15.2 Examples of Multiattributed Decisions

To provide perspective and motivation for studying the subject of multiattributed decision making, two realistic examples are offered that will introduce other topics to follow.

A common situation encountered by a new engineering graduate is that of selecting his or her first permanent professional job. Suppose that Mary Jones, a 22-year-old graduate engineer, is fortunate enough to have four acceptable job offers in writing. A choice among the four offers must be made within the next four weeks or else they become void. She is a bit perplexed concerning which offer to accept, but she decides to base her choice on these four important factors or attributes (not necessarily listed in their order of importance to her): (1) social climate of the town in which she will be working, (2) the opportunity for outdoor sports, (3) starting salary, and (4) potential for promotion and career advancement. Mary Jones next forms a table and fills it in with objective and subjective data relating to differences among the four offers. The completed table (or matrix) is shown as Table 15-1. Notice that several attributes are rated subjectively on a scale ranging from "poor" to "excellent."

It is not uncommon for monetary and nonmonetary data to be key ingredients in decision situations such as this rather elementary one. Take a minute or two to ponder which offer you would accept, given only the data in Table 15-1. Would starting salary dominate all other attributes, so that your choice would be the Apex

TABLE 15-1 Job Offer Selection Problem

Attributes	Alternatives (Offers and Locations)			
	Apex Corp., New York	Sycon, Inc., Los Angeles	Sigma Ltd., Macon, GA	McGraw-Wesley, Flagstaff, AZ
Social climate	Good	Good	Fair	Poor
Weather/sports	Poor	Excellent	Good	Very good
Starting salary (per annum)	$50,000	$45,000	$49,500	$46,500
Career advancement	Fair	Very good	Good	Excellent

TABLE 15-2 CAD Workstation Selection Problem

Attribute	Vendor A	Vendor B	Vendor C	Reference ("Do Nothing")
Cost of purchasing the system	$115,000	$338,950	$32,000	$0
Reduction in design time	60%	67%	50%	0
Flexibility	Excellent	Excellent	Good	Poor
Inventory control	Excellent	Excellent	Excellent	Poor
Quality	Excellent	Excellent	Good	Fair
Market share	Excellent	Excellent	Good	Fair
Machine utilization	Excellent	Excellent	Good	Poor

Corporation in New York? Would you try to trade off poor social climate against excellent career advancement in Flagstaff to make the McGraw-Wesley offer your top choice?

Many decision problems in industry can be reduced to matrix form similar to that of the foregoing job selection example. To illustrate the wide applicability of such a tabular summary of data, consider a second example involving the choice of a computer-aided design (CAD) workstation by an architectural engineering firm. The data are summarized in Table 15-2. Three vendors and "do nothing" compose the list of feasible alternatives (choices) in this decision problem, and a total of seven attributes is judged sufficient for purposes of discriminating among the alternatives. Aside from the question of which workstation to select, other significant questions come to mind in multiattributed decision making: (1) How are the attributes chosen in the first place? (2) Who makes the subjective judgments regarding nonmonetary attributes such as "quality" or "operating flexibility"? (3) What response is required—a partitioning of alternatives or a rank ordering of alternatives, for instance? Several simple, though workable and credible, models for selecting among alternatives such as those in Tables 15-1 and 15-2 are described in this chapter.

15.3 Choice of Attributes

The choice of attributes by which to judge alternative designs, systems, products, processes, and so on is one of the most important tasks in multiattribute decision analysis. (The most important task, of course, is to identify the feasible alternatives from which to select.) It has been observed that the articulation of attributes for a particular decision can, in some cases, shed enough light on the problem to make the final choice obvious to all involved.

Consider again the data in Tables 15-1 and 15-2. These general observations regarding the attributes used to discriminate among alternatives can immediately be made: (1) each attribute distinguishes at least two alternatives—in no case should identical values for an attribute apply to all alternatives; (2) each attribute captures a unique dimension or facet of the decision problem (i.e., attributes are independent and nonredundant); (3) all attributes, in a collective sense, are assumed to be sufficient for the purpose of selecting the best alternative; and (4) differences in values assigned to each attribute are presumed to be meaningful in distinguishing among feasible alternatives.

In practice, selection of a set of attributes is usually the result of group consensus, and it is clearly a subjective process. The final list of attributes, both monetary and nonmonetary, is therefore heavily influenced by the decision problem at hand, as well as by an intuitive feel for which attributes will or will not pinpoint relevant differences among feasible alternatives. If too many attributes are chosen, the analysis will become unwieldy and difficult to manage. Too few attributes, on the other hand, will limit discrimination among alternatives. Again, judgment is required to decide what number is too few or too many. If some attributes in the final list lack specificity or cannot be quantified, it will be necessary to subdivide them into lower-level attributes that can be measured.

To illustrate these points, we might consider adding an attribute called "cost of operating and maintaining the system" in Table 15–2 to capture a vital dimension of the CAD system's life-cycle cost. The attribute "flexibility" should perhaps be subdivided into two other, more specific attributes such as "ability to interface with computer-aided manufacturing equipment" (such as numerically controlled machine tools) and "ability to create and analyze solid geometry representations of engineering design concepts." Finally, it would be constructive to aggregate two attributes in Table 15-2 namely, "quality" and "market share." Because there is no difference in the values assigned to these two attributes across the four alternatives, they could be combined into a single attribute, possibly named "achievement of greater market share through quality improvements."

15.4 Selection of a Measurement Scale

Identifying feasible alternatives and appropriate attributes represents a large portion of the work associated with a multiattributed decision analysis. The next task is to develop metrics, or measurement scales, that permit various states of each attribute to be represented. For example, in Table 15-1, "dollars" was an obvious choice for the metric of starting salary. A subjective assessment of career advancement was made on a metric having five gradations that ranged from "poor" to "excellent." The gradations were "poor," "fair," "good," "very good," and "excellent." In many problems, the metric is simply the scale upon which a physical measurement is made. For instance, anticipated noise pollution for various routings of an urban highway project might be a relevant attribute whose metric is "decibels."

15.5 Dimensionality of the Problem

If you refer once again to Table 15-1, you will observe that there are two basic ways to process the information presented there. First, you could attempt to collapse each job offer into a single metric, or dimension. For instance, all attributes could somehow be forced into their dollar equivalents or they could be reduced to a *utility equivalent* ranging from, say, 0 to 100. Assigning a dollar value to good career advancement may not be too difficult, but how about placing a dollar value on a poor versus an excellent social climate? Similarly, translating all job offer data to a scale of worth or utility that ranges from 0 to 100 may not be plausible to most individuals. This first way of dealing with the data of Table 15-1 is called *single-dimensioned analysis*. (The dimension corresponds to the number of metrics used to represent the attributes that discriminate among alternatives.)

Collapsing all information into a single dimension is popular in practice because many analysts believe that a complex problem can be made computationally tractable in this manner. In fact, several useful models presented later are single-dimensioned. Such models are termed *compensatory* because changes in the values of a particular attribute can be offset by, or traded off against, opposing changes in another attribute.

The second basic way to process information in Table 15-1 is to retain the individuality of the attributes as the best alternative is being determined. That is, there is no attempt to collapse attributes into a common scale. This is referred to as *full-dimensioned analysis* of the multiattribute problem. For example, if r^* attributes have been chosen to characterize the alternatives under consideration, the predicted values for all r^* attributes are considered in the choice. If a metric is common to more than one attribute, as in Table 15-1, we have an intermediate-dimensioned problem that is analyzed with the same models as a full-dimensioned problem would be. Several of these models are illustrated in the next section, and they are often most helpful in eliminating inferior alternatives from the analysis. We refer to these models as *noncompensatory* because trade-offs among attributes are not permissible. Thus, comparisons of alternatives must be judged on an attribute-by-attribute basis.

15.6 Noncompensatory Models

In this section, we shall examine four noncompensatory models for making a choice when multiple attributes are present. They are (1) dominance, (2) satisficing, (3) disjunctive resolution, and (4) lexicography. In each model an attempt is made to select the best alternative in view of the full dimensionality of the problem. Example 15-1 is presented after the description of these models and will be utilized to illustrate each.

15.6.1 Dominance

Dominance is a useful screening method for eliminating inferior alternatives from the analysis. When one alternative is better than another with respect to all attributes, there is no problem in deciding between them. In this case, the first alternative *dominates* the second one. By comparing each possible pair of alternatives to determine whether the attribute values for one are at least as good as those for the other, it may be possible to eliminate one or more candidates from further consideration or even to select the single alternative that is clearly superior to all the others. Usually it will not be possible to select the best alternative based on dominance.

15.6.2 Satisficing

Satisficing, sometimes referred to as the *method of feasible ranges*, requires the establishment of minimum or maximum acceptable values (the standard) for each attribute. Alternatives having one or more attribute values that fall outside the acceptable limits are excluded from further consideration.

The upper and lower bounds of these ranges establish two fictitious alternatives against which maximum and minimum performance expectations of feasible alternatives can be defined. By bounding the permissible values of attributes from two sides (or from one), information processing requirements are substantially reduced. Restrictions on the domain of acceptable attribute values serve to make the evaluation problem more manageable.

Satisficing is more difficult to use than dominance because of the minimum acceptable attribute values that must be determined. Furthermore, satisficing is usually employed to evaluate feasible alternatives in more detail and to reduce the number being considered rather than to make a final choice. The satisficing principle is frequently used in practice when *satisfactory* performance on each attribute, rather than *optimal* performance, is good enough for decision-making purposes.

15.6.3 Disjunctive Resolution

The disjunctive method is similar to satisficing in that it relies on comparing the attributes of each alternative to the standard. The difference is that the disjunctive method evaluates each alternative on the best value achieved for any attribute. If an alternative has *just one* attribute that meets or exceeds the standard, that alternative is kept. In satisficing, *all* attributes must meet or exceed the standard in order for an alternative to be kept in the feasible set.

15.6.4 Lexicography

This model is particularly suitable for decision situations in which a single attribute is judged to be more important than all other attributes. A final choice *might* be based solely on the most acceptable value for this attribute. Comparing alternatives with respect to one attribute reduces the decision problem to a single dimension (i.e., the measurement scale of the predominant attribute). The alternative having the highest value for the most important attribute is then chosen. However, when two or more alternatives have identical values for the most important attribute, the second most important attribute must be specified and used to break the deadlock. If ties continue to occur, the analyst examines the next most important attribute until a single alternative is chosen or until all alternatives have been evaluated.

Lexicography requires that the importance of each attribute be specified to determine the order in which attributes are to be considered. If a selection is made by using one or a few of the attributes, lexicography does not take into account all the collected data. Lexicography does not require comparability across attributes, but it does process information in its original metric.

EXAMPLE 15-1

Mary Jones, the engineering graduate whose job offers were given in Table 15-1, has decided, based on comprehensive reasoning, to accept the Sigma position in Macon, Georgia. (Problem 15-8 will provide insight into why this choice was made.) Having moved to Macon, several other important multiattribute problems now face Mary Jones. Among them are (1) renting an apartment versus purchasing a small house, (2) what type of automobile or truck to purchase, and (3) whom to select for long-overdue dental work.

In this example, we shall consider the selection of a *dentist* as a means of illustrating noncompensatory (full-dimensioned) and compensatory (single-dimensioned) models for analyzing multiattribute decision problems.

After calling many dentists in the Yellow Pages, Mary finds that there are only four who are accepting new patients. They are Dr. Molar, Dr. Feelgood, Dr. Whoops, and Dr. Pepper. The alternatives are clear to Mary, and she has decided that her objectives in selecting a dentist are to obtain high-quality dental care at a reasonable cost with minimum disruption to her schedule and little (or no) pain involved. In this regard, Mary adopts these attributes to assist in gathering data and making her final choice: (1) reputation of the dentist, (2) cost per hour of dental work, (3) available office hours each week, (4) travel distance, and (5) method of anesthesia. Notice that these attributes are more or less independent in that the value of one attribute cannot be predicted by knowing the value of any other attribute.

Mary collects data by interviewing the receptionist in each dental office, talking with local townspeople, calling the Georgia Dental Association, and so on. A summary of information gathered by Mary is presented in Table 15-3.

We are now asked to determine whether a dentist can be selected by using (a) dominance, (b) satisficing, (c) disjunctive resolution, and (d) lexicography.

TABLE 15-3 Summary Information for Choice of a Dentist

Attribute	Alternatives			
	Dr. Molar	Dr. Feelgood	Dr. Whoops	Dr. Pepper
Cost ($/hr)	$50	$80	$20	$40
Method of anesthesia[a]	Novocaine	Acupuncture	Hypnosis	Laughing Gas
Driving distance (mi)	15	20	5	30
Weekly office hours	40	25	40	40
Quality of work	Excellent	Fair	Poor	Good

☐ Best value ☐ Worst value

[a] Mary has decided that novocaine > laughing gas > acupuncture > hypnosis, where $a > b$ means that a is preferred to b.

TABLE 15-4 Check for Dominance Among Alternatives

	Paired Comparison					
Attribute	Molar vs. Feelgood	Molar vs. Whoops	Molar vs. Pepper	Feelgood vs. Whoops	Feelgood vs. Pepper	Whoops vs. Pepper
Cost	Better	Worse	Worse	Worse	Worse	Better
Anesthesia	Better	Better	Better	Better	Worse	Worse
Distance	Better	Worse	Better	Worse	Better	Better
Office hours	Better	Equal	Equal	Worse	Worse	Equal
Quality	Better	Better	Better	Better	Worse	Worse
Dominance?	Yes	No	No	No	No	No

SOLUTION

(a) To check for dominance in Table 15-3, pairwise comparisons of each dentist's set of attributes must be inspected. There will be $4(3)/2 = 6$ pairwise comparisons necessary for the four dentists, and they are shown in Table 15-4. It is clear from Table 15-4 that Dr. Molar dominates Dr. Feelgood, so Dr. Feelgood will be dropped from further consideration. With dominance, it is not possible for Mary to select the best dentist.

(b) To illustrate the satisficing model, acceptable limits (feasible ranges) must be established for each attribute. After considerable thought, Mary comes up with the feasible ranges given in Table 15-5

   Comparison of attribute values for each dentist against the feasible range reveals that Dr. Whoops uses a less desirable type of anesthesia (hypnosis < acupuncture), and his quality rating is also not acceptable (poor < good). Thus, Dr. Whoops joins Dr. Feelgood on Mary's list of rejects. Notice that satisficing, by itself, did not produce the best alternative.

TABLE 15-5 Feasible Ranges for Satisficing

Attribute	Minimum Acceptable Value	Maximum Acceptable Value	Unacceptable Alternative
Cost	—	$60	None (Dr. Feelgood already eliminated)
Anesthesia	Acupuncture	—	Dr. Whoops
Distance (miles)	—	30	None
Office hours	30	40	None (Dr. Feelgood already eliminated)
Quality	Good	Excellent	Dr. Whoops

TABLE 15-6 Ordinal Ranking of Dentists' Attributes

A. Results of Paired Comparisons

Cost > anesthesia	(Cost is more important than anesthesia)
Quality > cost	(Quality is more important than cost)
Cost > distance	(Cost is more important than distance)
Cost > office hours	(Cost is more important than office hours)
Anesthesia > distance	(Anesthesia is more important than distance)
Anesthesia > office hours	(Anesthesia is more important than office hours)
Quality > anesthesia	(Quality is more important than anesthesia)
Office Hours > distance	(Office hours are more important than distance)
Quality > distance	(Quality is more important than distance)
Quality > office hours	(Quality is more important than office hours)

B. Attribute	Number of times on left of > (= Ordinal ranking)
Cost	3
Anesthesia	2
Distance	0
Office hours	1
Quality	4

(c) By applying the feasible ranges in Table 15-5 to the disjunctive resolution model, all dentists would be acceptable because each has at least one attribute value that meets or exceeds the minimum expectation. For instance, Dr. Whoops scores acceptably on three out of five attributes and Dr. Feelgood passes two out of five minimum expectations. Clearly, this model does not discriminate well among the four candidates.

(d) Many models, including lexicography, require that all attributes first be ranked in order of importance. Perhaps the easiest way to obtain a consistent ordinal ranking is to make paired comparisons between each possible attribute combination.* This is illustrated in Table 15-6. Each attribute can be ranked according to the number of times it appears on the left-hand side of the comparison when the preferred attribute is placed on the left as shown. In this case, the ranking is found to be quality > cost > anesthesia > office hours > distance.

Table 15-7 illustrates the application of lexicography to the ordinal ranking developed in Table 15-6. The final choice would be Dr. Molar because quality is the top-ranked attribute and Molar's quality rating is the best of all. If Dr. Pepper's work quality had also been rated as excellent, the choice would be made on the basis of cost. This would have resulted in the selection of Dr. Pepper. Therefore, lexicography does allow the best dentist to be chosen by Mary.

---

*An ordinal ranking is simply an ordering of attributes from most preferred to least preferred.

TABLE 15-7	Application of Lexicography	
Attribute	Rank[a]	Alternative Rank[b]
Cost	3	Whoops > Pepper > Molar > Feelgood
Anesthesia	2	Molar > Pepper > Feelgood > Whoops
Office hours	1	Molar = Whoops = Pepper > Feelgood
Distance	0	Whoops > Molar > Feelgood > Pepper
Quality	4	Molar > Pepper > Feelgood > Whoops

[a] Rank of 4 = most important, rank of 0 = least important.

[b] Selection is based on the highest ranked attribute (Whoops and Feelgood included only to illustrate the full procedure).

15.7 Compensatory Models

The basic principle behind all compensatory models, which involve a single dimension, is that the values for all attributes must be converted to a common measurement scale such as *dollars* or *utiles*.* When this is done, it is possible to construct an overall dollar index or utility index for each alternative. The form of the function used to calculate the index can vary widely. For example, the converted attribute values may be added together, they may be weighted and then added, or they may be sequentially multiplied. Regardless of the functional form, the end result is that good performance in one attribute can *compensate* for poor performance in another. This allows trade-offs among attributes to be made during the process of selecting the best alternative. Because lexicography involves no trade-offs, it was classified as a full-dimensional model in Section 15.6.4.

In this section, we examine three compensatory models for evaluating multi-attribute decision problems. The models are (1) nondimensional scaling, (2) the Hurwicz procedure, and (3) the additive weighting technique. Each model will be illustrated using the data of Example 15-1.

15.7.1 Nondimensional Scaling

A popular way to standardize attribute values is to convert them to nondimensional form. There are two important points to consider when doing this. First, the nondimensional values should all have a common range, such as 0 to 1 or 0 to 100. Without this constraint, the dimensionless attributes will contain implicit weighting factors. Second, all of the dimensionless attributes should follow the same trend with respect to desirability; the most preferred values should be either all small or all large. This is necessary in order to have a believable overall scale for selecting the best alternative.

Nondimensional scaling can be illustrated with the data of Example 15-1. As shown in Table 15-8, the preceding constraints may require that different procedures be used to *nondimensionalize* each attribute. For example, a cost-related at-

---

* A utile is a dimensionless unit of worth.

TABLE 15-8	Nondimensional Scaling for Example 15-1		
Attribute	Value	Rating Procedure	Dimensionless value
Cost	$20	$(80 - \text{cost})/60$	1.0
	40		0.67
	50		0.50
	80		0.0
Anesthesia	Hypnosis	$(\text{Relative rank}^a - 1)/3$	0.0
	Acupuncture		0.33
	Laughing gas		0.67
	Novocaine		1.0
Distance	5	$(30 - \text{distance})/25$	1.0
	15		0.60
	20		0.40
	30		0.0
Office Hours	25	$(\text{Office hours} - 25)/15$	0.0
	40		1.0
Quality	Poor	$(\text{Relative rank}^a - 1)/3$	0.0
	Fair		0.33
	Good		0.67
	Excellent		1.0

[a] Scale of 1 to 4, is used, 4 being the best (from Table 15-3).

tribute is best when it is low, but office hours are best when they are high. The goal should be to devise a nondimensionalizing procedure that rates each attribute in terms of its fractional accomplishment of the best attainable value. Table 15-3, the original table of information for Example 15-1, is restated in dimensionless terms in Table 15-9. The general procedure for converting the original data in Table 15-3 for a particular attribute to its dimensionless rating is

$$\text{Rating} = \frac{\text{worst outcome} - \text{outcome being made dimensionless}}{\text{worst outcome} - \text{best outcome}}. \quad (15\text{-}1)$$

Equation (15-1) applies when large numerical values, such as dollars or driving distance are considered to be *undesirable*. However, when large numerical values

TABLE 15-9	Nondimensional Data for Example 15-1			
Attribute	Dr. Molar	Dr. Feelgood	Dr. Whoops	Dr. Pepper
Cost	0.50	0.0	1.0	0.67
Method of anesthesia	1.0	0.33	0.0	0.67
Driving distance	0.60	0.40	1.0	0.0
Weekly office hours	1.0	0.0	1.0	1.0
Quality of work	1.0	0.33	0.0	0.67

are considered to be *desirable* (e.g., a rating with "4" as best and "1" as worst), the relationship for converting original data to their dimensionless ratings is

$$\text{Rating} = \frac{\text{outcome being made dimensionless} - \text{worst outcome}}{\text{best outcome} - \text{worst outcome}}. \quad (15\text{-}2)$$

If all the attributes in Table 15-9 are of equal importance, a score for each dentist could be found by merely summing the nondimensional values in each column. The results would be Dr. Molar = 4.10, Dr. Feelgood = 1.06, Dr. Whoops = 3.00, and Dr. Pepper = 3.01. Presumably, Dr. Molar would be the best choice in this case.

15.7.2 The Hurwicz Procedure

Nondimensional attribute values may be utilized in a variety of ways. The most pessimistic approach is to assume that each alternative is only as good as its lowest performing attribute. The goal would then be to pick the alternative with the most favorable value of its worst attribute (i.e., the *max*imum value of its *min*imum attribute). The left-hand column of Table 15-10 illustrates this for the data of Example 15-1, and Dr. Molar would be chosen by this procedure, which is called the *maximin rule*.

At the other extreme, one could be very optimistic and choose the alternative with the most favorable value of its best attribute (i.e., the *max*imum value of its *max*imum attribute). This rule, termed *maximax*, is illustrated on the righthand side of Table 15-10. Ties with either the maximin or maximax rule can be resolved by considering the second worst or best attribute, respectively, and so on until only one alternative remains. In Table 15-10 the maximax rule leads to Dr. Pepper as the preferred choice.

The *Hurwicz procedure* provides a means of reaching an intermediate level between the pessimism of maximin and the optimism of maximax. It is based on an index of optimism, α, which is chosen to reflect the decision maker's relative attitude. For example, α could be set equal to 0 for pure pessimism and equal to 1 for pure optimism. Values between 0 and 1 would reflect intermediate attitudes.

TABLE 15-10 Maximin and Maximax Rules Applied to Nondimensional Data

Alternative	Value of Worst Attribute (Table 15-9)	Value of Best Attribute (Table 15-9)	Value of Second-Best Attribute[a]
Dr. Molar	0.50	1.0	0.60
Dr. Feelgood	0.0	0.40	0.33
Dr. Whoops	0.0	1.0	0.0
Dr. Pepper	0.0	1.0	0.67

[a] When alternatives have more than one attribute with the maximum value, the second-best attribute can be selected in two different ways: (1) The maximum value can simply be repeated for alternatives where it occurs more than once, or (2) the next highest value may be selected instead. The latter method was used in this table.

TABLE 15-11 The Hurwicz Procedure Applied to Example 15-1			
Alternative	Value of Worst Attribute (Table 15-9)	Value of Best Attribute (Table 15-9)	Weighted Sum[a]
Dr. Molar	0.50	1.0	0.75
Dr. Feelgood	0.0	0.40	0.20
Dr. Whoops	0.0	1.0	0.50
Dr. Pepper	0.0	1.0	0.50

[a]Weighted sum for each alternative = α (value of best attribute) + $(1 - \alpha)$ (value of worst attribute), where $\alpha = 0.50$.

The optimism index is then used to weight the outcomes of maximin and maximax. The best alternative is chosen on the basis of the weighted sum.

Table 15-11 illustrates the Hurwicz procedure for $\alpha = 0.5$. The value of α can be varied as shown in Figure 15-1 to analyze the sensitivity of the selection of Dr. Molar, who is judged best in Table 15-11. With the Hurwicz procedure, Dr. Molar dominates all the other candidates.

An important criticism of these methods is that there is no attempt to include differential importance weightings among attributes. Up to this point, attributes have been equally weighted. Comparisons are made only on the basis of best and worst values, which usually represent different attributes from one alternative to another. This leads to some exaggerated comparisons in Mary's choice of a dentist. A good example of such an extreme comparison would be the equal ranking of Drs. Whoops and Pepper in terms of their lowest performing attributes. (See Table 15-9.) Dr. Pepper's poorest attribute is driving distance, while Dr. Whoops is poorest in quality of work and method of anesthesia. As a prospective patient, Mary would probably be much more concerned with the quality of work and min-

Figure 15-1

Sensitivity of Selection by the Hurwicz Procedure to Changes in α

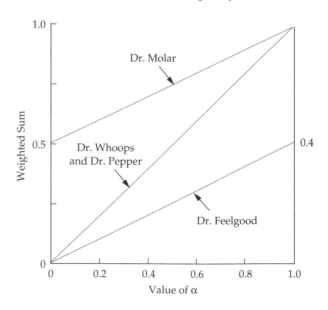

imizing of pain than with driving distance. Also, the Hurwicz procedure does not allow for making trade-offs among attributes.

15.7.3 The Additive Weighting Technique

Additive weighting provides for the direct use of nondimensional attributes such as those in Table 15-9 and the results of ordinal ranking as illustrated in Table 15-6. The procedure involves development of weights for attributes (based on ordinal rankings) that can be multiplied by the appropriate nondimensional attribute values to produce a *partial contribution* to the overall score for a particular alternative. When the partial contributions of all attributes are summed, the resulting set of alternative scores can be used to compare alternatives directly.

Attribute weights should be determined in two steps following the completion of ordinal ranking. First, relative weights are assigned to each attribute according to its ordinal ranking. The simplest procedure is to use rankings of 1, 2, 3, . . . , based on position, with higher numbers signifying greater importance; but one might also include subjective considerations by using uneven spacing in some cases. For instance, in a case where there are four attributes, two of which are much more important than the others, the top two may be rated as 7 and 5 instead of 3 and 4. The second step is to normalize the relative ranking numbers. This can be done by dividing each ranking number by the sum of all the rankings. Table 15-12 summarizes these steps for Example 15-1 and demonstrates how the overall score for each alternative is determined.

Additive weighting is probably the most popular single-dimensional method because it includes both the performance ratings and the importance weights of each attribute when evaluating alternatives. Furthermore, it produces recommendations that tend to agree with the intuitive feel of the decision maker concerning the best alternative. Perhaps its biggest advantage is that nondimensionalizing data and weighting attributes are separated into two distinct steps. This reduces confusion and allows for precise definition of each of these contributions. From Table 15-12, is apparent that the additive weighting score for Dr. Molar (0.84) makes him the top choice as Mary's dentist.

EXAMPLE 15-2

To illustrate an application of the additive weighting technique, consider a decision problem involving the choice of material for the wing spans of a new commercial aircraft. Suppose the General Aviation aircraft company has reduced its selection of wing span material to two alternatives generally viewed as superior to other options. Now the task facing engineers is to recommend the better material.

The first alternative is aluminum alloy and the second is a composite (an epoxy-resin reinforced by fibers of boron). In Table 15-13, the additive weighting technique was utilized to determine the relative worth (W) of the alternatives, with 100 representing the best performance possible. The costs (C) of materials for each alternative were also estimated, resulting in Table 15-14. In view of the information given, which material should be selected for the wing spans?

TABLE 15-12 The Additive Weighting Technique Applied to Example 15-1

| | Calculation of Weighting Factors | | Calculation of Scores for Each Alternative[b] | | | | | | | |
| | Step 1: | Step 2: | Dr. Molar | | Dr. Feelgood | | Dr. Whoops | | Dr. Pepper | |
Attribute	Relative Rank[a]	Normalized Weight (A)	(B)	(A) × (B)	(B)	(A) × (B)	(B)	(A) × (B)	(B)	(A) × (B)
Cost	4	4/15 = 0.27	0.50	0.14	0.00	0.00	1.00	0.27	0.67	0.18
Anesthesia	3	3/15 = 0.20	1.00	0.20	0.33	0.07	0.00	0.00	0.67	0.13
Distance	1	1/15 = 0.07	0.60	0.04	0.40	0.03	1.00	0.07	0.00	0.00
Office hours	2	2/15 = 0.13	1.00	0.13	0.00	0.00	1.00	0.13	1.00	0.13
Quality	5	5/15 = 0.33	1.00	0.33	0.33	0.11	0.00	0.00	0.67	0.22
	Sum = 15	Sum = 1.00		Sum = 0.84		Sum = 0.21		Sum = 0.47		Sum = 0.66

[a] Based on Table 15-6, relative rank = ordinal ranking + 1. A rank of 5 is best.
[b] Data in Column B are from Table 15-9.

TABLE 15-13 Multiattribute Analysis of Worth of Material

		Aluminum Alloy		Composite	
Attributes	Attribute Weights	Performance	Weighted Value	Performance	Weighted Value
Corrosion Resistance	0.15	50	7.5	90	13.5
Fatigue Resistance	0.20	80	16.0	70	14.0
Weight	0.45	50	22.5	100	45.0
Strength	0.20	30	6.0	90	18.0
Worth (W)			52.0		90.5[a]

[a] A higher score is preferred; the maximum score is 100.

TABLE 15-14 Cost Estimates for Materials

Alternative	Cost
Aluminum Alloy	$1,000,000
Composites	$1,200,000

SOLUTION

In this exercise, the minimum attainable cost and the maximum allowable cost selected by the evaluation team were $500,000 and $1,500,000, respectively. Furthermore, the cost factors in Table 15-15 result from Equation 15-1 by assuming that dollars are linearly scaled between $500,000 and $1,500,000.

With this information, the team then computed a value index of worth divided by cost (W/C) for both alternatives (Table 15-15). The final assessment reveals that the aluminum alloy with value index of 1.04 is acceptable, but the better material is composites with an *incremental* W/C of $|38.5 \div -20| = 1.92 (\geq 0)$.* Therefore, the composites alternative should be recommended.

TABLE 15-15 Computing Value Index

Alternative	Worth (W)	Cost Factor (C)[a]	Value Index = W/C
Aluminum Alloy	52.0	50.0	1.04
Composites	90.5	30.0	1.92[b]

[a] For aluminum, $100 \times \left[\frac{\$1,500,000 - \$1,000,000}{\$1,500,000 - \$500,000} \right] = 50.0$.

For composites, $100 \times \left[\frac{\$1,500,000 - \$1,200,000}{\$1,500,000 - \$500,000} \right] = 30.0$.

[b] Ratio of incremental W divided by incremental C.

*The absolute value of the ratio is correct because a large material cost has a small cost factor. (See Equation 15-1.) If the composites cost factor had been larger than 50, composites would have *dominated* aluminum alloy (i.e., there would have been no trade-off between W and C).

Companion Web Site (http://www.prenhall.com/sullivan_engineering/): Dealing with multiple attributes in a complex problem may seem like a juggling act to many engineers. Visit the Web site to view an example of a multiattributed decision analysis of alternative designs for mercury switches. This example shows an application of the *additive weighting technique* that considers the attributes of life-cycle cost, environmental effects, safety, and ease of use.

15.8 Summary

Several methods have been described for dealing with multiattribute decisions. Some key points are as follows:

1. When it is desired to maximize a single criterion of choice, such as PW, the evaluation of multiple alternatives is relatively straightforward.
2. For any decision, the objectives, available alternatives, and important attributes must be clearly defined at the beginning. Construction of a decision matrix such as that in Table 15-1 helps to systematize this process.
3. Decision making can become quite convoluted when multiple objectives and attributes must be included in an engineering economy study.
4. Multiattribute models can be classified as either multidimensional or single-dimensional. Multidimensional techniques analyze the attributes in terms of their original metrics. Single-dimensional techniques reduce attribute measurements to a common measurement scale.
5. Multidimensional, or noncompensatory, models are most useful for the initial screening of alternatives. In some instances, they can be used to make a final selection, but this usually involves a high degree of subjectivity. Of the multidimensional methods discussed, dominance is probably the least selective, while satisficing is probably the most selective.
6. Single-dimensional, or compensatory, models are useful for making a final choice among alternatives. The additive weighting technique allows superb performance in some attributes to compensate for poor performance in others.
7. When dealing with multiattribute problems that have many attributes and alternatives to be considered, it is advisable to apply a combination of several models in sequence for the purpose of reducing the selection process to a manageable activity.

15.9 References

Canada, J., and W. Sullivan. *Economic and Multiattribute Evaluation of Advanced Manufacturing Systems* (Englewood Cliffs, NJ, Prentice-Hall, Inc., 1989).

Cochrane, J. L., and M. Zeleny. *Multiple Criteria Decision Making*, Columbia, S.C., University of South Carolina Press, 1973.

Falkner, C., and S. Benhajla. "Multi-Attribute Decision Models in the Justification of CIM Systems," *The Engineering Economist*, vol. 35, no. 2, Winter 1990, pp. 91–114.

FRAZELLE, E. "Suggested Techniques Enable Multi-Criteria Evaluation of Material Handling Alternatives," *Industrial Engineering*, vol. 17, no. 2, February 1985, pp. 42–48.

HUANG, P., and P. GHANDFOROUSH. "Procedures Given for Evaluating, Selecting Robots," *Industrial Engineering*, vol. 16, no. 4, April 1984, pp. 44–48.

MACCRIMMON, K. R. "Decision Making Among Multiple Attribute Alternatives: A Survey and Consolidated Approach," Memo RM-4823-ARPA. Rand Corporation, December 1968.

SAATY, T. "Decision Making, Scaling, and Number Crunching," *Decision Sciences*, vol. 20, no. 2, Spring 1989, pp. 404–409.

SAATY, T. "Priority Setting in Complex Problems," *IEEE Transactions on Engineering Management*, vol. EM-30, no. 3, August 1983, pp. 140–155.

WEBER, STEPHEN F. "Automation: Decision Support Software for Automated Manufacturing Investments," No. N1ST1R89-4116. Washington, D.C.: U.S. Department of Commerce, August 1989.

ZELENY, M. *Multiple Criteria Decision Making* (New York: McGraw-Hill, 1982).

15.10 Problems

The number in parentheses () that follows each problem refers to the section from which the problem is taken.

15-1. Suppose that you have received your bachelor's degree, wish to earn a master's degree, and are attempting to decide in which graduate school to enroll. Your age, background, undergraduate area of emphasis, monetary situation, and so on are all valid inputs for your decision. Define six attributes you would use in the selection of a graduate school and rank them in order of importance. Assign approximate weights to the attributes by using a method discussed in this chapter. Be prepared to defend your position. (15.3, 15.7)

15-2. List two advantages and two disadvantages of noncompensatory models for dealing with multiattributed decision problems. Do the same for compensatory models. (15.6, 15.7)

15-3. Discuss ways in which satisficing and the Hurwicz procedure might be used in group decision-making exercises. (15.6, 15.7)

15-4. Discuss some of the difficulties of developing nonlinear functions for nondimensional scaling of qualitative (subjective) data. (15.7)

15-5. Given the matrix of outcomes in Table P15-5 for alternatives and attributes (with higher numbers being better), show what you can conclude using the following methods: (15.6)

TABLE P15-5 Matrix of Outcomes for Problem P15-5

	Alternative				
Attribute	1	2	3	Ideal	(Minimum Acceptable)
A	60	75	90	100	70
B	7	7	8	10	6
C	Poor	Excellent	Fair	Excellent	Good
D	7	8	8	10	6

TABLE P15-6 Data for Problem P15-6

	Alternative				
Attribute	Vendor I	Vendor II	Vendor III	Retain Existing System	Worst Acceptable Value
A. Reduction in throughput time	75%	60%	84%	—	50%
B. Flexibility	Good	Excellent	Good	Poor	Good
C. Reliability	Excellent	Good	Very good	—	Good
D. Quality	Good	Excellent	Excellent	Fair	Good
E. Cost of system (PW of life-cycle cost)	$270,000	$310,000	$214,000	$0	$350,000

Pairwise comparisons:

1. $A < B$ 6. $B > D$
2. $A = C$ 7. $B < E$
3. $A < D$ 8. $C < D$
4. $A < E$ 9. $C < E$
5. $B < C$ 10. $D < E$

a. Satisficing;
b. Dominance;
c. Lexicography, with rank order of attributes $D > C > B > A$.

15-6. With references to the data provided in Table P15-6, recommend the preferred alternative by using (a) dominance, (b) satisficing, (c) disjunctive resolution, and (d) lexicography. (15.6)

15-7. Three large industrial centrifuge designs are being considered for a new chemical plant.

   a. By using the data in Table P15-7, recommend a preferred design with each method that was discussed in this chapter for dealing with nonmonetary attributes.

   b. How would you modify your analysis if two or more of the attributes were found to be dependent (e.g., maintenance and product quality) (15.6, 15.7)

15-8. Mary Jones utilized the additive weighting technique to select a job with Sigma Ltd. in Macon, Georgia. The importance weights she placed on the four attributes in Table 15-1 were social climate = 1.00, starting salary = 0.50, career advancement = 0.33, and weather/sports = 0.25. Nondimensional values given to her ratings in Table 15-1 were excellent = 1.00, very good = 0.70, good = 0.40, fair = 0.25, and poor = 0.10.

TABLE P15-7 Data for Problem P15-7

		Design			
Attribute	Weight	A	B	C	Feasible Range
Initial cost	0.25	$140,000	$180,000	$100,000	$80,000-$180,000
Maintenance	0.10	Good	Excellent	Fair	Fair–excellent
Safety	0.15	Not known	Good	Excellent	Good–excellent
Reliability	0.20	98%	99%	94%	94-99%
Product quality	0.30	Good	Excellent	Good	Fair-excellent

TABLE P15-9 Data for Problem P15-9		
Attributes	Alignment A	Alignment B
Monetary		
Land	$4,044,662	$4,390,000
Bridges	10,134,000	8,701,000
Pavement	4,112,500	4,462,500
Grade and drainage	7,050,000	7,650,000
Erosion control	470,000	510,000
Clearing and replanting	188,000	204,000
Total	$25,999,162	$25,917,500
Miscellaneous		
Route length	4.7 miles	5.1 miles
Maintenance	Moderate (6)	High (3)
Noise pollution	Very good (6)	Good (5)
Cost savings (on gas)	Excellent	Poor
Access ability to another major roadway	U.S. Highway 41	None
Impact on wildlife	Little	Little
Relocation of residences	2	3
Road condition	Flat	Hilly

a. Normalize Mary's importance weights.

b. Develop nondimensional values for the starting salary attribute.

c. Use the results of (a) and (b) in a decision matrix to see if Mary's choice was consistent with the results obtained from the additive weighting technique. (15.7)

15-9. Two highway alignments have been proposed for access to a new manufacturing plant. Based on the data and information given in Table P15-9, a comparison of the two alignments can be made. The better one must be chosen as the proposed highway alignment that will be the connector between an interstate highway and the proposed site. The following attributes and data are to be taken into consideration:

a. grades, curve, and terrain;

b. rivers, creeks, lakes, and sinkholes;

c. roadway crossings;

d. route length;

e. churches, cemetaries, and residential areas;

f. noise, air, and water pollution.

Use any model presented in this chapter to recommend a highway alignment. Show all work. (15.6, 15.7)

15-10. Use all the multiattributed models of this chapter to decide which of the following job offers you should accept. Attempt to develop the data where question marks appear for your particular situation in view of the attributes shown. (See Table P15-10.) (15.6, 15.7)

15-11.

a. Use the additive weighting technique to select one of the three used automobiles for which data are given in Table P15-11. State your assumptions regarding miles driven each year, life of the automobile (how long you would keep it), market (resale) value at end of life, interest cost, price of fuel, cost of annual maintenance, and other subjectively based determinations. (15.7)

b. Use the data developed in (a) and the Hurwicz procedure with $\alpha = 0.70$ to select the automobile you should buy. Do your answers in (a) and (b) agree? Explain why they should (or should not) agree. (15.7)

TABLE P15-10 Job Offer Data for Problem P15-10

Attribute	Job Offer 1	Job Offer 2	Job Offer 3	Weight	Feasible Range
Location	Phoenix	Buffalo	Raleigh	?	—
Annual salary	$46,000	$47,500	$43,000	?	?
Proximity to relatives	?	?	?	?	?
Quality of leisure time	?	?	?	?	?
Promotion potential	Fair	Excellent	Excellent	?	?
Commuting time/day	1 hr	1 1/2 hr	1/2 hr	?	?
Fringe benefits	Excellent	Very good	Good	?	?
Type of work	Factory	Hospital	Government	?	?

15-12. You have volunteered to serve as a judge in a Midwestern contest to select Sunshine, the most wholesome pig in the world. Your assessments of the four finalists for each of the attributes used to distinguish among semifinalists are shown in Table P15-12.

a. Use dominance, feasible ranges, lexicography, and additive weighting to select your winner. Develop your own feasible ranges and weights for the attributes. (15.6, 15.7)

b. If there were two other judges, discuss how the final selection of this year's Sunshine might be made. (15.7)

15-13. You have decided to buy a new compact automobile and are willing to spend a maximum of $20,000 from your savings account. (Money not spent will remain in the account, where it earns effective interest of 12% per year). The choice has been narrowed to three cars having attribute values shown in Table P15-13.

Use four methods for dealing with nonmonetary attributes (dominance, feasible ranges, lexicography, and additive weighting) and determine whether a selection can be made with each. You will need to develop additional data that reflect your preferences. (15,6, 15.7)

TABLE P15-11 Data for the Three Used Automobiles for Problem P15-11

Attribute	Alternative Domestic 1	Alternative Domestic 2	Alternative Foreign
Price	$8,400	$10,000	$9,300
Gas mileage	25 mpg	30 mpg	35 mpg
Type of fuel	Gasoline	Gasoline	Diesel
Comfort	Very good	Excellent	Excellent
Aesthetic appeal	5 out of 10	7 out of 10	9 out of 10
Number of passengers	4	6	4
Ease of servicing	Excellent	Very good	Good
Performance on road	Fair	Very good	Very good
Stereo system	Poor	Good	Excellent
Ease of cleaning upholstery	Excellent	Very good	Poor
Luggage space	Very good	Excellent	Poor

TABLE P15-12 Assessment of Four Finalists for Problem P15-12

Attribute	Contestant			
	I	II	III	IV
Facial quality	Cute but plump	Sad eyes, great snout	Big lips, small ears	A real killer!
Poise[a]	10	8	8	3
Body tone[a]	5	10	7	8
Weight (lb)	400	325	300	380
Coloring	Brown	Spotted, black and white	Gray	Brown and white
Disposition	Friendly	Tranquil	Easily excited	Sour

[a] Data scaled from 1 to 10, with 10 being the highest possible rating.

TABLE P15-13 Attributes for Four Cars for Problem P15-13

Attribute	Alternative			Feasible Range
	Domestic 1	Domestic 2	Foreign	
Negotiated price	$18,400	$20,000	$19,300	$0-$20,000
Gas mileage (average)	25 mpg	30 mpg	35 mpg	20-50 mpg
Type of fuel	Gasoline	Gasoline	Diesel	Gasoline or diesel
Comfort	Very good	Excellent	Fair	Fair–excellent
Aesthetic appeal	4 out of 10	8 out of 10	9 out of 10	4-10
Number of passengers	6	6	4	2-6
Ease of servicing	Excellent	Very good	Good	Fair–excellent
Performance on road	Fair	Excellent	Very good	Fair–excellent

15-14. The additive weighting models a decision tool that aggregates information from different independent criteria to arrive at an overall score for each course of action being evaluated. The alternative with the highest score is preferred.

The general form of the model is

$$V_j = \sum_{i=1}^{n} w_i x_{ij},$$

where

V_j = the score of the jth alternative

w_i = the weight assigned to the ith decision attribute ($1 \le i \le n$)

x_{ij} = the rating assigned to the ith attribute, which reflects the performance of alternative j relative to maximum attainment of the attribute.

Consider Table P15-14 in view of these definitions and determine the value of each "?" shown. (15.7)

TABLE P15-14   Data for Problem P15-14

| i | w_i | Rank | Decision Factor | | Alternative j | |
					(1) Keep Existing Machine Tool	(2) Purchase a New Machine Tool
1	1.0	1	Annual cost of ownership (capital recovery cost)	Rank	?	?
				x_{ij}	1.0	0.7
2	?	4	Flexibility in types of jobs scheduled	Rank	2	1
				x_{ij}	0.8	1.0
3	0.8	2	Ease of training and operation	Rank	1	2
				x_{ij}	?	0.5
4	0.7	?	Time savings per part produced	Rank	2	1
				x_{ij}	0.7	1.0
				V_j	2.69	2.30
				V_j (normalized)	1.00	?

*A*ppendixes

*A*ccounting and Its Relationship to Engineering Economy

▼ A.1 Introduction

Engineering economy studies are made for the purpose of determining whether capital should be invested in a project or whether it should be used differently than it presently is being used. These studies always deal, at least for one of the alternatives being considered, with something that currently is not being done. Such studies provide information upon which investment and managerial decisions about future operations can be based. Thus, the engineer doing an economic analysis might be termed an *alternatives fortune teller.*

After a decision to invest capital in a project has been made and the capital has been invested, those who supply and manage the capital want to know the financial results. Therefore, procedures are established so that financial events relating to the investment can be recorded and summarized and financial productivity determined. At the same time, through the use of proper financial information, controls can be established and used to help guide the venture toward the desired financial goals. Financial accounting and cost accounting are the procedures that provide these necessary services in a business organization. Accounting studies thus are concerned with *past* and *current* financial events. Thus, the accountant might be termed a *financial historian.*

The accountant is somewhat like a data recorder in a scientific experiment. Such a recorder reads the pertinent gauges and meters and records all the essential data during the course of an experiment. From these, it is possible to determine the results of the experiment and to prepare a report. Similarly, the accountant records all significant financial events connected with an investment, and from these data he or she can determine what the results have been and can prepare financial reports. By taking cognizance of what is happening during the course of an experiment and making suitable corrections—thereby gaining more information and better results from the experiment—engineers and managers must rely on accounting reports to make corrective decisions in order to improve the current and future financial performance of the business.

Accounting is generally a source of much of the past financial data needed to make estimates of future financial conditions. Accounting is also a prime source of data for *postmortem,* or after-the-fact, analyses that might be made regarding how well an investment project has turned out compared to the results that were predicted in the engineering economy study.

A proper understanding of the origins and meaning of accounting data is needed in order to use or not use those data properly in making projections into the future and in comparing actual versus predicted results.

A.2 Accounting Fundamentals

Accounting is often referred to as the language of business. Engineers should make serious efforts to learn about a firm's accounting practice so that they can better communicate with top management. This section contains an extremely brief and simplified exposition of the elements of financial accounting in recording and summarizing transactions affecting the finances of the enterprise. These fundamentals apply to any entity (such as an individual or a corporation) called here a *firm.*

All accounting is based on the *fundamental accounting equation,* which is

$$\text{Assets} = \text{liabilities} + \text{owners' equity}, \tag{A-1}$$

where *assets* are those things of monetary value that the firm possesses, *liabilities* are those things of monetary value that the firm owes, and *owners' equity* is the worth of what the firm owes to its stockholders (also referred to as *equities, net worth,* etc.). For example, typical accounts in each term of Equation (A-1) are as follows:

Asset Accounts	=	Liability Accounts	+	Owner's Equity Accounts
Cash		Short-term debt		
Receivables		Payables		Capital stock
Inventories		Long-term debt		
Equipment				Retained earnings
Buildings				(income retained
Land				in the firm)

The fundamental accounting equation defines the format of the *balance sheet,* which is one of the two most common accounting statements and which shows the financial position of the firm at any given point in time.

Another important, and rather obvious, accounting relationship is

$$\text{Revenues} - \text{expenses} = \text{profit (or loss).} \tag{A-2}$$

This relationship defines the format of the *income statement* (also commonly known as a *profit-and-loss statement*), which summarizes the revenue and expense results of operations *over a period of time.* Equation (A-1) can be expanded to take account of profit defined in Equation (A-2):

$$\text{Assets} = \text{liabilities} + (\text{beginning owners' equity} + \text{revenue} - \text{expenses}). \quad \text{(A-3)}$$

Profit is the increase in money value (not to be confused with cash) that results from a firm's operations and is available for distribution to stockholders. It therefore represents the return on owners' invested capital.

A useful analogy is that a balance sheet is like a snapshot of the firm at an instant in time, whereas an income statement is a summarized moving picture of the firm over an interval of time. It is also useful to note that revenue serves to increase owners' interests in a firm, but an expense serves to decrease the owners' equity amount for a firm.

To illustrate the workings of accounts in reflecting the decisions and actions of a firm, suppose that an individual decides to undertake an investment opportunity and the following sequence of events occurs over a period of one year:

1. Organize XYZ firm and invest $3,000 cash as capital.
2. Purchase equipment for a total cost of $2,000 by paying cash.
3. Borrow $1,500 through a note to the bank.
4. Manufacture year's supply of inventory through the following:

   (a) Pay $1,200 cash for labor.

   (b) Incur $400 accounts payable for material.

   (c) Recognize the partial loss in value (depreciation) of the equipment amounting to $500.

5. Sell on credit all goods produced for year, 1,000 units at $3 each. Recognize that the accounting cost of these goods is $2,100, resulting in an increase in equity (through profits) of $900.
6. Collect $2,200 of accounts receivable.
7. Pay $300 of accounts payable and $1,000 of bank note.

A simplified version of the accounting entries recording the same information in a format that reflects the effects on the fundamental accounting equation (with a "+" denoting an increase and a "−" denoting a decrease) is shown in Figure A-1. A summary of results is shown in Figure A-2.

It should be noted that the profit for a period serves to increase the value of the owners' equity in the firm by that amount. Also, it is significant that the net cash flow from operation of $700 (= $2,200 − $1,200 − $300) is not the same as profit. This amount was recognized in transaction 4c, in which capital consumption (depreciation) for equipment of $500 was declared. Depreciation serves to convert part of an asset into an expense, which is then reflected in a firm's profits, as seen in Equation (A-2). Thus, the profit was $900, or $200 more than the net cash flow. For our purposes, revenue is recognized when it is earned, and expenses are recognized when they are incurred.

One important and potentially misleading indicator of after-the-fact financial performance that can be obtained from Figure A-2 is "annual rate of return." If the

	Account	Transaction							Balances at End of Year
		1	2	3	4	5	6	7	
Assets	Cash	+$3,000	−$2,000	+$1,500	−$1,200		+$2,200	−$1,300	+$2,200
	Accounts receivable					+$3,000	−2,200		+800
	Inventory				+2,100	−2,100			0
	Equipment		+2,000		−500				+1,500
equals									
Liabilities	Accounts payable				+400			−300	+100
	Bank note			+1,500				−1,000	+500
plus									
Owners' equity	Equity	+3,000				+900			+3,900

Assets balance: $4,500

Liabilities plus Owners' equity balance: $4,500

Figure A-1 Accounting Effects of Transactions: XYZ Firm

XYZ Firm Balance Sheet
As of Dec. 31, 1997

Assets		Liabilities and Owners' Equity	
Cash	$2,200	Bank note	$500
Accounts receivable	800	Accounts payable	100
Equipment	1,500	Equity	3,900
Total	$4,500	Total	$4,500

XYZ Firm Income Statement
for Year Ending Dec. 31, 1997

			Cash Flow
Operating revenues (Sales)		$3,000	$2,200
Operating costs (Inventory depleted)			
Labor	$1,200		−1,200
Material	400		−300
Depreciation	500		0
		$2,100	
Net income (profits)		900	$700

Figure A-2 **Balance Sheet and Income Statement Resulting from Transactions Shown in Figure A-1**

invested capital is taken to be the owners' (equity) investment, the annual rate of return at the end of this particular year is $900/$3,900 = 23%.

Financial statements are usually most meaningful if figures are shown for two or more years (or other reporting periods such as quarters or months) or for two or more individuals or firms. Such comparative figures can be used to reflect trends or financial indications that are useful in enabling investors and management to determine the effectiveness of investments *after* they have been made.

A.3 Cost Accounting

Cost accounting, or management accounting, is a phase of accounting that is of particular importance in engineering economic analysis because it is concerned principally with decision making and control in a firm. Consequently, cost accounting is the source of much of the cost data needed in making engineering economy studies. Modern cost accounting may satisfy any or all of the following objectives:

1. Determination of the actual cost of products or services;

2. Provision of a rational basis for pricing goods or services;
3. Provision of a means for allocating and controlling expenditures;
4. Provision of information on which operating decisions may be based and by means of which operating decisions may be evaluated.

Although the basic objectives of cost accounting are simple, the exact determination of costs usually is not. As a result, some of the procedures used are arbitrary devices that make it possible to obtain reasonably accurate answers for most cases but that may contain a considerable percentage of error in other cases, particularly with respect to the actual cash flow involved.

A.4 The Elements of Cost

One of the first problems in cost accounting is that of determining the elements of cost that arise in the production of a product or the rendering of a service. A study of how these costs occur gives an indication of the accounting procedure that must be established to give satisfactory cost information. Also, an understanding of the procedure that is used to account for these costs makes it possible to use them more intelligently.

From an engineering and managerial viewpoint in manufacturing enterprises, it is common to consider the general elements of cost to be *direct materials*, *direct labor*, and *overhead*. Such terms as *burden* and *indirect costs* are often used synonymously with *overhead*, and overhead costs are often divided into several subcategories.

Ordinarily, the materials that can be conveniently and economically charged directly to the cost of the product are called *direct materials*. Several guiding principles are used when we decide whether a material is classified as a direct material. In general, direct materials should be readily measurable, be of the same quantity in identical products, and be used in economically significant amounts. Those materials that do not meet these criteria are classified as *indirect materials* and are a part of the charges for overhead. For example, the exact amount of glue and sandpaper used in making a chair would be difficult to determine. Still more difficult would be the measurement of the exact amount of coal that was used to produce the steam that generated the electricity that was used to heat the glue. Some reasonable line must be drawn beyond which no attempt is made to measure directly the material that is used for each unit of production.

Labor costs also are ordinarily divided into *direct* and *indirect* categories. Direct labor costs are those that can be conveniently and easily charged to the product or service in question. Other labor costs, such as for supervisors, material handlers, and design engineers, are charged as indirect labor and are thus included as part of overhead costs. It is often imperative to know what is included in direct labor and direct material cost data before attempting to use them in engineering economy studies.

In addition to indirect materials and indirect labor, there are numerous other cost items that must be incurred in the production of products or the rendering of services. Property taxes must be paid; accounting and personnel departments must be maintained; buildings and equipment must be purchased and maintained;

supervision must be provided. It is essential that these necessary *overhead* costs be attached to each unit produced in proper proportion to the benefits received. Proper allocation of these overhead costs is not easy, and some factual, yet reasonably simple, method of allocation must be used.

As might be expected where solutions attempt to meet conflicting requirements such as exist in overhead-cost allocation, the resulting procedures are empirical approximations that are accurate in some cases and less accurate in others.*

There are many methods of allocating overhead costs among the products or services produced. The most commonly used methods involve allocation in proportion to direct labor cost, direct labor hours, direct materials cost, sum of direct labor and direct materials cost, or machine hours. In these methods, it is necessary to estimate what the total overhead costs will be if standard costs are being determined. Accordingly, *total overhead costs are customarily associated with a certain level of production,* which is an important condition that should always be remembered when dealing with unit-cost data. These costs can be correct only for the conditions for which they were determined.

To illustrate one method of allocation of overhead costs, consider the method that assumes that overhead is incurred in direct proportion to the cost of direct labor used. With this method the overhead rate (overhead per dollar of direct labor) and the overhead cost per unit would respectively be

$$\text{Overhead rate} = \frac{\text{total overhead in dollars for period}}{\text{direct labor in dollars for period}}$$

$$\text{Overhead cost/unit} = \text{overhead rate} \times \text{direct labor cost/unit}. \qquad \text{(A-4)}$$

Suppose that for a future period (say, a quarter), the total overhead cost is expected to be \$100,000 and the total direct labor cost is expected to be \$50,000. From this, the overhead rate = \$100,000/\$50,000 = \$2 per dollar of direct labor cost. Suppose further that for a given unit of production (or job) the direct labor cost is expected to be \$60. From Equation (A-4), the overhead cost for the unit of production would be \$60 × 2 = \$120.

This method obviously is simple and easy to apply. In many cases, it gives quite satisfactory results. However, in many other instances, it gives only very approximate results because some items of overhead, such as depreciation and taxes, have very little relationship to labor costs. Quite different total costs may be obtained for the same product when different procedures are used for the allocation of overhead costs. The magnitude of the difference will depend on the extent to which each method produces or fails to produce results that realistically capture the facts.

---

*Section A.7 discusses a relatively recent methodology, called *activity-based cost management,* for avoiding badly distorted cost estimates caused by traditional overhead allocations.

A.5 Cost Accounting Example

This relatively simple example involves a job-order system in which costs are assigned to work by job number. Schematically, this process is illustrated in the following diagram:

Costs are assigned to jobs in this manner:

1. Raw materials attach to jobs via material requisitions.
2. Direct labor attaches to jobs via direct labor tickets.
3. Overhead cannot be attached to jobs directly, but must have an allocation procedure that relates it to one of the resource factors, such as direct labor, which is already accumulated by the job.

Consider how an order for 100 tennis rackets accumulates costs at the Bowling Sporting Good Company:

Job #161	100 tennis rackets
Labor rate	$7 per hour
Leather	50 yards at $2 per yard
Gut	300 yards at $0.50 per yard
Graphite	180 pounds at $3 per pound
Labor hours for the job	200 hours
Total annual factory overhead costs	$600,000
Total annual direct labor hours	200,000 hours

The three major costs are now attached to the job. Direct labor and material expenses are straightforward:

Job #161		
Direct labor	$200 \times \$7 =$	$1,400
Direct material	leather: $50 \times \$2 =$	100
	gut: $300 \times \$0.5 =$	150
	graphite: $180 \times \$3 =$	540
Prime costs (direct labor + direct materials)		$2,190

Notice that this cost is not the total cost. We must somehow find a way to attach (allocate) factory costs that cannot be directly identified to the job, but are nevertheless involved in producing the 100 rackets. Costs such as the power to run the graphite molding machine, the depreciation on this machine, the depreciation of the factory building, and the supervisor's salary constitute overhead for this company. These overhead costs are part of the cost structure of the 100 rackets but cannot be directly traced to the job. For instance, do we really know how much machine obsolescence is attributable to the 100 rackets? Probably not. Therefore, we must allocate these overhead costs to the 100 rackets using the overhead rate determined as follows:

$$\text{Overhead rate} = \frac{\$600,000}{200,000} = \$3 \text{ per direct labor hour.}$$

This means that $600 ($3 × 200) of the total annual overhead cost of $600,000 would be allocated to Job #161. Thus, the total cost of Job #161 would be

Direct labor	$1,400
Direct materials	790
Factory overhead	600
	$2,790

The cost of manufacturing each racket is thus $27.90. If selling expenses and administrative expenses are allocated as 40% of the cost of goods sold, the total expense of a tennis racket becomes 1.4($27.90) = $39.06.

A.6 The Use of Accounting Costs in Engineering Economy Studies

When we recognize that accounting costs are linked to a definite set of conditions and that they are the result of certain arbitrary decisions concerning the allocation of overhead costs, it is apparent that they should not be used without modification in cases where the conditions are different from those for which they were determined. Engineering economy studies invariably deal with situations that now are *not* being done. Thus, ordinary accounting costs normally cannot be used without modification in these economy studies. However, if we understand how the accounting costs were determined, we should be able to break them down into their component elements, and then we often find that these cost elements will supply much of the cost information that is needed for an engineering economy study. Thus, an understanding of the basic objectives and procedures of cost accounting will enable the engineer doing an economic analysis to make best use of available cost information and to avoid needless work and serious mistakes.

It should not be assumed that the figures contained in accounting reports are absolutely correct and indicative, even though they have been prepared with the utmost care by highly professional accountants, because accounting procedures often must include certain assumptions that are based on subjective judgment or the current tax laws. For example, the years of life on which depreciation expense for a particular asset is based has to be determined or assumed and the estimate may turn out to have caused unrealistic depreciation expenses and book values in accounting reports. Also, there are many accepted practices in accounting that may provide unrealistic information for management control purposes. For example, the net book value of an asset is generally declared in the balance sheet at the original price (cost basis) minus any accumulated depreciation, even though it may be recognized that the true value of the asset at a particular time is far above or below this reported book value.

A.7 Bringing Cost Management Up to Date*

Traditional accounting systems allocate overhead costs using a volume-oriented base, such as direct labor hours or direct material dollars. The cost allocation bases of direct labor or production quantity were designed primarily for inventory valuation. As a consequence, traditional cost accounting methods are fully effective only when direct labor (or direct materials) is the dominant cause of cost.

Although traditional standard cost systems were effective in the past, changes in manufacturing technologies (such as the just-in-time philosophy, robotics, CAD, and flexible manufacturing systems) have made traditional cost models somewhat obsolete. Rapid technological advancement has resulted in the restructuring of manufacturing cost patterns (e.g., the direct labor and inventory components of total cost are decreasing, while those of technology depreciation, engineering, and data processing are increasing). Due to the changing nature of these cost components, existing cost accounting systems and cost management practices do not adequately support the objectives of advanced manufacturing. In fact, the direct labor component of product cost now accounts for as little as 5% of product cost, and overhead accounts for as much as 500%. In automated environments, volume-based allocations distort product costs because the measures used to allocate overhead do not cause the costs. As a result, product cost distortion occurs due to high overhead rates that are inflated by many costs that should be directly traceable to the product rather than arbitrarily allocated on the basis of volume. Several components of a product's cost that should be traced to the product include hidden overhead costs such as material movement, order processing, process planning, rework, warranty maintenance, production planning and control, and quality assurance.

Assume that a company uses a traditional cost accounting system that applies overhead based on direct labor dollars (Figure A-3). The product cost is computed as $550 with a sales price of $660, resulting in a reported net profit of $110 per unit.

The manufacture of the product involves a significant number of automated processes, however. Automated production requires a heavy investment in depreciation, software, and maintenance support. An analysis of these costs and others traced to the product results in a completely different financial outcome (Figure A-4). The new product cost after the analysis was $925. With a sales price of $660, the company was losing $265 on each unit produced.

A primary reason for this distortion is that the overhead rate is inflated by potentially traceable direct costs. The inflated overhead cost is then allocated to the

---

*Excerpted from J. A. Brimson, "Bringing Cost Management Up to Date," *Manufacturing Engineering*, vol. 102, no. 12, June 1988, pp. 49–51. Reproduced with permission of the Society of Manufacturing Engineers, Dearborn, MI

Traditional Cost Accounting		
Sales price		$660
Direct labor	$50	
Direct material	300	
Overhead	200	
Total production cost		550
Net profit		$110

Figure A-3 A Traditional Cost Accounting System
Applies Overhead Based on Direct Labor Dollars

products using a direct labor base. For this allocation method to be correct, there must be a complementary relationship between labor and technology. In other words, reported product cost is often based on a choice of accounting methods that do not mirror the manufacturing process.

A.7.1 Activity-Based Costing Systems

Activity-based cost management systems track hidden overhead costs to the specific activities that cause them and thus provide a more reliable product cost.

New Cost Accounting		
Sales price		$660
Cost		
Directly traceable		
Direct labor	$50	
Direct material	300	
Technology	200	
Scrap and Rework	50	
Imputed cost		
Raw materials inventory	20	
WIP inventory	60	
Other direct cost	90	
	770	
Nontraceable overhead	155	
Total cost		925
Net loss		($265)

Figure A-4 New Cost Management Techniques, Which Consider
the Ramifications of Automation, Produce Different Financial Results

Four key concepts differentiate activity-based costing from traditional costing systems, allowing activity-based systems to provide more accurate product cost data:

I. *Activity accounting.* In an activity-based system, the cost of the product is the sum of all costs required to manufacture and deliver the product. The activities a firm pursues consume its resources, and resource availability and usage create costs.

Activity accounting decomposes an organization into an activity structure that provides a cause-and-effect rationale for how fundamental objectives and their associated activities create costs and result in outputs. According to Brimson, an effective activity accounting system uses the following approach:*

A. Determine the fundamental activities that must be accomplished to satisfy a firm's objectives. Activities permit identification of how a company deploys its resources to achieve its basic aims.

B. Determine the causal relationships that permit outputs (performance) to be attributed to inputs (resources). A large number of these relationships will be based on non-volume-related measures such as the number of parts in a new design.

C. Ascertain the output of an activity in terms of a measure of *activity volume* through which costs of a business process vary most directly (e.g., number of machine setups required for a complex design).

D. Trace activities to products (or other objects) and determine how much of each activity's output is dedicated to them. A cost structure, known as a *bill of activities,* is used to describe each product's pattern of activity consumption.

E. Determine critical success factors by which activities of the enterprise can be aligned with stated strategic objectives. This step indicates how effectively desired performance is being accomplished through activities undertaken by a firm.

F. Take action, using a continuous improvement philosophy, on the productivity opportunities identified in Steps A–E. Because activity cost is the ratio of resources consumed by an activity to the measured output of the activity, a means for evaluating effectiveness and efficiency (i.e., productivity) is available to managers. Various alternatives for making desired changes in activity patterns, through investment or organizational means, can now be realistically appraised.

II. *Cost drivers.* A cost driver is an event that affects the cost/performance of a group of related activities. Familiar cost drivers include number of machine setups, number of engineering change notices, and number of purchase orders. Cost drivers reflect the demands placed on activities at both the activity and product levels. By controlling the cost driver, unnecessary costs can be eliminated, resulting in improved product cost.

---

*J. A. Brimson, *Activity Accounting: An Activity-Based Costing Approach* (New York: John Wiley & Sons, 1991).

III. *Direct traceability.* Direct traceability involves attributing costs to those products or processes that consume resources. Many hidden overhead costs can be effectively traced to products, thus providing a more accurate product cost.

IV. *Nonvalue-added costs.* In manufacturing processes, customers may perceive that certain activities add no value to the product. Through identification of cost drivers, a firm can pinpoint these unnecessary costs. Activity-based cost systems identify and place a cost on the activities performed (value-adding and nonvalue-adding) so that management can determine desired changes in resource requirements for each activity. In contrast, traditional cost systems accumulate costs by budgetary line items and by functions.

These four basic concepts are embodied in activity-based costing systems and lead to more accurate costing information. In addition, activity-based costing systems provide more flexibility than conventional costing systems because they produce a variety of cost figures useful for technology accounting, product costing, and life-cycle analysis. In addition, these cost figures can be applied in making various special decisions, including inventory valuation, budgeting/forecasting, product line analysis, make/buy decisions, and design to cost.

A.7.2 Example of Activity-Based Costing

The objective of this section is to show an example of how activity-based costing (ABC) can be used to capture more accurately the estimated cost of manufacturing a product.

In general terms, an ABC system identifies and then classifies the major activities of a facility's production process into four main "base" groups: *unit-level, batch-level, product-level,* and *facility-level* activities. The ABC approach assumes that not all overhead resources are consumed in proportion to the number of units produced. ABC introduces these hierarchical levels to ensure that the final estimate of product cost mirrors the manufacturing process as closely as possible.

Unit-level costs are those costs that can be directly apportioned to volume. These costs may include such activities as direct labor hour costs, directly attributable material costs, and costs per machine operating hour.

Batch-level costs are those costs that can be directly apportioned to the particular batch run. For this type of cost, certain activities are consumed in direct proportion to the number of batch runs for each product. These batch-level costs may include setup, ordering, material handling, and transportation costs.

Product-level costs are those costs that can be directly apportioned to the product, which assumes that certain activities are consumed to develop or permit production of different products. These product-level costs may include such activities as research and development (R&D), parts and material acquisition and inventory costs, technical administration, and specialized preproduction safety and manufacturing training.

Facility-level costs cause problems in an ABC environment because these costs are associated with the sustainment of a general manufacturing process. These facility-level costs may include such activities as travel costs, directors' fees, and general administration and can include a large segment of the estimated product cost.

EXAMPLE A-1

This example is constructed such that the attributes of ABC are clearly exhibited. Perhaps more important than representing the "advantages" of ABC, it is an illustration of the differences between ABC and traditional volume-based absorption costing (VBC).

The attributes that make ABC most productive and valuable are essentially diversity of products and high support costs and overheads. Certainly these characteristics are common to most markets today, and the trend is increasingly moving in that direction. Consider this simplified yet realistic example of ABC versus VBC:

Exhibit A-1. This exhibit shows the basic business scenario. It gives the detailed production budgets, volumes, and costs. The second part of the exhibit shows the

	Production: Scheduled Data for Year				
	Baseball	Glove	Bat	Pitching Machine	Totals
Budget units of production	20,000	10,000	5,000	200	35,200
Material cost per unit	$0.45	$5.00	$0.75	$2,000.00	n/a
Direct labor hours per unit	0.05	2.00	0.1	50.0	n/a
Direct labor cost per hour	$5.00	$5.00	$5.00	$5.00	$5.00
Machine hours per unit	0.1	0.1	0.2	100.0	24,000
Parts required per unit	3	4	1	250	n/a
Production orders (total budgeted)	500	25	100	100	725
Production set-ups (total budgeted)	1,000	100	100	100	1,300
Number of shipments	400	250	25	100	775

	Activity Conversion Costs			
			Cost Driver	
	Cost Driver	Costs	Total Units	Rate
Material handling	# of moves	$50,000	155,000	$0.32
Production planning dept.	# of prod. orders	40,000	725	$55.17
Set-up indirect labor	# of set-ups	25,000	1,300	$19.23
Depreciation on machinery	Machine hours	725,000	24,000	$30.21
Quality & finishing	Dir. labor hours	150,000	31,500	$4.76
Shipping department	# of shipments	100,000	775	$129.03
Total indirect costs		**$1,090,000**		
Indirect costs per direct labor hour		**$34.60** {*Traditional Volume-Based Costing*}		

Exhibit A-1 Business Scenario for a Small Firm

ABC "cost driver" calculations (there are six) and the VBC traditional overhead allocation of $34.60 per direct labor hour.

Note that the model shows a capital intensive business. Depreciation represents about 67% of total indirect costs.

The model has only four products. More diversity would better highlight the ABC advantages but would add unwanted complexity to the model. The products are very different in terms of production and market. The pitching machine requires much machine time, whereas the other products use hardly any. Direct labor is an insignificant factor for the ball and bat, but very significant for the glove. (The glove requires 40 times the labor of the ball and 20 times the labor of the bat.)

Exhibit A-2. Traditional Volume-Based Costing—All indirect costs are allocated based on the overall rate of $34.60 per direct labor hour.

The glove with relatively high direct labor content appears to be a loser. It sells for $57.00, but is assigned costs totaling $84.21—*a loss of $27.21 per unit!* Note that all other products seem to have very healthy margins. *Management would certainly have to consider dropping the glove from production and marketing.*

Exhibit A-3. Activity-Based Costing—Now we see completely different results in terms of margins and profitability. The glove is now the only profitable product with the others being losers!

The four products in Exhibits A-1 through A-3 were designed to show differences between ABC and VBC.

	Traditional Volume-Based Product Costs				
	Baseball	Glove	Bat	Pitching Machine	Totals
Units produced	20,000	10,000	5,000	200	35,200
Direct material costs	$9,000	$50,000	$3,750	$400,000	$462,750
Direct labor costs	5,000	100,000	2,500	50,000	157,500
Overhead allocations ⟶ (based on D.L. @ $34.60/hr.)	34,603	692,063	17,302	346,032	1,090,000
Total product costs	$48,603	$842,063	$23,552	$796,032	$1,710,250
Per unit costs					
Direct costs	$0.70	$15.00	$1.25	$2,250.00	
Overhead	1.73	69.21	3.46	1,730.16	
Total cost per unit	$2.43	$84.21	$4.71	$3,980.16	
Selling price	$4.45	$57.00	$10.00	$5,000.00	
Volume-based margin	$2.02	$(27.21)	$5.29	$1,019.84	
	45%	−48% LOSS	53%	20%	

Exhibit A-2 **Traditional Volume-Based Costing for Four Products**

			Activity-Based Product Costs		
		Baseball	Glove	Bat	Pitching Machine	Totals
Units produced		20,000	10,000	5,000	200	35,200
Direct material costs		$9,000	$50,000	$3,750	$400,000	$462,750
Direct labor costs		5,000	100,000	2,500	50,000	157,500
Overhead costs:	*ABC rate*					
Material handling	$0.32	19,355	12,903	1,613	16,129	50,000
Production planning dept.	$55.17	27,586	1,379	5,517	5,517	40,000
Set-up indirect labor	$19.23	19,231	1,923	1,923	1,923	25,000
Depreciation on machinery	$30.21	60,417	30,208	30,208	604,167	725,000
Quality & finishing	$4.76	4,762	95,238	2,381	47,619	150,000
Shipping department	$129.03	51,613	32,258	3,226	12,903	100,000
Total product costs		**$196,964**	**$323,910**	**$51,118**	**$1,138,258**	**$1,710,250**
Per unit costs:						
Direct costs		$0.70	$15.00	$1.25	$2,250.00	
Overhead		9.15	17.39	8.97	3,441.29	
Total cost per unit		*$9.85*	*$32.39*	*$10.22*	*$5,691.29*	
Selling price		$4.45	$57.00	$10.00	$5,000.00	
Activity-based margin		$(5.40)	$24.61	$(0.22)	$(691.29)	
		−121%	43%	−2%	−14%	
		LOSS		*LOSS*	*LOSS*	

Exhibit A-3 Activity-Based Costing for Four Products

Out of total costs of $1,710,250 direct labor represents only $157,500, less than 10%. In this case, direct labor used in production varied significantly in proportion to total product costs. The glove had the highest proportional costs. Not surprisingly, using VBC allocation methods on labor loaded up the indirect costs on the glove, which looked like a terrible product. However, production statistics in Exhibit A-1 show that the glove hardly uses any indirect resources, less than all the others. Setting up a glove manufacturing shop outside the company would require little indirect support and would probably be a profitable business.

Conversely, the pitching machine uses about 83% of the machinery's capacity, and the machinery is costly. Using direct labor to allocate indirect costs results in allocation of only $1,730 of indirect costs to the pitching machine—hardly near the 83% of costs on just machinery depreciation alone! Allocation based on machine hours of usage doubles the indirect cost assignment to the pitching machine, which, again, is logical because most of the indirect costs are machinery and the pitching machine seems to use most of the machinery.

When the products are costed on a resource-usage basis (ABC), only the glove seems to be profitable, which is a much more accurate reflection of the real costs of doing business.

This information is excellent feedback for both production and marketing. The pitching machine is only slightly unprofitable. Perhaps it should command a better price. Is it unique and a better product than the competition? Is it perhaps priced too low, based on the prior cost information (VBC) that indicated a 20% margin? Can it be redesigned to save some costs?

The bat is only a 2% margin loser. Redesign of production should be considered. A slight change in any production factor will make the bat profitable.

The baseball uses 1,000 setups. Setups should be re-examined. Perhaps we will end up with more setups, but with reduced costs. Serious redesign is needed here; otherwise this product just may not be good for the business.

ABC offers product profitability insights in areas that might otherwise cloud product costing. For strategic purposes, the following two factors need to be considered:

1. Many of the costs are fixed. Elimination of a product line may not reduce costs in total. In the short term, it may be best to keep some products even if they show fully costed losses. For long-term investment decisions, the full (ABC) costs are more relevant.
2. As indicated with the discussion of setups, ABC costs may give improper direction if followed "blindly." High setup costs assigned by a cost driver just means you have high setup costs. The way to reduce the setup costs may not be to reduce the number of setups, but it may be to redesign the setup process. (Perhaps you end up increasing the number of setups?)

ABC helps in identifying potential areas to pursue; it does not necessarily give the cost reduction answers; it just gives the direction to take in pursuing these difficult questions.

The graphs presented in Figure A-5 show the differing effects that ABC and VBC have on unit costs as overhead levels are varied. Overhead costs were varied by using the following levels: $0, $100,000, $500,000, $725,000, $1,000,000, and $2,000,000. All other factors remained exactly the same.

Note that at low overheads (around $100,000), the pitching machine and glove are indifferent to costing method.

The most significant point of Figure A-5 is that as overhead increases, the cost methodology becomes increasingly important. Absolute differences in unit product costs increase with higher overhead. Importantly, the rate of increase in unit cost differences (differences in slopes) increases in all cases. The rate of increase in slope differences was least for the baseball. Proportionately, VBC costs (based on direct labor) for the baseball was closest to ABC cost assignments as the levels of overhead increased.

Also, note that as overhead goes from $0 to $2,000,000 the overhead assignment from ABC remains roughly constant for the glove. Under VBC the overhead assignment increases significantly.

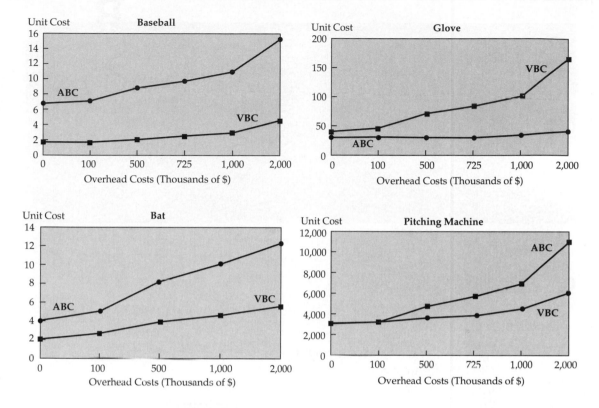

Notes: —All Products except the Glove show higher overhead costs for ABC than VBC.

—The glove has relatively constant assignment of overhead costs under ABC.

—The pitching machine actually has a transition where at zero overhead ABC has slightly lower unit cost, than at $100,000 of overhead ABC is slightly higher. Differences increase at increasing rate thereafter.

—The glove, the bat, and the pitching machine all have increasing differences in slopes of ABC and VBC cost. The baseball slope differences increase at a much lower rate.

Figure A-5 **Unit Cost Charts**

B

*Abbreviations and Notation**

Chapter 2

C_F	total fixed cost
C_V	total variable cost
c_v	variable cost per unit
C_T	total cost
C_U	average unit cost
D	demand for a product or service in units
D^*	optimal demand or production volume that maximizes profit
D'	breakeven point
$\hat{D}$	demand or production volume that will produce maximum revenue

Chapter 3

A	equal and uniform end-of-period cash flows (or equivalent end-of-period values)
APR	annual percentage rate (nominal interest)
A_1	end of period 1 cash flow in a geometric sequence of cash flows
$\overline{A}$	an amount of money flowing uniformly and continuously over a specified period of time
BOY	beginning of year
EOY	end of year
F	a future equivalent sum of money
$\overline{f}$	a geometric change from one time period to the next in cash flows or equivalent values

---

*Listed by the chapter in which they first appear.

G	an arithmetic (i.e., uniform) change from one period to the next in cash flows or equivalent values
$\underline{I}$	total interest earned or paid (simple interest)
i	effective interest rate per interest period
i_{CR}	an interest rate called the *convenience rate*
k	an index for time periods
P	principal amount of a loan; a present equivalent sum of money
M	number of compounding periods per year
N	number of interest periods
r	nominal interest rate per period (usually a year)
$\underline{r}$	a nominal interest rate that is continuously compounded

Chapter 4

AW($i\%$)	equivalent uniform annual worth, computed at $i\%$ interest, of one or more cash flows
CR($i\%$)	equivalent annual cost of capital recovery, computed at $i\%$ interest
$\underline{E}$	equivalent annual expenses
ϵ	external reinvestment rate
ERR	external rate of return
FW($i\%$)	future equivalent worth, calculated at $i\%$ interest, of one or more cash flows
I	initial investment for a project
IRR	internal rate of return, also designated $i'\%$
MARR	minimum attractive rate of return
N	the length of the study period (usually years)
O&M	equivalent annual operating and maintenance expenses
PW($i\%$)	present equivalent worth, computed at $i\%$ interest, of one or more cash flows
$\underline{R}$	equivalent annual revenues (or savings)
S	salvage (market) value at the end of the study period
θ	payback period
θ'	discounted payback period
V_N	value (price) of a bond N periods prior to redemption
Z	face value of a bond

Chapter 5

$A \rightarrow B$	increment (or incremental net cash flow) between Alternative A and Alternative B (read: A to B)
$\Delta(B - A)$	incremental net cash flow (difference) calculated from the cash flow of Alternative B minus the cash flow of Alternative A (read: delta B minus A)

Chapter 6

ACRS	accelerated cost recovery system
ADS	alternative depreciation system
ATCF	after-tax cash flow
B	cost basis
BTCF	before-tax cash flow
BV	book value of an asset
d	depreciation deduction
d^*	cumulative depreciation over a specified period of time
E	annual expense
GDS	general depreciation system
Int	interest expense
MACRS	modified accelerated cost recovery system
MV	market value of an asset; the price that a buyer will pay for a particular type of property
N	useful life of an asset (ADR life)
NIAT	net income after taxes
NIBT	net income before taxes
p	property classification for MACRS depreciation
$\underline{R}$	ratio of depreciation in a particular year to book value at the beginning of the same year
R	gross annual revenue
r_k	ACRS or MACRS depreciation rate (a decimal)
SV_N	salvage value of an asset at the end of useful life
T	income taxes
t	effective income-tax rate

Chapter 7

$\bar{I}_n$	an unweighted or a weighted index number dependent on the calculation
K	the number of input resource units needed to produce the first output unit
u	the output unit number
X	cost-capacity factor
Z_u	the number of input resource units needed to produce output unit number u

Chapter 8

A\$	actual (current) dollars
b	base time period index
e_j	total price escalation (or deescalation) rate for good or service j
e'_j	differential price inflation (or deflation) rate for good or service j

f_e	annual devaluation rate (rate of annual change in the exchange rate) between the currency of a foreign country and the U.S. dollar
f	general inflation rate
i_c	combined (nominal) interest rate; also called the *market interest rate*
i_{fc}	rate of return in terms of a combined (market) interest rate relative to the currency of a foreign country
i_r	real interest rate
i_{us}	rate of return in terms of a combined (market) interest rate relative to U.S. dollars
R$	real (constant) dollars

Chapter 9

EUAC	equivalent uniform annual cost
TC_k	total (marginal) cost for year k

Chapter 10

EW	equivalent worth (annual, present, or future)

Chapter 11

$B-C$	benefit–cost ratio
$\underline{B}$	equivalent uniform annual benefits of a proposed project
CR	capital recovery amount (a cost)
I	initial investment
O&M	equivalent uniform annual operating and maintenance expenses

Chapter 12

C	recurring annual expenses
CC	carrying charges
D_B	book depreciation
D_T	depreciation taken for income-tax purposes
e_a	return on equity (no inflation)
e_a'	return on equity (inflation-adjusted)
i_b	cost of borrowed capital (no inflation)
i_b'	cost of borrowed capital (inflation-adjusted)
K_a	weighted after-tax cost of capital (no inflation)
K_a'	weighted after-tax cost of capital (inflation-adjusted)
λ	fraction of total capitalization represented by borrowed money, also called the debt ratio
RR_k	annual revenue requirement in year k
$\overline{RR}$	levelized revenue requirement
T	income taxes paid

t	effective income tax rate
UI	unrecovered investment

Chapter 13

$E(X)$	mean of a random variable
EVPI	expected value of perfect information
$f(x)$	probability density function of a continuous random variable
$F(x)$	cumulative distribution function of a continuous random variable
$p(x)$	probability mass function of a discrete random variable
$p(x_i)$	probability that a discrete random variable takes on the value x_i
$P(x)$	cumulative distribution function of a discrete random variable
$Pr\{\cdots\}$	probability of the described event occurring
SD(X)	standard deviation of a random variable
$V(X)$	variance of a random variable

Chapter 14

B^*	present worth of an investment opportunity in a specified budgeting period
c	cash outlay for expenses in capital allocation problems
C_k	maximum cash outlay permissible in period k
Div	cash dividends (after taxes)
e_a	annual rate of return to owners of a firm (stockholders)
g	annual growth rate for the value of common stock and other equity interests
L	before-tax lease expense
l	after-tax lease expense
m	number of mutually exclusive projects being considered
X	binary decision variable ($= 0$ or 1) in capital allocation problems

Chapter 15

α	Hurwicz index of optimism
r^*	dimensionality of a multiattribute decision problem

Interest and Annuity Tables for Discrete Compounding

For various values of i from $\frac{1}{4}$% to 25%

$$i = \text{effective interest rate per period (usually one year)}$$
$$N = \text{number of compounding periods}$$

$$(F/P, i\%, N) = (1 + i)^N \qquad\qquad (A/F, i\%, N) = \frac{i}{(1 + i)^N - 1}$$

$$(P/F, i\%, N) = \frac{1}{(1 + i)^N} \qquad\qquad (A/P, i\%, N) = \frac{i(1 + i)^N}{(1 + i)^N - 1}$$

$$(F/A, i\%, N) = \frac{(1 + i)^N - 1}{i} \qquad (P/G, i\%, N) = \frac{1}{i}\left[\frac{(1 + i)^N - 1}{i(1 + i)^N} - \frac{N}{(1 + i)^N}\right]$$

$$(P/A, i\%, N) = \frac{(1 + i)^N - 1}{i(1 + i)^N} \qquad (A/G, i\%, N) = \frac{1}{i} - \frac{N}{(1 + i)^N - 1}$$

TABLE C-1 Discrete Compounding; $i = \frac{1}{4}\%$

| | Single Payment | | Uniform Series | | | | | Uniform Gradient | | |
| | Compound Amount Factor | Present Worth Factor | Compound Amount Factor | Present Worth Factor | Sinking Fund Factor | Capital Recovery Factor | | Gradient Present Worth Factor | Gradient Uniform Series Factor | |
N	To Find F Given P F/P	To Find P Given F P/F	To Find F Given A F/A	To Find P Given A P/A	To Find A Given F A/F	To Find A Given P A/P		To Find P Given G P/G	To Find A Given G A/G	N
1	1.0025	0.9975	1.0000	0.9975	1.0000	1.0025		0.000	0.0000	1
2	1.0050	0.9950	2.0025	1.9925	0.4994	0.5019		0.995	0.4994	2
3	1.0075	0.9925	3.0075	2.9851	0.3325	0.3350		2.980	0.9983	3
4	1.0100	0.9901	4.0150	3.9751	0.2491	0.2516		5.950	1.4969	4
5	1.0126	0.9876	5.0251	4.9627	0.1990	0.2015		9.901	1.9950	5
6	1.0151	0.9851	6.0376	5.9478	0.1656	0.1681		14.826	2.4927	6
7	1.0176	0.9827	7.0527	6.9305	0.1418	0.1443		20.722	2.9900	7
8	1.0202	0.9802	8.0704	7.9107	0.1239	0.1264		27.584	3.4869	8
9	1.0227	0.9778	9.0905	8.8885	0.1100	0.1125		35.406	3.9834	9
10	1.0253	0.9753	10.1133	9.8639	0.0989	0.1014		44.184	4.4794	10
11	1.0278	0.9729	11.1385	10.8368	0.0898	0.0923		53.913	4.9750	11
12	1.0304	0.9705	12.1664	11.8073	0.0822	0.0847		64.589	5.4702	12
13	1.0330	0.9681	13.1968	12.7753	0.0758	0.0783		76.205	5.9650	13
14	1.0356	0.9656	14.2298	13.7410	0.0703	0.0728		88.759	6.4594	14
15	1.0382	0.9632	15.2654	14.7042	0.0655	0.0680		102.244	6.9534	15
16	1.0408	0.9608	16.3035	15.6650	0.0613	0.0638		116.657	7.4469	16
17	1.0434	0.9584	17.3443	16.6235	0.0577	0.0602		131.992	7.9401	17
18	1.0460	0.9561	18.3876	17.5795	0.0544	0.0569		148.245	8.4328	18
19	1.0486	0.9537	19.4336	18.5332	0.0515	0.0540		165.411	8.9251	19
20	1.0512	0.9513	20.4822	19.4845	0.0488	0.0513		183.485	9.4170	20
21	1.0538	0.9489	21.5334	20.4334	0.0464	0.0489		202.463	9.9085	21
22	1.0565	0.9466	22.5872	21.3800	0.0443	0.0468		222.341	10.3995	22
23	1.0591	0.9442	23.6437	22.3241	0.0423	0.0448		243.113	10.8901	23
24	1.0618	0.9418	24.7028	23.2660	0.0405	0.0430		264.775	11.3804	24
25	1.0644	0.9395	25.7646	24.2055	0.0388	0.0413		287.323	11.8702	25
30	1.0778	0.9278	31.1133	28.8679	0.0321	0.0346		413.185	14.3130	30
36	1.0941	0.9140	37.6206	34.3865	0.0266	0.0291		592.499	17.2306	36
40	1.1050	0.9050	42.0132	38.0199	0.0238	0.0263		728.740	19.1673	40
48	1.1273	0.8871	50.9312	45.1787	0.0196	0.0221		1040.055	23.0209	48
60	1.1616	0.8609	64.6467	55.6524	0.0155	0.0180		1600.085	28.7514	60
72	1.1969	0.8355	78.7794	65.8169	0.0127	0.0152		2265.557	34.4221	72
84	1.2334	0.8108	93.3419	75.6813	0.0107	0.0132		3029.759	40.0331	84
100	1.2836	0.7790	113.4500	88.3825	0.0088	0.0113		4191.242	47.4216	100
∞				400.0000		0.0025				∞

TABLE C-2 Discrete Compounding; $i = \frac{1}{2}\%$

	Single Payment		Uniform Series				Uniform Gradient		
	Compound Amount Factor	Present Worth Factor	Compound Amount Factor	Present Worth Factor	Sinking Fund Factor	Capital Recovery Factor	Gradient Present Worth Factor	Gradient Uniform Series Factor	
	To Find F Given P	To Find P Given F	To Find F Given A	To Find P Given A	To Find A Given F	To Find A Given P	To Find P Given G	To Find A Given G	
N	F/P	P/F	F/A	P/A	A/F	A/P	P/G	A/G	N
1	1.0050	0.9950	1.0000	0.9950	1.0000	1.0050	0.000	0.0000	1
2	1.0100	0.9901	2.0050	1.9851	0.4988	0.5038	0.990	0.4988	2
3	1.0151	0.9851	3.0150	2.9702	0.3317	0.3367	2.960	0.9967	3
4	1.0202	0.9802	4.0301	3.9505	0.2481	0.2531	5.901	1.4938	4
5	1.0253	0.9754	5.0503	4.9259	0.1980	0.2030	9.803	1.9900	5
6	1.0304	0.9705	6.0755	5.8964	0.1646	0.1696	14.655	2.4855	6
7	1.0355	0.9657	7.1059	6.8621	0.1407	0.1457	20.449	2.9801	7
8	1.0407	0.9609	8.1414	7.8230	0.1228	0.1278	27.176	3.4738	8
9	1.0459	0.9561	9.1821	8.7791	0.1089	0.1139	34.824	3.9668	9
10	1.0511	0.9513	10.2280	9.7304	0.0978	0.1028	43.387	4.4589	10
11	1.0564	0.9466	11.2792	10.6770	0.0887	0.0937	52.853	4.9501	11
12	1.0617	0.9419	12.3356	11.6189	0.0811	0.0861	63.214	5.4406	12
13	1.0670	0.9372	13.3972	12.5562	0.0746	0.0796	74.460	5.9302	13
14	1.0723	0.9326	14.4642	13.4887	0.0691	0.0741	86.584	6.4190	14
15	1.0777	0.9279	15.5365	14.4166	0.0644	0.0694	99.574	6.9069	15
16	1.0831	0.9233	16.6142	15.3399	0.0602	0.0652	113.424	7.3940	16
17	1.0885	0.9187	17.6973	16.2586	0.0565	0.0615	128.123	7.8803	17
18	1.0939	0.9141	18.7858	17.1728	0.0532	0.0582	143.663	8.3658	18
19	1.0994	0.9096	19.8797	18.0824	0.0503	0.0553	160.036	8.8504	19
20	1.1049	0.9051	20.9791	18.9874	0.0477	0.0527	177.232	9.3342	20
21	1.1104	0.9006	22.0840	19.8880	0.0453	0.0503	195.243	9.8172	21
22	1.1160	0.8961	23.1944	20.7841	0.0431	0.0481	214.061	10.2993	22
23	1.1216	0.8916	24.3104	21.6757	0.0411	0.0461	233.677	10.7806	23
24	1.1272	0.8872	25.4320	22.5629	0.0393	0.0443	254.082	11.2611	24
25	1.1328	0.8828	26.5591	23.4456	0.0377	0.0427	275.269	11.7407	25
30	1.1614	0.8610	32.2800	27.7941	0.0310	0.0360	392.632	14.1265	30
36	1.1967	0.8356	39.3361	32.8710	0.0254	0.0304	557.560	16.9621	36
40	1.2208	0.8191	44.1588	36.1722	0.0226	0.0276	681.335	18.8359	40
48	1.2705	0.7871	54.0978	42.5803	0.0185	0.0235	959.919	22.5437	48
60	1.3489	0.7414	69.7700	51.7256	0.0143	0.0193	1448.646	28.0064	60
72	1.4320	0.6983	86.4089	60.3395	0.0116	0.0166	2012.348	33.3504	72
84	1.5204	0.6577	104.0739	68.4530	0.0096	0.0146	2640.664	38.5763	84
100	1.6467	0.6073	129.3337	78.5426	0.0077	0.0127	3562.793	45.3613	100
∞				200.0000		0.0050			∞

TABLE C-3 Discrete Compounding; $i = \frac{3}{4}\%$

	Single Payment		Uniform Series				Uniform Gradient	
	Compound Amount Factor	Present Worth Factor	Compound Amount Factor	Present Worth Factor	Sinking Fund Factor	Capital Recovery Factor	Gradient Present Worth Factor	Gradient Uniform Series Factor
	To Find F Given P	To Find P Given F	To Find F Given A	To Find P Given A	To Find A Given F	To Find A Given P	To Find P Given G	To Find A Given G
N	F/P	P/F	F/A	P/A	A/F	A/P	P/G	A/G
1	1.0075	0.9926	1.0000	0.9926	1.0000	1.0075	0.000	0.0000
2	1.0151	0.9852	2.0075	1.9777	0.4981	0.5056	0.985	0.4981
3	1.0227	0.9778	3.0226	2.9556	0.3308	0.3383	2.941	0.9950
4	1.0303	0.9706	4.0452	3.9261	0.2472	0.2547	5.853	1.4907
5	1.0381	0.9633	5.0756	4.8894	0.1970	0.2045	9.706	1.9851
6	1.0459	0.9562	6.1136	5.8456	0.1636	0.1711	14.487	2.4782
7	1.0537	0.9490	7.1595	6.7946	0.1397	0.1472	20.181	2.9701
8	1.0616	0.9420	8.2132	7.7366	0.1218	0.1293	26.775	3.4608
9	1.0696	0.9350	9.2748	8.6716	0.1078	0.1153	34.254	3.9502
10	1.0776	0.9280	10.3443	9.5996	0.0967	0.1042	42.606	4.4384
11	1.0857	0.9211	11.4219	10.5207	0.0876	0.0951	51.817	4.9253
12	1.0938	0.9142	12.5076	11.4349	0.0800	0.0875	61.874	5.4110
13	1.1020	0.9074	13.6014	12.3423	0.0735	0.0810	72.763	5.8954
14	1.1103	0.9007	14.7034	13.2430	0.0680	0.0755	84.472	6.3786
15	1.1186	0.8940	15.8137	14.1370	0.0632	0.0707	96.988	6.8606
16	1.1270	0.8873	16.9323	15.0243	0.0591	0.0666	110.297	7.3413
17	1.1354	0.8807	18.0593	15.9050	0.0554	0.0629	124.389	7.8207
18	1.1440	0.8742	19.1947	16.7792	0.0521	0.0596	139.249	8.2989
19	1.1525	0.8676	20.3387	17.6468	0.0492	0.0567	154.867	8.7759
20	1.1612	0.8612	21.4912	18.5080	0.0465	0.0540	171.230	9.2516
21	1.1699	0.8548	22.6524	19.3628	0.0441	0.0516	188.325	9.7261
22	1.1787	0.8484	23.8223	20.2112	0.0420	0.0495	206.142	10.1994
23	1.1875	0.8421	25.0010	21.0533	0.0400	0.0475	224.668	10.6714
24	1.1964	0.8358	26.1885	21.8891	0.0382	0.0457	243.892	11.1422
25	1.2054	0.8296	27.3849	22.7188	0.0365	0.0440	263.803	11.6117
30	1.2513	0.7992	33.5029	26.7751	0.0298	0.0373	373.263	13.9407
36	1.3086	0.7641	41.1527	31.4468	0.0243	0.0318	524.992	16.6946
40	1.3483	0.7416	46.4464	34.4469	0.0215	0.0290	637.469	18.5058
48	1.4314	0.6986	57.5207	40.1848	0.0174	0.0249	886.840	22.0691
60	1.5657	0.6387	75.4241	48.1734	0.0133	0.0208	1313.519	27.2665
72	1.7126	0.5839	95.0070	55.4768	0.0105	0.0180	1791.246	32.2882
84	1.8732	0.5338	116.4269	62.1540	0.0086	0.0161	2308.128	37.1357
100	2.1111	0.4737	148.1445	70.1746	0.0068	0.0143	3040.745	43.3311
∞				133.3333		0.0075		

TABLE C-4 Discrete Compounding; $i = 1\%$

	Single Payment		Uniform Series					Uniform Gradient		
	Compound Amount Factor	Present Worth Factor	Compound Amount Factor	Present Worth Factor	Sinking Fund Factor	Capital Recovery Factor		Gradient Present Worth Factor	Gradient Uniform Series Factor	
N	To Find F Given P F/P	To Find P Given F P/F	To Find F Given A F/A	To Find P Given A P/A	To Find A Given F A/F	To Find A Given P A/P		To Find P Given G P/G	To Find A Given G A/G	N
1	1.0100	0.9901	1.0000	0.9901	1.0000	1.0100		0.000	0.0000	1
2	1.0201	0.9803	2.0100	1.9704	0.4975	0.5075		0.980	0.4975	2
3	1.0303	0.9706	3.0301	2.9410	0.3300	0.3400		2.922	0.9934	3
4	1.0406	0.9610	4.0604	3.9020	0.2463	0.2563		5.804	1.4876	4
5	1.0510	0.9515	5.1010	4.8534	0.1960	0.2060		9.610	1.9801	5
6	1.0615	0.9420	6.1520	5.7955	0.1625	0.1725		14.321	2.4710	6
7	1.0721	0.9327	7.2135	6.7282	0.1386	0.1486		19.917	2.9602	7
8	1.0829	0.9235	8.2857	7.6517	0.1207	0.1307		26.381	3.3478	8
9	1.0937	0.9143	9.3685	8.5660	0.1067	0.1167		33.696	3.9337	9
10	1.1046	0.9053	10.4622	9.4713	0.0956	0.1056		41.844	4.4179	10
11	1.1157	0.8963	11.5668	10.3676	0.0865	0.0965		50.807	4.9005	11
12	1.1268	0.8874	12.6825	11.2551	0.0788	0.0888		60.569	5.3815	12
13	1.1381	0.8787	13.8093	12.1337	0.0724	0.0824		71.113	5.8607	13
14	1.1495	0.8700	14.9474	13.0037	0.0669	0.0769		82.422	6.3384	14
15	1.1610	0.8613	16.0969	13.8651	0.0621	0.0721		94.481	6.8143	15
16	1.1726	0.8528	17.2579	14.7179	0.0579	0.0679		107.273	7.2886	16
17	1.1843	0.8444	18.4304	15.5623	0.0543	0.0643		120.783	7.7613	17
18	1.1961	0.8360	19.6147	16.3983	0.0510	0.0610		134.996	8.2323	18
19	1.2081	0.8277	20.8109	17.2260	0.0481	0.0581		149.895	8.7017	19
20	1.2202	0.8195	22.0190	18.0456	0.0454	0.0554		165.466	9.1694	20
21	1.2324	0.8114	23.2392	18.8570	0.0430	0.0530		181.695	9.6354	21
22	1.2447	0.8034	24.4716	19.6604	0.0409	0.0509		198.566	10.0998	22
23	1.2572	0.7954	25.7163	20.4558	0.0389	0.0489		216.066	10.5626	23
24	1.2697	0.7876	26.9734	21.2434	0.0371	0.0471		234.180	11.0237	24
25	1.2824	0.7798	28.2432	22.0232	0.0354	0.0454		252.895	11.4831	25
30	1.3478	0.7419	34.7849	25.8077	0.0287	0.0387		355.002	13.7557	30
36	1.4308	0.6989	43.0769	30.1075	0.0232	0.0332		494.621	16.4285	36
40	1.4889	0.6717	48.8863	32.8346	0.0205	0.0305		596.856	18.1776	40
48	1.6122	0.6203	61.2226	37.9740	0.0163	0.0263		820.146	21.5976	48
60	1.8167	0.5504	81.6697	44.9550	0.0122	0.0222		1192.806	26.5333	60
72	2.0471	0.4885	104.7099	51.1504	0.0096	0.0196		1597.867	31.2386	72
84	2.3067	0.4335	130.6723	56.6485	0.0077	0.0177		2023.315	35.7170	84
100	2.7048	0.3697	170.4814	63.0289	0.0059	0.0159		2605.776	41.3426	100
∞				100.0000		0.0100				∞

TABLE C-5 Discrete Compounding; $i = 2\%$

| | Single Payment | | Uniform Series | | | | | | Uniform Gradient | | |
| | Compound Amount Factor | Present Worth Factor | Compound Amount Factor | Present Worth Factor | Sinking Fund Factor | Capital Recovery Factor | | | Gradient Present Worth Factor | Gradient Uniform Series Factor | |
N	To Find F Given P F/P	To Find P Given F P/F	To Find F Given A F/A	To Find P Given A P/A	To Find A Given F A/F	To Find A Given P A/P			To Find P Given G P/G	To Find A Given G A/G	N
1	1.0200	0.9804	1.0000	0.9804	1.0000	1.0200			0.000	0.0000	1
2	1.0404	0.9612	2.0200	1.9416	0.4950	0.5150			0.961	0.4950	2
3	1.0612	0.9423	3.0604	2.8839	0.3268	0.3468			2.846	0.9868	3
4	1.0824	0.9238	4.1216	3.8077	0.2426	0.2626			5.617	1.4752	4
5	1.1041	0.9057	5.2040	4.7135	0.1922	0.2122			9.240	1.9604	5
6	1.1262	0.8880	6.3081	5.6014	0.1585	0.1785			13.680	2.4423	6
7	1.1487	0.8706	7.4343	6.4720	0.1345	0.1545			18.904	2.9208	7
8	1.1717	0.8535	8.5830	7.3255	0.1165	0.1365			24.878	3.3961	8
9	1.1951	0.8368	9.7546	8.1622	0.1025	0.1225			31.572	3.8681	9
10	1.2190	0.8203	10.9497	8.9826	0.0913	0.1113			38.955	4.3367	10
11	1.2434	0.8043	12.1687	9.7868	0.0822	0.1022			46.998	4.8021	11
12	1.2682	0.7885	13.4121	10.5753	0.0746	0.0946			55.671	5.2642	12
13	1.2936	0.7730	14.6803	11.3484	0.0681	0.0881			64.948	5.7231	13
14	1.3195	0.7579	15.9739	12.1062	0.0626	0.0826			74.800	6.1786	14
15	1.3459	0.7430	17.2934	12.8493	0.0578	0.0778			85.202	6.6309	15
16	1.3728	0.7284	18.6393	13.5777	0.0537	0.0737			96.129	7.0799	16
17	1.4002	0.7142	20.0121	14.2919	0.0500	0.0700			107.555	7.5256	17
18	1.4282	0.7002	21.4123	14.9920	0.0467	0.0667			119.458	7.9681	18
19	1.4568	0.6864	22.8406	15.6785	0.0438	0.0638			131.814	8.4073	19
20	1.4859	0.6730	24.2974	16.3514	0.0412	0.0612			144.600	8.8433	20
21	1.5157	0.6598	25.7833	17.0112	0.0388	0.0588			157.796	9.2760	21
22	1.5460	0.6468	27.2990	17.6580	0.0366	0.0566			171.380	9.7055	22
23	1.5769	0.6342	28.8450	18.2922	0.0347	0.0547			185.331	10.1317	23
24	1.6084	0.6217	30.4219	18.9139	0.0329	0.0529			199.631	10.5547	24
25	1.6406	0.6095	32.0303	19.5235	0.0312	0.0512			214.259	10.9745	25
30	1.8114	0.5521	40.5681	22.3965	0.0246	0.0446			291.716	13.0251	30
36	2.0399	0.4902	51.9944	25.4888	0.0192	0.0392			392.041	15.3809	36
40	2.2080	0.4529	60.4020	27.3555	0.0166	0.0366			461.993	16.8885	40
48	2.5871	0.3865	79.3535	30.6731	0.0126	0.0326			605.966	19.7556	48
60	3.2810	0.3048	114.0515	34.7609	0.0088	0.0288			823.698	23.6961	60
72	4.1611	0.2403	158.0570	37.9841	0.0063	0.0263			1034.056	27.2234	72
84	5.2773	0.1895	213.8666	40.5255	0.0047	0.0247			1230.419	30.3616	84
100	7.2446	0.1380	312.2323	43.0984	0.0032	0.0232			1464.753	33.9863	100
∞				50.0000		0.0200					∞

TABLE C-6 Discrete Compounding; i = 3%

| | Single Payment | | Uniform Series | | | | Uniform Gradient | | |
| | Compound Amount Factor | Present Worth Factor | Compound Amount Factor | Present Worth Factor | Sinking Fund Factor | Capital Recovery Factor | Gradient Present Worth Factor | Gradient Uniform Series Factor | |
N	To Find F Given P F/P	To Find P Given F P/F	To Find F Given A F/A	To Find P Given A P/A	To Find A Given F A/F	To Find A Given P A/P	To Find P Given G P/G	To Find A Given G A/G	N
1	1.0300	0.9709	1.0000	0.9709	1.0000	1.0300	0.000	0.0000	1
2	1.0609	0.9426	2.0300	1.9135	0.4926	0.5226	0.943	0.4926	2
3	1.0927	0.9151	3.0909	2.8286	0.3235	0.3535	2.773	0.9803	3
4	1.1255	0.8885	4.1836	3.7171	0.2390	0.2690	5.438	1.4631	4
5	1.1593	0.8626	5.3091	4.5797	0.1884	0.2184	8.889	1.9409	5
6	1.1941	0.8375	6.4684	5.4172	0.1546	0.1846	13.076	2.4138	6
7	1.2299	0.8131	7.6625	6.2303	0.1305	0.1605	17.955	2.8819	7
8	1.2668	0.7894	8.8923	7.0197	0.1125	0.1425	23.481	3.3450	8
9	1.3048	0.7664	10.1591	7.7861	0.0984	0.1284	29.612	3.8032	9
10	1.3439	0.7441	11.4639	8.5302	0.0872	0.1172	36.309	4.2565	10
11	1.3842	0.7224	12.8078	9.2526	0.0781	0.1081	43.533	4.7049	11
12	1.4258	0.7014	14.1920	9.9540	0.0705	0.1005	51.248	5.1485	12
13	1.4685	0.6810	15.6178	10.6350	0.0640	0.0940	59.420	5.5872	13
14	1.5126	0.6611	17.0863	11.2961	0.0585	0.0885	68.014	6.0210	14
15	1.5580	0.6419	18.5989	11.9379	0.0538	0.0838	77.000	6.4500	15
16	1.6047	0.6232	20.1569	12.5611	0.0496	0.0796	86.348	6.8742	16
17	1.6528	0.6050	21.7616	13.1661	0.0460	0.0760	96.028	7.2936	17
18	1.7024	0.5874	23.4144	13.7535	0.0427	0.0727	106.014	7.7081	18
19	1.7535	0.5703	25.1169	14.3238	0.0398	0.0698	116.279	8.1179	19
20	1.8061	0.5537	26.8704	14.8775	0.0372	0.0672	126.799	8.5229	20
21	1.8603	0.5375	28.6765	15.4150	0.0349	0.0649	137.550	8.9231	21
22	1.9161	0.5219	30.5368	15.9369	0.0327	0.0627	148.509	9.3186	22
23	1.9736	0.5067	32.4529	16.4436	0.0308	0.0608	159.657	9.7093	23
24	2.0328	0.4919	34.4265	16.9355	0.0290	0.0590	170.971	10.0954	24
25	2.0938	0.4776	36.4593	17.4131	0.0274	0.0574	182.434	10.4768	25
30	2.4273	0.4120	47.5754	19.6004	0.0210	0.0510	241.361	12.3141	30
35	2.8139	0.3554	60.4621	21.4872	0.0165	0.0465	301.627	14.0375	35
40	3.2620	0.3066	75.4012	23.1148	0.0133	0.0433	361.750	15.6502	40
45	3.7816	0.2644	92.7199	24.5187	0.0108	0.0408	420.633	17.1556	45
50	4.3839	0.2281	112.7969	25.7298	0.0089	0.0389	477.480	18.5575	50
60	5.8916	0.1697	163.0534	27.6756	0.0061	0.0361	583.053	21.0674	60
80	10.6409	0.0940	321.3630	30.2008	0.0031	0.0331	756.087	25.0353	80
100	19.2186	0.0520	607.2877	31.5989	0.0016	0.0316	879.854	27.8444	100
∞				33.3333		0.0300			∞

TABLE C-7 Discrete Compounding; i = 4%

	Single Payment		Uniform Series				Uniform Gradient		
	Compound Amount Factor	Present Worth Factor	Compound Amount Factor	Present Worth Factor	Sinking Fund Factor	Capital Recovery Factor	Gradient Present Worth Factor	Gradient Uniform Series Factor	
	To Find F Given P	To Find P Given F	To Find F Given A	To Find P Given A	To Find A Given F	To Find A Given P	To Find P Given G	To Find A Given G	
N	F/P	P/F	F/A	P/A	A/F	A/P	P/G	A/G	N
1	1.0400	0.9615	1.0000	0.9615	1.0000	1.0400	0.000	0.0000	1
2	1.0816	0.9246	2.0400	1.8861	0.4902	0.5302	0.925	0.4902	2
3	1.1249	0.8890	3.1216	2.7751	0.3203	0.3603	2.703	0.9739	3
4	1.1699	0.8548	4.2465	3.6299	0.2355	0.2755	5.267	1.4510	4
5	1.2167	0.8219	5.4163	4.4518	0.1846	0.2246	8.555	1.9216	5
6	1.2653	0.7903	6.6330	5.2421	0.1508	0.1908	12.506	2.3857	6
7	1.3159	0.7599	7.8983	6.0021	0.1266	0.1666	17.066	2.8433	7
8	1.3686	0.7307	9.2142	6.7327	0.1085	0.1485	22.181	3.2944	8
9	1.4233	0.7026	10.5828	7.4353	0.0945	0.1345	27.801	3.7391	9
10	1.4802	0.6756	12.0061	8.1109	0.0833	0.1233	33.881	4.1773	10
11	1.5395	0.6496	13.4864	8.7605	0.0741	0.1141	40.377	4.6090	11
12	1.6010	0.6246	15.0258	9.3851	0.0666	0.1066	47.248	5.0343	12
13	1.6651	0.6006	16.6268	9.9856	0.0601	0.1001	54.455	5.4533	13
14	1.7317	0.5775	18.2919	10.5631	0.0547	0.0947	61.962	5.8659	14
15	1.8009	0.5553	20.0236	11.1184	0.0499	0.0899	69.736	6.2721	15
16	1.8730	0.5339	21.8245	11.6523	0.0458	0.0858	77.744	6.6720	16
17	1.9479	0.5134	23.6975	12.1657	0.0422	0.0822	85.958	7.0656	17
18	2.0258	0.4936	25.6454	12.6593	0.0390	0.0790	94.350	7.4530	18
19	2.1068	0.4746	27.6712	13.1339	0.0361	0.0761	102.893	7.8342	19
20	2.1911	0.4564	29.7781	13.5903	0.0336	0.0736	111.565	8.2091	20
21	2.2788	0.4388	31.9692	14.0292	0.0313	0.0713	120.341	8.5779	21
22	2.3699	0.4220	34.2480	14.4511	0.0292	0.0692	129.202	8.9407	22
23	2.4647	0.4057	36.6179	14.8568	0.0273	0.0673	138.128	9.2973	23
24	2.5633	0.3901	39.0826	15.2470	0.0256	0.0656	147.101	9.6479	24
25	2.6658	0.3751	41.6459	15.6221	0.0240	0.0640	156.104	9.9925	25
30	3.2434	0.3083	56.0849	17.2920	0.0178	0.0578	201.062	11.6274	30
35	3.9461	0.2534	73.6522	18.6646	0.0136	0.0536	244.877	13.1198	35
40	4.8010	0.2083	95.0255	19.7928	0.0105	0.0505	286.530	14.4765	40
45	5.8412	0.1712	121.0294	20.7200	0.0083	0.0483	325.403	15.7047	45
50	7.1067	0.1407	152.6671	21.4822	0.0066	0.0466	361.164	16.8122	50
60	10.5196	0.0951	237.9907	22.6235	0.0042	0.0442	422.997	18.6972	60
80	23.0498	0.0434	551.2450	23.9154	0.0018	0.0418	511.116	21.3718	80
100	50.5049	0.0198	1237.6237	24.5050	0.0008	0.0408	563.125	22.9800	100
∞				25.0000		0.0400			∞

TABLE C-8 Discrete Compounding; $i = 5\%$

| | Single Payment | | Uniform Series | | | | | Uniform Gradient | |
| | Compound Amount Factor | Present Worth Factor | Compound Amount Factor | Present Worth Factor | Sinking Fund Factor | Capital Recovery Factor | Gradient Present Worth Factor | Gradient Uniform Series Factor | |
N	To Find F Given P F/P	To Find P Given F P/F	To Find F Given A F/A	To Find P Given A P/A	To Find A Given F A/F	To Find A Given P A/P	To Find P Given G P/G	To Find A Given G A/G	N
1	1.0500	0.9524	1.0000	0.9524	1.0000	1.0500	0.000	0.0000	1
2	1.1025	0.9070	2.0500	1.8594	0.4878	0.5378	0.907	0.4878	2
3	1.1576	0.8638	3.1525	2.7232	0.3172	0.3672	2.635	0.9675	3
4	1.2155	0.8227	4.3101	3.5460	0.2320	0.2820	5.103	1.4391	4
5	1.2763	0.7835	5.5256	4.3295	0.1810	0.2310	8.237	1.9025	5
6	1.3401	0.7462	6.8019	5.0757	0.1470	0.1970	11.968	2.3579	6
7	1.4071	0.7107	8.1420	5.7864	0.1228	0.1728	16.232	2.8052	7
8	1.4775	0.6768	9.5491	6.4632	0.1047	0.1547	20.970	3.2445	8
9	1.5513	0.6446	11.0266	7.1078	0.0907	0.1407	26.127	3.6758	9
10	1.6289	0.6139	12.5779	7.7217	0.0795	0.1295	31.652	4.0991	10
11	1.7103	0.5847	14.2068	8.3064	0.0704	0.1204	37.499	4.5144	11
12	1.7959	0.5568	15.9171	8.8633	0.0628	0.1128	43.624	4.9219	12
13	1.8856	0.5303	17.7130	9.3936	0.0565	0.1065	49.988	5.3215	13
14	1.9799	0.5051	19.5986	9.8986	0.0510	0.1010	56.554	5.7133	14
15	2.0789	0.4810	21.5786	10.3797	0.0463	0.0963	63.288	6.0973	15
16	2.1829	0.4581	23.6575	10.8378	0.0423	0.0923	70.160	6.4736	16
17	2.2920	0.4363	25.8404	11.2741	0.0387	0.0887	77.141	6.8423	17
18	2.4066	0.4155	28.1324	11.6896	0.0355	0.0855	84.204	7.2034	18
19	2.5270	0.3957	30.5390	12.0853	0.0327	0.0827	91.328	7.5569	19
20	2.6533	0.3769	33.0660	12.4622	0.0302	0.0802	98.488	7.9030	20
21	2.7860	0.3589	35.7193	12.8212	0.0280	0.0780	105.667	8.2416	21
22	2.9253	0.3418	38.5052	13.1630	0.0260	0.0760	112.846	8.5730	22
23	3.0715	0.3256	41.4305	13.4886	0.0241	0.0741	120.009	8.8971	23
24	3.2251	0.3101	44.5020	13.7986	0.0225	0.0725	127.140	9.2140	24
25	3.3864	0.2953	47.7271	14.0939	0.0210	0.0710	134.228	9.5238	25
30	4.3219	0.2314	66.4388	15.3725	0.0151	0.0651	168.623	10.9691	30
35	5.5160	0.1813	90.3203	16.3742	0.0111	0.0611	200.581	12.2498	35
40	7.0400	0.1420	120.7998	17.1591	0.0083	0.0583	229.545	13.3775	40
45	8.9850	0.1113	159.7002	17.7741	0.0063	0.0563	255.315	14.3644	45
50	11.4674	0.0872	209.3480	18.2559	0.0048	0.0548	277.915	15.2233	50
60	18.6792	0.0535	353.5837	18.9293	0.0028	0.0528	314.343	16.6062	60
80	49.5614	0.0202	971.2288	19.5965	0.0010	0.0510	359.646	18.3526	80
100	131.5013	0.0076	2610.0252	19.8479	0.0004	0.0504	381.749	19.2337	100
∞				20.0000		0.0500			∞

TABLE C-9 Discrete Compounding; *i* = 6%

	Single Payment		Uniform Series					Uniform Gradient	
	Compound Amount Factor	Present Worth Factor	Compound Amount Factor	Present Worth Factor	Sinking Fund Factor	Capital Recovery Factor		Gradient Present Worth Factor	Gradient Uniform Series Factor
	To Find F Given P	To Find P Given F	To Find F Given A	To Find P Given A	To Find A Given F	To Find A Given P		To Find P Given G	To Find A Given G
N	F/P	P/F	F/A	P/A	A/F	A/P		P/G	A/G	N
1	1.0600	0.9434	1.0000	0.9434	1.0000	1.0600		0.000	0.0000	1
2	1.1236	0.8900	2.0600	1.8334	0.4854	0.5454		0.890	0.4854	2
3	1.1910	0.8396	3.1836	2.6730	0.3141	0.3741		2.569	0.9612	3
4	1.2625	0.7921	4.3746	3.4651	0.2286	0.2886		4.946	1.4272	4
5	1.3382	0.7473	5.6371	4.2124	0.1774	0.2374		7.935	1.8836	5
6	1.4185	0.7050	6.9753	4.9173	0.1434	0.2034		11.459	2.3304	6
7	1.5036	0.6651	8.3938	5.5824	0.1191	0.1791		15.450	2.7676	7
8	1.5938	0.6274	9.8975	6.2098	0.1010	0.1610		19.842	3.1952	8
9	1.6895	0.5919	11.4913	6.8017	0.0870	0.1470		24.577	3.6133	9
10	1.7908	0.5584	13.1808	7.3601	0.0759	0.1359		29.602	4.0220	10
11	1.8983	0.5268	14.9716	7.8869	0.0668	0.1268		34.870	4.4213	11
12	2.0122	0.4970	16.8699	8.3838	0.0593	0.1193		40.337	4.8113	12
13	2.1329	0.4688	18.8821	8.8527	0.0530	0.1130		45.963	5.1920	13
14	2.2609	0.4423	21.0151	9.2950	0.0476	0.1076		51.713	5.5635	14
15	2.3966	0.4173	23.2760	9.7122	0.0430	0.1030		57.555	5.9260	15
16	2.5404	0.3936	25.6725	10.1059	0.0390	0.0990		63.459	6.2794	16
17	2.6928	0.3714	28.2129	10.4773	0.0354	0.0954		69.401	6.6240	17
18	2.8543	0.3503	30.9057	10.8276	0.0324	0.0924		75.357	6.9597	18
19	3.0256	0.3305	33.7600	11.1581	0.0296	0.0896		81.306	7.2867	19
20	3.2071	0.3118	36.7856	11.4699	0.0272	0.0872		87.230	7.6051	20
21	3.3996	0.2942	39.9927	11.7641	0.0250	0.0850		93.114	7.9151	21
22	3.6035	0.2775	43.3923	12.0416	0.0230	0.0830		98.941	8.2166	22
23	3.8197	0.2618	46.9958	12.3034	0.0213	0.0813		104.701	8.5099	23
24	4.0489	0.2470	50.8156	12.5504	0.0197	0.0797		110.381	8.7951	24
25	4.2919	0.2330	54.8645	12.7834	0.0182	0.0782		115.973	9.0722	25
30	5.7435	0.1741	79.0582	13.7648	0.0126	0.0726		142.359	10.3422	30
35	7.6861	0.1301	111.4348	14.4982	0.0090	0.0690		165.743	11.4319	35
40	10.2857	0.0972	154.7620	15.0463	0.0065	0.0665		185.957	12.3590	40
45	13.7646	0.0727	212.7435	15.4558	0.0047	0.0647		203.110	13.1413	45
50	18.4202	0.0543	290.3359	15.7619	0.0034	0.0634		217.457	13.7964	50
60	32.9877	0.0303	533.1282	16.1614	0.0019	0.0619		239.043	14.7909	60
80	105.7960	0.0095	1746.5999	16.5091	0.0006	0.0606		262.549	15.9033	80
100	339.3021	0.0029	5638.3681	16.6175	0.0002	0.0602		272.047	16.3711	100
∞				16.6667		0.0600				∞

TABLE C-10 Discrete Compounding; $i = 7\%$

	Single Payment		Uniform Series						Uniform Gradient		
	Compound Amount Factor	Present Worth Factor	Compound Amount Factor	Present Worth Factor	Sinking Fund Factor	Capital Recovery Factor			Gradient Present Worth Factor	Gradient Uniform Series Factor	
	To Find F Given P	To Find P Given F	To Find F Given A	To Find P Given A	To Find A Given F	To Find A Given P			To Find P Given G	To Find A Given G	
N	F/P	P/F	F/A	P/A	A/F	A/P			P/G	A/G	N
1	1.0700	0.9346	1.0000	0.9346	1.0000	1.0700			0.000	0.0000	1
2	1.1449	0.8734	2.0700	1.8080	0.4831	0.5531			0.873	0.4831	2
3	1.2250	0.8163	3.2149	2.6243	0.3111	0.3811			2.506	0.9549	3
4	1.3108	0.7629	4.4399	3.3872	0.2252	0.2952			4.795	1.4155	4
5	1.4026	0.7130	5.7507	4.1002	0.1739	0.2439			7.647	1.8650	5
6	1.5007	0.6663	7.1533	4.7665	0.1398	0.2098			10.978	2.3032	6
7	1.6058	0.6227	8.6540	5.3893	0.1156	0.1856			14.715	2.7304	7
8	1.7182	0.5820	10.2598	5.9713	0.0975	0.1675			18.789	3.1465	8
9	1.8385	0.5439	11.9780	6.5152	0.0835	0.1535			23.140	3.5517	9
10	1.9672	0.5083	13.8164	7.0236	0.0724	0.1424			27.716	3.9461	10
11	2.1049	0.4751	15.7836	7.4987	0.0634	0.1334			32.467	4.3296	11
12	2.2522	0.4440	17.8885	7.9427	0.0559	0.1259			37.351	4.7025	12
13	2.4098	0.4150	20.1406	8.3577	0.0497	0.1197			42.330	5.0648	13
14	2.5785	0.3878	22.5505	8.7455	0.0443	0.1143			47.372	5.4167	14
15	2.7590	0.3624	25.1290	9.1079	0.0398	0.1098			52.446	5.7583	15
16	2.9522	0.3387	27.8881	9.4466	0.0359	0.1059			57.527	6.0897	16
17	3.1588	0.3166	30.8402	9.7632	0.0324	0.1024			62.592	6.4110	17
18	3.3799	0.2959	33.9990	10.0591	0.0294	0.0994			67.622	6.7225	18
19	3.6165	0.2765	37.3790	10.3356	0.0268	0.0968			72.599	7.0242	19
20	3.8697	0.2584	40.9955	10.5940	0.0244	0.0944			77.509	7.3163	20
21	4.1406	0.2415	44.8652	10.8355	0.0223	0.0923			82.339	7.5990	21
22	4.4304	0.2257	49.0057	11.0612	0.0204	0.0904			87.079	7.8725	22
23	4.7405	0.2109	53.4361	11.2722	0.0187	0.0887			91.720	8.1369	23
24	5.0724	0.1971	58.1767	11.4693	0.0172	0.0872			96.255	8.3923	24
25	5.4274	0.1842	63.2490	11.6536	0.0158	0.0858			100.677	8.6391	25
30	7.6123	0.1314	94.4608	12.4090	0.0106	0.0806			120.972	9.7487	30
35	10.6766	0.0937	138.2369	12.9477	0.0072	0.0772			138.135	10.6687	35
40	14.9745	0.0668	199.6351	13.3317	0.0050	0.0750			152.293	11.4233	40
45	21.0023	0.0476	285.7495	13.6055	0.0035	0.0735			163.756	12.0360	45
50	29.4570	0.0339	406.5289	13.8007	0.0025	0.0725			172.905	12.5287	50
60	57.9464	0.0173	813.5204	14.0392	0.0012	0.0712			185.768	13.2321	60
80	224.2344	0.0045	3189.0627	14.2220	0.0003	0.0703			198.075	13.9273	80
100	867.7163	0.0012	12381.6618	14.2693	0.0001	0.0701			202.200	14.1703	100
∞				14.2857		0.0700					∞

TABLE C-11 Discrete Compounding; $i = 8\%$

	Single Payment		Uniform Series						Uniform Gradient	
	Compound Amount Factor	Present Worth Factor	Compound Amount Factor	Present Worth Factor	Sinking Fund Factor	Capital Recovery Factor			Gradient Present Worth Factor	Gradient Uniform Series Factor
	To Find F Given P	To Find P Given F	To Find F Given A	To Find P Given A	To Find A Given F	To Find A Given P			To Find P Given G	To Find A Given G
N	F/P	P/F	F/A	P/A	A/F	A/P			P/G	A/G	N
1	1.0800	0.9259	1.0000	0.9259	1.0000	1.0800			0.000	0.0000	1
2	1.1664	0.8573	2.0800	1.7833	0.4808	0.5608			0.857	0.4808	2
3	1.2597	0.7938	3.2464	2.5771	0.3080	0.3880			2.445	0.9487	3
4	1.3605	0.7350	4.5061	3.3121	0.2219	0.3019			4.650	1.4040	4
5	1.4693	0.6806	5.8666	3.9927	0.1705	0.2505			7.372	1.8465	5
6	1.5869	0.6302	7.3359	4.6229	0.1363	0.2163			10.523	2.2763	6
7	1.7138	0.5835	8.9228	5.2064	0.1121	0.1921			14.024	2.6937	7
8	1.8509	0.5403	10.6366	5.7466	0.0940	0.1740			17.806	3.0985	8
9	1.9990	0.5002	12.4876	6.2469	0.0801	0.1601			21.808	3.4910	9
10	2.1589	0.4632	14.4866	6.7101	0.0690	0.1490			25.977	3.8713	10
11	2.3316	0.4289	16.6455	7.1390	0.0601	0.1401			30.266	4.2395	11
12	2.5182	0.3971	18.9771	7.5361	0.0527	0.1327			34.634	4.5957	12
13	2.7196	0.3677	21.4953	7.9038	0.0465	0.1265			39.046	4.9402	13
14	2.9372	0.3405	24.2149	8.2442	0.0413	0.1213			43.472	5.2731	14
15	3.1722	0.3152	27.1521	8.5595	0.0368	0.1168			47.886	5.5945	15
16	3.4259	0.2919	30.3243	8.8514	0.0330	0.1130			52.264	5.9046	16
17	3.7000	0.2703	33.7502	9.1216	0.0296	0.1096			56.588	6.2037	17
18	3.9960	0.2502	37.4502	9.3719	0.0267	0.1067			60.843	6.4920	18
19	4.3157	0.2317	41.4463	9.6036	0.0241	0.1041			65.013	6.7697	19
20	4.6610	0.2145	45.7620	9.8181	0.0219	0.1019			69.090	7.0369	20
21	5.0338	0.1987	50.4229	10.0168	0.0198	0.0998			73.063	7.2940	21
22	5.4365	0.1839	55.4568	10.2007	0.0180	0.0980			76.926	7.5412	22
23	5.8715	0.1703	60.8933	10.3711	0.0164	0.0964			80.673	7.7786	23
24	6.3412	0.1577	66.7648	10.5288	0.0150	0.0950			84.300	8.0066	24
25	6.8485	0.1460	73.1059	10.6748	0.0137	0.0937			87.804	8.2254	25
30	10.0627	0.0994	113.2832	11.2578	0.0088	0.0888			103.456	9.1897	30
35	14.7853	0.0676	172.3168	11.6546	0.0058	0.0858			116.092	9.9611	35
40	21.7245	0.0460	259.0565	11.9246	0.0039	0.0839			126.042	10.5699	40
45	31.9204	0.0313	386.5056	12.1084	0.0026	0.0826			133.733	11.0447	45
50	46.9016	0.0213	573.7702	12.2335	0.0017	0.0817			139.593	11.4107	50
60	101.2571	0.0099	1253.2133	12.3766	0.0008	0.0808			147.300	11.9015	60
80	471.9548	0.0021	5886.9354	12.4735	0.0002	0.0802			153.800	12.3301	80
100	2199.7613	0.0005	27484.5157	12.4943	*a*	0.0800			155.611	12.4545	100
∞				12.5000		0.0800					∞

a Less than 0.0001.

TABLE C-12 Discrete Compounding; i = 9%

| | Single Payment | | Uniform Series | | | | Uniform Gradient | | |
| | Compound Amount Factor | Present Worth Factor | Compound Amount Factor | Present Worth Factor | Sinking Fund Factor | Capital Recovery Factor | Gradient Present Worth Factor | Gradient Uniform Series Factor | |
N	To Find F Given P F/P	To Find P Given F P/F	To Find F Given A F/A	To Find P Given A P/A	To Find A Given F A/F	To Find A Given P A/P	To Find P Given G P/G	To Find A Given G A/G	N
1	1.0900	0.9174	1.0000	0.9174	1.0000	1.0900	0.000	0.0000	1
2	1.1881	0.8417	2.0900	1.7591	0.4785	0.5685	0.842	0.4785	2
3	1.2950	0.7722	3.2781	2.5313	0.3051	0.3951	2.386	0.9426	3
4	1.4116	0.7084	4.5731	3.2397	0.2187	0.3087	4.511	1.3925	4
5	1.5386	0.6499	5.9847	3.8897	0.1671	0.2571	7.111	1.8282	5
6	1.6771	0.5963	7.5233	4.4859	0.1329	0.2229	10.092	2.2498	6
7	1.8280	0.5470	9.2004	5.0330	0.1087	0.1987	13.375	2.6574	7
8	1.9926	0.5019	11.0285	5.5348	0.0907	0.1807	16.888	3.0512	8
9	2.1719	0.4604	13.0210	5.9952	0.0768	0.1668	20.571	3.4312	9
10	2.3674	0.4224	15.1929	6.4177	0.0658	0.1558	24.373	3.7978	10
11	2.5804	0.3875	17.5603	6.8052	0.0569	0.1469	28.248	4.1510	11
12	2.8127	0.3555	20.1407	7.1607	0.0497	0.1397	32.159	4.4910	12
13	3.0658	0.3262	22.9534	7.4869	0.0436	0.1336	36.073	4.8182	13
14	3.3417	0.2992	26.0192	7.7862	0.0384	0.1284	39.963	5.1326	14
15	3.6425	0.2745	29.3609	8.0607	0.0341	0.1241	43.807	5.4346	15
16	3.9703	0.2519	33.0034	8.3126	0.0303	0.1203	47.585	5.7245	16
17	4.3276	0.2311	36.9737	8.5436	0.0270	0.1170	51.282	6.0024	17
18	4.7171	0.2120	41.3013	8.7556	0.0242	0.1142	54.886	6.2687	18
19	5.1417	0.1945	46.0185	8.9501	0.0217	0.1117	58.387	6.5236	19
20	5.6044	0.1784	51.1601	9.1285	0.0195	0.1095	61.777	6.7674	20
21	6.1088	0.1637	56.7645	9.2922	0.0176	0.1076	65.051	7.0006	21
22	6.6586	0.1502	62.8733	9.4424	0.0159	0.1059	68.205	7.2232	22
23	7.2579	0.1378	69.5319	9.5802	0.0144	0.1044	71.236	7.4357	23
24	7.9111	0.1264	76.7898	9.7066	0.0130	0.1030	74.143	7.6384	24
25	8.6231	0.1160	84.7009	9.8226	0.0118	0.1018	76.927	7.8316	25
30	13.2677	0.0754	136.3075	10.2737	0.0073	0.0973	89.028	8.6657	30
35	20.4140	0.0490	215.7108	10.5668	0.0046	0.0946	98.359	9.3083	35
40	31.4094	0.0318	337.8824	10.7574	0.0030	0.0930	105.376	9.7957	40
45	48.3273	0.0207	525.8587	10.8812	0.0019	0.0919	110.556	10.1603	45
50	74.3575	0.0134	815.0836	10.9617	0.0012	0.0912	114.325	10.4295	50
60	176.0313	0.0057	1944.7921	11.0480	0.0005	0.0905	118.968	10.7683	60
80	986.5517	0.0010	10950.5741	11.0998	0.0001	0.0901	122.431	11.0299	80
100	5529.0408	0.0002	61422.6755	11.1091	a	0.0900	123.234	11.0930	100
∞				11.1111		0.0900			∞

a Less than 0.0001.

TABLE C-13 Discrete Compounding; $i = 10\%$

	Single Payment		Uniform Series					Uniform Gradient		
	Compound Amount Factor	Present Worth Factor	Compound Amount Factor	Present Worth Factor	Sinking Fund Factor	Capital Recovery Factor		Gradient Present Worth Factor	Gradient Uniform Series Factor	
	To Find F Given P	To Find P Given F	To Find F Given A	To Find P Given A	To Find A Given F	To Find A Given P		To Find P Given G	To Find A Given G	
N	F/P	P/F	F/A	P/A	A/F	A/P		P/G	A/G	N
1	1.1000	0.9091	1.0000	0.9091	1.0000	1.1000		0.000	0.0000	1
2	1.2100	0.8264	2.1000	1.7355	0.4762	0.5762		0.826	0.4762	2
3	1.3310	0.7513	3.3100	2.4869	0.3021	0.4021		2.329	0.9366	3
4	1.4641	0.6830	4.6410	3.1699	0.2155	0.3155		4.378	1.3812	4
5	1.6105	0.6209	6.1051	3.7908	0.1638	0.2638		6.862	1.8101	5
6	1.7716	0.5645	7.7156	4.3553	0.1296	0.2296		9.684	2.2236	6
7	1.9487	0.5132	9.4872	4.8684	0.1054	0.2054		12.763	2.6216	7
8	2.1436	0.4665	11.4359	5.3349	0.0874	0.1874		16.029	3.0045	8
9	2.3579	0.4241	13.5795	5.7590	0.0736	0.1736		19.422	3.3724	9
10	2.5937	0.3855	15.9374	6.1446	0.0627	0.1627		22.891	3.7255	10
11	2.8531	0.3505	18.5312	6.4951	0.0540	0.1540		26.396	4.0641	11
12	3.1384	0.3186	21.3843	6.8137	0.0468	0.1468		29.901	4.3884	12
13	3.4523	0.2897	24.5227	7.1034	0.0408	0.1408		33.377	4.6988	13
14	3.7975	0.2633	27.9750	7.3667	0.0357	0.1357		36.801	4.9955	14
15	4.1772	0.2394	31.7725	7.6061	0.0315	0.1315		40.152	5.2789	15
16	4.5950	0.2176	35.9497	7.8237	0.0278	0.1278		43.416	5.5493	16
17	5.0545	0.1978	40.5447	8.0216	0.0247	0.1247		46.582	5.8071	17
18	5.5599	0.1799	45.5992	8.2014	0.0219	0.1219		49.640	6.0526	18
19	6.1159	0.1635	51.1591	8.3649	0.0195	0.1195		52.583	6.2861	19
20	6.7275	0.1486	57.2750	8.5136	0.0175	0.1175		55.407	6.5081	20
21	7.4002	0.1351	64.0025	8.6487	0.0156	0.1156		58.110	6.7189	21
22	8.1403	0.1228	71.4027	8.7715	0.0140	0.1140		60.689	6.9189	22
23	8.9543	0.1117	79.5430	8.8832	0.0126	0.1126		63.146	7.1085	23
24	9.8497	0.1015	88.4973	8.9847	0.0113	0.1113		65.481	7.2881	24
25	10.8347	0.0923	98.3471	9.0770	0.0102	0.1102		67.696	7.4580	25
30	17.4494	0.0573	164.4940	9.4269	0.0061	0.1061		77.077	8.1762	30
35	28.1024	0.0356	271.0244	9.6442	0.0037	0.1037		83.987	8.7086	35
40	45.2593	0.0221	442.5926	9.7791	0.0023	0.1023		88.953	9.0962	40
45	72.8905	0.0137	718.9048	9.8628	0.0014	0.1014		92.454	9.3740	45
50	117.3909	0.0085	1163.9085	9.9148	0.0009	0.1009		94.889	9.5704	50
60	304.4816	0.0033	3034.8164	9.9672	0.0003	0.1003		97.701	9.8023	60
80	2048.4002	0.0005	20474.0021	9.9951	a	0.1000		99.561	9.9609	80
100	13780.6123	0.0001	137796.1234	9.9993	a	0.1000		99.920	9.9927	100
∞				10.0000		0.1000				∞

a Less than 0.0001.

TABLE C-14 Discrete Compounding; $i = 12\%$

	Single Payment		Uniform Series				Uniform Gradient		
	Compound Amount Factor	Present Worth Factor	Compound Amount Factor	Present Worth Factor	Sinking Fund Factor	Capital Recovery Factor	Gradient Present Worth Factor	Gradient Uniform Series Factor	
	To Find F Given P	To Find P Given F	To Find F Given A	To Find P Given A	To Find A Given F	To Find A Given P	To Find P Given G	To Find A Given G	
N	F/P	P/F	F/A	P/A	A/F	A/P	P/G	A/G	N
1	1.1200	0.8929	1.0000	0.8929	1.0000	1.1200	0.000	0.0000	1
2	1.2544	0.7972	2.1200	1.6901	0.4717	0.5917	0.797	0.4717	2
3	1.4049	0.7118	3.3744	2.4018	0.2963	0.4163	2.221	0.9246	3
4	1.5735	0.6355	4.7793	3.0373	0.2092	0.3292	4.127	1.3589	4
5	1.7623	0.5674	6.3528	3.5048	0.1574	0.2774	6.397	1.7746	5
6	1.9738	0.5066	8.1152	4.1114	0.1232	0.2432	8.930	2.1720	6
7	2.2107	0.4523	10.0890	4.5638	0.0991	0.2191	11.644	2.5515	7
8	2.4760	0.4039	12.2997	4.9676	0.0813	0.2013	14.471	2.9131	8
9	2.7731	0.3606	14.7757	5.3282	0.0677	0.1877	17.356	3.2574	9
10	3.1058	0.3220	17.5487	5.6502	0.0570	0.1770	20.254	3.5847	10
11	3.4785	0.2875	20.6546	5.9377	0.0484	0.1684	23.129	3.8953	11
12	3.8960	0.2567	24.1331	6.1944	0.0414	0.1614	25.952	4.1897	12
13	4.3635	0.2292	28.0291	6.4235	0.0357	0.1557	28.702	4.4683	13
14	4.8871	0.2046	32.3926	6.6282	0.0309	0.1509	31.362	4.7317	14
15	5.4736	0.1827	37.2797	6.8109	0.0268	0.1468	33.920	4.9803	15
16	6.1304	0.1631	42.7533	6.9740	0.0234	0.1434	36.367	5.2147	16
17	6.8660	0.1456	48.8837	7.1196	0.0205	0.1405	38.697	5.4353	17
18	7.6900	0.1300	55.7497	7.2497	0.0179	0.1379	40.908	5.6427	18
19	8.6128	0.1161	63.4397	7.3658	0.0158	0.1358	42.998	5.8375	19
20	9.6463	0.1037	72.0524	7.4694	0.0139	0.1339	44.968	6.0202	20
21	10.8038	0.0926	81.6987	7.5620	0.0122	0.1322	46.819	6.1913	21
22	12.1003	0.0826	92.5026	7.6446	0.0108	0.1308	48.554	6.3514	22
23	13.5523	0.0738	104.6029	7.7184	0.0096	0.1296	50.178	6.5010	23
24	15.1786	0.0659	118.1552	7.7843	0.0085	0.1285	51.693	6.6406	24
25	17.0001	0.0588	133.3339	7.8431	0.0075	0.1275	53.105	6.7708	25
30	29.9599	0.0334	241.3327	8.0552	0.0041	0.1241	58.782	7.2974	30
35	52.7996	0.0189	431.6635	8.1755	0.0023	0.1223	62.605	7.6577	35
40	93.0510	0.0107	767.0914	8.2433	0.0013	0.1213	65.116	7.8988	40
45	163.9876	0.0061	1358.2300	8.2825	0.0007	0.1207	66.734	8.0572	45
50	289.0022	0.0035	2400.0182	8.3045	0.0004	0.1204	67.762	8.1597	50
60	897.5969	0.0011	7471.6411	8.3240	0.0001	0.1201	68.810	8.2664	60
80	8658.4831	0.0001	72145.6925	8.3324	^{a}	0.1200	69.359	8.3241	80
100	83522.2657	^{a}	696010.5477	8.3332	^{a}	0.1200	69.434	8.3321	100
∞				8.3333		0.1200			∞

^{a} Less than 0.0001.

TABLE C-15 Discrete Compounding; $i = 15\%$

	Single Payment		Uniform Series				Uniform Gradient	
	Compound Amount Factor	Present Worth Factor	Compound Amount Factor	Present Worth Factor	Sinking Fund Factor	Capital Recovery Factor	Gradient Present Worth Factor	Gradient Uniform Series Factor
	To Find F Given P	To Find P Given F	To Find F Given A	To Find P Given A	To Find A Given F	To Find A Given P	To Find P Given G	To Find A Given G
N	F/P	P/F	F/A	P/A	A/F	A/P	P/G	A/G
1	1.1500	0.8696	1.0000	0.8696	1.0000	1.1500	0.000	0.0000	1
2	1.3225	0.7561	2.1500	1.6257	0.4651	0.6151	0.756	0.4651	2
3	1.5209	0.6575	3.4725	2.2832	0.2880	0.4380	2.071	0.9071	3
4	1.7490	0.5718	4.9934	2.8550	0.2003	0.3503	3.786	1.3263	4
5	2.0114	0.4972	6.7424	3.3522	0.1483	0.2983	5.775	1.7228	5
6	2.3131	0.4323	8.7537	3.7845	0.1142	0.2642	7.937	2.0972	6
7	2.6600	0.3759	11.0668	4.1604	0.0904	0.2404	10.192	2.4498	7
8	3.0590	0.3269	13.7268	4.4873	0.0729	0.2229	12.481	2.7813	8
9	3.5179	0.2843	16.7858	4.7716	0.0596	0.2096	14.755	3.0922	9
10	4.0456	0.2472	20.3037	5.0188	0.0493	0.1993	16.980	3.3832	10
11	4.6524	0.2149	24.3493	5.2337	0.0411	0.1911	19.129	3.6549	11
12	5.3503	0.1869	29.0017	5.4206	0.0345	0.1845	21.185	3.9082	12
13	6.1528	0.1625	34.3519	5.5831	0.0291	0.1791	23.135	4.1438	13
14	7.0757	0.1413	40.5047	5.7245	0.0247	0.1747	24.973	4.3624	14
15	8.1371	0.1229	47.5804	5.8474	0.0210	0.1710	26.693	4.5650	15
16	9.3576	0.1069	55.7175	5.9542	0.0179	0.1679	28.296	4.7522	16
17	10.7613	0.0929	65.0751	6.0472	0.0154	0.1654	29.783	4.9251	17
18	12.3755	0.0808	75.8364	6.1280	0.0132	0.1632	31.157	5.0843	18
19	14.2318	0.0703	88.2118	6.1982	0.0113	0.1613	32.421	5.2307	19
20	16.3665	0.0611	102.4436	6.2593	0.0098	0.1598	33.582	5.3651	20
21	18.8215	0.0531	118.8101	6.3125	0.0084	0.1584	34.645	5.4883	21
22	21.6447	0.0462	137.6316	6.3587	0.0073	0.1573	35.615	5.6010	22
23	24.8915	0.0402	159.2764	6.3988	0.0063	0.1563	36.499	5.7040	23
24	28.6252	0.0349	184.1678	6.4338	0.0054	0.1554	37.302	5.7979	24
25	32.9190	0.0304	212.7930	6.4641	0.0047	0.1547	38.031	5.8834	25
30	66.2118	0.0151	434.7451	6.5660	0.0023	0.1523	40.753	6.2066	30
35	133.1755	0.0075	881.1702	6.6166	0.0011	0.1511	42.359	6.4019	35
40	267.8635	0.0037	1779.0903	6.6418	0.0006	0.1506	43.283	6.5168	40
45	538.7693	0.0019	3585.1285	6.6543	0.0003	0.1503	43.805	6.5830	45
50	1083.6574	0.0009	7217.7163	6.6605	0.0001	0.1501	44.096	6.6205	50
60	4383.9987	0.0002	29219.9916	6.6651	a	0.1500	44.343	6.6530	60
80	71750.8794	a	478332.5293	6.6666	a	0.1500	44.436	6.6656	80
100	1174313.4507	a	7828749.6713	6.6667	a	0.1500	44.444	6.6666	100
∞				6.6667		0.1500			∞

a Less than 0.0001.

TABLE C-16 Discrete Compounding; $i = 18\%$

	Single Payment		Uniform Series				Uniform Gradient		
	Compound Amount Factor	Present Worth Factor	Compound Amount Factor	Present Worth Factor	Sinking Fund Factor	Capital Recovery Factor	Gradient Present Worth Factor	Gradient Uniform Series Factor	
	To Find F Given P	To Find P Given F	To Find F Given A	To Find P Given A	To Find A Given F	To Find A Given P	To Find P Given G	To Find A Given G	
N	F/P	P/F	F/A	P/A	A/F	A/P	P/G	A/G	N
1	1.1800	0.8475	1.0000	0.8475	1.0000	1.1800	0.000	0.0000	1
2	1.3924	0.7182	2.1800	1.5655	0.4587	0.6387	0.718	0.4587	2
3	1.6430	0.6086	3.5724	2.1743	0.2799	0.4599	1.935	0.8902	3
4	1.9388	0.5158	5.2154	2.6901	0.1917	0.3717	3.483	1.2947	4
5	2.2878	0.4371	7.1542	3.1272	0.1398	0.3198	5.231	1.6728	5
6	2.6996	0.3704	9.4420	3.4976	0.1059	0.2859	7.083	2.0252	6
7	3.1855	0.3139	12.1415	3.8115	0.0824	0.2624	8.967	2.3526	7
8	3.7589	0.2660	15.3270	4.0776	0.0652	0.2452	10.829	2.6558	8
9	4.4355	0.2255	19.0859	4.3030	0.0524	0.2324	12.633	2.9358	9
10	5.2338	0.1911	23.5213	4.4941	0.0425	0.2225	14.353	3.1936	10
11	6.1759	0.1619	28.7551	4.6560	0.0348	0.2148	15.972	3.4303	11
12	7.2876	0.1372	34.9311	4.7932	0.0286	0.2086	17.481	3.6470	12
13	8.5994	0.1163	42.2187	4.9095	0.0237	0.2037	18.877	3.8449	13
14	10.1472	0.0985	50.8180	5.0081	0.0197	0.1997	20.158	4.0250	14
15	11.9737	0.0835	60.9653	5.0916	0.0164	0.1964	21.327	4.1887	15
16	14.1290	0.0708	72.9390	5.1624	0.0137	0.1937	22.389	4.3369	16
17	16.6722	0.0600	87.0680	5.2223	0.0115	0.1915	23.348	4.4708	17
18	19.6733	0.0508	103.7403	5.2732	0.0096	0.1896	24.212	4.5916	18
19	23.2144	0.0431	123.4135	5.3162	0.0081	0.1881	24.988	4.7003	19
20	27.3930	0.0365	146.6280	5.3527	0.0068	0.1868	25.681	4.7978	20
21	32.3238	0.0309	174.0210	5.3837	0.0057	0.1857	26.300	4.8851	21
22	38.1421	0.0262	206.3448	5.4099	0.0048	0.1848	26.851	4.9632	22
23	45.0076	0.0222	244.4868	5.4321	0.0041	0.1841	27.339	5.0329	23
24	53.1090	0.0188	289.4945	5.4509	0.0035	0.1835	27.773	5.0950	24
25	62.6686	0.0160	342.6035	5.4669	0.0029	0.1829	28.156	5.1502	25
30	143.3706	0.0070	790.9480	5.5168	0.0013	0.1813	29.486	5.3448	30
35	327.9973	0.0030	1816.6516	5.5386	0.0006	0.1806	30.177	5.4485	35
40	750.3783	0.0013	4163.2130	5.5482	0.0002	0.1802	30.527	5.5022	40
45	1716.6839	0.0006	9531.5771	5.5523	0.0001	0.1801	30.701	5.5293	45
50	3927.3569	0.0003	21813.0937	5.5541	a	0.1800	30.786	5.5428	50
60	20555.1400	a	114189.6665	5.5553	a	0.1800	30.847	5.5526	60
80	563067.6604	a	3128148.1133	5.5555	a	0.1800	30.863	5.5554	80
∞				5.5556		0.1800			∞

^{a} Less than 0.0001.

TABLE C-17 Discrete Compounding; $i = 20\%$

	Single Payment		Uniform Series						Uniform Gradient		
	Compound Amount Factor	Present Worth Factor	Compound Amount Factor	Present Worth Factor	Sinking Fund Factor	Capital Recovery Factor	Gradient Present Worth Factor	Gradient Uniform Series Factor			
	To Find F Given P	To Find P Given F	To Find F Given A	To Find P Given A	To Find A Given F	To Find A Given P	To Find P Given G	To Find A Given G			
N	F/P	P/F	F/A	P/A	A/F	A/P	P/G	A/G			N
1	1.2000	0.8333	1.0000	0.8333	1.0000	1.2000	0.000	0.0000			1
2	1.4400	0.6944	2.2000	1.5278	0.4545	0.6545	0.694	0.4545			2
3	1.7280	0.5787	3.6400	2.1065	0.2747	0.4747	1.852	0.8791			3
4	2.0736	0.4823	5.3680	2.5887	0.1863	0.3863	3.299	1.2742			4
5	2.4883	0.4019	7.4416	2.9906	0.1344	0.3344	4.906	1.6405			5
6	2.9860	0.3349	9.9299	3.3255	0.1007	0.3007	6.581	1.9788			6
7	3.5832	0.2791	12.9159	3.6046	0.0774	0.2774	8.255	2.2902			7
8	4.2998	0.2326	16.4991	3.8372	0.0606	0.2606	9.883	2.5756			8
9	5.1598	0.1938	20.7989	4.0310	0.0481	0.2481	11.434	2.8364			9
10	6.1917	0.1615	25.9587	4.1925	0.0385	0.2385	12.887	3.0739			10
11	7.4301	0.1346	32.1504	4.3271	0.0311	0.2311	14.233	3.2893			11
12	8.9161	0.1122	39.5805	4.4392	0.0253	0.2253	15.467	3.4841			12
13	10.6993	0.0935	48.4966	4.5327	0.0206	0.2206	16.588	3.6597			13
14	12.8392	0.0779	59.1959	4.6106	0.0169	0.2169	17.601	3.8175			14
15	15.4070	0.0649	72.0351	4.6755	0.0139	0.2139	18.510	3.9588			15
16	18.4884	0.0541	87.4421	4.7296	0.0114	0.2114	19.321	4.0851			16
17	22.1861	0.0451	105.9306	4.7746	0.0094	0.2094	20.042	4.1976			17
18	26.6233	0.0376	128.1167	4.8122	0.0078	0.2078	20.681	4.2975			18
19	31.9480	0.0313	154.7400	4.8435	0.0065	0.2065	21.244	4.3861			19
20	38.3376	0.0261	186.6880	4.8696	0.0054	0.2054	21.740	4.4643			20
21	46.0051	0.0217	225.0256	4.8913	0.0044	0.2044	22.174	4.5334			21
22	55.2061	0.0181	271.0307	4.9094	0.0037	0.2037	22.555	4.5941			22
23	66.2474	0.0151	326.2369	4.9245	0.0031	0.2031	22.887	4.6475			23
24	79.4968	0.0126	392.4842	4.9371	0.0025	0.2025	23.176	4.6943			24
25	95.3962	0.0105	471.9811	4.9476	0.0021	0.2021	23.428	4.7352			25
30	237.3763	0.0042	1181.8816	4.9789	0.0008	0.2008	24.263	4.8731			30
35	590.6682	0.0017	2948.3411	4.9915	0.0003	0.2003	24.661	4.9406			35
40	1469.7716	0.0007	7343.8578	4.9966	0.0001	0.2001	24.847	4.9728			40
45	3657.2620	0.0003	18281.3099	4.9986	0.0001	0.2001	24.932	4.9877			45
50	9100.4382	0.0001	45497.1908	4.9995	a	0.2000	24.970	4.9945			50
60	56347.5144	a	281732.5718	4.9999	a	0.2000	24.994	4.9989			60
80	2160228.4620	a	10801137.3101	5.0000	a	0.2000	25.000	5.0000			80
∞				5.0000		0.2000					∞

a Less than 0.0001.

TABLE C-18 Discrete Compounding; $i = 25\%$

| | Single Payment | | Uniform Series | | | | Uniform Gradient | | |
| | Compound Amount Factor | Present Worth Factor | Compound Amount Factor | Present Worth Factor | Sinking Fund Factor | Capital Recovery Factor | Gradient Present Worth Factor | Gradient Uniform Series Factor | |
N	To Find F Given P F/P	To Find P Given F P/F	To Find F Given A F/A	To Find P Given A P/A	To Find A Given F A/F	To Find A Given P A/P	To Find P Given G P/G	To Find A Given G A/G	N
1	1.2500	0.8000	1.0000	0.8000	1.0000	1.2500	0.000	0.0000	1
2	1.5625	0.6400	2.2500	1.4400	0.4444	0.6944	0.640	0.4444	2
3	1.9531	0.5120	3.8125	1.9520	0.2623	0.5123	1.664	0.8525	3
4	2.4414	0.4096	5.7656	2.3616	0.1734	0.4234	2.893	1.2249	4
5	3.0518	0.3277	8.2070	2.6893	0.1218	0.3718	4.204	1.5631	5
6	3.8147	0.2621	11.2588	2.9514	0.0888	0.3388	5.514	1.8683	6
7	4.7684	0.2097	15.0735	3.1611	0.0663	0.3163	6.773	2.1424	7
8	5.9605	0.1678	19.8419	3.3289	0.0504	0.3004	7.947	2.3872	8
9	7.4506	0.1342	25.8023	3.4631	0.0388	0.2888	9.021	2.6048	9
10	9.3132	0.1074	33.2529	3.5705	0.0301	0.2801	9.987	2.7971	10
11	11.6415	0.0859	42.5661	3.6564	0.0235	0.2735	10.846	2.9663	11
12	14.5519	0.0687	54.2077	3.7251	0.0184	0.2684	11.602	3.1145	12
13	18.1899	0.0550	68.7596	3.7801	0.0145	0.2645	12.262	3.2437	13
14	22.7374	0.0440	86.9495	3.8241	0.0115	0.2615	12.833	3.3559	14
15	28.4217	0.0352	109.6868	3.8593	0.0091	0.2591	13.326	3.4530	15
16	35.5271	0.0281	138.1085	3.8874	0.0072	0.2572	13.748	3.5366	16
17	44.4089	0.0225	173.6357	3.9099	0.0058	0.2558	14.109	3.6084	17
18	55.5112	0.0180	218.0446	3.9279	0.0046	0.2546	14.415	3.6698	18
19	69.3889	0.0144	273.5558	3.9424	0.0037	0.2537	14.674	3.7222	19
20	86.7362	0.0115	342.9447	3.9539	0.0029	0.2529	14.893	3.7667	20
21	108.4202	0.0092	429.6809	3.9631	0.0023	0.2523	15.078	3.8045	21
22	135.5253	0.0074	538.1011	3.9705	0.0019	0.2519	15.233	3.8365	22
23	169.4066	0.0059	673.6264	3.9764	0.0015	0.2515	15.363	3.8634	23
24	211.7582	0.0047	843.0329	3.9811	0.0012	0.2512	15.471	3.8861	24
25	264.6978	0.0038	1054.7912	3.9849	0.0009	0.2509	15.562	3.9052	25
30	807.7936	0.0012	3227.1743	3.9950	0.0003	0.2503	15.832	3.9628	30
35	2465.1903	0.0004	9856.7613	3.9984	0.0001	0.2501	15.937	3.9858	35
40	7523.1638	0.0001	30088.6554	3.9995	a	0.2500	15.977	3.9947	40
45	22958.8740	a	91831.4962	3.9993	a	0.2500	15.992	3.9980	45
50	70064.9232	a	280255.6929	3.9999	a	0.2500	15.997	3.9993	50
60	652530.4468	a	2610117.7872	4.0000	a	0.2500	16.000	3.9999	60
∞		a		4.0000	a	0.2500			∞

*a*Less than 0.0001.

I nterest and Annuity Tables for Continuous Compounding

For various values of r from 5% to 20%

r = nominal interest rate per period, compounded continuously

N = number of compounding periods

$$(F/P, \underline{r}\%, N) = e^{rN} \qquad (P/A, \underline{r}\%, N) = \frac{e^{rN} - 1}{e^{rN}(e^r - 1)}$$

$$(P/F, \underline{r}\%, N) = e^{-rN} = \frac{1}{e^{rN}} \qquad (F/\overline{A}, \underline{r}\%, N) = \frac{e^{rN} - 1}{r}$$

$$(F/A, \underline{r}\%, N) = \frac{e^{rN} - 1}{e^r - 1} \qquad (P/\overline{A}, \underline{r}\%, N) = \frac{e^{rN} - 1}{re^{rN}}$$

TABLE D-1 Continuous Compounding; $r = 8\%$

	Discrete Flows				Continuous Flows		
	Single Payment		Uniform Series		Uniform Series		
	Compound Amount Factor	Present Worth Factor	Compound Amount Factor	Present Worth Factor	Compound Amount Factor	Present Worth Factor	
	To Find F Given P	To Find P Given F	To Find F Given $\overline{A}$	To Find P Given $\overline{A}$	To Find F Given $\overline{A}$	To Find P Given $\overline{A}$	
N	F/P	P/F	$F/\overline{A}$	$P/\overline{A}$	$F/\overline{A}$	$P/\overline{A}$	N
1	1.0833	0.9231	1.0000	0.9231	1.0411	0.9610	1
2	1.1735	0.8521	2.0833	1.7753	2.1689	1.8482	2
3	1.2712	0.7866	3.2568	2.5619	3.3906	2.6672	3
4	1.3771	0.7261	4.5280	3.2880	4.7141	3.4231	4
5	1.4918	0.6703	5.9052	3.9584	6.1478	4.1210	5
6	1.6161	0.6188	7.3970	4.5771	7.7009	4.7652	6
7	1.7507	0.5712	9.0131	5.1483	9.3834	5.3599	7
8	1.8965	0.5273	10.7637	5.6756	11.2060	5.9088	8
9	2.0544	0.4868	12.6602	6.1624	13.1804	6.4156	9
10	2.2255	0.4493	14.7147	6.6117	15.3193	6.8834	10
11	2.4109	0.4148	16.9402	7.0265	17.6362	7.3152	11
12	2.6117	0.3829	19.3511	7.4094	20.1462	7.7138	12
13	2.8292	0.3535	21.9628	7.7629	22.8652	8.0818	13
14	3.0649	0.3263	24.7920	8.0891	25.8107	8.4215	14
15	3.3201	0.3012	27.8569	8.3903	29.0015	8.7351	15
16	3.5966	0.2780	31.1770	8.6684	32.4580	9.0245	16
17	3.8962	0.2567	34.7736	8.9250	36.2024	9.2917	17
18	4.2207	0.2369	38.6698	9.1620	40.2587	9.5384	18
19	4.5722	0.2187	42.8905	9.3807	44.6528	9.7661	19
20	4.9530	0.2019	47.4627	9.5826	49.4129	9.9763	20
21	5.3656	0.1864	52.4158	9.7689	54.5694	10.1703	21
22	5.8124	0.1720	57.7813	9.9410	60.1555	10.3494	22
23	6.2965	0.1588	63.5938	10.0998	66.2067	10.5148	23
24	6.8120	0.1466	69.8903	10.2464	72.7620	10.6674	24
25	7.3891	0.1353	76.7113	10.3817	79.8632	10.8083	25
26	8.0045	0.1249	84.1003	10.5067	87.5559	10.9384	26
27	8.6711	0.1153	92.1048	10.6220	95.8892	11.0584	27
28	9.3933	0.1065	100.776	10.7285	104.917	11.1693	28
29	10.1757	0.0983	110.169	10.8267	114.696	11.2716	29
30	11.0232	0.0907	120.345	10.9174	125.290	11.3660	30
35	16.4446	0.0608	185.439	11.2765	193.058	11.7399	35
40	24.5325	0.0408	282.547	11.5172	294.157	11.9905	40
45	36.5982	0.0273	427.416	11.6786	444.978	12.1585	45
50	54.5982	0.0183	643.535	11.7868	669.977	12.2711	50
55	81.4509	0.0123	965.947	11.8593	1005.64	12.3465	55
60	121.510	0.0082	1446.93	11.9079	1506.38	12.3971	60
65	181.272	0.0055	2164.47	11.9404	2253.40	12.4310	65
70	270.426	0.0037	3234.91	11.9623	3367.83	12.4538	70
75	403.429	0.0025	4831.83	11.9769	5030.36	12.4690	75
80	601.845	0.0017	7214.15	11.9867	7510.56	12.4792	80
85	897.847	0.0011	10768.1	11.9933	11210.6	12.4861	85
90	1339.43	0.0007	16070.1	11.9977	16730.4	12.4907	90
95	1998.20	0.0005	23979.7	12.0007	24964.9	12.4937	95
100	2980.96	0.0003	35779.3	12.0026	37249.5	12.4958	100

TABLE D-2 Continuous Compounding; $r = 10\%$

	Discrete Flows				Continuous Flows		
	Single Payment		Uniform Series		Uniform Series		
	Compound Amount Factor	Present Worth Factor	Compound Amount Factor	Present Worth Factor	Compound Amount Factor	Present Worth Factor	
	To Find F Given P	To Find P Given F	To Find F Given $\overline{A}$	To Find P Given $\overline{A}$	To Find F Given $\overline{A}$	To Find P Given $\overline{A}$	
N	F/P	P/F	F/A	P/A	$F/\overline{A}$	$P/\overline{A}$	N
1	1.1052	0.9048	1.0000	0.9048	1.0517	0.9516	1
2	1.2214	0.8187	2.1052	1.7236	2.2140	1.8127	2
3	1.3499	0.7408	3.3266	2.4644	3.4986	2.5918	3
4	1.4918	0.6703	4.6764	3.1347	4.9182	3.2968	4
5	1.6487	0.6065	6.1683	3.7412	6.4872	3.9347	5
6	1.8221	0.5488	7.8170	4.2900	8.2212	4.5119	6
7	2.0138	0.4966	9.6391	4.7866	10.1375	5.0341	7
8	2.2255	0.4493	11.6528	5.2360	12.2554	5.5067	8
9	2.4596	0.4066	13.8784	5.6425	14.5960	5.9343	9
10	2.7183	0.3679	16.3380	6.0104	17.1828	6.3212	10
11	3.0042	0.3329	19.0563	6.3433	20.0417	6.6713	11
12	3.3201	0.3012	22.0604	6.6445	23.2012	6.9881	12
13	3.6693	0.2725	25.3806	6.9170	26.6930	7.2747	13
14	4.0552	0.2466	29.0499	7.1636	30.5520	7.5340	14
15	4.4817	0.2231	33.1051	7.3867	34.8169	7.7687	15
16	4.9530	0.2019	37.5867	7.5886	39.5303	7.9810	16
17	5.4739	0.1827	42.5398	7.7713	44.7395	8.1732	17
18	6.0496	0.1653	48.0137	7.9366	50.4965	8.3470	18
19	6.6859	0.1496	54.0634	8.0862	56.8589	8.5043	19
20	7.3891	0.1353	60.7493	8.2215	63.8906	8.6466	20
21	8.1662	0.1225	68.1383	8.3440	71.6617	8.7754	21
22	9.0250	0.1108	76.3045	8.4548	80.2501	8.8920	22
23	9.9742	0.1003	85.3295	8.5550	89.7418	8.9974	23
24	11.0232	0.0907	95.3037	8.6458	100.232	9.0928	24
25	12.1825	0.0821	106.327	8.7278	111.825	9.1791	25
26	13.4637	0.0743	118.509	8.8021	124.637	9.2573	26
27	14.8797	0.0672	131.973	8.8693	138.797	9.3279	27
28	16.4446	0.0608	146.853	8.9301	154.446	9.3919	28
29	18.1741	0.0550	163.298	8.9852	171.741	9.4498	29
30	20.0855	0.0498	181.472	9.0349	190.855	9.5021	30
35	33.1155	0.0302	305.364	9.2212	321.154	9.6980	35
40	54.5981	0.0183	509.629	9.3342	535.982	9.8168	40
45	90.0171	0.0111	846.404	9.4027	890.171	9.8889	45
50	148.413	0.0067	1401.65	9.4443	1474.13	9.9326	50
55	244.692	0.0041	2317.10	9.4695	2436.92	9.9591	55
60	403.429	0.0025	3826.43	9.4848	4024.29	9.9752	60
65	665.142	0.0015	6314.88	9.4940	6641.42	9.9850	65
70	1096.63	0.0009	10417.6	9.4997	10956.3	9.9909	70
75	1808.04	0.0006	17182.0	9.5031	18070.7	9.9945	75
80	2980.96	0.0003	28334.4	9.5051	29799.6	9.9966	80
85	4914.77	0.0002	46721.7	9.5064	49137.7	9.9980	85
90	8103.08	0.0001	77037.3	9.5072	81020.8	9.9988	90
95	13359.7	a	127019.0	9.5076	133587.0	9.9993	95
100	22026.5	a	209425.0	9.5079	220255.0	9.9995	100

a Less than 0.0001.

TABLE D-3 Continuous Compounding; $r = 20\%$

	Discrete Flows				Continuous Flows		
	Single Payment		Uniform Series		Uniform Series		
	Compound Amount Factor	Present Worth Factor	Compound Amount Factor	Present Worth Factor	Compound Amount Factor	Present Worth Factor	
N	To Find F Given P F/P	To Find P Given F P/F	To Find F Given $\overline{A}$ F/A	To Find P Given $\overline{A}$ P/A	To Find F Given $\overline{A}$ $F/\overline{A}$	To Find P Given $\overline{A}$ $P/\overline{A}$	N
1	1.2214	0.8187	1.0000	0.8187	1.1070	0.9063	1
2	1.4918	0.6703	2.2214	1.4891	2.4591	1.6484	2
3	1.8221	0.5488	3.7132	2.0379	4.1106	2.2559	3
4	2.2255	0.4493	5.5353	2.4872	6.1277	2.7534	4
5	2.7183	0.3679	7.7609	2.8551	8.5914	3.1606	5
6	3.3201	0.3012	10.4792	3.1563	11.6006	3.4940	6
7	4.0552	0.2466	13.7993	3.4029	15.2760	3.7670	7
8	4.9530	0.2019	17.8545	3.6048	19.7652	3.9905	8
9	6.0496	0.1653	22.8075	3.7701	25.2482	4.1735	9
10	7.3891	0.1353	28.8572	3.9054	31.9453	4.3233	10
11	9.0250	0.1108	36.2462	4.0162	40.1251	4.4460	11
12	11.0232	0.0907	45.2712	4.1069	50.1159	4.5464	12
13	13.4637	0.0743	56.2944	4.1812	62.3187	4.6286	13
14	16.4446	0.0608	69.7581	4.2420	77.2232	4.6959	14
15	20.0855	0.0498	86.2028	4.2918	95.4277	4.7511	15
16	24.5325	0.0408	106.288	4.3325	117.633	4.7962	16
17	29.9641	0.0334	130.821	4.3659	144.820	4.8331	17
18	36.5982	0.0273	160.785	4.3932	177.991	4.8634	18
19	44.7012	0.0224	197.383	4.4156	218.506	4.8881	19
20	54.5981	0.0183	242.084	4.4339	267.991	4.9084	20
21	66.6863	0.0150	296.682	4.4489	328.432	4.9250	21
22	81.4509	0.0123	363.369	4.4612	402.254	4.9386	22
23	99.4843	0.0101	444.820	4.4713	492.422	4.9497	23
24	121.510	0.0082	544.304	4.4795	602.552	4.9589	24
25	148.413	0.0067	665.814	4.4862	737.066	4.9663	25
26	181.272	0.0055	814.227	4.4917	901.361	4.9724	26
27	221.406	0.0045	995.500	4.4963	1102.03	4.9774	27
28	270.426	0.0037	1216.91	4.5000	1347.13	4.9815	28
29	330.299	0.0030	1487.33	4.5030	1646.50	4.9849	29
30	403.429	0.0025	1817.63	4.5055	2012.14	4.9876	30
35	1096.63	0.0009	4948.60	4.5125	5478.17	4.9954	35
40	2980.96	0.0003	13459.4	4.5151	14899.8	4.9983	40
45	8103.08	0.0001	36594.3	4.5161	40510.4	4.9994	45
50	22026.5	a	99481.4	4.5165	110127.0	4.9998	50
55	59874.1	a	270426.0	4.5166	299366.0	4.9999	55
60	162755.0	a	735103.0	4.5166	813769.0	5.0000	60

a Less than 0.0001.

Standardized Normal Distribution Function

The standardized normal distribution function cumulates the normal density function from minus infinity to the standardized value in question, $Z = (X - \mu)/\sigma$. Interested readers are referred to any introductory statistics textbook for an in-depth discussion of the use of the standardized normal distribution function illustrated on page 555.

TABLE E-1   Areas under the Normal Curve[a]

z	0.00	0.01	0.02	0.03	0.04	0.05	0.06	0.07	0.08	0.09
−3.4	0.0003	0.0003	0.0003	0.0003	0.0003	0.0003	0.0003	0.0003	0.0003	0.0002
−3.3	0.0005	0.0005	0.0005	0.0004	0.0004	0.0004	0.0004	0.0004	0.0004	0.0003
−3.2	0.0007	0.0007	0.0006	0.0006	0.0006	0.0006	0.0006	0.0005	0.0005	0.0005
−3.1	0.0010	0.0009	0.0009	0.0009	0.0008	0.0008	0.0008	0.0007	0.0007	0.0007
−3.0	0.0013	0.0013	0.0013	0.0012	0.0012	0.0011	0.0011	0.0011	0.0010	0.0010
−2.9	0.0019	0.0018	0.0017	0.0017	0.0016	0.0016	0.0015	0.0015	0.0014	0.0014
−2.8	0.0026	0.0025	0.0024	0.0023	0.0023	0.0022	0.0021	0.0021	0.0020	0.0019
−2.7	0.0035	0.0034	0.0033	0.0032	0.0031	0.0030	0.0029	0.0028	0.0027	0.0026
−2.6	0.0047	0.0045	0.0044	0.0043	0.0041	0.0040	0.0039	0.0038	0.0037	0.0036
−2.5	0.0062	0.0060	0.0059	0.0057	0.0055	0.0054	0.0052	0.0051	0.0049	0.0048
−2.4	0.0082	0.0080	0.0078	0.0075	0.0073	0.0071	0.0069	0.0068	0.0066	0.0064
−2.3	0.0107	0.0104	0.0103	0.0099	0.0096	0.0094	0.0091	0.0089	0.0087	0.0084
−2.2	0.0139	0.0136	0.0132	0.0129	0.0125	0.0122	0.0119	0.0116	0.0113	0.0110
−2.1	0.0179	0.0174	0.0170	0.0166	0.0162	0.0158	0.0154	0.0150	0.0146	0.0143
−2.0	0.0228	0.0222	0.0217	0.0212	0.0207	0.0202	0.0197	0.0192	0.0118	0.0183
−1.9	0.0287	0.0281	0.0274	0.0268	0.0262	0.0256	0.0250	0.0244	0.0239	0.0233
−1.8	0.0359	0.0352	0.0344	0.0336	0.0329	0.0322	0.0314	0.0307	0.0301	0.0294
−1.7	0.0446	0.0436	0.0427	0.0418	0.0409	0.0401	0.0392	0.0384	0.0375	0.0367
−1.6	0.0548	0.0537	0.0526	0.0516	0.0505	0.0495	0.0485	0.0475	0.0465	0.0455
−1.5	0.0668	0.0655	0.0643	0.0630	0.0618	0.0606	0.0594	0.0582	0.0571	0.0559
−1.4	0.0808	0.0793	0.0778	0.0764	0.0749	0.0735	0.0722	0.0708	0.0694	0.0681
−1.3	0.0968	0.0951	0.0934	0.0918	0.0901	0.0885	0.0869	0.0853	0.0838	0.0823
−1.2	0.1151	0.1131	0.1112	0.1093	0.1075	0.1056	0.1038	0.1020	0.1003	0.0985
−1.1	0.1357	0.1335	0.1314	0.1292	0.1271	0.1251	0.1230	0.1210	0.1190	0.1170
−1.0	0.1587	0.1562	0.1539	0.1515	0.1492	0.1469	0.1446	0.1423	0.1401	0.1379
−0.9	0.1841	0.1841	0.1788	0.1762	0.1736	0.1711	0.1685	0.1660	0.1635	0.1611
−0.8	0.2119	0.2090	0.2061	0.2033	0.2005	0.1977	0.1949	0.1922	0.1894	0.1867
−0.7	0.2420	0.2389	0.2358	0.2327	0.2296	0.2266	0.2236	0.2206	0.2177	0.2148
−0.6	0.2743	0.2709	0.2676	0.2643	0.2611	0.2578	0.2546	0.2514	0.2483	0.2451
−0.5	0.3085	0.3050	0.3015	0.2981	0.2946	0.2912	0.2877	0.2843	0.2810	0.2776
−0.4	0.3446	0.3409	0.3372	0.3336	0.3300	0.3264	0.3228	0.3192	0.3156	0.3121
−0.3	0.3821	0.3783	0.3745	0.3707	0.3669	0.3632	0.3594	0.3557	0.3520	0.3483
−0.2	0.4207	0.4168	0.4129	0.4090	0.4052	0.4013	0.3974	0.3936	0.3897	0.3859
−0.1	0.4602	0.4562	0.4522	0.4483	0.4443	0.4404	0.4364	0.4325	0.4286	0.4247
−0.0	0.5000	0.4960	0.4920	0.4880	0.4840	0.4801	0.4761	0.4721	0.4681	0.4641

[a] From R. E. Walpole and R. H. Myers, *Probability and Statistics for Engineers and Scientists*, 2nd ed. (New York: Macmillan, 1978), p. 513.

(continued)

TABLE E-1		Areas under the Normal Curve		(cont'd.)					
z	0.00	0.01	0.02	0.03	0.04	0.05	0.06	0.07	0.08	0.09
0.0	0.5000	0.5040	0.5080	0.5120	0.5160	0.5199	0.5239	0.5279	0.5319	0.5359
0.1	0.5398	0.5438	0.5478	0.5517	0.5557	0.5596	0.5636	0.5675	0.5714	0.5753
0.2	0.5793	0.5832	0.5871	0.5910	0.5948	0.5987	0.6026	0.6064	0.6103	0.6141
0.3	0.6179	0.6217	0.6255	0.6293	0.6331	0.6368	0.6406	0.6443	0.6480	0.6517
0.4	0.6554	0.6591	0.6628	0.6664	0.6700	0.6736	0.6772	0.6808	0.6844	0.6879
0.5	0.6915	0.6950	0.6985	0.7019	0.7054	0.7088	0.7123	0.7157	0.7190	0.7224
0.6	0.7257	0.7291	0.7324	0.7357	0.7389	0.7422	0.7454	0.7486	0.7517	0.7549
0.7	0.7580	0.7611	0.7642	0.7673	0.7704	0.7734	0.7764	0.7794	0.7823	0.7852
0.8	0.7881	0.7910	0.7939	0.7967	0.7995	0.8023	0.8051	0.8078	0.8106	0.8133
0.9	0.8159	0.8186	0.8212	0.8238	0.8264	0.8289	0.8315	0.8340	0.8365	0.8389
1.0	0.8413	0.8438	0.8461	0.8485	0.8508	0.8531	0.8554	0.8577	0.8599	0.8621
1.1	0.8643	0.8665	0.8686	0.8708	0.8729	0.8749	0.8770	0.8790	0.8810	0.8830
1.2	0.8849	0.8869	0.8888	0.8907	0.8925	0.8944	0.8962	0.8980	0.8997	0.9015
1.3	0.9032	0.9049	0.9066	0.9082	0.9099	0.9115	0.9131	0.9147	0.9162	0.9177
1.4	0.9192	0.9207	0.9222	0.9236	0.9251	0.9265	0.9278	0.9292	0.9306	0.9319
1.5	0.9332	0.9345	0.9357	0.9370	0.9382	0.9394	0.9406	0.9418	0.9429	0.9441
1.6	0.9452	0.9463	0.9474	0.9484	0.9495	0.9505	0.9515	0.9525	0.9535	0.9545
1.7	0.9554	0.9564	0.9573	0.9582	0.9591	0.9599	0.9608	0.9616	0.9625	0.9633
1.8	0.9641	0.9649	0.9656	0.9664	0.9671	0.9678	0.9686	0.9693	0.9699	0.9706
1.9	0.9713	0.9719	0.9726	0.9732	0.9738	0.9744	0.9750	0.9756	0.9761	0.9767
2.0	0.9772	0.9778	0.9783	0.9788	0.9793	0.9798	0.9803	0.9808	0.9812	0.9817
2.1	0.9821	0.9826	0.9830	0.9834	0.9838	0.9842	0.9846	0.9850	0.9854	0.9857
2.2	0.9861	0.9864	0.9868	0.9871	0.9875	0.9878	0.9881	0.9884	0.9887	0.9890
2.3	0.9893	0.9896	0.9898	0.9901	0.9904	0.9906	0.9909	0.9911	0.9913	0.9916
2.4	0.9918	0.9920	0.9922	0.9925	0.9927	0.9929	0.9931	0.9932	0.9934	0.9936
2.5	0.9938	0.9940	0.9941	0.9943	0.9945	0.9946	0.9948	0.9949	0.9951	0.9952
2.6	0.9953	0.9955	0.9956	0.9957	0.9959	0.9960	0.9961	0.9962	0.9963	0.9964
2.7	0.9965	0.9966	0.9967	0.9968	0.9969	0.9970	0.9971	0.9972	0.9973	0.9974
2.8	0.9974	0.9975	0.9976	0 9977	0.9977	0.9978	0.9979	0.9979	0.9980	0.9981
2.9	0.9981	0.9982	0.9982	0.9983	0.9984	0.9984	0.9985	0.9985	0.9986	0.9986
3.0	0.9987	0.9987	0.9987	0.9988	0.9988	0.9989	0.9989	0.9989	0.9990	0.9990
3.1	0.9990	0.9991	0.9991	0.9991	0.9992	0.9992	0.9992	0.9992	0.9993	0.9993
3.2	0.9993	0.9993	0.9994	0.9994	0.9994	0.9994	0.9994	0.9995	0.9995	0.9995
3.3	0.9995	0.9995	0.9995	0.9996	0.9996	0.9996	0.9996	0.9996	0.9996	0.9997
3.4	0.9997	0.9997	0.9997	0.9997	0.9997	0.9997	0.9997	0.9997	0.9997	0.9998

F

Selected References

AMERICAN TELEPHONE AND TELEGRAPH COMPANY, Engineering Department. *Engineering Economy,* 3rd ed. (New York: American Telephone and Telegraph Co., 1977).

AU, T., and T. P. AU. *Engineering Economics for Capital Investment Analysis,* 2nd ed. (Boston: Allyn and Bacon, 1992).

BARISH, N. N., and S. KAPLAN. *Economic Analysis for Engineering and Managerial Decision Making* (New York: McGraw-Hill 1978).

BIERMAN, H., JR., and S. SMIDT. *The Capital Budgeting Decision,* 8th ed. (New York: Macmillan, 1993).

BLANK, L. T., and A. J. TARQUIN. *Engineering Economy,* 5th ed. (New York: McGraw-Hill, 2001).

BRIMSON, J. A. *Activity Accounting: An Activity-Based Approach* (New York: John Wiley & Sons, 1991).

BUSSEY, L. E., and T. G. ESCHENBACH. *The Economic Analysis of Industrial Projects,* 2nd ed. (Upper Saddle River, NJ: Prentice Hall, 1992).

CAMPEN, J. T. *Benefit, Cost, and Beyond* (Cambridge, MA: Ballinger Publishing Company, 1986).

CANADA, J. R., and W. G. SULLIVAN. *Economic and Multiattribute Analysis of Advanced Manufacturing Systems* (Upper Saddle River, NJ: Prentice Hall, 1989).

CANADA, J. R., W. G. SULLIVAN, and J. A. WHITE. *Capital Investment Decision Analysis for Engineering and Management* (Upper Saddle River, NJ: Prentice Hall, 1996).

CLARK, J. J., T. J. HINDELANG, and R. E. PRITCHARD. *Capital Budgeting: Planning and Control of Capital Expenditures* (Upper Saddle River, NJ: Prentice Hall, 1979).

COCHRANE, J. L., and M. ZELENY. *Multiple Criteria Decision Making* (Columbia, SC: University of South Carolina, 1973).

COLLIER, C. A., and W. B. LEDBETTER. *Engineering Cost Analysis,* 2nd ed. (New York: Harper & Row, 1987).

DELAMARE, R. F. *Manufacturing Systems Economics* (East Sussex, England: Holt Reinhart & Winston, 1982).

Engineering Economist, The. A quarterly journal jointly published by the Engineering Economy Division of the American Society for Engineering Education and the Institute of Industrial Engineers. Published by IIE, Norcross, GA.

Engineering News-Record. Published monthly by McGraw-Hill, New York.

ENGLISH, J. M., ed. *Cost Effectiveness: Economic Evaluation of Engineering Systems* (New York: John Wiley & Sons, 1968).

ESCHENBACH, T. G. *Engineering Economy: Applying Theory to Practice* (Chicago: Richard D. Irwin, 1995).

FABRYCKY, W. J., G. J. THUESEN, and D. VERMA. *Economic Decision Analysis*, 3rd ed. (Upper Saddle River, NJ: Prentice Hall, 1998).

FLEISCHER, G. A. *Introduction to Engineering Economy* (Boston: PWS Publishing Company, 1994).

FLEISCHER, G. A. *Risk and Uncertainty: Non-Deterministic Decision Making in Engineering Economy* (Norcross, GA: Institute of Industrial Engineers, Publication EE-75-1, 1975).

GOICOECHEA, A., D. R. HANSEN, and L. DUCKSTEIN. *Multiobjective Decision Analysis with Engineering and Business Applications* (New York: John Wiley & Sons, 1982).

GRANT, E. L., W. G. IRESON, and R. S. LEAVENWORTH. *Principles of Engineering Economy*, 8th ed. (New York: John Wiley & Sons, 1990).

HAPPEL, J., and D. JORDAN. *Chemical Process Economics*, 2nd ed. (New York: Marcel Dekker, 1975).

HULL, J. C. *The Evaluation of Risk in Business Investment* (New York: Pergamon Press, 1980).

Industrial Engineering. A monthly magazine published by the Institute of Industrial Engineers, Norcross, GA.

Internal Revenue Service Publication 534. *Depreciation.* U.S. Government Printing Office, revised periodically.

JELEN, F. C., and J. H. BLACK. *Cost and Optimization Engineering*, 3rd ed. (New York: McGraw-Hill, 1991).

JONES, B. W. *Inflation in Engineering Economic Analysis* (New York: John Wiley & Sons, 1982).

KAHL, A. L., and W. F. RENTZ. *Spreadsheet Applications in Engineering Economics* (St. Paul, MN West Publishing Company, 1992).

KAPLAN, R. S., and R. COOPER. *The Design of Cost Management Systems* (Upper Saddle River, NJ: Prentice Hall, 1999).

KEENEY, R. L., and H. RAIFFA. *Decisions with Multiple Objectives: Preferences and Value Trade-offs* (New York: John Wiley & Sons, 1976).

KLEINFELD, IRA H. *Engineering and Managerial Economics* (New York: Holt, Rinehart & Winston, 1986).

LASSER, J. K. *Your Income Tax* [New York: Simon & Schuster (see latest edition)].

MACHINERY AND ALLIED PRODUCTS INSTITUTE. *MAPI Replacement Manual.* Washington, DC: Machinery and Allied Products Institute, 1950.

MALLIK, A. K. *Engineering Economy with Computer Applications* (Mahomet, IL: Engineering Technology, 1979).

MAO, J. *Quantitative Analysis of Financial Decisions* (New York: Macmillan, 1969).

MATTHEWS, L. M. *Estimating Manufacturing Costs: A Practical Guide for Managers and Estimators* (New York: McGraw-Hill, 1983).

MAYER, R. R. *Capital Expenditure Analysis for Managers and Engineers* (Prospect Heights, IL: Waveland Press, 1978).

MERRETT, A. J., and A. SYKES. *The Finance and Analysis of Capital Projects* (New York: John Wiley & Sons, 1963).

MISHAN, E. J. *Cost-Benefit Analysis* (New York: Praeger Publishers, 1976).

MORRIS, W. T. *Decision Analysis* (Columbus, OH: Grid, 1977).

MORRIS, W. T. *Engineering Economic Analysis* (Reston, VA: Reston Publishing, 1976).

NEWNAN, D. G., J. P. LAVELLE, and T. G. ESCHENBACH. *Engineering Economic Analysis*, 8th ed. (San Jose, CA: Engineering Press, 2001).

OAKFORD, R. V. *Capital Budgeting: A Quantitative Evaluation of Investment Alternatives* (New York: John Wiley & Sons, 1970).

OSTWALD, P. F. *Cost Estimating for Engineering and Management*, 3rd ed. (Upper Saddle River, NJ: Prentice Hall, 1992).

PARK, C. S. *Contemporary Engineering Economics* (Upper Saddle River, NJ: Prentice Hall, 2002).

PARK, C. S., and G. P. SHARP-BETTE. *Advanced Engineering Economics* (New York: John Wiley & Sons, 1990).

PARK, W. R., and D. E. JACKSON. *Cost Engineering Analysis: A Guide to Economic Evaluation of Engineering Projects*, 2nd ed. (New York: John Wiley & Sons, 1984).

PETERS, M. S., and K. D. TIMMERHAUS. *Plant Design and Economics for Chemical Engineers*, 4th ed. (New York: McGraw-Hill, 1991).

PORTER, M. E. *Competitive Strategy: Techniques for Analyzing Industries and Competitors* (New York: The Free Press, 1980).

RIGGS, J. L., D. D. BEDWORTH, and S. V. RANDHAWA. *Engineering Economics*, 4th ed. (New York: McGraw-Hill, 1996).

ROSE, L. M. *Engineering Investment Decisions: Planning Under Uncertainty* (Amsterdam: Elsevier, 1976).

SMITH, G. W. *Engineering Economy: The Analysis of Capital Expenditures*, 4th ed. (Ames, IO: Iowa State University Press, 1987).

STEINER, H. M. *Engineering Economic Principles* (New York: McGraw-Hill, 1992).

STERMOLE, F. J., and J. M. STERMOLE. *Economic Evaluation and Investment Decision Methods*, 6th ed. (Golden, CO: Investment Evaluations Corp., 1987).

STEWART, R. D. *Cost Estimating* (New York: John Wiley & Sons, 1982).

STEWART, R. D., and R. M. WYSKIDA, eds. *Cost Estimators' Reference Manual* (New York: John Wiley & Sons, 1987).

SULLIVAN, W. G., and W. W. CLAYCOMBE. *Fundamentals of Forecasting* (Reston, VA: Reston Publishing, 1977).

TAYLOR, G. A. *Managerial and Engineering Economy*, 3rd ed. (New York: Van Nostrand Reinhold, 1980).

TERBORGH, G. *Business Investment Management* (Washington, DC: Machinery and Allied Products Institute, 1967).

THUESEN, G. J., and W. J. FABRYCKY. *Engineering Economy*, 9th ed. (Upper Saddle River, NJ: Prentice Hall, 2001).

VANHORNE, J. C. *Financial Management and Policy*, 8th ed. (Upper Saddle River, NJ: Prentice Hall, 1989).

WEINGARTNER, H. M. *Mathematical Programming and the Analysis of Capital Budgeting Problems* (Upper Saddle River, NJ: Prentice Hall, 1975).

WELLINGTON, A. M. *The Economic Theory of Railway Location* (New York: John Wiley & Sons, 1887).

WHITE, J. A., K. E. CASE, D. B. PRATT, and M. H. AGEE. *Principles of Engineering Economic Analysis*, 4th ed. (New York: John Wiley & Sons, 1998).

WOODS, D. R. *Financial Decision Making in the Process Industry* (Upper Saddle River, NJ: Prentice Hall, 1975).

ZELENY, M. *Multiple Criteria Decision Making* (New York: McGraw-Hill, 1982).

G

Answers to Problems

CHAPTER 2

2-6 **a.** $D^* = 2,425$ circuit boards/month
b. Maximum Profit = \$75,612.50/month
c. $D'_1 = 480.6 \approx 481$ circuit boards/month
d. $D'_2 = 4,369.4 \approx 4,369$ circuit boards/month
e. Range of profitable demand is 481 to 4,369 circuit boards per month

2-7 **a.** $D^* = 227.27$ units per year
b. $\frac{d^2(\text{Profit})}{dD^2} = -4.4 < 0$; profit maximized

2-8 $D^* = 240$ units/month; Maximum profit = \$4,960 per month
$D'_1 = 17.3$, or 18 units/month;
$D'_2 = 462.7$, or 462 units/month

2-9 **a.** Even though $D^* = 300$ units/month is the optimal demand, the company would incur a loss at this production volume. Do not produce the new product.

2-10 **a.** $D^* = 4,685$ units/month; Maximum profit = \$197,461.25/month
b. $D'_1 = 2,698$ units/month
$D'_2 = 6,672$ units/month
Range of profitable demand is 2,698 units to 6,672 units per month

2-11 **a.** $D^* = 200$ units/month
$\frac{d^2(\text{Profit/Loss})}{dD^2} = -0.4 < 0$; therefore D^* is a maximum
b. Maximum profit = \$7,000/month
c. Profitable range is $D = 13$ to 387 units per month

2-12 **a.** $D^* = 10$ units
b. $\frac{d^2\text{TP}}{dD^2} = -60$; at $D = D^*$, $\frac{d^2\text{TP}}{dD^2} = -60 < 0$; maximum profit
c. $D = 15$ units

2-13 **a.** The solid waste site should be located at Site B.
b. $X = 15.64$ megawatts; $\frac{d^2\text{Profit}}{dX^2} = -0.94 < 0$; profit maximized

2-14 $D' = 3,112$ pumps/month; 22.75% reduction in the breakeven point

2-15 **a.** From graph, $D' = \$8,400,000$ per year
b. D' is lowered by 12%
c. $D' = \$8,400,000$ per year ∴ no change

2-16 **a.** $D^* = 50$ units/month
b. $\frac{d^2(\text{Profit})}{dD^2} = \frac{-10,000}{D^3} < 0$ for $D > 1$; Maximum profit

2-17 **a.** $D^* = 2$ units/week; $\frac{d^2(\text{Profit})}{dD^2} = -90 < 0$ ∴ maximum profit
b. Total profit = \$180/week

2-18 **a.** $D' = 40,000$ units/year

b. Profit (90% of capacity) = $2,500,000/year

c. Profit (100% of capacity) = $2,800,000/year

2-19 a. $X^* = 0.0305$ meters

b. $\frac{d^2(\text{TAC})}{dX^2} = \frac{3.6}{X^{5/2}} > 0$, for $X > 0$; X^* is a minimum

c. Cost of the extra insulation (a directly varying cost) is being traded-off against the value of reduction in lost heat (an indirectly varying cost)

2-20 Harvest crop in 2.5 weeks to receive maximum revenue = $6,125

2-21 a. 500 castings/hr

b. Preference changes from 500 castings/hr to 100 castings/hr for a 42% increase in total hourly production costs

2-22 Velocity = 10.25 mph

2-23 $v^* = 44.7$ mph

2-24 a. Select Machine B ($5.32/part)

b. Select Machine A ($3.09/part)

2-25 a. Speed B (Cost/piece = $0.104/piece)

b. Extra tooling and operating expenses are traded off for extra output (production)

2-26 Select tool steel to minimize overall cost per piece ($0.71/piece)

2-27 Speed C (Cost/piece = $0.0195/piece)

2-28 a. Purchase item (cost = $8.50/item)

b. Manufacture item (cost = $8.65/item)

2-29 Brass–Copper alloy should be selected to save $28.25 over the life cycle of each radiator

2-30 a. Select Operation 1 to maximize profit (Profit = $4,640/day)

b. Increased production for Operation 1 is being traded off for increased tool changing time (downtime), and the balance is favorable for Operation 1 to Operation 2

2-31 a. Either machine will produce the required 30,000 nondefective units/3-months

b. Select machine A (Cost/nondefective unit = $6.46/unit)

2-32 a. Select Design B (cost = $0.333/unit)

b. Savings of Design B over Design A are $0.04065/unit

2-33 Answer depends on assumptions made

2-34 a. Lathe: $200 for 40 units; A-S machine: $97 for 40 units

2-35 Select Method 1 (Profit = $10,974,000)

2-36 Select Method B (Profit/ounce = $76.50)

2-37 (a) False; (b) False; (c) True; (d) True; (e) True; (f) False; (g) True; (h) True; (i) False; (j) True; (k) True; (l) False; (m) True; (n) False; (o) True (p) True; (q) False

2-38 a. $\lambda^* = (C_I/C_R t)^{1/2}$

b. $\frac{d^2C}{d\lambda^2} = \frac{2C_I}{\lambda^3} > 0$ for $\lambda > 0$; minimum life-cycle cost

c. Investment cost versus total repair costs

2-39 Select Process 1 (Profit = $2,640.00/day)

2-40 (a) $X = 2,500$ miles (b) $X = 1,100$ miles (c) there are 2 breakeven points

CHAPTER 3

3-1 $\underline{I} = \$4,250$

3-2 $\underline{I} = \$7,560$

3-3 Select c

3-4 $200 (years 1–4); $100 (years 5–8); total interest paid = $1,200

3-5 Total interest paid = $1,823.07. Difference of $623.07 is due to compounding

3-6 a. Total interest paid = $2,400

b. Year 3 principal payment = $2,070.60; total interest paid = $1,660.60

3-7 a. $P_1 = \$3,141$; $P_3 = \$3,529.54$

b. Amount left to repay at beginning of year 3 = $3,529.54

c. Amount is less because part of the principal is repaid each year

3-8 $A = \$2,925$

3-9 $A = \$497$

3-10 $F = \$3,215.40$

3-11 $A = \$5,548$

3-12 Total interest = $7,200; total interest paid in Problem 3–11 = $7,740

3-13 $A = \$184$

3-14 $i = 14\%$

3-15 b. $F_7 = \$1,754,102.16$ (section 3.8); $F_7 = \$1,516,600$ (section 3.14); difference in F_7 amounts is due to rounding the interest factor values

3-17 $A = \$3,397.50$

3-18 $P = \$73,748.40$

3-19 $P \leq \$3,280.16$

3-20 $A = \$4,417$

3-21 $F_4 = \$124,966$

3-22 a. $N \approx 8$ years **b.** $i = 15.11\%$
c. $P = \$720.96$ **d.** $A = \$277.40$

3-23 Rule of 72, $N \approx 7.2$ years; Exact solution, $N = 7.2725$ years

3-24 c. $P_0 = 8.3333A$

3-25 I $= \$1,477.50$; II $= \$342.78$; III $= \$110.25$; IV $= \$783.63$; V $= \$1,000$

3-26 $F_4 = \$13,490$, select D

3-27 $A = \$55.74$

3-28 $N = 49$ years

3-29 a. $N \approx 5$ years
b. $P_2 = \$656.04$

3-30 $A = -\$681.86$

3-31 $i = 4.94\%$ per year

3-32 $P = \$33,511.70$

3-33 $i \approx 7\%$ per year

3-34 $P_0 = \$14,171.62$

3-35 $Z = \$3,848.15$

3-36 $F_{12} = \$3,269.12$

3-37 $A_2 = \$189.68$

3-38 $A_1 = \$1993.67$; $A_2 = \$1543.50$

3-39 $i'/\text{year} = 11.55\%$ per year

3-40 $P_0 = \$433.28$

3-41 $F_5 = \$664.99$

3-42 $W = \$714.25$

3-43 $Z = \$63.09$

3-44 $A = \$1,417.16$

3-45 $P_0 = -\$165,104$

3-46 $F_{12} = \$8,198.11$

3-47 $N = 11$ years

3-48 $P_0 = \$471.20$; $A = \$90.52$

3-49 $A = \$1,203.69$

3-50 $P_0(\text{rental income}) = \$8,288.56 > \$8,000 = P_0$ (investment), therefore, it is a good investment

3-51 $Z = \$608.21$

3-52 $Z = \$1,256.05$

3-53 a. $i = 7.86\%$
b. $N = 6.1$ periods; if an integer value of N is desired, choose $N = 7$ periods
c. $F = \$93,841.30$
d. $G = \$466.34$

3-54 $P_0 = \$820.12$

3-55 $K = \$1,034.25$

3-56 $P_0 = \$100(P/A, 10\%, 4) + \$100(P/G, 10\%, 8)$

3-57 $F = \$3,500.14$

3-58 $A = \$124.34$

3-59 $A = \$437.14$

3-60 $A = \$593.10$

3-61 $P_0 = \$24,678.64$; install insulation

3-62 $Q = \$435.75$

3-63 $N = 8$ years

3-64 $B = \$13,370.26$

3-65 $P_0 = \$721,285$

3-66 $P_0 = \$4,672.61$

3-67 $A = \$2,790.83$

3-68 a. $P_0 = \$61,217.76$
b. $A = \$12,323.13$
c. $A_0 = \$9,345.79$

3-69 $P_0 = \$9,191.97$

3-70 $P_0 = \$23,853.74$

3-71 $\$5,573.90$

3-72 a. $i_{CR} = 2\%$ per year
b. $P_0 = \$36,204.86$
c. $P_0 = \$29,896 + 34.22G$
d. $G = \$184.36$

3-74 $F = \$28,226.38$

3-75 a. $i = 10.25\%$
b. $i = 10.38\%$
c. $i = 10.51\%$

3-76 $A = \$1,430$; select D

3-77 a. $A = \$249.99$
b. $A = \$22,742.33$

3-78 a. $A = \$1,696.00/\text{month}$; $i/\text{year} = 6.17\%$
b. $i/\text{qtr} = 1.51\%$

3-79 $P_0 = \$10,847.43$

3-80 $P_0 = \$4,729.87$

3-81 a. $P = \$91,276.00$
b. $P = \$93,820.50$
c. $P = \$93,363.50$

3-82 $F = \$24,465$

3-83 $N = 30$ months

3-84 $F = \$1,402.63$

3-85 $N \approx 70$ months

3-86 $P_0 = \$14,579$; select C

3-87 **a.** i/year $= 8.24\%$; $F = \$6,340.50$
b. i/6 months $= 4.04\%$; $F = \$2,655.84$

3-88 $P_0 = \$11,359$

3-89 $P_0 = \$1,824.21$

3-90 $A = \$312$; select C

3-91 $r = 17.56\%$

3-92 $A = \$557.25$

3-93 $F = \$17,303.19$

3-94 $P_0 = \$4,653.33$

3-95 $F_4 = \$11,109.06$

3-96 **a.** False　**b.** False　**c.** False
d. True　**e.** False　**f.** True
g. False　**h.** False　**i.** False

3-97 **a.** $A = \$543.67$
b. $P = \$7,409.40$
c. $F = \$3,668.30$
d. $F = \$2,054.40$

3-98 $Z = \$1,421.67$

3-99 $F_{18} = \$42,207$

3-100 $P_0 = \$767.43$

3-101 $r = 8.35\%$

3-102 $A = \$1,320.66$

3-103 **a.** $P = \$13,094.20$
b. $r = 9.19\%$

3-104 **a.** $F = \$362,944$
b. $A = \$60,386$

3-105 **a.** $F = \$526,217$
b. $r = 10\%$, $F = \$133,965$
c. $N = 16.38$ (or 17) years

3-106 $N = 5$ years

3-107 **a.** $P = \$3,296,800$
b. $P = \$40,260.60$
c. $r = 20\%$/year $= 5\%$/quarter; $P = \$7,408$

3-108 Difference $= \$1,269.00$

3-109 **a.** True
b. True
c. False
d. False
e. False
f. (i) (F/P,i%,N) (ii) (P/G,i%,N)

CHAPTER 4

4-1 No. A higher MARR reduces the present worth of future cash inflows created by savings (reductions) in annual operating costs. The initial investment (at time 0) is unaffected, so higher MARRs reduce the price that a company should be willing to pay for this equipment.

4-2 **a.** PW $= \$82,082.78 > 0$, proposal is acceptable
b. IRR $= \$15.48\% > 12\%$, proposal is acceptable
c. Simple payback period $= 5$ years

4-3 **a.** PW $= -\$13,423.57$
b. CR $= \$1,828$ (same for all three equations)

4-4 **a.** PW $= \$2,911.60$; $FW = \$5,855.60$; $AW = \$868.70$
b. IRR $= 27.2\%$, yes, since the IRR $> MARR$, the project is acceptable
c. ERR $= 21\%$

4-5 Buy the ranch (PW $= \$1,185.80 > 0$)

4-6 PW $= \$3,526.50 > 0$; the company should invest in the new product line

4-7 **b.** $V_N = \$750.77$

4-8 $P_0 = \$6,693.37$

4-9 i/year $= 10.88\%$

4-10 **a.** $V_N = \$7,688.96$
b. $A = \$150,892.90$

4-11 $C = \$702.21$

4-12 $i = 7.5\%$ every six months

4-13 $N \approx 24$ months

4-14 PW $= \$5,671.40$

4-15 $A = \$4,490$/year

4-16 I $=$ capital investment $= \$25,058.39$

4-17 $A = \$3,102.45$

4-18 FW $= \$14,580.72 \geq 0$, recommend process R
AW $= \$620 \geq 0$, recommend process R

4-19 AW $= -\$577.81 < 0$; not a good investment

4-20 Capital Recovery Amounts:
Year 1 $= \$250$, Year 2 $= \$240$, Year 3 $= \$230$, Year 4 $= \$120$

4-21 a. As $i \to \infty$, the PW approaches $-\$3,000$
  b. $\theta' = 6$ years
  c. PW(0%) $= -\$1,000$; AW(0%) $= -\$166.70$

4-22 Total cost/unit $= \$4.32/$unit; Selling price/unit $= \$5.18/$unit

4-23 APR $= 22.8\%$ compounded monthly

4-24 b. i/year $= 22.4\%$ per year

4-25 $i = 40.9\%$ per year

4-26 $i = 51.1\%$ per year

4-27 $i' \approx 14.7\%$, and you want to start a personal savings program as soon as possible because delay requires a higher rate of return, increased annual amount, or both to reach a set goal

4-28 i/year $= 26.7\%$ per year

4-29 $X = \$19,778$

4-30 IRR $= 21.5\% >$ MARR; the product line appears profitable

4-31 a. 15.2%
  b. 18.8%
  c. 21.5%
  d. 20%

4-32 a. IRR $= 14.1\%$
  b. IRR $- 0\%$
  c. i/year $= 20\%$ per year

4-33 $i'\% = 12.3\%$

4-36 $i = 8.6\%$ per year

4-37 a. IRR $= 10\%$
  b. FW(12%) $= -\$27,070.25$
  c. ERR $= 10.74\%$

4-39 $i = 1.24\%$ per year

4-40 PW $= \$630.43 \geq 0$, project is acceptable
  FW $= \$1,677.14 \geq 0$, project is acceptable
  AW $= \$151.56 \geq 0$, project is acceptable
  IRR $= 24.9\% \geq 15\%$, project is acceptable
  Simple payback period $= 4$ years $= \theta <$ 5 years, project is acceptable
  Discounted payback period $= 6$ years $= \theta' \leq 6$ years, project is acceptable
  ERR $= 20.9\% \geq 15\%$, project is acceptable

4-41 a. $\theta' = 6$ years
  b. $i'\% = 29.4\%$ per year

4-42 a.

N years	P (= affordable price)
5	$328,403.80
6	$373,572.20
7	$413,908.10
8	$449,911.30
9	$482,062.20
10	$510,772.00

  b. $\theta \approx 4$ years

4-43 a. PW(15%) $= \$185.95 \geq 0$; financially profitable
  b. $\theta = 5$ years
  c. $\theta' = 5$ years

4-44 a. IRR $= 24.7\% >$ MARR, project is profitable
  b. $\theta = 4$ years > 3 years; project is not acceptable

4-45 IRR $= 4.9\%$ and 31.2% per year
  ERR $= 7.6\%$ per year

4-46 b. ERR $= 25.9\%$

4-47 a. $i'\% = 1/2\%$ and 28.8% per year
  b. ERR $> 20\%$; project is acceptable

4-48 a. In all three cases, IRR $= 15.3\%$; this is true for both EOY 0 and EOY 4 as reference points in time
  b. Select the third case to maximize PW(10%); however, the PW(IRR $= 15.3\%$) would be zero for all three situations

4-49 a. $N \approx 348$ months (past age 62), he should probably start drawing social security payments at age 65
  b. Take Social Security starting at age 62
  c. If your uncle expects to live past age 75, deferring social security payments to the regular age of 65 years is preferred at $i = 0\%$ per month

4-50 a. AW $= \$1,828$, it is a good investment
  b. IRR $= 15.3\%, \theta = 4$ years, $\theta' = 5$ years
  c. Other factors include sales price of reworked units, life of the machine, the company's reputation, and demand for the product

CHAPTER 5

5-1 **a.** Alternative II
  b. Alternative II
  c. Rule 1; the net annual revenues vary among the alternatives

5-2 **a.** Select Design 3 (AW = $141.10)
  b. Select Design 3 (FW = $2,886.16)

5-3 Select Design D3 (PW = −$5,233,268.80)

5-4 Select the apartment house (AW = $32,016)

5-5 **a.** 1;
  b. 4;
  c. 5;
  d. 2

5-6 **a.** Select product 2 (PW = $12,897)
  b. IRR_{Δ} = 10.4%, select product 2

5-7 Select C

5-8 i'_{Δ}% = 13.7% < 15% so keep Alternative B as best

5-9 Select Design 3

5-10 Select Design B

5-11 Select A1; because it maximizes the AW value (and A2 is not economically justified at demand = 91,000 units/year).

5-12 **a.** Select Design C (AW = −$25,781)
  b. Select Design C

5-13 Select Design A (concrete pavement)

5-14 Both methods select motor B (PW = −$3,470.54)

5-15 Select Motor A (AW = −$3,606.43)

5-16 Assume repeatability; select Alternative C (AW = $60.00)

5-17 Select Process S (AW = $1,639.84)

5-18 Select machine D2 (Equivalent annual cost = $25,116)

5-19 **a.** B
  b. B
  c. B
  d. Leasing crane A is *not* preferred to the selected alternative (B), but would be preferred to the purchase of crane A

5-20 **a.** Standard light bulb is less expensive by $0.44 per year

5-21 **a.** Select Machine A (AW = $227)
  b. Select Machine A (PW = $1,139)
  c. Select Machine A (i'_{Δ}% = 0 < 15%)

5-22 **a.** Select Alternative 1, AW = $4,552
  b. Select Alternative 1, AW = $4,552
  c. Select Alternative 2, FW = $47,179

5-23 L1 is the preferred choice

5-24 **a.** Select Alternative E1 (AW = −$16,990)
  b. Select Alternative E2 (AW = −$19,256)

5-25 Select Plan A (CW = −$66,500)

5-26 Select Design D1 (CW = −$147,000)

5-27 Select Bridge Design L (CW = −$378,733)

5-28 **a.** CW = $34,591
  b. N = 80 years

5-29 Select Alternative E2 (PW = −$273,100)

5-30 **a.** Select Alternative D1, (FW = $87,722)
  b. Select Alternative D1, (IRR_{Δ} = 9.13%)
  c. Because MARR = 12% > IRR_{Δ} = 9.13%

5-31 Recommend Design ER2

5-32 Select projects A1 and C1 for investment

5-33 Of the 29 feasible combinations, MEC 25 maximizes present worth at i = 10%; proposals A1, B1, and C1 should be implemented; the remaining $200,000 is assumed to be invested elsewhere in the firm at MARR = 10%

5-34 Invest in projects X and Y (MEC #4) for investment (PW = $17,520)

5-35 **a.** MEC 0,1,2,3
  b. PW_{MEC2}(12%) = $28,713

5-36 Select mutually exclusive combination 2 (projects A and B1) based on the present worths

5-37 Select Alternative S1 (CW = −$150,927)

5-38 A 50-story building should be recommended

5-39 Produce ice cream in quart containers

5-40 Plot of PW(Δ) versus i% shows a unique IRR_{Δ} at i = 16.9%

5-41 X = $1,147,790 every five years

5-43 Select option II to continue the project (PW = $43,792)

CHAPTER 6

6-6 **a.** $d_2 = \$6,000$
b. $d_2 = \$7,143$
c. $d_2 = \$11,200$
d. $d_2 = \$5,000$

6-7 **a.** 5
b. 3
c. 4

6-9 Basis = $120,000

1. (d) $19,200
2. (a) $96,000
3. (c) $12,885
4. (b) 7 years
5. (a) $17,148
6. (b) $52,476
7. (b) $7,494

6-10 **a.** $B = \$17,200$
b. $6,480

6-11 **a.** $d_3 = \$3,428.57;\ BV_5 = \$42,857.15$
b. $d_3 = \$5,485.71;\ BV_5 = \$32,571.43$
c. $d_3 = \$6,297.38;\ BV_5 = \$27,759.86$
d. $d_3 = \$10,494.00;\ BV_5 = \$13,386.00$
e. $d_3 = \$4,285.71;\ BV_5 = \$40,714.30$

6-12 **a.** 5 years
b. $d_4 = \$17,280$
c. $BV_4 = \$25,920$ (beginning of year 5)

6-13 **a.** GDS:

$d_1 = \$60,000;\ d_4 = \$34,560$
$d_2 = \$96,000;\ d_5 = \$34,560$
$d_3 = \$57,600;\ d_6 = \$17,280$

ADS: $d_1 = d_7 = \$25,000$
$d_2 = d_3 = \cdots = d_6 = \$50,000$

b. $\text{PW}_{\text{GDS}} = \$221,431.15;\ \text{PW}_{\text{ADS}} = \$194,566.30;\ \text{PW}_\Delta = \$26,864.85$

6-14 **b.** PW(SL) = $294,941; PW(DB) = $319,538; PW(GDS) = $360,720

6-15 **a.** $d_3 = \$21,984$
b. $BV_4 = \$19,786$
c. $d_4^* = \$70,015$

6-16 $d_4 = \$2,000;\ BV_4 = \$11,000$

6-17 **a.** cost depletion (this year) = $280,000
b. percentage depletion allowance = $236,250; acceptable

6-18 **a.** cost depletion = $200,000 (maximum)
b. percentage depletion allowance = $43,000

6-19 adjusted unit cost for depletion allowance = $0.371/MCF

6-20 **a.** $18,850
b. $71,150
c. $130,000

6-21 **a.** Taxable Income = $1,700,000; Income Tax = $578,000
b. NIAT = $1,122,000
c. ATCF = $2,322,000

6-22 **a.** (b)
b. (c)
c. (d)
d. (e)
e. (d)
f. (c)
g. (a)

6-23 $t = 37.96\%;\ t = 41.92\%$

6-24 Income taxes (with venture) = $11,250
Income taxes (without venture) = $9,250

6-25 $i\% = 4.4\%/6$ months, $r = 8.8\%$/year, $i = 8.99\%$/year (effective interest)

6-26 Assume repeatability; select Alternative A: plastic $(\text{AW}_A = -\$1,184)$

6-27 If leasing cost < $8,733 per year, lease the tanks; otherwise, purchase the tanks

6-28 Assume repeatability; select Fixture X $(\text{AW} = -\$4,989)$

6-29 IRR = 22.2%

6-30 Select machine B (AW = $6,678, PW = $60,000)

6-31 Select Design S1 (AW = −$290)

6-32 **a.** Select Method M2 (PW = −$54,211)
b. Select Method M2 (AW = −$9,180)

6-33 Added annual expense can be as high as $1,774 and the IRR will still exceed 10%

6-34 **a.** MARR (before taxes) = 25%
b. $\text{MV}_8 - \text{BV}_8 = \$10,000$
c. Do not purchase the new machine $[(\text{PW}_{\text{ATCF}}(15\%) = -\$25,082 < 0)]$

6-35 $N = 6$ years

6-36 **a.** Select A $(\text{AW}_A = -\$7,883)$
b. Select A $(\text{PW}_A = -\$26,426)$
c. Select A based on incremental IRR analysis

6-37 $864,135/year

6-38 **a.** PW = −$171,592
b. $AW = -\$37,115$

6-39 a. IRR = 75.3%;
b. IRR = 79.4%;
c. IRR = 129.3%

6-40 a. PW(10%) = $66,150
b. Yes

6-41 PW(10%) = $17,208

6-42 AW(12%) = $3,468 for both ATCF and EVA

6-43 No, not a profitable investment
PW(MARR$_{AT}$) = −$3,561.43 < 0

6-44 ATCF for EOY 6–10 = $64,000

6-45 a. ATCF for EOY 1–10 = $6,700,000
b. PW(12%) = −$2,143,660

6-46 Accept Quotation II

CHAPTER 7

7-4 C$_{2005}$ = $262,780

7-5 $\bar{I}_{2004}$ = 153.5

7-6 a. $\bar{I}_R$ = 154.9; $\bar{I}_C$ = 203.4
b. C$_C$ = $412,710

7-7 a. $\bar{I}_{2004}$ = 176
b. $\bar{I}_{2000}$ = 144.5
C$_{2004}$ = $791,696

7-8 Cost = $354,879

7-9 Difference between high and low estimates is 59.8%; therefore; the estimate should not be expected to be more accurate than about +50%

7-10 Cost = $12,641,919

7-11 a. Cost = $6,300/year

7-12 a. Cost = $229,707
b. Cost = $127,512

7-13 Cost = $345,914

7-14 C$_{2006}$ = $11,541

7-15 a. Z$_8$ = 108 hr; Z$_{50}$ = 94.3 hr
b. C$_5$ = 117.5 hr

7-16 a. K = 19.7 hr
b. Z$_{1000}$ = 3.9 hr

7-17 31.4% reduction in overhead costs

7-18 Total cost = $228,678

7-19 a. $\bar{I}_{B1} = \frac{198.6}{127.3} = 1.56$ or $\underline{156\%}$
$\bar{I}_{B2} = \frac{192.0}{125.5} = 1.53$ or $\underline{153\%}$
b. $3,351,600

7-20 a. Cost = 50,631 + 51.5x (at x = 23,000 ft^2, cost = $1,235,131)
b. SE = 59,730, R = 0.9765

7-21 a. $y = 31.813 + 0.279x$
b. $R = 0.99$
c. $y = 101.56

7-22 Unit selling price = $248.00

7-23 Unit selling price = $31.50/widget

7-24 a. Maximum profit = $26.04; profit margin = 6.61%
b. Target cost @ 15% profit margin = $365.22 (cannot be achieved)

7-25 X = 4,497 units

7-26 a. Total manufacturing cost = $2,284.94; unit selling price = $2,741.92
b. Target cost = $2,000

7-28 Total cost = $2,239,046

7-29 s = 0.9 (90% learning curve)

7-30 a. Unit selling price = $0.445 per unit
b. Target cost = $0.435 per unit
c. The $0.01 reduction in unit cost can possibly be achieved by renegotiating the cost of production material

7-31 1997 estimate of $320,274,240 for the plant

CHAPTER 8

8-1 a. PW(i$_c$) = $6,082
b. PW(i$_r$) = $8,111

8-2 $N \approx 18$ years

8-3 Select situation 2 (FW$_5$ = $4,000)

8-4 Alternative B has least negative PW at time 0 (PW = −$369,070)

8-5 Alternative 1 (PW = $10,000)

8-6 $31,746; $32,069; $31,564; and $30,361

8-7 f = 2.77%

8-8 FW$_{10}$ = $19,231

8-9 a. Total Capital Investment = $24,230,790
b. PW(12%) = −$28,584,440
c. AW(4%) = −$3,524,460

8-11 P$_0$(A$) = $43,755

8-12 N = 5 years

8-13 IRR$_r$ = 1.85% per year

8-14 a. FW (in A$) = $144,105
b. AW (in R$) = $44,932

8-16 Product A: $\bar{f} = 6.00\%$ per year; Product B: $\bar{f} = 8.33\%$ per year

8-17 a. ATCF = −$50,640 (EOY 1); −$38,904 (EOY 2); −$33,194 (EOY 3); −$33,514 (EOY 4); −$33,865 (EOY 5); −$34,252 (EOY 6)

b. $i_c = 15\%$ per year; PW = −$146,084; EUAC = $38,595

8-18 Annual Revenue = $305,286

8-19 PW(18%) = −$12,234

8-20 Year 1 savings (in A$) must be $11,875 per motor to break even

8-21 a. AW(A$) = −$1,859

b. AW(R$) = −$1,309

8-22 $IRR_{A \to B} < 0\%$; select Machine A

8-23 X = $393,790,000 (for both a and b)

8-24 Select Design 2 (PW = $5,789.86)

8-25 PW = −$359,665

8-25 $i_c = 26\%$

8-27 a. $356,557

8-28 Select Purchase alternative (FW$_6$ = −$1,823,920)

8-29 a. 6.08 units of X

b. 6.91 units of X

8-30 a. $i_{fc} = 36.08\%$ per year

b. $i_{fc} = 18.44\%$ per year

8-31 a. PW(18%) = −$19,635, not acceptable

b. $IRR_{fc} = 28.0\%$/year

c. $IRR_{US} = 14.29\% < 18\%$; not acceptable

8-32 $i_{US} = 22.7\%$,
$PW(22.7\%) = \$70,583,300 > 0$
Yes, this project will meet the company's economic decision criteria.

8-33 $39,836 could be spent for software with a 3-year upgrade agreement (i.e., Option 1)

8-34 This is intended to be a tailor-made exercise (at the discretion of the instructor).

CHAPTER 9

9-1 Keep the old lift truck (PW$_{Defender}$ = −$23,331 > PW$_{Challenger}$ = −$24,247)

9-2 Keep the old car based on incremental analysis

9-3 Replace the old crane (AW$_{Challenger}$ = −$3,678 > AW$_{Defender}$ = −$4,952)

9-4 a. $N = 3$ years

b. $N = 1$ year

c. $N = 4$ years

9-5 $N = 3$ years

9-6 Replace defender after two years

9-7 a. New improved machine should replace present machine immediately

b. Keep the present machine for two more years

9-8 $N = 6$ years

9-9 Defender should be retained because the AW over its useful life has the least negative value (−$15,382)

9-10 Keep the old machine

9-11 Reinforce the existing bridge

9-12 Centrifuge should be retained for three years before abandonment

9-13 $N = 5$ years

9-14 Total after-tax investment in the defender is $13,938

9-15 a. $N_{defender} = 1$ year; $N_{challenger} = 3$ years

b. $N = 2$ years

c. Assumptions: Infinite analysis period with repetitive cycles of replacement with challenger (every three years) starting at the end of the second year

9-16 Keep the defender (PW$_{ATCF}$ = −$3,677)

9-17 b. Keep the defender, since ERR$_\Delta$ < MARR

9-18 Keep the defender, since $i'_\Delta \approx 1.36\% <$ MARR (12%)

9-19 Select the challenger (PW$_{ATCF}$ = −$1,440,423)

9-20 Keep the defender (AW$_{ATCF}$ = −$10,507)

9-21 Select the defender (PW$_{ATCF}$ = −$70,875)

9-22 Select the challenger (PW$_{ATCF}$ = −$46,793)

9-23 Lease the challenger ($i'_\Delta \approx 4.5\% <$ MARR)

9-24 $I = \$93,939$

9-25 Relocate exisiting transformers ($CW = -\$4,239$)

CHAPTER 10

10-4 Net Annual Worth is most sensitive to changes in the net annual cash flow; however, the decision to invest in the project is relatively insensitive to changes in the specified range.

10-5 **a.** $MV_2 = \$2,050$
b. $N = 7.3$ years

10-6 **a.** $H = 867$ hr/year
b. Select XYZ brand motor (AW = $-\$17,987$)

10-7 **a.** The preferred plan is relatively insensitive to errors in estimating the annual disbursements.
b. The preferred plan is relatively insensitive to errors in estimating the MARR.

10-8 Electrical energy cost = 1.88 cents/kW-hr

10-9 X = \$933,953 in annual revenues per year

10-10 **a.** The project is economically attractive (AW = \$232,625)
b. The decision is most sensitive to changes in the occupancy rate

10-14 Optimal insulation thickness is seven inches

10-15 IRR $\geq$ 10% when EOY 3 repair cost $\leq$ \$6,872

10-16 Optimal height of the proposed building is four floors over the specified range of the MARR; however, unless the MARR is less than 17%, the proposed building would not be profitable

10-17 X = 362,500,000 Btu per year

10-18 **a.** Optimistic: AW = \$23,330; Most likely: AW = \$14,325; Pessimistic: AW = $-\$9,184$

10-19 **a.** Optimistic: AW = \$23,192; Most likely: AW = \$14,984; Pessimistic AW = $-\$7,552$

10-20 Build the four-lane bridge now (PW = $-\$350,000$)

10-21 Pump A is the better investment

10-22 Select the ABC brand motor (AW = $-\$9,831$)

10-24 **a.** Select Alternative B (PW = $-\$79,065,532$)
b. Select Alternative B (PW = $-\$60,788,379$), 23.1% less expensive due to cotermination at EOY 10
c. Alternative B is not reversed

CHAPTER 11

11-1 Select B

11-2 Select Alternative C

11-3 Recommend F

11-4 No alternative is recommended

11-5 Select Alternative B

11-6 **a.** Projects B,C, and E are acceptable for funding
b. $B > C > E(> D > A)$
c. Select project B

11-7 **a.** All three plans (A, B, and C) should be selected ($B-C$ ratios > 1)
c. A constant amount subtracted from the denominator and numerator of the $B-C$ ratio does not appreciably affect the recomputed ratio

11-8 Select RS-511 ($\Delta B/\Delta C = 1.03$)

11-9 **a.** Maximum benefit—choose levees
b. Minimum cost—choose no flood control
c. Maximum ($B-C$)—choose the small dam
d. Largest investment having incremental $B-C$ ratio larger than 1; choose the small dam
e. Largest $B-C$ ratio—choose the small dam (which is coincidentally the correct choice). The correct choice based on incremental analysis would be to select the small dam as seen in part (d)

11-10 **a.** Option B should not be recommended
b. Select option B

11-11 Route B is the least objectionable

11-12 If an option must be chosen, build the flood-control dam

11-13 **a.** Failure to consider the time value of money
b. More economical to retain the steel pier (AW = $-\$29,332$)

11-14 The toll bridge should be constructed ($B-C = 1.28$)

11-15 **a.** CW(10%) = $3,639,750
   b. $B-C = 1.14$
   Select initial design (described in Problem 11-14) should be selected ($\Delta B/\Delta C = -0.16 < 1$)

11-16 Construct the levee ($\Delta B/\Delta C = 1.17 > 1$)

11-17 **a.** Recommend constructing the tunnel ($B-C = 1.16 > 1$)
   b. X = $706,053

11-18 **a.** Select Design 3
   b. Select Design 3

11-19 Select Design B

11-20 Select Alternative A

CHAPTER 12

12-8 Select Alternative B, $RR = \$17,498$

12-9 Wait three years to build ($\overline{RR}_2 = \$142,524 < \overline{RR}_1 = \$159,638$)

12-10 Company must be able to purchase power for 18.25 mils per kilowatt-hour

12-12 **a.** ATCF$_5$ (in A$) = −$1,641
   b. T_5 (in A$) = $3,543

12-13 **a.** $RR_3 = \$12,878.11$
   b. Increase in RR_3 of $8.33

12-14 Build geothermal generating plant ($RR = \$525,088$)

12-15 Select Alternative B (levelized revenue requirement = $22,677)

12-16 Alternative A is marginally preferred to Alternative B ($\overline{RR}_A = \$145,056$; $\overline{RR}_B = \$145,338$)

12-17 Select Alternative 2 (levelized revenue requirement = $7,107)

CHAPTER 13

13-1 $\Pr(PW > 0) = 0.10$

13-2 E(PW : 4-lanes now) = −$3,500,000 and E(PW : 2 now + 2 later) = −$3,839,500, ∴ build the 4-lane bridge now

13-3 An interest rate of $i = 15\%$ will not reverse the initial decision to build the four-lane bridge now; the two-lane bridge would be preferred for interest rates greater than 16.78%

13-4 E(concrete) = 1,350 cubic yards, V(concrete) = 66,500 (cubic yards)^2; SD(concrete) = 258 cubic yards

13-5 $SD(R) = \$24.06$

13-6 Implement Design E; E(PW) = −$239,414

13-7 Recommend Alternative T to minimize total annual cost in both cases

13-8 Do not install lift; E(PW) = −$85,142 < 0

13-9 SD(PW) = $79,005; Recommend that the lift be installed since E(PW) = $115,848 > 0 and one SD ($79,005) is only 68% of $E(PW)$

13-10 Yes; $E(PW_{AT}) = \$33,386 > 0$

13-11 **a.** V(PW) = $8,606.78 \times 10^6 (\$)^2$, SD(PW) = $92,773
   b. $\Pr\{PW < 0\} = 0.1$; Recommend purchase of equipment since E(PW) = $114,862 is favorable; SD(PW) = $92,773 is less than the E(PW); and $\Pr\{PW < 0\} = 0.1$ is low.

13-12 **a.** E(PW) = $175; V(PW) = $28.04 \times 10^6 (\$)^2$; SD(PW) = $5,295
   b. $\Pr(PW \geq 0) = 0.68$

13-13 **a.** V(PW) = $1,097.8 \times 10^6 (\$)^2$, SD(PW) = $33,133
   b. $\Pr(PW > 0) = 0.57$
   E(AW)$_{R\$} = \$1,866$; project appears questionable since the E(PW) is positive but the SD(PW) is approximately two times the expected value. Also the $\Pr\{PW > 0\} = 0.57$ is only moderately attractive.

13-14 E(PW) = −$7,599; V(PW) = $404.74 \times 10^6 (\$)^2$, SD(PW) = $20,118; $\Pr(PW \leq 0) = 0.70$

13-15 $\Pr\{X \geq 171\} = 0.7881$

13-16 $\Pr(AW < \$1,700) = 0.5871$

13-17 E(PW) = $2,477 V(PW) = $1,096,863(\$)^2$ $\Pr(PW > 0) = 0.9911$

13-18 **a.** E(PW) = $1,354; V(PW) = $1,639,240(\$)^2$
   b. $\Pr(PW \geq 0) = 0.8554$

c. Yes; if PW (at i = MARR) > 0 then the IRR > MARR. Therefore, it is correct to conclude that Pr{IRR $\geq$ MARR = Pr{PW $\geq$ 0}.

13-19 Since E(AW) < 0, the lift should not be installed.

13-20 b. E(PW) = $84,280 > 0, recommend purchase of equipment

13-21 Alternative 1

13-22 Build the culvert, AW(culvert) = −$7,687 versus cost per year due to mudslide of −$10,000

13-23 Estimated value of test information = $15,891

13-24 Select new design, E(PW) = $20,225; EVPI = $9,567

13-25 $PW_{New\ Product}$ = $62,165, $PW_{Do\ Nothing}$ = $0; select new product

13-26 Select Alternative 2, E(PW) = $61,839; EVPI = $9,089

13-27 E(Value of Survey) = $3,162

CHAPTER 14

14-6 A = $302,500

14-9 e_a = 12.5%

14-11 C_B = 6.62% per year

14-12 Combination 1: EUAC(15%) = $6,264, purchase the truck
Combination 2: EUAC(15%) = $6,731, lease the truck

14-13 Recommend leasing the equipment (AW = −$1,800)

14-14 a. Lease the machine, AW = −$500
b. Lease the machine as long as the annual lease rate is $\leq$ $1,410

14-15 Objective function value = $219,887

14-16 Objective function value = $4,478

14-17 Objective function value = $2.47

14-18 Objective function value = $8,822

14-19 Ex. 14-1 updated, C_L = 5.28%
Ex. 14-2 updated, C_B = 3.29%
Ex. 14-3 updated, **a.** Stock price = $20.63/share **b.** e_a = 12%
Ex. 14-4 updated, e_p = 6.9%, WACC = 8.09% per year

CHAPTER 15

15-5 a. Alternative 2
b. Alternative 2
c. Alternative 2

15-6 a. No alternatives are removed from consideration
b. Remove "Retain Existing System" from consideration
c. All alternatives still available ("Retain" already eliminated) pass because all options are acceptable in at least one attribute.
d. Select Vendor III

15-7 Dominance—no alternatives eliminated
Satisficing—Alternative A eliminated
Lexicography—no alternatives eliminated
Hurwicz procedure—Alternative A eliminated
Additive weighting—Alternative B selected

15-8 a. See Table G15-8(b)
b. See Table G15-8(b)
c. See Table G15-8(c)
Using lexicography we conclude that social climate is the most important attribute, and additive weighting also selects Apex.

TABLE G15-8a

Attribute	Relative Rank	Normalized Rank
Social Climate	1.00	1/2.08 = 0.481
Starting Salary	0.50	0.5/2.08 = 0.240
Career Adv.	0.33	0.33/2.08 = 0.159
Weather/Sports	0.25	0.25/2.08 = 0.120
	2.08	

TABLE G15-8b

Attribute	Alternatives			
	Apex (N.Y.)	Sycon (L.A.)	Sigma (GA.)	Mc-Graw-Wesley (AZ.)
Starting Salary	$35,000	$30,000	$34,500	$31,500
Dimensionless				
Equivalent (DE)	1.0	0.0	0.9	0.3

$$DE = \frac{\text{Worst Outcome–Outcome Being Made Dimensionless}}{\text{Worst Outcome–Best Outcome}}$$

TABLE G15-8c

Attribute	Normalized Weight	Apex	Sycon	Sigma	Mc-Graw Wesley
Social Climate	0.48	1×0.48	1×0.48	0.5×0.48	0×0.48
Starting Salary	0.24	1×0.24	1×0.24	0.9×0.24	0.3×0.25
Career Adv.	0.16	0×0.16	0×0.16	0.6×0.16	1×0.16
Weather Sports	0.12	0×0.12	0×0.12	0.33×0.12	0.67×0.12
	Sum	0.72	0.63	0.59	0.31

15-9 Alternative A

15-10 Dominance—no offers eliminated
Satisficing—eliminate Offers 1 and 2, accept Offer 3
Disjunctive Resolution—no offers eliminated
Lexicography—accept Offer 2
Non-dimensional scaling—accept Offer 3
Additive weighting—accept Offer 3

15-11 The solution involves subjective factors which will vary from one student to another.

15-12 a. Dominance—no alternatives eliminated
Feasible ranges—eliminate contestants 1 and IV
Lexicography—select contestant II
Additive weighting—select contestant I

15-13 Dominance—no selection
Feasible Ranges—no selection
Lexicography—Domestic 2
Additive weighting—Domestic 2

15-14 X_{4j} 0.7 (keep tool)
1.0 (purchase new)

Index